Introduction to Finite Element Analysis for Engineers

Now in its second edition, *Introduction to Finite Element Analysis for Engineers* is an essential introduction to FEA as a method to solve differential equations. With many practical examples focusing on both solid mechanics and fluid mechanics, it includes problems for both applications.

Using a structure of classes of differential equations, the book also includes MATLAB® codes and aims to build a comprehensive understanding of FEA and its applications in modern engineering. New chapters present finite-element models of a system of partial differential equations in two or more independent variables typified by problems in theory of elasticity and plates. Chapter ten presents the finite element method for a nonlinear Mindlin-Reissner plate, and panel flutter is included as a typical example of fluid-structure interactions. The book demonstrates the power and versatility of FEA as a tool with a large number of examples of practical engineering problems. These problems range from those which can be solved without a computer, to those requiring MATLAB® or Python.

With applications in civil, mechanical, aerospace, and biomedical engineering, the textbook is ideal for senior undergraduate and first-year graduate students and also aligns with mathematics courses.

Introduction to Finite Element Analysis for Engineers

Second Edition

Saad A. Ragab and Hassan E. Fayed

CRC Press
Taylor & Francis Group
Boca Raton London New York

CRC Press is an imprint of the
Taylor & Francis Group, an **informa** business

Designed cover image: Saad A. Ragab and Hassan E. Fayed

Second edition published 2025
by CRC Press
2385 NW Executive Center Drive, Suite 320, Boca Raton FL 33431

and by CRC Press
4 Park Square, Milton Park, Abingdon, Oxon, OX14 4RN

CRC Press is an imprint of Taylor & Francis Group, LLC

First edition published by CRC Press 2017

ISBN: 978-1-032-34629-8 (hbk)
ISBN: 978-1-032-34640-3 (pbk)
ISBN: 978-1-003-32315-0 (ebk)

DOI: 10.1201/9781003323150

Typeset in Nimbus Roman font
by KnowledgeWorks Global Ltd.

Ragab: To the Memory of my Mother and Father,
my Sisters and Brothers

Fayed: To my family in the United States and Egypt

Contents

Preface

Our primary aim of the second edition of *Introduction to Finite Element Analysis for Engineers* is to include topics of interest to graduate students in engineering mechanics and related fields. Three new chapters are added: Plane elasticity, linear analysis of plates, and geometrically nonlinear thick plates with applications to fluid-structure interactions. The first seven chapters of the first edition are unchanged except for corrections of typos and the addition of new examples and problems.

Computational approaches to solving engineering problems have become essential analysis and design tools for engineers. The finite-element method is at the center of modern computer analysis techniques. Before embarking on using massive finite-element commercial software, engineers need to know how finite-element models are derived from the basic principles that are usually expressed as differential or integral statements. Having a strong mathematical foundation of the finite-element method, engineering students will be better prepared to tackle complex problems. This textbook, *Introduction to Finite Element Analysis for Engineers,* has evolved from the first author's lecture notes for finite-element courses that were taught in the department of Engineering Science and Mechanics (now Biomedical Engineering and Mechanics) at Virginia Tech for the past 20 years. The book serves as an introduction to the finite-element method, and presents it as a numerical technique for solving differential equations that describe problems in civil, mechanical, aerospace, and biomedical engineering. It enables engineering students to formulate finite-element models and solve practical problems and analyze the results. Although commercial finite-element software is not used in this book, it explains the mathematical foundation underpinning such software. By mastering the techniques presented in this book, students will be better prepared to use commercial software as practicing engineers. The book is intended for senior undergraduate or first-year graduate students in engineering and related disciplines. Thus, mathematical rigor is not compromised but presented at a level consistent with the anticipated mathematics background required in most engineering curricula. The power and versatility of the finite-element method are demonstrated by a large number of examples and exercises of practical engineering problems. Some examples are simple enough so that they can be solved completely in a reasonable time without the use of a computer. Other problems require the use of a PC and programming software such as MATLAB®, Mathematica, or Python. The MATLAB® codes for representative examples solved in the book can be downloaded from the publisher's website.

The ability to verify computational results is an especially important facet in using the finite-element method and numerical techniques in general. This book emphasizes the importance of verifying finite-element results by providing comparisons to exact analytic solutions when possible. Excellent agreement between computational results and exact analytic solutions of complex problems strengthens students' confidence in the finite-element method. The book also emphasizes conducting mesh refinement studies so that students become aware of discretization errors, and develop healthy skepticism of physically implausible answers.

After the first introductory chapter, each of the following six chapters deals with a certain type of differential equation: Second-order ordinary differential equations (ODE), fourth-order ODE, elliptic partial differential equations (PDE), parabolic PDE, hyperbolic PDE, and differential eigenvalue problems. The first chapter is a review of basic mathematical concepts, definitions, and methods in linear algebra and differential equations. For Chapters 2 through 7, each chapter starts with a statement of a standard problem given by a differential equation and associated boundary and/or initial conditions. The weak form of the stated problem (also called the principle of virtual displacements) is derived and used to develop finite-element models.

Chapter 2 deals with second-order ODE. Almost all concepts and definitions of the finite-element method are introduced in that chapter including the method of weighted residuals, weak form, trial solutions, test functions, classification of boundary conditions (essential, natural, mixed), element shape functions, element equations and assembly. Elastic deformation of bars and steady heat conduction in fins and slabs are the main applications. Biomedical engineering applications include oxygen diffusion and consumption in cells and tissues and pulsatile blood flow in arteries. Chapter 3 deals with fourth-order differential equations; essentially it is the Euler–Bernoulli beam equation. The higher-order equation increases the level of complexity of the element shape functions, and introduces shape functions that satisfy the continuity of the trial solution and its first spatial derivative. In addition to beam problems, the finite-element method is extended to plane frames and trusses. The connection between the weak form of the equations of equilibrium and the principle of virtual displacements in solid mechanics is elucidated in that chapter. The principle is derived for a 3D continuum in static equilibrium with the assumption of infinitesimally small displacements. Then, it is used to give a better understanding of the equivalence of distributed forces and nodal concentrated forces and moments. That chapter also presents the principle of minimum total potential energy, and demonstrates its use for deriving finite-element beam equations.

Chapter 4 deals with elliptic partial differential equations in two space dimensions. Isoparametric elements and numerical integration of element matrices are covered in that chapter. Applications include steady heat conduction, torsion of non-circular sections, laminar viscous flow in ducts, and potential flow around two-dimensional airfoils with lift calculations. Finite-element models are developed for oxygen diffusion and consumption in a Krogh capillary-tissue cylinder in cylindrical coordinates. Linear and nonlinear (Michaelis–Menten) reaction kinetics are covered. Chapter 5 deals with parabolic equations in one space dimension and time. The finite-element method is used for the space operator and the finite-difference method is used for time integration. Applications include transient heat transfer in fins and slabs. Finite-element models for two biomedical engineering problems are presented. First, Pennes heat equation is used to analyze thermal injuries of skin tissues due to flash fire and heat exchange with hot surfaces. Second, transient oxygen diffusion and consumption in cylindrical tissue and spherical cells are analyzed with linear and nonlinear (Michaelis–Menten) reaction kinetics. Hyperbolic equations in one space dimension and time are considered in Chapter 6. The finite-element method is used in space and the Newmark method is used for direct time integration. The space operator is either second-order or fourth-order derivative. Waves on strings, and free and forced vibrations of bars and beams are analyzed. Chapter 7 deals with differential eigenvalue problems. The finite-element method is used to compute natural frequencies of bars, beams, and frames. Hydrodynamic stability theory is a rich source of differential eigenvalue equations; the Orr–Sommerfeld equation is a well-known example. Finite-element treatment for that equation is presented and applied to Poiseuille channel flow.

Chapters eight, nine, and ten deal with a system of partial differential equations in two or more independent variables; hence there are multi-degrees-of-freedom at each node. Chapter eight presents a finite-element analysis of the plane elasticity of linear isotropic materials including uncoupled thermoelasticity. Chapter nine is on thin and moderately thick plates. Both Kirchhoff–Love and Reissner–Mindlin plates are presented with examples for static deflection and free vibrations of plates in vacuum and submerged in heavy fluids. Chapter ten presents the finite-element method for geometrically nonlinear Reissner–Mindlin plate. Applications include buckling of plates due to in-plane and thermal loading, and nonlinear free vibrations. Examples of fluid-structure interactions include vibrations of submerged plates, hydroelastic panel divergence, and panel flutter and determination of limit cycle oscillations amplitudes and frequencies. As previously mentioned, great emphasis is placed on assessment of the accuracy of the finite-element method by comparison with exact analytic solution if available. MATLAB® codes developed for the examples can be easily modified to solve exercises at the end of each chapter.

In teaching the finite-element method to senior students, we set the learning objectives to be:

1. Develop a clear understanding of the concepts of the weighted residuals and weak form of a differential equation, including defining the physical meaning of primary and secondary variables and essential and natural boundary conditions.

2. Apply the weak form of a differential equation to an element and develop the appropriate element matrices. Assemble element matrices to develop a multi-element model and apply boundary conditions.

3. Be able to determine the consistent equivalent nodal representation of a distributed effect; replace distributed load by concentrated forces and moments at the element's nodes.

4. Develop the weak forms of linear partial differential equations (elliptic, parabolic, and hyperbolic) in one dependent variable and two independent variables.

5. Develop element shape functions and element matrices for triangular and rectangular elements.

6. Solve 2D Poisson's equation with applications to problems in engineering sciences and mechanics (heat transfer, viscous flow and solid mechanics).
7. Solve transient diffusion problems (e.g., heat transfer in a fin, transient Poiseuille and Couette flows) in one space dimension.
8. Approximate natural frequencies of bars, beams, and frames.
9. Verify finite-element results by comparison with exact analytic solutions, and conduct mesh independence studies.
10. Recognize and exploit the versatility of the finite-element method for simultaneously solving problems in different disciplines such as solid and fluid mechanics, heat and mass transfer, etc.

Similarly, we set the learning objectives for the graduate course to be:

1. Derive the weak form of differential equations, and construct appropriate basis functions.
2. Derive Lagrange and Hermite element shape functions for one- and two-dimensional problems.
3. Develop finite-element models using the weak form or the principle of virtual displacements for elasticity problems.
4. Use numerical integration to evaluate element matrices for isoparametric elements.
5. Solve potential flow problems and determine lift over 2D airfoils of arbitrary profiles.
6. Determine stress concentration factors and thermal stresses for plane elasticity problems.
7. Determine static deflections of plates of arbitrary shape and various boundary conditions.
8. Compute natural frequencies of plates in vacuum.

Senior students have been able to easily read the first edition of the book, follow mathematical derivations, and accomplish the above learning objectives. Graduate students also read a draft of Chapters 8 and 9, and have been able to accomplish related objectives. We felt that a second edition of the book is a chance to share the new chapters so that students and practicing engineers could benefit from them.

The authors would like to thank Professor Slimane Adjerid of the Mathematics Department at Virginia Tech for helpful discussions on the finite-element method. We thank Dr. Youssef Bichiou for reading parts of the book. The first author thanks all senior and graduate students who took Introduction to the Finite Element courses, made corrections, and offered supportive comments. He also thanks his sons, Ahmad Ragab, for reading the entire manuscript of the first edition and making corrections, and Amr Ragab, for discussions on biological and chemical problems. The authors also thank Ms. Katya Porter, Ms. Nicola Sharpe, Mr. Jonathan Plant, Ms. Claudia Kisielewicz, Ms. Karen Simon, and Ms. Shatakshi Singh of the editorial staff of Taylor and Francis for their help in the production of both editions of the book.

Author's Biography

Saad A. Ragab held the position of Professor in the Department of Biomedical Engineering & Mechanics at Virginia Tech until May 2022. Presently, he is Professor Emeritus in the same department. Dr. Ragab's fields of interest and research include computational fluid dynamics, hydrodynamic stability, optimization, multiphase fluid flow, and fluid-structure interactions. Dr. Ragab received his B.S. and M.S. degrees in aeronautical engineering, Cairo University, Egypt, and earned his Ph.D. in Engineering Mechanics in 1979 from Virginia Tech, Blacksburg, Virginia. He has received several awards for excellence in teaching, and certificates of recognition from the APS and ASME.

Hassan E. Fayed is the founder and CEO of Narmer-EngSim, a Research and Development Company based in Virginia. He received his B.S. and M.S. degrees in Mechanical Engineering from Zagazig University in Egypt, and his Ph.D. in Engineering Mechanics in 2013 from Virginia Tech. Dr. Fayed's specialities include multiphase flow physics and numerical simulations, hydro-cyclone technology and applications in the mining and oil and gas industries.

1 Introduction

1.1 COMPUTATIONAL SCIENCES AND MECHANICS

Analytical methods for exact or approximate closed-form solutions and laboratory experiments have been long established for solving engineering problems. These methods are indispensable for an in-depth understanding of basic physical phenomena, but are not adequate for design of complex systems and solving multi-physics problems. Very often, engineers have to make drastic simplifying assumptions in the problem geometry, material properties, and operating conditions just to be able to obtain analytical solutions. Experimental measurements are usually too expensive and time-consuming but are necessary for characterization of new materials and discovery or study of new phenomena.

Computational approaches have evolved over the years to meet the demands for accurate solutions of complex and multidisciplinary engineering problems. The versatility and power of computational (numerical) solutions lead to the introduction of computational physics and mechanics in undergraduate engineering curricula. The availability of affordable computers and advancements in programming software (MATLAB®, *Mathematica*, PYTHON, etc.) enable engineering students to learn advanced computational techniques and apply them for solving complex physics and mechanics problems. For example, a personal computer with 16 GB of RAM, 256 GB SSD, and 6th GEN Intel Core i5 or i7 processor is standard equipment in the hands of incoming undergraduates at Virginia Tech in 2016. This constitutes significant computing resources, adequate for solving many complex problems in physics and mechanics.

Continuum mechanics is founded on the conservation of mass, the principles of linear and angular momenta, and the first and second laws of thermodynamics. Mathematical formulations of those principles are a set of partial differential equations or integral statements for kinematic (displacement or velocity) and kinetic (stresses and energy) variables. Constitutive relations (e.g., Hooke's law and Fourier's law) characterize materials and provide relations among the kinematic and kinetic variables, thereby closing the set of equations. Boundary conditions in space and initial conditions in time must also be specified to obtain a unique solution. Various numerical methods have been developed to deal with certain classes of problems in each discipline of mechanics. Popular among these methods are panel methods (or boundary integral methods), finite difference, finite volume, and finite element methods. Here, we present the basic features of these methods.

Panel Methods

Panel methods are common for solving low-speed aerodynamic (e.g., [59]) and ship hydrodynamic problems (e.g., [77]), and their application to ship hull optimization is demonstrated in [90]. Panel methods belong to a wider class of methods known

DOI: 10.1201/9781003323150-1

as boundary-integral methods. Instead of solving for field variables, fundamental solutions are superimposed to satisfy the differential equation, and the boundary condition provides an integral equation for the strength of the fundamental solutions on the bounding surface. Thus, a three-dimensional problem is reduced to a two-dimensional problem. Similarly, a problem in the plane, which is two-dimensional, is reduced to a one-dimensional problem on a line. The surface or line is divided (discretized) into panels, and the unknown strength is defined over the panel in terms of its values at key points. The integral equation is then replaced by a system of algebraic equations for those values,

$$[A]\{U\} = \{b\}. \tag{1.1}$$

Finite-Difference Method

The basic feature of the finite-difference method is that derivatives of a function are approximated by differences of the functions at discrete points, thereby a differential equation is replaced by an algebraic system. If the differential equation is linear, the system will be linear as given by Equation (1.1). For nonlinear equations, iterative methods can be used to solve the system, which also involves multiple solutions of linear systems as Equation (1.1). The finite-difference method is very efficient for domains with simple geometry, and has been applied in heat transfer, fluid and solid mechanics. With coordinate transformation, the method can also be applied to simple configurations with curved boundaries (e.g., [51, 115]). An example of finite-difference method is given in Section 1.3.2.

Finite-Volume Method

The finite-volume method uses integral statements of the basic laws. The domain of interest is divided into small elements. Volume averages of different variables are defined, and the integral statements are reduced to algebraic systems that may be linear or nonlinear. Eventually, one has to solve systems such as Equation (1.1) perhaps many times. With the introduction of unstructured meshes, the method can be applied to complex geometries (e.g., [51, 115]).

Finite-Element Method–A Semi-Analytical Solution

Finite-element models are developed by applying certain procedures to integral statements. For example, Hamilton's principle, the principles of virtual displacements and minimum total potential energy are integral statements often used in dynamics and solid mechanics to formulate finite-element models. For problems defined by differential equations and boundary conditions, the method of weighted residuals can be used to generate integral statements. In those statements, there are two sets of variables: the first is the unknown set of variables such as displacements, velocity, pressure, and temperature that we want to determine. The second is a set of variations, virtual (generalized) displacements, or test (weight) functions. Roughly speaking, field variables (the first set) that satisfy the basic laws also make the integral statements valid for all admissible selections of the second set of variables. Thus, we assume a trial solution for the first set of unknowns $Q(\vec{r},t)$ in terms of selected functions of space (basis function $\Phi(\vec{r})$) and unknown coefficients $\{U(t)\}$,

$$Q(\vec{r},t) = [\Phi(\vec{r})]\{U(t)\}. \tag{1.2}$$

Then, we specify the second set of variables in terms of predefined functions of space $\Psi(\vec{r})$ and *known* coefficients $\{D\}$,

$$W(\vec{r},t) = [\Psi(\vec{r})]\{D\}. \tag{1.3}$$

We substitute these expressions into the integral statement, and since it has to hold for arbitrary $\{D\}$, it provides equations (differential in time, or algebraic for static problems) for the unknown coefficients. A typical form of the equations for a linear dynamic or transient problem is

$$[M]\{\ddot{U}\} + [C]\{\dot{U}\} + [A]\{U\} = \{F\} + \{R\} \tag{1.4}$$

where $[M]$, $[C]$, and $[A]$ are called mass, damping, and coefficient (stiffness) matrices, respectively, $\{F\}$ is a forcing vector, and $\{R\}$ is a vector of (secondary) variables that results from terms in the integral statement evaluated on the outer boundary of the problem domain. Its value is decided by the boundary conditions.

The major part of computational work is the computation of matrices and solving system Equation (1.4) for $\{U\}$. However, we emphasize that the finite-element solution is given by Equation (1.2), not just by $\{U\}$ or its values at discrete points in the field. As such, the finite-element method provides us with a well-defined procedure for evaluating the solution at any point in the domain. For that reason, we call it a semi-analytical solution rather than a totally numerical solution.

In the actual implementation of the method, the domain of interest is divided into contiguous non-overlapping elements, which makes it easy and natural, to deal with complex geometries of non-similar boundaries. Equally important is that the division of the domain into small elements facilitates the construction of trial solutions and virtual displacements (or test functions) that need to be constructed over the element domain only instead of the entire domain. The integral statement written for the element is then used to generate element equations that take the same general form as Equation (1.4),

$$[M^e]\{\ddot{U}^e\} + [C^e]\{\dot{U}^e\} + [A^e]\{U^e\} = \{F^e\} + \{R^e\}. \tag{1.5}$$

Now, the matrices and vectors are properties of the elements. Because boundary conditions are only known for the entire problem domain rather than for each element, we must assemble all elements into a global system. The assembly of matrices and arrays gives the global system Equation (1.4) as if the problem domain was never divided into elements.

In this book, we use the weak form of the differential equation or the principle of virtual displacements. Moreover, we use the Galerkin method for selecting the test (weight or virtual displacements) functions, and hence

$$[\Psi(\vec{r})] = [\Phi(\vec{r})]. \tag{1.6}$$

1.2 BRIEF MATHEMATICAL BACKGROUND: LINEAR ALGEBRA

1.2.1 VECTORS

A geometric vector is a mathematical representation of a physical quantity that has a magnitude and a direction. The force, momentum, and acceleration are familiar vector quantities. A vector quantity is denoted by writing a letter with a small arrow above the letter ($\vec{v}$). The vector may be specified by its projections on coordinate axes. In Cartesian coordinates, the position vector can be written as

$$\vec{r} = r_x\hat{i} + r_y\hat{j} + r_z\hat{k} \tag{1.7}$$

where $\hat{i}$, $\hat{j}$, and $\hat{k}$ are unit vectors in directions of the x, y, and z coordinates, respectively. This position vector can be written by listing its components in a row (i.e., vector row) as

$$\{r\} = \begin{bmatrix} r_x & r_y & r_z \end{bmatrix} \tag{1.8}$$

or in a column (i.e., vector column) as

$$\{r\} = \begin{bmatrix} r_x \\ r_y \\ r_z \end{bmatrix}. \tag{1.9}$$

Vector Addition

The two vectors $\vec{r} = r_x\hat{i} + r_y\hat{j} + r_z\hat{k}$ and $\vec{s} = s_x\hat{i} + s_y\hat{j} + s_z\hat{k}$ can be added (subtracted) by adding (subtracting) the respective components

$$\vec{r} \pm \vec{s} = (r_x \pm s_x)\hat{i} + (r_y \pm s_y)\hat{j} + (r_z \pm s_z)\hat{k}. \tag{1.10}$$

Vector Multiplication by a Scalar

The vector $\vec{r}$ can be multiplied by a scalar α, where each component of this vector is multiplied by the scalar quantity as

$$\alpha\vec{r} = \alpha r_x\hat{i} + \alpha r_y\hat{j} + \alpha r_z\hat{k}. \tag{1.11}$$

Multiplication of Two Vectors

The two vectors $\vec{r}$ and $\vec{s}$ can be multiplied in two different ways; *dot product* and *cross product*. The dot product of two vectors is a scalar quantity while the cross product is a new vector that is normal to the plane containing the two multiplied vectors $\vec{r}$ and $\vec{s}$. The dot product of these two vectors is given as

$$\vec{r} \cdot \vec{s} = r_x s_x + r_y s_y + r_z s_z \tag{1.12}$$

that is a scalar quantity.

The cross product of the two vectors $\vec{r}$ and $\vec{s}$ produces a third vector $\vec{w}$ normal to the plane made by the two multiplied vectors. The cross product of these two vectors is given by

$$\vec{r}\times\vec{s} = (r_y s_z - r_z s_y)\hat{\imath} - (r_x s_z - r_z s_x)\hat{\jmath}. + (r_x s_y - r_y s_x)\hat{k}. \tag{1.13}$$

It can be written as a determinant,

$$\vec{r}\times\vec{s} = \begin{vmatrix} \hat{\imath} & \hat{\jmath} & \hat{k} \\ r_x & r_y & r_z \\ s_x & s_y & s_z \end{vmatrix}.$$

Index Notation and Summation Convention

In index notation, the components of a geometric vector $\vec{r}$ can be written as r_i, for $i = 1,2,3$.

A repeated index indicates summation over the range of the index. For example, the dot product of two vectors in Equation (1.12) can be written as

$$\begin{aligned} \vec{r}\cdot\vec{s} &= r_1 s_1 + r_2 s_2 + r_3 s_3 \\ &= r_i s_i. \end{aligned} \tag{1.14}$$

Kronecker Delta δ_{ij} and Permutation Symbol ε_{ijk}

The Kronecker delta is defined such that $\delta_{ij} = 1$ if the index i is equal to the index j, and $\delta_{ij} = 0$ if $i \neq j$. Thus, $\delta_{11} = \delta_{22} = \delta_{33} = 1$ but $\delta_{12} = \delta_{13} = \delta_{23} = \delta_{21} = \delta_{31} = \delta_{32} = 0$. Note that $\delta_{jj} = \delta_{11} + \delta_{22} + \delta_{33} = 3$ because the index j is repeated.

The permutation symbol is defined such that $\varepsilon_{ijk} = 0$ if $i = j$, $j = k$, or $i = k$ (i.e., if any two indexes are equal). The value of $\varepsilon_{ijk} = 1$ if the three indexes are in the order of $1,2,3$ or $2,3,1$ or $3,1,2$, while the value of $\varepsilon_{ijk} = -1$ if the three indexes are in the order of $1,3,2$ or $3,2,1$ or $2,1,3$. The cross product of two vectors can be written in index notation using the permutation symbol. Let $\vec{v} = \vec{r}\times\vec{s}$, then in index notation we have

$$v_i = \varepsilon_{jik} r_k s_j. \tag{1.15}$$

1.2.2 MATRICES

A matrix is a two-dimensional array of data in the form of rows and columns. The dimension (or size) of the matrix is given by the number of rows and columns. An $(m \times n)$ matrix has m rows and n columns as given by

$$A = \begin{bmatrix} a_{11} & a_{12} & a_{13} & \dots & a_{1n} \\ a_{21} & a_{22} & a_{23} & \dots & a_{2n} \\ \dots & \dots & \dots & \dots & \dots \\ a_{m1} & a_{m2} & a_{m3} & \dots & a_{mn} \end{bmatrix}.$$

The matrix $[A]$ can be viewed as a set of row or column vectors. The row vector is a special case of a matrix where $m = 1$,

$$\{r\} = \begin{bmatrix} r_1 & r_2 & r_3 \ldots & r_n \end{bmatrix} \tag{1.16}$$

and a column vector is a special matrix which has $n = 1$,

$$\{c\} = \begin{bmatrix} c_1 \\ c_2 \\ \ldots \\ c_m \end{bmatrix}. \tag{1.17}$$

The entries of the matrix $[A]$ can be written using index notation as a_{ij} where $i = 1,2,..,m$ and $j = 1,2,3,..,n$.

The matrix $[A]$ can be multiplied by a scalar quantity α. Let $[B] = \alpha[A]$, then

$$b_{ij=} = \alpha a_{ij}. \tag{1.18}$$

Addition of Matrices

Two matrices can be added or subtracted only if they have the same dimensions (i.e., sizes). This means that matrix $[A]$ and matrix $[B]$ can be added or subtracted if they have the same number of rows and columns. Let $[C] = [A] \pm [B]$. Then the entries of $[C]$ are

$$c_{ij} = a_{ij} \pm b_{ij} \tag{1.19}$$

where the resulting matrix $[C]$ has the same size of matrices $[A]$ and $[B]$.

Transpose of a Matrix

The transpose of a matrix $[A]$ is obtained by turning the rows into columns, which is the same process as switching the columns into rows. Let $[B] = [A]^T$, then

$$b_{ij} = a_{ji}. \tag{1.20}$$

If the dimension of $[A]$ is $(m \times n)$, then its transpose is of dimension $(n \times m)$. For example, the transpose of a 2×3 matrix

$$A = \begin{bmatrix} a_{11} & a_{12} & a_{13} \\ a_{21} & a_{22} & a_{23} \end{bmatrix}$$

is a matrix $[A]^T$ with dimensions 3×2 given as

$$[A]^T = \begin{bmatrix} a_{11} & a_{21} \\ a_{12} & a_{22} \\ a_{13} & a_{23} \end{bmatrix}.$$

Matrix Multiplication

The multiplication of two matrices $[A]$ and $[B]$ is only defined when the number of columns of $[A]$ is equal to the number of rows of $[B]$ without any restrictions on the number of columns and rows of matrices $[A]$ and $[B]$, respectively. The resultant matrix $[C]$ has the same number of rows of matrix $[A]$ and number of columns of matrix $[B]$. Therefore, if matrix $[A]$ has dimensions $m \times n$, then matrix $[B]$ must have the dimensions $n \times p$. The multiplication of these two matrices produces a matrix $[C]$ that has dimensions of $m \times p$. As an example, the multiplication of the matrix

$$[A] = \begin{bmatrix} a_{11} & a_{12} \\ a_{21} & a_{22} \\ a_{31} & a_{32} \end{bmatrix}$$

and matrix

$$[B] = \begin{bmatrix} b_{11} & b_{12} & b_{13} \\ b_{21} & b_{22} & b_{23} \end{bmatrix}$$

is a matrix $[C]$ that is written as

$$[C] = \begin{bmatrix} a_{11} & a_{12} \\ a_{21} & a_{22} \\ a_{31} & a_{32} \end{bmatrix} \begin{bmatrix} b_{11} & b_{12} & b_{13} \\ b_{21} & b_{22} & b_{23} \end{bmatrix}$$

$$[C] = \begin{bmatrix} a_{11} \times b_{11} + a_{12} \times b_{21} & a_{11} \times b_{12} + a_{12} \times b_{22} & a_{11} \times b_{13} + a_{12} \times b_{23} \\ a_{21} \times b_{11} + a_{22} \times b_{21} & a_{21} \times b_{12} + a_{21} \times b_{22} & a_{21} \times b_{13} + a_{22} \times b_{23} \\ a_{31} \times b_{11} + a_{32} \times b_{21} & a_{31} \times b_{12} + a_{31} \times b_{22} & a_{31} \times b_{13} + a_{32} \times b_{23} \end{bmatrix},$$

which is a 3×3 matrix.

The multiplication of two matrices $[A]$ and $[B]$

$$[C] = [A][B]$$

is given in the index form as

$$c_{ij} = a_{ik} b_{kj}. \tag{1.21}$$

Note that index k is repeated, so it is summed over the range 1 to n, where n is the number of columns of $[A]$ that is equal to the number of rows of $[B]$.

The transposition of products follows the rules

$$([A][B])^T = [B]^T [A]^T \tag{1.22}$$

$$\begin{aligned} ([A][B][C])^T &= ([B][C])^T [A]^T \\ &= [C]^T [B]^T [A]^T. \end{aligned} \tag{1.23}$$

The product of a $(1 \times n)$ row vector $\{r\}$ and a $(n \times 1)$ column vector $\{c\}$ produces a scalar

$$w = \{r\}\{c\}.$$

The scalar w is equal to its transpose, hence we can also write w as

$$w = \{c\}^T \{r\}^T.$$

Special Matrices

There are matrices with a special data structure. The following is a list of these matrices with a short description.

Square Matrix

A matrix $[A]$ is said to be a square matrix if the number of rows equals the number of columns. This matrix is of the dimensions of $n \times n$. In such a matrix, the data or elements along the diagonal such as a_{11}, a_{22} ... and a_{nn} are called *diagonal elements*. The elements a_{ij} where the index j is larger than the index i are called *super-diagonal* while the elements that have index j smaller than the index i are called *sub-diagonal*.

Diagonal Matrix

A square matrix is called diagonal if all *off-diagonal* elements are zeros and *diagonal* elements are non-zero. This matrix is denoted by $[D]$.

Upper-Triangular Matrix

The matrix $[A]$ is denoted as an *upper-triangular* matrix if the *sub-diagonal* elements are zeros and the *super-diagonal* elements are non-zero.

Lower-Triangular Matrix

The matrix $[A]$ is denoted as a *lower-triangular* matrix if the *super-diagonal* elements are zeros and the *sub-diagonal* elements are non-zero.

Identity Matrix

The identity matrix is denoted by $[I]$ where all *diagonal* elements are ones and all *off-diagonal* elements are zeros. The multiplication of any other matrix with the identity matrix produces the same matrix.

Null Matrix

A matrix is denoted as a *null* matrix if all elements are zeros.

Symmetric and Hermitian Matrices

A square real matrix $[A]$ is symmetric if the matrix is equal to its transpose

$$a_{ij} = a_{ji}.$$

A complex-valued matrix $[A]$ is said to be Hermitian if the matrix is equal to the complex conjugate of its transpose.

$$a_{ij} = a_{ji}^*.$$

Skew-Symmetric Matrix

A matrix $[A]$ is a skew-symmetric matrix if the matrix is equal to the negative of its transpose, $[A] = -[A]^T$.

Determinant of a Matrix

The determinant of a matrix is defined for square matrices. For a 2×2 matrix

$$[A] = \begin{bmatrix} a_{11} & a_{12} \\ a_{21} & a_{22} \end{bmatrix}$$

the determinant is

$$|A| = a_{11}a_{22} - a_{12}a_{21}. \tag{1.24}$$

For a 3×3 matrix

$$[A] = \begin{bmatrix} a_{11} & a_{12} & a_{13} \\ a_{21} & a_{22} & a_{23} \\ a_{31} & a_{32} & a_{33} \end{bmatrix}$$

the determinant is

$$|A| = a_{11}\begin{vmatrix} a_{22} & a_{23} \\ a_{32} & a_{33} \end{vmatrix} - a_{12}\begin{vmatrix} a_{21} & a_{23} \\ a_{31} & a_{33} \end{vmatrix} + a_{13}\begin{vmatrix} a_{21} & a_{22} \\ a_{31} & a_{32} \end{vmatrix} \tag{1.25}$$

$$|A| = a_{11}(a_{22}a_{33} - a_{23}a_{32}) - a_{12}(a_{21}a_{33} - a_{23}a_{31}) + a_{13}(a_{21}a_{32} - a_{22}a_{31}).$$

For a 4×4 matrix

$$[A] = \begin{bmatrix} a_{11} & a_{12} & a_{13} & a_{14} \\ a_{21} & a_{22} & a_{23} & a_{24} \\ a_{31} & a_{32} & a_{33} & a_{34} \\ a_{41} & a_{42} & a_{43} & a_{44} \end{bmatrix}$$

the determinant is

$$\begin{aligned} |A| = a_{11}&\begin{vmatrix} a_{22} & a_{23} & a_{24} \\ a_{32} & a_{33} & a_{34} \\ a_{42} & a_{43} & a_{44} \end{vmatrix} - a_{12}\begin{vmatrix} a_{21} & a_{23} & a_{24} \\ a_{31} & a_{33} & a_{34} \\ a_{41} & a_{43} & a_{44} \end{vmatrix} \\ +a_{13}&\begin{vmatrix} a_{21} & a_{22} & a_{24} \\ a_{31} & a_{32} & a_{34} \\ a_{41} & a_{42} & a_{44} \end{vmatrix} - a_{14}\begin{vmatrix} a_{21} & a_{22} & a_{23} \\ a_{31} & a_{32} & a_{33} \\ a_{41} & a_{42} & a_{43} \end{vmatrix}. \end{aligned} \tag{1.26}$$

The 3×3 determinants in the above equation are evaluated using Equation (1.25).

Inverse of a Matrix

The inverse of a matrix $[A]$ is a matrix $[B]$ such that the multiplication of these two matrices results in the *identity* matrix as

$$[A][B] = I$$

or

$$[B] = [A]^{-1}.$$

The entries of $[B]$ in index notation are

$$b_{ij} = \frac{1}{2|A|}\varepsilon_{jkl}\varepsilon_{imn}a_{km}a_{ln}. \tag{1.27}$$

If the determinant of a matrix is zero, $|A| = 0$, the matrix is said to be singular and its inverse does not exist.

A System of Linear Algebraic Equations

Let $\{x\}$ and $\{b\}$ be $(n \times 1)$ column vectors, and $[A]$ be $(n \times n)$ square matrix. A system of linear equations

$$\begin{aligned} a_{11}x_1 + a_{11}x_2 + \cdots + a_{1n}x_n &= b_1 \\ a_{21}x_1 + a_{22}x_2 + \cdots + a_{2n}x_n &= b_2 \\ &\cdots \\ a_{n1}x_1 + a_{n2}x_1 + \cdots + a_{nn}x_n &= b_n \end{aligned} \tag{1.28}$$

can be written in matrix form

$$[A]\{x\} = \{b\}. \tag{1.29}$$

If $[A]$ is not singular, we multiply both sides by the inverse of $[A]$, and obtain the solution for $\{x\}$

$$\{x\} = [A]^{-1}\{b\}. \tag{1.30}$$

Eigenvalues and Eigenvectors

The eigenvalues and eigenvectors of a matrix have special importance in engineering science and mechanics. As an example in vibration problems, the eigenvalues represent the natural frequencies of the vibrating system and eigenvectors are the modes of oscillations. Another example can be found in the strength of material where the eigenvalues are the principle stresses of the stress tensor and the eigenvectors are the principle directions of the associated stresses. Also, the eigenvalues of a matrix provide important information about the convergence of iterative methods for solving systems of equations such as Jacobi and Gauss–Seidel methods.

For a scalar λ and a square matrix $[A]$ of dimension $n \times n$, we consider the system

$$[A]\{u\} = \lambda\{u\} \tag{1.31}$$

for the unknown column vector $\{u\}$ of dimension $n \times 1$. We write the system as

$$[A - \lambda I]\{u\} = \{0\}. \tag{1.32}$$

It is obvious that the trivial solution $\{u\} = \{0\}$ satisfies Equation (1.32) for any value of λ. We are interested in non-trivial solutions of the system. Such solutions exist only if the parameter λ takes certain values that make the coefficient matrix $[A - \lambda I]$ singular; that is to say, the determinant of the coefficient matrix is zero,

$$|A - \lambda I| = 0. \tag{1.33}$$

The specific values of λ are called the eigenvalues of the matrix $[A]$. Equation (1.33) is called the *characteristic equation.* This equation yields a polynomial of order n equal to the dimension of matrix $[A]$, which has n roots. There are n eigenvalues λ_i, $i = 1, 2, ..., n$, and for each λ_i there is a solution $\{u\}_i$, which is the corresponding eigenvector. Once the roots (i.e., eigenvalues λ_i) are known, the corresponding eigenvectors are determined by direct substitution in Equation (1.31). For small matrix $[A]$, the eigenvalues λ can be determined directly by calculating the determinant and solving an nth-order polynomial. However, for large matrices, it will be difficult to solve this polynomial. Therefore, numerical methods such as the *power method* and *QR factorization* are recommended to find λ.

Eigenvalues of a Hermitian Matrix

The eigenvalues of a Hermitian matrix are all real. A real symmetric matrix is a special case of a Hermitian matrix.

Positive Definite Matrix

A real symmetric matrix whose eigenvalues are strictly positive is called a positive definite matrix.

In MATLAB, the eigenvalues of a matrix $[A]$ can be found by implementing a built-in function $eig(A)$ that performs QR-factorization. If only the eigenvalues are needed, then the following built-in function is used:

$$\lambda = eig(A) \tag{1.34}$$

where λ is a vector containing the eigenvalues of matrix $[A]$. To determine the eigenvalues and the eigenvectors, the following built-in function is used

$$[V, \lambda] = eig(A). \tag{1.35}$$

1.3 BRIEF MATHEMATICAL BACKGROUND: PARTIAL DIFFERENTIAL EQUATIONS

Conservation of mass, the principles of linear and angular momenta, and the first and second laws of thermodynamics are the fundamental laws of continuum mechanics. Mathematical formulation of the basic principles give rise to partial differential equations in terms of kinematic variables (velocity or displacement), thermodynamic variables (temperature, pressure, density), and sources of momentum and energy (forces and heat flux). Multiple physical phenomena (convection, diffusion, wave propagation, equilibrium) can take place simultaneously in complex problems, but with practical simplifying assumptions a dominant phenomenon gets emphasized and isolated. Simplified models highlight the mathematical character of a physical phenomenon and help develop methods of solution and formulate boundary and initial conditions.

We consider a linear second-order partial differential equation for the unknown function u of two independent variables (x,y),

$$p_{11}\frac{\partial^2 u}{\partial x^2}+2p_{12}\frac{\partial^2 u}{\partial x\partial y}+p_{22}\frac{\partial^2 u}{\partial y^2}+p_{10}\frac{\partial u}{\partial x}+p_{02}\frac{\partial u}{\partial y}+p_{00}u+f=0 \tag{1.36}$$

where $p_{11}, p_{12}, p_{22}, p_{10}, p_{02}, p_{00}$, and f are given functions of x and y. Equation (1.36) is classified as *elliptic*, *parabolic* or *hyperbolic* if the term $(p_{12}^2 - p_{11}p_{22})$ is negative, zero, or positive, respectively (e.g., [40]). We note that only the coefficients of the highest derivatives of u (the first three terms) determine the type of the equation. The standard forms of the three types are

Elliptic:
$$\frac{\partial^2 u}{\partial x^2}+\frac{\partial^2 u}{\partial y^2}=0 \tag{1.37}$$

Parabolic:
$$\frac{\partial u}{\partial t}-\frac{\partial^2 u}{\partial x^2}=0 \tag{1.38}$$

Hyperbolic:
$$\frac{\partial^2 u}{\partial t^2}-\frac{\partial^2 u}{\partial x^2}=0. \tag{1.39}$$

These are Laplace's equation, the heat, and the wave equations, respectively. In the second and third equations, we changed the independent y to t, which usually denotes time. Next, we formulate and solve problems in solid mechanics, transient heat transfer, and elastic waves as typical examples of elliptic, parabolic, and hyperbolic equations.

1.3.1 ELLIPTIC EQUATIONS: POISSON'S EQUATION

The equation

$$\nabla^2 u = f \tag{1.40}$$

where f is a prescribed function, is the well-known Poisson's equation. It is a typical elliptic equation. In two dimensions, it reads

$$\frac{\partial^2 u}{\partial x^2}+\frac{\partial^2 u}{\partial y^2}=f(x,y). \tag{1.41}$$

Poisson's equation has wide applications in physics and engineering sciences including torsion of prismatic bars of arbitrary cross-section, laminar viscous flow in ducts,

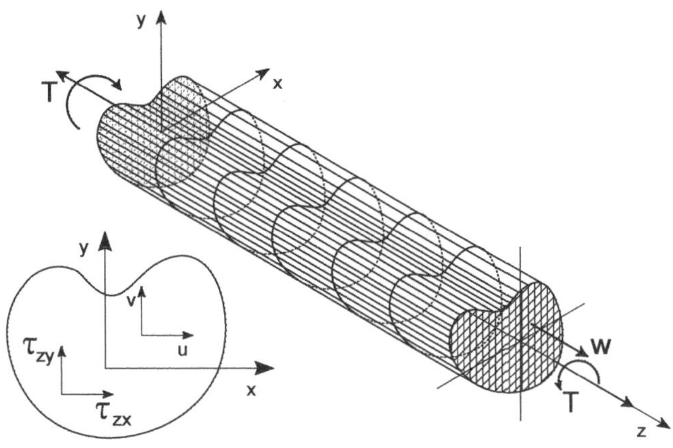

Figure 1.1 A non-circular prismatic bar twisted by two equal and opposite torques T. Displacements (u, v, w), and shear stresses (τ_{zx}, τ_{zy}).

heat conduction in solids, flow in porous media, and electrostatics, among other applications. The case of $f = 0$ is Laplace's equation, which describes potential flow just to name an example. Here, we consider the torsion problem. We want to determine the stresses in an elastic prismatic bar of an arbitrary cross-section twisted by two equal and opposite torques applied at its ends as shown in Figure 1.1. Boresi [14] presented approaches for solving the problem for solid sections (simply connected domains) and tubular sections (multiply connected domains); here we consider only solid sections. The twist angle per unit length is denoted by θ, and the shear modulus of the material is G. The twist axis is an axis parallel to the prism generatrix, which is taken as the z-axis, and the cross-section is in the x-y plane. Boresi [14] showed that if the twist axis is transferred to another axis parallel to the z-axis, the displacement field differs by a rigid body displacement, which has no effects on the stresses and torque. The Cartesian components of displacements are denoted by (u, v, w). The displacement in the x-y plane is a rigid body rotation about the z-axis; that is

$$u(x,y,z) = -\theta zy, \quad v(x,y,z) = \theta zx. \tag{1.42}$$

A non-circular section also warps, so there is a nonuniform displacement in the z-direction $w(x,y)$ assumed to be a function of x and y only. The linear strain tensor is

$$\varepsilon_{ij} = \frac{1}{2}\left(\frac{\partial u_i}{\partial x_j} + \frac{\partial u_j}{\partial x_i}\right). \tag{1.43}$$

For the assumed displacement field, we find the components of strains

$$\begin{gathered} \varepsilon_{xx} = \varepsilon_{yy} = \varepsilon_{zz} = 0, \quad \gamma_{xy} \equiv 2\varepsilon_{xy} = 0 \\ \gamma_{zx} \equiv 2\varepsilon_{zx} = \frac{\partial w}{\partial x} - \theta y, \quad \gamma_{zy} \equiv 2\varepsilon_{zy} = \frac{\partial w}{\partial y} + \theta x. \end{gathered} \tag{1.44}$$

For a linear elastic isotropic material the constitutive equation is

$$\sigma_{ij} = \lambda \varepsilon_{kk}\delta_{ij} + 2\mu\varepsilon_{ij} \tag{1.45}$$

where ε_{ij} is the linear strain tensor, and λ and $\mu = G$ are the Lamé elastic constants. In terms of Young's modulus E and Poisson's ratio ν, we have

$$\lambda = \frac{\nu E}{(1+\nu)(1-2\nu)}, \; G = \frac{E}{2(1+\nu)}. \tag{1.46}$$

The normal stresses are

$$\sigma_{xx} = \sigma_{yy} = \sigma_{zz} = 0 \tag{1.47}$$

and shear stresses are

$$\tau_{xy} = 0, \; \tau_{zx} = G\left(\frac{\partial w}{\partial x} - \theta y\right), \; \tau_{zy} = G\left(\frac{\partial w}{\partial y} + \theta x\right). \tag{1.48}$$

The equations of equilibrium are

$$\begin{aligned} \frac{\partial \sigma_{xx}}{\partial x} + \frac{\partial \tau_{yx}}{\partial y} + \frac{\partial \tau_{zx}}{\partial z} + f_x &= 0 \\ \frac{\partial \tau_{xy}}{\partial x} + \frac{\partial \sigma_{yy}}{\partial y} + \frac{\partial \tau_{zy}}{\partial z} + f_y &= 0 \\ \frac{\partial \tau_{xz}}{\partial x} + \frac{\partial \tau_{yz}}{\partial y} + \frac{\partial \sigma_{zz}}{\partial z} + f_z &= 0. \end{aligned} \tag{1.49}$$

We neglect the body forces (f_x, f_y, f_z), and invoke Equations (1.47) and (1.48),

$$\begin{aligned} \frac{\partial \tau_{zx}}{\partial z} &= 0 \\ \frac{\partial \tau_{zy}}{\partial z} &= 0 \\ \frac{\partial \tau_{xz}}{\partial x} + \frac{\partial \tau_{yz}}{\partial y} &= 0. \end{aligned} \tag{1.50}$$

The first two equations of equilibrium are readily satisfied by the assumed form of displacement field and associated strains and stresses Equations (1.47) and (1.48), and the third equation of equilibrium becomes an equation for the warping function,

$$\frac{\partial^2 w}{\partial x^2} + \frac{\partial^2 w}{\partial y^2} = 0. \tag{1.51}$$

Boundary conditions can be formulated for w (e.g., Boresi [14]). However, the introduction of the Prandtl stress function $\phi(x,y)$ offers a simpler formulation. The non-zero shear stresses are defined as

$$\tau_{zx} = \frac{\partial \phi}{\partial y}, \; \tau_{zy} = -\frac{\partial \phi}{\partial x}. \tag{1.52}$$

With these definitions, the third equation of equilibrium is automatically satisfied. Relations between w and ϕ follow from Equations (1.48) and (1.52),

$$\begin{aligned} \frac{\partial \phi}{\partial y} &= G\left(\frac{\partial w}{\partial x} - \theta y\right) \\ \frac{\partial \phi}{\partial x} &= -G\left(\frac{\partial w}{\partial y} + \theta x\right). \end{aligned} \tag{1.53}$$

The consistency condition

$$\frac{\partial^2 w}{\partial x \partial y} = \frac{\partial^2 w}{\partial y \partial x} \tag{1.54}$$

provides the Poisson's equation for Prandtl stress function ϕ

$$-\left(\frac{\partial^2 \phi}{\partial x^2} + \frac{\partial^2 \phi}{\partial y^2}\right) = 2G\theta. \tag{1.55}$$

The surface traction (stress vector) on the bounding surface of the prismatic bar is zero. Let $\hat{n}$ be a unit normal to the bar surface. We consider the shear stresses in the cross-section (which is normal to the z-axis) and contains the unit vector $\hat{n}$. At the point where $\hat{n}$ intersects the cross-section boundary, the shear stress component in the direction of $\hat{n}$ must be zero; in other words, the shear stress vector $\vec{\tau}$ is tangent to the boundary. The unit normal can be written as

$$\begin{aligned} \hat{n} &= cos(\hat{n},x)\hat{i} + cos(\hat{n},y)\hat{j} \\ &= \frac{dy}{ds}\hat{i} - \frac{dx}{ds}\hat{j} \end{aligned} \tag{1.56}$$

where s is an arc length measured along the cross-section boundary. The boundary condition is

$$\begin{aligned} \vec{\tau}\cdot\hat{n} &= 0 \\ \tau_{zx}\frac{dy}{ds} - \tau_{zy}\frac{dx}{ds} &= 0 \\ \frac{\partial \phi}{\partial y}\frac{dy}{ds} + \frac{\partial \phi}{\partial x}\frac{dx}{ds} &= 0 \\ \frac{d\phi}{ds} &= 0. \end{aligned} \tag{1.57}$$

Thus, Prandtl stress function is constant on the boundary; $\phi = \phi_o$. We set $\phi = 0$ on the boundary since adding a constant will have no effects on the stresses. We solve Poisson's equation for ϕ, and determine stresses on the cross section from Equation (1.52). The warping function $w(x,y)$ can be determined by integrating the equations

$$\begin{aligned} \frac{\partial w}{\partial x} &= \theta y + \frac{1}{G}\frac{\partial \phi}{\partial y} \\ \frac{\partial w}{\partial y} &= -\left(\theta x + \frac{1}{G}\frac{\partial \phi}{\partial x}\right). \end{aligned} \tag{1.58}$$

Integrating the shear stresses on the cross-section, we get

$$\begin{aligned} F_x &= \int\int \tau_{zx} dxdy \\ &= \int\int \frac{\partial \phi}{\partial y} dxdy \\ &= 0 \\ F_y &= \int\int \tau_{zy} dxdy \\ &= \int\int -\frac{\partial \phi}{\partial x} dxdy \\ &= 0 \end{aligned} \tag{1.59}$$

because ϕ is constant on the boundary. The moment of the distributed stresses about the z-axis is the torque T,

$$\begin{aligned} T &= \int\int (-y\tau_{zx} + x\tau_{zy}) dxdy \\ &= \int\int \left(-y\frac{\partial \phi}{\partial y} - x\frac{\partial \phi}{\partial x}\right) dxdy \\ &= 2\int\int (\phi - \phi_o) dxdy \end{aligned} \tag{1.60}$$

where ϕ_o the value of ϕ on the boundary, which we take to be zero. Thus, the torque is

$$T = 2\int \phi dA. \tag{1.61}$$

Example: Torsion of a Rectangular Cross-Section

We consider a rectangular section of dimensions $2a$ and $2b$. Solution of Poisson's equation is determined by the method of separation of variables (e.g., [17]). Taking the origin at the rectangle center, and defining nondimensional coordinates $\xi = x/a$, and $\eta = y/b$, we find the Prandtl stress function

$$\frac{\phi(\xi,\eta)}{G\theta a^2} = 1 - \xi^2 - 4\sum_{m=0}^{\infty} \frac{(-1)^m cosh\left(k_m \frac{b}{a}\eta\right) cos(k_m\xi)}{k_m^3 cosh(k_m \frac{b}{a})} \tag{1.62}$$

where $k_m = (2m+1)\pi/2$. The shear stresses are

$$\frac{\tau_{zx}(\xi,\eta)}{G\theta a} = -4\sum_{m=0}^{\infty} \frac{(-1)^m sinh\left(k_m \frac{b}{a}\eta\right) cos(k_m\xi)}{k_m^2 cosh(k_m \frac{b}{a})} \tag{1.63}$$

$$\frac{\tau_{zy}(\xi,\eta)}{G\theta a} = 2\xi - 4\sum_{m=0}^{\infty} \frac{(-1)^m cosh\left(k_m \frac{b}{a}\eta\right) sin(k_m\xi)}{k_m^2 cosh(k_m \frac{b}{a})}. \tag{1.64}$$

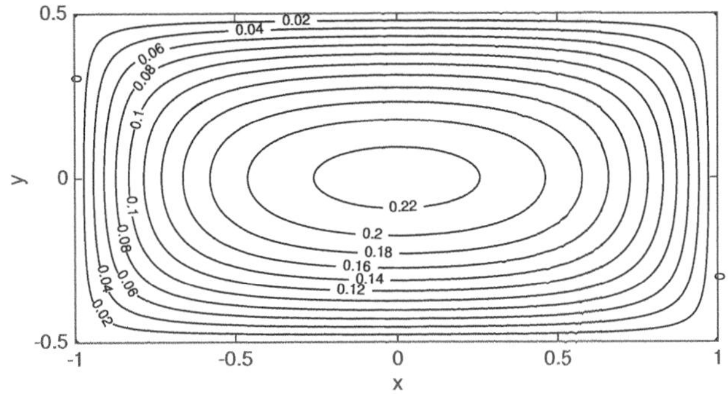

Figure 1.2 Torsion of rectangular section. Contours of Prandtl stress function ϕ.

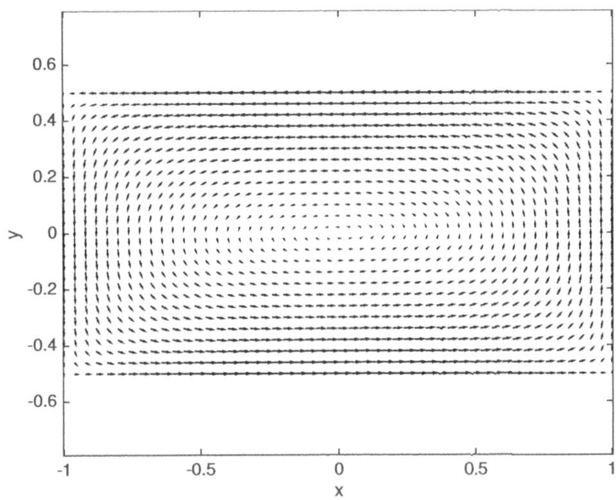

Figure 1.3 Torsion of rectangular section. Shear stress vectors on cross-section.

The warping function is

$$\frac{w(\xi,\eta)}{\theta a^2} = \frac{b}{a}\xi\eta - 4\sum_{m=0}^{\infty}\frac{(-1)^m sinh\left(k_m\frac{b}{a}\eta\right) sin(k_m\xi)}{k_m^3 cosh(k_m\frac{b}{a})} \tag{1.65}$$

and the torque T is given by

$$\frac{T}{G\theta a^4} = 16\left[\frac{b}{3a} - 2\sum_{m=0}^{\infty}\frac{tanh(k_m\frac{b}{a})}{k_m^5}\right]. \tag{1.66}$$

We evaluated the different quantities given by Equations (1.62)-(1.66) for a rectangular section of dimension $2a$ and $2b = a$. Contours of Prandtl stress function are shown in Figure 1.2; we note the contours double symmetry and that $\phi_{max} = 0.227743664$ at the center. Shear stress vectors in the section are plotted in Figure 1.3; they are equivalent to a couple of moment $T = 0.457363354G\theta a^4$. Contours of shear stresses τ_{zx}

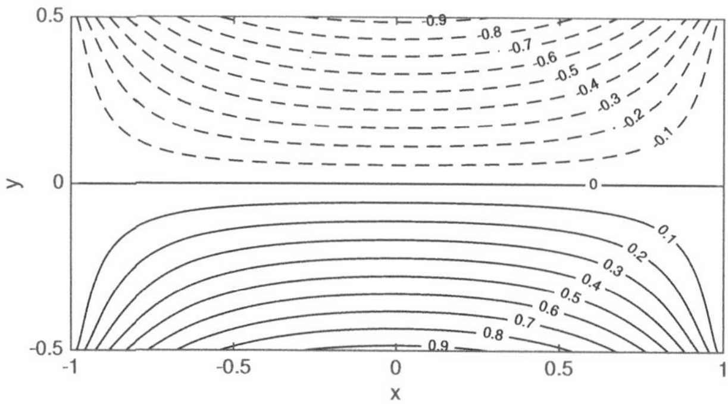

Figure 1.4 Torsion of a rectangular section. Contours of the shear stress component τ_{zx}.

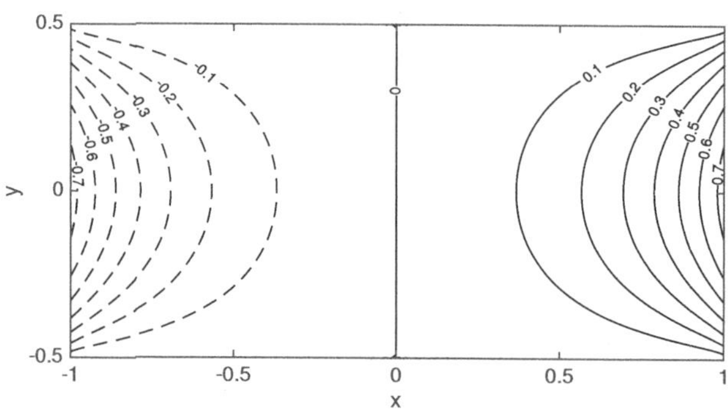

Figure 1.5 Torsion of a rectangular section. Contours of the shear stress component τ_{zy}.

and τ_{zy} are shown in Figures 1.4 and 1.5, respectively. The shear is maximum at the midpoint of the long side of the rectangle. Warping function w (displacement in the z-direction) contours are plotted in Figure 1.6. With the torque in the counterclockwise direction (torque vector is in the z-direction), the displacement is positive in the second and fourth quadrants, and negative in the other two quadrants.

1.3.2 PARABOLIC EQUATIONS: HEAT EQUATION

We start with a simple case of one-dimensional heat conduction in which the temperature field T varies only in one space direction x and time t. The heat flux q_x is defined as the rate of heat flow in the x-direction per unit area; the area being normal to the x-axis as shown in Figure 1.7. We assume q_x to be proportional to the local temperature gradient $\partial T/\partial x$,

$$q_x = -k\frac{\partial T}{\partial x} \tag{1.67}$$

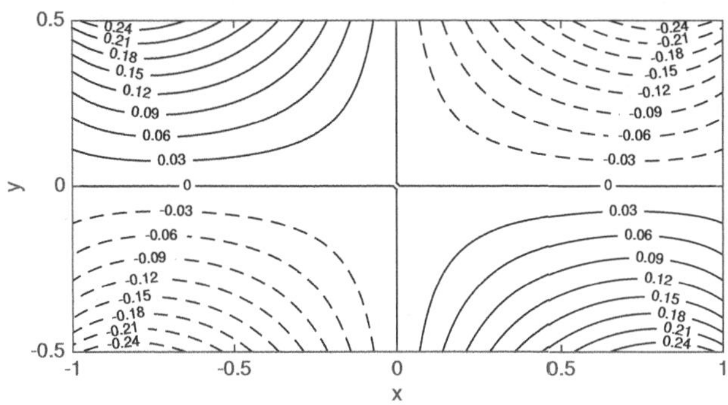

Figure 1.6 Torsion of a rectangular section. Warping function, $w(x,y)$.

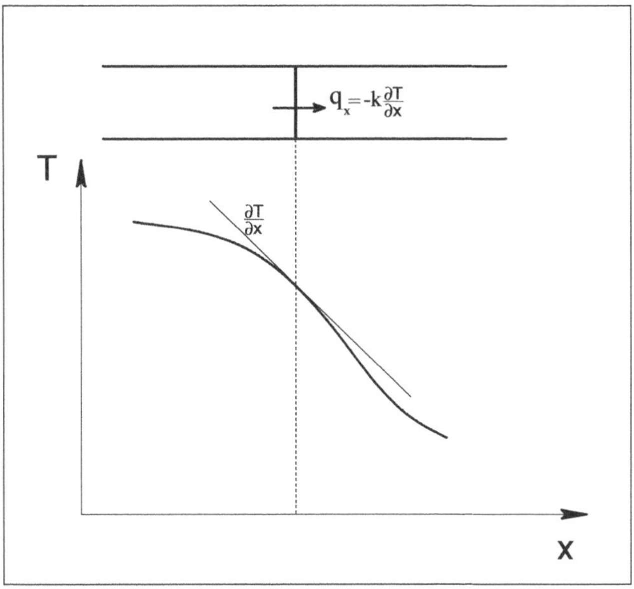

Figure 1.7 Heat flux vector q_x in one-dimensional heat conduction.

where k is the thermal conductivity. Equation (1.67) is known as Fourier's law of heat conduction, it is a constitutive equation that is valid for solids, liquids, and gases. The dimensions of q_x are [energy/(area.time)] = $[MLT^{-2}L/(L^2T)] = [MT^{-3}]$ and its unit is W/m^2. The dimensions of k are $[k] = [MT^{-3}/(\theta/L)] = [MLT^{-3}/\theta]$, where $[\theta]$ is the dimension of temperature. The unit for k is W/m $^\circ K$. The thermal conductivity depends on the atomic and molecular structure of matter. It varies with temperature and, for liquids and gases, may also depend on pressure. Typical values of k (W/m$^\circ K$) at 300 $^\circ K$ are: 237, 401, and 317 for pure aluminum, copper, and gold, respectively. And values of k are 0.613, 0.0196, and 0.0263 for saturated water (0.03531 bar), saturated steam (0.03531 bar), and air (1 atm), respectively.

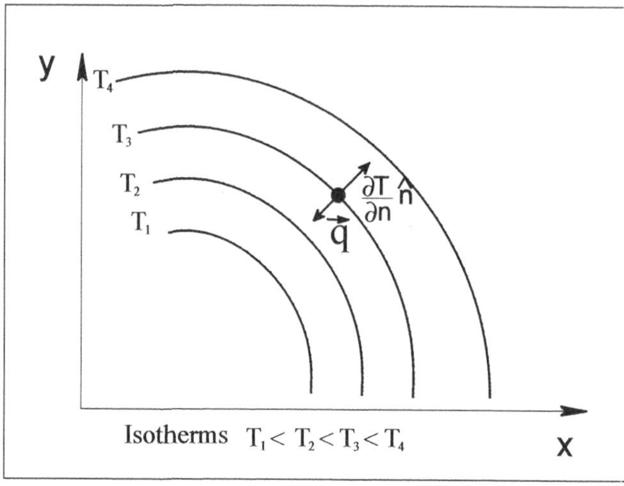

Figure 1.8 In three dimensions the heat flux vector $\vec{q}$ is opposite to the local temperature gradient.

In three dimensions the heat flux vector $\vec{q}$ is locally normal to the isotherms, which are surfaces of constant temperature as shown in Figure 1.8. For isotropic materials, we generalize Equation (1.67) to

$$\vec{q} = -k\frac{\partial T}{\partial n}\hat{n} \tag{1.68}$$

where $\hat{n}$ is a unit vector normal to isotherms and n is the distance measured along $\hat{n}$. Thus, the heat flux vector is opposite to the local temperature gradient ($grad\ T = \partial T/\partial n\ \hat{n}$) if k is positive, as required by the second law of thermodynamics. Another form of Equation (1.68) is

$$\vec{q} = -k\nabla T \tag{1.69}$$

where ∇T in rectangular Cartesian coordinates is defined by

$$\nabla T = \frac{\partial T}{\partial x}\hat{i} + \frac{\partial T}{\partial y}\hat{j} + \frac{\partial T}{\partial z}\hat{k}. \tag{1.70}$$

The Cartesian components of the heat flux vector are $q_x = -k\ \partial T/\partial x$, $q_y = -k\ \partial T/\partial y$, and $q_z = -k\ \partial T/\partial z$. The heat flow over an element of area dA identified by a unit normal $\hat{m}$ is $d\dot{Q} = \vec{q}\cdot\hat{m}dA$ as shown in Figure 1.9. Note that $\vec{q}\cdot\hat{m}$ is the projection of the heat flux vector onto the unit normal $\hat{m}$. If dA is tangential to an isotherm, the heat flow $d\dot{Q}$ will have maximum magnitude, whereas if it is normal to an isotherm, the flow will be zero.

In order to determine the temperature distribution in a continuum, we need to formulate the principle of conservation of energy (first law of thermodynamics). Here, we consider a special case of a medium with negligible mean (bulk) motion. For an

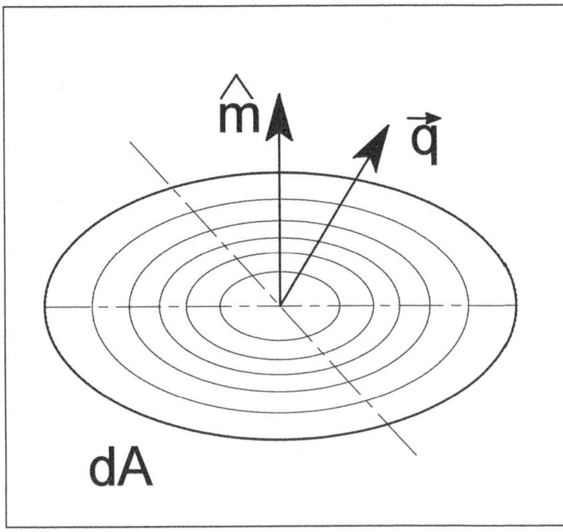

Figure 1.9 Heat flux over an element of area $d\dot{Q} = \vec{q} \cdot \hat{m} dA$.

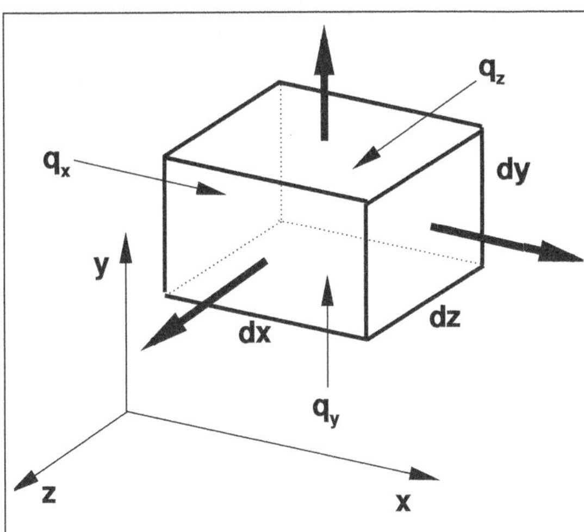

Figure 1.10 Heat flux over an element of volume $dV = dxdydz$.

infinitesimal element with dimensions (dx, dy, dz) as shown in Figure 1.10, the net rate of heat transferred by conduction out of volume is

$$d\dot{Q} = (q_x + \frac{\partial q_x}{\partial x} dx) dydz - q_x dydz$$
$$+ (q_y + \frac{\partial q_y}{\partial y} dy) dxdz - q_y dxdz$$

$$
\begin{aligned}
&+(q_z+\frac{\partial q_z}{\partial z}dz)dxdy-q_zdxdy\\
&=(\frac{\partial q_x}{\partial x}+\frac{\partial q_y}{\partial y}+\frac{\partial q_z}{\partial z})dxdydz\\
&=\nabla\cdot\vec{q}dV
\end{aligned}
$$

where $dV = dxdydz$ is the element volume. Thermal energy within the element may also be generated by electrical, chemical, or nuclear processes. Let f denote the rate of energy generated per unit volume. The conservation of energy principle states that the rate of increase of internal energy of the matter within the volume is equal to the rate of energy generated minus the rate of heat conducted out of the volume.

$$\rho C_p dV\frac{\partial T}{\partial t} = fdV - \nabla\cdot\vec{q}dV \tag{1.71}$$

where ρ is the density and C_p is the specific heat. Expressed per unit volume, this equation becomes

$$\rho C_p\frac{\partial T}{\partial t} = -\nabla\cdot\vec{q}+f. \tag{1.72}$$

Using Fourier's law Equation (1.69), we obtain the heat diffusion equation

$$\rho C_p\frac{\partial T}{\partial t} = \nabla\cdot(k\nabla T)+f \tag{1.73}$$

and in Cartesian coordinates we have

$$\rho C_p\frac{\partial T}{\partial t} = \frac{\partial}{\partial x}(k\frac{\partial T}{\partial x})+\frac{\partial}{\partial y}(k\frac{\partial T}{\partial y})+\frac{\partial}{\partial z}(k\frac{\partial T}{\partial z})+f. \tag{1.74}$$

If k is uniform throughout the domain of interest, we get

$$\frac{\partial T}{\partial t} = \alpha(\frac{\partial^2 T}{\partial x^2}+\frac{\partial^2 T}{\partial y^2}+\frac{\partial^2 T}{\partial z^2})+\frac{f}{\rho C_p}. \tag{1.75}$$

where $\alpha = k/\rho C_p$ is thermal diffusivity. It has the dimensions of $[L^2/T]$ and its unit is m^2/s. For heat conduction in one dimension, the heat equation is simplified to

$$\frac{\partial T}{\partial t} = \alpha\frac{\partial^2 T}{\partial x^2}+\frac{f}{\rho C_p} \tag{1.76}$$

with the understanding that the energy source f depends on x and t only.

First, we present exact analytic solutions to the one-dimensional unsteady (transient) heat equation. Then, we present a brief description of the finite-difference method for solving Equation (1.76). This will help introduce the concepts of consistency, stability, and convergence of numerical methods for PDEs.

Example: Exact Analytic Solutions

We want to solve the equation

$$\frac{\partial T}{\partial t} = \alpha\frac{\partial^2 T}{\partial x^2} \tag{1.77}$$

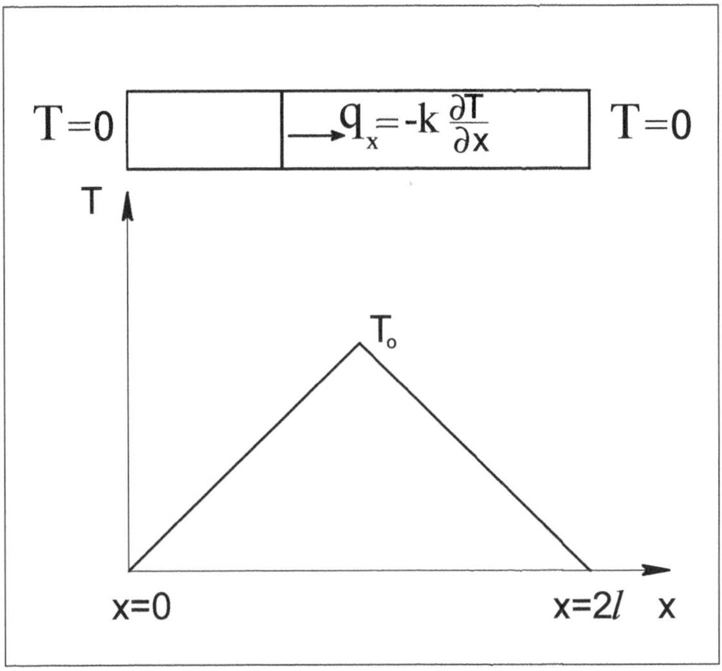

Figure 1.11 Initial temperature distribution in a slab of width 2ℓ.

for the temperature T in the domain $0 < x < 2\ell$, subjected to boundary conditions $T = 0$ at the two faces $x = 0$ and $x = 2\ell$. The initial ($t = 0$) temperature distribution is

$$T(x,0) \equiv \phi(x) = \left\{ \begin{array}{c} T_o\, x/\ell : 0 \leq x \leq \ell \\ T_o(2\ell - x)/\ell : \ell \leq x \leq 2\ell \end{array} \right\} \tag{1.78}$$

where T_o is a constant, as shown in Figure 1.11. The objective is to determine $T(x,t)$ for $t > 0$. We try separation of variables

$$T = P(t)R(x)$$

and substitute into Equation (1.77),

$$\dot{P}R = \alpha PR''$$

$$\frac{\dot{P}}{\alpha P} = \frac{R''}{R} = -\lambda^2$$

where λ is a constant. We obtain two equations, for P

$$\dot{P} + \alpha\lambda^2 P = 0$$

and for R

$$R'' + \lambda^2 R = 0.$$

The solutions are

$$P(t) = Ae^{-\alpha\lambda^2 t}$$

$$R(x) = C\sin\lambda x + D\cos\lambda x.$$

Apply boundary conditions, at $x = 0$ we have $T = 0$ for all t, hence

$$R(0) = 0 \Rightarrow D = 0.$$

Similarly, at $x = 2\ell$ we have $T = 0$, hence

$$R(2\ell) = 0 \Rightarrow C\sin 2\lambda\ell = 0 \Rightarrow 2\lambda\ell = n\pi \qquad n = 1,2,3,\ldots \text{ because } C \neq 0.$$

$$T = \sum_{n=1}^{\infty} C_n exp(-\alpha\frac{n^2\pi^2 t}{4\ell^2})\sin(\frac{n\pi x}{2\ell}). \tag{1.79}$$

To determine the coefficients C_n we apply the initial condition Equation (1.78),

$$\phi(x) = \sum_{n=1}^{\infty} C_n \sin(\frac{n\pi x}{2\ell}). \tag{1.80}$$

We need to represent the function $\phi(x)$ in the interval $0 \leq x \leq 2\ell$ by a Fourier sine series.

Fourier Series

A "nice" function $f(x)$ defined in the interval $-L \leq x \leq L$ and is periodic elsewhere with period $2L$ can be represented by a Fourier series

$$f(x) = \frac{a_o}{2} + \sum_{m=1}^{\infty} (a_m \cos\frac{m\pi x}{L} + b_m \sin\frac{m\pi x}{L}) \tag{1.81}$$

where

$$a_m = \frac{1}{L}\int_{-L}^{L} f(x)\cos\frac{m\pi x}{L}dx \quad m = 0,1,2,\ldots \tag{1.82}$$

$$b_m = \frac{1}{L}\int_{-L}^{L} f(x)\sin\frac{m\pi x}{L}dx \quad m = 1,2,\ldots. \tag{1.83}$$

For our problem, we take $L = 2\ell$ and define $f(x)$ by

$$f(x) = \begin{cases} -\phi(-x) : -L < x < 0 \\ \phi(x) : 0 \leq x \leq L \end{cases}$$

The function $f(x)$ is defined elsewhere to be periodic with period $2L$ as depicted in Figure 1.12. Now that $f(x)$ is an odd function $[f(-x) = -f(x)]$ it can be expanded into a Fourier sine series ($a_m = 0,\ m = 0,1,2,\ldots$) with coefficients

$$b_m = \frac{2}{L}\int_{0}^{L} f(x)\sin\frac{m\pi x}{L}dx \quad m = 1,2,3,\ldots \tag{1.84}$$

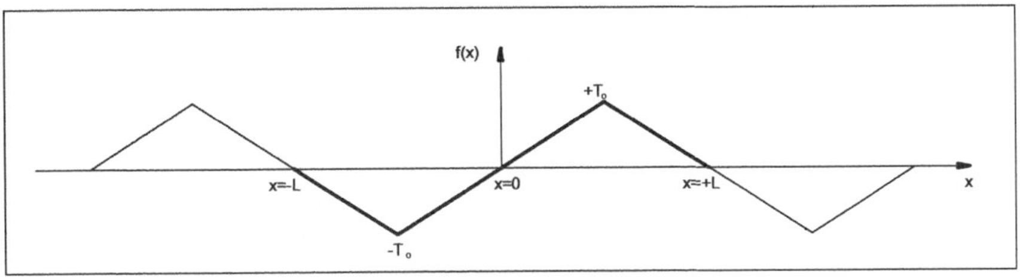

Figure 1.12 Initial temperature distribution extended into an odd periodic function.

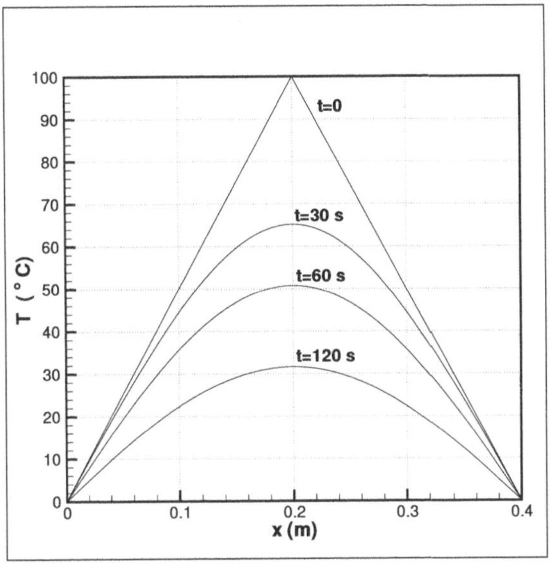

Figure 1.13 Exact solution to cooling of a golden slab, $\alpha = 1.2732 \times 10^{-4}$ m$^2/s$.

For the given initial condition $\phi(x)$, Equation (1.84) provides the coefficients C_n in Equation (1.80),

$$C_n = \frac{8T_o}{\pi^2}\frac{\sin(n\pi/2)}{n^2} \qquad n = 1,2,3,... \tag{1.85}$$

The final solution for $T(x,t)$ is

$$T(x,t) = \frac{8T_o}{\pi^2}\sum_{n=1}^{\infty}\frac{(-1)^{n+1}}{(2n-1)^2}exp[-\alpha(2n-1)^2\pi^2\, t/L^2]\sin[(2n-1)\pi\, x/L]. \tag{1.86}$$

This solution is plotted in Figure 1.13 for a material with $\alpha = 1.2732 \times 10^{-4}$ m$^2/s$ (gold) and $T_o = 100°C$.

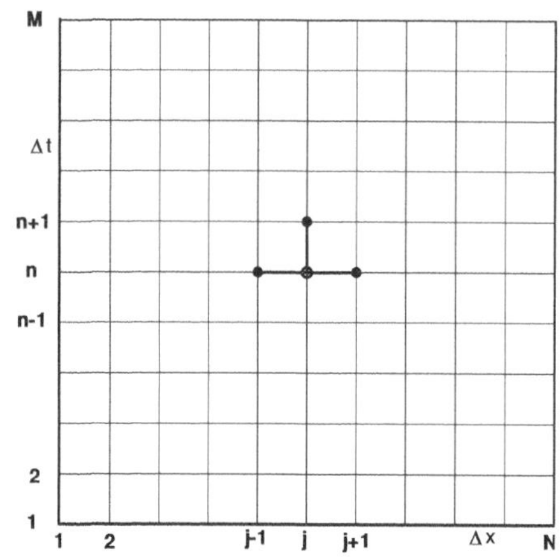

Figure 1.14 Finite-difference module for the simple explicit method.

Finite-Difference Method

The analytic solution to the heat equation gives the temperature field $T(x,t)$ as a continuous function of x and t that can be evaluated at any point in the domain of interest in the xt-plane. Let's cover that domain by a uniform mesh (or grid) as shown in Figure 1.14 where the grid points are $(x_j = j\Delta x,\ t_n = n\Delta t)$, where Δx and Δt are the step sizes in the x- and t-directions, respectively. Each node in the grid is identified by an ordered pair of integers (j,n). A finite-difference method aims at determining numerical values for T at the discrete set of points specified by the grid (x_j,t_n); that is to say, to determine $T(x_j,t_n) \equiv T_j^n$ for all values of j and n in the grid. Afterward, suitable interpolations can be used to find T at intermediate points not specified by the grid, if desired.

Simple Explicit Method

For fixed $x = x_j$ we expand T_j^{n+1} in a Taylor series in time about point (j,n),

$$T_j^{n+1} = T_j^n + (\frac{\partial T}{\partial t})_j^n \Delta t + \frac{1}{2}(\frac{\partial^2 T}{\partial t^2})_j^n (\Delta t)^2 + 0[(\Delta t)^3], \tag{1.87}$$

If the time step is small enough, we approximate $\partial T/\partial t$ at the grid point (j,n) by

$$(\frac{\partial T}{\partial t})_j^n = \frac{T_j^{n+1} - T_j^n}{\Delta t}. \tag{1.88}$$

The neglected terms constitute the truncation error whose leading order term is proportional to Δt. Therefore, approximation Equation (1.88) is termed first-order in Δt. It is also called forward in time since the derivative at time level n uses the temperature at the forward level $(n+1)$. Similarly, at fixed time $t = t_n$, we write a

Taylor series expansion in x for T^n_{j+1} and T^n_{j-1}:

$$T^n_{j+1} = T^n_j + (\frac{\partial T}{\partial x})^n_j \Delta x + \frac{1}{2}(\frac{\partial^2 T}{\partial x^2})^n_j (\Delta x)^2 + \frac{1}{6}(\frac{\partial^3 T}{\partial x^3})^n_j (\Delta x)^3 + 0[(\Delta x)^4] \quad (1.89)$$

$$T^n_{j-1} = T^n_j - (\frac{\partial T}{\partial x})^n_j \Delta x + \frac{1}{2}(\frac{\partial^2 T}{\partial x^2})^n_j (\Delta x)^2 - \frac{1}{6}(\frac{\partial^3 T}{\partial x^3})^n_j (\Delta x)^3 + 0[(\Delta x)^4]. \quad (1.90)$$

If the spatial step Δx is small enough, an approximation for $\partial^2 T/\partial x^2$ at the grid point (j,n) is obtained by adding the above two equations

$$(\frac{\partial^2 T}{\partial x^2})^n_j = \frac{T^n_{j+1} - 2T^n_j + T^n_{j-1}}{(\Delta x)^2}. \quad (1.91)$$

The leading term in the truncation error is proportional to $(\Delta x)^2$, and hence approximation Equation (1.91) is called second-order in Δx. Because of the equal weights given to T^n_{j+1} and T^n_{j-1}, this is a central difference formula.

We write the heat equation at grid point (j,n)

$$(\frac{\partial T}{\partial t})^n_j - \alpha(\frac{\partial^2 T}{\partial x^2})^n_j = 0. \quad (1.92)$$

Next, we substitute approximations Equations (1.88) and (1.91) into Equation (1.92) and obtain

$$\frac{T^{n+1}_j - T^n_j}{\Delta t} - \alpha\frac{T^n_{j+1} - 2T^n_j + T^n_{j-1}}{(\Delta x)^2} = 0. \quad (1.93)$$

Rearranging terms, we get

$$T^{n+1}_j = T^n_j + r(T^n_{j+1} - 2T^n_j + T^n_{j-1}) \quad (1.94)$$

where $r = \alpha\Delta t/(\Delta x)^2$. Equation (1.94) is an explicit algorithm because the temperature at one grid point $(j,n+1)$ on the left is given explicitly in terms of temperatures at a previous time level n.

Example

A slab of thickness $L = 0.06$ m is divided into six steps each of size $\Delta x = 0.01$ m as shown in Figure 1.15. The initial $(t = 0)$ temperature at the seven nodes is also shown in the figure. The thermal diffusivity $\alpha = 10^{-4}\text{m}^2/\text{s}$. The faces $x = 0$ and $x = 0.06$ m are kept at zero temperature for $t > 0$. We want to determine the temperature distribution at $t = 6.4$ s using (a) $\Delta t = 0.4$s and (b) $\Delta t = 0.8$ s.

Solution

(a) We compute $r = \alpha\Delta t/(\Delta x)^2 = 10^{-4}(0.4)/(0.01)^2 = 0.4$, and use Equation (1.94),

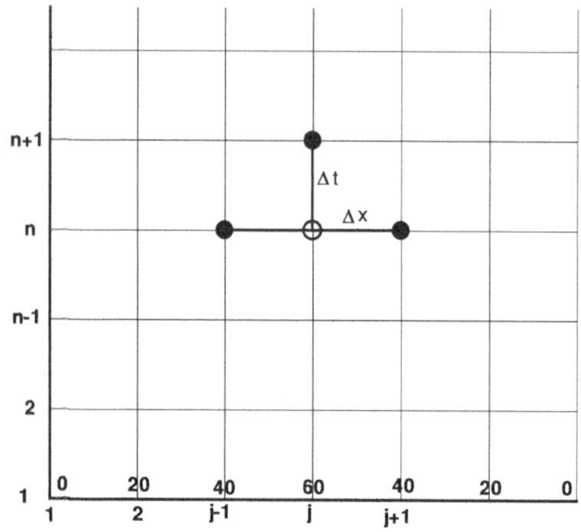

Figure 1.15 Initial temperature distribution.

$$
\begin{aligned}
&t = 0, \quad n = 1 \;\; T_j^1 = (0, 20, 40, 60, 40, 20, 0) \\
&t = 0.4\text{s}, \; n = 2 \;\; T_j^2 = (0, 20, 40, 44, 40, 20, 0) \\
&t = 0.8\text{s}, \; n = 3 \;\; T_j^3 = (0, 20, 33.6, 40.8, 33.6, 20, 0) \\
&t = 1.2\text{s}, \; n = 4 \;\; T_j^4 = (0, 17.4, 31.0, 35.0, 31.0, 17.4, 0) \\
&t = 3.2\text{s}, \; n = 9 \;\; T_j^9 = (0, 10.1, 17.4, 20.1, 17.4, 10.1, 0) \\
&t = 6.4\text{s}, \; n = 17 \;\; T_j^{17} = (0, 4.06, 7.02, 8.11, 7.02, 4.06, 0).
\end{aligned}
$$

(b) We compute $r = \alpha \Delta t / (\Delta x)^2 = 10^{-4}(0.8)/(0.01)^2 = 0.8$, and use Equation (1.94),

$$
\begin{aligned}
&t = 0, \quad n = 1 \;\; T_j^1 = (0, 20, 40, 60, 40, 20, 0) \\
&t = 0.8\text{s}, \; n = 2, \;\; T_j^2 = (0, 20, 40, 28, 40, 20, 0) \\
&t = 1.6\text{s}, \; n = 3 \;\; T_j^3 = (0, 20, 14.4, 47.2, 14.2, 20, 0) \\
&t = 2.4\text{s}, \; n = 4 \;\; T_j^4 = (0, -0.480, 45.1, -5.28, 45.1, -0.480, 0) \\
&t = 3.2\text{s}, \; n = 5 \;\; T_j^5 = (0, 36.4, -31.7, 75.4, -31.7, 36.4, 0) \\
&t = 6.4\text{s}, \; n = 8 \;\; T_j^8 = (0, 435, -741, 871, -741, 435, 0).
\end{aligned}
$$

It is clear that for $\Delta t = 0.8$s the results do not make sense.

Consistency, Stability, and Convergence

Three important concepts in the analysis of finite-difference methods for initial value problems are consistency, stability, and convergence.

Consistency

First, we discuss consistency. A finite-difference equation (FDE) is consistent with a partial differential equation (PDE) if in the limit of infinitesimally small step sizes ($\Delta t \to 0$ and $\Delta x \to 0$) the FDE reduces to the PDE. The difference between the PDE and FDE is called the truncation error (TE). For the heat equation

$$\text{PDE}: \; \frac{\partial T}{\partial t} - \alpha \frac{\partial^2 T}{\partial x^2} = 0. \tag{1.95}$$

Consider the simple explicit method,

$$\text{FDE}: \; \frac{T_j^{n+1} - T_j^n}{\Delta t} - \alpha \frac{T_{j+1}^n - 2T_j^n + T_{j-1}^n}{(\Delta x)^2} = 0 \tag{1.96}$$

$$\text{TE} = \text{PDE} - \text{FDE} = \left(\frac{\partial T}{\partial t} - \alpha \frac{\partial^2 T}{\partial x^2}\right)_j^n - \left[\frac{T_j^{n+1} - T_j^n}{\Delta t} - \alpha \frac{T_{j+1}^n - 2T_j^n + T_{j-1}^n}{(\Delta x)^2}\right]. \tag{1.97}$$

To find the order of the truncation error, we expand T_j^{n+1}, T_{j+1}^n, and T_{j-1}^n in Taylor series about point (j,n).

$$T_j^{n+1} = T_j^n + \left(\frac{\partial T}{\partial t}\right)_j^n \Delta t + \frac{1}{2}\left(\frac{\partial^2 T}{\partial t^2}\right)_j^n (\Delta t)^2 + O[(\Delta t)^3] \tag{1.98}$$

$$\begin{aligned} T_{j+1}^n = T_j^n &+ \left(\frac{\partial T}{\partial x}\right)_j^n \Delta x + \frac{1}{2}\left(\frac{\partial^2 T}{\partial x^2}\right)_j^n (\Delta x)^2 + \frac{1}{6}\left(\frac{\partial^3 T}{\partial x^3}\right)_j^n (\Delta x)^3 \\ &+ \frac{1}{24}\left(\frac{\partial^4 T}{\partial x^4}\right)_j^n (\Delta x)^4 + O[(\Delta x)^5] \end{aligned} \tag{1.99}$$

$$\begin{aligned} T_{j-1}^n = T_j^n &- \left(\frac{\partial T}{\partial x}\right)_j^n \Delta x + \frac{1}{2}\left(\frac{\partial^2 T}{\partial x^2}\right)_j^n (\Delta x)^2 - \frac{1}{6}\left(\frac{\partial^3 T}{\partial x^3}\right)_j^n (\Delta x)^3 \\ &+ \frac{1}{24}\left(\frac{\partial^4 T}{\partial x^4}\right)_j^n (\Delta x)^4 + O[(\Delta x)^5]. \end{aligned} \tag{1.100}$$

Substituting Equations (1.98)–(1.100) into Equation (1.97), we obtain

$$\text{TE} = -\left(\frac{\partial^2 T}{\partial t^2}\right)_j^n \frac{\Delta t}{2} + \alpha \left(\frac{\partial^4 T}{\partial x^4}\right)_j^n \frac{(\Delta x)^2}{12} + \ldots. \tag{1.101}$$

The leading order terms are $O(\Delta t)$ and $O(\Delta x^2)$. Thus, the simple explicit method is first order in Δt and second order in Δx. The FDE given by Equation (1.96) is consistent with the PDE given by Equation (1.95) because as $\Delta t \to 0$ and $\Delta x \to 0$ the TE given by Equation (1.101) vanishes.

The Du Fort-Frankel finite-difference scheme of the heat equation is a well-known example of an inconsistent scheme. Central difference is used for the time derivative,

$$\left(\frac{\partial T}{\partial t}\right)_j^n = \frac{u_j^{n+1} - u_j^{n-1}}{2\Delta t}. \tag{1.102}$$

It is second-order in Δt. If the space derivative is also central

$$(\frac{\partial^2 T}{\partial x^2})_j^n = \frac{T_{j+1}^n - 2T_j^n + T_{j-1}^n}{(\Delta x)^2} \tag{1.103}$$

we obtain the scheme

$$\frac{T_j^{n+1} - T_j^{n-1}}{2\Delta t} - \alpha \frac{T_{j+1}^n - 2T_j^n + T_{j-1}^n}{(\Delta x)^2} = 0 \tag{1.104}$$

which can be shown to be unconditionally unstable. However, it can be stabilized by substituting

$$2T_j^n = T_j^{n+1} + T_j^{n-1} + O[(\Delta t^2)] \tag{1.105}$$

into Equation (1.104), and the result is the Du Fort-Frankel scheme,

$$\text{FDE}: \frac{T_j^{n+1} - T_j^{n-1}}{2\Delta t} - \alpha \frac{T_{j+1}^n + T_{j-1}^n - T_j^{n+1} - T_j^{n-1}}{(\Delta x)^2} = 0. \tag{1.106}$$

This is an explicit unconditionally stable method. Unfortunately, it is inconsistent with the heat equation. The truncation error is

$$\text{TE} = -(\frac{\partial^3 T}{\partial t^3})_j^n \frac{\Delta t^2}{6} + \alpha(\frac{\partial^4 T}{\partial x^4})_j^n \frac{\Delta x^2}{12} - \alpha(\frac{\partial^2 T}{\partial t^2})_j^n (\frac{\Delta t}{\Delta x})^2 + \ldots. \tag{1.107}$$

The first and second terms vanish as $\Delta t \to 0$ and $\Delta x \to 0$, but the third term may not. If the mesh is refined (decrease Δt and Δx) such that the ratio $\Delta t/\Delta x = \beta$ is held fixed, then the FDE will be consistent with the equation

$$\frac{\partial T}{\partial t} + \alpha\beta^2 \frac{\partial^2 T}{\partial t^2} = \alpha \frac{\partial^2 T}{\partial x^2} \tag{1.108}$$

which is a hyperbolic equation. The inconsistency has arisen because we substituted an expression, Equation (1.105), for T_j^n that is $O[(\Delta t)^2]$ into Equation (1.104) and divided by $(\Delta x)^2$.

Stability Analysis

Stability is a property of finite-difference equations; it describes how the solution to those equations behave as the solution is marched in time (increasing n). For initial-boundary-value problems, the essence of stability is that there should be a limit to the extent to which any component of an initial solution can be amplified in the numerical procedure [103].

Before we present stability analysis, let's first consider the heat equation with periodic boundary conditions.

$$\frac{\partial T}{\partial t} = \alpha \frac{\partial^2 T}{\partial x^2}. \tag{1.109}$$

Initial conditions:

$$T(x,0) = f(x), \quad 0 \leq x \leq L. \tag{1.110}$$

We replace the actual boundary conditions at $x = 0$ and $x = L$ by the requirement that the solution be periodic in x; that is

$$T(0,t) = T(L,t), \quad t > 0. \tag{1.111}$$

We also assume that the initial condition Equation (1.110) is consistent with Equation (1.111); that is, $f(0) = f(L)$. We try the separation of variables,

$$\begin{aligned}
&T = P(t) \cdot R(x) \\
&\dot{P}R = \alpha P R'' \\
&\frac{\dot{P}}{\alpha P} = \frac{R''}{R} = -\lambda^2 \\
&\dot{P} = -\alpha \lambda^2 P \\
&P = E e^{-\alpha \lambda^2 t} \\
&R'' + \lambda^2 R = 0 \\
&R = A \cos \lambda x + B \sin \lambda x.
\end{aligned}$$

Periodic boundary condition Equation (1.111) can be satisfied if we choose λ to be

$$\lambda = 2\pi k/L, \quad k = 0, 1, 2, \ldots.$$

The solution is

$$T(x,t) = \frac{A_o}{2} + \sum_{k=1}^{\infty} e^{-\alpha(\frac{2\pi k}{L})^2 t}\left(A_k \cos \frac{2\pi k x}{L} + B_k \sin \frac{2\pi k x}{L}\right). \tag{1.112}$$

Now, we apply the initial condition,

$$f(x) = \frac{A_o}{2} + \sum_{k=1}^{\infty} A_k \cos \frac{2\pi k x}{L} + B_k \sin \frac{2\pi k x}{L}. \tag{1.113}$$

It is clear that A_k and B_k are the Fourier coefficients of the Fourier series expansion of $f(x)$. Introducing the notation $i = \sqrt{-1}$, and using the complex form of Fourier series, we write Equation (1.112) as

$$T(x,t) = \sum_{k=-\infty}^{\infty} C_k e^{-\alpha(\frac{2\pi k}{L})^2 t} e^{i\frac{2\pi k x}{L}} \tag{1.114}$$

where

$$C_k = \frac{1}{L} \int_0^L f(x) e^{-i\frac{2\pi k x}{L}} dx \tag{1.115}$$

The term $C_k e^{-\alpha(\frac{2\pi k}{L})^2 t} e^{i\frac{2\pi kx}{L}}$ is called a Fourier mode; its complex amplitude is $C_k e^{-\alpha(\frac{2\pi k}{L})^2 t}$, which decays with time for all values of k. Its initial value (C_k) as given by Equation (1.115) depends only on the initial condition; $f(x)$.

von Neumann Stability Analysis

We divide the interval $0 \leq x \leq L$ into N equally spaced steps, $\Delta x = L/N$, and denote the grid nodes by $x_j = j\Delta x$, $j = 0, 1, ..., N-1$. In the von Neumann stability analysis (also called Fourier analysis) of finite-difference methods, the solution to the finite-difference equations at any time level n, T_j^n, is assumed to be periodic in j with period N; that is to say ($T_{j+mN} = T_j$, for any integer m). The actual boundary conditions at $x = 0$ and $x = L$ are ignored. The solution is then expanded into a (discrete) Fourier transform in the space index j,

$$T_j^n = \sum_{k=-\frac{N}{2}}^{\frac{N}{2}-1} \hat{T}_k^n e^{i\frac{2\pi kj}{N}}, \quad j = 0, 1, 2, ..., N-1 \tag{1.116}$$

where

$$\hat{T}_k^n = \frac{1}{N} \sum_{\ell=0}^{N-1} T_\ell^n e^{-i\frac{2\pi k\ell}{N}}, \quad k = -\frac{N}{2}, -\frac{N}{2}+1, ..., \frac{N}{2}-1. \tag{1.117}$$

Note that the amplitudes $\hat{T}_k^n$ are functions of the time level n. The ratio

$$g = \frac{\hat{T}_k^{n+1}}{\hat{T}_k^n}$$

is called the amplification factor of the kth Fourier mode. A necessary condition for stability of a finite-difference method applied to the initial boundary value problem Equations (1.109)–(1.111) is that the magnitude of the amplification factor $|g|$ satisfies

$$|g| \leq 1 \tag{1.118}$$

for all Fourier modes, i.e., all values of k. Let

$$\phi_k = \frac{2\pi k}{N} \tag{1.119}$$

and because k varies between $-\frac{N}{2}$ and $+\frac{N}{2}$, hence ϕ_k varies between $-\pi$ and π.

Stability of the Simple Explicit Method

Because the problem under investigation is linear, we can consider the amplification of one Fourier mode at a time. There is no interactions among the different modes.

$$\frac{T_j^{n+1} - T_j^n}{\Delta t} = \frac{\alpha}{(\Delta x)^2}(T_{j+1}^n - 2T_j^n + T_{j-1}^n). \tag{1.120}$$

Consider the amplification of the kth Fourier mode

$$T_j^n = \hat{T}_k^n e^{i\phi_k j} \tag{1.121}$$

$$(\hat{T}_k^{n+1} - \hat{T}_k^n)e^{i\phi_k j} = \frac{\alpha \Delta t}{(\Delta x)^2}\hat{T}_k^n[e^{i\phi_k(j+1)} - 2e^{i\phi_k j} + e^{i\phi_k(j-1)}]. \tag{1.122}$$

Let

$$r = \frac{\alpha \Delta t}{(\Delta x)^2}$$
$$g = \hat{T}_k^{n+1}/\hat{T}_k^n.$$

We write Equation (1.122) as

$$\begin{aligned} g - 1 &= r(e^{i\phi_k} - 2 + e^{-i\phi_k}) \\ g &= 1 + r(2\cos\phi_k - 2) \\ &= 1 + 2r(\cos\phi_k - 1). \end{aligned} \tag{1.123}$$

We use $cos2\theta = 1 - 2\sin^2\theta$, drop subscript k, and obtain

$$g = 1 - 4r\sin^2(\frac{\phi}{2}).$$

For stability we must have

$$|1 - 4r\sin^2(\frac{\phi}{2})| \leq 1.$$

$$-1 \leq 1 - 4r\sin^2(\frac{\phi}{2}) \leq 1$$

The right inequality demands that $r = \frac{\alpha\Delta t}{\Delta x^2} \geq 0$. This is satisfied for all Δt and Δx. The left inequality demands

$$r\sin^2(\frac{\phi}{2}) \leq \frac{1}{2}$$

for all values of $-\pi \leq \phi \leq \pi$ (this covers all Fourier modes). The simple explicit method is stable if

$$0 \leq r \leq \frac{1}{2}$$

or

$$0 \leq \frac{\alpha \Delta t}{(\Delta x)^2} \leq \frac{1}{2}. \tag{1.124}$$

This condition puts limitations on the permissible values of Δt and Δx for a given α. Thus, the simple explicit method is said to be conditionally stable. In the finite-difference example, we used $r = 0.4$ for which the method is stable, but for $r = 0.8$ the method is unstable and results did not make sense.

Convergence

Let's assume that we are able to find the exact solution to the given initial-boundary-value problem in the domain of interest in the xt-plane. An example of such a solution is given by Equation (1.86). We can evaluate this exact solution at the nodes of the grid that we intend to use for the finite-difference method. At node (j,n), we denote this solution by $\tilde{T}_j^n$. Using a finite-difference method such as the simple explicit method Equation (1.94) or Du Fort-Frankel method Equation (1.106), and applying the same initial and boundary conditions as with the exact solution, we can also determine a finite-difference solution at the same nodes in the xt-plane, which we denote by T_j^n. For finite values of Δx and Δt, we do not expect the finite-difference solution T_j^n to be identical to the exact solution $\tilde{T}_j^n$; the solution error is $SE(\Delta x, \Delta t) = \tilde{T}_j^n - T_j^n$. Convergence means that the $SE \Rightarrow 0$ as the grid is refined, i.e., as $(\Delta x, \Delta t) \Rightarrow 0$.

We recall that consistency concerns the truncation error, which is the difference between the differential equation and the finite-difference equation, whereas convergence concerns the difference between the solutions to those equations. The explicit finite-difference example in this section shows that a method such as the simple explicit method, which is consistent with the heat equation, failed to provide a meaningful solution when the stability condition $r \leq \frac{1}{2}$ was violated. Therefore, the consistency of a finite-difference method is not sufficient to ensure convergence. The following theorem shows that both consistency and stability are needed for convergence.

Lax's Equivalence Theorem

Richtmyer and Morton [103] state: "*Given a well posed initial value problem and a finite-difference approximation to it that satisfies the consistency condition, stability is necessary and sufficient condition for convergence.*"

Implicit Methods

Finite-difference methods that are implicit in time offer better stability characteristics. We first consider an ordinary differential equation

$$\frac{d^2u}{dx^2} + \frac{1}{4}\pi^2 u = 0,$$

with boundary conditions: $u(0) = 1$ and $u(1) = 0$. The exact solution is $u = \cos\frac{\pi}{2}x$. To solve this problem using a finite-difference method, we discretize the interval $0 \leq x \leq 1$ into five equally spaced segments, $\Delta x = 0.2$. The nodes are denoted by x_j and the values of u at these nodes are denoted by u_j, $j = 1,2,...,6$. At any node j, we use the second-order central difference

$$\left(\frac{d^2u}{dx^2}\right)_j = \frac{u_{j+1} - 2u_j + u_{j-1}}{(\Delta x)^2}.$$

The differential equation at node j is

$$\left(\frac{d^2u}{dx^2}\right)_j + \frac{1}{4}\pi^2 u_j = 0,$$

$$u_{j+1} - 2u_j + u_{j-1} + \frac{1}{4}\pi^2(\Delta x)^2 u_j = 0$$

rearrange:

$$u_{j-1} + \left(-2 + \frac{\pi^2}{4}(\Delta x)^2\right) u_j + u_{j+1} = 0.$$

We can write this equation for $j = 2$, 3, 4, and 5.

For $j = 1$ and $j = 6$, we have the boundary conditions

$$u_1 = 1 \quad \text{and} \quad u_6 = 0.$$

Write the six equations in matrix form

$$\begin{bmatrix} 1 & 0 & 0 & 0 & 0 & 0 \\ 1 & b & 1 & 0 & 0 & 0 \\ 0 & 1 & b & 1 & 0 & 0 \\ 0 & 0 & 1 & b & 1 & 0 \\ 0 & 0 & 0 & 1 & b & 1 \\ 0 & 0 & 0 & 0 & 0 & 1 \end{bmatrix} \begin{bmatrix} u_1 \\ u_2 \\ u_3 \\ u_4 \\ u_5 \\ u_6 \end{bmatrix} = \begin{bmatrix} 1 \\ 0 \\ 0 \\ 0 \\ 0 \\ 0 \end{bmatrix}$$

where $b = -2 + \frac{\pi^2}{4}(\Delta x)^2$.

The next step is to solve the above system of linear algebraic equations for u_j, for $j = 1, 2, ..., 6$. The coefficient matrix has a special form. In each equation, all coefficients are zero except the diagonal, the one below the diagonal and the one above the diagonal. Such a matrix is called a tridiagonal matrix. In its general form, a tridiagonal system has the form

$$\begin{bmatrix} b_1 & c_1 & 0 & 0 & \dots & & & & & 0 \\ a_2 & b_2 & c_2 & 0 & 0 & 0 & \dots & & & \\ 0 & a_3 & b_3 & c_3 & 0 & 0 & \dots & & & \\ 0 & 0 & a_4 & b_4 & c_4 & 0 & 0 & \dots & & \\ \vdots & & & & & & & & & \\ \vdots & & 0 & 0 & a_j & b_j & c_j & 0 & 0 & \\ \vdots & & & & & & & & & 0 \\ 0 & & & & & & 0 & a_{N-1} & b_{N-1} & c_{N-1} \\ 0 & & & & & & 0 & 0 & a_N & b_N \end{bmatrix} \begin{Bmatrix} u_1 \\ u_2 \\ u_3 \\ \vdots \\ u_{j-1} \\ u_j \\ u_{j+1} \\ \vdots \\ u_{N-1} \\ u_N \end{Bmatrix} = \begin{Bmatrix} d_1 \\ d_2 \\ d_3 \\ \\ d_j \\ \\ d_{N-1} \\ d_N \end{Bmatrix}$$

An efficient method for solving this system exists and is known as the Thomas algorithm. It is an LU decomposition method.

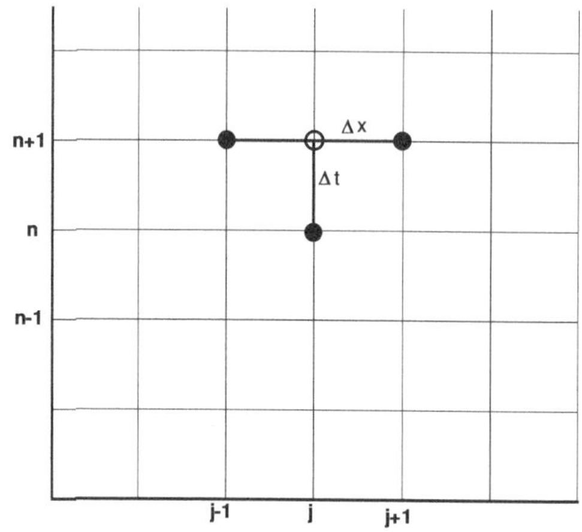

Figure 1.16 Finite-difference stencil for a simple implicit method.

Simple Implicit Method

Now, we consider the heat equation:

$$\frac{\partial T}{\partial t} = \alpha \frac{\partial^2 T}{\partial x^2}$$

where $\alpha = k/\rho C_p$ is thermal diffusivity, with boundary conditions:

$$T(0,t) = F_1(t) \text{ and } T(L,t) = F_2(t)$$

and an initial condition $T(x,0) = f(x)$. We write the heat equation at the node $(j, n+1)$:

$$\left(\frac{\partial T}{\partial t}\right)_j^{n+1} = \alpha \left(\frac{\partial^2 T}{\partial x^2}\right)_j^{n+1}.$$

The finite-difference stencil of the simple implicit method is shown in Figure 1.16. The time derivative is approximated by a first-order backward formula

$$\left(\frac{\partial T}{\partial t}\right)_j^{n+1} = \frac{T_j^{n+1} - T_j^n}{\Delta t}$$

and the space derivative is approximated by a second-order central difference formula,

$$\left(\frac{\partial^2 T}{\partial x^2}\right)_j^{n+1} = \frac{T_{j+1}^{n+1} - 2T_j^{n+1} + T_{j-1}^{n+1}}{(\Delta x)^2}.$$

The finite-difference equation is

$$\frac{T_j^{n+1} - T_j^n}{\Delta t} = \frac{\alpha}{(\Delta x)^2}(T_{j+1}^{n+1} - 2T_j^{n+1} + T_{j-1}^{n+1}).$$

Rearranging, we obtain

$$T_{j-1}^{n+1} + \left(-2 - \frac{(\Delta x)^2}{\alpha \Delta t}\right) T_j^{n+1} + T_{j+1}^{n+1} = -\frac{(\Delta x)^2}{\alpha \Delta t} T_j^n.$$

We write this equation for nodes $j = 2, 3, ..., N-1$ and obtain $N-2$ equations for the unknowns T_j^{n+1} for $j = 1, 2, 3, ..., N$. To close the system we need to add the boundary conditions, at $j = 1$, $T_1^{n+1} = F_1^{n+1}$ (Given) and at $j = N$, $T_N^{n+1} = F_2^{n+1}$ (Given). We drop superscript $(n+1)$, and write the system

$$\begin{bmatrix} 1 & 0 & 0 & 0 & \ldots & & & & & \\ 1 & b & 1 & 0 & 0 & & & & & \\ 0 & 1 & b & 1 & 0 & \ldots & & & & \\ & & & & & & & & & \\ & & 0 & 1 & b & 1 & 0 & & & \\ & & & & & & & & & \\ & & & & & 0 & 0 & 1 & b & 1 \\ & & & & & & & 0 & 0 & 1 \end{bmatrix} \begin{Bmatrix} T_1 \\ T_2 \\ T_3 \\ \\ T_{j-1} \\ T_j \\ T_{j+1} \\ \\ T_{N-1} \\ T_N \end{Bmatrix} = \begin{Bmatrix} d_1 \\ d_2 \\ d_3 \\ \\ d_j \\ \\ d_{N-1} \\ d_N \end{Bmatrix}$$

where

$$b = (-2 - (\Delta x)^2/\alpha \Delta t),\ d_j = -\frac{(\Delta x)^2}{\alpha \Delta t} T_j^n,$$

$$\text{for } j = 2, ..., N-1, \text{ and } d_1 = F_1^{n+1},\ d_N = F_2^{n+1}.$$

Knowing the temperature distribution at time level n, we solve this system for the temperature distribution at the new time level $(n+1)$. Starting with the initial conditions (level $n = 1$) we march in time. At each time level we need to solve a tridiagonal system of equations. Note that in this example the coefficient matrix is constant (does not vary with time); therefore, only the right-hand side d_j needs to be recomputed at each time level. This implicit method is unconditionally stable according to von Neumann analysis. But it is first-order accurate in Δt. For engineering calculations, first-order accuracy is usually not sufficient. We need a method that is at least second-order in Δt. To represent $(\partial T/\partial t)_j^{n+1}$ to second-order accuracy, we consider the Taylor series expansions of T_j^n and T_j^{n-1} about the node $(j, n+1)$,

$$T_j^n = T_j^{n+1} - \left(\frac{\partial T}{\partial t}\right)_j^{n+1} \Delta t + \frac{1}{2}\left(\frac{\partial^2 T}{\partial t^2}\right)_j^{n+1} (\Delta t)^2 + 0[(\Delta t)^3]$$

$$T_j^{n-1} = T_j^{n+1} - \left(\frac{\partial T}{\partial t}\right)_j^{n+1} (2\Delta t) + \frac{1}{2}\left(\frac{\partial^2 T}{\partial t^2}\right)_j^{n+1} (2\Delta t)^2 + 0[(\Delta t)^3].$$

To eliminate the $0[(\Delta t)^2]$ term, we multiply the first equation by 4 and subtract the second,

$$4T_j^n - T_j^{n-1} = 3T_j^{n+1} - 2\left(\frac{\partial T}{\partial t}\right)_j^{n+1} \Delta t + 0[(\Delta t)^3].$$

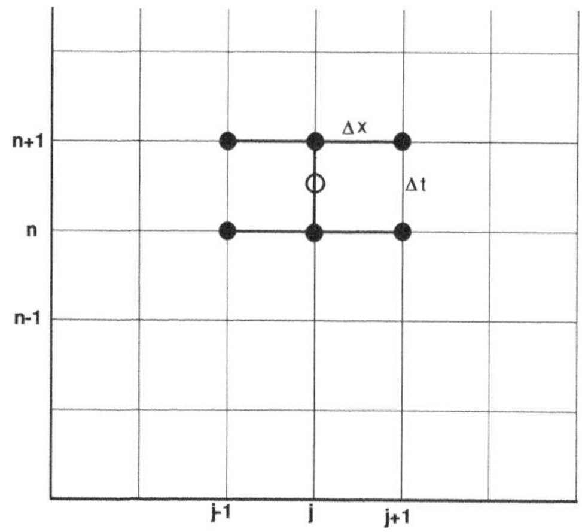

Figure 1.17 Finite-difference module for Crank–Nicolson method.

Rearranging, we get

$$\left(\frac{\partial T}{\partial t}\right)_j^{n+1} = \frac{3T_j^{n+1} - 4T_j^n + T_j^{n-1}}{2\Delta t} + 0[(\Delta t)^2].$$

This is called the three-point backward difference formula. Using this expression in the heat equation, we obtain the same system as with the simple implicit method except that

$$b = \left(-2 - \frac{3(\Delta x)^2}{2\alpha\Delta t}\right)$$

$$d_j = -\frac{(\Delta x)^2}{2\alpha\Delta t}(T_j^n - T_j^{n-1}).$$

This method can be used for $t \geq 2\Delta t$, but it cannot be used for level $t = \Delta t$.

Crank–Nicolson Method

The finite-difference stencil for the Crank–Nicolson method is shown in Figure 1.17. We apply the heat equation at the mid point $(j, n+\frac{1}{2})$,

$$\left(\frac{\partial T}{\partial t}\right)_j^{n+\frac{1}{2}} = \alpha\left(\frac{\partial^2 T}{\partial x^2}\right)_j^{n+\frac{1}{2}} \tag{1.125}$$

and use the central difference formula for both time and space derivatives,

$$\left(\frac{\partial T}{\partial t}\right)_j^{n+\frac{1}{2}} = \frac{T_j^{n+1} - T_j^n}{\Delta t} + 0[(\Delta t)^2] \tag{1.126}$$

$$
\begin{aligned}
\left(\frac{\partial^2 T}{\partial x^2}\right)_j^{n+\frac{1}{2}} &= \frac{1}{2}\left[\left(\frac{\partial^2 T}{\partial x^2}\right)_j^{n+1} + \left(\frac{\partial^2 T}{\partial x^2}\right)_j^{n}\right] + 0[(\Delta t)^2] \\
\left(\frac{\partial^2 T}{\partial x^2}\right)_j^{n+1} &= \frac{1}{(\Delta x)^2}(T_{j+1}^{n+1} - 2T_j^{n+1} + T_{j-1}^{n+1}) + 0[(\Delta x)^2] \\
\left(\frac{\partial^2 T}{\partial x^2}\right)_j^{n} &= \frac{1}{(\Delta x)^2}(T_{j+1}^{n} - 2T_j^{n} + T_{j-1}^{n}) + 0[(\Delta x)^2].
\end{aligned} \tag{1.127}
$$

Substituting from Equations (1.126) and (1.127) into Equation (1.125), we get

$$
\frac{T_j^{n+1} - T_j^n}{\Delta t} = \frac{\alpha}{2(\Delta x)^2}[(T_{j+1}^{n+1} - 2T_j^{n+1} + T_{j-1}^{n+1}) + (T_{j+1}^{n} - 2T_j^{n} + T_{j-1}^{n})]. \tag{1.128}
$$

The Crank–Nicolson scheme is

$$
T_{j-1} + \left(-2 - \frac{2(\Delta x)^2}{\alpha \Delta t}\right) T_j + T_{j+1} = d_j \tag{1.129}
$$

where

$$
d_j = -\frac{2(\Delta x)^2}{\alpha \Delta t} T_j^n - (T_{j+1}^{n} - 2T_j^{n} + T_{j-1}^{n}), \text{ for } j = 2, 3, ..., N-1. \tag{1.130}
$$

Boundary Conditions

$$
\begin{aligned}
j = 1 \quad & T_1^{n+1} = F_1^{n+1} \quad \text{(Given)} \\
j = N \quad & T_N^{n+1} = F_2^{n+1} \quad \text{(Given)}
\end{aligned}
$$

We solve a tridiagonal system at each time step n. This method is second-order accurate in both Δt and Δx. It is also unconditionally stable according to von Neumann's analysis.

von Neumann Stability of Crank–Nicolson Method

Crank–Nicolson scheme is written as

$$
T_j^{n+1} - T_j^n = \frac{\alpha \Delta t}{2(\Delta x)^2}[(T_{j+1}^{n+1} - 2T_j^{n+1} + T_{j-1}^{n+1}) + (T_{j+1}^{n} - 2T_j^{n} + T_{j-1}^{n})]. \tag{1.131}
$$

We consider the amplification of the kth Fourier mode

$$
T_j^n = \hat{T}_k^n e^{i\phi j}. \tag{1.132}
$$

Let $r = \alpha \Delta t/(\Delta x)^2$, and substituting into Equation (1.131), we get

$$
\hat{T}_n^{n+1} - \hat{T}_n^n = r[\hat{T}_j^{n+1}(\cos\phi - 1) + \hat{T}_k^n(\cos\phi - 1)].
$$

Dividing by $\hat{T}_k^n$, we get the amplification factor

$$g-1=r[g(\cos\phi-1)+\cos\phi-1]$$
$$g[1+r(1-\cos\phi)]=1-r(1-\cos\phi)$$
$$g=\frac{1-r(1-\cos\phi)}{1+r(1-\cos\phi)}$$
$$g=\frac{1-2r\sin^2(\frac{\phi}{2})}{1+2r\sin^2(\frac{\phi}{2})}. \tag{1.133}$$

It is clear that $|g|\leq 1$ for all values of ϕ without any limitations on r (of course for $r\geq 0$). The Crank–Nicolson method is unconditionally stable according to von Neumann analysis.

1.3.3 HYPERBOLIC EQUATIONS: WAVE EQUATION

The standard form of a hyperbolic equation in one space dimension x and time t is

$$\frac{\partial^2 u}{\partial t^2}=C^2\frac{\partial^2 u}{\partial x^2}. \tag{1.134}$$

It describes waves of different physical origins where C is the speed of propagation of waves in the x-direction. For transverse waves on taut strings, u is the deflection and $C^2=T/\rho\prime$ where T is tension and $\rho\prime$ is density per unit length. For longitudinal waves in slender rods, u is axial displacement and $C^2=E/\rho$ where E is Young's modulus and ρ is density per unit volume. For acoustic waves in gases, u may be the pressure fluctuations and $C^2=\gamma RT$ where T is the absolute temperature, R is the gas constant, and γ is the specific heat ratio.

Transverse Waves on Strings

We consider a flexible string (no resistance to bending) stretched along the x-axis with tension T. We give the string a small transverse displacement and velocity and study the resulting motion. We make the assumptions: (a) neglect changes in tension due to the extra stretching due to deflection, thus tension T remains tangential to the deflected string and is independent of x, (b) material points move normaly to the x-axis and longitudinal motion is negligible, (c) deflection is small so that the slope magnitude $|\partial v/\partial x|<<1$, and (d) neglect gravity and air resistance.

We draw the free-body diagram of a small element as shown in Figure 1.18 and write the equation of motion,

$$ma_y=\sum F_y$$
$$\rho A\Delta x\frac{\partial^2 v}{\partial t^2}=Tsin(\theta+\frac{\partial\theta}{\partial x}\Delta x)-Tsin\theta \tag{1.135}$$

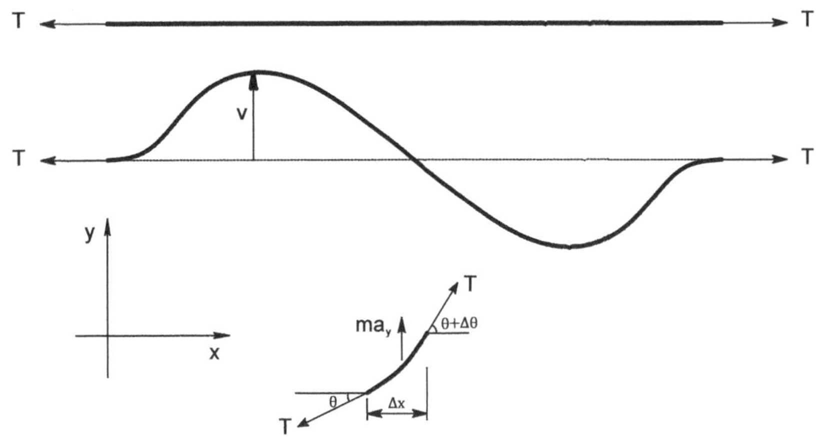

Figure 1.18 Transverse vibration of a taut string.

where $tan\theta = \partial v/\partial x$, and A is the cross-sectional area of string. We let $\rho\prime = \rho A$, and for small angles $sin\theta \approx tan\theta \approx \theta \approx \partial v/\partial x$, we get

$$\rho\prime\Delta x\frac{\partial^2 v}{\partial t^2} = T(\theta + \frac{\partial \theta}{\partial x}\Delta x) - T\theta$$

$$\rho\prime\Delta x\frac{\partial^2 v}{\partial t^2} = T\frac{\partial^2 v}{\partial x^2}\Delta x$$

$$\frac{\partial^2 v}{\partial t^2} = C^2\frac{\partial^2 v}{\partial x^2} \tag{1.136}$$

where $C^2 = T/\rho\prime$. The dimension $[C^2] = [force/mass/length] = [MLT^{-2}/ML^{-1}] = [L^2T^{-2}]$. Thus, the dimension of C is $[LT^{-1}]$, and its unit is m/s. As we shall see, C is the speed of propagation of transverse waves on the string.

Right and Left Running Waves

We define a variable $\xi = x - Ct$, and consider a function $F(\xi)$ as sketched in Figure 1.19, where F is zero everywhere except in the domain $0 \leq \xi \leq L$. We choose values of $Ct = L/2$, L, $3L/2$, and $2L$ and sketch the function F against x in Figure 1.20. It is evident from the sketch that $F(x - Ct)$ represents a wave that propagates in the positive x-direction without a change in shape. It is called a right-running wave. We follow a particular point on the wave, such as point a, by giving ξ a fixed value ξ_a. Therefore, $x - Ct = \xi_a$, and hence $dx/dt = C$ (x and t are independent variables but when we follow a particular point, x becomes dependent on t). This shows that the speed of propagation is given by the constant C. A function $G(\eta)$ where $\eta = x + Ct$ represents a wave that propagates in the negative x-direction without a change in shape. It is called a left-running wave, and its speed is $-C$.

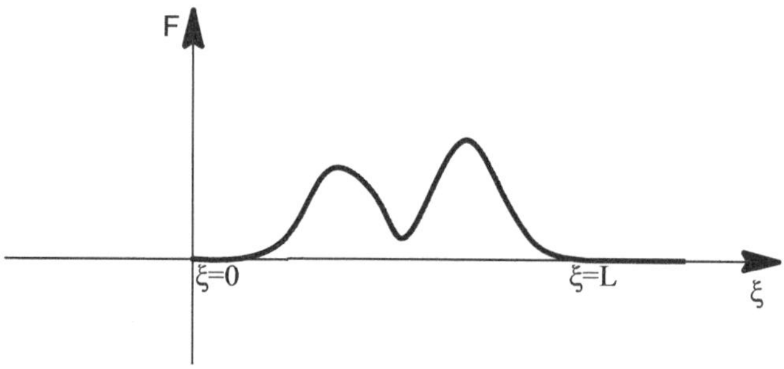

Figure 1.19 Arbitrary function of ξ.

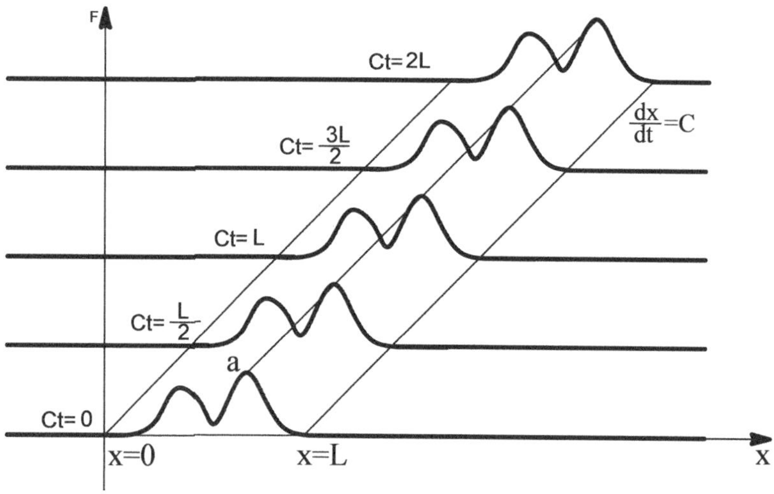

Figure 1.20 Right-running wave.

Next, we develop the general solution to the wave equation Equation (1.136),

$$\left(\frac{\partial^2}{\partial t^2} - C^2\frac{\partial^2}{\partial x^2}\right)v = 0 \tag{1.137}$$

which we factor as

$$\left(\frac{\partial}{\partial t} - C\frac{\partial}{\partial x}\right)\left(\frac{\partial}{\partial t} + C\frac{\partial}{\partial x}\right)v = 0. \tag{1.138}$$

We introduce two new independent variables

$$\begin{aligned}\xi &= x - Ct \\ \eta &= x + Ct.\end{aligned} \tag{1.139}$$

To change variables from (x,t) to (ξ,η), we use the chain rule

$$\begin{aligned}
\frac{\partial}{\partial x} &= \frac{\partial}{\partial \xi}\frac{\partial \xi}{\partial x} + \frac{\partial}{\partial \eta}\frac{\partial \eta}{\partial x} \\
\frac{\partial}{\partial x} &= \frac{\partial}{\partial \xi} + \frac{\partial}{\partial \eta} \\
\frac{\partial}{\partial t} &= \frac{\partial}{\partial \xi}\frac{\partial \xi}{\partial t} + \frac{\partial}{\partial \eta}\frac{\partial \eta}{\partial t} \\
\frac{\partial}{\partial t} &= -C\frac{\partial}{\partial \xi} + C\frac{\partial}{\partial \eta}.
\end{aligned} \tag{1.140}$$

The combined time and space operators become

$$\begin{aligned}
\frac{\partial}{\partial t} - C\frac{\partial}{\partial x} &= -2C\frac{\partial}{\partial \xi} \\
\frac{\partial}{\partial t} + C\frac{\partial}{\partial x} &= 2C\frac{\partial}{\partial \eta}
\end{aligned} \tag{1.141}$$

and we write Equation (1.138) as

$$\frac{\partial^2 v}{\partial \xi \partial \eta} = 0. \tag{1.142}$$

We integrate Equation (1.142) with respect to ξ, and obtain

$$\frac{\partial v}{\partial \eta} = g(\eta) \tag{1.143}$$

then we integrate with respect to η,

$$v = F(\xi) + G(\eta) \tag{1.144}$$

where F and G are arbitrary functions of ξ and η, respectively. In terms of x and t, the solution is

$$v(x,t) = F(x - Ct) + G(x + Ct). \tag{1.145}$$

Therefore, the general solution to the wave equation is the superposition of two waves, a right-running wave $F(x-Ct)$, and a left-running wave $G(x+Ct)$. The forms of the waves are determined by the initial and boundary conditions. We first consider the case of an infinite string and specify initial conditions at $t = 0$ by

$$v(x,0) = \phi(x) \qquad -\infty < x < \infty \tag{1.146}$$

$$\left(\frac{\partial v}{\partial t}\right)_{t=0} = \psi(x) \qquad -\infty < x < \infty. \tag{1.147}$$

That is to say that the string is deflected in the form given by $\phi(x)$, and then imparted a transverse velocity given by $\psi(x)$. Evaluating Equation (1.145) at $t = 0$ and using Equation (1.146), we get

$$F(x)+G(x) = \phi(x). \tag{1.148}$$

Differentiating Equation (1.145) with respect to t, we get

$$\frac{\partial v}{\partial t} = -CF\prime(x-Ct)+CG\prime(x+Ct) \tag{1.149}$$

where the ($\prime$) means derivative with respect to the argument of the function, and hence

$$-CF\prime(x)+CG\prime(x) = \psi(x). \tag{1.150}$$

This is an ordinary differential equation in x, and upon integration we get

$$-CF(x)+CG(x) = \int_0^x \psi(\beta)d\beta + BC \tag{1.151}$$

where B is an arbitrary constant. From Equations (1.148) and (1.151), we find

$$F(x) = \frac{1}{2}\phi(x) - \frac{1}{2C}\int_0^x \psi(\beta)d\beta - \frac{1}{2}B \tag{1.152}$$

$$G(x) = \frac{1}{2}\phi(x) + \frac{1}{2C}\int_0^x \psi(\beta)d\beta + \frac{1}{2}B. \tag{1.153}$$

D'Alembert's Solution

The general solution ($v = F(x-Ct)+G(x+Ct)$) is

$$v(x,t) = \frac{1}{2}\phi(x-Ct) + \frac{1}{2}\phi(x+Ct) + \frac{1}{2C}\int_{x-Ct}^{x+Ct} \psi(\beta)d\beta. \tag{1.154}$$

Example

We determine the deflection of a string for the initial conditions

$$\begin{aligned} v(x,0) &= a_o(b^2-x^2) && -b<x<b \\ &= 0 && |x|>b \end{aligned} \tag{1.155}$$

$$\left(\frac{\partial v}{\partial t}\right)_{t=0} = 0 \qquad -\infty < x < \infty. \tag{1.156}$$

Using the Heaviside step function ($H(\zeta) = 1$ if $\zeta > 1$, and $H(\zeta) = 0$ if $\zeta < 0$), we write the initial conditions as

$$\begin{aligned} \phi(x) &= a_o(b^2-x^2)[H(x+b)-H(x-b)] \\ \psi(x) &= 0 \end{aligned} \tag{1.157}$$

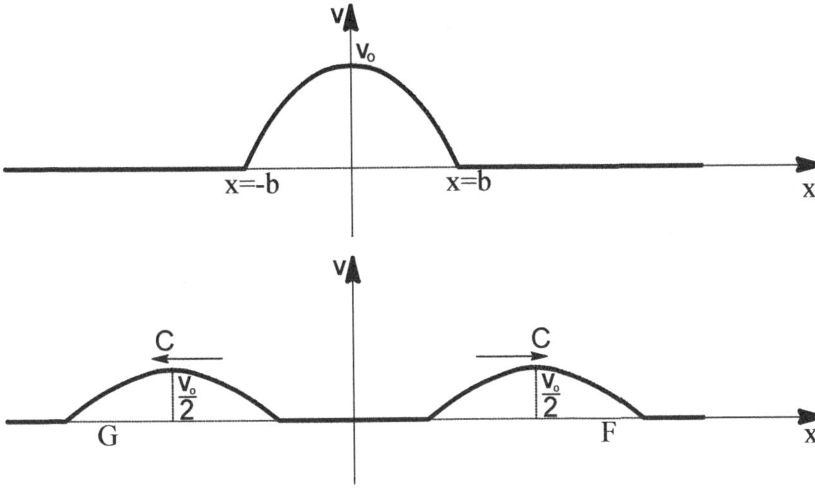

Figure 1.21 The initial deflection splits into two waves, one propagates to the right and the other to the left of the initial position, $v_o = a_o b^2$.

and we obtain the solution Equation (1.154),

$$\begin{aligned} v(x,t) &= \frac{1}{2}a_o[b^2-(x-Ct)^2][H(x-Ct+b)-H(x-Ct-b)] \\ &+\frac{1}{2}a_o[b^2-(x+Ct)^2][H(x+Ct+b)-H(x+Ct-b)]. \end{aligned} \tag{1.158}$$

The initial deflection is divided into two equal waves, one half propagates to the right of initial position and the other half propagates to the left as shown in Figure 1.21.

If the initial conditions are

$$\begin{aligned} \phi(x) &= a_o(b^2-x^2)[H(x+b)-H(x-b)] \\ \psi(x) &= 2a_oCx[H(x+b)-H(x-b)] \end{aligned} \tag{1.159}$$

we can obtain a solution that is a right-running wave only,

$$v(x,t) = a_o[b^2-(x-Ct)^2][H(x-Ct+b)-H(x-Ct-b)]. \tag{1.160}$$

Characteristic Lines of the Wave Equation

We consider a right-running wave

$$v = F(x-Ct). \tag{1.161}$$

For all points on a straight line in the *xt*-plane

$$x - Ct = \gamma^+ \tag{1.162}$$

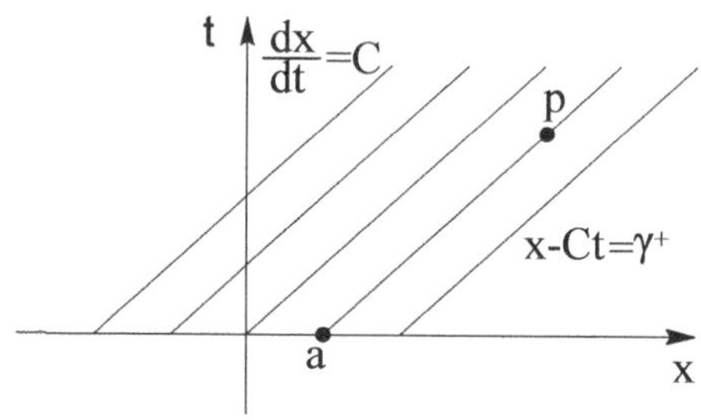

Figure 1.22 The γ^+ characteristic lines, slope $dx/dt = C$. For a right-running wave, solution at point p is $v_p = v_a$.

the function F is constant. By giving the constant γ^+ different values, we generate a family of parallel straight lines of slope $dx/dt = C$ in the xt-plane as shown in Figure 1.22. The function F may change if we move in the xt-plane across those lines, but it remains constant if we stay on one line. These lines are called γ^+-family of characteristic lines of the wave equation. Given the initial distribution of $v(x)$ on the x-axis at time $t = 0$, we can use characteristic lines to construct the solution of a right-running wave $v(x,t)$ at a later time $t > 0$. The value of v at any point p in the xt-plane is equal to v at the point where the γ^+ characteristic line through p intersects the x-axis. This process is demonstrated in Figure 1.22, we get $v_p = v_a$. Now, we consider a left-running wave

$$v = G(x + Ct). \tag{1.163}$$

For all points on a straight line in the xt-plane

$$x + Ct = \gamma^- \tag{1.164}$$

the function G is constant. By giving the constant γ^- different values, we generate a family of parallel straight lines of slope $dx/dt = -C$ in the xt-plane as shown in Figure 1.23. The function G may change if we move in the xt-plane across those lines, but it remains constant if we stay on one line. These lines are called the γ^--family of characteristic lines of the wave equation. Given the initial distribution of $v(x)$ on the x-axis at time $t = 0$, we can use characteristic lines to construct the solution of a left-running wave $v(x,t)$ at a later time $t > 0$. The value of v at any point p in the xt-plane is equal to v at the point where the γ^- characteristic line through p intersects the x-axis. This process is demonstrated in Figure 1.24, we get $v_p = v_b$.

Method of Characteristics

The general solution to the wave equation is a superposition of a right-running wave $F(x - Ct)$ and a left-running wave $G(x + Ct)$. Their forms are determined by the initial and boundary conditions. For the initial conditions given by Equations (1.146) and (1.147), the functions $F(x)$ and $G(x)$ are given by Equations (1.152) and (1.153),

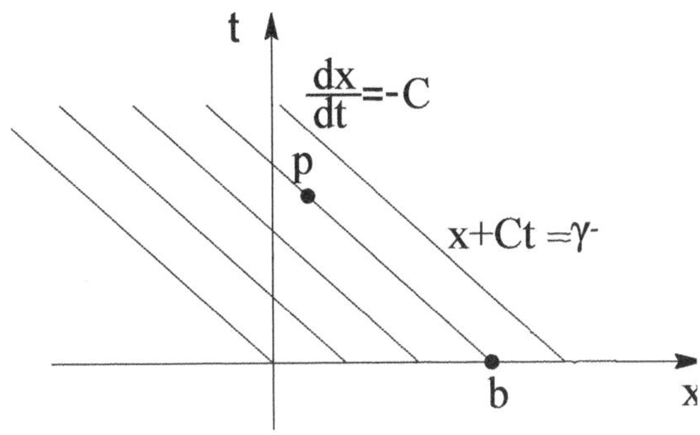

Figure 1.23 The γ^- characteristic lines, slope $dx/dt = -C$. For a left-running wave, solution at point p is $v_p = v_b$.

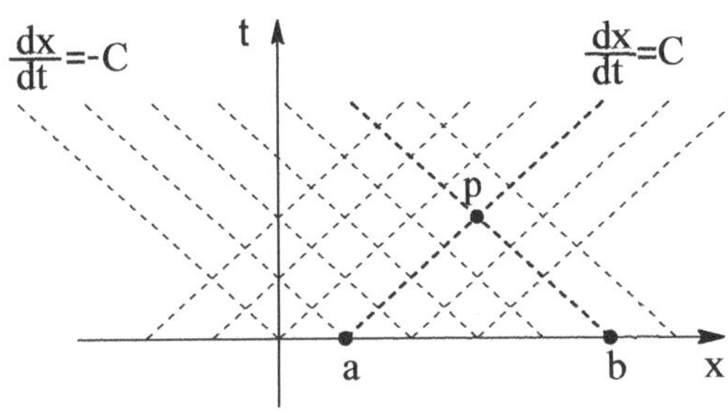

Figure 1.24 Method of characteristics, solution to the wave equation at point p is $v_p = F(x_a) + G(x_b)$.

respectively. Now, to find v at any point p in the xt-plane, we construct the γ^+ and γ^- characteristic lines through point p as shown in Figure 1.24. Let x_a and x_b be the points where the γ^+ and γ^- characteristic lines intersect the x-axis, respectively. The solution at p is

$$v_p = F(x_a) + G(x_b). \tag{1.165}$$

Domain of Dependence and Range of Influence

The solution to the wave equation in the domain bounded by the lines $a-b$, $b-p$, and $p-a$ is uniquely determined by the initial data ϕ and ψ on $a-b$ as shown in Figure 1.25. Initial data outside $a-b$ have no effects on the solution within that domain. Thus, $a-b$ is called the domain of dependence of point p. Initial data on a segment $a-b$ influence the solution at all points in the range bounded by $a-b$, and γ^- and γ^+ characteristic lines at a and b as shown in Figure 1.26. This is the range of influence of the initial data on $a-b$.

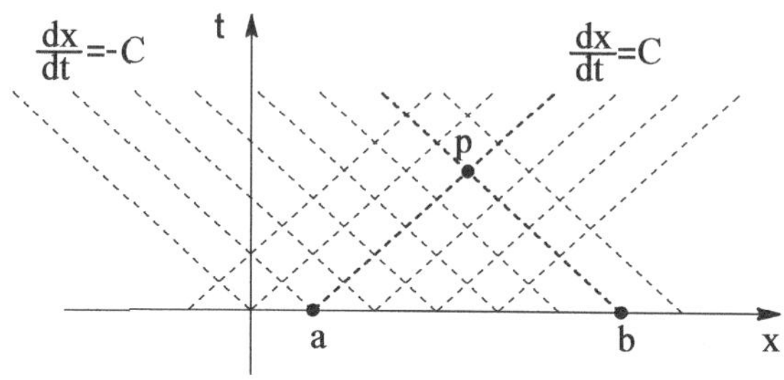

Figure 1.25 Domain of dependence, the solution to the wave equation at any point in region abp depends on the initial data on $a-b$ only.

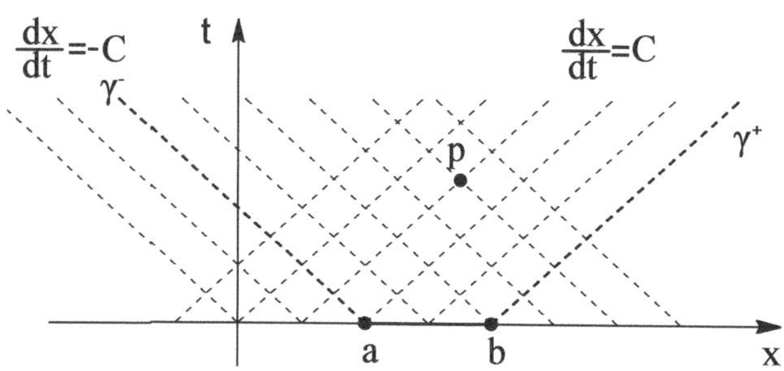

Figure 1.26 Range of influence, initial data on $a-b$ influence the solution in the range bounded by γ^- at a and γ^+ at b.

Example: Reflection of a Right-Running Wave From a Fixed End

A string is stretched and fixed at the points $x = 0$ and $x = 3$. It is given an initial deflection v and velocity $\partial v/\partial t$

$$\begin{aligned} v(x,0) &= 1+cos[\pi(x-1)] && 0 \leq x \leq 2 \\ &= 0 && \text{otherwise} \end{aligned} \tag{1.166}$$

$$\begin{aligned} \dot{v}(x,0) &= \pi sin[\pi(x-1)] && 0 \leq x \leq 2 \\ &= 0 && \text{otherwise} \end{aligned} \tag{1.167}$$

that create a right-running wave ($F \neq 0$, and $G = 0$). The boundary condition at the fixed end $x = 3$ is $v = 0$. We use the method of characteristics to determine the reflected wave. We select nine points on the incident wave as shown in Figure 1.27, and draw the corresponding γ^+ characteristic lines initiated on the x-axis as shown in Figure 1.28. We assume a nondimensional wave speed $C = 1$. The γ^+ line at point 1 intersects the fixed end at point a, where a γ^- characteristic line emanates (we may

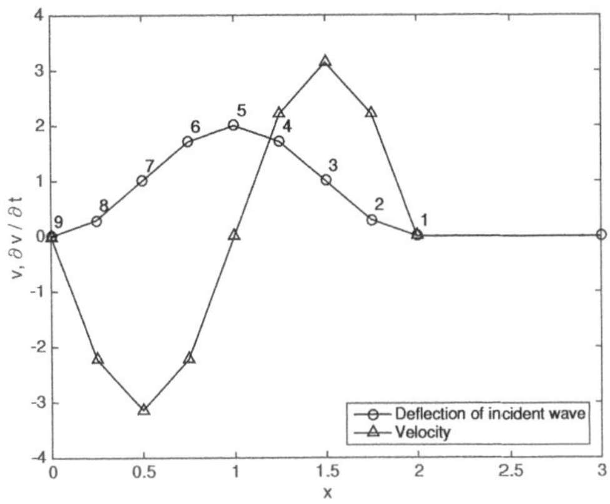

Figure 1.27 Initial deflection and velocity in a right-running wave on taut string.

consider it to be the reflection of the γ^+ line). The nine γ^+ lines reflect at the fixed end ($x=3$) at the points a, b, ..., and i. In the right-running wave, we have $v=F(x-Ct)$. We use the initial deflection to find F_1, F_2, ..., and F_9 on the 9 γ^+ characteristic lines. The values of G on the reflected γ^- lines are determined by the boundary condition at $x=3$. For example, at point d, $v_d = F_d + G_d = 0$, and hence $G_d = -F_d$. Since $F_d = F_4$, we get $G_d = -F_4$. Similarly, $G_a = -F_a$, $G_b = -F_b$, ..., and $G_i = -F_i$. Those values of G remain constant on the γ^- characteristic lines emanating at $x=3$. At time $t=3$, the reflected wave occupies the domain $1 \leq x \leq 3$, and is given by $v=G$ at points 1, 2, 3, ... , and 9. At these points there are also γ^+ characteristic lines but the values of F on them are zero; they are not shown in Figure 1.28. The reflected wave at time $t=3$ is shown in Figure 1.29; the deflection is inverted.

Example: Reflection of a Right-Running Wave from a Free End

We consider a right-running longitudinal wave on a bar that is fixed at $x=0$ and free at $x=3$. The initial displacement and strain are

$$\begin{aligned} u(x,0) &= 1+cos[\pi(x-1)] \quad 0 \leq x \leq 2 \\ &= 0 \quad \text{otherwise} \end{aligned} \tag{1.168}$$

$$\begin{aligned} \partial u/\partial x &= -\pi sin[\pi(x-1)] \quad 0 \leq x \leq 2 \\ &= 0 \quad \text{otherwise.} \end{aligned} \tag{1.169}$$

As shown in Figure 1.30, the forward half of the wave (points 1, 2, 3, and 4) is in compression (negative strain), and the rear half (points 9, 8, 7, and 6) is under tension (positive strain). The characteristic lines are given in Figure 1.28.

At a free end ($x=x_o$), the boundary condition demands that the stress (or strain) must be zero. Since the solution is $u=F(x-Ct)+G(x+Ct)$, we get $F\prime(x-Ct)+G\prime(x+Ct)=0$ at $x=x_o$, where the $\prime$ indicates differentiation with respect to the

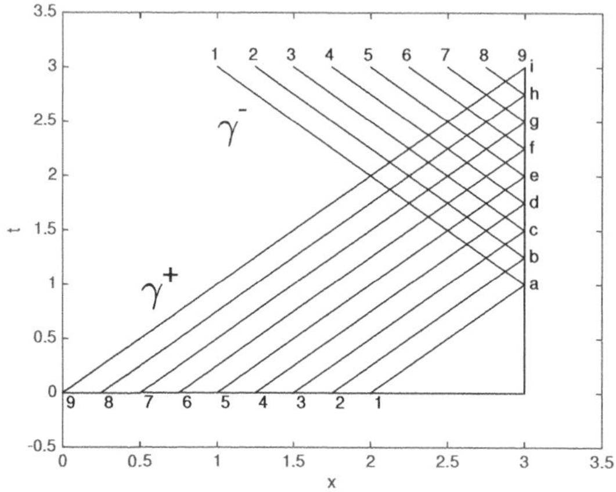

Figure 1.28 Reflection of a right-running wave. γ^+ characteristic lines initiated on the initial data line ($t = 0$), and γ^- characteristic initiated at the fixed end $x = 3$. Other characteristic lines are not shown.

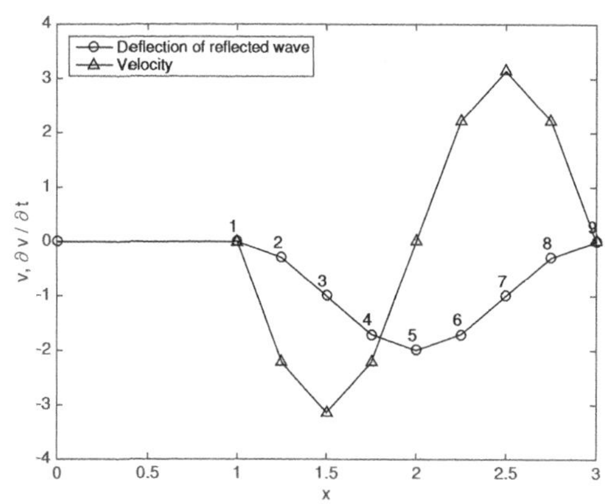

Figure 1.29 Reflected wave at time $t = 3$.

argument. Thus, we have $F\prime(x_o - Ct) + G\prime(x_o + Ct) = 0$, and integrating with respect to time we get, $-F(x_o - Ct) + G(x_o + Ct) = B$, where B is a constant. Thus, we get $G(x_o + Ct) = F(x_o - Ct) + B$. But $G(x_o) = 0$, thus $B = -F(x_o)$, and in the present example $B = 0$. Hence at the free end, we find $G = F$. The values of G on the reflected γ^- characteristic lines are determined by the boundary condition at $x = 3$. The free end condition gives $G_a = F_a$, $G_b = F_b$, ..., and $G_i = F_i$. Those values of G remain constant on the γ^- characteristic lines emanating at $x = 3$. At time $t = 3$, the reflected wave occupies the domain $1 \leq x \leq 3$, and is given by $u = G$ at points 1, 2, 3, .. , and 9. At these points there are also γ^+ characteristic lines but the values of F on them are zero; they are not shown in Figure 1.28. The displacement and strain (or stress)

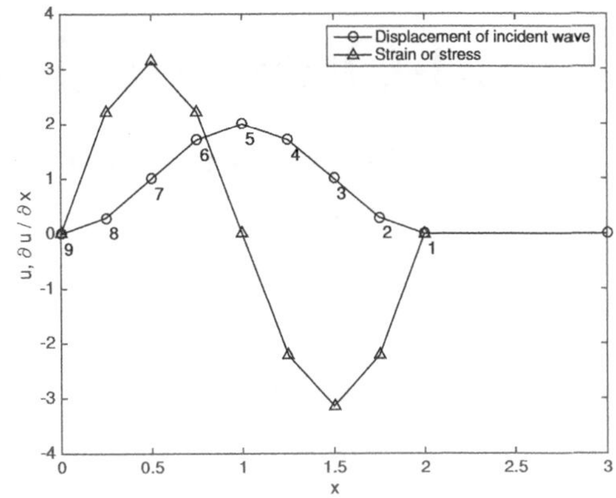

Figure 1.30 Initial longitudinal displacement u in a right-running wave in a bar, and the strain $\partial u/\partial x$. The stress is $E\partial u/\partial x$.

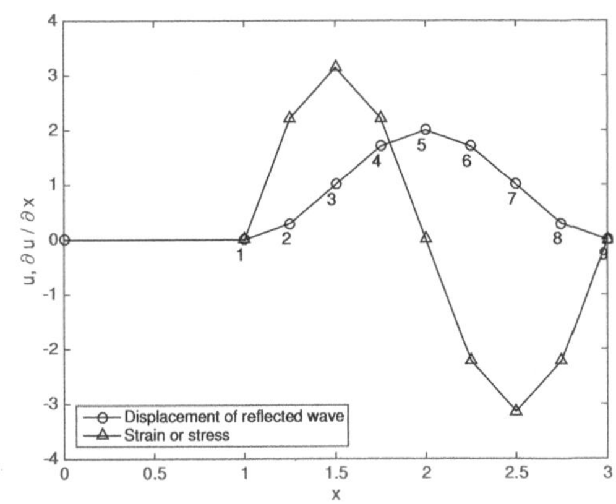

Figure 1.31 Longitudinal displacement u in a reflected left-running wave in a bar, and the strain $\partial u/\partial x$. The stress is $E\partial u/\partial x$.

in the (reflected) left-running wave at time $t = 3$ are shown in Figure 1.31. We note that the forward half (points 1, 2, 3, and 4) is now in tension (positive strain) whereas the rear half (points 6, 7, 8, and 9) is in compression (negative strain). Thus, at a free end, the stress of the reflected wave is of the opposite sign to that of the incident wave. A compression wave is reflected as a tension wave, and vice versa. Also, at a free end, the displacement $u = F + G = 2F$ is double that of the incident wave.

For reflection at a fixed end ($x = x_o$), we find $G(x_o + Ct) = -F(x_o - Ct)$. Letting $\theta = x_o + Ct$, we write $G(\theta) = -F(2x_o - \theta)$, hence $G(x + Ct) = -F(2x_o - x - Ct)$. The displacement is $u = F + G$ becomes $u = F(x - Ct) - F(2x_o - x - Ct)$ and strain (or stress) is $\partial u/\partial x = F'(x - Ct) + F'(2x_o - x - Ct)$. At the fixed end, we find $\partial u/\partial x = 2F'(x_o - Ct)$, thus stress is double the stress in the incident wave.

2 Second-Order Ordinary Differential Equations

In this chapter, we develop finite-element models for a class of boundary-value problems governed by a second-order ordinary differential equation (*ODE*). We consider a standard form of the differential equation, which we call the *Model Problem*. We first present the method of weighted residuals and then the weak form (also called the variational formulation) which is used for all problems in this book. Development of the finite-element method for a second-order *ODE* serves the purpose of introducing mathematical concepts and definitions that extend to more complex *ODE* as well as to a system of partial differential equations. Derivation of weak forms, definitions of primary and secondary variables, classification of boundary conditions, element shape functions, matrices, equations, and assembly of elements into a global system are common to all equations presented in this book. This chapter includes applications in solid and fluid mechanics (axial deformation of bars, and laminar channel flow), heat conduction in cooling fins and slabs, and mass diffusing (oxygen consumption in tissues.)

2.1 MODEL PROBLEM

A model boundary-value problem is given by the second-order ordinary differential equation

$$-\frac{d}{dx}[p(x)\frac{du}{dx}]+q(x)u-f(x)=0 \qquad a<x<b \tag{2.1}$$

and boundary conditions

$$\begin{aligned} x &= a: & \gamma_1\frac{du}{dx}+\beta_1 u=\alpha_1 \\ x &= b: & \gamma_2\frac{du}{dx}+\beta_2 u=\alpha_2 \end{aligned} \tag{2.2}$$

where p, q, and f are given functions of x, and α_i, β_i, and γ_i are given constants. We want to determine the unknown function $u(x)$. This boundary-value problem is encountered in a variety of engineering science and mechanics applications. Such applications include axial deformation of a slender rod (bar) and conduction heat transfer in one-dimensional fins or slabs. Consideration of these two problems will help us associate definite physical or mechanical entities with mathematical terms and definitions that arise during the formulation of the finite-element method.

We rewrite the model equation in operator form

$$\mathscr{L}(u)-f(x)=0 \qquad a<x<b \tag{2.3}$$

DOI: 10.1201/9781003323150-2

where $\mathscr{L}(u)$ is a linear operator defined by

$$\mathscr{L}(u) \equiv -\frac{d}{dx}[p(x)\frac{du}{dx}] + q(x)u. \tag{2.4}$$

We consider two functions $u(x)$ and $v(x)$, and the apply the operator to their sum,

$$\begin{aligned}
\mathscr{L}(u+v) &= -\frac{d}{dx}[p(x)\frac{d}{dx}(u+v)] + q(x)(u+v) \\
&= -\frac{d}{dx}[p(x)\frac{du}{dx}] + q(x)u - \frac{d}{dx}[p(x)\frac{dv}{dx}] + q(x)v \\
&= \mathscr{L}(u) + \mathscr{L}(v).
\end{aligned} \tag{2.5}$$

For an arbitrary constant C, we get

$$\mathscr{L}(Cu) = C\mathscr{L}(u). \tag{2.6}$$

For arbitrary constants C_1 and C_2, we get

$$\begin{aligned}
\mathscr{L}(C_1u + C_2v) &= \mathscr{L}(C_1u) + \mathscr{L}(C_2v) \\
&= C_1\mathscr{L}(u) + C_2\mathscr{L}(v)
\end{aligned} \tag{2.7}$$

and as a result, we also have for N functions $u_j(x)$ and constants C_j,

$$\mathscr{L}\left(\sum_{j=1}^{N} C_j u_j\right) = \sum_{j=1}^{N} C_j \mathscr{L}(u_j). \tag{2.8}$$

We note that the boundary conditions Equation (2.2) are non-homogeneous because α_i are not zero. We can make the boundary conditions homogeneous by the substitution

$$u = g + v \tag{2.9}$$

where g is a function that we construct to satisfy the conditions

$$\begin{aligned}
x &= a: & \gamma_1\frac{dg}{dx} + \beta_1 g &= \alpha_1 \\
x &= b: & \gamma_2\frac{dg}{dx} + \beta_2 g &= \alpha_2
\end{aligned} \tag{2.10}$$

and hence v satisfies the equation,

$$\mathscr{L}(v) - \tilde{f}(x) = 0 \qquad a < x < b \tag{2.11}$$

where

$$\tilde{f}(x) = f(x) - \mathscr{L}(g) \tag{2.12}$$

with homogeneous boundary conditions,

$$\begin{aligned}
x &= a: & \gamma_1\frac{dv}{dx} + \beta_1 v &= 0 \\
x &= b: & \gamma_2\frac{dv}{dx} + \beta_2 v &= 0
\end{aligned} \tag{2.13}$$

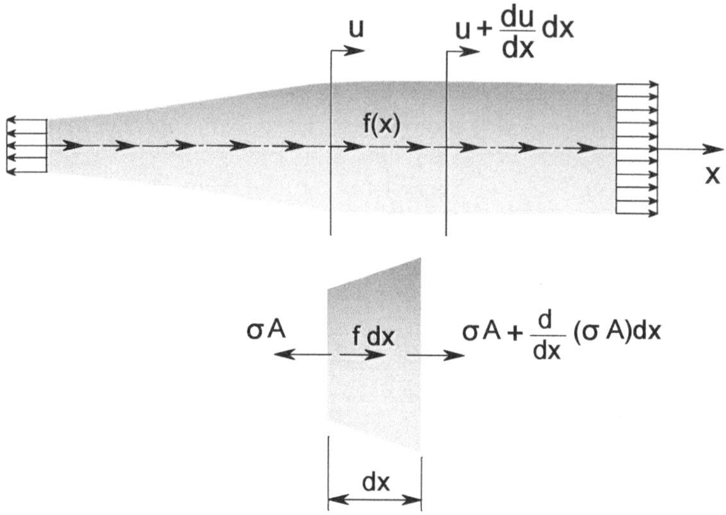

Figure 2.1 Axial deformation of a bar (slender rod). $u(x)$ is displacement and $f(x)$ is distributed axial force per unit length. Shown also is the free-body diagram of a small element of the bar, where σ is normal stress and A is the cross-sectional area.

2.1.1 AXIAL DEFORMATION OF A BAR

We consider the axial deformation of a bar (also called a slender rod). We let $A(x)$ denote the cross-sectional area of the bar which is loaded axially by a body force per unit length $f(x)$ as shown in Figure 2.1. Also, we let $u(x)$ denote the axial displacement of the section located initially at x. The normal strain ε and stress σ are given by

$$\varepsilon(x) = \frac{du}{dx} \tag{2.14}$$

$$\sigma(x) = E(x)\frac{du}{dx} \tag{2.15}$$

where E is Young's modulus, which may vary with x for a non-homogeneous material or a composite bar. The internal axial force at section x is then

$$R(x) = EA\frac{du}{dx}. \tag{2.16}$$

To determine the governing equation for the displacement field $u(x)$, we write the equation of equilibrium ($\sum F_x = 0$) for an infinitesimally small element as shown in Figure 2.1, and obtain

$$-\frac{d}{dx}[EA\frac{du}{dx}] - f(x) = 0. \tag{2.17}$$

A comparison of Equation (2.17) with the model equation Equation (2.1) shows that $p(x) = EA$, $q = 0$, and that f is the distributed axial force per unit axial length. The

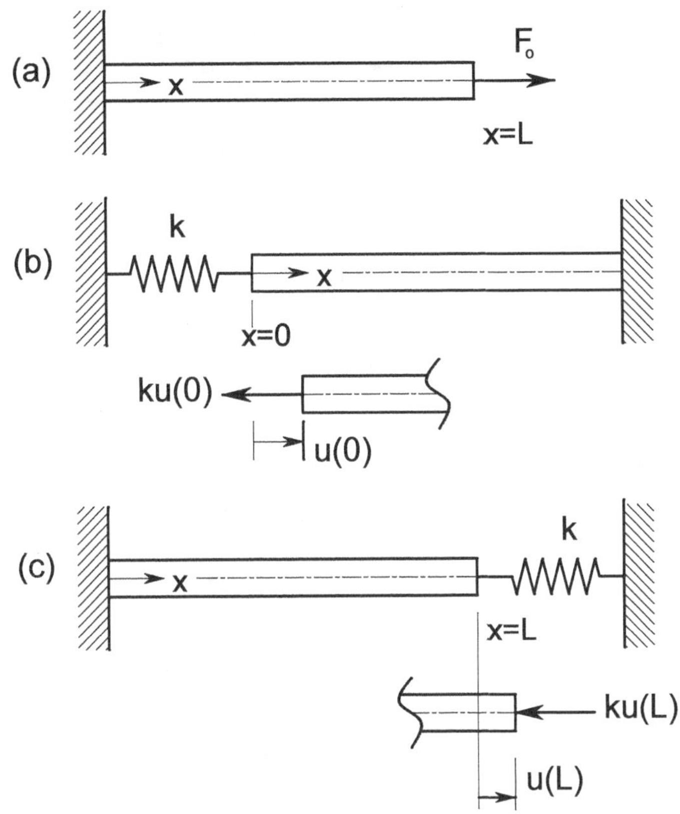

Figure 2.2 Boundary conditions for axial deformation problems, (a) concentrated force at a tip, and (b) and (c) one end is restrained by a linear spring.

method by which the bar is supported at its ends imposes certain boundary conditions on the displacement $u(x)$ or the axial force R. For example, we consider a bar of length L, which is fixed at $x=0$ and is axially loaded by a force F_o at the other end as shown in Figure 2.2a. The boundary conditions are

$$\begin{aligned} x &= 0: & u &= 0 \\ x &= L: & EA\frac{du}{dx} &= F_o. \end{aligned} \tag{2.18}$$

We consider another example of a bar which is restrained at $x=0$ by a linear spring of stiffness k and is fixed at the other end as shown in Figure 2.2b. The boundary conditions are

$$\begin{aligned} x &= 0: & -EA\frac{du}{dx}+ku &= 0 \\ x &= L: & u &= 0. \end{aligned} \tag{2.19}$$

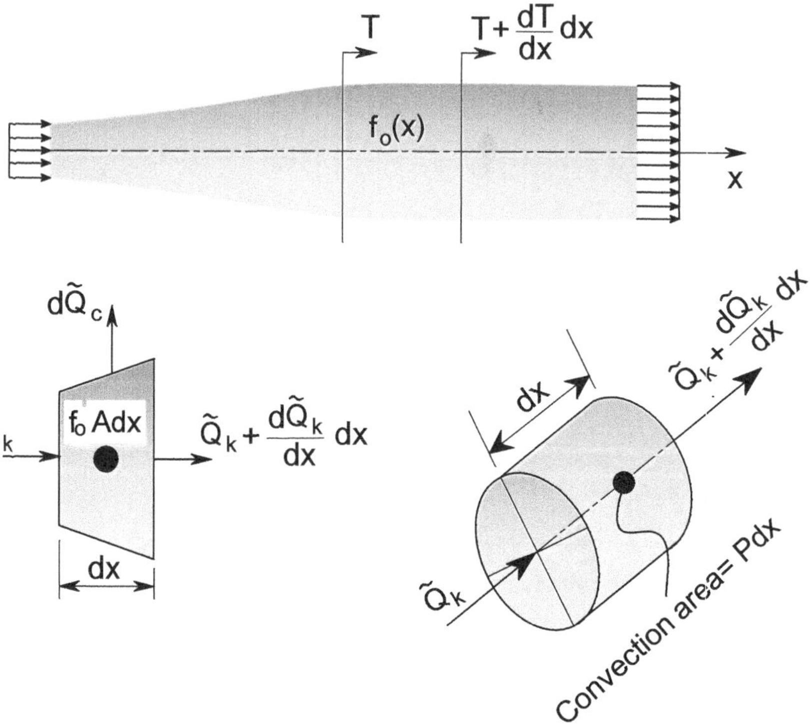

Figure 2.3 Heat conduction in a fin. $\tilde{Q}_k$ is heat transferred by conduction on the cross-section, and $d\tilde{Q}_c$ is heat transferred by convection on the bounding surface Pdx. P is the perimeter of the cross-section.

For this case both the displacement u and strain du/dx at the end $x = 0$ are unknown. Alternatively, if the end $x = L$ is restrained by a spring as shown in Figure 2.2c the boundary conditions are

$$
\begin{aligned}
x &= 0: & u &= 0 \\
x &= L: & EA\frac{du}{dx} + ku &= 0.
\end{aligned}
\tag{2.20}
$$

All of the above boundary conditions are special cases of the boundary conditions Equation (2.2).

2.1.2 ONE-DIMENSIONAL HEAT CONDUCTION IN SOLIDS

A second application that can be described by the model problem is conduction heat transfer in heating/cooling fins and slabs as shown in Figure 2.3. The temperature distribution, $T(x)$, in the fin, is governed by the energy equation (first law of thermodynamics). We isolate a small element of length dx as shown in Figure 2.3 and enforce energy balance. The rate of heat transfer by conduction is given by Fourier's law,

$$\widetilde{Q}_k = -kA\frac{dT}{dx} \tag{2.21}$$

where k is the thermal conductivity of the material and $A(x)$ is the cross-sectional area of the fin. The net heat transfer by conduction out of the element is

$$d\widetilde{Q}_k = \frac{d}{dx}\left(-kA\frac{dT}{dx}\right)dx. \tag{2.22}$$

Heat is also transferred by convection through the outer surface to the surrounding fluid, which is given by Newton's law of cooling,

$$d\widetilde{Q}_c = \tilde{h}(T - T_\infty)P(x)dx \tag{2.23}$$

where $\tilde{h}$ is the coefficient of convection heat transfer, T_∞ is the ambient temperature of the cooling fluid, and $P(x)$ is the perimeter of the fin cross-section at location x. Internal energy may also be generated within the element (due to chemical reaction or electric current) at the rate

$$d\widetilde{Q}_g = f_o A dx \tag{2.24}$$

where f_o is the rate of internal energy generation per unit volume. For steady conditions, energy balance gives

$$-d\widetilde{Q}_k - d\widetilde{Q}_c + d\widetilde{Q}_g = 0. \tag{2.25}$$

Substituting Equations (2.22)–(2.24) into Equation (2.25), we get

$$-\frac{d}{dx}[kA\frac{dT}{dx}] + \tilde{h}PT - (f_oA + \tilde{h}T_\infty P) = 0. \tag{2.26}$$

Again, a comparison with the model equation, Equation (2.1), shows that $p = kA$, $q = \tilde{h}P$, and $f = f_oA + \tilde{h}T_\infty P$. The thermal treatment at the fin's ends determines the boundary conditions to be imposed on the temperature field. At the end of the fin, we may specify the temperature or the rate of heat transfer by conduction. For example, if the ends of the fin are kept at given temperatures, the boundary conditions are

$$\begin{aligned} x &= 0: \quad && T = T_o \\ x &= L: \quad && T = T_L \end{aligned} \tag{2.27}$$

In this case, the rate of heat flow at the ends is unknown. Or, if the end at $x = L$ is insulated, the boundary conditions are

$$\begin{aligned} x &= 0: \quad && T = T_o \\ x &= L: \quad && kA\frac{dT}{dx} = 0 \end{aligned} \tag{2.28}$$

and in this case, the temperature at the end $x = L$ is unknown. A convection boundary condition may also be applied at the fin's end. For example, if end $x = L$ is subjected to convection, the boundary conditions are

$$\begin{aligned} x &= 0: & T &= T_o \\ x &= L: & kA\frac{dT}{dx} + \tilde{h}AT &= \tilde{h}AT_{\infty}. \end{aligned} \tag{2.29}$$

Here, both the temperature T and heat flow $-kAdT/dx$ at the end $x = L$ are unknown.

2.2 A MOTIVATIONAL EXAMPLE

The method of weighted residuals and the weak form (also called the variational formulation) are the main methods for generating finite-element models for boundary-value problems. To demonstrate the basic idea of the method of weighted residuals, we consider a special case of the model problem for which $p = 1$, $q = 1$, and $f = x^2$,

$$-\frac{d}{dx}[\frac{du}{dx}] + u - x^2 = 0 \qquad 0 < x < 1 \tag{2.30}$$

and boundary conditions

$$\begin{aligned} x &= 0: & u &= 1 \\ x &= 1: & \frac{du}{dx} &= -2. \end{aligned} \tag{2.31}$$

A solution to this problem is a function $u(x)$ that satisfies the differential equation at all points $0 < x < 1$, and also satisfies the two boundary conditions. The function $u(x)$ and its first derivative must be continuous, and its second derivative must also exist. The boundary conditions are non-homogeneous. To render them homogeneous, we make the substitution

$$u = g + v \tag{2.32}$$

and construct the function g such that

$$\begin{aligned} x &= 0: & g &= 1 \\ x &= 1: & \frac{dg}{dx} &= -2 \end{aligned} \tag{2.33}$$

and impose homogeneous boundary conditions on v

$$\begin{aligned} x &= 0: & v &= 0 \\ x &= 1: & \frac{dv}{dx} &= 0. \end{aligned} \tag{2.34}$$

In this example, a linear function is sufficient for g; hence we assume

$$g(x) = a_o + b_o x \tag{2.35}$$

and determine the constants a_o and b_o by enforcing the conditions Equation (2.33), we obtain

$$\begin{aligned} a_o &= 1 \\ b_o &= -2 \end{aligned} \tag{2.36}$$

and hence

$$g(x) = 1 - 2x. \tag{2.37}$$

One can assume a *trial solution* for v of the form

$$v(x) \approx V_N(x) = \sum_{j=1}^{N} C_j \phi_j(x) \tag{2.38}$$

where the functions $\phi_j, j = 1, ..., N$ are to be constructed such that the trial solution satisfies the two homogeneous boundary conditions for any values of the unknown coefficients $C_j, j = 1, ..., N$. To that end, we impose the conditions:

$$\begin{aligned} x = 0: &\quad \phi_j = 0 &\quad j = 1,2,...,N \\ x = 1: &\quad \frac{d\phi_j}{dx} = 0 &\quad j = 1,2,..,N. \end{aligned} \tag{2.39}$$

The functions ϕ_j may be assumed to be polynomials in x. They must be linearly independent functions. Simply stated, linear independence means that for a set of functions, no function can be a linear superposition of the remaining functions. For simplicity, we take $N = 2$, and construct ϕ_1 and ϕ_2. In view of conditions Equation (2.39), a linear function in x for ϕ_1 will be trivial, thus the lowest-order polynomial is the second order,

$$\phi_1 = a_1 + b_1 x + c_1 x^2 \tag{2.40}$$

We determine the coefficients a_1, b_1, and c_1 by applying the boundary conditions, Equation (2.39), which gives

$$\begin{aligned} a_1 &= 0 \\ b_1 + 2c_1 &= 0. \end{aligned} \tag{2.41}$$

The second equation shows that $b_1 = -2c_1$, and hence

$$\phi_1(x) = c_1(-2x + x^2) \tag{2.42}$$

The function ϕ_1 in the trial solution is multiplied by, as of now, an unknown coefficient C_1. Thus, we take $c_1 = 1$ in Equation (2.42), and consider

$$\phi_1(x) = -2x + x^2. \tag{2.43}$$

For ϕ_2 we have to consider a polynomial of order greater than 2. However, we should not skip consecutive powers in x, and hence we choose

$$\phi_2 = a_2 + b_2 x + c_2 x^2 + d_2 x^3. \tag{2.44}$$

The boundary conditions Equation (2.39) give

$$\begin{aligned} a_2 &= 0, \\ b_2 + 2c_2 + 3d_2 &= 0. \end{aligned} \tag{2.45}$$

The second condition can be satisfied by different choices for b_2, c_2, and d_2. But we must not choose $d_2 = 0$, because this will make ϕ_2 linearly dependent on ϕ_1. Here we take $c_2 = 0$, $d_2 = 1$, and deduce $b_2 = -3$. Hence,

$$\phi_2(x) = -3x + x^3. \tag{2.46}$$

The trial solution for the function u

$$u(x) \approx U_N(x) = g(x) + \sum_{j=1}^{N} C_j \phi_j(x) \tag{2.47}$$

becomes

$$u(x) \approx U_2(x) = 1 - 2x + C_1(-2x + x^2) + C_2(-3x + x^3). \tag{2.48}$$

It is easy to check that the trial solution Equation (2.48) satisfies the boundary conditions Equation (2.31) for any values for the coefficients C_1 and C_2. It remains to determine these coefficients that make the trial solution satisfy the differential equation in the domain $0 < x < 1$. However, this may not be possible unless the exact solution happens to be a third-order polynomial as implied by the trial solution.

The Residual

The residual, $R(x;C_j)$ is a measure of the error in satisfying the differential equation. Upon substituting the trial solution, Equation (2.47), into the differential equation Equation (2.1), we obtain the residual

$$R(x;C_j) = -\frac{d}{dx}[p(x)\frac{dU_N}{dx}] + q(x)U_N - f(x) \tag{2.49}$$

and for the example at hand, we get

$$R(x;C_1,C_2) = C_2x^3 + (-1 + C_1)x^2 + (-2 - 9C_2 - 2C_1)x + (-2C_1 + 1). \tag{2.50}$$

It is clear that we cannot find values for C_1 and C_2 that make the residual identically zero at every point in the domain $0 < x < 1$ as this requires that the coefficient of every power of x to be zero independently. Therefore, the trial solution with two parameters C_1 and C_2 cannot be an exact solution to our example. There are different strategies that we can follow to determine these two parameters that make the trial solution an *approximate* solution in some sense.

Since we are seeking values for two coefficients C_1 and C_2, we can demand that the residual be exactly zero at two arbitrary points; say $x = 1/3$ and $x = 2/3$, which are called *collocation points*. This will provide two algebraic equations that we can solve for the unknown coefficients. This method is known as the *collocation* method. Although the differential equation is exactly satisfied at the two collocation points, the trial solution is not necessarily equal to the exact solution at those points.

Another approach for finding the coefficients C_1 and C_2 is to minimize the integral of the square of the residual function, which is given by

$$E(C_1, C_2) = \int_0^1 R^2(x; C_1, C_2)dx \tag{2.51}$$

The necessary conditions for minimizing E are

$$\frac{\partial E}{\partial C_1} = 0 \tag{2.52}$$

$$\frac{\partial E}{\partial C_2} = 0 \tag{2.53}$$

hence, we have

$$\int_0^1 \frac{\partial R}{\partial C_1} R(x; C_1, C_2)dx = 0 \tag{2.54}$$

$$\int_0^1 \frac{\partial R}{\partial C_2} R(x; C_1, C_2)dx = 0. \tag{2.55}$$

Again we solve two algebraic equations for C_1 and C_2. This approach is known as the *least-squares method.* A more general approach is the method of *Weighted Residuals.*

Weighted Residuals

Weighted residuals are obtained by multiplying the residual function $R(x; C_j)$ by an arbitrary weight (or test) function $w(x)$ and integrating the result over the problem domain

$$R_w(C_1, C_2) = \int_0^1 w(x)R(x; C_1, C_2)dx. \tag{2.56}$$

We generate two independent equations by selecting two linearly independent weight (or test) functions w_1 and w_2, and insist that the weighted residuals vanish.

$$\int_0^1 w_1(x)R(x; C_1, C_2)dx = 0 \tag{2.57}$$

$$\int_0^1 w_2(x)R(x; C_1, C_2)dx = 0. \tag{2.58}$$

One may take $w_1 = 1$ and $w_2 = x$. Other admissible choices are $w_1 = \phi_1(x)$ and $w_2 = \phi_2(x)$. The approach of selecting the weight (test) functions to be the same as the trial solution basis functions is called the Galerkin method.

2.3 THE METHOD OF WEIGHTED RESIDUALS

We write the model equation (2.1) in operator form

$$\mathscr{L}(u) - f(x) = 0 \qquad a < x < b \tag{2.59}$$

where $\mathscr{L}(u)$ is defined by

$$\mathscr{L}(u) \equiv -\frac{d}{dx}[p(x)\frac{du}{dx}] + q(x)u. \tag{2.60}$$

Trial Solution

The trial solution for the function u is written as

$$u(x) \approx U_N(x) = g(x) + \sum_{j=1}^{N} C_j \phi_j(x). \tag{2.61}$$

The function g satisfies the two conditions:

$$\begin{aligned} x &= a: \qquad \gamma_1 \frac{dg}{dx} + \beta_1 g = \alpha_1 \\ x &= b: \qquad \gamma_2 \frac{dg}{dx} + \beta_2 g = \alpha_2. \end{aligned} \tag{2.62}$$

Each of the functions ϕ_j satisfies the homogeneous form of the boundary conditions

$$\begin{aligned} x &= a: \qquad \gamma_1 \frac{d\phi_j}{dx} + \beta_1 \phi_j = 0 \\ x &= b: \qquad \gamma_2 \frac{d\phi_j}{dx} + \beta_2 \phi_j = 0. \end{aligned} \tag{2.63}$$

Because the function g and the N basis functions ϕ_j satisfy the boundary conditions Equation (2.62) and Equation (2.63), the trial solution $U_N(x)$ given by Equation (2.61) satisfies the boundary conditions on $u(x)$ for any values of the N coefficients C_j. Note that the function g is not needed (choose $g = 0$) if the two boundary conditions are homogeneous ($\alpha_1 = 0$ and $\alpha_2 = 0$).

Residual Function

For the trial solution U_N, the residual function is

$$R(x; C_j) = \mathscr{L}(U_N) - f(x). \tag{2.64}$$

The operator $\mathscr{L}$ is a linear operator. Thus, for functions $F(x)$ and $G(x)$, and constants A_1 and A_2, it is easy to show that

$$\begin{aligned} \mathscr{L}(F+G) &= \mathscr{L}(F) + \mathscr{L}(G) \\ \mathscr{L}(A_1 F) &= A_1 \mathscr{L}(F) \\ \mathscr{L}(A_1 F + A_2 G) &= A_1 \mathscr{L}(F) + A_2 \mathscr{L}(G). \end{aligned} \tag{2.65}$$

Using the linearity properties of the operator $\mathscr{L}$, we obtain

$$R(x;C_j) = \mathscr{L}(g) + \sum_{j=1}^{N} C_j \mathscr{L}(\phi_j) - f(x). \tag{2.66}$$

The residual function $R(x,C_j)$ is a function of x and all the unknown coefficients C_j, $j = 1,2,...,N$. The weighted residuals are obtained by multiplying the residual function $R(x,C_j)$ by weight (test) functions $w_i(x)$ that we select, and then integrating the result over the problem domain. Equating the weighted residuals to zero, we obtain the statement of weighted residuals,

$$\int_a^b w_i(x)R(x;C_j)dx = 0, \qquad i = 1,2,...,N. \tag{2.67}$$

Since we have N unknown coefficients, we apply the weighted residuals statement for N linearly independent weight (test) functions $w_i(x)$, $i = 1,2,...,N$. Substituting Equation (2.66) into Equation (2.67), we get

$$\sum_{j=1}^{N} \left[\int_a^b w_i \mathscr{L}(\phi_j) dx \right] C_j = \int_a^b w_i \left[f(x) - \mathscr{L}(g) \right] dx, \qquad i = 1,2,...,N. \tag{2.68}$$

This is a linear system of algebraic equations to be solved for the coefficients C_j, $j = 1,2,...N$. In matrix form the system is

$$[A]\{C\} = \{\tilde{b}\} \tag{2.69}$$

where the entries of matrix $[A]$ are

$$a_{ij} = \int_a^b w_i \mathscr{L}(\phi_j) dx \tag{2.70}$$

and

$$\tilde{b}_i = \int_a^b w_i \left[f(x) - \mathscr{L}(g) \right] dx. \tag{2.71}$$

In summary, the method of weighted residuals involves the following steps: (1) construct the function g, (2) construct N basis functions ϕ_i and select N weight functions $w(x)$, (3) evaluate the integrals Equation (2.70) and Equation (2.71), and (4) solve a linear system Equation (2.69). An approximate $N-$term solution to the boundary-value problem is given by Equation (2.61).

The Galerkin Method

The hallmark of the Galerkin method is that the weight (test) functions are linear superposition of the basis functions ϕ_j constructed in the trial solution. Since the trial solution has N unknown coefficients and N corresponding basis functions ϕ_j, we can construct N linearly independent weight (test) functions, each of the general form

$$w(x) = \sum_{j=1}^{N} d_j \phi_j \tag{2.72}$$

where d_j are arbitrary constants that we select. It suffices to take

$$w_i(x) = \phi_i, \qquad i = 1,2,...,N \tag{2.73}$$

and hence the linear system coefficients are

$$a_{ij} = \int_a^b \phi_i \mathscr{L}(\phi_j) dx. \tag{2.74}$$

and

$$\tilde{b}_i = \int_a^b \phi_i \left[f(x) - \mathscr{L}(g)\right] dx. \tag{2.75}$$

Least-Squares Method

The least-squares method corresponds to a weighted residual method with weight (test) functions given by

$$w_i = \frac{\partial R}{\partial C_i}. \tag{2.76}$$

Using Equation (2.66), we get

$$w_i = \mathscr{L}(\phi_i) \tag{2.77}$$

and hence the corresponding linear system coefficients are

$$a_{ij} = \int_a^b \mathscr{L}(\phi_i)\mathscr{L}(\phi_j) dx \tag{2.78}$$

which is a symmetric matrix ($a_{ij} = a_{ji}$), and

$$\tilde{b}_i = \int_a^b \mathscr{L}(\phi_i) \left[f(x) - \mathscr{L}(g)\right] dx. \tag{2.79}$$

Collocation Method

By taking $w_i(x) = \delta(x - x_i)$, where $\delta(x)$ is the Dirac delta function, for N collocation points $a < x_i < b$, $i = 1,2,...,N$, we get coefficients

$$a_{ij} = \left[\mathscr{L}(\phi_j)\right]_{x=x_i} \tag{2.80}$$

$$\tilde{b}_i = \left[f(x) - \mathscr{L}(g)\right]_{x=x_i}. \tag{2.81}$$

Example 2.1. Weighted Residuals

Given $p(x) = 1 + x$, $q(x) = e^x$ and $f(x) = -(1+x)e^x + e^{2x}$, consider the boundary value problem

$$-\frac{d}{dx}(p\frac{du}{dx}) + qu = f \quad \text{on} \quad 0 < x < 1$$

with boundary conditions

$$x = 0: \ u = 2$$

$$x = 1: \ p\frac{du}{dx} + 3u = 3 + 5e.$$

We wish to use a weighted-residuals method to determine a two-term ($N = 2$) approximate solution

$$U_2 = g(x) + C_1\phi_1(x) + C_2\phi_2(x).$$

(i) Select admissible polynomials for g and ϕ_j, and determine the coefficients C_j by the Galerkin method.

(ii) Plot the pointwise error defined by $|u - U_2|$ against $0 \leq x \leq 1$, where $u = 1 + e^x$ is the exact solution.

(iii) Determine the *Supmetric* of the error, which is the maximum of the pointwise error.

(iv) Determine the L_2 norm of the error, which is defined by

$$\|u - U_2\|_o = \left[\int_0^1 |u - U_2|^2 \, dx\right]^{1/2}.$$

(v) Compute and plot the residual function $R(x; C_1, C_2)$.

Solution

Assume a linear function for $g(x)$,

$$g(x) = a_o + b_o x. \tag{2.82}$$

Noting that at the boundary $x = 1$ the function $p(x) = 1 + x = 2$, we determine the constants a_o and b_o by enforcing the conditions:

$$\begin{aligned} x &= 0: & g &= 2 \\ x &= 1: & 2\frac{dg}{dx} + 3g &= 3 + 5e \end{aligned} \tag{2.83}$$

thus

$$\begin{aligned} a_o &= 2 \\ 2b_o + 3(a_o + b_o) &= 3 + 5e \end{aligned} \tag{2.84}$$

and hence

$$g(x) = 2 + (-3/5 + e)x \tag{2.85}$$

The lowest-order polynomial for ϕ_1 is second order.

$$\phi_1 = a_1 + b_1 x + c_1 x^2. \tag{2.86}$$

We determine the coefficients a_1, b_1, and c_1 by applying the boundary conditions:

$$\begin{aligned} x &= 0: & \phi_1 &= 0 \\ x &= 1: & 2\frac{d\phi_1}{dx} + 3\phi_1 &= 0 \end{aligned} \tag{2.87}$$

thus

$$\begin{aligned} a_1 &= 0 \\ 2(b_1+2c_1)+3(a_1+b_1+c_1) &= 0. \end{aligned} \tag{2.88}$$

The second equation shows that $b_1 = -7c_1/5$. We set $c_1 = 5$ and $b_1 = -7$, hence

$$\phi_1(x) = -7x+5x^2. \tag{2.89}$$

For ϕ_2 we have to consider a polynomial of order greater than two, hence we choose

$$\phi_2 = a_2+b_2x+c_2x^2+d_2x^3 \tag{2.90}$$

The boundary conditions give

$$\begin{aligned} x &= 0: & \phi_2 &= 0 \\ x &= 1: & 2\frac{d\phi_2}{dx}+3\phi_2 &= 0. \end{aligned} \tag{2.91}$$

thus

$$\begin{aligned} a_2 &= 0 \\ 2(b_2+2c_2+3d_2)+3(a_2+b_2+c_2+d_2) &= 0. \end{aligned} \tag{2.92}$$

The second condition can be satisfied by different choices for b_2, c_2, and d_2. But we must not choose $d_2 = 0$ because this will make ϕ_2 linearly dependent on ϕ_1. Here we take $c_2 = 0$, $d_2 = 5$, and deduce $b_2 = -9$. Hence,

$$\phi_2(x) = -9x+5x^3. \tag{2.93}$$

These functions are plotted in Figure 2.4a. The two-term trial solution ($N = 2$)

$$u(x) \approx U_2(x) = g(x)+C_1\phi_1(x)+C_2\phi_2(x) \tag{2.94}$$

becomes

$$u(x) \approx U_2(x) = 2+(-3/5+e)x+C_1(-7x+5x^2)+C_2(-9x+5x^3). \tag{2.95}$$

For the Galerkin method, we use Equations (2.74) and (2.75) and obtain the system

$$\begin{bmatrix} 34.5353 & 63.2783 \\ 63.2783 & 119.490 \end{bmatrix} \begin{Bmatrix} C_1 \\ C_2 \end{Bmatrix} = \begin{Bmatrix} 6.54160 \\ 12.1973 \end{Bmatrix}. \tag{2.96}$$

Solving this system, we get $C_1 = 0.0802570$, and $C_2 = 0.0595764$. A comparison between the exact solution and the approximate Galerkin solution is shown in Figure 2.4b, and the pointwise error is shown in Figure 2.4c. The maximum of the pointwise error is the supmetric error = 1.25017×10^{-3}, and the L_2 norm is 8.53116×10^{-4}.

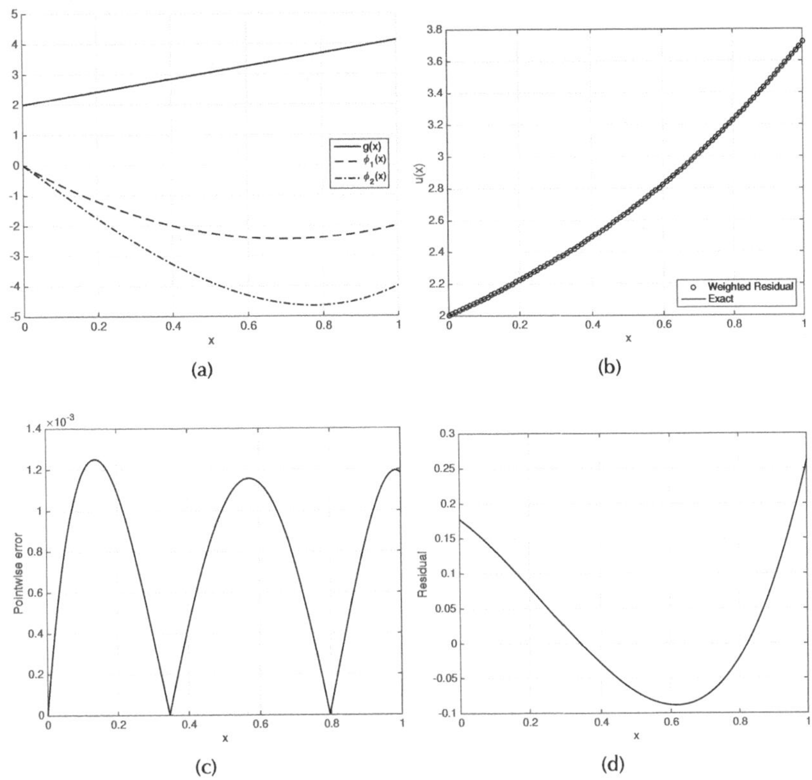

Figure 2.4 Example 2.1. (a) The functions g, ϕ_1, and ϕ_2. (b) A comparison between the two-term Galerkin solution and the exact solution. (c) Pointwise error. (d) Residual function.

These measures show the difference between the exact solution and the approximate solution. The residual function $R(x;C_1,C_2)$ is a measure of the error in satisfying the differential equation; this is shown in Figure 2.4d. The differential equation is exactly satisfied at two points only where the residual is zero. The error in the solution, u is zero at $x = 0$ by the construction of the trial solution, but it is not necessarily zero at the other boundary which involves the derivative of u. Even though the trial solution satisfies the two boundary conditions, the differential equation is not exactly satisfied at the boundary points, hence the residual is not zero at $x = 0$ or $x = 1$.

2.4 WEAK FORM

The formulation of boundary-value problems as given by Equations (2.1) and (2.2) or the weighted residual statement Equation (2.67) are called strong formulations

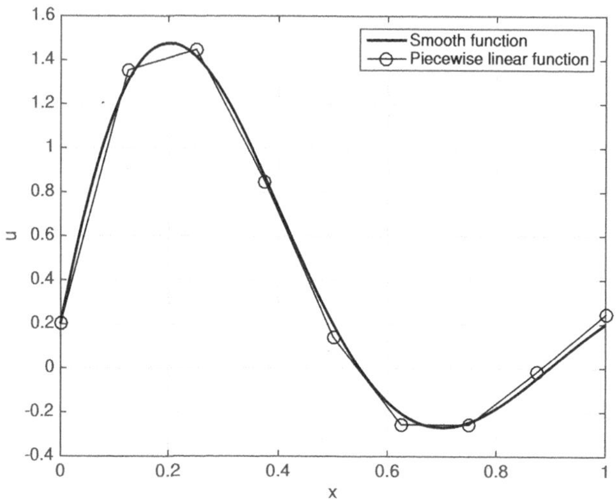

Figure 2.5 A smooth function $u(x)$, and a piecewise linear approximation.

because of the required continuity of the trial solutions. Another formulation of the problem with less stringent continuity conditions on the unknown function $u(x)$ is possible; it is known as the weak form or variational formulation. The exact solution to the model problem is a smooth function $u(x)$ that is continuous with continuous first derivative. Such a function is sketched in Figure 2.5 along with a piecewise linear function that appears to be a good approximation to the exact solution. However, the piecewise linear function is not a genuine solution to the second-order differential equation because its first derivative is discontinuous and the second derivative is zero everywhere except at the nodes where it is undefined. Thus, such a function is not admissible as a solution to the strong form that demands continuity of the first derivative. As we shall see, the weak form, because it has less stringent requirements on the continuity of the unknown function, admits the piecewise linear approximation as a *generalized* or *weak* solution. Furthermore, the weak form is a natural framework for problems with discontinuous data. We consider the axial deformation of a bar whose cross-sectional area A or modulus E has a discontinuity at some section. In this case, the axial force $EAdu/dx$ must have the same value on both sides of that section. Since EA is discontinuous, then du/dx must also be discontinuous at that section. Similarly, for heat conduction in a composite wall made of materials with different thermal conductivity k, the heat flux $-kdT/dx$ must be continuous at the interface. But because k is discontinuous, so is dT/dx. Such discontinuities in the slope of the unknown displacement or temperature are legitimate and must be represented by the trial solution.

To derive the weak form, we multiply Equation (2.1) by a test (or weight) function $w(x)$ and integrate over the problem domain,

$$\int_a^b w\left[-\frac{d}{dx}\left(p\frac{du}{dx}\right)+qu-f\right]dx=0. \tag{2.97}$$

Next, integrate the first term by parts

$$\left(-wp\frac{du}{dx}\right)_{x=a}^{x=b}+\int_a^b\left[\frac{dw}{dx}p\frac{du}{dx}+wqu\right]dx-\int_a^b wf\,dx=0. \tag{2.98}$$

Definition of Primary and Secondary Variables

The first term is called the boundary term, and based on this term we define secondary and primary variables. Reddy [97] defined the secondary variable to be the coefficient of the test function in the boundary term.

$$R=p\frac{du}{dx}. \tag{2.99}$$

The primary variable is u. Introducing the notations

$$R_a=-\left(p\frac{du}{dx}\right)_{x=a} \tag{2.100}$$

and

$$R_b=\left(p\frac{du}{dx}\right)_{x=b} \tag{2.101}$$

into Equation (2.98), we get

$$-w(b)R_b-w(a)R_a+\int_a^b\left[\frac{dw}{dx}p\frac{du}{dx}+wqu\right]dx-\int_a^b wf\,dx=0. \tag{2.102}$$

Classification of Boundary Conditions

To complete the derivation of the weak form, we need to apply the boundary conditions. First, we classify the given boundary conditions. A boundary condition may be classified as one of three types: essential, natural, or mixed. If a boundary condition specifies a primary variable, it is called an essential boundary condition (EBC). If a boundary condition specifies a secondary variable, it is called a natural boundary condition (NBC). A boundary condition that gives a relationship between the primary and secondary variables on the boundary is called a mixed boundary condition (MBC). Thus, an essential boundary condition means that u is given on the boundary, and a natural boundary condition means that R is given on the boundary. A mixed boundary condition means that neither u nor R is given on the boundary but we have an equation that can be used to express R in terms of u on the boundary.

Now we are in a position to make decisions regarding the boundary terms in Equation (2.102). At an NBC, the specified value of the secondary variable R is substituted into the corresponding boundary term in Equation (2.102). At an MBC, we use the boundary condition to express the secondary variable R in terms of the primary variable u and the result is substituted into the corresponding boundary term in Equation (2.102). Finally, at an EBC we require that the selected test (weight) function to be zero at the corresponding boundary point. For example, if the boundary condition at $x = a$ is an EBC, then we choose $w(x)$ such that $w(a) = 0$, and if the boundary condition at $x = b$ is an EBC, then $w(b) = 0$. By choosing a $w(x)$ that is zero at an EBC, we remove from the weak form the unknown secondary variable at an EBC. Therefore, the weak form depends on the type of boundary conditions given by the problem statement.

Examples of EBC are given by the first condition of Equation (2.18), the second condition of Equation (2.19), the first condition of Equation (2.20), the two conditions of Equation (2.27), the first condition of Equation (2.28), the first condition of Equation (2.29), the first condition of Equation (2.31), and the condition at $x = 0$ of Example 2.1. Examples of NBC are given by the second condition of Equation (2.18), the second condition of Equation (2.28), and the second condition of Equation (2.31). Examples of MBC are given by the first condition of Equation (2.19), the second condition of Equation (2.20), the second condition of Equation (2.29), and the condition at $x = 1$ of Example 2.1.

We note that if the primary variable is specified at the boundary, the corresponding secondary variable is unknown and will be determined as part of the solution. And if the secondary variable is specified on the boundary, the corresponding primary variable is unknown and will be determined as part of the solution. At an MBC both the primary and secondary variables are unknown. Also, in the present formulation of boundary-value problems, we cannot specify both the primary and secondary variables at the same boundary point; only one of them can be specified and the other will be determined as part of the solution.

Considering Example 2.1, we see that the boundary condition at $x = 0$ is an EBC, while the condition at $x = 1$ is an MBC. Thus, we set $w(0) = 0$. The MBC gives

$$x = 1: \; p\frac{du}{dx} = -3u + 3 + 5e$$

which can be written as

$$R_b = -3u + 3 + 5e$$

We substitute this value of R_b into Equation (2.102), and obtain the weak form for Example 2.1

$$\int_0^1 \left[\frac{dw}{dx}p\frac{du}{dx} + wqu\right] dx + 3w(1)u(1) = \int_0^1 wf dx + w(1)(3 + 5e). \qquad (2.103)$$

A solution to the weak form is a continuous function $u(x)$ that satisfies the EBC $x = 0: u = 2$, has a non-zero first derivative, and satisfies Equation (2.103) for all test (weight) functions, $w(x)$ that are continuous, have first derivative and satisfy the homogeneous form of the EBC ($w(0) = 0$).

A solution to the strong form of Example 2.1 is also a solution to its weak form Equation (2.103). It can be shown that a twice differentiable solution to the weak form is also a solution to the strong form. However, a solution to the weak form need not be twice differentiable; only the first derivative needs to exist. Such a solution is not a genuine solution to the strong form and it is called a weak or generalized solution.

The weak form simplifies the finite-element analysis of multi-dimensional problems. It also provides a straightforward treatment of problems with discontinuous coefficients as in the case of axial deformation of stepped bars, for which EA is discontinuous, or heat conduction in composite walls for which kA is also discontinuous.

To find an approximate solution to the weak form of a problem, we write it in operator form

$$\mathtt{A}(w,u) = \mathtt{F}(w) \tag{2.104}$$

where, for the example at hand, we have

$$\mathtt{A}(w,u) = \int_0^1 \left[\frac{dw}{dx}p\frac{du}{dx} + wqu\right] dx + 3w(1)u(1) \tag{2.105}$$

and

$$\mathtt{F}(w) = \int_0^1 wf dx + w(1)(3+5e). \tag{2.106}$$

We note that $\mathtt{A}(w,u)$ is a bilinear functional in w and u, and $\mathtt{F}(w)$ is a linear functional in w. For functions $F(x)$ and $G(x)$, and a constant C, it is easy to show that

$$\begin{aligned}
\mathtt{A}(w,F+G) &= \mathtt{A}(w,F)+\mathtt{A}(w,G)\\
\mathtt{A}(F+G,u) &= \mathtt{A}(F,u)+\mathtt{A}(G,u)\\
\mathtt{A}(w,Cu) &= C\mathtt{A}(w,u)\\
\mathtt{A}(Cw,u) &= C\mathtt{A}(w,u)\\
\mathtt{F}(F+G) &= \mathtt{F}(F)+\mathtt{F}(G)\\
\mathtt{F}(Cw) &= C\mathtt{F}(w).
\end{aligned} \tag{2.107}$$

We assume a trial solution for the function u in the general form

$$u(x) \approx U_N(x) = g(x) + \sum_{j=1}^{N} C_j\phi_j(x). \tag{2.108}$$

The trial solution must satisfy the EBC(s), if any, but it need not satisfy NBC or MBC because these conditions are already included in the weak form and will be approximately satisfied. We substitute selected test functions (w_i, $i = 1,2,...,N$) and the assumed trial solution into the weak form,

$$\mathtt{A}(w_i, g + \sum_{j=1}^{N} C_j\phi_j) = \mathtt{F}(w_i) \qquad i = 1,2,...,N. \tag{2.109}$$

Using the linearity properties of A, we obtain

$$\sum_{j=1}^{N} \mathtt{A}(w_i,\phi_j)C_j = \mathtt{F}(w_i) - \mathtt{A}(w_i,g) \qquad i = 1,2,...,N. \tag{2.110}$$

This is a linear system of algebraic equations for the coefficients C_j.

Example 2.2. Weak Form

Determine a three-term Galerkin approximation

$$U_3 = g(x) + C_1\phi_1(x) + C_2\phi_2(x) + C_3\phi_3(x)$$

to the solution of the weak form of the boundary value problem specified in Example 2.1. Construct $g(x)$ so that it satisfies only the non-homogeneous essential boundary condition(s), and ϕ_j to be the lowest order polynomials admissible by the weak form. Determine the L_2 norm of the error and compare it with that of the weighted residual Galerkin method found in Example 2.1.

Solution

The function g needs to satisfy the EBC at $x = 0$, that is $g(0) = 2$. There are no other requirements. Thus, the simplest choice is $g(x) = 2$. The functions ϕ_j must have at least first derivative and satisfy the homogeneous form of the EBC, $\phi_j(0) = 0$. And of course, they have to be linearly independent. The lowest order polynomials are $\phi_1 = x$, $\phi_2 = x^2$, and $\phi_3 = x^3$. The basis functions are plotted in Figure 2.6a. The three-term approximation is

$$U_3 = 2 + C_1x + C_2x^2 + C_3x^3.$$

The weak form is given by Equation (2.103), which we write here in operator form

$$\mathtt{A}(w,u) = \mathtt{F}(w) \tag{2.111}$$

where

$$\mathtt{A}(w,u) = \int_0^1 \left[\frac{dw}{dx}p\frac{du}{dx} + wqu\right]dx + 3w(1)u(1) \tag{2.112}$$

and

$$\mathtt{F}(w) = \int_0^1 wf dx + w(1)(3+5e). \tag{2.113}$$

For this example, the linear system Equation (2.110) is

$$[A]\{C\} = \{\tilde{b}\} \tag{2.114}$$

where

$$a_{ij} = \int_0^1 \left[\frac{dw_i}{dx}p\frac{d\phi_j}{dx} + w_iq\phi_j\right]dx + 3w_i(1)\phi_j(1) \tag{2.115}$$

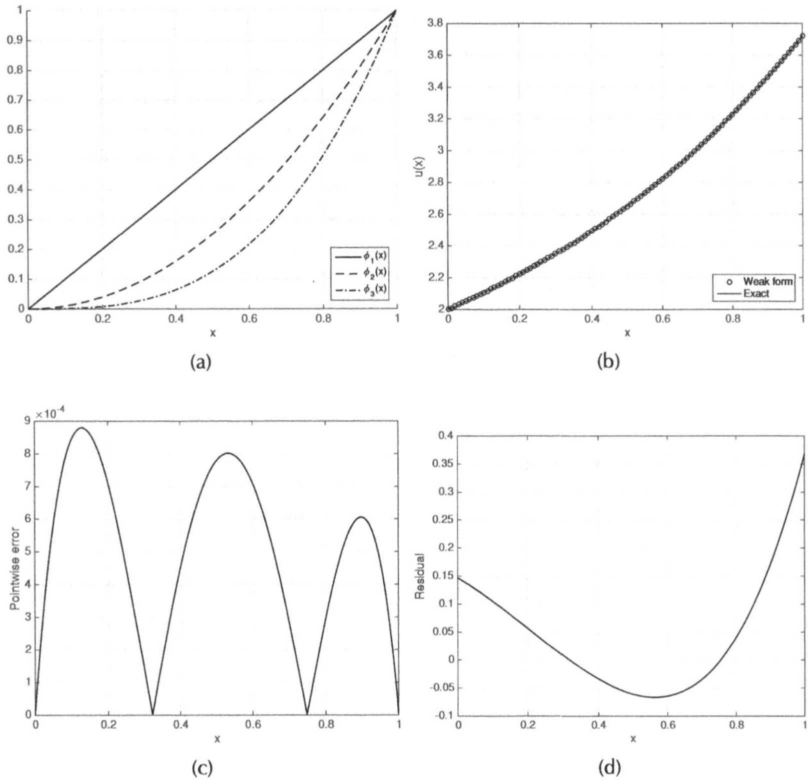

Figure 2.6 Example 2.2. (a) Basis functions ϕ_1, ϕ_2, and ϕ_3. Note that $g = 2$. (b) A comparison between a three-term weak-form Galerkin solution and an exact solution. (c) Pointwise error. (d) Residual function.

and

$$\tilde{b}_i = \int_0^1 w_i f dx + w_i(1)(3+5e) - \mathtt{A}(w_i, g) \tag{2.116}$$

In the Galerkin method, we take

$$w_i(x) = \phi_i(x) \qquad i = 1, ..., N \tag{2.117}$$

and substitute into Equations (2.115) and (2.116) to obtain

$$a_{ij} \equiv \mathtt{A}(\phi_i, \phi_j) = \int_0^1 \left[\frac{d\phi_i}{dx} p \frac{d\phi_j}{dx} + \phi_i q \phi_j \right] dx + 3\phi_i(1)\phi_j(1) \tag{2.118}$$

and

$$\tilde{b}_i \equiv \mathtt{F}(\phi_i) - \mathtt{A}(\phi_i, g) = \int_0^1 \phi_i f dx + \phi_i(1)(3+5e) - \mathtt{A}(\phi_i, g). \tag{2.119}$$

We evaluate the integrals and obtain the system

$$\begin{bmatrix} 5.21828 & 5.23010 & 5.21454 \\ 5.23010 & 5.79787 & 6.09560 \\ 5.21454 & 6.09560 & 6.64468 \end{bmatrix} \begin{Bmatrix} C_1 \\ C_2 \\ C_3 \end{Bmatrix} = \begin{Bmatrix} 8.9704 \\ 9.4704 \\ 9.7352 \end{Bmatrix}. \quad (2.120)$$

Solving this system, we get $C_1 = 1.01540$, $C_2 = 0.41907$, and $C_3 = 0.28382$. A comparison between the exact solution and the approximate Galerkin solution is shown in Figure 2.6b, and the pointwise error is shown in Figure 2.6c. The supmetric error is 8.7904×10^{-4} and L_2 norm is 5.5854×10^{-4}. A comparison between the two-term solution to the strong form obtained in Example 2.1 and the three-term solution to the weak form shows that the latter is more accurate. The higher accuracy comes at a higher effort of formulating and solving three equations as compared to only two equations for the strong form. However, a two-term solution to the weak form for this example will be much less accurate than the two-term solution to the strong form. We also note that a trial solution to the strong form is constructed such that all boundary conditions are satisfied, but the trial solution to the weak form is constructed such that only essential boundary conditions are satisfied whereas other types of boundary conditions need not be exactly satisfied. The residual function $R(x;C_1,C_2,C_3)$ is a measure of the error in satisfying the differential equation; this is shown in Figure 2.6d. The differential equation is exactly satisfied at two points only where the residual is zero.

2.5 GLOBAL BASIS FUNCTIONS AND MATRIX FORMULATION

In this section, we develop the finite-element method for the model problem using functions that are defined over the entire problem domain; they are called global basis functions. Those functions are used to represent the trial solution and test (weight) functions. The model boundary-value problem considered in this section is given by the second-order ordinary differential equation

$$-\frac{d}{dx}[p(x)\frac{du}{dx}] + q(x)u - f(x) = 0 \qquad a < x < b \quad (2.121)$$

and natural boundary conditions

$$\begin{aligned} x &= a: & -p\frac{du}{dx} &= R_a \\ x &= b: & p\frac{du}{dx} &= R_b \end{aligned} \quad (2.122)$$

where p, q, and f are given functions of x, and R_a, and R_b are given constants, which are the secondary variables at the end points. Other boundary conditions will be analyzed afterward.

It is important to note that for the special case of $q = 0$ and these boundary conditions, the values of R_a, R_b, and f must satisfy a compatibility (solvability) condition. Integrating Equation (2.121) over the domain and substituting the boundary conditions, we obtain $\int_a^b f(x)dx + R_a + R_b = 0$. Therefore, the given R_a, R_b, and $f(x)$ must satisfy this condition. Furthermore, the solution is not unique within an additive constant, that is to say, if $u(x)$ is a solution then $u(x)+C$ is also a solution, where C is an arbitrary constant. For the axially loaded bar, the condition means that the sum of forces (distributed plus end loads) must be zero.

The weak form is

$$\int_a^b \left[\frac{dw}{dx}p(x)\frac{du}{dx}+wq(x)u\right]dx-\int_a^b wf(x)dx-w(b)R_b-w(a)R_a=0. \tag{2.123}$$

The domain $a \leq x \leq b$ is divided into contiguous non-overlapping elements by using N global nodes. The coordinates of the nodes are denoted by x_i, with the first node $x_1 = a$ and the last node $x_N = b$. A trial solution is assumed

$$u(x) \approx U(x) = \sum_{j=1}^{N} C_j\phi_j(x) \tag{2.124}$$

where ϕ_j $(j = 1,2,...,N)$ are N global basis functions. These functions are constructed such that

$$\phi_j(x_i) = \delta_{ij} \tag{2.125}$$

where δ_{ij} is the Kronecker delta ($\delta_{ij} = 1$ if $i = j$ and $\delta_{ij} = 0$ otherwise). Substituting $x = x_i$ in the trial solution Equation (2.124), we find $C_i = U_i$, where $U_i = U(x_i)$. These are the primary variables at the global nodes. The trial solution now is written as

$$u(x) \approx U(x) = \sum_{j=1}^{N} U_j\phi_j(x). \tag{2.126}$$

To write $U(x)$ in matrix form, we define the arrays

$$\{\phi(x)\} = \{\phi_1(x)\ \phi_2(x) \ldots \phi_j(x) \ldots \phi_N(x)\} \tag{2.127}$$

$$\{U\} = \left\{\begin{matrix} U_1 \\ U_2 \\ \cdot \\ \cdot \\ \cdot \\ U_j \\ \cdot \\ \cdot \\ \cdot \\ U_N \end{matrix}\right\}. \tag{2.128}$$

In matrix form, the trial solution is

$$U(x) = \{\phi(x)\}\{U\} \tag{2.129}$$

as can be verified

$$U(x) = \{\phi_1(x)\ \phi_2(x) \dots \phi_j(x) \dots \phi_N(x)\} \begin{Bmatrix} U_1 \\ U_2 \\ \cdot \\ \cdot \\ \cdot \\ U_j \\ \cdot \\ \cdot \\ \cdot \\ U_N \end{Bmatrix}$$

$$= \phi_1(x)U_1 + \phi_2(x)U_2 + \dots + \phi_j(x)U_j + \dots + \phi_N(x)U_N$$

$$= \sum_{j=1}^{N} U_j \phi_j(x). \tag{2.130}$$

We note that the dependence of the trial solution on x is given by the array $\{\phi(x)\}$, whereas the array $\{U\}$ on the right does not depend on x; these are values of the primary variable at the global nodes. Therefore, the trial solution gradient is

$$\frac{dU(x)}{dx} = \{\frac{d\phi_1(x)}{dx}\ \frac{d\phi_2(x)}{dx} \dots \frac{d\phi_j(x)}{dx} \dots \frac{d\phi_N(x)}{dx}\} \begin{Bmatrix} U_1 \\ U_2 \\ \cdot \\ \cdot \\ \cdot \\ U_j \\ \cdot \\ \cdot \\ \cdot \\ U_N \end{Bmatrix}$$

$$= \frac{d\{\phi(x)\}}{dx}\{U\}. \tag{2.131}$$

Galerkin

In the Galerkin method, the test (or weight) function is taken as a linear superposition of the global basis functions.

$$w(x) = \sum_{i=1}^{N} d_i \phi_i(x) \tag{2.132}$$

where d_i $(i = 1,2,\ldots N)$ is an array of (arbitrary) constants that are the values of the test function at the global nodes. We define the array

$$\{d\} = \begin{Bmatrix} d_1 \\ d_2 \\ \cdot \\ \cdot \\ \cdot \\ d_i \\ \cdot \\ \cdot \\ \cdot \\ d_N \end{Bmatrix} \tag{2.133}$$

and write $w(x)$ in matrix form

$$w(x) = \{\phi(x)\}\{d\} \tag{2.134}$$

or

$$w(x) = \{d\}^T\{\phi(x)\}^T. \tag{2.135}$$

Note that $w(x) = \phi_1(x)$ is obtained if we choose $d_1 = 1$ and $d_i = 0$ for $i \neq 1$. Similarly, $w(x) = \phi_2(x)$ is obtained if we choose $d_2 = 1$ and $d_i = 0$ if $i \neq 2$, and so on. The gradient of $w(x)$ in matrix form is

$$\frac{dw(x)}{dx} = \{d\}^T \frac{d\{\phi(x)\}^T}{dx}. \tag{2.136}$$

Once we decided on the trial solution and test function, we used the weak form to obtain a system of algebraic equations for the array of unknown primary variables at the global nodes. To that end, we substitute the trial solution Equation (2.129) and the test function Equation (2.135) into the weak form Equation (2.123).

$$\int_a^b \left[\{d\}^T \frac{d\{\phi(x)\}^T}{dx} p(x) \frac{d\{\phi(x)\}}{dx}\{U\} + \{d\}^T\{\phi(x)\}^T q(x)\{\phi(x)\}\{U\}\right] dx$$
$$- \int_a^b \{d\}^T\{\phi(x)\}^T f(x)dx - \{d\}^T\{\phi(x_b)\}^T R_b - \{d\}^T\{\phi(x_a)\}^T R_a = 0. \tag{2.137}$$

Recall that $\{d\}$ and $\{U\}$ are independent of x, we factor them out,

$$\{d\}^T \left\{ \int_a^b \left[\frac{d\{\phi(x)\}^T}{dx} p(x) \frac{d\{\phi(x)\}}{dx} + \{\phi(x)\}^T q(x)\{\phi(x)\}\right] dx\{U\} \right.$$
$$\left. - \int_a^b \{\phi(x)\}^T f(x)dx - \{\phi(x_b)\}^T R_b - \{\phi(x_a)\}^T R_a \right\} = 0. \tag{2.138}$$

Define the coefficient matrix

$$[A] = \int_a^b \left[\frac{d\{\phi(x)\}^T}{dx} p(x) \frac{d\{\phi(x)\}}{dx} + \{\phi(x)\}^T q(x) \{\phi(x)\} \right] dx \tag{2.139}$$

and forcing vector

$$\begin{aligned}\{\tilde{b}\} &= \int_a^b \{\phi(x)\}^T f(x)dx + \{\phi(x_b)\}^T R_b + \{\phi(x_a)\}^T R_a \\ &= \{F\} + \{R\}\end{aligned} \tag{2.140}$$

where

$$\{F\} = \int_a^b \{\phi(x)\}^T f(x)dx \tag{2.141}$$

$$\{R\} = \begin{Bmatrix} R_a \\ 0 \\ . \\ . \\ . \\ 0 \\ . \\ . \\ 0 \\ R_b \end{Bmatrix}. \tag{2.142}$$

We write Equation (2.138) as

$$\{d\}^T \left[[A]\{U\} - \{\tilde{b}\} \right] = 0. \tag{2.143}$$

Because $\{d\}$ is arbitrary, the bracket must be zero, and we obtain the desired system of equations

$$[A]\{U\} = \{\tilde{b}\}. \tag{2.144}$$

The ith equation of the system can be obtained from Equation (2.143) by choosing all entries of $\{d\}$ to be zero except the entry in the ith row to be 1.

In index notation, the entries of the coefficient matrix are

$$a_{ij} = \int_a^b \left[\frac{d\phi_i(x)}{dx} p(x) \frac{d\phi_j(x)}{dx} + \phi_i(x) q(x) \phi_j(x) \right] dx \tag{2.145}$$

and forcing vector is

$$\tilde{b}_i = \int_a^b \phi_i(x) f(x)dx \qquad \text{if} \quad 2 \le i \le N-1 \tag{2.146}$$

$$\tilde{b}_1 = \int_a^b \phi_1(x) f(x)dx + R_a \tag{2.147}$$

$$\tilde{b}_N = \int_a^b \phi_N(x) f(x) dx + R_b. \tag{2.148}$$

The coefficient matrix can be written as the sum of two matrices $a_{ij} = k_{ij} + m_{ij}$, where

$$k_{ij} = \int_a^b \frac{d\phi_i(x)}{dx} p(x) \frac{d\phi_j(x)}{dx} dx \tag{2.149}$$

which is called the stiffness matrix, and

$$m_{ij} = \int_a^b \phi_i(x) q(x) \phi_j(x) dx \tag{2.150}$$

which is called the mass matrix. Also, we write it as

$$[A] = [K] + [M]. \tag{2.151}$$

The global system in expanded form is

$$\begin{bmatrix} a_{11} & a_{12} & \cdots & a_{1,N-1} & a_{1,N} \\ a_{21} & a_{22} & \cdots & a_{2,N-1} & a_{2,N} \\ a_{31} & a_{32} & \cdots & a_{3,N-1} & a_{3,N} \\ \cdots & \cdots & \cdots & \cdots & \cdots \\ a_{N-2,1} & a_{N-2,2} & \cdots & a_{N-2,N-1} & a_{N-2,N} \\ a_{N-1,1} & a_{N-1,2} & \cdots & a_{N-1,N-1} & a_{N-1,N} \\ a_{N,1} & a_{N,2} & \cdots & a_{N,N-1} & a_{N,N} \end{bmatrix} \begin{Bmatrix} U_1 \\ U_2 \\ U_3 \\ \cdot \\ U_{N-2} \\ U_{N-1} \\ U_N \end{Bmatrix} = \begin{Bmatrix} \tilde{b}_1 \\ \tilde{b}_2 \\ \tilde{b}_3 \\ \cdot \\ \tilde{b}_{N-2} \\ \tilde{b}_{N-1} \\ \tilde{b}_N \end{Bmatrix}. \tag{2.152}$$

We solve this system for U_j, $j = 1, 2, ..., N$, and the approximate solution is given by Equation (2.126) that can be evaluated at any point $a \leq x \leq b$.

Piecewise Linear Global Basis Functions

To demonstrate the use of global basis functions, we consider a simple case of piecewise linear global basis functions. The problem domain $a \leq x \leq b$ is divided into equal segments (elements) of length, h,

$$h = \frac{b-a}{N-1} \tag{2.153}$$

where N is the number of global nodes whose coordinates are

$$x_j = a + (j-1)h, \qquad j = 1, 2, ..., N. \tag{2.154}$$

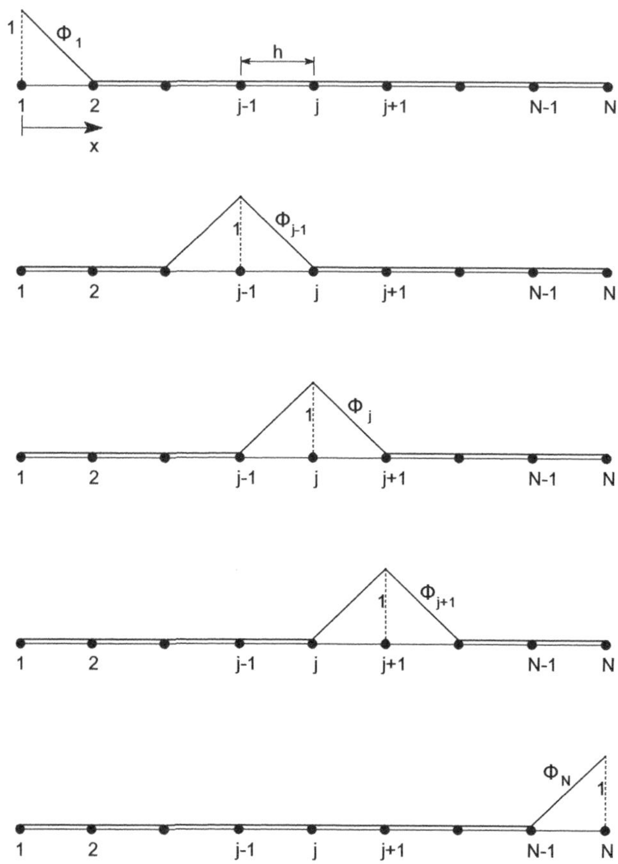

Figure 2.7 Piecewise linear global basis functions.

There are N global basis functions (Φ_j , $j = 1, \cdots, N$), and each one is associated with a global node as shown in Figure 2.7. These functions are linearly independent as can be seen from the plot of the functions at three consecutive nodes $j-1$, j, and $j+1$ shown in Figure 2.8. None of the functions can be obtained as a linear superposition of the remaining functions. The expressions for these functions become simpler when written in terms of a scaled local coordinate ξ defined by

$$\xi = \frac{x - x_j}{h} \tag{2.155}$$

and shown in Figure 2.9. For the jth node, the global basis function is

$$\phi_j(\xi) = \begin{cases} 1+\xi & \text{if } -1 \leq \xi \leq 0 \\ 1-\xi & \text{if } 0 \leq \xi \leq 1 \\ 0 & \text{if } |\xi| > 1. \end{cases} \tag{2.156}$$

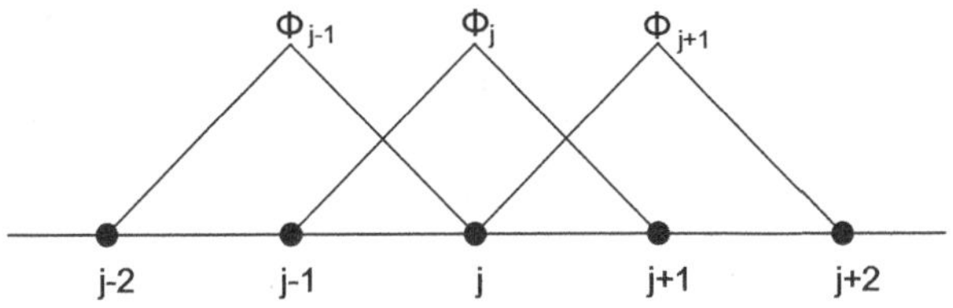

Figure 2.8 Compact support of piecewise linear global basis functions, ϕ_j is zero everywhere except in the domain between nodes $j-1$ and $j+1$.

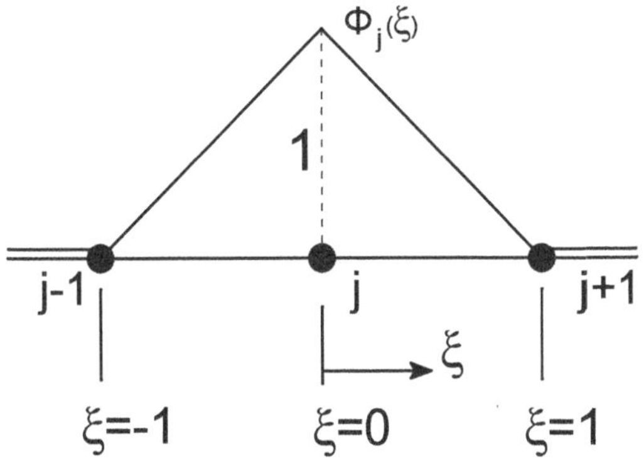

Figure 2.9 Scaled local coordinate, ξ.

The functions at the end nodes $j = 1$ and $j = N$ are of a special form as shown in Figure 2.7 and given by

$$\phi_1(\xi) = \begin{cases} 1-\xi & \text{if } 0 \leq \xi \leq 1 \\ 0 & \text{if } \xi > 1 \end{cases} \tag{2.157}$$

$$\phi_N(\xi) = \begin{cases} 1+\xi & \text{if } -1 \leq \xi \leq 0 \\ 0 & \text{if } \xi < -1. \end{cases} \tag{2.158}$$

Substituting the expressions of ϕ_j into Equation (2.145), we determine the coefficient matrix entries a_{ij} as follows. We note that because of the compact support of the functions ϕ_j, all elements a_{ij} are zero except for the diagonal terms a_{ii}, and the terms below and above the diagonal $a_{i,i-1}$ and $a_{i,i+1}$.

$$a_{ii} = \int_a^b \left[\frac{d\phi_i(x)}{dx} p(x) \frac{d\phi_i(x)}{dx} + \phi_i(x)q(x)\phi_i(x) \right] dx. \tag{2.159}$$

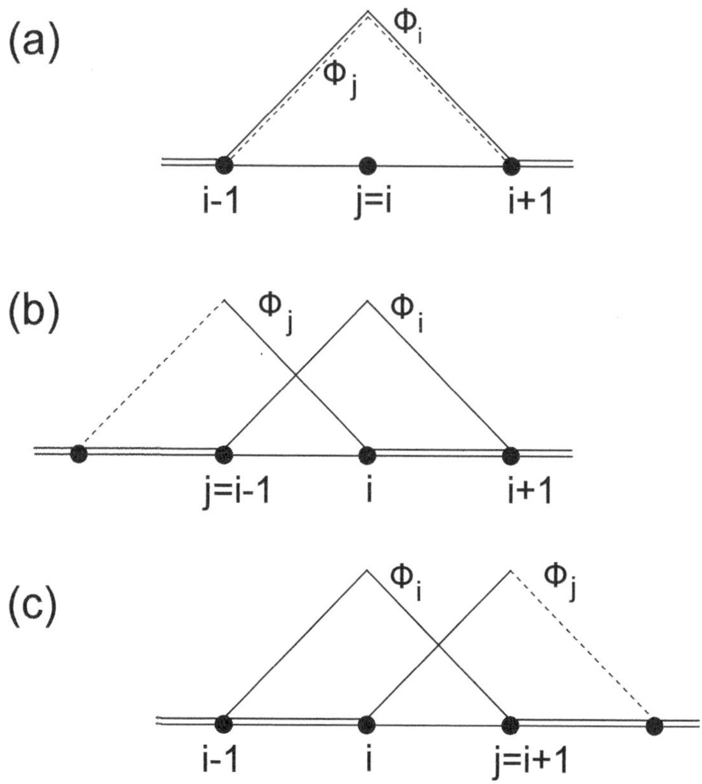

Figure 2.10 Domain of non-zero contributions to the matrix elements a_{ij}, (a) a_{ii}, domain $-1 \leq \xi \leq 1$; (b) $a_{i,i-1}$, domain $-1 \leq \xi \leq 0$; and (c) $a_{i,i+1}$, domain $0 \leq \xi \leq 1$,.

The domain of non-zero contributions to the integrals is shown in Figure 2.10. Changing variables from x to ξ and noting that $dx = hd\xi$, we obtain

$$a_{ii} = \int_{-1}^{0} \left[\frac{1}{h} p(\xi) \frac{1}{h} + (1+\xi) q(\xi)(1+\xi)\right] d\xi$$
$$+ \int_{0}^{1} \left[\left(-\frac{1}{h}\right) p(\xi) \left(-\frac{1}{h}\right) + (1-\xi) q(\xi)(1-\xi)\right] d\xi. \tag{2.160}$$

We take a special case of constant $p(\xi) = p_o$, and $q(\xi) = q_o$, and evaluate the integrals

$$a_{ii} = 2\frac{p_o}{h} + \frac{2}{3} q_o h \qquad \text{for } i = 2, 3, ..., N-1. \tag{2.161}$$

Similarly, we obtain

$$a_{i,i-1} = \int_{-1}^{0} \left[\frac{1}{h} p(\xi)(-\frac{1}{h}) + (1+\xi) q(\xi)(-\xi)\right] d\xi \tag{2.162}$$

$$a_{i,i-1} = -\frac{p_o}{h} + \frac{1}{6}q_o h \qquad \text{for } i = 2,3,...,N-1 \tag{2.163}$$

and

$$a_{i,i+1} = \int_0^1 \left[(-\frac{1}{h})p(\xi)\frac{1}{h} + (1-\xi)q(\xi)(\xi) \right] d\xi \tag{2.164}$$

$$a_{i,i+1} = -\frac{p_o}{h} + \frac{1}{6}q_o h \qquad \text{for } i = 2,3,...,N-1. \tag{2.165}$$

For $i = 1$, we use ϕ_1 and ϕ_2, and get

$$a_{11} = \frac{p_o}{h} + \frac{1}{3}q_o h \tag{2.166}$$

$$a_{12} = -\frac{p_o}{h} + \frac{1}{6}q_o h. \tag{2.167}$$

For $i = N$, we use ϕ_N and ϕ_{N-1}, and get

$$a_{NN} = \frac{p_o}{h} + \frac{1}{3}q_o h \tag{2.168}$$

$$a_{N,N-1} = -\frac{p_o}{h} + \frac{1}{6}q_o h. \tag{2.169}$$

The global stiffness and mass matrices for this special case of piecewise linear global basis functions and constant p and q are given by

$$[K] = \frac{p_o}{h}\begin{bmatrix} 1 & -1 & 0 & 0 & \cdots & 0 & 0 & 0 \\ -1 & 2 & -1 & 0 & \cdots & 0 & 0 & 0 \\ 0 & -1 & 2 & -1 & \cdots & 0 & 0 & 0 \\ \cdots & \cdots & \cdots & \cdots & \cdots & \cdots & \cdots & \cdots \\ 0 & 0 & 0 & 0 & \cdots & -1 & 2 & -1 \\ 0 & 0 & 0 & 0 & \cdots & 0 & -1 & 1 \end{bmatrix} \tag{2.170}$$

$$[M] = \frac{q_o h}{6}\begin{bmatrix} 2 & 1 & 0 & 0 & \cdots & 0 & 0 & 0 \\ 1 & 4 & 1 & 0 & \cdots & 0 & 0 & 0 \\ 0 & 1 & 4 & 1 & \cdots & 0 & 0 & 0 \\ \cdots & \cdots & \cdots & \cdots & \cdots & \cdots & \cdots & \cdots \\ 0 & 0 & 0 & 0 & \cdots & 1 & 4 & 1 \\ 0 & 0 & 0 & 0 & \cdots & 0 & 1 & 2 \end{bmatrix}. \tag{2.171}$$

The coefficient matrix is given by

$$[A] = [K] + [M]. \tag{2.172}$$

Next, we find the vector $\tilde{b}_i$ by evaluating the integrals in Equation (2.146).

$$\tilde{b}_i = \int_{-1}^0 (1+\xi)f(\xi)d\xi + \int_0^1 (1-\xi)f(\xi)d\xi. \tag{2.173}$$

We take a special case of constant $f(\xi) = f_o$, and evaluate the integrals

$$\tilde{b}_i = f_o h \qquad \text{for } i = 2,3,...,N-1. \tag{2.174}$$

For the end nodes we have

$$\tilde{b}_1 = \frac{1}{2} f_o h + R_a \tag{2.175}$$

$$\tilde{b}_N = \frac{1}{2} f_o h + R_b. \tag{2.176}$$

Finally, we solve the algebraic linear system of N equations

$$[A]\{U\} = \{\tilde{b}\}. \tag{2.177}$$

This provides the values of primary variables U_i at the global nodes and the solution varies linearly between two consecutive global nodes.

Other Types of Boundary Conditions

Thus far we assumed that the two boundary conditions are natural boundary conditions where the secondary variables R_a and R_b are specified at the end points. Other boundary conditions can be easily applied by modifying the first and the last equations of the algebraic system Equation (2.144) without repeating the entire derivation.

First, we consider a case of two EBC where the primary variable u is specified at the boundary points. In this case, $U_1 = u_a$ and $U_N = u_b$ are known whereas the secondary variables R_a and R_b are unknown. There are two ways to modify the system of equations given by Equation (2.144). The first method is to replace the first equation in the system by

$$U_1 = u_a. \tag{2.178}$$

Thus, the entries in the coefficient matrix in the first row become $a_{1j} = \delta_{1j}$ and the right-hand side becomes $\tilde{b}_1 = u_a$. Similarly, the Nth equation is replaced by

$$U_N = u_b. \tag{2.179}$$

Accordingly, the entries in the coefficient matrix in the last row become $a_{jN} = \delta_{jN}$ and the right-hand side becomes $\tilde{b}_N = u_b$. This method is straightforward but it destroys the symmetry of the matrix $[A]$ if it is originally symmetric. To keep the symmetry, in addition to the above-mentioned modifications, we also move the known terms that involve u_a and u_b from the left side to the right side of the system of equations. Hence, we replace the right-hand side with $\tilde{b}_i - a_{i1}u_a - a_{iN}u_b$, and then set $a_{i1} = a_{iN} = 0$ in the coefficient matrix for $i = 2, 3, ..., N-1$. A reduced system of

$(N-2)$ equations can then be obtained by dropping the first and last equations.

$$\begin{bmatrix} a_{22} & a_{23} & \cdots & a_{2,N-3} & a_{2,N-1} \\ a_{32} & a_{33} & \cdots & a_{3,N-3} & a_{3,N-1} \\ \cdots & \cdots & \cdots & \cdots & \cdots \\ a_{N-2,2} & a_{N-2,3} & \cdots & a_{N-2,N-3} & a_{N-2,N-1} \\ a_{N-1,2} & a_{N-1,3} & \cdots & a_{N-1,N-3} & a_{N-1,N-1} \end{bmatrix} \begin{Bmatrix} U_2 \\ U_3 \\ \cdot \\ U_{N-2} \\ U_{N-1} \end{Bmatrix} = \begin{Bmatrix} \tilde{b}_2 - a_{21}u_a - a_{2,N}u_b \\ \tilde{b}_3 - a_{31}u_a - a_{3,N}u_b \\ \cdot \\ \tilde{b}_{N-2} - a_{N-2,1}u_a - a_{N-2,N}u_b \\ \tilde{b}_{N-1} - a_{N-1,1}u_a - a_{N-1,N}u_b \end{Bmatrix}. \tag{2.180}$$

The unknowns are U_i for $i = 2, 3, ..., N-1$.

After solving for the primary variables, we determine the unknown secondary variables at the boundary points by using the first and last equations of the original system before modification.

$$R_a = -f_1 + \sum_{j=1}^{N} a_{1j}U_j \tag{2.181}$$

$$R_b = -f_N + \sum_{j=1}^{N} a_{Nj}U_j. \tag{2.182}$$

Next, we consider a case of two MBC where the boundary condition gives a relationship between the unknown primary variable u and the secondary variable R.

$$R_a + \beta_a U_1 = \alpha_a \tag{2.183}$$

$$R_b + \beta_b U_N = \alpha_b. \tag{2.184}$$

From these two equations we obtain $R_a = \alpha_a - \beta_a U_1$ and $R_b = \alpha_b - \beta_b U_N$. Substituting these values into Equations (2.147) and (2.148), we get

$$\tilde{b}_1 = \int_a^b \phi_1(x) f dx + \alpha_a - \beta_a U_1 \tag{2.185}$$

$$\tilde{b}_N = \int_a^b \phi_N(x) f dx + \alpha_b - \beta_b U_N. \tag{2.186}$$

However, U_1 and U_N are unknown, and we have to move them from the right-hand side to the left-hand side of the first and last equations of the system Equation (2.144). Thus, the first entry in the matrix coefficient in the first row is replaced by $a_{11} + \beta_a$ and the Nth entry in the last row is replaced by $a_{NN} + \beta_b$. The right-hand side entries become

$$\tilde{b}_1 = \int_a^b \phi_1(x) f(x) dx + \alpha_a \tag{2.187}$$

$$\tilde{b}_N = \int_a^b \phi_N(x) f(x) dx + \alpha_b \tag{2.188}$$

$$\begin{bmatrix} a_{11}+\beta_a & a_{12} & \cdots & a_{1,N-1} & a_{1,N} \\ a_{21} & a_{22} & \cdots & a_{2,N-1} & a_{2,N} \\ a_{31} & a_{32} & \cdots & a_{3,N-1} & a_{3,N} \\ \cdots & \cdots & \cdots & \cdots & \cdots \\ a_{N-2,1} & a_{N-2,2} & \cdots & a_{N-2,N-1} & a_{N-2,N} \\ a_{N-1,1} & a_{N-1,2} & \cdots & a_{N-1,N-1} & a_{N-1,N} \\ a_{N,1} & a_{N,2} & \cdots & a_{N,N-1} & a_{N,N}+\beta_b \end{bmatrix} \begin{Bmatrix} U_1 \\ U_2 \\ U_3 \\ \cdot \\ U_{N-2} \\ U_{N-1} \\ U_N \end{Bmatrix} = \begin{Bmatrix} \tilde{b}_1 \\ \tilde{b}_2 \\ \tilde{b}_3 \\ \cdot \\ \tilde{b}_{N-2} \\ \tilde{b}_{N-1} \\ \tilde{b}_N \end{Bmatrix} \tag{2.189}$$

where $\tilde{b}_1$ and $\tilde{b}_N$ are given by Equations (2.187) and (2.188), respectively. After solving for the primary variables, we determine the secondary variables at the boundary from

$$R_a = \alpha_a - \beta_a U_1 \tag{2.190}$$

$$R_b = \alpha_b - \beta_b U_N. \tag{2.191}$$

Modifications of the system of equations for any other combination of boundary conditions can be done following similar steps.

2.6 FINITE ELEMENT FORMULATION USING ELEMENT SHAPE FUNCTIONS

Formulation of the finite element method using global basis functions as presented in Section 2.5 establishes a strong connection between the weighted residual method or the weak form and the finite element method. However, the construction of global basis functions and evaluations of integrals are not practical for complex geometries particularly for multidimensional problems. A more practical approach is to divide the problem domain into elements (**Mesh Generation**), and evaluate the integrals over each element domain (**Element Matrices**), then add or assemble the contribution of all elements into a global system (**Assemble Elements**). When the integrals are evaluated over an element, only parts of the global basis functions that fall in the element domain are needed. We call these parts the *element shape functions*. In practice, each element is defined by a set of nodes, and the trial solution over the element domain is expressed in terms of the primary variables at the element nodes (**Approximations and Element Shape Functions**). In other words, the process is reversed in that we construct the element shape functions first, and then piece them together to obtain the global basis functions, if needed.

Boundary conditions are specified on the boundary of the whole domain, and no boundary conditions are available for any of the elements except for an element whose boundary falls on the global boundary. The weak form before application of boundary conditions, such as Equation (2.123), is valid for each element provided that the integrals are evaluated over the element domain only and the secondary variables are evaluated at the element boundary points (**Element Equations**). Since there are no boundary conditions to be enforced at the element level, those secondary variables remain unknown. Upon assembly of the elements, the sum of the unknown secondary variables at an interior node must be zero or specified by external effects (**Balance of secondary variables** [97]). Secondary variables at nodes on the boundary of the whole domain will be decided by the problem boundary conditions (**Apply Boundary Conditions**). Solving the assembled system provides the primary variables at all global nodes (**Solve Global System**). With primary variables now known at the element nodes, we can use the element equations to determine the secondary variables at the element nodes, if desired. Furthermore, we can evaluate the solution at any point in the element using the element shape functions (**Post Processing**).

2.6.1 LINEAR ELEMENT

Mesh Generation and Elements Connectivity

We divide the problem domain into a number (ne) of non-overlapping elements by inserting global nodes. Each element (e) is identified by a set of global nodes. A linear element is identified by two nodes, a quadratic element by three nodes, cubic element by four nodes, and so on. In this section, we develop a finite element model for the linear element. For a domain $0 \leq x \leq 1$, Figure 2.11 shows a mesh of three linear elements and the corresponding four global nodes. The solution over each element varies linearly with x. The mesh of three elements is disassembled as shown in Figure 2.12, and local nodes and local primary variables are defined for each element. Now, each element has two local nodes labeled 1 and 2 with global coordinates x_1^e and x_2^e, where superscript e refers to element number e. The corresponding primary variables at the element nodes are also labeled U_1^e and U_2^e. We call them the degrees of freedom of the element. Note that node 1 of element e is the same as node 2 of element $e-1$, hence $U_2^{e-1} = U_1^e$, and both are equal to the value of the primary variable (global degree of freedom) at the corresponding global node.

Connectivity Matrix $CM(ne,n)$

The connectivity matrix identifies the local nodes of the elements with global nodes. Here, ne is the number of elements, and n is number of nodes per element. Each local node of an element corresponds to a global node. The connectivity matrix gives the global nodes that correspond to the element local nodes. The i-th row in the connectivity matrix gives the global nodes of the i-th element. For the mesh shown in Figure 2.11, element number 1 is given by global nodes 1 and 2, and element number 2 is identified by global nodes 2 and 3, and element number 3 has global

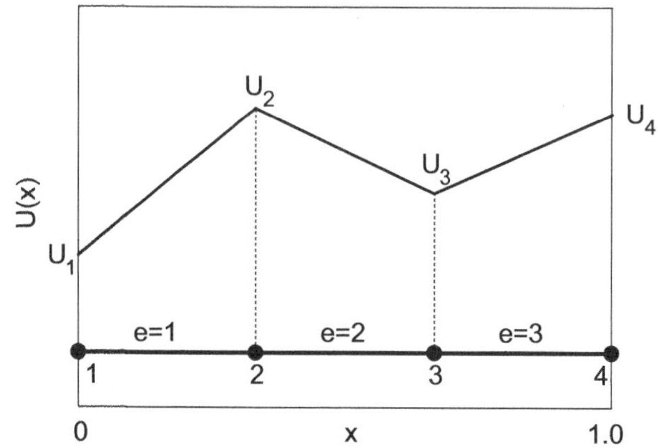

Figure 2.11 A mesh of three linear elements, and four global nodes.

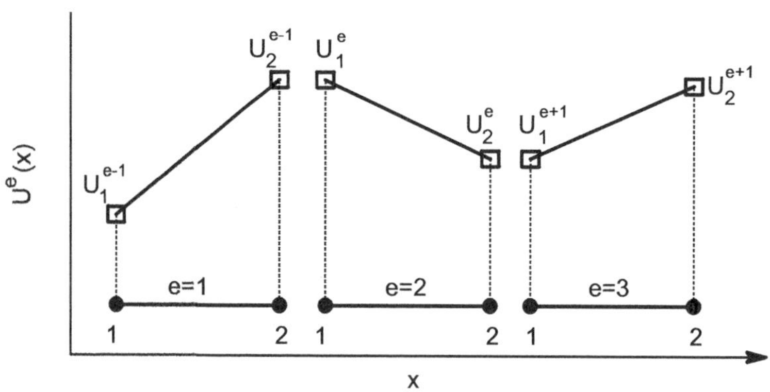

Figure 2.12 A mesh of three disassembled linear elements, each element has two local nodes.

nodes 3 and 4. Therefore, the connectivity matrix for this mesh is

$$[CM] = \begin{bmatrix} 1 & 2 \\ 2 & 3 \\ 3 & 4 \end{bmatrix}. \tag{2.192}$$

In this step, we should also compute the global coordinates of all nodes in the mesh, and store them in an array ($x_j, j = 1, 2, ..., N$), where N is the number of global nodes.

Element Equations and Matrices $[A^e], \{F^e\}, \{R^e\}$

We consider the linear element, which is defined by two nodes with global coordinates x_1^e and x_2^e. Recalling Equation (2.123), we write the weak form for element (e)

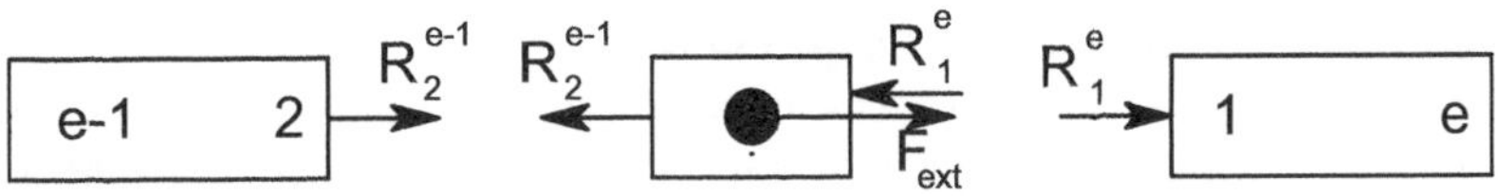

Figure 2.13 Free-body diagram of the common node between two elements. F_{ext} is an externally applied concentrated force at the node.

as

$$\int_{x_1^e}^{x_2^e} \left[\frac{dw}{dx} p \frac{du}{dx} + wqu \right] dx = \int_{x_1^e}^{x_2^e} wf dx + w(x_1^e) R_1^e + w(x_2^e) R_2^e \tag{2.193}$$

where

$$\begin{aligned} R_1^e &= -\left(p \frac{du}{dx} \right)_{x_1^e} \\ R_2^e &= \left(p \frac{du}{dx} \right)_{x_2^e} \end{aligned} \tag{2.194}$$

are the secondary variables at the element's nodes. Note that they are defined in terms of the derivative of the unknown function $u(x)$, thus we have to consider them as unknowns. There are no boundary conditions to be imposed at the element level. Boundary conditions are only known at the first node of the first element $x_1^1 = 0$ and the last node of the last element $x_2^{ne} = 1$. These conditions will be applied to the global system that we will obtain by assembling all elements. At inter-element nodes, secondary variables satisfy [97],

$$\begin{aligned} R_2^{e-1} + R_1^e &= 0 \\ R_2^e + R_1^{e+1} &= 0. \end{aligned} \tag{2.195}$$

However, in some applications externally applied concentrated sources of secondary variables (e.g., concentrated forces at nodes, not accounted for by the distributed axial load $f(x)$), will contribute to the right-hand side of Equation (2.195). For example, consider the free-body diagram of the common node between two adjacent elements $e-1$ and e as shown in Figure 2.13. An axial force F_{ext}, that is not accounted for by the distributed load $f(x)$, is applied at the common node. Equilibrium of the free-body diagram gives

$$R_2^{e-1} + R_1^e = F_{ext}. \tag{2.196}$$

For later use, we write the weak form Equation (2.193) in operator form

$$\mathtt{A}^e(w,u) = \mathtt{F}^e(w). \tag{2.197}$$

Approximation and Element Shape Functions

Considering the two-node linear element, we approximate the solution $u(x)$ to the weak form over the element domain by a linear function that interpolates the nodal values of the primary variable, U_1^e and U_2^e.

$$u^e(x) \approx U^e(x) = c_1 + c_2 x. \tag{2.198}$$

To determine the coefficients c_1 and c_2, we use

$$\begin{aligned} x = x_1^e : \qquad & U^e = U_1^e \\ x = x_2^e : \qquad & U^e = U_2^e. \end{aligned} \tag{2.199}$$

This approach is inconvenient. Derivation of the shape functions is much easier in terms of a scaled local coordinate, ξ, which is related to the global coordinate x by

$$\xi = (x - x_1^e)/h_e \qquad x_1^e \leq x \leq x_2^e \tag{2.200}$$

where $h_e = x_2^e - x_1^e$, is the element length. Note that the element domain $x_1^e \leq x \leq x_2^e$ is mapped onto $0 \leq \xi \leq 1$ by

$$x = x_1^e + h_e \xi \qquad 0 \leq \xi \leq 1 \tag{2.201}$$

and $dx = h_e d\xi$. Now we write the approximate solution in terms of ξ,

$$U^e(\xi) = c_1 + c_2 \xi. \tag{2.202}$$

To determine the coefficients c_1 and c_2, we use

$$\begin{aligned} \xi &= 0 : \qquad & U^e = U_1^e \\ \xi &= 1 : \qquad & U^e = U_2^e \end{aligned} \tag{2.203}$$

and obtain

$$\begin{aligned} c_1 &= U_1^e \\ c_1 + c_2 &= U_2^e. \end{aligned} \tag{2.204}$$

The solution is $c_1 = U_1^e$ and $c_2 = U_2^e - U_1^e$. Substituting into Equation (2.202), we rewrite the approximation as

$$U^e(\xi) = (1 - \xi)U_1^e + \xi U_2^e. \tag{2.205}$$

The shape functions are given by the coefficients of U_1^e and U_2^e,

$$\begin{aligned} \psi_1(\xi) &= 1 - \xi \\ \psi_2(\xi) &= \xi \end{aligned} \tag{2.206}$$

and are shown in Figure 2.14. Note that ψ_1 is one at node 1 and zero at node 2. And ψ_2 is zero at node 1 and one at node 2. Both functions vary linearly with ξ. The

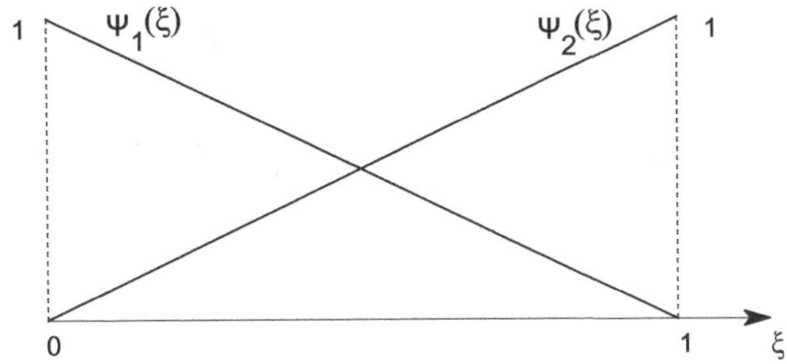

Figure 2.14 Element shape functions for the linear element.

piecewise linear global basis functions ϕ_j shown in Figure 2.7 can be constructed from the element shape functions.

Galerkin Finite Element Model

To obtain the element matrices, we choose two linearly independent weight (test) functions and apply the weak form to the approximate solution $U^e(x)$. In the Galerkin method, the weight (test) functions are the same as the shape functions, $w_1^e = \psi_1$ and $w_2^e = \psi_2$. Let's write the weak form for the element as

$$\mathrm{A}^e(\psi_i, U^e(x)) = \mathrm{F}^e(\psi_i) \qquad i = 1,2. \tag{2.207}$$

Substituting the approximate solution, we get

$$\mathrm{A}^e(\psi_i, \psi_1 U_1^e + \psi_2 U_2^e) = \mathrm{F}^e(\psi_i) \qquad i = 1,2. \tag{2.208}$$

Because A^e is bilinear in its two argument, we obtain

$$\mathrm{A}^e(\psi_i, \psi_1)U_1^e + \mathrm{A}^e(\psi_i, \psi_2)U_2^e = \mathrm{F}^e(\psi_i) \qquad i = 1,2 \tag{2.209}$$

where

$$\mathrm{A}^e(\psi_i, \psi_j) = \int_{x_1^e}^{x_2^e} [\frac{d\psi_i}{dx} p(x) \frac{d\psi_j}{dx} + \psi_i q(x) \psi_j] dx \tag{2.210}$$

$$\mathrm{F}^e(\psi_i) = \int_{x_1^e}^{x_2^e} \psi_i f dx + \psi_i(x_1^e) R_1^e + \psi_i(x_2^e) R_2^e. \tag{2.211}$$

Recalling the properties of the shape functions, $\psi_i(x_j^e) = \delta_{ij}$, we write the finite-element model in matrix form

$$\begin{bmatrix} a_{11}^e & a_{12}^e \\ a_{21}^e & a_{22}^e \end{bmatrix} \begin{Bmatrix} U_1^e \\ U_2^e \end{Bmatrix} = \begin{Bmatrix} F_1^e \\ F_2^e \end{Bmatrix} + \begin{Bmatrix} R_1^e \\ R_2^e \end{Bmatrix} \tag{2.212}$$

$$a_{ij}^e = \int_{x_1^e}^{x_2^e} [\frac{d\psi_i}{dx} p(x) \frac{d\psi_j}{dx} + \psi_i q(x) \psi_j] dx. \tag{2.213}$$

The matrix resulting from the integration of the first term that depends on the function $p(x)$ is called the element stiffness matrix,

$$k_{ij}^e = \int_{x_1^e}^{x_2^e} \frac{d\psi_i}{dx} p(x) \frac{d\psi_j}{dx} dx \tag{2.214}$$

whereas the matrix obtained by integration of the second term that depends on $q(x)$ is called the element mass matrix.

$$m_{ij}^e = \int_{x_1^e}^{x_2^e} \psi_i q(x) \psi_j dx. \tag{2.215}$$

The forcing vector is

$$F_i^e = \int_{x_1^e}^{x_2^e} \psi_i f(x) dx. \tag{2.216}$$

To evaluate these integrals, we need to specify the functions $p(x), q(x)$, and $f(x)$. For simplicity, we consider a special case where these functions are element-wise constant $p = p_e$, $q = q_e$ and $f = f_e$. Evaluating the integral, we obtain

$$\begin{bmatrix} a_{11} & a_{12} \\ a_{21} & a_{22} \end{bmatrix} = \frac{p_e}{h_e} \begin{bmatrix} 1 & -1 \\ -1 & 1 \end{bmatrix} + \frac{q_e h_e}{6} \begin{bmatrix} 2 & 1 \\ 1 & 2 \end{bmatrix}. \tag{2.217}$$

The first matrix on the right is the stiffness matrix of the linear element,

$$[K^e] = \frac{p_e}{h_e} \begin{bmatrix} 1 & -1 \\ -1 & 1 \end{bmatrix} \tag{2.218}$$

and the second matrix is the mass matrix,

$$[M^e] = \frac{q_e h_e}{6} \begin{bmatrix} 2 & 1 \\ 1 & 2 \end{bmatrix} \tag{2.219}$$

and the forcing vector is

$$\begin{Bmatrix} F_1^e \\ F_2^e \end{Bmatrix} = \frac{f_e h_e}{2} \begin{Bmatrix} 1 \\ 1 \end{Bmatrix}. \tag{2.220}$$

It is important to note that the above matrices do not apply if the functions p, q, and f vary with x. For each element, we have two equations given by Equation (2.212) that contain four unknowns: two primary variables U_1^e and U_2^e and two secondary variables R_1^e and R_2^e.

Assemble Elements and Obtain Global System

To understand the assembly process, we rewrite the element equations in terms of global variables. For the mesh of three elements, we have for the first element

$$\begin{bmatrix} a_{11}^1 & a_{12}^1 & 0 & 0 \\ a_{21}^1 & a_{22}^1 & 0 & 0 \\ 0 & 0 & 0 & 0 \\ 0 & 0 & 0 & 0 \end{bmatrix} \begin{Bmatrix} U_1 \\ U_2 \\ U_3 \\ U_4 \end{Bmatrix} = \begin{Bmatrix} F_1^1 \\ F_2^1 \\ 0 \\ 0 \end{Bmatrix} + \begin{Bmatrix} R_1^1 \\ R_2^1 \\ 0 \\ 0 \end{Bmatrix}, \tag{2.221}$$

for the second element we get

$$\begin{bmatrix} 0 & 0 & 0 & 0 \\ 0 & a_{11}^2 & a_{12}^2 & 0 \\ 0 & a_{21}^2 & a_{22}^2 & 0 \\ 0 & 0 & 0 & 0 \end{bmatrix} \begin{Bmatrix} U_1 \\ U_2 \\ U_3 \\ U_4 \end{Bmatrix} = \begin{Bmatrix} 0 \\ F_1^2 \\ F_2^2 \\ 0 \end{Bmatrix} + \begin{Bmatrix} 0 \\ R_1^2 \\ R_2^2 \\ 0 \end{Bmatrix}, \tag{2.222}$$

and for the third element we have

$$\begin{bmatrix} 0 & 0 & 0 & 0 \\ 0 & 0 & 0 & 0 \\ 0 & 0 & a_{11}^3 & a_{12}^3 \\ 0 & 0 & a_{21}^3 & a_{22}^3 \end{bmatrix} \begin{Bmatrix} U_1 \\ U_2 \\ U_3 \\ U_4 \end{Bmatrix} = \begin{Bmatrix} 0 \\ 0 \\ F_1^3 \\ F_2^3 \end{Bmatrix} + \begin{Bmatrix} 0 \\ 0 \\ R_1^3 \\ R_2^3 \end{Bmatrix}. \tag{2.223}$$

The global system is obtained by adding the above systems of equations,

$$\begin{bmatrix} a_{11}^1 & a_{12}^1 & 0 & 0 \\ a_{21}^1 & a_{22}^1+a_{11}^2 & a_{12}^2 & 0 \\ 0 & a_{21}^2 & a_{22}^2+a_{11}^3 & a_{12}^3 \\ 0 & 0 & a_{21}^3 & a_{22}^3 \end{bmatrix} \begin{Bmatrix} U_1 \\ U_2 \\ U_3 \\ U_4 \end{Bmatrix} = \begin{Bmatrix} F_1^1 \\ F_2^1+F_1^2 \\ F_2^2+F_1^3 \\ F_2^3 \end{Bmatrix} + \begin{Bmatrix} R_1^1 \\ R_2^1+R_1^2 \\ R_2^2+R_1^3 \\ R_2^3 \end{Bmatrix}. \tag{2.224}$$

Secondary variables at inter-element nodes [see (2.195)] satisfy

$$\begin{aligned} R_2^1+R_1^2 &= 0 \\ R_2^2+R_1^3 &= 0 \end{aligned} \tag{2.225}$$

and hence the global system before applying the boundary conditions becomes

$$\begin{bmatrix} a_{11}^1 & a_{12}^1 & 0 & 0 \\ a_{21}^1 & a_{22}^1+a_{11}^2 & a_{12}^2 & 0 \\ 0 & a_{21}^2 & a_{22}^2+a_{11}^3 & a_{12}^3 \\ 0 & 0 & a_{21}^3 & a_{22}^3 \end{bmatrix} \begin{Bmatrix} U_1 \\ U_2 \\ U_3 \\ U_4 \end{Bmatrix} = \begin{Bmatrix} F_1^1 \\ F_2^1+F_1^2 \\ F_2^2+F_1^3 \\ F_2^3 \end{Bmatrix} + \begin{Bmatrix} R_1^1 \\ 0 \\ 0 \\ R_2^3 \end{Bmatrix}. \tag{2.226}$$

Apply Boundary Conditions and Solve Global System

In our analysis so far, we have not made any reference to the boundary conditions specified in the problem statement. Thus, the obtained global system is applicable to a variety of boundary conditions. The system is composed of four equations, one equation for each global node. However, there are six unknowns: four primary variables, one for each global node, and two secondary variables, one at the first node R_1^1 and the other at the last node R_2^3. The boundary conditions provide two additional equations to close the system.

We consider a case of two EBCs that are given by

$$\begin{aligned} x &= a: \qquad & U_1 = u_a \\ x &= b: \qquad & U_4 = u_b. \end{aligned} \tag{2.227}$$

There are different ways of combining these conditions with the global system. The simplest way, not necessarily the most efficient way, is to replace the first and last equations in the global system by the above equations, and obtain

$$\begin{bmatrix} 1 & 0 & 0 & 0 \\ a_{21}^1 & a_{22}^1 + a_{11}^2 & a_{12}^2 & 0 \\ 0 & a_{21}^2 & a_{22}^2 + a_{11}^3 & a_{12}^3 \\ 0 & 0 & 0 & 1 \end{bmatrix} \begin{Bmatrix} U_1 \\ U_2 \\ U_3 \\ U_4 \end{Bmatrix} = \begin{Bmatrix} u_a \\ F_2^1 + F_1^2 \\ F_2^2 + F_1^3 \\ u_b \end{Bmatrix}. \tag{2.228}$$

We solve this system of linear algebraic equations for the primary variables. The secondary variables at the boundary nodes are determined by the first and fourth equations of the original global system Equation (2.226) before applying the boundary conditions.

$$\begin{aligned} R_1^1 &= a_{11}^1 U_1 + a_{12}^1 U_2 - F_1^1 \\ R_2^3 &= a_{21}^3 U_3 + a_{22}^3 U_4 - F_2^3. \end{aligned} \tag{2.229}$$

Another method for applying the boundary conditions that preserves the symmetry of the original global system is

$$\begin{bmatrix} 1 & 0 & 0 & 0 \\ 0 & a_{22}^1 + a_{11}^2 & a_{12}^2 & 0 \\ 0 & a_{21}^2 & a_{22}^2 + a_{11}^3 & 0 \\ 0 & 0 & 0 & 1 \end{bmatrix} \begin{Bmatrix} U_1 \\ U_2 \\ U_3 \\ U_4 \end{Bmatrix} = \begin{Bmatrix} u_a \\ F_2^1 + F_1^2 - a_{21}^1 u_a \\ F_2^2 + F_1^3 - a_{21}^3 u_b \\ u_b \end{Bmatrix}. \tag{2.230}$$

Yet, a third method is to obtain a reduced system for the unknown primary variables. Here U_2 and U_3 are unknown; therefore, the reduced system is given by the second and third equations of the global system.

$$\begin{bmatrix} a_{22}^1 + a_{11}^2 & a_{12}^2 \\ a_{21}^2 & a_{22}^2 + a_{11}^3 \end{bmatrix} \begin{Bmatrix} U_2 \\ U_3 \end{Bmatrix} = \begin{Bmatrix} F_2^1 + F_1^2 - a_{21}^1 u_a \\ F_2^2 + F_1^3 - a_{21}^3 u_b \end{Bmatrix}. \tag{2.231}$$

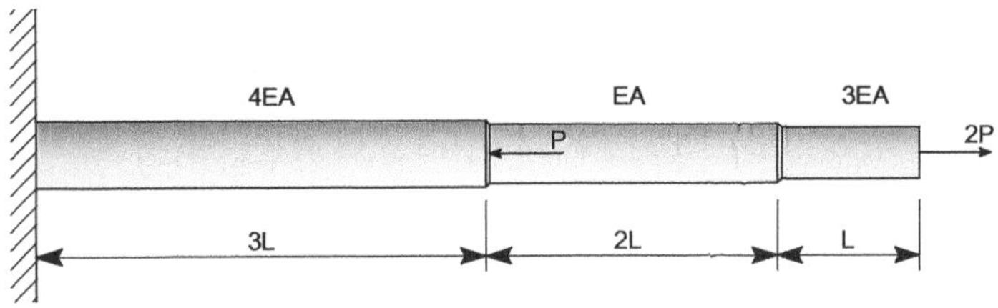

Figure 2.15 Example 2.3, stepped bar.

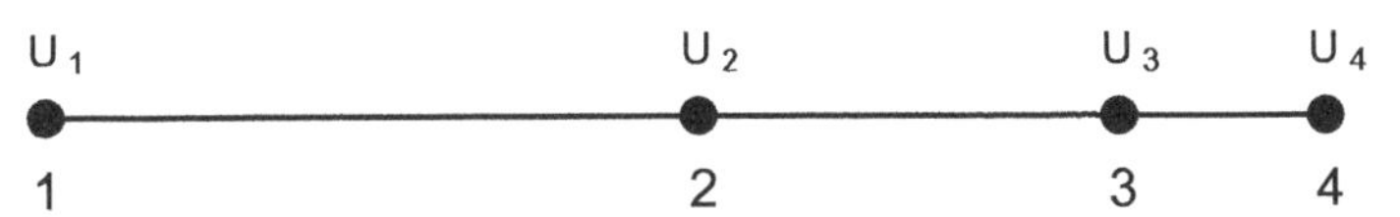

Figure 2.16 Four global nodes and corresponding degrees of freedom.

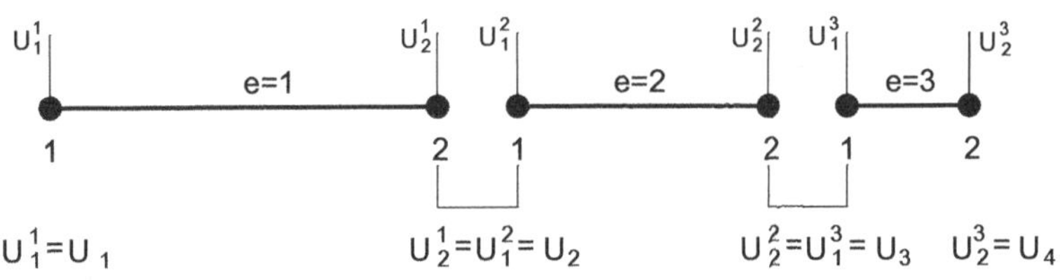

Figure 2.17 Three linear elements, element local nodes, and degrees of freedom.

Example 2.3. Axial Deformation of a Stepped Bar

We want to use the finite element method to determine the displacement field $u(x)$ in the bar shown in Figure 2.15 due to the externally applied concentrated forces P and $2P$. The bar is composed of three segments of different stiffness and lengths as shown. Determine the reaction at the fixed end and the displacement at a distance 3.5 L from that end.

Solution

We model each segment by a linear element. There are four global nodes as shown in Figure 2.16. The displacements at the global nodes are denoted by the array $\{U_1\, U_2\, U_3\, U_4\}^T$. The elements are disassembled in Figure 2.17. Next, we write the element equations for each of the three elements. In this example, there is no distributed axial force $f(x) = 0$, and $q(x) = 0$. The function $p(x)$ is uniform over each element but it varies from one element to another. For the first element $p_e = 4EA$, for the second element $p_e = EA$, and for the third element $p_e = 3EA$. The element lengths are $h_e = 3L, 2L$, and L. Here, EA and L are some reference values. The element matrices $[A^e]$ and $\{F^e\}$ are given by Equations (2.217) and (2.220), respectively. The element equations are Equation (2.212).

For element #1, we get

$$\frac{4EA}{3L}\begin{bmatrix} 1 & -1 \\ -1 & 1 \end{bmatrix}\begin{Bmatrix} U_1^1 \\ U_2^1 \end{Bmatrix} = \begin{Bmatrix} 0 \\ 0 \end{Bmatrix} + \begin{Bmatrix} R_1^1 \\ R_2^1 \end{Bmatrix} \tag{2.232}$$

and for element #2, we get

$$\frac{EA}{2L}\begin{bmatrix} 1 & -1 \\ -1 & 1 \end{bmatrix}\begin{Bmatrix} U_1^2 \\ U_2^2 \end{Bmatrix} = \begin{Bmatrix} 0 \\ 0 \end{Bmatrix} + \begin{Bmatrix} R_1^2 \\ R_2^2 \end{Bmatrix}. \tag{2.233}$$

and for element #3, we have

$$\frac{3EA}{L}\begin{bmatrix} 1 & -1 \\ -1 & 1 \end{bmatrix}\begin{Bmatrix} U_1^3 \\ U_2^3 \end{Bmatrix} = \begin{Bmatrix} 0 \\ 0 \end{Bmatrix} + \begin{Bmatrix} R_1^3 \\ R_2^3 \end{Bmatrix}. \tag{2.234}$$

Next, we assemble the elements into a global system. Since there are four global nodes and each node has one degree of freedom (the value of a primary variable), the global system is four equations. Noting that $U_1^1 = U_1$, $U_2^1 = U_1^2 = U_2$, $U_2^2 = U_1^3 = U_3$, and $U_2^3 = U_4$, we obtain

$$\frac{EA}{6L}\begin{bmatrix} 8 & -8 & 0 & 0 \\ -8 & 11 & -3 & 0 \\ 0 & -3 & 21 & -18 \\ 0 & 0 & -18 & 18 \end{bmatrix}\begin{Bmatrix} U_1 \\ U_2 \\ U_3 \\ U_4 \end{Bmatrix} = \begin{Bmatrix} R_1^1 \\ R_2^1 + R_1^2 \\ R_2^2 + R_1^3 \\ R_2^3 \end{Bmatrix}. \tag{2.235}$$

In the axial deformation problem, the secondary variable $R = EAdu/dx$ is the axial force. By considering a free-body diagram of a small piece about a common node between two elements, we can show that the secondary variables at the two interior nodes 2 and 3 satisfy

$$\begin{aligned} R_2^1 + R_1^2 &= -P \\ R_2^2 + R_1^3 &= 0. \end{aligned} \tag{2.236}$$

Thus, the global system before enforcing the boundary conditions is

$$\frac{EA}{6L}\begin{bmatrix} 8 & -8 & 0 & 0 \\ -8 & 11 & -3 & 0 \\ 0 & -3 & 21 & -18 \\ 0 & 0 & -18 & 18 \end{bmatrix}\begin{Bmatrix} U_1 \\ U_2 \\ U_3 \\ U_4 \end{Bmatrix} = \begin{Bmatrix} R_1^1 \\ -P \\ 0 \\ R_2^3 \end{Bmatrix}. \tag{2.237}$$

There are four equations in six unknowns. The boundary conditions provide two additional equations. The bar is fixed at global node # 1, hence $U_1 = 0$, which is an essential boundary condition. Note that the corresponding secondary variable R_1^1 is unknown. At the other end, the secondary variable being the axial force is given by the externally applied force 2P, hence $R_2^3 = 2P$.

$$\frac{EA}{6L}\begin{bmatrix} 8 & -8 & 0 & 0 \\ -8 & 11 & -3 & 0 \\ 0 & -3 & 21 & -18 \\ 0 & 0 & -18 & 18 \end{bmatrix}\begin{Bmatrix} U_1 = 0 \\ U_2 \\ U_3 \\ U_4 \end{Bmatrix} = \begin{Bmatrix} R_1^1 \\ -P \\ 0 \\ 2P \end{Bmatrix}. \tag{2.238}$$

Knowing U_1, we obtain a reduced system for the unknown primary variables as

$$\frac{EA}{6L}\begin{bmatrix} 11 & -3 & 0 \\ -3 & 21 & -18 \\ 0 & -18 & 18 \end{bmatrix}\begin{Bmatrix} U_2 \\ U_3 \\ U_4 \end{Bmatrix} = \begin{Bmatrix} -P \\ 0 \\ 2P \end{Bmatrix} + \frac{EA}{6L}U_1\begin{Bmatrix} 8 \\ 0 \\ 0 \end{Bmatrix}. \tag{2.239}$$

In this example, the second term on the right is zero because $U_1 = 0$, but it is kept here as a reminder that it needs to be included for problems when the specified U_1 is not zero. Solving the system, we get

$$\begin{Bmatrix} U_2 \\ U_3 \\ U_4 \end{Bmatrix} = \frac{PL}{12EA}\begin{Bmatrix} 9 \\ 57 \\ 65 \end{Bmatrix}. \tag{2.240}$$

The unknown secondary variable R_1^1 can now be determined from the first equation of the system Equation (2.237),

$$\begin{aligned} R_1^1 &= \frac{EA}{6L}(-8U_2) \\ &= -P. \end{aligned} \tag{2.241}$$

R_1^1 is the reaction at the fixed end as can be verified by the static equilibrium of the entire bar. Finite element analysis provides approximate analytic expressions that can be used to evaluate displacements and strains at any point in the bar. Measuring x from the fixed end, we see that $x = 3.5L$ falls in the domain of element # 2. We calculate $\xi = (3.5L - 3L)/2L = 0.25$ and use Equation (2.205)

$$\begin{aligned} U^e(\xi) &= (1-\xi)U_1^e + \xi U_2^e \\ U(x=3.5L) &= (1-0.25)U_1^2 + 0.25U_2^2 \\ &= 0.75\frac{9PL}{12EA} + 0.25\frac{57PL}{12EA} \\ &= \frac{7PL}{4EA}. \end{aligned} \tag{2.242}$$

Example 2.4. Axial Deformation of a Stepped Bar—Mixed Boundary Condition

The right end of the bar of the previous example is now attached to a linear spring of stiffness k that is not stretched before application of the force P. We wish to determine the axial displacement field.

Solution

The finite element model given by Equation (2.237) applies and is repeated here

$$\frac{EA}{6L}\begin{bmatrix} 8 & -8 & 0 & 0 \\ -8 & 11 & -3 & 0 \\ 0 & -3 & 21 & -18 \\ 0 & 0 & -18 & 18 \end{bmatrix}\begin{Bmatrix} U_1 \\ U_2 \\ U_3 \\ U_4 \end{Bmatrix} = \begin{Bmatrix} R_1^1 \\ -P \\ 0 \\ R_2^3 \end{Bmatrix}. \tag{2.243}$$

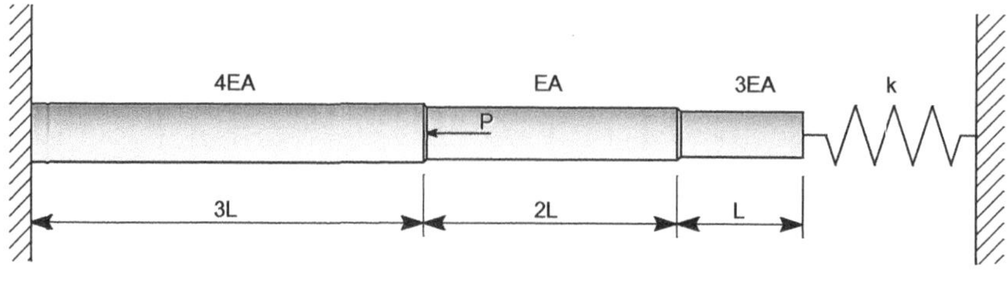

Figure 2.18 Example 2.4, stepped bar restrained by a linear spring at one end.

In this example both U_4 and R_2^3 are unknown, but they are related by the mixed boundary condition [see Equation (2.20)]

$$R_2^3 + kU_4 = 0. \tag{2.244}$$

We substitute $R_2^3 = -kU_4$ into the last equation of system Equation (2.243), rearrange terms, and impose the condition $U_1 = 0$ to obtain

$$\frac{EA}{6L}\begin{bmatrix} 8 & -8 & 0 & 0 \\ -8 & 11 & -3 & 0 \\ 0 & -3 & 21 & -18 \\ 0 & 0 & -18 & 18+\frac{6kL}{EA} \end{bmatrix}\begin{Bmatrix} U_1 = 0 \\ U_2 \\ U_3 \\ U_4 \end{Bmatrix} = \begin{Bmatrix} R_1^1 \\ -P \\ 0 \\ 0 \end{Bmatrix}. \tag{2.245}$$

The reduced system is

$$\frac{EA}{6L}\begin{bmatrix} 11 & -3 & 0 \\ -3 & 21 & -18 \\ 0 & -18 & 18+\frac{6kL}{EA} \end{bmatrix}\begin{Bmatrix} U_2 \\ U_3 \\ U_4 \end{Bmatrix} = \begin{Bmatrix} -P \\ 0 \\ 0 \end{Bmatrix}. \tag{2.246}$$

Solution provides the unknown nodal displacements (U_2, U_3 and U_4), and the reaction at the fixed end (R_1^1) follows from the first equation of the global system. The force exerted by the linear spring is $R_2^3 = -kU_4$.

Remark 1: In these two examples the finite element analysis using linear elements for each segment of the stepped bar gives the exact solution to the axial deformation problem. This is because for a bar of uniform EA with no distributed load $f(x) = 0$, the displacement varies linearly with x, and the linear element is an exact representation of the displacement field.

Remark 2: In the analysis of stepped bars it is recommended that we introduce a global node and start a new element wherever there is a discontinuity in the property EA or a concentrated force is applied. This recommendation also applies in the analysis of heat conduction in fins or slabs with discontinuous property kA.

2.6.2 QUADRATIC ELEMENT

Approximation and Element Shape Functions

A quadratic element is defined by three nodes with global coordinates (x_1^e, x_2^e, x_3^e).

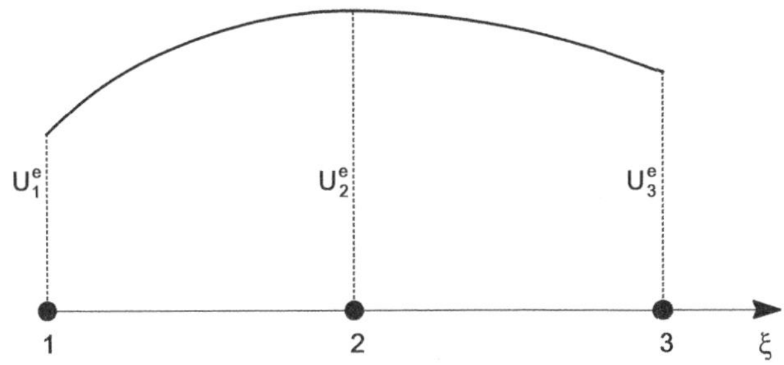

Figure 2.19 Quadratic element, local degrees of freedom U_1^e, U_2^e, and U_3^e.

The array of primary variables at the element's nodes is denoted by $\{U_1^e \, U_2^e \, U_3^e\}^T$, as shown in Figure 2.19. We derive the shape functions in terms of the local coordinate ξ

$$\xi = (x - x_1^e)/h_e \qquad x_1^e \le x \le x_3^e \tag{2.247}$$

where $h_e = x_3^e - x_1^e$, is the element length. Note that the element domain $x_1^e \le x \le x_3^e$ is mapped onto $0 \le \xi \le 1$. It follows that

$$x = x_1^e + h_e \xi \tag{2.248}$$

and $dx = h_e d\xi$.
The approximate solution is a quadratic function in ξ,

$$U^e(\xi) = c_1 + c_2\xi + c_3\xi^2. \tag{2.249}$$

Next, we determine the coefficients c_1, c_2, c_3, in terms of the primary variables at the element's nodes. We assume that the interior node number 2 is the middle point of the element. We apply the conditions

$$\begin{aligned} \xi &= 0: & U^e &= U_1^e \\ \xi &= 1/2: & U^e &= U_2^e \\ \xi &= 1: & U^e &= U_3^e \end{aligned} \tag{2.250}$$

and obtain

$$\begin{bmatrix} 1 & 0 & 0 \\ 1 & 1/2 & 1/4 \\ 1 & 1 & 1 \end{bmatrix} \begin{Bmatrix} c_1 \\ c_2 \\ c_3 \end{Bmatrix} = \begin{Bmatrix} U_1^e \\ U_2^e \\ U_3^e \end{Bmatrix}. \tag{2.251}$$

Solution

$$\begin{aligned} c_1 &= U_1^e \\ c_2 &= -3U_1^e + 4U_2^e - U_3^e \\ c_3 &= 2U_1^e - 4U_2^e + 2U_3^e. \end{aligned} \tag{2.252}$$

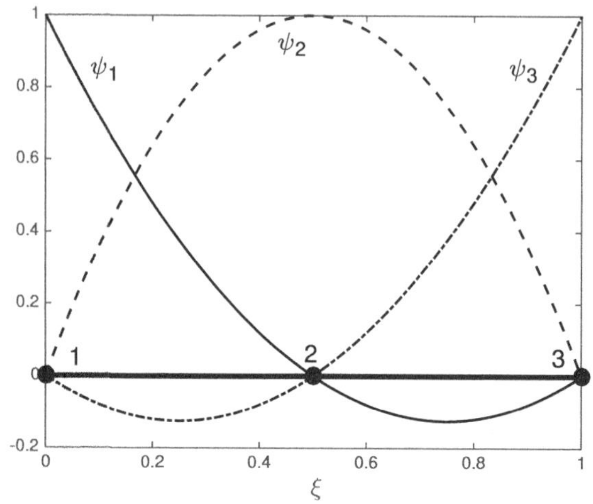

Figure 2.20 Shape functions of a quadratic element.

We substitute these coefficients into the approximate solution Equation (2.249),

$$U^e(\xi) = U_1^e + (-3U_1^e + 4U_2^e - U_3^e)\xi + (2U_1^e - 4U_2^e + 2U_3^e)\xi^2. \tag{2.253}$$

Rearranging, we obtain

$$U^e(\xi) = (1-\xi)(1-2\xi)U_1^e + 4\xi(1-\xi)U_2^e + \xi(2\xi-1)U_3^e. \tag{2.254}$$

The element shape functions are given by the coefficients of U_1^e, U_2^e, and U_3^e,

$$\begin{aligned} \psi_1 &= (1-\xi)(1-2\xi) \\ \psi_2 &= 4\xi(1-\xi) \\ \psi_3 &= \xi(2\xi-1). \end{aligned} \tag{2.255}$$

Note that ψ_1 is one at node 1 and zero at nodes 2 and 3. And ψ_2 is one at node 2 and zero at nodes 1 and 3. And ψ_3 is one at node 3 and zero at nodes 1 and 2. That is to say $\psi_i(x_j^e) = \delta_{ij}$. It is easy to show that

$$\psi_1(\xi) + \psi_2(\xi) + \psi_3(\xi) = 1 \qquad 0 \leq \xi \leq 1. \tag{2.256}$$

The shape functions are plotted in Figure 2.20. From the element shape functions we can construct global basis functions. The basis function of a global node that coincides with the central node of an element is ψ_2 over the element domain and zero everywhere else. The global basis function for a node that coincide with an end node such as node 1 or 3 is composed of two piecewise quadratic segments. Over the element to the left of the node it is given by ψ_3, and over the element to the right it is given by ψ_1, otherwise it is zero everywhere else. The global basis functions for a mesh of two quadratic elements are shown in Figure 2.21. The trial solution over the

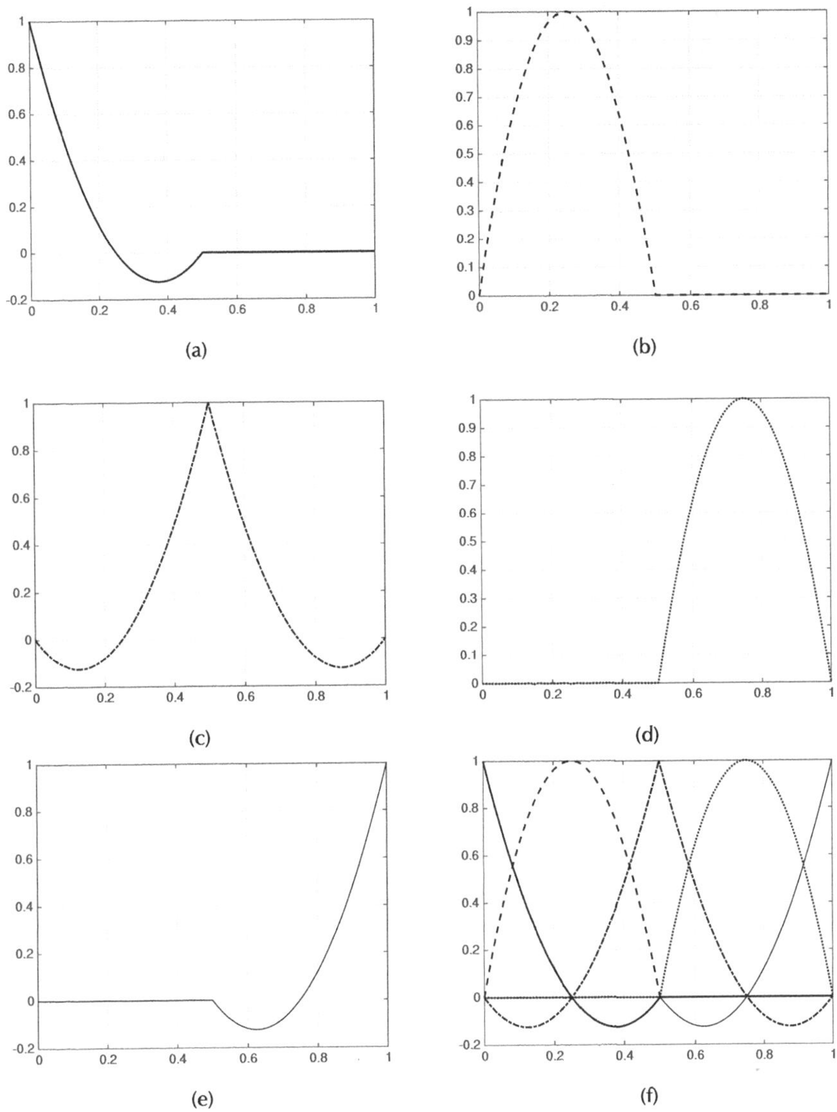

Figure 2.21 Global basis functions for a mesh of two quadratic elements. (a) ϕ_1, (b) ϕ_2, (c) ϕ_3, (d) ϕ_4, (e) ϕ_5, and (f) five global basis functions.

entire domain can be written in terms of the global basis functions, and subsequently used to derive the global system of equation as described in Section 2.5. However, the procedure we adopt is to use the element shape functions to determine element equations, and then assemble them into a global system.

Element Equations and Matrices

To obtain element matrices for the quadratic element, we choose three linearly independent weight (test) functions and apply the weak form to the approximate solution $U^e(x)$. In the Galerkin method, the weight (test) functions are the same as the shape functions, $w_1^e = \psi_1$, $w_2^e = \psi_2$, and $w_3^e = \psi_3$. Let's write the weak form for the element as

$$\mathtt{A}^e(\psi_i, U^e(x)) = \mathtt{F}^e(\psi_i), \qquad i = 1, 2, 3. \tag{2.257}$$

Substituting the approximate solution, we get

$$\mathtt{A}^e(\psi_i, \psi_1 U_1^e + \psi_2 U_2^e + \psi_3 U_3^e) = \mathtt{F}^e(\psi_i), \qquad i = 1, 2, 3. \tag{2.258}$$

Because $\mathtt{A}^e$ is bilinear in its two argument, we obtain,

$$\mathtt{A}^e(\psi_i, \psi_1)U_1^e + \mathtt{A}^e(\psi_i, \psi_2)U_2^e + \mathtt{A}^e(\psi_i, \psi_3)U_3^e = \mathtt{F}^e(\psi_i), \qquad i = 1, 2, 3 \tag{2.259}$$

where

$$\mathtt{A}^e(\psi_i, \psi_j) = \int_{x_1^e}^{x_3^e} [\frac{d\psi_i}{dx} p(x) \frac{d\psi_j}{dx} + \psi_i q(x) \psi_j] dx \tag{2.260}$$

$$\mathtt{F}^e(\psi_i) = \int_{x_1^e}^{x_3^e} \psi_i f dx + \psi_i(x_1^e) R_1^e + \psi_i(x_3^e) R_3^e. \tag{2.261}$$

Secondary variables at interior nodes, such as node 2 of the quadratic element, represent internal effects (e.g., forces), hence they dot appear in the element equations. This is demonstrated by Equation (2.261) that includes only secondary variables at the end nodes of the element R_1^e and R_3^e. Recalling the properties of the shape functions, $\psi_i(x_j^e) = \delta_{ij}$, we write the finite-element model in matrix form

$$\begin{bmatrix} a_{11}^e & a_{12}^e & a_{13}^e \\ a_{21}^e & a_{22}^e & a_{23}^e \\ a_{31}^e & a_{32}^e & a_{33}^e \end{bmatrix} \begin{Bmatrix} U_1^e \\ U_2^e \\ U_3^e \end{Bmatrix} = \begin{Bmatrix} F_1^e \\ F_2^e \\ F_3^e \end{Bmatrix} + \begin{Bmatrix} R_1^e \\ 0 \\ R_3^e \end{Bmatrix} \tag{2.262}$$

where the entries of the coefficient matrix are

$$a_{ij}^e = \int_{x_1^e}^{x_3^e} [\frac{d\psi_i}{dx} p \frac{d\psi_j}{dx} + \psi_i q \psi_j] dx \tag{2.263}$$

and the forcing vector is

$$F_i^e = \int_{x_1^e}^{x_3^e} \psi_i f dx. \tag{2.264}$$

Changing variables, $x = x_1^e + h_e \xi$, and $dx = h_e d\xi$, we obtain

$$a_{ij}^e = \frac{1}{h_e} \int_0^1 \frac{d\psi_i}{d\xi} p \frac{d\psi_j}{d\xi} d\xi + h_e \int_0^1 \psi_i q \psi_j d\xi \tag{2.265}$$

$$F_i^e = h_e \int_0^1 \psi_i f d\xi. \tag{2.266}$$

To evaluate these integrals, we need to specify the functions $p(x), q(x)$, and $f(x)$. For simplicity, we consider a special case where these functions are element-wise constant $p = p_e$, $q = q_e$, and $f = f_e$.

$$\begin{bmatrix} a_{11}^e & a_{12}^e & a_{13}^e \\ a_{21}^e & a_{22}^e & a_{23}^e \\ a_{31}^e & a_{32}^e & a_{33}^e \end{bmatrix} = \frac{p_e}{3h_e}\begin{bmatrix} 7 & -8 & 1 \\ -8 & 16 & -8 \\ 1 & -8 & 7 \end{bmatrix} + \frac{q_e h_e}{30}\begin{bmatrix} 4 & 2 & -1 \\ 2 & 16 & 2 \\ -1 & 2 & 4 \end{bmatrix}. \tag{2.267}$$

The first matrix on the right is the stiffness matrix for the quadratic element,

$$[K^e] = \frac{p_e}{3h_e}\begin{bmatrix} 7 & -8 & 1 \\ -8 & 16 & -8 \\ 1 & -8 & 7 \end{bmatrix} \tag{2.268}$$

and the second matrix is the mass matrix,

$$[M^e] = \frac{q_e h_e}{30}\begin{bmatrix} 4 & 2 & -1 \\ 2 & 16 & 2 \\ -1 & 2 & 4 \end{bmatrix}. \tag{2.269}$$

The forcing vector is

$$\begin{Bmatrix} F_1^e \\ F_2^e \\ F_3^e \end{Bmatrix} = \frac{f_e h_e}{6}\begin{Bmatrix} 1 \\ 4 \\ 1 \end{Bmatrix}. \tag{2.270}$$

It is important to note that the matrices given by Equations (2.268)–(2.270) do not apply if the functions p, q, and f vary with x.

Assemble Elements and Obtain Global System

We consider a mesh of two quadratic elements defined by five global nodes. Each element has three equations. We rewrite the element equations in terms of global variables. For the first element, we have

$$\begin{bmatrix} a_{11}^1 & a_{12}^1 & a_{13}^1 & 0 & 0 \\ a_{21}^1 & a_{22}^1 & a_{23}^1 & 0 & 0 \\ a_{31}^1 & a_{32}^1 & a_{33}^1 & 0 & 0 \\ 0 & 0 & 0 & 0 & 0 \\ 0 & 0 & 0 & 0 & 0 \end{bmatrix}\begin{Bmatrix} U_1 \\ U_2 \\ U_3 \\ U_4 \\ U_5 \end{Bmatrix} = \begin{Bmatrix} F_1^1 \\ F_2^1 \\ F_3^1 \\ 0 \\ 0 \end{Bmatrix} + \begin{Bmatrix} R_1^1 \\ 0 \\ R_3^1 \\ 0 \\ 0 \end{Bmatrix} \tag{2.271}$$

and for the second element we get,

$$\begin{bmatrix} 0 & 0 & 0 & 0 & 0 \\ 0 & 0 & 0 & 0 & 0 \\ 0 & 0 & a_{11}^2 & a_{12}^2 & a_{13}^2 \\ 0 & 0 & a_{21}^2 & a_{22}^2 & a_{23}^2 \\ 0 & 0 & a_{31}^2 & a_{32}^2 & a_{33}^2 \end{bmatrix}\begin{Bmatrix} U_1 \\ U_2 \\ U_3 \\ U_4 \\ U_5 \end{Bmatrix} = \begin{Bmatrix} 0 \\ 0 \\ F_1^2 \\ F_2^2 \\ F_3^2 \end{Bmatrix} + \begin{Bmatrix} 0 \\ 0 \\ R_1^2 \\ 0 \\ R_3^2 \end{Bmatrix}. \tag{2.272}$$

The global system is obtained by adding the above systems of equations,

$$\begin{bmatrix} a_{11}^1 & a_{12}^1 & a_{13}^1 & 0 & 0 \\ a_{21}^1 & a_{22}^1 & a_{23}^1 & 0 & 0 \\ a_{31}^1 & a_{32}^1 & a_{33}^1 + a_{11}^2 & a_{12}^2 & a_{13}^2 \\ 0 & 0 & a_{21}^2 & a_{22}^2 & a_{23}^2 \\ 0 & 0 & a_{31}^2 & a_{32}^2 & a_{33}^2 \end{bmatrix} \begin{Bmatrix} U_1 \\ U_2 \\ U_3 \\ U_4 \\ U_5 \end{Bmatrix} = \begin{Bmatrix} F_1^1 \\ F_2^1 \\ F_3^1 + F_1^2 \\ F_2^2 \\ F_3^2 \end{Bmatrix} + \begin{Bmatrix} R_1^1 \\ 0 \\ R_3^1 + R_1^2 \\ 0 \\ R_3^2 \end{Bmatrix}. \tag{2.273}$$

Secondary variables at inter-element nodes satisfy

$$R_3^1 + R_1^2 = 0 \tag{2.274}$$

and hence the global system before applying the boundary conditions becomes

$$\begin{bmatrix} a_{11}^1 & a_{12}^1 & a_{13}^1 & 0 & 0 \\ a_{21}^1 & a_{22}^1 & a_{23}^1 & 0 & 0 \\ a_{31}^1 & a_{32}^1 & a_{33}^1 + a_{11}^2 & a_{12}^2 & a_{13}^2 \\ 0 & 0 & a_{21}^2 & a_{22}^2 & a_{23}^2 \\ 0 & 0 & a_{31}^2 & a_{32}^2 & a_{33}^2 \end{bmatrix} \begin{Bmatrix} U_1 \\ U_2 \\ U_3 \\ U_4 \\ U_5 \end{Bmatrix} = \begin{Bmatrix} F_1^1 \\ F_2^1 \\ F_3^1 + F_1^2 \\ F_2^2 \\ F_3^2 \end{Bmatrix} + \begin{Bmatrix} R_1^1 \\ 0 \\ 0 \\ 0 \\ R_3^2 \end{Bmatrix}. \tag{2.275}$$

Apply Boundary Conditions and Solve Global System

In our analysis so far, we have not made any reference to the boundary conditions specified in the problem statement. Thus, the obtained global system is applicable to a variety of boundary conditions. The system is composed of five equations; one equation for each global node. However, there are seven unknowns: five primary variables, one for each global node, and two secondary variables, one at the first node R_1^1 and the other at the last node R_3^2. The boundary conditions provide two additional equations to close the system.

We consider a case of two EBCs, which are given by

$$\begin{aligned} x &= a \qquad & U_1 &= u_a \\ x &= b \qquad & U_5 &= u_b. \end{aligned} \tag{2.276}$$

There are different ways of combining these conditions with the global system. The simplest way, not necessarily the most efficient way, is to replace the first and the last equations in the global system by the above equations, and obtain

$$\begin{bmatrix} 1 & 0 & 0 & 0 & 0 \\ a_{21}^1 & a_{22}^1 & a_{23}^1 & 0 & 0 \\ a_{31}^1 & a_{32}^1 & a_{33}^1 + a_{11}^2 & a_{12}^2 & a_{13}^2 \\ 0 & 0 & a_{21}^2 & a_{22}^2 & a_{23}^2 \\ 0 & 0 & 0 & 0 & 1 \end{bmatrix} \begin{Bmatrix} U_1 \\ U_2 \\ U_3 \\ U_4 \\ U_5 \end{Bmatrix} = \begin{Bmatrix} u_a \\ F_2^1 \\ F_3^1 + F_1^2 \\ F_2^2 \\ u_b \end{Bmatrix}. \tag{2.277}$$

We solve this system of linear algebraic equations for the primary variables. The secondary variables at the boundary nodes are determined by the first and fifth equations of the original global system (before applying the boundary conditions) Equation (2.275).

$$\begin{aligned} R_1^1 &= a_{11}^1 U_1 + a_{12}^1 U_2 + a_{13}^1 U_3 - F_1^1 \\ R_3^2 &= a_{31}^2 U_3 + a_{32}^2 U_4 + a_{33}^2 U_5 - F_3^2. \end{aligned} \tag{2.278}$$

Example 2.5. Mesh Refinement and Error Measures

Given $p(x) = 1$, $q(x) = 1$, and $f(x) = 2$, consider the boundary value problem

$$-\frac{d}{dx}(p\frac{du}{dx}) + qu = f \quad \text{on} \quad 0 < x < 1 \tag{2.279}$$

with boundary conditions

$$\begin{aligned} x = 0: &\quad \frac{du}{dx} = 0 \\ x = 1: &\quad u = 1. \end{aligned} \tag{2.280}$$

We want to solve this problem by the finite-element method and assess the accuracy of the solution as we refine the mesh (h-refinement). Divide the the problem domain into ne quadratic elements, where each element is defined by three nodes. For a given mesh (ne), we define the error at global node j by $\varepsilon_j = |u(x_j) - U_j|$, where $u(x_j)$ is the exact solution evaluated at global node j. And we also define the maximum error by $\varepsilon_{max} = max(\varepsilon_j)$. Obtain solutions on different meshes by using $ne = 2, 4, 8, 16, 32$, and 64. Plot the maximum error ε_{max} against the number of elements ne on a log-log scale, and establish the order of accuracy of the quadratic element.

Solution

We developed a MATLAB code to solve this problem. The maximum error at the global nodes is depicted in Figure 2.22. On a log-log scale, the error drops as ne^{-4}. The error at the nodes is proportional to h_e^4.

Example 2.6. Equation with Variable $f(x)$

Consider the two-point boundary value problem

$$-\frac{d}{dx}\left(\frac{du}{dx}\right) - u + \frac{x}{\pi} = 0, \qquad 0 < x < \pi \tag{2.281}$$

with boundary conditions

$$\begin{aligned} x &= 0: \quad u = 0 \\ x &= \pi: \quad \frac{du}{dx} + u = \frac{1}{\pi}. \end{aligned} \tag{2.282}$$

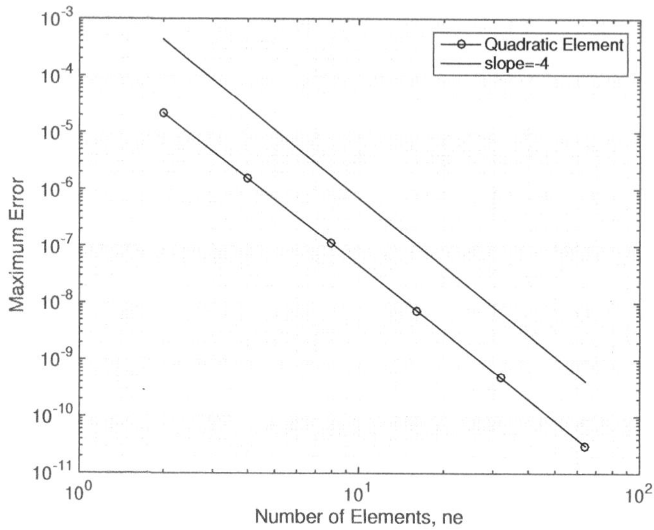

Figure 2.22 Example 2.5, maximum error at global nodes using *ne* quadratic elements, as the mesh is refined the error drops as ne^{-4}.

Use two quadratic elements and compare the finite-element solution to the exact solution for the primary variable at the nodes and the secondary variable at $x = \pi$.

Solution

For this problem we have $p(x) = 1$, $q(x) = -1$, and $f(x) = -x/\pi$. Because p and q are uniform over the element, we use the coefficient matrix as given by Equation (2.267) with $h_e = \pi/2$, and obtain

$$\begin{bmatrix} a_{11}^e & a_{12}^e & a_{13}^e \\ a_{21}^e & a_{22}^e & a_{23}^e \\ a_{31}^e & a_{32}^e & a_{33}^e \end{bmatrix} = \frac{2}{3\pi}\begin{bmatrix} 7 & -8 & 1 \\ -8 & 16 & -8 \\ 1 & -8 & 7 \end{bmatrix} - \frac{\pi}{60}\begin{bmatrix} 4 & 2 & -1 \\ 2 & 16 & 2 \\ -1 & 2 & 4 \end{bmatrix}, \quad (2.283)$$

$$[A^e] = \begin{bmatrix} 1.2760 & -1.8024 & 0.2646 \\ -1.8024 & 2.5575 & -1.8024 \\ 0.2646 & -1.8024 & 1.2760 \end{bmatrix}. \quad (2.284)$$

We note that Equation (2.270) should not be used for the forcing term because $f(x)$ is not uniform over the element, instead we have to evaluate the integrals Equation (2.266).

$$F_i^e = h_e \int_0^1 \psi_i(\xi) f(\xi) d\xi. \quad (2.285)$$

For $f(x) = -x/\pi$, and $x = (x_1^e + h_e\xi)$, we have $f(\xi) = -(x_1^e + \pi\xi/2)/\pi$. The shape functions $\psi_i(\xi)$ are given by Equation (2.255). For the first element $x_1^e = 0$ and for the second element $x_1^e = \pi/2$.

$$\begin{aligned}
F_1^1 &= \frac{\pi}{2}\int_0^1 (1-\xi)(1-2\xi)(-\xi/2)d\xi \\
&= 0 \\
F_2^1 &= \frac{\pi}{2}\int_0^1 4\xi(1-\xi)(-\xi/2)d\xi \\
&= -\frac{\pi}{12} \\
F_3^1 &= \frac{\pi}{2}\int_0^1 \xi(2\xi-1)(-\xi/2)d\xi \\
&= -\frac{\pi}{24}
\end{aligned} \tag{2.286}$$

$$\{F^1\} = -\frac{\pi}{24}\begin{Bmatrix} 0 \\ 2 \\ 1 \end{Bmatrix} \tag{2.287}$$

$$\begin{aligned}
F_1^2 &= \frac{\pi}{2}\int_0^1 (1-\xi)(1-2\xi)[-(\frac{1}{2}+\frac{1}{2}\xi)]d\xi \\
&= -\frac{\pi}{24} \\
F_2^2 &= \frac{\pi}{2}\int_0^1 4\xi(1-\xi)[-(\frac{1}{2}+\frac{1}{2}\xi)]d\xi \\
&= -\frac{\pi}{4} \\
F_3^2 &= \frac{\pi}{2}\int_0^1 \xi(2\xi-1)[-(\frac{1}{2}+\frac{1}{2}\xi)]d\xi \\
&= -\frac{\pi}{12}
\end{aligned} \tag{2.288}$$

$$\{F^2\} = -\frac{\pi}{24}\begin{Bmatrix} 1 \\ 6 \\ 2 \end{Bmatrix}. \tag{2.289}$$

Now we assemble the two elements into a global system.

$$\begin{bmatrix} 1.2760 & -1.8024 & 0.2646 & 0 & 0 \\ -1.8024 & 2.5575 & -1.8024 & 0 & 0 \\ 0.2646 & -1.8024 & 1.2760+1.2760 & -1.8024 & 0.2646 \\ 0 & 0 & -1.8024 & 2.5575 & -1.8024 \\ 0 & 0 & 0.2646 & -1.8024 & 1.2760 \end{bmatrix}\begin{Bmatrix} U_1 \\ U_2 \\ U_3 \\ U_4 \\ U_5 \end{Bmatrix} =$$

$$-\frac{\pi}{24}\begin{Bmatrix} 0 \\ 2 \\ 1+1 \\ 6 \\ 2 \end{Bmatrix} + \begin{Bmatrix} R_1^1 \\ 0 \\ R_3^1 + R_1^2 \\ 0 \\ R_3^2 \end{Bmatrix}. \tag{2.290}$$

The secondary variables at the common nodes between the two elements satisfy

$$R_3^1 + R_1^2 = 0. \tag{2.291}$$

The boundary condition at $x = 0$ gives $U_1 = 0$, and the boundary condition at $x = \pi$ gives $R_3^2 + U_5 = 1/\pi$, and hence $R_3^2 = 1/\pi - U_5$. The reduced system for the unknown primary variables is

$$\begin{bmatrix} 2.5575 & -1.8024 & 0 & 0 \\ -1.8024 & 2.5520 & -1.8024 & 0.2646 \\ 0 & -1.8024 & 2.5575 & -1.8024 \\ 0 & 0.2646 & -1.8024 & 2.2760 \end{bmatrix} \begin{Bmatrix} U_2 \\ U_3 \\ U_4 \\ U_5 \end{Bmatrix} = -\frac{\pi}{24} \begin{Bmatrix} 2 \\ 2 \\ 6 \\ 2 - \frac{24}{\pi^2} \end{Bmatrix}. \tag{2.292}$$

Solving this system, we get for the two quadratic element solution,

$$\{U\} = \begin{Bmatrix} 0 \\ 0.9588 \\ 1.5058 \\ 1.4669 \\ 1.0114 \end{Bmatrix}. \tag{2.293}$$

The secondary variable at the boundary $x = \pi$ is $R_3^2 = 1/\pi - U_5 = -1.0114 + 1/\pi$. The exact solution $u(x) = x/\pi + sinx$ evaluated at the global nodes gives

$$\{u_e\} = \begin{Bmatrix} 0 \\ 0.9571 \\ 1.5 \\ 1.4571 \\ 1 \end{Bmatrix} \tag{2.294}$$

and the secondary variable at the boundary $x = \pi$ is $du/dx = -1 + 1/\pi$.

2.6.3 HIGHER-ORDER ELEMENTS AND LAGRANGE INTERPOLATION FUNCTIONS

When using linear or quadratic elements, we can improve the accuracy of the finite-element solution simply by increasing the number of elements or mesh refinement (called h-refinement). Alternatively, we can also obtain higher accuracy by using higher-order polynomials to approximate the solution over each element (called p-refinement). If the number of nodes per element is n, the approximation polynomial is of order $n-1$. Let x_i^e, and U_i^e $(i = 1, ..., n)$ denote the global coordinates of the nodes of element e, and the corresponding primary variables at the element's nodes. We introduce the local coordinate ξ

$$\xi = (x - x_1^e)/h_e \qquad\qquad x_1^e \le x \le x_n^e \tag{2.295}$$

where $h_e = x_n^e - x_1^e$, is the element length. Note that the element domain $x_1^e \leq x \leq x_n^e$ is mapped onto $0 \leq \xi \leq 1$. It follows that

$$x = x_1^e + h_e \xi \tag{2.296}$$

and $dx = h_e d\xi$. The approximate solution is a polynomial of degree $n-1$ in ξ,

$$U^e(\xi) = \sum_{j=1}^{n} U_j^e \psi_j(\xi). \tag{2.297}$$

The shape functions $\psi_j(\xi)$ are the Lagrange polynomials of degree $n-1$.

$$\psi_j(\xi) = \prod_{m=1, m\neq j}^{n} \frac{\xi - \xi_m}{\xi_j - \xi_m} \qquad 0 \leq \xi \leq 1 \tag{2.298}$$

Note that ψ_i satisfies $\psi_i(\xi_j) = \delta_{ij}$. It is easy to show that

$$\sum_{j=1}^{n} \psi_j(\xi) = 1 \qquad 0 \leq \xi \leq 1. \tag{2.299}$$

The shape functions of the linear and quadratic elements can be obtained as special cases corresponding to $n = 2$ and $n = 3$, respectively. A cubic element $n = 4$ has four local nodes and four shape functions each is a third-order polynomial. For equally spaced nodes, we have $\xi_1 = 0$, $\xi_2 = 1/3$, $\xi_3 = 2/3$, and $\xi_4 = 1$. The shape functions of the Lagrange cubic element are

$$\begin{aligned}
\psi_1(\xi) &= \frac{\xi - \xi_2}{\xi_1 - \xi_2} \cdot \frac{\xi - \xi_3}{\xi_1 - \xi_3} \cdot \frac{\xi - \xi_4}{\xi_1 - \xi_4} \\
&= -\frac{9}{2}(\xi - 1/3)(\xi - 2/3)(\xi - 1) \\
\psi_2(\xi) &= \frac{\xi - \xi_1}{\xi_2 - \xi_1} \cdot \frac{\xi - \xi_3}{\xi_2 - \xi_3} \cdot \frac{\xi - \xi_4}{\xi_2 - \xi_4} \\
&= \frac{27}{2}(\xi)(\xi - 2/3)(\xi - 1) \\
\psi_3(\xi) &= \frac{\xi - \xi_1}{\xi_3 - \xi_1} \cdot \frac{\xi - \xi_2}{\xi_3 - \xi_2} \cdot \frac{\xi - \xi_4}{\xi_3 - \xi_4} \\
&= -\frac{27}{2}(\xi)(\xi - 1/3)(\xi - 1) \\
\psi_4(\xi) &= \frac{\xi - \xi_1}{\xi_4 - \xi_1} \cdot \frac{\xi - \xi_2}{\xi_4 - \xi_2} \cdot \frac{\xi - \xi_3}{\xi_4 - \xi_3} \\
&= \frac{9}{2}(\xi)(\xi - 1/3)(\xi - 2/3).
\end{aligned} \tag{2.300}$$

Using the weak form, we derive the element equations for an element with n nodes. For each element, we obtain a system of n linear algebraic equations of the general form:

$$[A^e]\{U^e\} = \{F^e\} + \{R^e\} \tag{2.301}$$

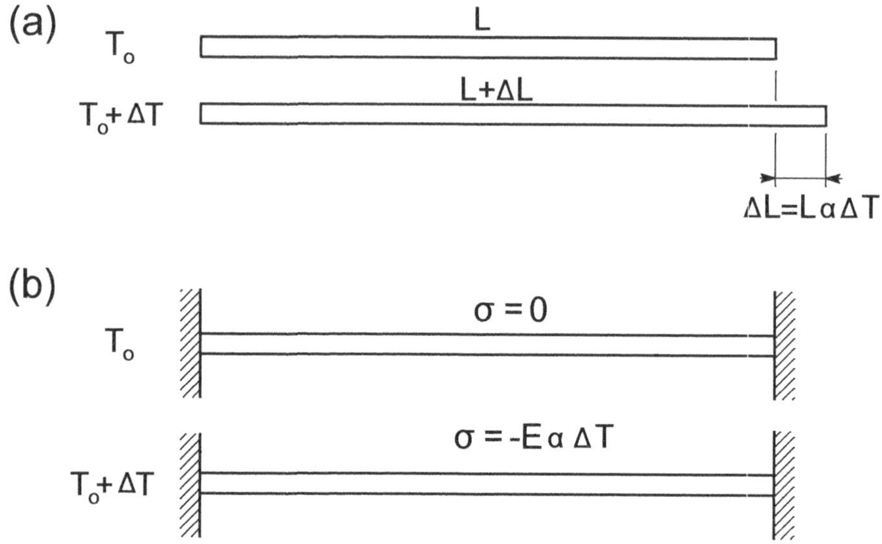

Figure 2.23 Axial deformation of a bar (slender rod) due to temperature change.

The entries for the coefficient matrix $[A^e]$ are

$$a_{ij}^e = \frac{1}{h_e}\int_0^1 \frac{d\psi_i}{d\xi}p\frac{d\psi_j}{d\xi}d\xi + h_e\int_0^1 \psi_i q \psi_j d\xi \tag{2.302}$$

and the forcing vector $\{F^e\}$ is

$$F_i^e = h_e\int_0^1 \psi_i f d\xi. \tag{2.303}$$

The secondary variable array is $\{R^e\}^T = \{R_1^e, 0, ...0, R_n^e\}^T$, where R_1^e and R_n^e are the secondary variables at the first and last nodes of the element, respectively.

2.7 THERMOELASTIC EFFECTS IN ONE DIMENSION

We consider a slender rod (bar) of length L, cross-sectional area A, at uniform temperature T_o. If the temperature is increased to $T_o + \Delta T$ and the bar is unrestrained, it will expand, as shown in Figure 2.23a. The increase in length is $\Delta L = L\alpha\Delta T$, where α is the coefficient of thermal expansion. In this case, there is a thermal strain $\varepsilon_T = \alpha\Delta T$, but no thermal stress. However, if the bar is rigidly constrained as shown in Figure 2.23b, then uniform temperature increase produces no change in length, and hence no thermal strain, but there will be (compressive) thermal stress $\sigma_T = -E\alpha\Delta T$. The equation of equilibrium of an infinitesimally small element of a bar under the combined effects of external axial loading and temperature change is

$$-\frac{d}{dx}(\sigma A) - f(x) = 0 \tag{2.304}$$

where the normal stress is now given by a modified Hook's law,

$$\sigma = E[\varepsilon - \alpha(T - T_o)] \tag{2.305}$$

and $\varepsilon = du/dx$ is the actual strain, $T(x)$ is the current temperature and T_o is the temperature for zero thermal stresses. We neglect the variation of E with temperature. Thus, we solve two problems in sequence. First, we determine the temperature field $T(x)$ by solving a heat conduction problem. Second, we solve for the displacement field $u(x)$ for the obtained temperature field.

The weak form of Equation (2.304) with σ given by Equation (2.305) for an element is

$$\int_{x_1^e}^{x_n^e} \frac{dw}{dx} EA \frac{du}{dx} dx = \int_{x_1^e}^{x_n^e} EA\alpha(T - T_o) \frac{dw}{dx} dx + \int_{x_1^e}^{x_n^e} wf dx + w(x_1^e)R_1^e + w(x_n^e)R_n^e \tag{2.306}$$

where the secondary variable is $R = \sigma A = EA[\varepsilon - \alpha(T - T_o)]$. Using the Galerkin method, we obtain the element equations,

$$[A^e]\{U^e\} = \{F_T^e\} + \{F^e\} + \{R^e\}. \tag{2.307}$$

The entries for the coefficient matrix $[A^e]$ are

$$a_{ij} = \frac{1}{h_e} \int_0^1 \frac{d\psi_i}{d\xi} EA \frac{d\psi_j}{d\xi} d\xi. \tag{2.308}$$

The entries of the forcing vector $\{F_T^e\}$ due to temperature changes are

$$(F_T^e)_i = \int_0^1 EA\alpha(T - T_o) \frac{d\psi_i}{d\xi} d\xi \tag{2.309}$$

it is called thermal loading. The entries of forcing vector $\{F^e\}$ due to the axial distributed load $f(x)$ are

$$F_i^e = h_e \int_0^1 \psi_i f d\xi. \tag{2.310}$$

The secondary variable array is $\{R^e\}^T = \{R_1^e, 0, ...0, R_n^e\}^T$, where R_1^e and R_n^e are the secondary variables at the first and last nodes of the element, respectively.

To evaluate the thermal loading vector, one needs to know the temperature $[T^e(x) - T_o^e]$ distribution over the element. Let's assume that the same type of element is used for both elasticity and thermal problems. Thus, the same shape functions are used to approximate both the temperature and displacement fields,

$$T^e(x) - T_o^e = \sum_{j=1}^{n} \psi_j(\xi)(T_j^e - T_{oj}^e). \tag{2.311}$$

The thermal loading vector becomes

$$(F_T^e)_i = \sum_{j=1}^{n} \left[\int_0^1 EA\alpha \frac{d\psi_i}{d\xi} \psi_j(\xi) d\xi \right] (T_j^e - T_{oj}^e) \tag{2.312}$$

or in matrix form, we have

$$\{F_T^e\} = [G^e]\{T^e - T_o^e\} \tag{2.313}$$

where the elements of the matrix $[G^e]$ are

$$g_{ij}^e = \int_0^1 EA\alpha \frac{d\psi_i}{d\xi} \psi_j(\xi) d\xi. \tag{2.314}$$

For element-wise uniform properties (E, A, and α), we obtain for the linear element,

$$[G^e] = \frac{(EA\alpha)^e}{2} \begin{bmatrix} -1 & -1 \\ 1 & 1 \end{bmatrix} \tag{2.315}$$

and for the quadratic element,

$$[G^e] = \frac{(EA\alpha)^e}{6} \begin{bmatrix} -3 & -4 & 1 \\ 4 & 0 & -4 \\ -1 & 4 & 3 \end{bmatrix}. \tag{2.316}$$

Example 2.7. Axial Deformation due to Temperature Variation

A uniform slender rod (Young's modulus E, coefficient of thermal expansion α, length h, and cross-sectional area A) is rigidly constrained and is unstressed at uniform temperature T_o. The temperature at its ends is changed such that at the end $x = 0$ the temperature is T_a and at the end $x = h$ the temperature is T_b. The rod surface is assumed to be insulated. Use the finite-element method to determine the displacement field and the internal stress at steady state.

Solution

Since there is no heat generation within the bar and the properties are uniform, the steady- state temperature varies linearly with x, hence the final temperature is

$$T(x) = T_a + (T_b - T_a)x/h. \tag{2.317}$$

We use one quadratic element, the shape functions $\psi_i(\xi)$ are given by Equation (2.255), and evaluate the thermal loading vector Equation (2.309).

$$(F_T)_i = \int_0^1 EA\alpha[T(x) - T_o]\frac{d\psi_i}{d\xi} d\xi. \tag{2.318}$$

Evaluating the integrals [or using Equations (2.313) and (2.316)], we obtain

$$\begin{aligned} (F_T^1)_1 &= -EA\alpha(5T_a + T_b - 6T_o)/6 \\ (F_T^1)_2 &= -2EA\alpha(T_b - T_a)/3 \\ (F_T^1)_3 &= EA\alpha(T_a + 5T_b - 6T_o)/6. \end{aligned} \tag{2.319}$$

Substituting $\{F_T^1\}$ into Equation (2.307), realizing that $\{F^e\} = 0$, and that the matrix $[A^e]$ is the stiffness matrix of a quadratic element with constant properties, we obtain the system

$$\frac{EA}{3h}\begin{bmatrix} 7 & -8 & 1 \\ -8 & 16 & -8 \\ 1 & -8 & 7 \end{bmatrix}\begin{Bmatrix} U_1 \\ U_2 \\ U_3 \end{Bmatrix} = \frac{EA\alpha}{6}\begin{Bmatrix} -(5T_a + T_b - 6T_o) \\ -4(T_b - T_a) \\ T_a + 5T_b - 6T_o \end{Bmatrix} + \begin{Bmatrix} R_1 \\ 0 \\ R_3 \end{Bmatrix}. \tag{2.320}$$

Applying the boundary conditions $U_1 = 0$ and $U_3 = 0$, we obtain U_2 from the second equation

$$U_2 = -\alpha h(T_b - T_a)/8 \tag{2.321}$$

and obtain the reactions from the first and third equations,

$$\begin{aligned} R_1 &= EA\alpha\left(\frac{T_a + T_b}{2} - T_o\right) \\ R_3 &= -EA\alpha\left(\frac{T_a + T_b}{2} - T_o\right). \end{aligned} \tag{2.322}$$

The two forces are equal but opposite as expected. The thermal stress is $\sigma_o = -R_1/A = R_3/A$,

$$\sigma_o = -E\alpha\left(\frac{T_a + T_b}{2} - T_o\right). \tag{2.323}$$

The bar is compressed if the average temperature $(T_a + T_b)/2$ is greater than the initial temperature T_o, otherwise it will be under tension. Since $U_1 = U_3 = 0$, the displacement field is $u(x) = U_2\psi_2(x)$.

2.8 NUMERICAL EVALUATIONS OF ELEMENT MATRICES

In practical problems with involved functions (p, q, and f), analytic expressions for the element matrices and vectors may not be available or are given by unwieldy expressions that take excessive computation time to evaluate. For such cases, we evaluate the integrals using numerical methods. Gauss–Legendre quadrature formula is a well-known method. All matrices involve integrals over the element domain and are of the general form

$$I = \int_{x_1^e}^{x_n^e} g(x)dx. \tag{2.324}$$

We introduce the transformation

$$x = \frac{1}{2}(x_n^e + x_1^e) + \frac{1}{2}(x_n^e - x_1^e)\zeta \tag{2.325}$$

and obtain

$$I = \frac{1}{2}(x_n^e - x_1^e)\int_{-1}^{+1} g(x)d\zeta. \tag{2.326}$$

Table 2.1
Gauss–Legendre Quadrature Base Points (ζ_m) and Weight Factors (ω_m)

M	m	ζ_m	ω_m
1	1	0	2
2	1	$-1/\sqrt{3}$	1
	2	$+1/\sqrt{3}$	1
3	1	$-\sqrt{3/5}$	5/9
	2	0	8/9
	3	$+\sqrt{3/5}$	5/9

Gauss–Legendre quadrature formula is

$$\int_{-1}^{+1} G(\zeta)d\zeta = \sum_{m=1}^{M} \omega_m G(\zeta_m). \tag{2.327}$$

The base points ζ_m and corresponding weight factors ω_m $(m = 1, 2, ..., M)$ are given in Table 2.1 for $M = 1, 2$, and 3. Parameters for higher-order quadrature formulas are given by Abramowitz and Stegun [3]. The M-point formula is an exact integration of a polynomial of degree less than or equal to $(2M - 1)$.

2.9 BIOMEDICAL ENGINEERING APPLICATIONS

2.9.1 REACTION-DIFFUSION: OXYGEN CONSUMPTION IN FLAT TISSUES

Dissolved oxygen transported by blood flow diffuses through capillary walls to be consumed in the surrounding tissue. The rate of metabolic consumption of oxygen in the tissue may be modeled by the Michaelis–Menten reaction (e.g., Fournier [38]),

$$B(C_{O_2}) = \frac{V_{max} C_{O_2}}{K_m + C_{O_2}} \tag{2.328}$$

where C_{O_2} is the oxygen concentration in the tissue [μmole/cm^3], V_{max} is the maximum reaction rate [μmole/sec/cm^3], and K_m is known as Michaelis constant [μmole/cm^3]. If the concentration is much greater than K_m, the reaction rate is maximum $B = V_{max}$, which is called a zero-order reaction since the rate is constant. For zero-order reaction, the rate of oxygen uptake is constant regardless of oxygen supply. If the concentration is much lower than K_m, the rate is linearly proportional to the concentration, $B = (V_{max}/K_m)C_{O_2}$, which is called a first-order reaction. Hypoxia is the term used to describe tissues where oxygen concentration drops below certain levels whereas anoxia means tissues with zero oxygen concentration. Because the zero-order rate is inaccurate at low concentration, the Michaelis–Menten reaction rate is needed for determining the hypoxia or anoxia conditions. In this section, we solve the problem of oxygen consumption in flat tissues. Spherical cells and cylindrical tissues are considered in the next section. The objective is to study the effects of the used oxygen uptake rate on the distribution of oxygen concentration in the tissue. The results show that first-order and Michaelis–Menten reaction rates predict hypoxic but not anoxic conditions. The zero-order reaction rate predicts anoxic regions, but the predictions are not valid because that rate is not applicable when the concentration is less than K_m.

We consider a flat tissue (a slab) of thickness $2b$, and solve for a steady state condition where the oxygen concentration does not change with time. The two sides of the tissue are subjected to convection, whereas the midplane is a plane of symmetry. The concentration is assumed to vary only across the tissue thickness. The concentration is then given by the mass reaction-diffusion equation

$$-\frac{d}{dx_1}\left[D_T\frac{dC_{O_2}}{dx_1}\right]+B(C_{O_2})=0 \tag{2.329}$$

where D_T is oxygen diffusivity in tissue [cm^2/sec]. The boundary conditions are

$$\begin{aligned} x_1 &= 0: & -D_T\frac{dC_{O_2}}{dx_1} &= \gamma_o(C_o-C_{O_2}) \\ x_1 &= b: & D_T\frac{dC_{O_2}}{dx_1} &= 0. \end{aligned} \tag{2.330}$$

The first condition is a convection boundary condition where γ_o is the convection coefficient and C_o is the oxygen concentration outside the boundary layer. The second condition is the symmetry condition.

Introducing nondimensional variables, $x = x_1/b$, $u = C_{O_2}/C_o$, and $\alpha = V_{max}b^2/(C_oD_T)$, $\beta = K_m/C_o$, and $\gamma = \gamma_o b/D_T$, we rewrite Equation (2.329) in nondimensional form

$$-\frac{d}{dx}\left(\frac{du}{dx}\right)+B(u)=0 \tag{2.331}$$

where

$$B(u)=\frac{\alpha u}{\beta+u}. \tag{2.332}$$

The boundary conditions are

$$\begin{aligned} x &= 0: & -\frac{du}{dx}+\gamma u &= \gamma \\ x &= 1: & \frac{du}{dx} &= 0. \end{aligned} \tag{2.333}$$

The concentration distribution within the tissue depends only on three nondimensional parameters, namely α, β, and γ.

Defining $y = du/dx$, we write Equation (2.331) as $-ydy+B(u)du=0$. For $B(u)$ given by Equation (2.332), we find an exact analytic solution to the problem given by Equations (2.331)–(2.333),

$$x = \frac{1}{\sqrt{2\alpha}} \int_{u}^{u_o} \frac{d\zeta}{\sqrt{\zeta - u_1 - \beta ln\left(\frac{\zeta+\beta}{u_1+\beta}\right)}} \tag{2.334}$$

where $u_o = u(0)$ and $u_1 = u(1)$ given implicitly by

$$\alpha = \frac{1}{2}\left[\int_{u_1}^{u_o} \frac{d\zeta}{\sqrt{\zeta - u_1 - \beta ln\left(\frac{\zeta+\beta}{u_1+\beta}\right)}}\right]^2 \tag{2.335}$$

$$\gamma = \frac{\sqrt{2\alpha}}{1-u_o}\sqrt{u_o - u_1 - \beta ln\left(\frac{u_o+\beta}{u_1+\beta}\right)}. \tag{2.336}$$

For the purpose of comparison of the finite element solution to the exact solution, we can prescribe values for β, u_o, and u_1, and use Equations (2.335) and (2.336) to find α and γ.

The finite element method can be applied directly to the boundary-value problem given by Equations (2.331)–(2.333) but the resulting system of algebraic equations will be nonlinear because the reaction rate $B(u)$ is nonlinear in the unknown concentration u. An iterative method can then be used to solve the nonlinear system. Another iterative approach that we adopt here is to quasi-linearize the nonlinear term before developing the finite element model. Let $v(x)$ be a known initial guess for $u(x)$. The nonlinear function $B(u)$ may be linearized as

$$B(u) = B(v) + \left[\frac{dB}{du}\right]_{u=v}(u-v). \tag{2.337}$$

This is linear in u since v is known. Quasi-linearization of $B(u)$ given by Equation (2.332) is

$$B(u) = \frac{\alpha v}{\beta + v} + \frac{\alpha\beta}{(\beta+v)^2}(u-v) \tag{2.338}$$

or,

$$B(u) = \frac{\alpha v^2}{(\beta+v)^2} + \frac{\alpha\beta u}{(\beta+v)^2}. \tag{2.339}$$

Substituting Equation (2.339) into Equation (2.331), we write it in the form of the model equation as

$$-\frac{d}{dx}\left(p(x)\frac{du}{dx}\right) + q(x)u - f(x) = 0 \tag{2.340}$$

where $p = 1$, $q(x) = \alpha\beta/[\beta + v(x)]^2$, and $f(x) = -\alpha v^2(x)/[\beta + v(x)]^2$.

The boundary conditions are given by Equation (2.333). For a given $v(x)$, we use the finite element method as presented in this chapter to solve this linear equation.

The solution is denoted by $u^*(x)$. A measure of the change $u^* - v$ is computed and the iterative process is terminated if the measure falls below some prescribed value. If not, we set $v(x) = u^*(x)$ and use the finite element method to solve the problem again and obtain a new $u^*(x)$. The iteration continues until the convergence criterion is met.

An initial guess might be the exact analytic solution to the linear problem assuming zero or first-order reaction rate denoted by u_{zo} and u_{fo}, respectively.

$$\begin{aligned} u_{zo}(x) &= -\alpha(1-\frac{1}{2}x)x+1-\frac{\alpha}{\gamma} && \text{if} \quad \alpha < z_o \\ u_{fo}(x) &= A_o cosh[\lambda(1-x)] && \text{if} \quad \alpha \geq z_o \end{aligned} \tag{2.341}$$

where $z_o = 2\gamma/(\gamma+2)$, $\lambda = \sqrt{\alpha/\beta}$ and $A_o = \gamma/(\lambda sinh\lambda + \gamma cosh\lambda)$.

Comparison Between Finite-Element Method and Exact Solution for Michaelis–Menten Kinetics

The finite-element method for the nonlinear Michaelis–Menten kinetics is validated by comparison with the exact analytic solution for two cases. For case 1 the oxygen concentrations at the boundary are specified $u_o = 0.6$ and $u_1 = 0.2$, and for case 2 they are $u_o = 0.9$ and $u_1 = 0.01$. For Michaelis constant of $\beta = 0.1$ and tissue half-thickness $b = 1$, we use Equations (2.335) and (2.336) to find $\alpha = 1.118179$ and $\gamma = 2.099192$ for case 1, and $\alpha = 5.101351$ and $\gamma = 26.131185$ for case 2. These values of α, β, and γ are used in the reaction rate Equation (2.332) and boundary conditions Equation (2.333) for the finite-element method, but the values of u_o and u_1 are not prescribed; instead, they are predicted by the finite-element solution. A mesh of 20 quadratic elements is used. As shown in Figure 2.24, excellent agreement is obtained between the finite-element solution and exact solution as given by Equation (2.334). The errors in the values of u_o and u_1 predicted by the finite element method are $O(10^{-9})$.

Example 2.8. Comparison of Michaelis–Menten Reaction Rate with the First- and Zero-order Reactions

We consider a case of $\alpha = 1.3, \beta = 0.1$, and $\gamma = 5$, and use 10 quadratic elements. The finite-element solution using the nonlinear reaction rate is compared with the solutions given by Equation (2.341) for zero- and first-order rates in Figure 2.25. We conclude that, in this particular case, it is important to consider the nonlinear reaction rate, and that neither of zero- or first-order rate is accurate.

The finite element method is then used to study the effects of the maximum reaction rate α of the nonlinear rate on the oxygen distribution within the tissue. We use 10 quadratic elements and $\beta = 0.1$, and $\gamma = 5$. The distributions for $\alpha = 0.5$, 1, 1.5, 2, and 4 are depicted in Figure 2.26. As the maximum reaction rate increases, the oxygen concentration near the tissue center decreases, and the tissue becomes hypoxic.

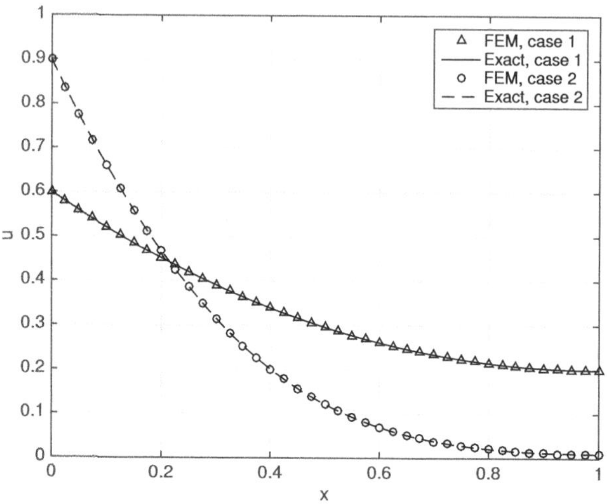

Figure 2.24 Comparison of finite-element method to the exact analytic solution for oxygen concentration in a flat tissue with Michaelis–Menten reaction rate. Case 1: $u_o = 0.6$, and $u_1 = 0.2$, and case 2: $u_o = 0.9$, and $u_1 = 0.01$. A mesh of 20 quadratic elements is used.

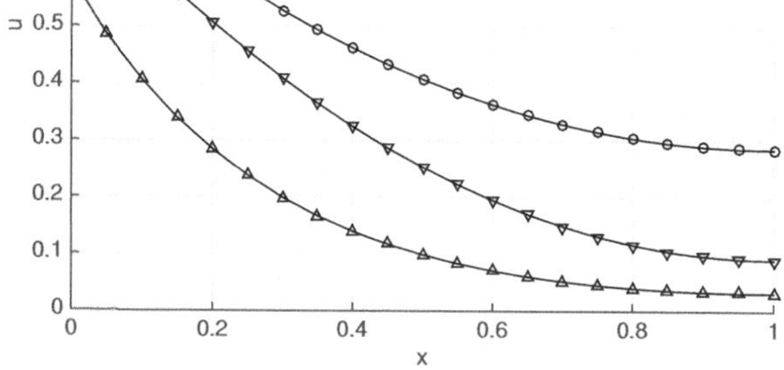

Figure 2.25 Comparison of Michaelis–Menten reaction rate with the first- and zero-order reactions. $\alpha = 1.3$, $\beta = 0.1$, and $\gamma = 5$. A mesh of 10 quadratic elements is used.

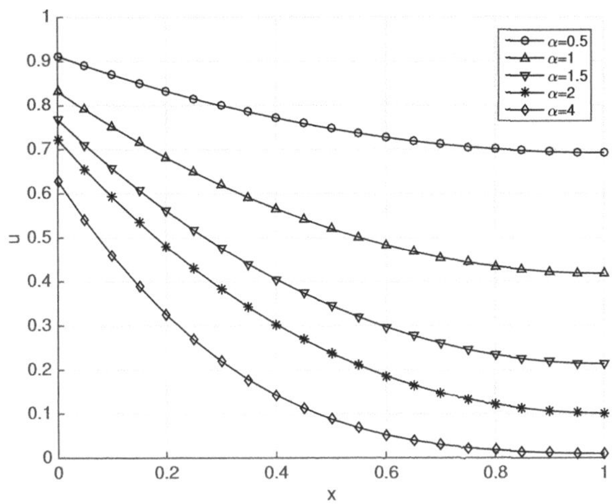

Figure 2.26 Effects of maximum reaction rate α of Michaelis–Menten kinetics on oxygen distribution in a flat tissue. $\beta = 0.1$, and $\gamma = 5$. A mesh of 10 quadratic elements is used.

2.9.2 REACTION-DIFFUSION: OXYGEN CONSUMPTION IN CYLINDRICAL TISSUES AND SPHERICAL CELLS

Oxygen diffusion and consumption in tissues!cylindrical tissues and spherical cells

In this section, we consider the problem of oxygen consumption in a cylindrical or spherical domain. In three-dimensional space, the mass flux vector is given by Fick's law:

$$\vec{J} = -D_T \nabla C_{O_2} \tag{2.342}$$

where C_{O_2} is oxygen concentration (mass per unit volume) and D_T is diffusivity coefficient (area/time). In the absence of convection, mass balance in an infinitesimally small volume gives the governing equation for the concentration

$$-\nabla \cdot (D_T \nabla C_{O_2}) + B(C_{O_2}) = 0 \tag{2.343}$$

where $B(C_{O_2})$ is the rate of mass consumption per unit volume. We consider a cylindrical or spherical domain and assume that the concentration varies only in the radial direction.

$$-\frac{1}{r^m}\frac{d}{dr}\left(D_T r^m \frac{dC_{O_2}}{dr}\right) + B(C_{O_2}) = 0 \tag{2.344}$$

where $m = 1$ for cylindrical coordinates, and $m = 2$ for spherical coordinates.

We introduce nondimensional variables $r^* = r/L_r$, $u = C_{O_2}/C_r$, $p = D_T/D_{Tr}$, $B^* = BL_r^2/C_r D_{Tr}$, where L_r, C_r, and D_{Tr} are reference length, concentration, and diffusivity, respectively. We drop the asterisk for simplicity of notation, and write the governing equation in nondimensional form as

$$-\frac{1}{r^m}\frac{d}{dr}\left(pr^m\frac{du}{dr}\right)+B(u)=0. \tag{2.345}$$

The rate of oxygen consumption is assumed to be a Michaelis–Menten reaction

$$B(u)=\frac{\alpha u}{\beta+u}. \tag{2.346}$$

The parameter α is the maximum reaction rate and β is the half saturation concentration. To derive a finite-element model, we first quasi-linearize $B(u)$ around a guessed solution $v(r)$ as we have done in Section 2.9.1.

$$B(u)=\frac{\alpha v^2}{(\beta+v)^2}+\frac{\alpha\beta u}{(\beta+v)^2}. \tag{2.347}$$

Now the equation takes the form

$$-\frac{1}{r^m}\frac{d}{dr}\left[p(r)r^m\frac{du}{dr}\right]+q(r)u-f(r)=0 \tag{2.348}$$

where $q(r)=\alpha\beta/[\beta+v(r)]^2$, and $f(r)=-\alpha v^2(r)/[\beta+v(r)]^2$.

Let us consider a cylindrical or a spherical shell of inner radius a and outer radius b (a and b are nondimensional), and specify the boundary conditions:

$$\begin{aligned} r &= a: \quad -pr^m\frac{du}{dr}+\beta_a u=\alpha_a \\ r &= b: \quad pr^m\frac{du}{dr}+\beta_b u=\alpha_b. \end{aligned} \tag{2.349}$$

To derive the weak form for the problem Equations (2.348)–(2.349), we multiply Equation (2.348) by a test (weight) function $w(r)$, and integrate over the domain volume. The element of volume is a cylindrical or spherical shell of radial thickness dr. For a cylindrical shell of length L_z the volume is $2\pi r dr L_z$, and for a spherical shell, it is $4\pi r^2 dr$. We assume a unit length in the z-direction for the cylindrical shell $L_z=1$, and write the element volume as $2^m\pi r^m dr$ for both shells. Furthermore, for simplicity, we drop the common constant factor $2^m\pi$.

$$\int_a^b -w(r)\frac{d}{dr}\left[p(r)r^m\frac{du}{dr}\right]dr+\int_a^b w(r)q(r)u(r)r^m dr-\int_a^b w(r)f(r)r^m dr=0. \tag{2.350}$$

Integrating the first term by parts, we obtain

$$\int_a^b\left[\frac{dw}{dr}p(r)r^m\frac{du}{dr}+w(r)q(r)u(r)r^m\right]dr=\int_a^b w(r)f(r)r^m dr \\ +w(a)R_a+w(b)R_b \tag{2.351}$$

where

$$\begin{aligned} R_a &= -\left(pr^m \frac{du}{dr}\right)_{r=a} \\ R_b &= \left(pr^m \frac{du}{dr}\right)_{r=b} \end{aligned} \tag{2.352}$$

are the secondary variables at the end points. For the boundary conditions Equation (2.349), the weak form becomes

$$\int_a^b \left[\frac{dw}{dr}p(r)r^m \frac{du}{dr}dr + w(r)q(r)u(r)r^m\right] dr + \beta_a w(a)u(a) + \beta_b w(b)u(b)$$
$$= \int_a^b w(r)f(r)r^m dr + \alpha_a w(a) + \alpha_b w(b). \tag{2.353}$$

To develop the finite element model, we divide the radial distance $a \le r \le b$ into elements, each element is defined by n nodes. The first node of a typical element (e) is r_1^e and its nth node is at r_n^e. The weak form for such an element follows from Equation (2.351),

$$\int_{r_1^e}^{r_n^e} \left[\frac{dw}{dr}p(r)r^m \frac{du}{dr}dr + w(r)q(r)u(r)r^m\right] dr$$
$$= \int_{r_1^e}^{r_n^e} w(r)f(r)r^m dr + w(r_1^e)R_1^e + w(r_n^e)R_n^e \tag{2.354}$$

where

$$\begin{aligned} R_1^e &= -\left(pr^m \frac{du}{dr}\right)_{r=r_1^e} \\ R_n^e &= \left(pr^m \frac{du}{dr}\right)_{r=r_n^e} \end{aligned} \tag{2.355}$$

are the secondary variables at the end nodes of the element, which are unknown since no boundary conditions are imposed for the typical element.

Approximation

An approximation of the primary variable over the element domain is written in terms of the element shape functions $\psi_i(r)$,

$$u(r) \approx U^e(r) = \sum_{j=1}^{n} U_j^e \psi_j(r). \tag{2.356}$$

To write $U^e(r)$ in matrix form, we define the arrays

$$\{\psi(r)\} = \{\psi_1(r)\ \psi_2(r) \ldots \psi_j(r) \ldots \psi_n(r)\} \tag{2.357}$$

$$\{U^e\} = \begin{Bmatrix} U_1^e \\ U_2^e \\ \cdot \\ \cdot \\ \cdot \\ U_j^e \\ \cdot \\ \cdot \\ \cdot \\ U_n^e \end{Bmatrix}. \tag{2.358}$$

In matrix form, the trial solution over element # e is

$$U^e(r) = \{\psi(r)\}\{U^e\}. \tag{2.359}$$

We note that the dependence of the trial solution on r is given by the array $\{\psi(r)\}$, whereas the array $\{U^e\}$ on the right-hand side does not depend on r; these are values of the primary variable at the element nodes.

Galerkin

In the Galerkin method, the test (or weight) function is taken as a linear superposition of the element shape functions.

$$W^e(r) = \sum_{i=1}^{n} d_i \psi_i(r) \tag{2.360}$$

where d_i ($i = 1, 2, \ldots n$) is an array of (arbitrary) constants that are the values of the test function at the element nodes. We define the array

$$\{d\} = \begin{Bmatrix} d_1 \\ d_2 \\ \cdot \\ \cdot \\ \cdot \\ d_i \\ \cdot \\ \cdot \\ \cdot \\ d_n \end{Bmatrix} \tag{2.361}$$

and write $W^e(r)$ in matrix form

$$W^e(r) = \{\psi(r)\}\{d\} \tag{2.362}$$

or

$$W^e(r) = \{d\}^T\{\psi(r)\}^T. \tag{2.363}$$

Note that the choice $W^e(r) = \psi_1(r)$ is obtained if $d_1 = 1$ and $d_i = 0$ if $i \neq 1$. Similarly, the choice $W^e(r) = \psi_2(x)$ is obtained if $d_2 = 1$ and $d_i = 0$ if $i \neq 2$, and so on.

Once we decided on the trial solution and test function, we use the weak form to obtain a system of algebraic equations for the array of unknown primary variables at the element nodes and the unknown secondary variables at the element end nodes. To that end, we substitute the trial solution Equation (2.359) and the test function Equation (2.363) into the weak form Equation (2.354).

$$\int_{r_1^e}^{r_n^e} \left[\{d\}^T \frac{d\{\psi(r)\}^T}{dr} p(r) \frac{d\{\psi(r)\}}{dr} \{U^e\} + \{d\}^T \{\psi(r)\}^T q(r) \{\psi(r)\}\{U^e\} \right] r^m dr$$
$$- \int_{r_1^e}^{r_n^e} \{d\}^T \{\psi(r)\}^T f(r) r^m dr - \{d\}^T \{\psi(r_n^e)\}^T R_n^e - \{d\}^T \{\psi(r_1^e)\}^T R_1^e = 0. \tag{2.364}$$

Recall that $\{d\}$ and $\{U^e\}$ are independent of r, we factor them out,

$$\{d\}^T \left\{ \int_{r_1^e}^{r_n^e} \left[\frac{d\{\psi(r)\}^T}{dr} p(r) \frac{d\{\psi(r)\}}{dr} + \{\psi(r)\}^T q(r) \{\psi(r)\} \right] r^m dr \{U^e\} \right.$$
$$\left. - \int_{r_1^e}^{r_n^e} \{\psi(r)\}^T f(r) r^m dr - \{\psi(r_n^e)\}^T R_n^e - \{\psi(r_1^e)\}^T R_1^e \right\} = 0. \tag{2.365}$$

Define the element coefficient matrix

$$[A^e] = \int_{r_1^e}^{r_n^e} \left[\frac{d\{\psi(r)\}^T}{dr} p(r) \frac{d\{\psi(r)\}}{dr} + \{\psi(r)\}^T q(r) \{\psi(r)\} \right] r^m dr \tag{2.366}$$

and element vector

$$\begin{aligned} \{\tilde{b}^e\} &= \int_{r_1^e}^{r_n^e} \{\psi(r)\}^T f(r) r^m dr + \{\psi(r_n^e)\}^T R_n^e + \{\psi(r_1^e)\}^T R_1^e \\ &= \{F^e\} + \{R^e\} \end{aligned} \tag{2.367}$$

where the forcing vector is

$$\{F^e\} = \int_{r_1^e}^{r_n^e} \{\psi(r)\}^T f(r) r^m dr \tag{2.368}$$

and the array of secondary variables is

$$\{R^e\} = \left\{ \begin{array}{c} R_1^e \\ 0 \\ \cdot \\ \cdot \\ \cdot \\ 0 \\ \cdot \\ \cdot \\ 0 \\ R_n^e \end{array} \right\}. \tag{2.369}$$

We write Equation (2.365) as

$$\{d\}^T \left[[A^e]\{U^e\} - \{\tilde{b}^e\}\right] = 0. \tag{2.370}$$

Because $\{d\}$ is arbitrary, the bracket must be zero, and we obtain the desired element equations

$$[A^e]\{U^e\} = \{\tilde{b}^e\}. \tag{2.371}$$

In index notation, the entries of the coefficient matrix are

$$a_{ij}^e = \int_{r_1^e}^{r_n^e} \left[\frac{d\psi_i(r)}{dr} p(r) \frac{d\psi_j(r)}{dr} + \psi_i(r) q(r) \psi_j(r)\right] r^m dr \tag{2.372}$$

and forcing vector is

$$\tilde{b}_i^e = \int_{r_1^e}^{r_n^e} \psi_i(r) f(r) r^m dr \qquad \text{if} \quad 2 \le i \le n-1 \tag{2.373}$$

and

$$\tilde{b}_1^e = \int_{r_1^e}^{r_n^e} \psi_1(r) f(r) r^m dr + R_1^e \tag{2.374}$$

$$\tilde{b}_n^e = \int_{r_1^e}^{r_n^e} \psi_n(r) f(r) r^m dr + R_n^e. \tag{2.375}$$

Example 2.9. Reaction-Diffusion in a Spherical Cell

We consider reaction-diffusion in a spherical cell of nondimensional radius $b = 1$ and diffusion coefficient $p = 1$, with Michaelis–Menten reaction Equation (2.346) with constant $\beta = 0.1$, and study the effect of α on the distribution of C_{O2} in the cell for $\alpha = 5, 10, 15$, and 20. Boundary conditions are

$$\begin{aligned} r &= 0: \quad du/dr = 0 \\ r &= 1: \quad u = 1. \end{aligned}$$

For a mesh of 10 quadratic elements, oxygen concentration is shown in Figure 2.27. Since the concentration is fixed on the cell outer boundary, as the rate of consumption increases the concentration decreases rapidly toward the cell center. The low level of oxygen concentration implies that the core of the cell suffers from hypoxia [38]. We conclude that according to Michaelis–Menten kinetics, the cell may become hypoxic but never anoxic.

To determine the anoxic radius r_o for a spherical cell, we apply the boundary conditions

$$\begin{aligned} r &= r_o: \; u = 0 \text{ and } du/dr = 0 \\ r &= 1: \; u = 1. \end{aligned} \tag{2.376}$$

For zero-order reaction $\beta = 0$, we solve the problem analytically and obtain the condition, $(3 - 2r_o)r_o^2 = 1 - 6/\alpha$. Thus, according to the zero-order kinetics if the reaction rate α is greater than 6, a core region $0 \le r \le r_o$ will be anoxic. However, this result conflicts with the results of the more accurate Michaelis–Menten kinetics.

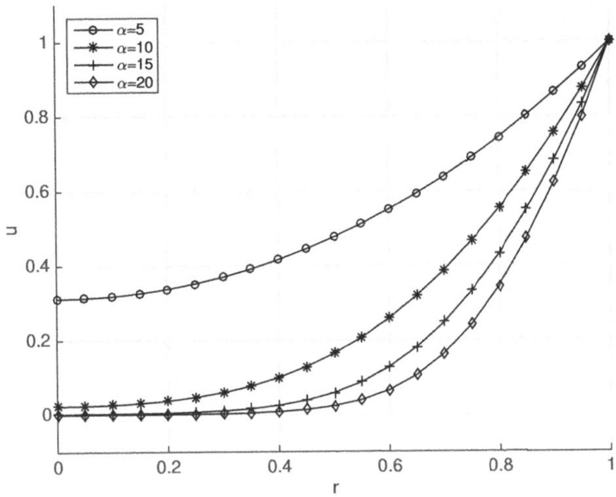

Figure 2.27 Effects of maximum reaction rate of Michaelis-Menten kinetics on oxygen concentration in a spherical cell. $\beta = 0.1$. A mesh of 10 quadratic elements is used.

2.9.3 PULSATILE BLOOD FLOW IN ARTERIES: WOMERSLEY PROBLEM

Pulsatile blood flow in an artery is modeled as an unsteady incompressible viscous flow in a long circular tube of radius b. Dynamic viscosity μ and density ρ are assumed to be constants. We set the continuity and momentum equations in cylindrical coordinates (r, θ, z), and assume the streamlines to be straight lines parallel to the tube axis, which is taken to be the z-axis. The flow is also assumed to be axisymmetric (i.e., no variation in θ, and $v_\theta = 0$). With these assumptions, we reduce the continuity and momentum equations to one equation for the velocity component $w(r,t)$ along the z-axis,

$$\rho \frac{\partial w}{\partial t} = -\frac{\partial p}{\partial z} + \frac{1}{r}\frac{\partial}{\partial r}\left(\mu r \frac{\partial w}{\partial r}\right) \tag{2.377}$$

where $-\partial p/\partial z$ is the pressure gradient; it is a function of time only. The boundary conditions are

$$\begin{aligned} r &= 0: \quad \frac{\partial w}{\partial r} = 0 \\ r &= b: \quad w = 0. \end{aligned} \tag{2.378}$$

McDonald [72] reported measured pressure gradient and volume flow rate in arteries of dogs. Using McDonald's data, Womersley [126] approximated the measured pressure gradient by a Fourier series. We denote the pressure gradient by a periodic function $g(t) = -\partial p/\partial z$ of period T. We divide the period T into N equally spaced segments each of size $\Delta t = T/N$; where N is an even number. Let $g_i = g(t_i)$ denote the value of the function g at time t_i. If we know the N values of the function g at the time points $0, \Delta t, 2\Delta t, ..., (N-1)\Delta t$ we can curve-fit the function $g(t)$ by a harmonic function in time:

$$-\frac{\partial p}{\partial z} = g(t) = \sum_{k=-N/2}^{N/2-1} C_k e^{ik\omega t} \tag{2.379}$$

where $\omega = 2\pi/T$. The pressure gradient is a real-valued function. There are N coefficients in the series of, which $N-2$ coefficients are complex. The coefficient C_{-k} is the complex conjugate of C_k for $k = 1, 2, ..., N/2-1$. The remaining two coefficients C_o and $C_{-N/2}$ are purely real. The values $(g_i, i = 0, 1, 2, ..., N-1)$ are measured pressure gradient such as those given in Figure 7 of McDonald's paper. To determine the complex coefficients C_k, we input the N values of g_i into the MATLAB fast Fourier transform (fft) function; the output is the complex coefficients C_k (note that MATLAB returns NC_k.)

Since Equation (2.377) is a linear differential equation, we solve it for one Fourier mode (i.e., one frequency $k\omega$, where k is an integer) at a time and then use superposition to find the total response. Thus, we take

$$-\frac{\partial p}{\partial z} = C_k e^{ik\omega t}. \tag{2.380}$$

Here, we are interested in the long-time response; therefore, we ignore the initial conditions and transient response, and assume the response to be harmonic in time with the same frequency $k\omega$ as the pressure gradient.

$$w(r,t) = W_k(r) e^{ik\omega t} \tag{2.381}$$

where $W_k(r)$ is a complex-valued function of r. Substituting Equations (2.380) and (2.381) into Equation (2.377), we get

$$-\frac{1}{r}\frac{d}{dr}\left(\mu r \frac{dW_k}{dr}\right) + ik\omega\rho W_k - C_k = 0. \tag{2.382}$$

Introduce nondimensional variables $u_k = W_k/(C_k b^2/\mu)$ and $r^* = r/b$, and rewrite Equation (2.382),

$$-\frac{1}{r^*}\frac{d}{dr^*}\left(r^* \frac{du_k}{dr^*}\right) + i\alpha_k^2 u_k - 1 = 0 \tag{2.383}$$

where $\alpha_k = b\sqrt{k\omega/\nu}$ is the Womersley number, and $\nu = \mu/\rho$ is the kinematic viscosity. The boundary conditions become

$$\begin{aligned} r^* &= 0: \quad \frac{du_k}{dr^*} = 0 \\ r^* &= 1: \quad u_k = 0. \end{aligned} \tag{2.384}$$

The problem specified by Equations (2.383) and (2.384) is the model problem in cylindrical coordinates where $p = 1$, $q = i\alpha_k^2$ and $f = 1$. We use the finite-element model of Section (2.9.2) to solve this problem for u_k. Solving this problem for all frequencies in the pressure gradient $k = 0, 1, 2, ..., N/2 - 1$, and $k = -N/2$, we obtain the finite-element approximation to the velocity $w(r,t)$,

$$w(r,t) = \sum_{k=-N/2}^{N/2-1} \frac{C_k b^2}{\mu} u_k(r^*) e^{ik\omega t}. \tag{2.385}$$

The exact analytic solution to the problem specified by Equations (2.383) and (2.384) is

$$u_k(r^*) = -\frac{i}{\alpha_k^2} \left[1 - \frac{J_o(\sqrt{-i}\alpha_k r^*)}{J_o(\sqrt{-i}\alpha_k)} \right] \tag{2.386}$$

where J_0 is the Bessel function of the first kind and order 0. With $w(r,t)$ given by Equation (2.385), the instantaneous volume flow rate $Q(t)$ is

$$\begin{aligned} Q(t) &= \int_0^b w(r,t) 2\pi r dr \\ &= 2\pi b^2 \sum_{k=-N/2}^{N/2-1} \frac{C_k b^2}{\mu} \hat{Q}_k e^{ik\omega t} \end{aligned} \tag{2.387}$$

where

$$\hat{Q}_k = \int_0^1 u_k(r^*) r^* dr^*. \tag{2.388}$$

For the finite-element method, the integral is evaluated for each element using the shape functions of the element, and for the exact analytic solution Equation (2.386) the result is

$$\hat{Q}_k = \frac{1}{\alpha_k^4} \left[\frac{Z_k^2}{2} - \frac{Z_k J_1(Z_k)}{J_o(Z_k)} \right] \tag{2.389}$$

where $Z_k = \alpha_k \sqrt{-i}$, and J_1 is the Bessel function of the first kind and order 1. Another quantity of interest is the instantaneous wall shear stress $\tau(t)$, which is calculated from

$$\begin{aligned} \tau(t) &= \mu \left[\frac{\partial w(r,t)}{\partial r} \right]_{r=b} \\ &= b \sum_{k=-N/2}^{N/2-1} C_k \left[\frac{du_k(r^*)}{dr^*} \right]_{r^*=1} e^{ik\omega t} \end{aligned} \tag{2.390}$$

For the finite-element method, the derivative $du_k(r^*)/dr^*$ at $r^* = 1$ is the secondary variable (R_k) at the node on the artery wall. After solving for the primary variables at all global nodes, the equation for the boundary node, before the application of the boundary conditions, is used to compute the unknown secondary variable; no

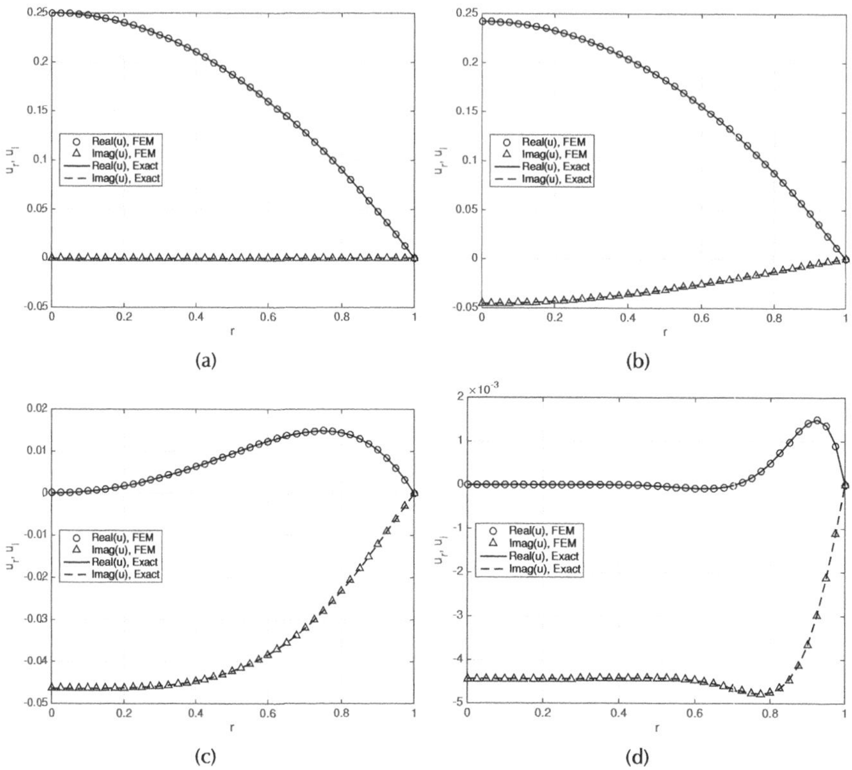

Figure 2.28 Comparison between finite-element method and exact solution for velocity profiles of pulsatile flow in a tube at different frequencies (Womersley number, α). A mesh of 20 quadratic elements is used. (a) $\alpha = 0.1$, (b) $\alpha = 1$, (c) $\alpha = 5$, and (d) $\alpha = 15$.

numerical differentiation is performed. For the exact analytic solution, the derivative is

$$\left[\frac{du_k(r^*)}{dr^*}\right]_{r^*=1} = -\frac{1}{\alpha_k\sqrt{-i}}\frac{J_1(Z_k)}{J_o(Z_k)}. \tag{2.391}$$

First, we investigate the basic features of velocity profile $u_k(r^*)$ for selected values of Womersley number α_k. The profiles are depicted in Figures 2.28a–2.28d for $\alpha_k = 0.1, 1, 5$, and 15 for both the finite-element results and exact analytic solution. As one anticipates, for very low frequency or $\alpha_k = 0.1$ the real part of the velocity profile (u_r) is close to the Poiseuille parabolic profile which prevails for steady flow in the tube (i.e., at zero frequency). The imaginary part (u_i) is nearly zero indicating that the velocity is in-phase with the pressure gradient. For low frequency, local acceleration $\rho\partial w/\partial t$ is negligible, and the pressure gradient is balanced by the net viscous force. As α_k or frequency increases the real part decreases near the tube axis while the imaginary part increases in magnitude but is negative. For high frequency or $\alpha_k = 15$, the real part is nearly zero for $(0 \le r^* < 0.7)$ while the imaginary part is dominant but negative. At high frequency, the local acceleration $\rho \partial w/\partial t$ balances the pressure gradient over most of the artery cross-section, and the negative imaginary part of velocity indicates that the velocity lags the pressure gradient by about $\pi/2$. Because of the no-slip boundary condition, viscous forces must participate in the force balance and the velocity magnitude may exceed its value on the centerline. The viscous layer near the wall is called the Stokes layer. It is also possible that for a short time during the period of oscillation, the velocity near the wall will be opposite to that on the centerline.

To demonstrate the total time response, we synthesized a pressure gradient pulse

$$-\frac{\partial p}{\partial z} = 55t^*(1-t^*)(0.22-t^*)(0.42-t^*)(0.81-t^*)E(t^*) \quad \text{for} \quad 0 \le t^* \le 1 \tag{2.392}$$

where $t^* = t/T$, and

$$\begin{aligned} E(t^*) \quad = \quad & (1+140e^{-36(0.32-t^*)^2})(1+2e^{-120(0.64-t^*)^2}) \times \\ & (1+80e^{-50(0.925-t^*)^2})/(1+10t^{*2}). \end{aligned} \tag{2.393}$$

The units for pressure gradient are (mmHg/cm). The pulse is plotted in Figure 2.29; it is similar to the pressure gradient measured by McDonald (Figure 7 of [72]). We also use the following data from [72]: The artery radius $b = 0.15$ cm, blood properties $\mu = 0.004$ Pa.s and $\rho = 1050$ kg/m^3, and pulse period $T = 1/3$ s. For $N = 64$, we use fft function in MATLAB to determine the coefficients C_k in Equation (2.379). A mesh of 20 quadratic elements is used. The finite-element results are compared to the exact analytic solution in Figure 2.29. For the exact solution, the volume flow rate is computed using Equations (2.387) and (2.389), and shear stress using Equation (2.391). Figure 2.29 shows that the volume flow rate lags the pressure gradient by a phase angle of about 25 degrees. Also, the wall shear lags the pressure gradient by approximately 12 degrees. The mean pressure gradient is 0.2118 mmHg/cm, and the mean flow rate and wall shear stress are 1.4034 ml/s and 2.1178 Pa, respectively. The maximum and minimum levels of the instantaneous volume flow rate and wall shear are significantly higher than those of the Poiseuille velocity profile corresponding to the mean pressure gradient. However, the net forward flow in the artery is due to the mean pressure gradient only, which is given by the coefficient C_o in the Fourier decomposition, and the average over a period of oscillation of the flow rate of any harmonic is zero.

2.10 PROBLEMS

Exact Analytic Solution

(1) An aluminum pin fin of diameter $d = 12$ mm, length $L = 72$ mm, and thermal conductivity $k = 204$ W m^{-1}°C^{-1} is exposed to ambient air at $T_\infty = 25$°C with a convection heat transfer coefficient $\tilde{h} = 300$ W m^{-2}°C^{-1}. The left end of the

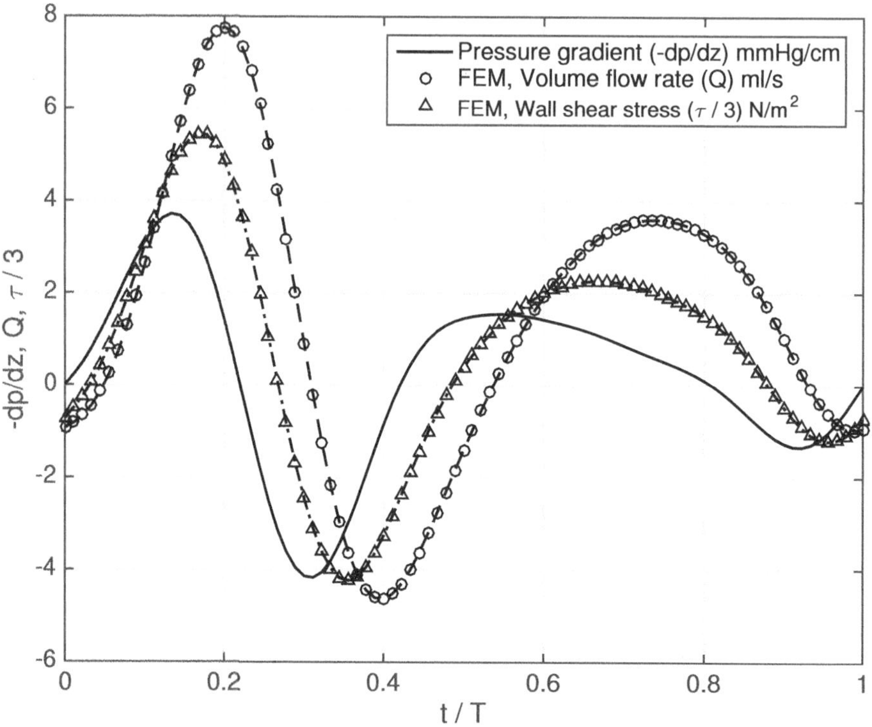

Figure 2.29 Time response of volume flow rate and wall shear in a pulsatile blood flow in an artery. Comparison between finite element results (symbols) and exact analytic solution (lines). The pressure gradient is given by Equation (2.392). A mesh of 20 quadratic elements is used.

fin is maintained at temperature $T_o = 275\,^\circ\mathrm{C}$. The governing equation of the problem is

$$-\frac{d^2\theta}{dx^2} + q\theta = 0 \qquad \text{on} \qquad 0 < x < L$$

where $\theta = T - T_\infty$ is the temperature, and q is given by

$$q = \frac{4\tilde{h}}{kd}.$$

The boundary conditions are

$$\theta(0) = T(0) - T_\infty = 250\,^\circ\mathrm{C}, \qquad \left(\frac{d\theta}{dx}\right)_{x=L} + \frac{\tilde{h}}{k}\theta(L) = 0.$$

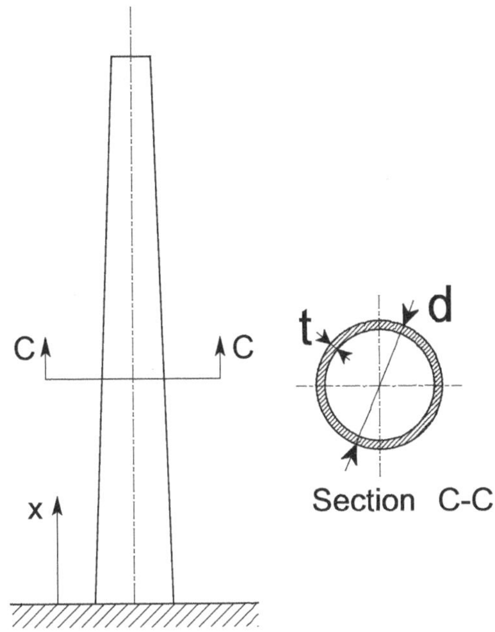

Figure 2.30 Tubular column.

(a) Derive the exact analytic solution for the temperature distribution in the fin.
Answer:

$$\theta(x) = \theta(0)\frac{[cosh\, b(L-x) + (\tilde{h}/bk) sinh\, b(L-x)]}{cosh\, bL + (\tilde{h}/bk) sinh\, bL}$$

where $b = \sqrt{q}$.

(b) Plot the temperature (primary variable) θ against x.

(c) Plot the rate of heat flow (the negative of secondary variable) $\tilde{Q}_k = -kAdT/dx$ against x.

(2) A slender rod (bar) of length L, cross-sectional area A, and Young's modulus E, is fixed at $x = 0$ and restrained by a linear spring at $x = L$; the spring constant is assumed to be $k = EA/2L$. The bar is subjected to axial force per unit length $f(x) = f_o sin(\pi x/L)$. Determine the displacement (primary variable), $u(x)$, and axial force (secondary variable), $R(x)$, at any section x. Plot the nondimensional displacement field $EAu(x)/f_oL^2$, and axial force $R(x)/f_oL$, for $0 \leq x/L \leq 1$. Determine the location of maximum displacement.

(3) For the purpose of assessing the error of the finite element method, we wish to determine the analytic solution for the axial displacement field due to gravity in a vertical tubular aluminum column, see Figure 2.30. The external diameter

$d(x)$ varies linearly with x

$$d(x) = d_b + (d_t - d_b)x/L$$

where the base diameter is $d_b = 400$ mm and the tip diameter is $d_t = 200$ mm, and $L = 8$ m is the column height. The tube thickness t is uniform, $t = 15$mm. Solve the differential equation of equilibrium and determine the exact solution in the form

$$u(x) = (\rho g L^2/2E)\left[\frac{1}{2}(x/L)^2 + \lambda x/L - (1+\lambda)^2 ln(1 + x/\lambda L)\right]$$

where $\lambda = (d_b - t)/(d_t - d_b)$. For aluminum, $E = 70$ GPa, $\rho = 2700$ kg m^{-3}, and $g = 9.81$ m s^{-2}. Plot the displacement field $u(x)$ for $0 \leq x \leq L$.

(4) An equation that governs oxygen concentration in a flat tissue (a slab) is

$$-\frac{d}{dx}\left(\frac{du}{dx}\right) + B(u) = 0 \tag{2.394}$$

where

$$B(u) = \frac{\alpha u}{\beta + u} \tag{2.395}$$

is the consumption rate assumed to be the Michaelis–Menten reaction. The boundary conditions are

$$\begin{aligned} x &= 0: \quad & -\frac{du}{dx} + \gamma u = \gamma \\ x &= 1: \quad & \frac{du}{dx} = 0. \end{aligned} \tag{2.396}$$

The concentration distribution within the tissue depends only on three nondimensional parameters, namely α, β, and γ.

(a) Determine an analytic solution in the limit of high concentration $u >> \beta$, hence $B = \alpha$ is a constant. The reaction in this case is called a zero-order reaction.

(b) Determine an analytic solution in the limit of low concentration $u << \beta$, hence $B = \alpha u/\beta$. The reaction in this case is called a first-order reaction.

(c) For $\alpha = 2$, $\beta = 0.1$, and $\gamma = 5$: Plot the oxygen concentration as a function of x for the two cases of zero- and first-order reactions, and calculate the rate of oxygen transport to tissue per unit area for each case.

(5) Uniform bar AB (length L, Young's modulus E, and cross-sectional area A) is fixed at A and bounded by a rigid wall near B as shown in Figure 2.31. Before application of uniform axial load $f(x) = f_o$, a gap δ existed between end B and the wall. Determine the reaction at A and the gap at B after the application of the load for three different values of f_o:

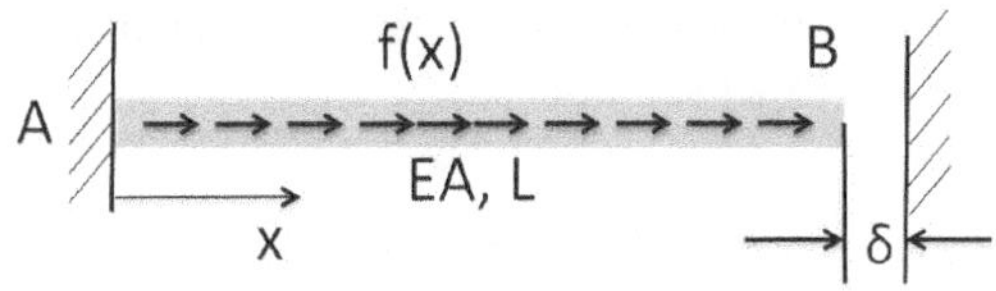

Figure 2.31 Axial deformation of a bar bounded by a nearby wall.

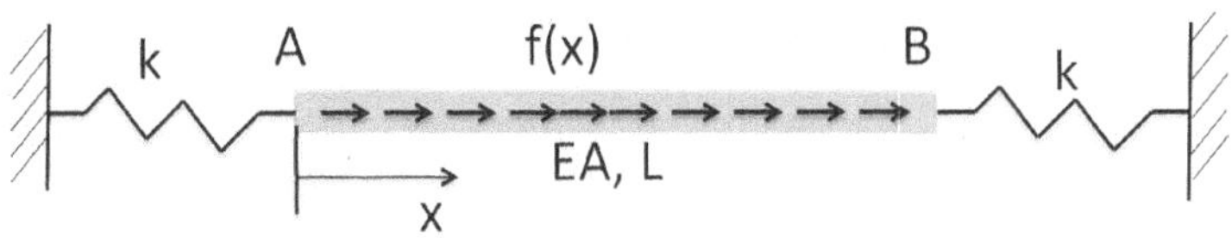

Figure 2.32 Axial deformation of a bar attached to two springs.

(a) $f_o = EA\delta/L^2$.

(b) $f_o = 2EA\delta/L^2$.

(c) $f_o = 3EA\delta/L^2$.

(6) Uniform bar AB (length L, Young's modulus E, and cross-sectional area A) is restrained by two linear springs of stiffness k at A and B as shown in Figure 2.32. Before the application of uniform axial load $f(x) = f_o$, the springs are undeformed. Determine the displacement $u(x)$ after application of the load f_o.

Weighted-Residual Method: Strong Form

(7) Given $p(x) = 1+x$, $q(x) = 1$ and $f(x) = -xcosh(x) - sinh(x)$, consider the boundary value problem

$$-\frac{d}{dx}(p\frac{du}{dx}) + qu = f \quad \text{on} \quad 0 < x < 1$$

with boundary conditions

$$\begin{aligned} x &= 0: \ u = 1 \\ x &= 1: \ p\frac{du}{dx} = 2sinh(1). \end{aligned}$$

We wish to use a weighted-residuals method to determine a three-term ($N = 3$) approximate solution

$$U_3 = g(x) + C_1\phi_1(x) + C_2\phi_2(x) + C_3\phi_3(x).$$

(a) Select admissible polynomials for g and ϕ_j, and determine the coefficients C_j by the Galerkin method.

(b) Plot the pointwise error defined by $\mid u - U_3 \mid$ against $0 \leq x \leq 1$, where $u = cosh(x)$ is the exact solution.

(c) Determine the L_2-norm of the error, which is defined by

$$\|u - U_3\|_o = \left[\int_0^1 \mid u - U_3 \mid^2 dx\right]^{1/2}.$$

(8) Change boundary conditions in Problem (7) to

$$\begin{aligned} x &= 0: \; p\frac{du}{dx} = 0 \\ x &= 1: \; u = cosh(1) \end{aligned}$$

and resolve the problem.

(9) Given $p(x) = 1$, $q(x) = \pi^2$ and $f(x) = \pi^2$, consider the boundary value problem

$$-\frac{d}{dx}(p\frac{du}{dx}) + qu = f \quad \text{on} \quad 0 < x < 1$$

with boundary conditions

$$\begin{aligned} x &= 0: \; -p\frac{du}{dx} = 1 \\ x &= 1: \; p\frac{du}{dx} = 3. \end{aligned}$$

We wish to use a weighted-residuals method to determine a three-term ($N = 3$) approximate solution

$$U_3 = g(x) + C_1\phi_1(x) + C_2\phi_2(x) + C_3\phi_3(x).$$

(a) Select admissible polynomials for g and ϕ_j, and determine the coefficients C_j by the Galerkin method.

(b) Plot the pointwise error defined by $\mid u - U_3 \mid$ against $0 \leq x \leq 1$, where u is the exact solution.

(c) Determine the L_2-norm of the error, which is defined by

$$\|u - U_3\|_o = \left[\int_0^1 \mid u - U_3 \mid^2 dx\right]^{1/2}$$

Partial answers:
$g = -x + 2x^2$, $\phi_1 = 1$, $\phi_i = -(i+1)x^i + ix^{i+1}$ for $i = 2, 3, \ldots$.

(10) A one-term solution to Problem (6) is proposed

$$u \approx U_1 = g(x) + C_1\phi_1.$$

Use lowest-order polynomials to construct the functions g and ϕ_1. Determine the constant C_1 by the Galerkin weighted-residuals method (strong form). Show that the solution is identical to the exact solution found in Problem (6).

Weak (or Variational) Form

(11) Using Equation (2.102), show that

$$\begin{aligned} R_a + R_b &= \int_a^b (qu - f)dx \\ (b-a)R_b &= \int_a^b p\frac{du}{dx}dx + \int_a^b (x-a)(qu-f)dx. \end{aligned} \tag{2.397}$$

Hint: use $w = 1$ and $w = x - a$.

(12) Given $p(x)$, $q(x)$, and $f(x)$, consider the boundary-value problem

$$-\frac{d}{dx}(p\frac{du}{dx}) + qu = f \quad \text{on} \quad 0 < x < 1$$

with boundary conditions

$$\begin{aligned} x &= 0: \; -p\frac{du}{dx} = -1 \\ x &= 1: \; u = 1. \end{aligned}$$

(a) Derive the weak form, and classify the boundary conditions as essential, natural, or mixed.

(b) Given $p(x) = 1$, $q(x) = 0$, and $f(x) = x$, we wish to determine a one-term Galerkin approximation

$$U_1 = g + C_1\phi_1$$

to the solution of the weak form. Using the lowest-order admissible polynomials, construct $g(x)$ and $\phi_1(x)$.

(c) Determine the coefficient C_1.

(13) Construct the weak (also called the variational) form for the boundary-value problem given in Problem (7). Classify the boundary conditions as essential, natural, or mixed.

(14) Determine a three-term Galerkin approximation

$$U_3 = g(x) + C_1\phi_1(x) + C_2\phi_2(x) + C_3\phi_3(x)$$

to the solution of the weak form constructed in Problem (13). Construct $g(x)$ so that it satisfies only the non-homogeneous essential boundary condition(s), and

ϕ_j to be the lowest order polynomials admissible by the weak form. Determine the L_2-norm of the error and compare it with that of the weighted residuals Galerkin method found in part (c) of Problem (7).

(15) An equation that governs oxygen concentration in a flat tissue (a slab) and boundary conditions are given in Problem (4).

(a) Consider the case of low concentration where the reaction is first order, $B = \alpha u/\beta$, and derive the weak form. Classify the boundary conditions. For $\alpha = 2$, $\beta = 0.1$, and $\gamma = 5$:

(b) Determine a two-term Galerkin approximation

$$U_2 = g(x) + C_1\phi_1(x) + C_2\phi_2(x)$$

to the solution of the weak form.

(c) Plot the oxygen concentration as a function of x for the exact solution and the two-term Galerkin approximation.

(16) Justify that the functions $g(x)$ and $\phi_j(x)$, $(j = 1, ..., 3)$ of the weighted-residuals Galerkin's method constructed in part (a) of Problem (7) are admissible by the weak form. Using these functions, determine a three-term Galerkin approximation

$$U_3 = g(x) + C_1\phi_1(x) + C_2\phi_2(x) + C_3\phi_3(x)$$

to the solution of the weak form constructed in Problem (13). Compare the coefficients C_j with those of the weighted residual Galerkin method found in part (a) of Problem (7).

(17) A viscous incompressible fluid between two parallel plates is driven by a streamwise pressure gradient, see Figure 2.33. The lower plate is stationary and the upper plate is sliding. The velocity $u(y)$ depends on y only and is governed by the $x-$component of the momentum equation written here in nondimensional form

$$-\frac{d^2u}{dy^2} - f_o = 0 \tag{2.398}$$

where f_o is a constant. The boundary conditions are the no-slip conditions at the walls

$$\begin{aligned} y = 0: &\qquad u = 0 \\ y = 1: &\qquad u = 1 \end{aligned} \tag{2.399}$$

(a) Derive the weak form of the problem, and classify the boundary conditions.

(b) A one-term ($N = 1$) trial solution is assumed for the weak form

$$U_1(y) = g(y) + C_1\phi_1(y)$$

Construct lowest-order admissible polynomials for g and ϕ_1.

(c) Use the Galerkin method to determine the coefficients C_1.

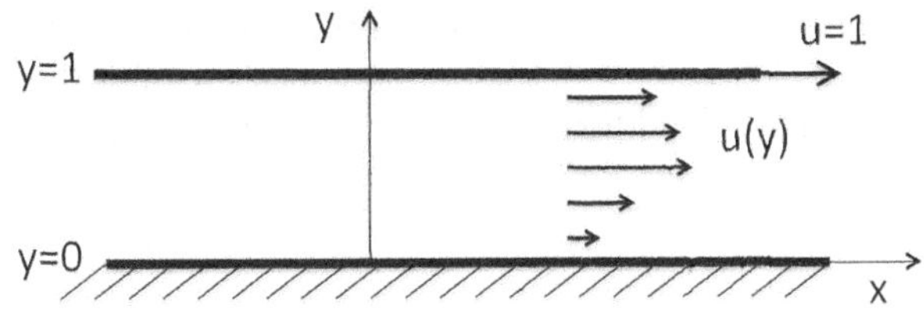

Figure 2.33 Viscous flow between two parallel plates.

(18) Derive the weak form for Problem (2), and determine a two-term Galerkin approximation

$$U_2 = g(x) + C_1\phi_1(x) + C_2\phi_2(x)$$

to its solution. Justify that $g(x) = 0$. Compare the nondimensional displacement (exact solution) found in Problem (2) to the two-term Galerkin approximate solution. Plot the pointwise error in the range $0 \leq x/L \leq 1$.
To simplify the analysis, introduce nondimensional variables, $v = EAu(x)/f_oL^2$ and $y = x/L$, and show that the equation of equilibrium becomes

$$-\frac{d^2v}{dy^2} - sin(\pi y) = 0 \quad \text{on} \quad 0 < y < 1$$

with boundary conditions

$$\begin{aligned} y &= 0: \; v = 0 \\ y &= 1: \; \frac{dv}{dy} + k^* v = 0 \end{aligned}$$

where $k^* = kL/EA = 1/2$.

(19) Derive the weak form for Problem (3) . The boundary conditions are

$$\begin{aligned} x &= 0: \; u = 0 \\ x &= L: \; EA\frac{du}{dx} = 0 \end{aligned}$$

Determine a two-term Galerkin approximation to the solution of the weak form.

$$U_2 = g(x) + C_1\phi_1(x) + C_2\phi_2(x).$$

Justify that $g(x) = 0$, and use lowest order polynomials for the basis functions ϕ_1 and ϕ_2. Compare the two-term weak solution to the exact solution found in Problem (3). Plot the pointwise error in displacement in the range $0 \leq x/L \leq 1$. The natural boundary condition at the column top $x = L$ is a result of the normal stress (and hence axial force) being zero at the free end, $\sigma = Edu/dx = 0$. Since the natural boundary condition is not exactly satisfied by the trial solution, it is instructive to investigate the effects of the number of terms N on the satisfaction of that boundary condition. Let us use the ratio of the normal stress at the top (σ_t) to the normal stress at the base (σ_b) as a measure of the error in satisfying the natural boundary condition. Determine that ratio σ_t/σ_b for $N = 2$, 3 and 4. What can you conclude?

(20) Given $p(x) = 1+x$, $q(x) = 1$ and $f(x) = sinx$, consider the boundary-value problem

$$-\frac{d}{dx}(p\frac{du}{dx})+qu=f \quad \text{on} \quad 0<x<1$$

with boundary conditions

$$x=0: \; -p\frac{du}{dx}+3u=1$$

$$x=1: \; u=1.$$

A two-term ($N = 2$) trial solution is assumed for this boundary-value problem:

$$U_2(x) = g(x)+C_1\phi_1(x)+C_2\phi_2(x).$$

Construct lowest-order admissible polynomials for g, ϕ_1, and ϕ_2 in the following two different cases:

(a) If the coefficients C_1 and C_2 are to be determined by using the method of weighted residuals (strong form).

(b) If the coefficients C_1 and C_2 are to be determined by solving the weak form of the problem.

(21) Derive the weak form of Problem (6), and classify the boundary conditions.

(a) A three-term solution to the weak form is proposed

$$u \approx U_3 = g(x)+C_1\phi_1+C_2\phi_2+C_3\phi_3.$$

Use lowest-order polynomials to construct the functions g and the basis functions ϕ_j. Determine the constants C_j by the Galerkin method, and show that the solution is identical to the exact solution found in Problem (6).

Partial answers: $g = 0$, $\phi_1 = 1$, $\phi_2 = x$, $\phi_3 = x^2$.

(b) Solve the weak form of Problem (6) by using the one-term trial solution constructed in Problem (10), and show that the solution is identical to the strong form solution.

Finite Element Method: 2nd Order ODE

(22) Given $p(x) = 1+x$, $q(x) = 1$ and $f(x) = x$, consider the boundary value problem

$$-\frac{d}{dx}(p\frac{du}{dx})+qu=f \quad \text{on} \quad 0<x<1$$

with boundary conditions

$$x = 0: \ u = 1$$
$$x = 1: \ p\frac{du}{dx} = 2.$$

We wish to determine an approximate solution by the Galerkin finite-element method using one quadratic element. The element equations before the application of the boundary conditions are written in the form:

$$\begin{bmatrix} 89/30 & -49/15 & 7/15 \\ -49/15 & a_{22} & -23/5 \\ 7/15 & -23/5 & 43/10 \end{bmatrix} \begin{Bmatrix} U_1 \\ U_2 \\ U_3 \end{Bmatrix} = \begin{Bmatrix} F_1 \\ 1/3 \\ 1/6 \end{Bmatrix} + \begin{Bmatrix} R_1 \\ 0 \\ R_3 \end{Bmatrix} \tag{2.400}$$

(a) Verify the given entries, and calculate a_{22} and F_1.

(b) Apply the boundary conditions and determine the unknown primary and secondary variables.

(23) Given $p(x) = 0.01, q(x) = 1$, and $f(x) = 1$, consider the boundary value problem

$$-\frac{d}{dx}(p\frac{du}{dx}) + qu = f \quad \text{on} \quad 0 < x < 1$$

with boundary conditions

$$x = 0: \ u = 0$$
$$x = 1: \ u = 1.$$

We want to solve this problem by the FEM and assess the accuracy of the solution as we refine the mesh (h-refinement). Divide the problem domain into *ne* linear elements. For a given mesh (*ne*), we define the error at global node j by $\varepsilon_j = |u(x_j) - U_j|$, where $u(x_j)$ is the exact solution evaluated at global node j. And we also define the maximum error by $\varepsilon_{max} = max(\varepsilon_j)$. Obtain solutions on different meshes by using $ne = 2, 4, 8, 16, 32, 64$, and 128. Plot the maximum error ε_{max} against the number of elements *ne* (a log-log scale is recommended). Comment on the accuracy of this method.
For very small p, a boundary layer develops at $x = 0$, and hence a finer mesh is needed there. A uniform mesh is inefficient in resolving the boundary layer [See Problem (26)].

(24) Given $p(x) = 1 + x$, $q(x) = 1$ and $f(x) = -x\cosh(x) - \sinh(x)$, consider the boundary value problem

$$-\frac{d}{dx}(p\frac{du}{dx}) + qu = f \quad \text{on} \quad 0 < x < 1$$

with boundary conditions

$$x = 0: \ u = 1$$
$$x = 1: \ p\frac{du}{dx} = 2\sinh(1).$$

Determine an approximate solution by the Galerkin finite-element method. Use a uniform mesh and quadratic elements. Note that element matrices based on element-wise constant p, q, and f are not appropriate for the present problem. We must account for the variations of p and f with x when computing the element matrices. For a mesh of three quadratic elements, provide:

(a) The element matrices (A^e and F^e) for the third element ($2/3 \leq x \leq 1$)

(b) The error at all nodes; recall that the exact solution is $u = cosh(x)$.

(c) A plot of the pointwise error over the domain of the third element only.

(25) Given $p(x) = 1$, $q(x) = 9$ and $f(x) = 3$, consider the boundary value problem

$$-\frac{d}{dx}(p\frac{du}{dx}) + qu = f \quad \text{on} \quad 0 < x < 1$$

with boundary conditions,

$$x = 0: \ \frac{du}{dx} = 2$$
$$x = 1: \ u = 0.$$

Divide the problem domain into two non-overlapping equal elements ($h = 1/2$), and approximate the solution on each element by a cubic polynomial. Note that each element will have equally spaced four nodes (this is called a cubic element). (a) Plot the shape functions for a cubic element. (b) Assume element-wise uniform p, q, and f, and derive the element matrices $[A^e]$ and $\{F^e\}$ for a cubic element. (c) Assemble the two elements and provide the global coefficient matrix $[A]$ and forcing vector $\{F\}$ before application of the boundary conditions. (d) Apply the boundary conditions and solve for the unknown nodal values of the primary variable. (e) Plot the finite-element solution $U(x)$ on the domain $x = [0, 1]$, and compare it with the exact solution. (f) Also determine the error at: (i) the global nodes, and (ii) the mid-point of each element. (g) Determine the error in the secondary variable at $x = 1$.

(26) Given $p(x) = 10^{-5}$, $q(x) = 1$ and $f(x) = 1$, consider the boundary value problem

$$-\frac{d}{dx}(p\frac{du}{dx}) + qu = f \quad \text{on} \quad 0 < x < 1$$

with boundary conditions

$$x = 0: \ u = 0$$
$$x = 1: \ u = 1$$

Because a small parameter p is multiplying the highest derivative in the differential equation, the solution exhibits a boundary-layer behavior. A boundary layer is a thin region near the boundary where the solution changes very rapidly from the prescribed value at the boundary. In the present problem, the boundary layer is at the end $x = 0$. To resolve the rapid change of $u(x)$ in the boundary layer, a fine mesh is needed near $x = 0$, where the element size should be of the order of 0.003 ($\sim \sqrt{p}$). Therefore, a uniform mesh over the entire domain is not adequate because the number of elements would be too big to handle numerically.

Determine an approximate solution by the Galerkin finite-element method. Use linear elements on a nonuniform mesh where the mesh should be fine near $x = 0$ and coarser near $x = 1$. For a mesh of 20 (or less) elements:

(a) Plot the FEM solution and the exact solution in the boundary layer, $0 \leq x \leq 0.05$.

(b) Print the error at the nodes.

Finite Element Method: Solid Mechanics, Heat Conduction, and Other Engineering Applications.

(27) Determine an approximate solution to the boundary value problem given in Problem (1) by the finite-element method. Use a uniform mesh of three linear elements and provide:

(a) The element matrices ($[A^e]$ and $\{F^e\}$) for each of the elements.

(b) The global (assembled) matrices before applying the boundary conditions ($[A]$ and $\{F\}$).

(c) The error in temperature (primary variable) at all nodes; recall that the exact solution is given in Problem (1).

(d) The error in the heat flow (secondary variable) at $x = 0$.

(28) Determine an approximate solution to the boundary value problem given in Problem (1) by the finite-element method. Use a uniform mesh of two quadratic elements and provide:

(a) The element matrices ($[A^e]$ and $\{F^e\}$) for each of the elements.

(b) The global (assembled) matrices before applying the boundary conditions ($[A]$ and $\{F\}$).

(c) The error in temperature (primary variable) at all nodes; recall that the exact solution is given in Problem (1).

(d) The error in the heat flow (secondary variable) at all nodes.

(29) Consider steady heat conduction in two tightly attached concrete slabs of thickness $h_1 = 1$ m and $h_2 = 0.6$ m as shown in Figure 2.34. In the curing process, internal energy is generated within the first slab at the rate of 72 W/m^3, but no energy is generated in the second slab. Thermal conductivity of concrete is 1.2 W m^{-1}°C^{-1}. The outer surface of the first slab is maintained at 10°C while the outer surface of the second is at 20°C. Temperature is assumed to be a function of x only.

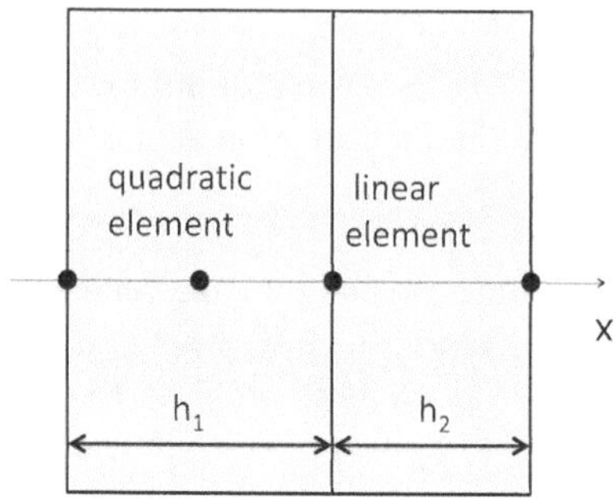

Figure 2.34 Composite concrete wall.

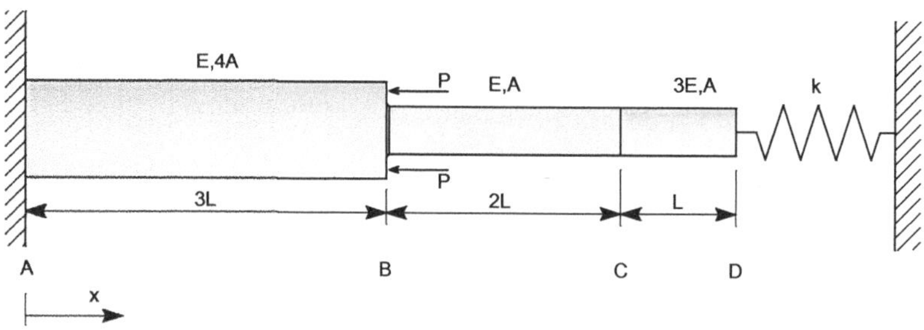

Figure 2.35 Stepped bar.

The composite slab is modeled by two elements, a quadratic element for the first slab and a linear element for the second. Determine:

(a) The global assembled system before applying the boundary conditions.

(b) The temperature at the interior nodes.

(c) Heat flux on the outer surfaces of the slabs.

(30) Use three linear elements to analyze the static axial deformation of a stepped bar with its right end supported by a linear spring as shown in Figure 2.35. The spring is undeformed before the application of the axial forces. The boundary condition at end D is

$$EA\frac{du}{dx}+ku=0 \qquad \text{at} \quad x=6L.$$

Take $k=3EA/L$. Find u_B, u_C, and u_D, and the reaction at A. Comment on the accuracy of the finite element solution.

(31) The finite element method is used to analyze the axial displacement field due to gravity in the vertical tubular aluminum column shown in Figure 2.30. For the

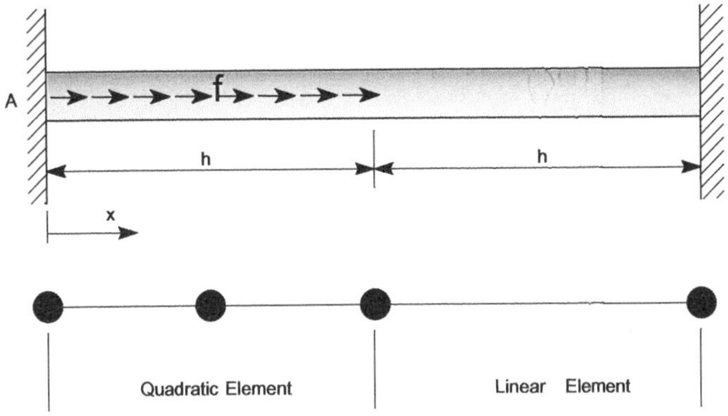

Figure 2.36 Axial deformation of a bar using elements of different order.

data of Problem (3), divide the column into ne linear elements, and plot the displacements U at the global nodes against x. Compare the finite element solution with the exact analytic solution [given in Problem (3)] for two cases $ne = 5$ and $ne = 10$. What is the percentage error in displacement at the tip?

Note that in this problem, $p = EA$ and $f = -\rho g A$ are not uniform over the element. New element matrices $[A^e]$ and $\{F^e\}$ need to be developed to account for the variation of EA and the axial load (weight per unit length) with x.

(32) Consider the axial deformation of a slender rod AB (length $= 2h$, cross-sectional area $= A$, and Young's modulus $= E$). The bar is fixed at both ends. Uniform axial distributed load f is externally applied to one-half of the bar as shown in Figure 2.36. The loaded half is modeled by a quadratic element, whereas the other half is modeled by a linear element. Provide:

(a) The global (assembled) system of equations before applying the boundary conditions.

(b) Given EA, h, and f, find the displacement at a distance $0.75h$ from end A.

(33) The equation of equilibrium of a small element of a circular shaft subjected to external torque is given by

$$-\frac{d}{dx}(GJ\frac{d\theta}{dx}) - f(x) = 0$$

where θ is the twist angle at any section x, $GJ(x)$ is the torsional rigidity of the cross-section, and $f(x)$ is externally applied distributed torque per unit axial length of the shaft. Consider a shaft of length $3h$ and uniform torsional rigidity GJ, which is fixed at $x = 0$ as shown in Figure 2.37. It is loaded by uniformly distributed torque per unit length f_o over the domain $0 \leq x \leq 2h$ and a concentrated torque $T_o = f_o h$ applied in the opposite direction at the end $x = 3h$.

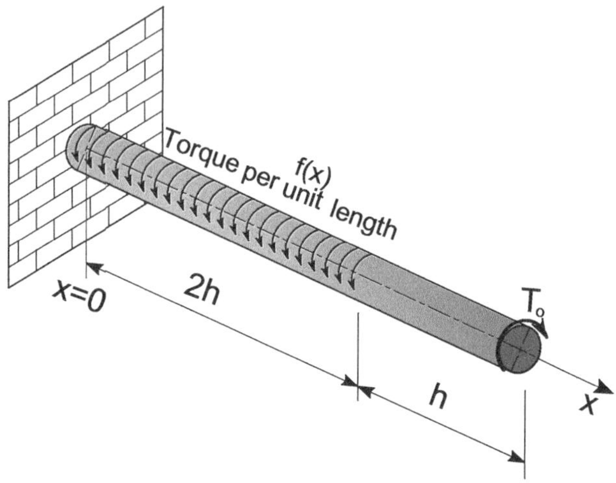

Figure 2.37 Torsion of a circular shaft

We want to determine the twist angle $\theta(x)$ by the finite element method. The shaft is modeled by two elements. A quadratic element is used for the domain $0 \leq x \leq 2h$ and a linear element for the domain $2h \leq x \leq 3h$. Determine the twist angle at $x = 0.75h$. Note that the primary variable is θ, and the secondary variable $GJd\theta/dx$ is the torque.

(34) Uniform bar (cross-sectional area A, Young's modulus E, thermal conductivity k, coefficient of thermal expansion α, and length $L = 2h$) is constrained by two rigid walls at $x = 0$ and $x = 2h$ as shown in Figure 2.38. There is no stress in the bar when its temperature is uniform and equals 0°C.

Thermal Analysis
Internal energy per unit volume is generated within the bar at a uniform rate of f_o while the end $x = 0$ is at 0°C and the end $x = L$ is at a higher temperature T_o. The outside boundary of the bar is insulated. At a steady state, the temperature $T(x)$ is governed by the energy equation

$$-\frac{d}{dx}(kA\frac{dT}{dx}) - f_oA = 0$$

(a) Use two linear elements each of length $h = L/2$, and determine the temperature at $x = h$. Draw a neat sketch of the temperature distribution in the bar for $0 \leq x \leq L$.

Stress Analysis
Because of the non-uniform temperature distribution, the equation of equilibrium gives

$$-\frac{d}{dx}(\sigma A) = 0$$

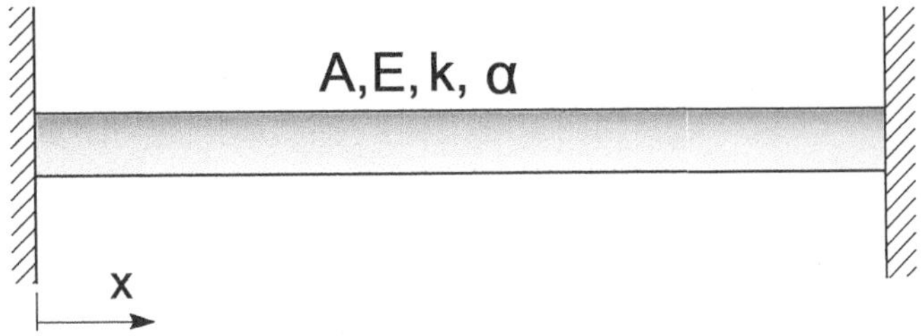

Figure 2.38 Axial deformation of a bar subjected to a change in temperature.

where

$$\sigma = E\frac{du}{dx} - E\alpha\Delta T$$

Note that in this problem $\Delta T = T(x)$.
(b) Use two linear elements and determine the displacement at $x = h$. Also, determine the normal stress in the bar.
In solving this problem, assume A, E, α, k, L, f_o, and T_o to be known.

(35) A rod of diameter $d = 25$ mm, length $L = 1/3$ m, and thermal conductivity $k = 400$ W m^{-1}°C^{-1} is exposed to ambient air at $T_\infty = 0$°C with a convection heat transfer coefficient $\tilde{h} = 225$ W m^{-2}°C^{-1}. One end of the rod ($x = 0$) is maintained at a temperature $T = 180$°C while the other end ($x = L$) is at a temperature $T = 0$°C. The governing equation for the temperature distribution in the rod is

$$-\frac{d^2T}{dx^2} + qT = 0 \quad \text{on} \quad 0 < x < L$$

where $q = 4\tilde{h}/kd = 90$ m^{-2}. Using one quadratic element, determine the temperature at $x = L/4$.

(36) Uniform bar is subjected to axially distributed load that varies according to $f(x) = f_o(x/L)^2$, where f_o is a constant with the dimensions of (force/length), and L is the bar length. The bar is fixed at one end and supported by a linear spring at the other end as shown in Figure 2.39. The spring is undeformed before the application of the axial load. We denote the cross-sectional area by A and Young's modulus by E, and take the spring constant as $k = 5EA/3L$. We wish to analyze the axial deformation in the bar by using one quadratic element.

(a) Replace the given axial load by equivalent concentrated forces at the element nodes, $\{F^e\}$.

(b) Find the force in the linear spring.

(37) The stepped bar shown in Figure 2.40 is composed of three elements with cross-sectional areas (3A) for element one, (2A) for element two, and (A) for element three. All elements have equal length (L) and Young's modulus (E). Using three linear elements, determine the displacement at the nodes of the middle element.

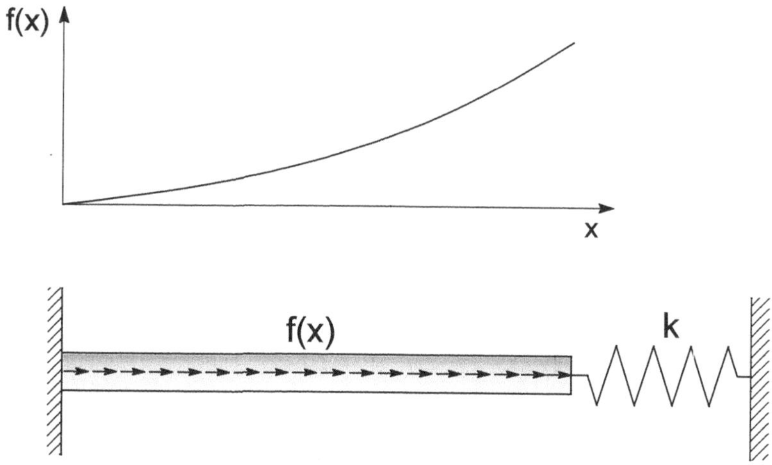

Figure 2.39 Axially loaded bar.

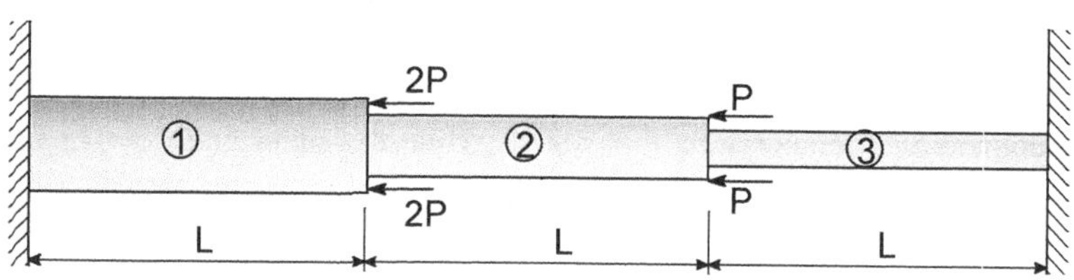

Figure 2.40 Stepped bar fixed at both ends.

(38) A rod of diameter $d = 50$ mm, length $L = 1/3$ m, and thermal conductivity $k = 200$ W m^{-1}°C^{-1} is exposed to ambient air at $T_\infty = 0$°C with a convection heat transfer coefficient $\tilde{h} = 225$ W m^{-2}°C^{-1}. One end of the rod ($x = 0$) is maintained at a temperature $T = 160$°C while the other end ($x = L$) is insulated ($dT/dx = 0$). The governing equation for the temperature distribution in the rod is

$$-\frac{d^2T}{dx^2} + qT = 0 \quad \text{on} \quad 0 < x < L$$

where $q = 4\tilde{h}/kd = 90$ m^{-2}. Using one quadratic element, determine the heat flux at $x = 0$.

(39) The stepped bar shown in Figure 2.41 is composed of two elements with cross-sectional areas ($2A$) for element one, and (A) for element two. The two elements have equal length (L) and Young's modulus (E). The right end is restrained by a linear spring of stiffness $k = 3EA/L$, which is undeformed before the application of the axial forces. Determine the displacement at the nodes of the element # 2.

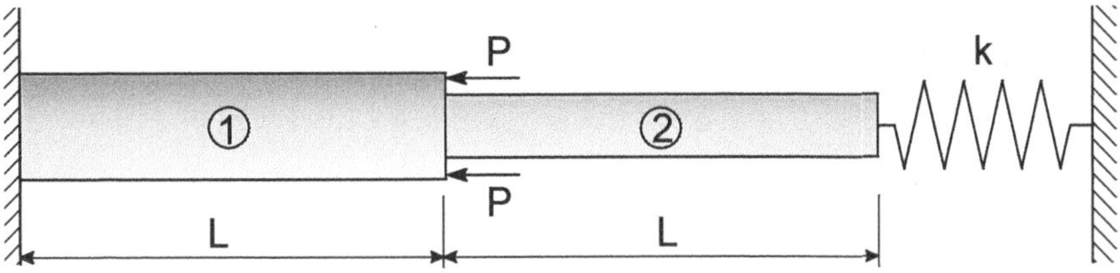

Figure 2.41 Stepped bar restrained by a linear spring.

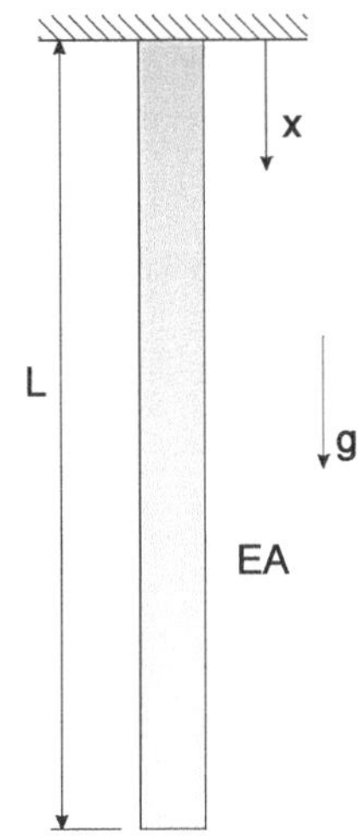

Figure 2.42 Displacement in a slender rod due to gravity.

(40) Uniform elastic bar (cross-sectional area A, length L, mass density ρ, and Young's modulus E) is fixed at one end in a vertical plane as shown in Figure 2.42. The body force due to gravity is represented by uniformly distributed axial force per unit length $f = \rho g A$, where g is the gravitational acceleration. Use the Galerkin finite-element method, and model the bar by one quadratic element. Determine the displacement at any section x. Show that the obtained finite-element solution is identical to the exact solution.

(41) Uniform elastic bar (cross-sectional area A, length h, and Young's modulus E) is fixed at one end and restrained by a linear spring ($k = EA/3h$), which is initially undeformed, at the other end. A linearly varying axial load is applied to the bar as shown in Figure 2.43. Use the Galerkin finite-element method, and model the bar by one quadratic element. Determine:

(a) The reaction at the fixed end.

(b) The displacement at $x = 3h/4$.

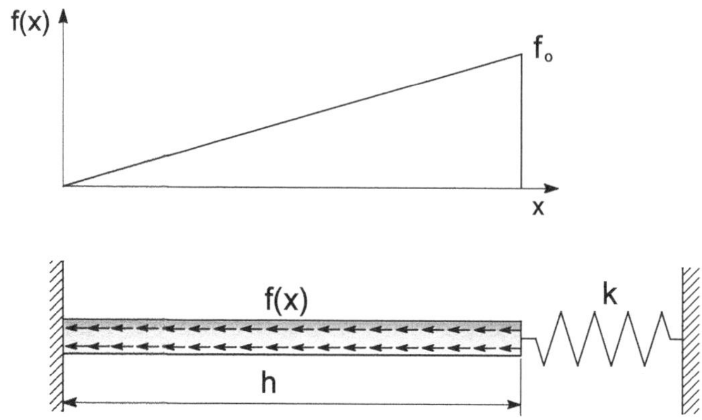

Figure 2.43 A bar subjected to linearly varying axial load and restrained by a linear spring.

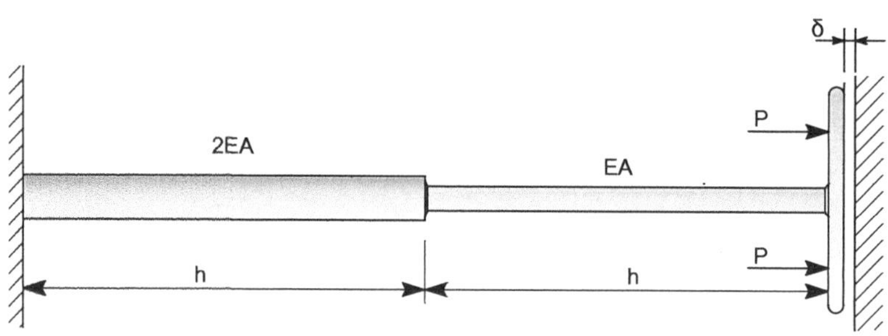

Figure 2.44 Stepped bar with an end plate.

(42) The stepped bar shown in Figure 2.44 is fixed at one end and is rigidly attached to a plate at the other end. Before application of the two axial forces, there is a gap δ between the plate and the wall. Model the stepped bar by two linear elements and neglect the deformation of the end plate. Given EA, h and δ, determine the minimum force P required to close the gap between the end plate and wall.

(43) An elastic bar (length h, and Young's modulus E) is fixed at one end and is subjected to an axial force P at the other end as shown in Figure 2.45. The bar has a variable cross-sectional area that is given by

$$A(x) = A_1(1-\xi) + A_2\xi$$

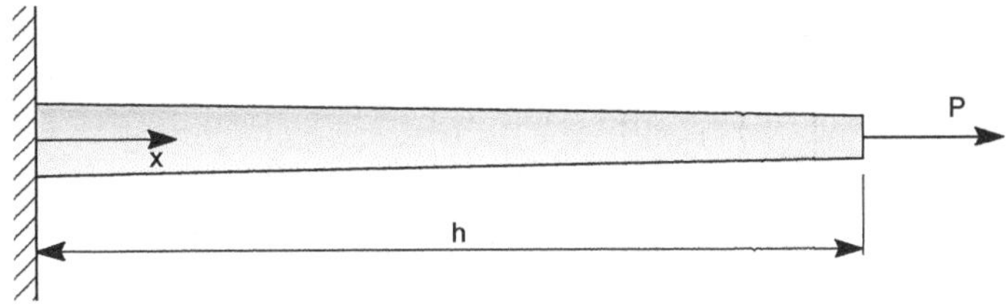

Figure 2.45 A bar with variable cross-sectional area.

where $\xi = x/h$. The bar is modeled by one quadratic element and all the entries of the stiffness matrix (k_{ij}) except k_{22} are given:

$$\begin{aligned}
k_{11} &= (E/6h)(11A_1 + 3A_2) \\
k_{12} &= k_{21} = -(E/6h)(12A_1 + 4A_2) \\
k_{13} &= k_{31} = (E/6h)(A_1 + A_2) \\
k_{23} &= k_{32} = -(E/6h)(4A_1 + 12A_2) \\
k_{33} &= (E/6h)(3A_1 + 11A_2)
\end{aligned}$$

(a) Verify the given entries and determine k_{22}.

(b) Determine the displacement at $x = h$.

(44) A wall is composed of three homogeneous layers, as shown in Figure 2.46. Thermal conductivity for aluminum and copper are $k = 204$ W m^{-1}°C^{-1}, and 400 W m^{-1}°C^{-1}, respectively. The outside temperature of the composite wall is 100°C on the left side and 300°C on the right side. The steady temperature distribution in the wall is governed by

$$-\frac{d}{dx}\left(k\frac{dT}{dx}\right) = 0$$

To analyze steady temperature distribution in the composite wall, three linear elements are used; one for each layer. Provide:

(a) The element matrices ($[K^e]$ and $\{F^e\}$) for each of the elements.

(b) The global (assembled) matrices before applying the boundary conditions ($[K]$ and $\{F\}$).

(c) The temperature at the middle of the copper layer.

(45) Consider the axial deformation of a slender rod (length = $3h$, cross-sectional area = A, and Young's modulus = E). The rod is fixed at $x = 0$ and attached to a linear spring of stiffness k at end $x = 3h$. Uniform axially distributed load f is externally applied in the range $h \le x \le 3h$ as shown in Figure 2.47. The spring is undeformed before the application of the load. The loaded part is modeled

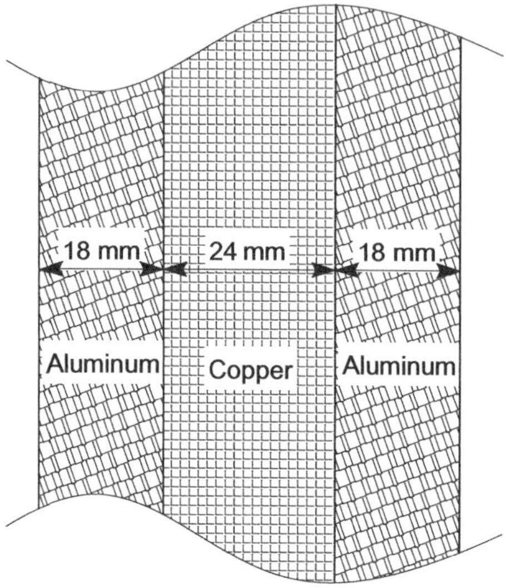

Figure 2.46 Steady heat conduction in a composite wall.

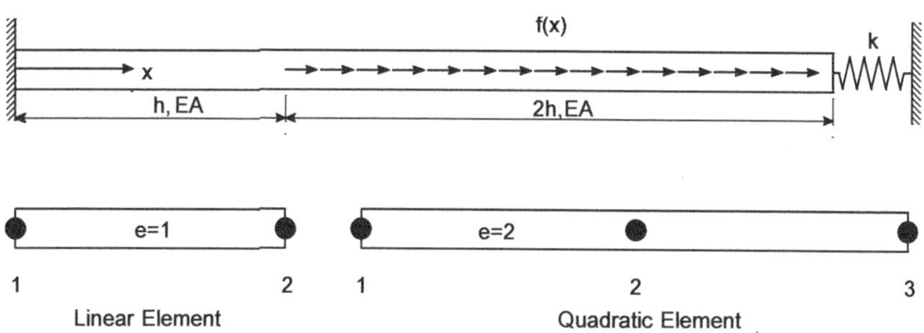

Figure 2.47 Axial deformation of a bar using linear and quadratic elements.

by a quadratic element, whereas the other part is modeled by a linear element. Provide:

(a) The global (assembled) system of equations before applying the boundary conditions.

(b) Given EA, h, f, and $k = EA/(2h)$ find the reaction at $x = 0$ and the displacement at a distance $x = 1.5h$.

(46) A viscous incompressible fluid between two parallel plates is driven by a streamwise pressure gradient, see Figure 2.33. The lower plate is stationary and

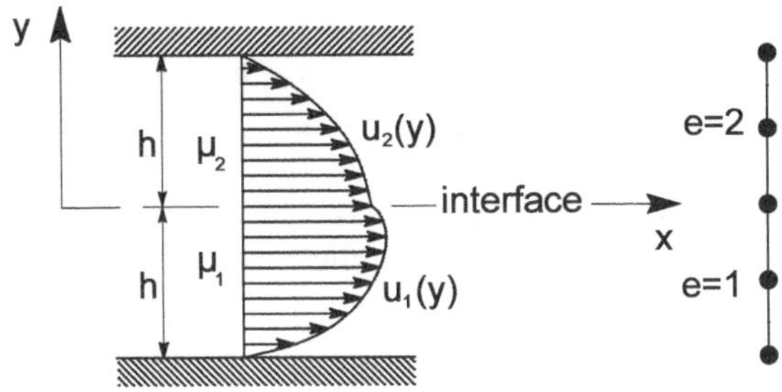

Figure 2.48 Viscous flow of two superimposed layers between parallel plates.

the upper plate is sliding. The velocity $u(y)$ depends on y only and is governed by the x–component of the momentum equation:

$$-\frac{d^2u}{dy^2} - f_o = 0 \tag{2.401}$$

where f_o is a constant. The boundary conditions are the no-slip conditions at the walls

$$\begin{aligned} y = 0: &\quad u = 0 \\ y = 1: &\quad u = 1. \end{aligned} \tag{2.402}$$

Determine an approximation to the velocity profile by the finite-element method. Use a uniform mesh of three linear elements and provide:

(a) The assembled global system before the application of boundary conditions.

(b) The velocity at the interior nodes.

(c) The shear stress on the lower and upper walls.

(47) Two liquid layers of equal thickness ($h = 1$) but different viscosity ($\mu_1 = 1$ and $\mu_2 = 3$) flow between two parallel plates, see Figure 2.48. The velocity $u(y)$ depends on y only and is governed by the x-component of the momentum equation:

$$-\frac{d}{dy}\left(\mu \frac{du}{dy}\right) = 1. \tag{2.403}$$

Use two quadratic elements (one element in each layer) to determine the velocity and shear stress at the interface between the two layers. The boundary conditions are the no-slip conditions at the walls ($u = 0$ *at* $y = \pm 1$). Note that the velocity and shear stress are continuous at the interface ($y = 0$).

(48) (a) Derive the weak form of the following boundary-value problem:

$$-\frac{d}{dx}[D(x)\frac{du}{dx}] + V(x)\frac{du}{dx} + q(x)u - f(x) = 0 \qquad 0 < x < L \tag{2.404}$$

and boundary conditions

$$\begin{aligned} x &= 0: & u &= u_o \\ x &= L: & u &= u_L. \end{aligned} \tag{2.405}$$

(b) Use the weak form to develop a Galerkin finite-element model for Equation (2.404). Use a linear element and write the element equations as

$$[A^e]\{U^e\} = \{F^e\} + \{R^e\} \tag{2.406}$$

and provide expressions for $[A^e]$ and $\{F^e\}$.

(c) Assume constant $D = D_o$, $V = V_o$, and $q = q_o$ evaluate the matrix $[A^e]$.

(d) The Peclet number is defined by $Pe = V_oL/D_o$; it is the ratio of convection to diffusion (we may also call it the Reynolds number). Take $q = 0$, $f = 0$, $L = 1$, and $u_L = 1$, and write the problem as

$$-\frac{d^2u}{dx^2} + Pe\frac{du}{dx} = 0 \qquad 0 < x < 1 \tag{2.407}$$

and boundary conditions

$$\begin{aligned} x &= 0: & u &= 0 \\ x &= 1: & u &= 1. \end{aligned} \tag{2.408}$$

Equation (2.407) is the steady convection-diffusion equation. For convection dominated problems (high Peclet or Reynolds number), the Galerkin finite-element model may develop artificial oscillations ("wiggles"). The purpose of this exercise is to investigate the effects of the cell Peclet number Pe_h $(= V_oh/D_o = Pe(h/L))$ on the quality of the Galerkin finite-element solution, where h is the element length. Use 10 linear elements $(h = 0.1)$, and compare the finite element solution to the exact solution for $Pe = 2, 5, 20$, and 50, which give $Pe_h = 0.2, 0.5, 2$, and 5. What effect does the cell Peclet number have on the artificial oscillations? Brooks and Hughes [18] presented upwind/Petrov-Galerkin finite-element methods that eliminate the oscillations and reduce artificial dissipation for convection-dominated flows.

(49) Derive the exact solution given by Equations (2.334)–(2.336) for oxygen concentration in a flat tissue with Michaelis–Menten kinetics.

3 Fourth-Order Ordinary Differential Equations

This chapter presents a finite-element model for fourth-order ordinary differential equations. In solid mechanics, a typical equation is the Euler–Bernoulli beam equation that governs the deflection of beams. The beam equation is then combined with the bar equation of Chapter 2 to analyze structures such as plane frames and trusses. Principles of virtual displacements and minimum total potential energy are frequently used to derive finite-element models in structural and solid mechanics. Starting with the equations of equilibrium of a continuum and boundary conditions, we derive the principle of virtual displacements, which is mathematically identical to the weak form of the problem. We then present the principle of minimum total potential energy and demonstrate its use for deriving finite-element models for bars and beams. In fluid mechanics, a well-known fourth-order equation is the Orr–Sommerfeld equation, which is encountered in hydrodynamic stability analysis of shear flows. However, it is a differential eigenvalue problem and will be presented in Chapter 7.

3.1 EULER–BERNOULLI BEAM THEORY

Beams are common structural members designed to support bending and shearing due to transverse loading (i.e., forces and moments normal to the member's axis). The free-body diagram of a segment of a straight beam is shown in Figure 3.1, where the x-axis passes through the centroid of the beam cross-section, which is symmetric about the x-y plane, see Figure 3.2. The beam is loaded by transverse distributed load $f(x)$; that is, a force per unit length. The beam can also be loaded by external moments about the z-axis. We are interested in determining the deflection $v(x)$, which is the displacement of the center line in the y-direction, and the internal shear force $Q(x)$ and bending moment $M(x)$.

We obtain a differential equation for $v(x)$ by writing the equations of equilibrium of an infinitesimally small element of the beam as shown in Figure 3.3 and invoking the Euler–Bernoulli beam theory assumptions; sections normal to the beam axis remain plane and normal after bending deformation (e.g., [11]).

The result is

$$\frac{d^2}{dx^2}[D\frac{d^2v}{dx^2}] - f(x) = 0 \qquad 0 < x < L \tag{3.1}$$

DOI: 10.1201/9781003323150-3

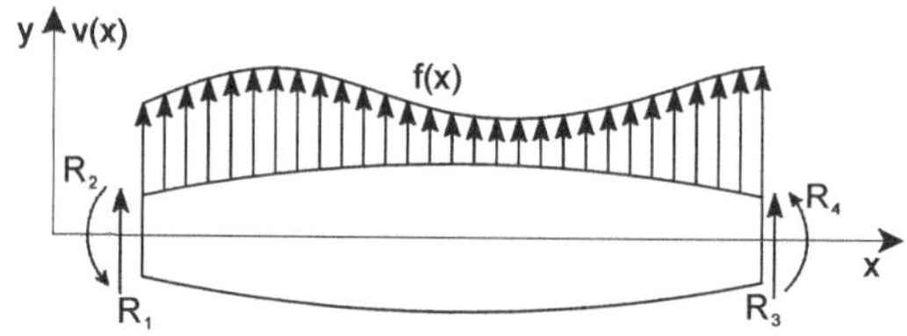

Figure 3.1 A beam subjected to transverse loading, $f(x)$ is force per unit length. The deflection $v(x)$ and $f(x)$ are positive upward. R_1 and R_3 are vertical forces, and R_2 and R_4 are moments about the z-axis; xyz is a right-handed system.

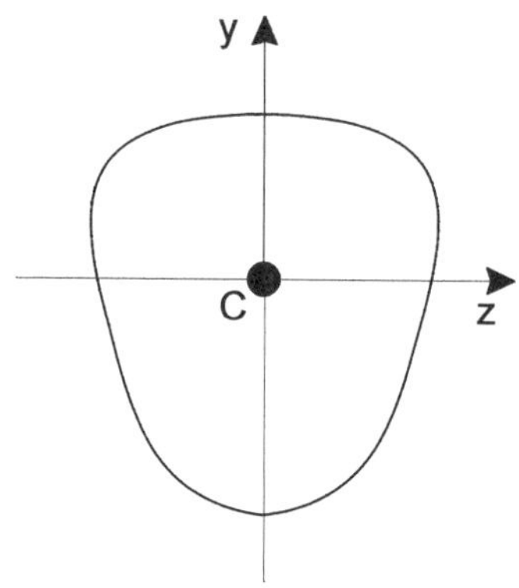

Figure 3.2 Beam cross-section with centroid at C, the x-axis is into the page, and the x-y plane is a plane of symmetry.

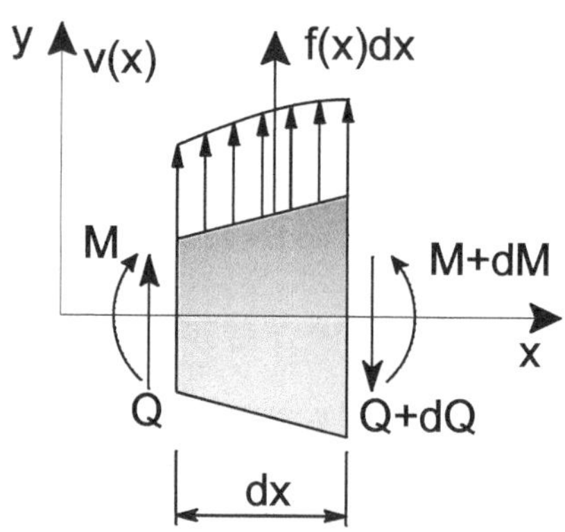

Figure 3.3 Free-body diagram of a small element of a loaded beam; $f(x)dx$ is the resultant of distributed force on the element, Q and M are shear force and bending moment statically equivalent to the shear and normal stresses distributed on the cross-section.

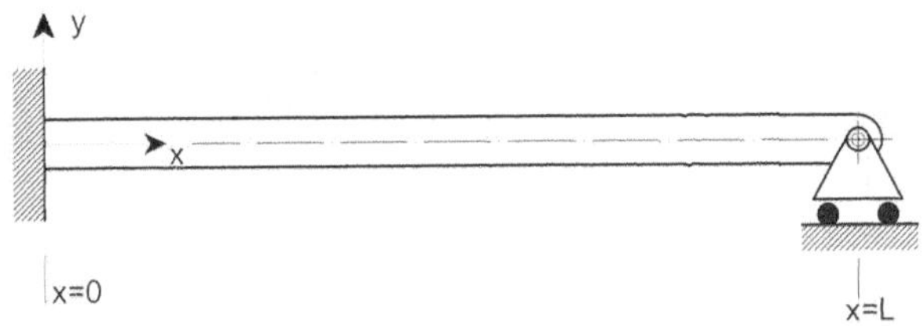

Figure 3.4 Beam fixed at $x = 0$ and simply supported at $x = L$. External loads are not shown.

where $D = EI$ is the flexural rigidity, which may vary with x. The shear force Q and bending moment M are given by

$$\begin{aligned} Q &= \frac{d}{dx}\left(D\frac{d^2v}{dx^2}\right) \\ M &= D\frac{d^2v}{dx^2}. \end{aligned} \tag{3.2}$$

The deflection curve $v(x)$ must satisfy the differential equation and boundary conditions at the supports. For example, Figure 3.4 shows a beam that is fixed at $x = 0$ and simply supported at $x = L$ (external loads are not shown). At the fixed end, the deflection v and slope dv/dx must be zero. At the simply supported end, the deflection is zero but the slope is unknown. If no externally concentrated moment is applied at $x = L$, then the bending moment will be zero there. Therefore, we have four boundary conditions:

$$\begin{aligned} x &= 0: \quad v = 0 \quad \text{and} \quad \frac{dv}{dx} = 0 \\ x &= L: \quad v = 0 \quad \text{and} \quad D\frac{d^2v}{dx^2} = 0. \end{aligned} \tag{3.3}$$

Another example is a cantilever beam attached to a linear spring at the free end as shown in Figure 3.5. To formulate the boundary conditions at the end $x = L$, we give that end a positive vertical displacement $v(L)$ as shown in Figure 3.6; and hence, the linear spring exerts a vertically downward force $F_s = kv(L)$, this is the shear force at that end. There is no externally applied moment at the free end, thus the four boundary conditions are

$$\begin{aligned} x &= 0: \ v = 0 \quad \text{and} \quad \frac{dv}{dx} = 0 \\ x &= L: \ \frac{d}{dx}\left(D\frac{d^2v}{dx^2}\right) = kv \quad \text{and} \quad D\frac{d^2v}{dx^2} = 0. \end{aligned} \tag{3.4}$$

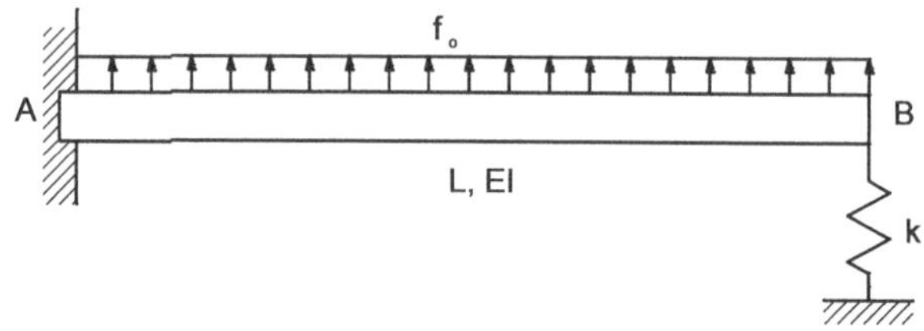

Figure 3.5 Cantilever attached to a linear spring at the tip.

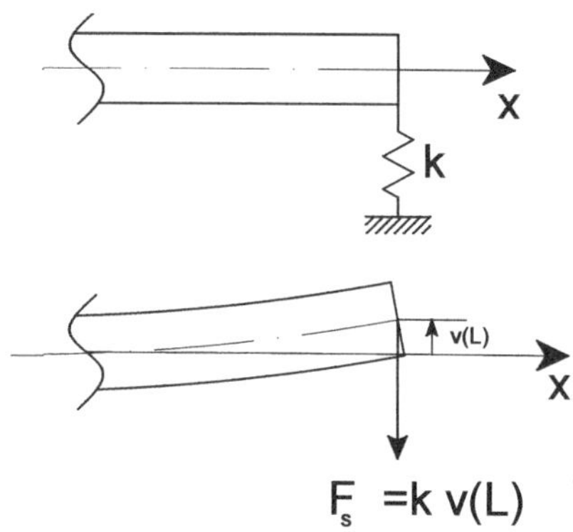

Figure 3.6 Boundary condition at an end attached to a linear spring.

3.2 WEAK FORM OF THE BEAM EQUATION

To derive the weak form of the beam equation, Equation (3.1), we multiply the equation by a test (weight) function $w(x)$ and integrate the result over the problem domain

$$\int_0^L w(x)\frac{d^2}{dx^2}\left[D\frac{d^2v}{dx^2}\right]dx - \int_0^L w(x)f(x)dx = 0. \tag{3.5}$$

Integration by parts of the first term yields

$$\left\{w(x)\frac{d}{dx}\left[D\frac{d^2v}{dx^2}\right]\right\}_{x=0}^{x=L} - \int_0^L \frac{dw(x)}{dx}\frac{d}{dx}\left[D\frac{d^2v}{dx^2}\right]dx - \int_0^L w(x)f(x)dx = 0. \tag{3.6}$$

Once again, we integrate the second term by parts and obtain

$$\left\{w(x)\frac{d}{dx}\left[D\frac{d^2v}{dx^2}\right] - \frac{dw(x)}{dx}D\frac{d^2v}{dx^2}\right\}_{x=0}^{x=L} + \int_0^L \frac{d^2w(x)}{dx^2}D\frac{d^2v}{dx^2}dx$$
$$- \int_0^L w(x)f(x)dx = 0. \tag{3.7}$$

Definition of Secondary and Primary Variables

The integration-by-parts twice produced two boundary terms, each term defines a pair of primary and secondary variables. The coefficients of w and dw/dx in the boundary terms are the secondary variables. We define the secondary variables by

$$\begin{aligned} R_1 &= \left[\frac{d}{dx}(D\frac{d^2v}{dx^2})\right]_{x=0} \\ R_2 &= -\left(D\frac{d^2v}{dx^2}\right)_{x=0} \\ R_3 &= -\left[\frac{d}{dx}(D\frac{d^2v}{dx^2})\right]_{x=L} \\ R_4 &= \left(D\frac{d^2v}{dx^2}\right)_{x=L}. \end{aligned} \tag{3.8}$$

It is clear that R_1 and $-R_2$ are the shear force and bending moment at the left end of the beam ($x = 0$), whereas $-R_3$ and R_4 are the shear force and bending moment at the right end of the beam ($x = L$). The use of the negative signs for R_2 and R_3 is a matter of convention. The primary variables are the deflection v and slope (or rotation about the z-axis) dv/dx. Thus, the deflection and shear force constitute a pair of primary and secondary variables, whereas the slope (or rotation) and bending moment constitute another pair of primary and secondary variables. The primary variables being displacement and slope refer to kinematic (geometric) variables, whereas the secondary variables being force and moment refer to kinetic variables. A vertical force does work through a vertical displacement. Thus, the shear force (vertical force) as a secondary variable is associated with the deflection (vertical displacement) as a primary variable. Similarly, a moment about the z-axis does work through rotation about the same axis. Hence, the bending moment as a secondary variable is associated with the slope (or rotation) as the corresponding primary variable. Using the definitions Equation (3.8), we rewrite Equation (3.7)

$$\begin{aligned} &\int_0^L \frac{d^2w(x)}{dx^2} D\frac{d^2v}{dx^2}dx - \int_0^L w(x)f(x)dx \\ &- \left[w(0)R_1 + \frac{dw(x)}{dx}|_{x=0} R_2 + w(L)R_3 + \frac{dw(x)}{dx}|_{x=L} R_4\right] = 0. \end{aligned} \tag{3.9}$$

Classification of Boundary Conditions

Boundary conditions are classified into three types: (1) essential (EBC), (2) natural (NBC), and (3) mixed (MBC). An EBC means that the value of a primary variable is specified at the boundary point while the corresponding secondary variable is unknown. An NBC means that the value of a secondary variable is given at the boundary while the corresponding primary variable is unknown. An MBC means that a relationship between a primary variable and the corresponding secondary variable is given at the boundary, but the values of both variables are unknown.

To complete the derivation of the weak form, we need to apply the boundary conditions. The weight (test) function $w(x)$ is an arbitrary function but its second derivative must exist as required by Equation (3.9). At an EBC that specifies the deflection v, we select the test function $w(x)$ such that $w = 0$ at the boundary. Similarly, at an EBC that specifies the slope dv/dx, we select $w(x)$ such that $dw/dx = 0$ at the boundary. As a result of this construction of $w(x)$, the corresponding unknown secondary variables do not appear in the weak form. At an NBC, the given value of the secondary variable is substituted into the boundary term in the weak form. Thus, the weak form includes the NBCs. At an MBC, both the primary and secondary variables are unknown. The given relationship between them is used to express the secondary variable in terms of the primary variable, and then the secondary variable is substituted in the boundary term in the weak form. Of course, the value of the primary variable that now appears in the weak form at the MBC is still unknown.

For the boundary conditions Equation (3.3), we have

$$\begin{aligned} x &= 0: w = 0 \quad \text{(EBC), and} \quad \frac{dw}{dx} = 0 \quad \text{(EBC)} \\ x &= L: w = 0 \quad \text{(EBC), and} \quad R_4 = 0 \quad \text{(NBC)} \end{aligned} \tag{3.10}$$

therefore, the weak form becomes

$$\int_0^L \frac{d^2w}{dx^2} D \frac{d^2v}{dx^2} dx = \int_0^L wf dx. \tag{3.11}$$

A solution to the weak form is a function $v(x)$ that satisfies Equation (3.11) for all weight (test) functions, $w(x)$. The function $v(x)$ must be continuous, with continuous first derivative and non-zero second derivative, and satisfy the prescribed EBCs. The function $w(x)$ must also have the same continuity properties as $v(x)$ and satisfy the homogeneous forms of EBC's (at $x = 0$: $w = 0$ and $dw/dx = 0$, and at $x = L$: $w = 0$.)

For the boundary conditions Equation (3.4), we have

$$\begin{aligned} x &= 0: w = 0 \quad \text{(EBC), and} \quad \frac{dw}{dx} = 0 \quad \text{(EBC)} \\ x &= L: -R_3 = kv \quad \text{(MBC), and} \quad R_4 = 0 \quad \text{(NBC)} \end{aligned} \tag{3.12}$$

therefore, the weak form becomes

$$\int_0^L \frac{d^2w}{dx^2} D \frac{d^2v}{dx^2} dx + w(L)kv(L) = \int_0^L wf dx. \tag{3.13}$$

In this case the function $v(x)$ satisfies the prescribed EBCs, and $w(x)$ satisfies the homogeneous forms of EBCs (at $x = 0$: $w = 0$ and $dw/dx = 0$).

Example 3.1.

Using the weak form, determine an approximation of the deflection of the beam shown in Figure 3.7, and determine the reactions. The beam carries a uniformly distributed load f_o.

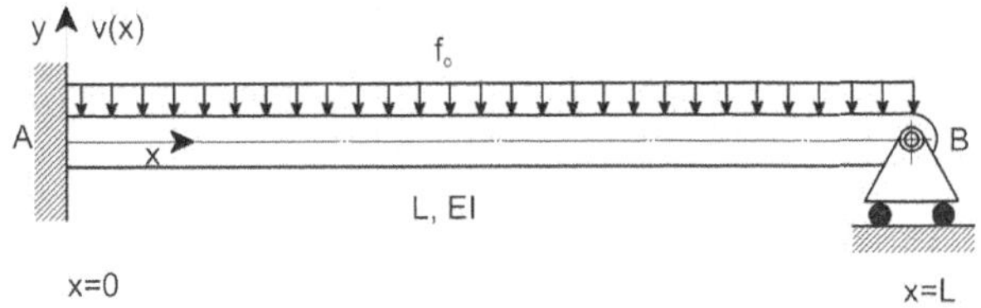

Figure 3.7 Beam fixed at $x = 0$ and simply supported at $x = L$.

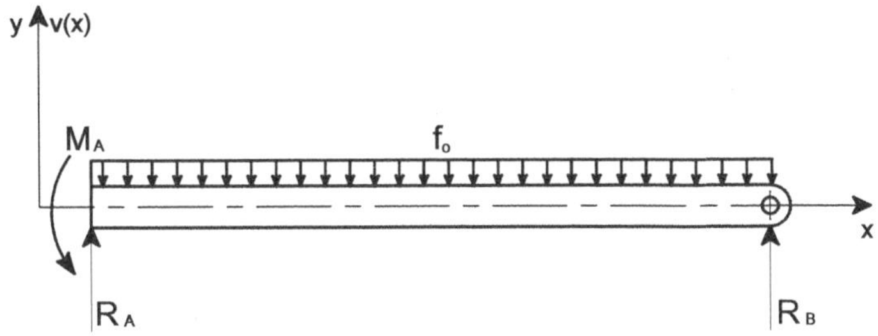

Figure 3.8 Free-body diagram of beam fixed at $x = 0$ and simply supported at $x = L$.

Solution

The free-body diagram of the beam is shown in Figure 3.8. The trial solution must at least satisfy the EBCs.

$$\begin{aligned} x &= 0: v = 0 \qquad \text{and} \qquad \frac{dv}{dx} = 0. \\ x &= L: v = 0. \end{aligned} \tag{3.14}$$

By inspection, the simplest trial solution is

$$v = Cx^2(L - x) \tag{3.15}$$

where C is an unknown constant. We select a weight function $w(x)$ that must be zero at the EBCs. Thus, we take

$$w = x^2(L - x). \tag{3.16}$$

Substituting the trial solution and weight function into the weak form Equation (3.11), we get

$$\begin{aligned} \int_0^L (2L - 6x)EIC(2L - 6x)dx &= \int_0^L x^2(L - x)(-f_o)dx \\ 4EIL^3C &= -\frac{1}{12}f_oL^4 \end{aligned} \tag{3.17}$$

hence $C = -f_o L/48EI$, and the approximate solution for deflection is

$$v(\xi) = -\frac{f_o L^4}{48EI}(1-\xi)\xi^2 \tag{3.18}$$

where $\xi = x/L$. The exact solution to the deflection is

$$v_{exact}(\xi) = -\frac{f_o L^4}{48EI}(2\xi^2 - 5\xi + 3)\xi^2. \tag{3.19}$$

We can use Equation (3.9) to determine the unknown reactions. In this example $R_1 = R_A$, $R_2 = M_A$, and $R_3 = R_B$ are unknown while $R_4 = M_B = 0$. With $v(x)$ given by the approximate solution Equation (3.18), we select three weight (test) functions $w_1 = 1$, $w_2 = x$, and $w_3 = x^2$, and use Equation (3.9) to obtain

$$\begin{aligned} R_A + R_B &= f_o L \\ M_A + R_B L &= \frac{1}{2} f_o L^2 \\ R_B L^2 &= \frac{3}{8} f_o L^3. \end{aligned} \tag{3.20}$$

The first two equations are the equations of equilibrium of a rigid body; sum of forces in the vertical direction is zero and sum of moments about the z-axis through the point $x = 0$ is zero. Later on, we will show that the weak form is nothing but the principle of virtual displacements. For beams, w is a virtual deflection, and the choices $w = 1$ and $w = x$ correspond to rigid-body virtual translation and rotation, respectively. The reactions thus obtained are $R_A = 5f_o L/8$, $R_B = 3f_o L/8$, and $M_A = f_o L^2/8$. These reactions are exact. Alternatively, we can determine the reactions by using the test functions

$$\begin{aligned} w_1 &= 1 - (x/L)^2 \\ w_2 &= (x/L)^2 \\ w_3 &= (x/L)(1 - x/L) \end{aligned} \tag{3.21}$$

in Equation (3.9). This is equivalent to applying the unit-displacement theorem (e.g., [89]).

3.3 FINITE ELEMENT METHOD: BEAM ELEMENT

Mesh Generation

The beam is divided into contiguous non-overlapping elements by introducing N global nodes; the number of elements is denoted by ne. Each element is defined by a set of nodes, the number of nodes per element is n. The mesh generation step provides the global coordinates of the nodes, and the connectivity (or identification) matrix of the elements.

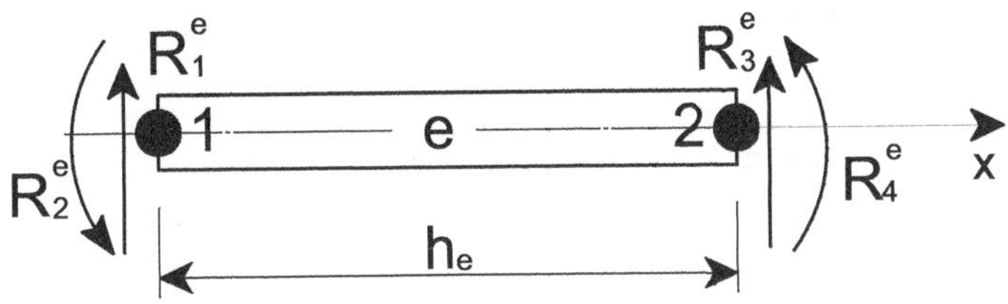

Figure 3.9 Secondary variables at the nodes of a Euler–Bernoulli beam element, R_1^e and R_3^e are forces, and R_2^e and R_4^e are moments.

Derivation of Element Matrices $[K^e]$, and $\{F^e\}$

We consider an element that is defined by two nodes ($n = 2$) with global coordinates x_1^e and x_2^e and refer to it as element (e). Following the same procedure leading to Equation (3.7), we obtain the weak form for element (e) as

$$\int_{x_1^e}^{x_2^e} \frac{d^2w}{dx^2} D \frac{d^2v}{dx^2} dx = \int_{x_1^e}^{x_2^e} wf dx$$

$$+w(x_1^e)R_1^e + \left(\frac{dw}{dx}|_{x_1^e}\right) R_2^e + w(x_2^e) R_3^e + \left(\frac{dw}{dx}|_{x_2^e}\right) R_4^e \tag{3.22}$$

where

$$\begin{aligned} R_1^e &= \left[\frac{d}{dx}\left(D\frac{d^2v}{dx^2}\right)\right]_{x_1^e} \\ R_2^e &= -\left(D\frac{d^2v}{dx^2}\right)_{x_1^e} \\ R_3^e &= -\left[\frac{d}{dx}\left(D\frac{d^2v}{dx^2}\right)\right]_{x_2^e} \\ R_4^e &= \left(D\frac{d^2v}{dx^2}\right)_{x_2^e} \end{aligned} \tag{3.23}$$

are the secondary variables at the element's nodes. Note that they are defined in terms of the derivatives of the unknown function $v(x)$, thus we have to consider them as unknowns. We note that R_1^e and R_2^e are the shear force (Q) and the negative of the bending moment ($-M$) at the first node of the element, while R_3^e and R_4^e are $-Q$ and M at the second node of the element. There are no boundary conditions to be imposed at the element level. Boundary conditions are known for the whole beam and will be applied on the assembled global system.

Figure 3.10 shows the free body diagram of a node that is common to two adjacent elements, say elements $e-1$ and e. Equilibrium of the node requires that secondary

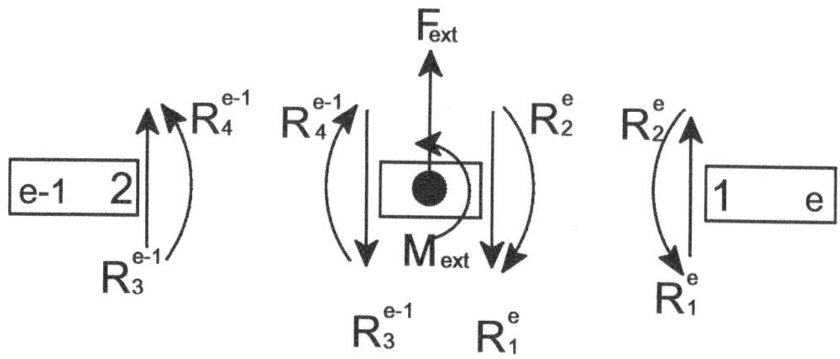

Figure 3.10 Free body diagram of a common node between two elements, F_{ext} and M_{ext} are externally applied force and moment at the common node between element $e-1$ and e.

variables satisfy

$$
\begin{aligned}
R_3^{e-1} + R_1^e &= F_{ext} \\
R_4^{e-1} + R_2^e &= M_{ext}
\end{aligned}
\tag{3.24}
$$

where F_{ext} and M_{ext} are externally applied concentrated force and moment at the common node. F_{ext} is positive when upward, and M_{ext} is positive when counterclockwise. These are externally applied loads not accounted for by the distributed load $f(x)$.

For later use, we write the weak form in the operator form

$$
\mathtt{A}(w,v) = \mathtt{F}(w) \tag{3.25}
$$

where

$$
\mathtt{A}(w,v) = \int_{x_1^e}^{x_2^e} \frac{d^2w}{dx^2} D \frac{d^2v}{dx^2} dx \tag{3.26}
$$

and

$$
\mathtt{F}(w) = \int_{x_1^e}^{x_2^e} wf dx + w(x_1^e)R_1^e + w(x_2^e)R_3^e + (\frac{dw}{dx}|_{x_1^e})R_2^e + (\frac{dw}{dx}|_{x_2^e})R_4^e. \tag{3.27}
$$

The operator $\mathtt{A}(w,v)$ is bilinear in its two arguments. It can be shown that

$$
\begin{aligned}
\mathtt{A}(w, v_1 + v_2) &= \mathtt{A}(w, v_1) + \mathtt{A}(w, v_2) \\
\mathtt{A}(w_1 + w_2, v) &= \mathtt{A}(w_1, v) + \mathtt{A}(w_2, v) \\
\mathtt{A}(w, cv) &= c\mathtt{A}(w, v).
\end{aligned}
\tag{3.28}
$$

Similarly, $\mathtt{F}(w)$ is linear in w.

Approximation and Hermite Cubic Shape Functions

We recall that in the case of a second-order differential equation, the weak form contains only the first derivative du/dx of the unknown function, and hence it requires

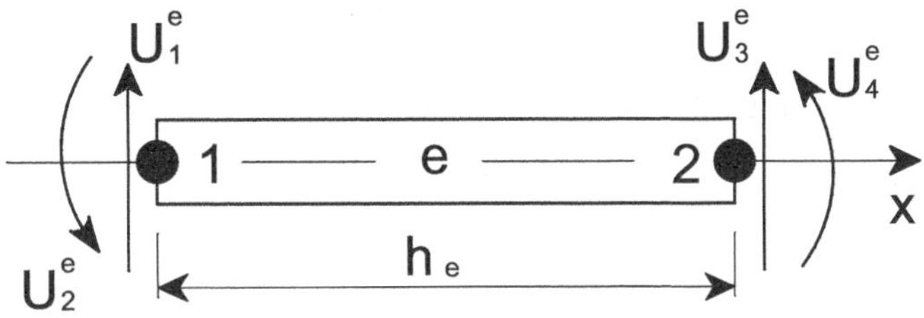

Figure 3.11 Primary variables at the nodes of a Euler–Bernoulli beam element. U_1^e and U_3^e are deflection (displacement), and U_2^e and U_4^e are rotations (dv/dx) about the z-axis.

continuity of the trial solution $u(x)$ only and the existence of its first derivative. Lagrange family of shape functions (linear, quadratic, cubic, etc.) allows us to satisfy these continuity requirements. In the present case of a fourth-order differential equation, the weak form contains the second derivative d^2v/dx^2, and hence it requires continuity of the trial solution $v(x)$ and its first derivative dv/dx. Lagrange interpolation (shape) functions are not admissible in this case because they do not allow us to enforce continuity of the slope at the inter-element nodes even if we use quadratic or cubic elements. For the case of the Lagrange interpolation functions, the approximation over the element domain is expressed in terms of the nodal values of the unknown function but not its derivative. To be able to satisfy continuity of the function and its slope between elements, we have to express the approximation over the element's domain in terms of the nodal values of the function v and its first derivative dv/dx. There are two degrees of freedom at each node, namely deflection and slope. For a two-node element, there are four degrees of freedom per element. The approximation $V^e(x)$ over the element's domain must be a cubic polynomial in x.

$$v^e(x) \approx V^e(x) = c_1 + c_2x + c_3x^2 + c_4x^3 \tag{3.29}$$

To determine the coefficients c_1 through c_4, we use the four conditions

$$\begin{aligned} x &= x_1^e: & V^e &= U_1^e \\ x &= x_1^e: & \frac{dV^e}{dx} &= U_2^e \\ x &= x_2^e: & V^e &= U_3^e \\ x &= x_2^e: & \frac{dV^e}{dx} &= U_4^e. \end{aligned} \tag{3.30}$$

We note that U_1^e and U_2^e are the deflection and slope at the first node, and U_3^e and U_4^e are the deflection and slope at the second node of element e. This approach is inconvenient. Derivation of the shape functions is much easier in terms of a scaled local coordinate, ξ, which is related to the global coordinate x by

$$\xi = (x - x_1^e)/h_e \qquad x_1^e \le x \le x_2^e \tag{3.31}$$

where $h_e = x_2^e - x_1^e$ is the element length. Note that the element's domain $x_1^e \le x \le x_2^e$ is mapped onto $0 \le \xi \le 1$. It follows that

$$x = x_1^e + h_e \xi \tag{3.32}$$

and $dx = h_e d\xi$, hence $dV^e/d\xi = h_e dV^e/dx$. Now we write the approximate solution in terms of ξ,

$$V^e(\xi) = c_1 + c_2\xi + c_3\xi^2 + c_4\xi^3 \tag{3.33}$$

$$\frac{dV^e(\xi)}{d\xi} = c_2 + 2c_3\xi + 3c_4\xi^2. \tag{3.34}$$

To determine the coefficients c_1 through c_4, we use

$$\begin{aligned}
\xi &= 0: & V^e &= U_1^e \\
\xi &= 0: & \frac{dV^e}{d\xi} &= h_e U_2^e \\
\xi &= 1: & V^e &= U_3^e \\
\xi &= 1: & \frac{dV^e}{d\xi} &= h_e U_4^e
\end{aligned} \tag{3.35}$$

and obtain

$$\begin{aligned}
c_1 &= U_1^e \\
c_2 &= h_e U_2^e \\
c_1 + c_2 + c_3 + c_4 &= U_3^e \\
c_2 + 2c_3 + 3c_4 &= h_e U_4^e.
\end{aligned} \tag{3.36}$$

The solution is

$$\begin{aligned}
c_1 &= U_1^e \\
c_2 &= h_e U_2^e \\
c_3 &= 3(U_3^e - U_1^e) - h_e(2U_2^e + U_4^e) \\
c_4 &= -2(U_3^e - U_1^e) + h_e(U_2^e + U_4^e).
\end{aligned} \tag{3.37}$$

We rewrite the approximation $V^e(\xi)$ in terms of the U^e's, which are the degrees of freedom of the element, and then rearrange into the form

$$V^e(\xi) = \psi_1(\xi)U_1^e + \psi_2(\xi)U_2^e + \psi_3(\xi)U_3^e + \psi_4(\xi)U_4^e. \tag{3.38}$$

The shape functions are given by,

$$\begin{aligned}
\psi_1 &= 1 - 3\xi^2 + 2\xi^3 \\
\psi_2 &= h_e\xi(1-\xi)^2 \\
\psi_3 &= \xi^2(3 - 2\xi) \\
\psi_4 &= -h_e\xi^2(1-\xi).
\end{aligned} \tag{3.39}$$

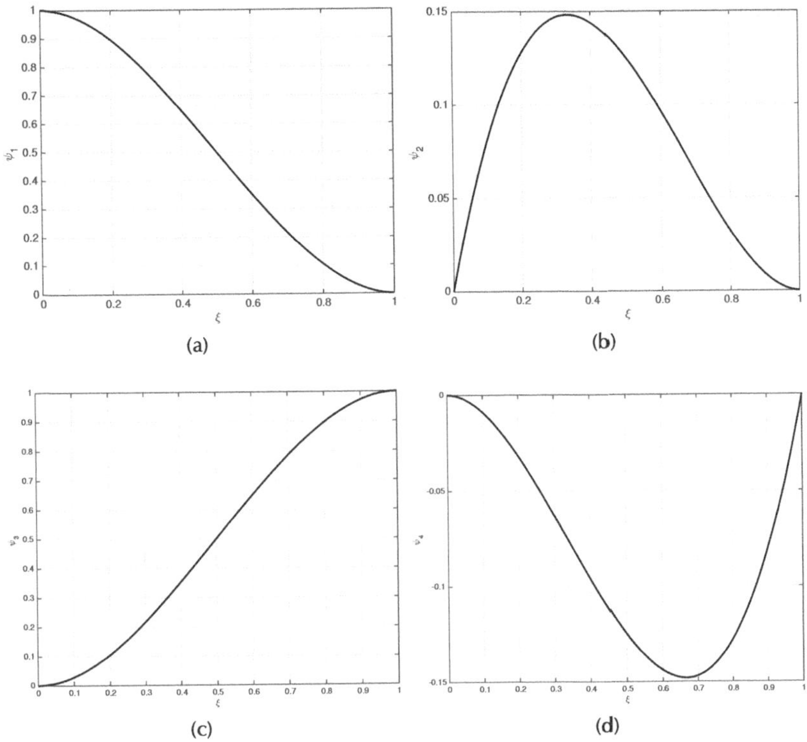

Figure 3.12 Hermite cubic polynomials, (a)$\psi_1(\xi)$, (b) $\psi_2(\xi)$ for $h_e = 1$, (c) $\psi_3(\xi)$, and (d) $\psi_4(\xi)$ for $h_e = 1$.

These functions are called the Hermite cubic interpolation polynomials, see Figures 3.12a–3.12d. They have the following properties. At node one ($\xi = 0$), all $\psi_j = 0$ except $\psi_1 = 1$. And all $d\psi_j/dx = 0$ except $d\psi_2/dx = 1$. At node two ($\xi = 1$), all $\psi_j = 0$ except $\psi_3 = 1$. And all $d\psi_j/dx = 0$ except $d\psi_4/dx = 1$. The values of a function and its derivative at the end points ($\xi = 0$, and $\xi = 1$) are four parameters; three parameters are zero and only one parameter is equal to 1.

The first derivative of these functions are

$$
\begin{aligned}
\frac{d\psi_1}{d\xi} &= -6\xi(1-\xi) \\
\frac{d\psi_2}{d\xi} &= h_e(1-4\xi+3\xi^2) \\
\frac{d\psi_3}{d\xi} &= 6\xi(1-\xi) \\
\frac{d\psi_4}{d\xi} &= h_e\xi(3\xi-2)
\end{aligned}
\tag{3.40}
$$

Derivatives with respect to x are

$$\begin{aligned} \frac{d\psi_i}{dx} &= \frac{1}{h_e}\frac{d\psi_i}{d\xi} \\ \frac{d^2\psi_i}{dx^2} &= \frac{1}{h_e^2}\frac{d^2\psi_i}{d\xi^2}. \end{aligned} \quad (3.41)$$

Element Matrices: The Galerkin Finite Element Model

To obtain the element matrices, we choose four linearly independent weight (test) functions and apply the weak form to the approximate solution $V^e(x)$. In the Galerkin method, the weight (test) functions are the same as the shape functions, $W_i^e = \psi_i$ for $i = 1,...,4$. Let's write the weak form for the element as

$$\mathtt{A}(\psi_i, V^e(x)) = \mathtt{F}(\psi_i) \qquad i = 1,...,4. \quad (3.42)$$

Substituting the approximate solution, we get

$$\mathtt{A}(\psi_i, \sum_{j=1}^{4} \psi_j U_j^e) = \mathtt{F}(\psi_i) \qquad i = 1,...,4. \quad (3.43)$$

Because $\mathtt{A}$ is bilinear, see Equation (3.28), we obtain,

$$\sum_{j=1}^{4} \mathtt{A}(\psi_i, \psi_j) U_j^e = \mathtt{F}(\psi_i) \qquad i = 1,...,4 \quad (3.44)$$

where

$$\mathtt{A}(\psi_i, \psi_j) = \int_{x_1^e}^{x_2^e} \frac{d^2\psi_i}{dx^2} D \frac{d^2\psi_j}{dx^2} dx \quad (3.45)$$

$$\mathtt{F}(\psi_i) = \int_{x_1^e}^{x_2^e} \psi_i f dx + \psi_i(x_1^e)R_1^e + (\frac{d\psi_i}{dx}|_{x_1^e})R_2^e + \psi_i(x_2^e)R_3^e + (\frac{d\psi_i}{dx}|_{x_2^e})R_4^e. \quad (3.46)$$

Recalling the properties of the shape functions, we write the finite-element model in matrix form

$$[K^e]\{U^e\} = \{F^e\} + \{R^e\} \quad (3.47)$$

where $[K^e]$ is the stiffness matrix, $\{F^e\}$ is the forcing vector that is equivalent to the distributed load f, and $\{R^e\}$ is the array of secondary variables at the element nodes. In expanded form, we have

$$\begin{bmatrix} k_{11}^e & k_{12}^e & k_{13}^e & k_{14}^e \\ k_{21}^e & k_{22}^e & k_{23}^e & k_{24}^e \\ k_{31}^e & k_{32}^e & k_{33}^e & k_{34}^e \\ k_{41}^e & k_{42}^e & k_{43}^e & k_{44}^e \end{bmatrix} \begin{Bmatrix} U_1^e \\ U_2^e \\ U_3^e \\ U_4^e \end{Bmatrix} = \begin{Bmatrix} F_1^e \\ F_2^e \\ F_3^e \\ F_4^e \end{Bmatrix} + \begin{Bmatrix} R_1^e \\ R_2^e \\ R_3^e \\ R_4^e \end{Bmatrix} \quad (3.48)$$

$$k_{ij}^e = \int_{x_1^e}^{x_2^e} \frac{d^2\psi_i}{dx^2} D \frac{d^2\psi_j}{dx^2} dx \quad (3.49)$$

$$F_i^e = \int_{x_1^e}^{x_2^e} \psi_i f dx. \tag{3.50}$$

In terms of the local coordinate ξ, we have

$$k_{ij}^e = \frac{1}{h_e^3} \int_0^1 \frac{d^2\psi_i}{d\xi^2} D \frac{d^2\psi_j}{d\xi^2} d\xi \tag{3.51}$$

$$F_i^e = h_e \int_0^1 \psi_i f d\xi. \tag{3.52}$$

To evaluate these integrals, we need to specify the functions $D(x) = EI$ and $f(x)$. For simplicity, we consider a special case where these functions are element-wise constant $D = (EI)_e$, and $f = f_e$. The beam element stiffness matrix is

$$[K^e] = \frac{2(EI)_e}{h_e^3} \begin{bmatrix} 6 & 3h_e & -6 & 3h_e \\ 3h_e & 2h_e^2 & -3h_e & h_e^2 \\ -6 & -3h_e & 6 & -3h_e \\ 3h_e & h_e^2 & -3h_e & 2h_e^2 \end{bmatrix} \tag{3.53}$$

and the forcing vector is

$$\{F^e\} = \frac{f_e h_e}{12} \begin{Bmatrix} 6 \\ h_e \\ 6 \\ -h_e \end{Bmatrix}. \tag{3.54}$$

The distributed load f_e is replaced by concentrated forces F_1^e and F_3^e and moments F_2^e and F_4^e at the element's nodes.

Assemble Elements and Obtain Global System

We consider a mesh of two elements as shown in Figure 3.13. There are three global nodes labeled 1, 2, and 3. Each node has two degrees of freedom, namely the beam deflection and slope at the node. Therefore, the global mesh of two elements has six degrees of freedom. The two elements are disassembled and the local degrees of freedom for each element are depicted in Figure 3.14. These are the primary variables at the nodes of each element. The secondary variables for each element are shown in Figure 3.15. For each element we have four equations that contain eight unknowns: four primary variables U_1^e through U_4^e and four secondary variables R_1^e through R_4^e. To understand the assembly process, we rewrite the element equations in terms of global primary variables. For a mesh of two elements, we have for the first element

$$\begin{bmatrix} k_{11}^1 & k_{12}^1 & k_{13}^1 & k_{14}^1 & 0 & 0 \\ k_{21}^1 & k_{22}^1 & k_{23}^1 & k_{24}^1 & 0 & 0 \\ k_{31}^1 & k_{32}^1 & k_{33}^1 & k_{34}^1 & 0 & 0 \\ k_{41}^1 & k_{42}^1 & k_{43}^1 & k_{44}^1 & 0 & 0 \\ 0 & 0 & 0 & 0 & 0 & 0 \\ 0 & 0 & 0 & 0 & 0 & 0 \end{bmatrix} \begin{Bmatrix} U_1 \\ U_2 \\ U_3 \\ U_4 \\ U_5 \\ U_6 \end{Bmatrix} = \begin{Bmatrix} F_1^1 \\ F_2^1 \\ F_3^1 \\ F_4^1 \\ 0 \\ 0 \end{Bmatrix} + \begin{Bmatrix} R_1^1 \\ R_2^1 \\ R_3^1 \\ R_4^1 \\ 0 \\ 0 \end{Bmatrix} \tag{3.55}$$

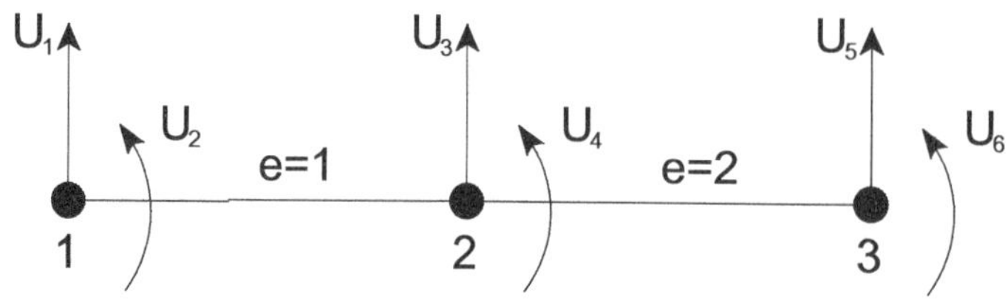

Figure 3.13 Global degrees of freedom for a mesh of two elements.

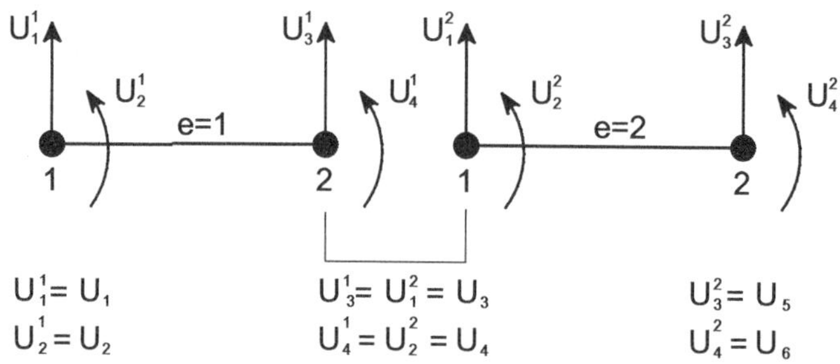

Figure 3.14 Primary variables (element's degrees of freedom) of isolated elements and their relation to global degrees of freedom.

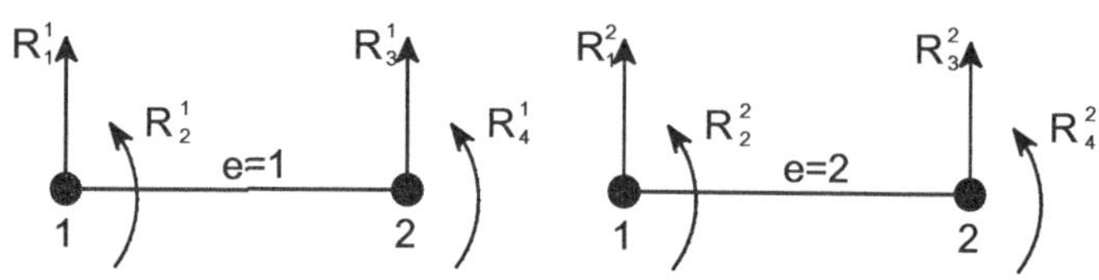

Figure 3.15 Secondary variables of isolated elements.

and for the second element

$$\begin{bmatrix} 0 & 0 & 0 & 0 & 0 & 0 \\ 0 & 0 & 0 & 0 & 0 & 0 \\ 0 & 0 & k_{11}^2 & k_{12}^2 & k_{13}^2 & k_{14}^2 \\ 0 & 0 & k_{21}^2 & k_{22}^2 & k_{23}^2 & k_{24}^2 \\ 0 & 0 & k_{31}^2 & k_{32}^2 & k_{33}^2 & k_{34}^2 \\ 0 & 0 & k_{41}^2 & k_{42}^2 & k_{43}^2 & k_{44}^2 \end{bmatrix} \begin{Bmatrix} U_1 \\ U_2 \\ U_3 \\ U_4 \\ U_5 \\ U_6 \end{Bmatrix} = \begin{Bmatrix} 0 \\ 0 \\ F_1^2 \\ F_2^2 \\ F_3^2 \\ F_4^2 \end{Bmatrix} + \begin{Bmatrix} 0 \\ 0 \\ R_1^2 \\ R_2^2 \\ R_3^2 \\ R_4^2 \end{Bmatrix}. \tag{3.56}$$

The global system is obtained by adding the above systems of equations,

$$\begin{bmatrix} k_{11}^1 & k_{12}^1 & k_{13}^1 & k_{14}^1 & 0 & 0 \\ k_{21}^1 & k_{22}^1 & k_{23}^1 & k_{24}^1 & 0 & 0 \\ k_{31}^1 & k_{32}^1 & k_{33}^1+k_{11}^2 & k_{34}^1+k_{12}^2 & k_{13}^2 & k_{14}^2 \\ k_{41}^1 & k_{42}^1 & k_{43}^1+k_{21}^2 & k_{44}^1+k_{22}^2 & k_{23}^2 & k_{24}^2 \\ 0 & 0 & k_{31}^2 & k_{32}^2 & k_{33}^2 & k_{34}^2 \\ 0 & 0 & k_{41}^2 & k_{42}^2 & k_{43}^2 & k_{44}^2 \end{bmatrix} \begin{Bmatrix} U_1 \\ U_2 \\ U_3 \\ U_4 \\ U_5 \\ U_6 \end{Bmatrix} = \begin{Bmatrix} F_1^1 \\ F_2^1 \\ F_3^1+F_1^2 \\ F_4^1+F_2^2 \\ F_3^2 \\ F_4^2 \end{Bmatrix} + \begin{Bmatrix} R_1^1 \\ R_2^1 \\ R_3^1+R_1^2 \\ R_4^1+R_2^2 \\ R_3^2 \\ R_4^2 \end{Bmatrix}. \tag{3.57}$$

Let's assume that at node 2, there are (given) externally applied force F_{ext} and moment M_{ext}. Balance of secondary variables at inter-element nodes, see Equation (3.24), requires

$$\begin{aligned} R_3^1+R_1^2 &= F_{ext} \\ R_4^1+R_2^2 &= M_{ext}. \end{aligned} \tag{3.58}$$

The global system becomes,

$$\begin{bmatrix} k_{11}^1 & k_{12}^1 & k_{13}^1 & k_{14}^1 & 0 & 0 \\ k_{21}^1 & k_{22}^1 & k_{23}^1 & k_{24}^1 & 0 & 0 \\ k_{31}^1 & k_{32}^1 & k_{33}^1+k_{11}^2 & k_{34}^1+k_{12}^2 & k_{13}^2 & k_{14}^2 \\ k_{41}^1 & k_{42}^1 & k_{43}^1+k_{21}^2 & k_{44}^1+k_{22}^2 & k_{23}^2 & k_{24}^2 \\ 0 & 0 & k_{31}^2 & k_{32}^2 & k_{33}^2 & k_{34}^2 \\ 0 & 0 & k_{41}^2 & k_{42}^2 & k_{43}^2 & k_{44}^2 \end{bmatrix} \begin{Bmatrix} U_1 \\ U_2 \\ U_3 \\ U_4 \\ U_5 \\ U_6 \end{Bmatrix} = \begin{Bmatrix} F_1^1 \\ F_2^1 \\ F_3^1+F_1^2 \\ F_4^1+F_2^2 \\ F_3^2 \\ F_4^2 \end{Bmatrix} + \begin{Bmatrix} R_1^1 \\ R_2^1 \\ F_{ext} \\ M_{ext} \\ R_3^2 \\ R_4^2 \end{Bmatrix}. \tag{3.59}$$

We note that the distributed load $f(x)$ is replaced by equivalent nodal forces and moments given by F_i^e, this is the first array on the right-hand side of Equation (3.59). We also note that F_{ext} and M_{ext} are externally applied concentrated force and moment at the common node, and as such they are not included in the distributed load.

Apply Boundary Conditions and Solve Global System

In our analysis so far, we have not made any reference to the boundary conditions specified in the problem statement. Thus, the obtained global system is applicable to a variety of boundary conditions. The system consists of six equations, two equations for each global node because there are two primary variables at each node. However, there are 10 unknowns: six primary variables, two for each global node; and four secondary variables, shear and moment at the first and last node. The boundary conditions provide four additional equations to close the system.

Let's consider the boundary conditions given in Equation (3.3) for the beam shown in Figure 3.4. We rewrite the EBCs as

$$U_1 = 0, \ U_2 = 0, \ U_5 = 0. \tag{3.60}$$

The NBC at $x = L$ gives $R_4^2 = 0$. There are different ways of combining these conditions with the global system. The simplest way, not necessarily the most efficient way, is to replace the first, second, and fifth equations in the global system by the above equations, and obtain

$$\begin{bmatrix} 1 & 0 & 0 & 0 & 0 & 0 \\ 0 & 1 & 0 & 0 & 0 & 0 \\ k_{31}^1 & k_{32}^1 & k_{33}^1 + k_{11}^2 & k_{34}^1 + k_{12}^2 & k_{13}^2 & k_{14}^2 \\ k_{41}^1 & k_{42}^1 & k_{43}^1 + k_{21}^2 & k_{44}^1 + k_{22}^2 & k_{23}^2 & k_{24}^2 \\ 0 & 0 & 0 & 0 & 1 & 0 \\ 0 & 0 & k_{41}^2 & k_{42}^2 & k_{43}^2 & k_{44}^2 \end{bmatrix} \begin{Bmatrix} U_1 \\ U_2 \\ U_3 \\ U_4 \\ U_5 \\ U_6 \end{Bmatrix}$$

$$= \begin{Bmatrix} 0 \\ 0 \\ F_3^1 + F_1^2 + F_{ext} \\ F_4^1 + F_2^2 + M_{ext} \\ 0 \\ F_4^2 \end{Bmatrix}. \tag{3.61}$$

Another method is to obtain a reduced system

$$\begin{bmatrix} k_{33}^1 + k_{11}^2 & k_{34}^1 + k_{12}^2 & k_{14}^2 \\ k_{43}^1 + k_{21}^2 & k_{44}^1 + k_{22}^2 & k_{24}^2 \\ k_{41}^2 & k_{42}^2 & k_{44}^2 \end{bmatrix} \begin{Bmatrix} U_3 \\ U_4 \\ U_6 \end{Bmatrix} = \begin{Bmatrix} F_3^1 + F_1^2 + F_{ext} \\ F_4^1 + F_2^2 + M_{ext} \\ F_4^2 \end{Bmatrix}. \tag{3.62}$$

We solve this system for the primary variables. Afterward, the secondary variables (the reactions) at the boundary nodes are determined by the first, second, and fifth equations of the original global system Equation (3.59), i.e., before applying the boundary conditions.

Example 3.2.

Determine the reactions at the supports of the loaded beam shown in Figure 3.16.

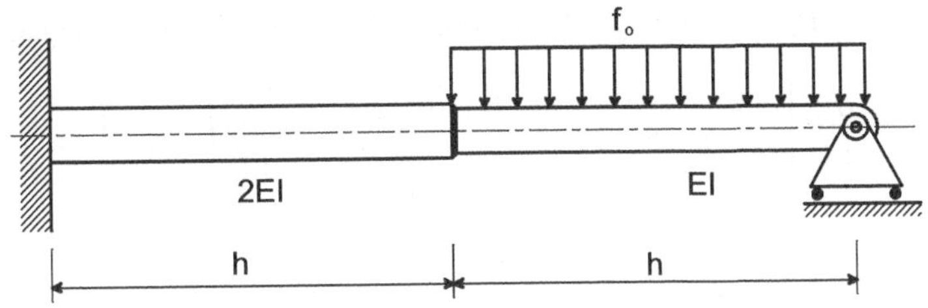

Figure 3.16 Example 3.2. Loaded beam.

Solution

We use two Euler–Bernoulli beam elements. For element # 1, $(EI)_e = 2EI$, $f_e = 0$, and $h_e = h$, and the element equations are

$$\frac{2(2EI)}{h^3}\begin{bmatrix} 6 & 3h & -6 & 3h \\ 3h & 2h^2 & -3h & h^2 \\ -6 & -3h & 6 & -3h \\ 3h & h^2 & -3h & 2h^2 \end{bmatrix}\begin{Bmatrix} U_1^1 \\ U_2^1 \\ U_3^1 \\ U_4^1 \end{Bmatrix} = \begin{Bmatrix} 0 \\ 0 \\ 0 \\ 0 \end{Bmatrix} + \begin{Bmatrix} R_1^1 \\ R_2^1 \\ R_3^1 \\ R_4^1 \end{Bmatrix}. \tag{3.63}$$

For element # 2, $(EI)_e = EI$, $f_e = -f_o$, and $h_e = h$, and the element equations are

$$\frac{2EI}{h^3}\begin{bmatrix} 6 & 3h & -6 & 3h \\ 3h & 2h^2 & -3h & h^2 \\ -6 & -3h & 6 & -3h \\ 3h & h^2 & -3h & 2h^2 \end{bmatrix}\begin{Bmatrix} U_1^2 \\ U_2^2 \\ U_3^2 \\ U_4^2 \end{Bmatrix} = -\frac{f_o h}{12}\begin{Bmatrix} 6 \\ h \\ 6 \\ -h \end{Bmatrix} + \begin{Bmatrix} R_1^2 \\ R_2^2 \\ R_3^2 \\ R_4^2 \end{Bmatrix}. \tag{3.64}$$

Assemble the two elements into a global system

$$\frac{2EI}{h^3}\begin{bmatrix} 12 & 6h & -12 & 6h & 0 & 0 \\ 6h & 4h^2 & -6h & 2h^2 & 0 & 0 \\ -12 & -6h & 12+6 & -6h+3h & -6 & 3h \\ 6h & 2h^2 & -6h+3h & 4h^2+2h^2 & -3h & h^2 \\ 0 & 0 & -6 & -3h & 6 & -3h \\ 0 & 0 & 3h & h^2 & -3h & 2h^2 \end{bmatrix}\begin{Bmatrix} U_1 \\ U_2 \\ U_3 \\ U_4 \\ U_5 \\ U_6 \end{Bmatrix}$$

$$= -\frac{f_o h}{12}\begin{Bmatrix} 0 \\ 0 \\ 6 \\ h \\ 6 \\ -h \end{Bmatrix} + \begin{Bmatrix} R_1^1 \\ R_2^1 \\ R_3^1 + R_1^2 \\ R_4^1 + R_2^2 \\ R_3^2 \\ R_4^2 \end{Bmatrix}. \tag{3.65}$$

Because there are no externally applied loads at node 2, we have $R_3^1 + R_1^2 = 0$ and $R_4^1 + R_2^2 = 0$. R_1^1, R_2^1, and R_3^2 are the unknown reactions. Apply boundary conditions

$U_1 = 0$, $U_2 = 0$, $U_5 = 0$, and $R_4^2 = 0$, and obtain the reduced system of equations,

$$\frac{2EI}{h^3}\begin{bmatrix} 18 & -3h & 3h \\ -3h & 6h^2 & h^2 \\ 3h & h^2 & 2h^2 \end{bmatrix}\begin{Bmatrix} U_3 \\ U_4 \\ U_6 \end{Bmatrix} = -\frac{f_o h}{12}\begin{Bmatrix} 6 \\ h \\ -h \end{Bmatrix}. \tag{3.66}$$

Solving this system, we get

$$U_3 = -\frac{8}{9}\frac{f_o h^4}{24EI}, \quad U_4 = -\frac{f_o h^3}{24EI}, \quad U_6 = \frac{7}{3}\frac{f_o h^3}{24EI}. \tag{3.67}$$

The reactions are the secondary variables, which we obtain from the first, the second, and the fifth equations of the system Equation (3.65). The equations are

$$\begin{aligned} \frac{2EI}{h^3}(-12U_3 + 6hU_4) &= R_1^1 \\ \frac{2EI}{h^3}(-6hU_3 + 2h^2U_4) &= R_2^1 \\ \frac{2EI}{h^3}(-6U_3 - 3hU_4 - 3hU_6) &= -\frac{6f_o h}{12} + R_3^2 \end{aligned} \tag{3.68}$$

hence,

$$R_1^1 = \frac{7}{18}f_o h, \quad R_2^1 = \frac{5}{18}f_o h^2, \quad R_3^2 = \frac{11}{18}f_o h. \tag{3.69}$$

The free-body diagram of the entire beam with reactions is shown in Figure 3.17a; the beam is in static equilibrium. For the present example, we can integrate Equation (3.1) analytically and obtain the exact solution as:

$$\begin{aligned} v(x) &= \frac{1}{6}c_1(x/h)^3 + \frac{1}{2}c_2(x/h)^2 \qquad \text{if} \quad 0 \le x \le h \\ v(x) &= b_o(x/h-1)^4 + \frac{1}{6}b_1(x/h-1)^3 + \frac{1}{2}b_2(x/h-1)^2 + b_3(x/h-1) + b_4 \\ &\qquad \text{if} \quad h \le x \le 2h \end{aligned} \tag{3.70}$$

where $b_o = -f_o h^4/24EI$, $c_1 = -(14/3)b_o$, $c_2 = (10/3)b_o$, $b_1 = -(28/3)b_o$, $b_2 = -(8/3)b_o$, $b_3 = b_o$, and $b_4 = (8/9)b_o$.

The finite-element solutions for the shear force, deflection, and bending moment diagrams are compared with the exact solutions in Figures 3.17b–3.17d, respectively. There is evident error in the deflection of the element that carries the uniformly distributed load because the trial solution is cubic (3rd order) whereas the exact solution is quartic (4th order). The error can be reduced by dividing the span carrying the distributed load into more elements.

The finite-element shear force and bending moment diagrams are computed from the definitions Equation (3.2). Since the trial solution $v(x)$ over the element is a 3rd-order polynomial, its second derivative (bending moment) is linear and its third

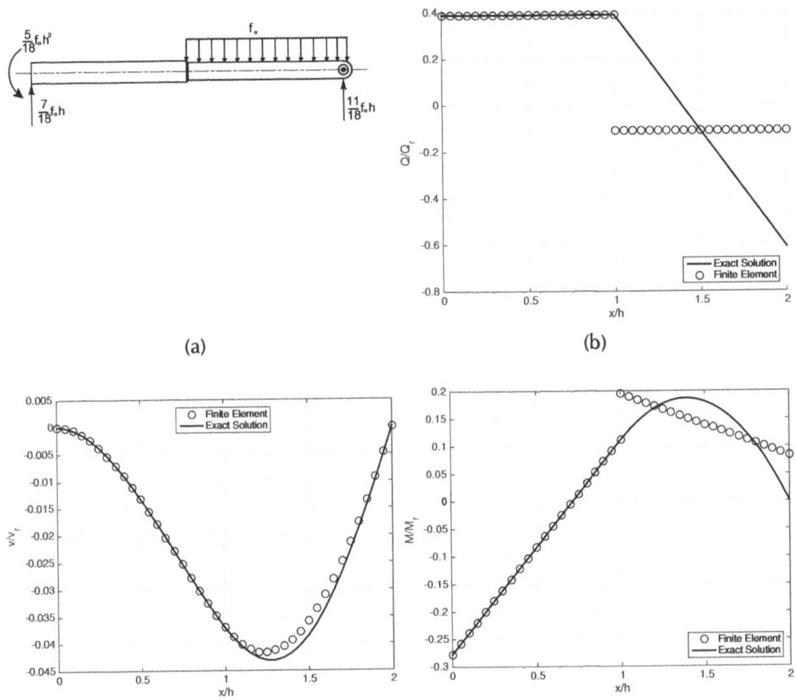

Figure 3.17 Comparison between finite element method using two elements and the exact solution. (a) Reactions, (b) shear force diagram, $Q_r = f_o h$, (c) deflection, $v_r = f_o h^4/EI$, and (d) bending moment diagram, $M_r = f_o h^2$.

derivative (shear) is constant. Therefore, computing shear and bending moment by differentiation of the trial solution is not recommended. Instead, we first use Equation (3.48) to determine the shear forces and bending moments (the secondary variables) at the element nodes; of course, this is to be done after the global system is solved for the unknown primary variables. Then we construct the shear and moment diagrams for each element knowing all the loads on it and that it is in equilibrium.

Three-node fifth-order beam element

A higher-order beam element can be constructed by increasing the number of degrees of freedom per element. An element defined by three nodes has six degrees of freedom; two degrees of freedom at each node (deflection and slope), as shown in Figure 3.18. The approximation over the element is written as

$$V^e(\xi) = \psi_1(\xi)U_1^e + \psi_2(\xi)U_2^e + \psi_3(\xi)U_3^e + \psi_4(\xi)U_4^e + \psi_5(\xi)U_5^e + \psi_6(\xi)U_6^e \tag{3.71}$$

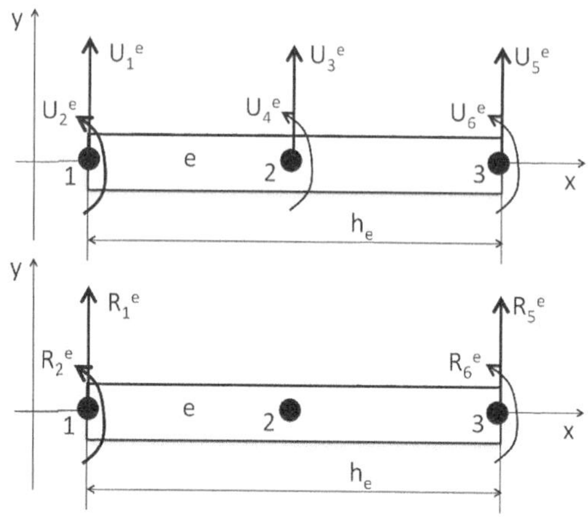

Figure 3.18 Quintic beam element. Three local nodes and six degrees of freedom; deflection and slope at each node. Node 2 is an interior node at midpoint.

where U_1^e, U_3^e, and U_5^e are the deflections at the element's nodes and U_2^e, U_4^e, and U_6^e are the slopes. The shape functions are fifth-order polynomials; we may call it a *quintic* element. If the interior node is the midpoint of the element, the shape functions are

$$\begin{aligned}
\psi_1 &= (1+6\xi)(2\xi-1)^2(\xi-1)^2 \\
\psi_2 &= h_e\xi(2\xi-1)^2(\xi-1)^2 \\
\psi_3 &= 16\xi^2(\xi-1)^2 \\
\psi_4 &= 8h_e(2\xi-1)\xi^2(\xi-1)^2 \\
\psi_5 &= (7-6\xi)\xi^2(2\xi-1)^2 \\
\psi_6 &= h_e(\xi-1)\xi^2(2\xi-1)^2
\end{aligned} \tag{3.72}$$

where h_e is the element length. For uniform $(EI)_e$ and distributed load f_e, the element stiffness matrix is

$$[K^e] = \frac{2(EI)_e}{35h_e^3}\begin{bmatrix} 2546 & 569h_e & -1792 & 960h_e & -754 & 121h_e \\ 569h_e & 166h_e^2 & -448h_e & 160h_e^2 & -121h_e & 19h_e^2 \\ -1792 & -448h_e & 3584 & 0 & -1792 & 448h_e \\ 960h_e & 160h_e^2 & 0 & 640h_e^2 & -960h_e & 160h_e^2 \\ -754 & -121h_e & -1792 & -960h_e & 2546 & -569h_e \\ 121h_e & 19h_e^2 & 448h_e & 160h_e^2 & -569h_e & 166h_e^2 \end{bmatrix} \tag{3.73}$$

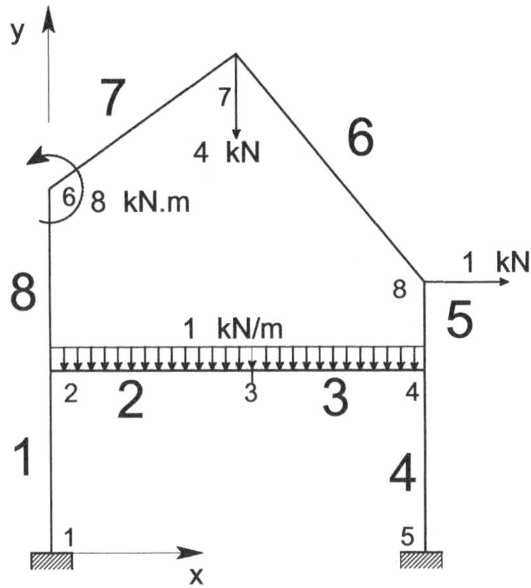

Figure 3.19 A typical frame and global coordinates (x, y). Eight global nodes and eight members.

and the forcing vector is

$$\{F^e\} = \frac{f_e h_e}{60} \begin{Bmatrix} 14 \\ h_e \\ 32 \\ 0 \\ 14 \\ -h_e \end{Bmatrix}. \tag{3.74}$$

3.4 PLANE FRAMES

An example of a plane frame is shown in Figure 3.19. All members of a frame are rigidly connected, so all members connected at a joint rotate through the same angle. A member in a plane frame can carry both axial and transverse loading. An isolated plane frame element with two nodes is shown in Figure 3.20. Local axes for the element are ($\bar{x}$, $\bar{y}$), where $\bar{x}$ is along the element from node 1 to node 2. There are three degrees of freedom at each node (primary variables). At node # 1, axial displacement, transverse deflection, and rotation about an axis that is parallel to the z-axis are denoted by $\bar{U}_1^e$, $\bar{U}_2^e$, and $\bar{U}_3^e$, respectively. And at node #2, they are denoted by $\bar{U}_4^e$, $\bar{U}_5^e$, and $\bar{U}_6^e$. The secondary variables at the nodes are the axial force, shear

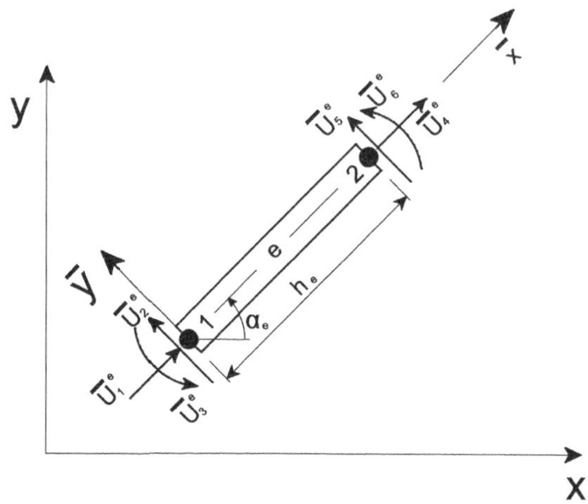

Figure 3.20 Global coordinates (x, y) and element local coordinates $(\bar{x}, \bar{y})$ for a plane frame element. Primary variables (degrees of freedom) at the nodes of a frame element: $\bar{U}_1^e$, $\bar{U}_2^e$, $\bar{U}_4^e$, and $\bar{U}_5^e$ are displacements, and $\bar{U}_3^e$ and $\bar{U}_6^e$ are rotations about an axis that is parallel to the z-axis.

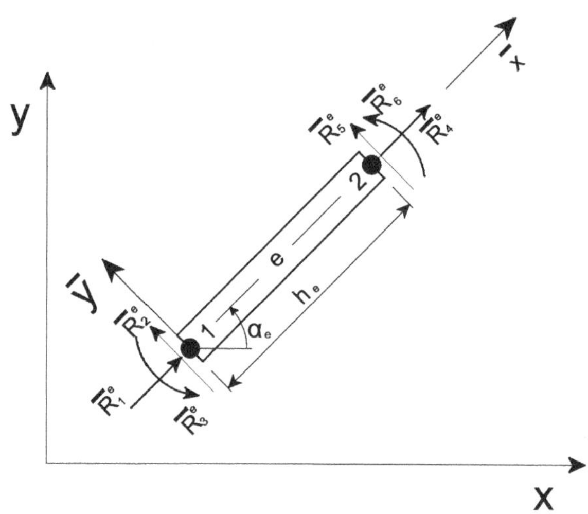

Figure 3.21 Secondary variables at the nodes of a frame element: $\bar{R}_1^e$, $\bar{R}_2^e$, $\bar{R}_4^e$, and $\bar{R}_5^e$ are forces, and $\bar{R}_3^e$ and $\bar{R}_6^e$ are moments.

force, and bending moment as shown in Figure 3.21. At node # 1, they are denoted by $\bar{R}_1^e$, $\bar{R}_2^e$, $\bar{R}_3^e$, respectively. And at node 2 they are $\bar{R}_4^e$, $\bar{R}_5^e$, and $\bar{R}_6^e$. We assume that displacements and strains to be very small, and neglect the coupling between axial displacements and transverse deflection and rotation. Thus, the axial displacements are related to the axial forces by the bar element equations derived in Chapter 2:

$$\frac{(EA)_e}{h_e}\begin{bmatrix} 1 & -1 \\ -1 & 1 \end{bmatrix}\begin{Bmatrix} \bar{U}_1^e \\ \bar{U}_4^e \end{Bmatrix} = \frac{f_e^a h_e}{2}\begin{Bmatrix} 1 \\ 1 \end{Bmatrix} + \begin{Bmatrix} \bar{R}_1^e \\ \bar{R}_4^e \end{Bmatrix}. \tag{3.75}$$

Here we assumed the member to be loaded with a uniform axial distributed load f_e^a (force per unit length). The transverse displacements (deflection) and rotations are related to the shear forces and bending moments by the Euler–Bernoulli beam element equations:

$$\frac{2(EI)_e}{h_e^3}\begin{bmatrix} 6 & 3h_e & -6 & 3h_e \\ 3h_e & 2h_e^2 & -3h_e & h_e^2 \\ -6 & -3h_e & 6 & -3h_e \\ 3h_e & h_e^2 & -3h_e & 2h_e^2 \end{bmatrix}\begin{Bmatrix} \bar{U}_2^e \\ \bar{U}_3^e \\ \bar{U}_5^e \\ \bar{U}_6^e \end{Bmatrix} = \frac{f_e^t h_e}{12}\begin{Bmatrix} 6 \\ h_e \\ 6 \\ -h_e \end{Bmatrix} + \begin{Bmatrix} \bar{R}_2^e \\ \bar{R}_3^e \\ \bar{R}_5^e \\ \bar{R}_6^e \end{Bmatrix}. \tag{3.76}$$

We also assumed the member to be loaded with a uniform transverse distributed load f_e^t (force per unit length). The frame element equations in local degrees of freedom are obtained by combining the bar and beam element equations in one system:

$$\frac{2(EI)_e}{h_e^3}\begin{bmatrix} \mu_e & 0 & 0 & -\mu_e & 0 & 0 \\ 0 & 6 & 3h_e & 0 & -6 & 3h_e \\ 0 & 3h_e & 2h_e^2 & 0 & -3h_e & h_e^2 \\ -\mu_e & 0 & 0 & \mu_e & 0 & 0 \\ 0 & -6 & -3h_e & 0 & 6 & -3h_e \\ 0 & 3h_e & h_e^2 & 0 & -3h_e & 2h_e^2 \end{bmatrix}\begin{Bmatrix} \bar{U}_1^e \\ \bar{U}_2^e \\ \bar{U}_3^e \\ \bar{U}_4^e \\ \bar{U}_5^e \\ \bar{U}_6^e \end{Bmatrix} = \frac{h_e}{12}\begin{Bmatrix} 6f_e^a \\ 6f_e^t \\ h_e f_e^t \\ 6f_e^a \\ 6f_e^t \\ -h_e f_e^t \end{Bmatrix} + \begin{Bmatrix} \bar{R}_1^e \\ \bar{R}_2^e \\ \bar{R}_3^e \\ \bar{R}_4^e \\ \bar{R}_5^e \\ \bar{R}_6^e \end{Bmatrix} \tag{3.77}$$

where $\mu_e = A_e h_e^2/2I_e$. These equations are now written as

$$[\bar{K}^e]\{\bar{U}^e\} = \{\bar{F}^e\} + \{\bar{R}^e\}. \tag{3.78}$$

For the local degrees of freedom, the stiffness matrix for a frame element is

$$[\bar{K}^e] = \frac{2(EI)_e}{h_e^3}\begin{bmatrix} \mu_e & 0 & 0 & -\mu_e & 0 & 0 \\ 0 & 6 & 3h_e & 0 & -6 & 3h_e \\ 0 & 3h_e & 2h_e^2 & 0 & -3h_e & h_e^2 \\ -\mu_e & 0 & 0 & \mu_e & 0 & 0 \\ 0 & -6 & -3h_e & 0 & 6 & -3h_e \\ 0 & 3h_e & h_e^2 & 0 & -3h_e & 2h_e^2 \end{bmatrix} \tag{3.79}$$

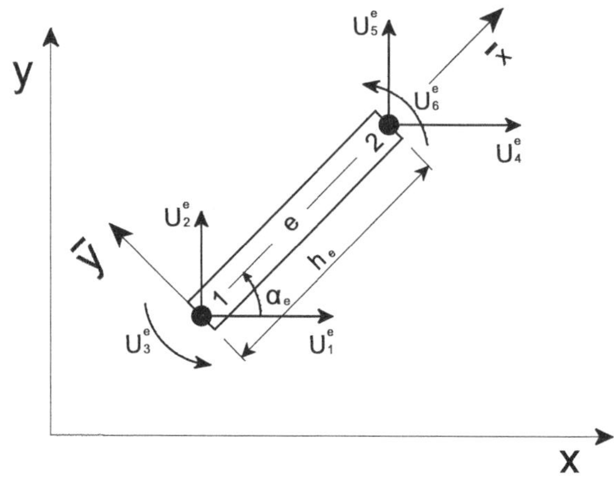

Figure 3.22 Primary variables (degrees of freedom) at the nodes of a frame element projected onto the global coordinate system. U_1^e, U_2^e, U_4^e, and U_5^e are displacements, and U_3^e and U_6^e are rotations.

and the forcing vector is

$$\{\bar{F}^e\} = \frac{h_e}{12}\left\{\begin{array}{c} 6f_e^a \\ 6f_e^t \\ h_e f_e^t \\ 6f_e^a \\ 6f_e^t \\ -h_e f_e^t \end{array}\right\}. \tag{3.80}$$

The primary and secondary variables in Equation (3.78) are projected onto the local element axes $\bar{x}$, $\bar{y}$. In general, each element has its own orientation relative to the global axes (x, y). To be able to assemble all elements of the frame, we need to transform the element equations from the local axes to the global axes so that the displacements and force components are projected onto the global coordinate system as shown in Figures 3.22 and 3.23.

Consider a vector $\vec{V}$ whose components are denoted by (u, v) in the global coordinate system (x, y), and by $(\bar{u}, \bar{v})$ in the local coordinate system $(\bar{x}, \bar{y})$ as shown in Figure 3.24. The angle between the two coordinate systems is α. It can be shown that $\bar{u} = ucos\alpha + vsin\alpha$, and $\bar{v} = vcos\alpha - usin\alpha$. Let $C = cos\alpha_e$ and $S = sin\alpha_e$. Since the joint rotation about the z-axis is the same for the local and global axes, we can write the transformation of the three degrees of freedom at one node between the two coordinate system as

$$\left\{\begin{array}{c} \bar{U}_1^e \\ \bar{U}_2^e \\ \bar{U}_3^e \end{array}\right\} = \left[\begin{array}{ccc} C & S & 0 \\ -S & C & 0 \\ 0 & 0 & 1 \end{array}\right]\left\{\begin{array}{c} U_1^e \\ U_2^e \\ U_3^e \end{array}\right\}. \tag{3.81}$$

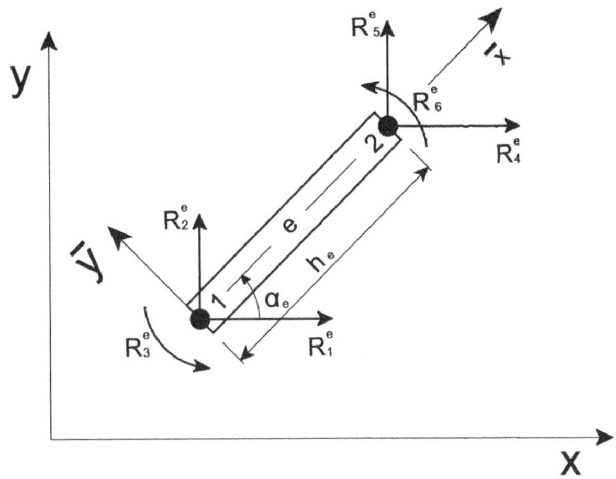

Figure 3.23 Secondary variables at the nodes of a frame element projected onto the global coordinate system. R_1^e, R_2^e, R_4^e, and R_5^e are forces, and R_3^e and R_6^e are moments.

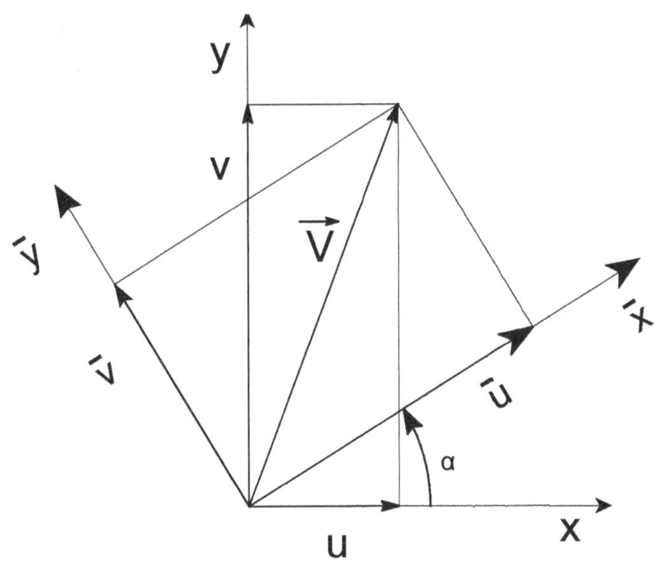

Figure 3.24 Displacement vector in local and global coordinates. The vector $\vec{V}$ is projected onto components (u, v) in the global system, and components $(\bar{u}, \bar{v})$ in the element's local coordinates.

The transformation of the element's six degrees of freedom is

$$\left\{\begin{array}{c} \bar{U}_1^e \\ \bar{U}_2^e \\ \bar{U}_3^e \\ \bar{U}_4^e \\ \bar{U}_5^e \\ \bar{U}_6^e \end{array}\right\} = \begin{bmatrix} C & S & 0 & 0 & 0 & 0 \\ -S & C & 0 & 0 & 0 & 0 \\ 0 & 0 & 1 & 0 & 0 & 0 \\ 0 & 0 & 0 & C & S & 0 \\ 0 & 0 & 0 & -S & C & 0 \\ 0 & 0 & 0 & 0 & 0 & 1 \end{bmatrix} \left\{\begin{array}{c} U_1^e \\ U_2^e \\ U_3^e \\ U_4^e \\ U_5^e \\ U_6^e \end{array}\right\} \tag{3.82}$$

or

$$\{\bar{U}^e\} = [T^e]\{U^e\} \tag{3.83}$$

where the transformation matrix is

$$[T^e] = \begin{bmatrix} C & S & 0 & 0 & 0 & 0 \\ -S & C & 0 & 0 & 0 & 0 \\ 0 & 0 & 1 & 0 & 0 & 0 \\ 0 & 0 & 0 & C & S & 0 \\ 0 & 0 & 0 & -S & C & 0 \\ 0 & 0 & 0 & 0 & 0 & 1 \end{bmatrix}. \tag{3.84}$$

The inverse of the transformation matrix $[T^e]$ is its transpose

$$[T^e]^{-1} = [T^e]^T \tag{3.85}$$

$$\{U^e\} = [T^e]^T \{\bar{U}^e\}. \tag{3.86}$$

Similarly, the forcing and secondary variable vectors can also be transformed

$$\{F^e\} = [T^e]^T \{\bar{F}^e\} \tag{3.87}$$

$$\{R^e\} = [T^e]^T \{\bar{R}^e\}. \tag{3.88}$$

Multiplying Equation (3.78) by $[T^e]^T$, and using Equation (3.83), we get

$$[T^e]^T [\bar{K}^e][T^e]\{U^e\} = [T^e]^T \{\bar{F}^e\} + [T^e]^T \{\bar{R}^e\}. \tag{3.89}$$

The element equations for a frame element are

$$[K^e]\{U^e\} = \{F^e\} + \{R^e\} \tag{3.90}$$

where the stiffness matrix is

$$[K^e] = [T^e]^T [\bar{K}^e][T^e] \tag{3.91}$$

and the force vector is

$$\{F^e\} = [T^e]^T \{\bar{F}^e\}. \tag{3.92}$$

Remarks:

(1) For a plane frame, each node has three degrees of freedom (primary variables); two displacement components and a rotation about the z-axis.

(2) At global node # j, the primary variables are x-displacement U_{3j-2}, y-displacement U_{3j-1}, and rotation about an axis that is parallel to the z-axis U_{3j}. These are shown in Figure 3.25.

(3) Distributed axial (f^a) and transverse loads (f^t) are prescribed as force per unit length in the element's local coordinates. f^a is positive along the $\bar{x}$-axis, and f^t is positive in the direction of $\bar{y}$-axis.

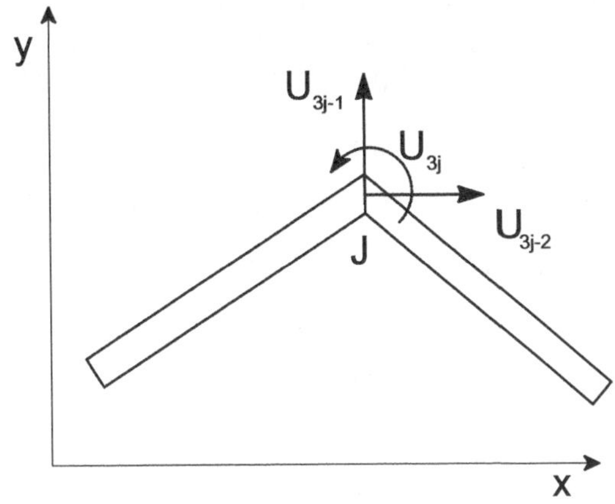

Figure 3.25 Primary variables at node number j of plane frame, U_{3j-2} is horizontal displacement, U_{3j-1} is vertical displacement, and U_{3j} is rotation about the z-axis.

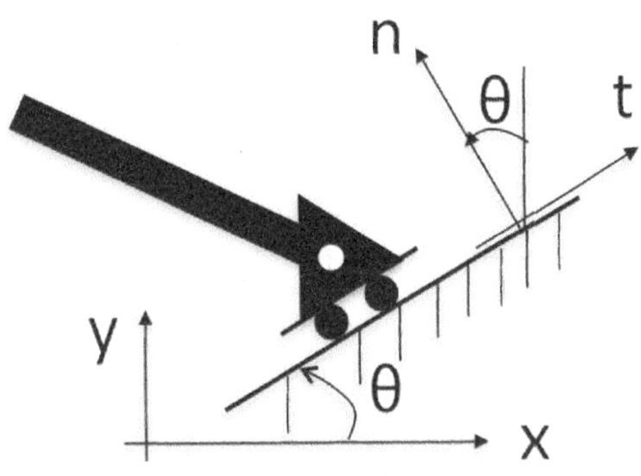

Figure 3.26 Roller supported by an inclined plane.

At a joint (node) that connects more than one element, the sums of secondary variables at that node are equal to the externally applied concentrated forces and moments that are not accounted for by the distributed loads. Externally applied concentrated forces and moments at the global joints (nodes) are specified by the x- and y-force components and moments about the z-axis, which is out of the page.

Conditions at a Roller Supported by an Inclined Plane

Application of the boundary conditions at a roller supported by an inclined plane relative to the global coordinate system requires special treatment. We assume that global node # i in a frame is supported by a roller on an inclined plane as shown in Figure 3.26. The plane makes an angle θ with the positive x-axis. The displacements in the global x- and y-directions are denoted by U_{3i-2} and U_{3i-1}, respectively, and the rotation about the z-axis by U_{3i}. The corresponding secondary variables are R_{3i-2}, R_{3i-1}, and R_{3i}. At the roller, the displacement normal to the inclined plane is zero, hence

$$-U_{3i-2} sin\theta + U_{3i-1} cos\theta = 0 \tag{3.93}$$

and the corresponding normal force (reaction at the roller)

$$R_n = -R_{3i-2} sin\theta + R_{3i-1} cos\theta \tag{3.94}$$

is unknown. The roller is free to slide parallel to the inclined plane; hence, the tangential force is zero

$$R_{3i-2} cos\theta + R_{3i-1} sin\theta = 0 \tag{3.95}$$

and the corresponding displacement parallel to the plane

$$U_t = U_{3i-2} cos\theta + U_{3i-1} sin\theta \tag{3.96}$$

is unknown. To implement the conditions given by Equations (3.93) and (3.95), we write the equations # $(3i-2)$ and $(3i-1)$ of the assembled global system,

$$\begin{aligned} a_{3i-2,1}U_1 + a_{3i-2,2}U_2 + \cdots + a_{3i-2,3i-2}U_{3i-2} + a_{3i-2,3i-1}U_{3i-1} + \cdots \\ + a_{3i-2,N}U_N = F_{3i-2} + R_{3i-2} \end{aligned} \tag{3.97}$$

$$\begin{aligned} a_{3i-1,1}U_1 + a_{3i-1,2}U_2 + \cdots + a_{3i-1,3i-2}U_{3i-2} + a_{3i-1,3i-1}U_{3i-1} + \cdots \\ + a_{3i-1,N}U_N = F_{3i-1} + R_{3i-1}. \end{aligned} \tag{3.98}$$

We multiply Equation (3.97) by $cos\theta$ and Equation (3.98) by $sin\theta$ and then add the two equations. Enforcing the condition given by Equation (3.95), we obtain

$$\begin{aligned} (a_{3i-2,1} cos\theta + a_{3i-1,1} sin\theta) U_1 + \cdots \\ + (a_{3i-2,3i-2} cos\theta + a_{3i-1,3i-2} sin\theta) U_{3i-2} \\ + (a_{3i-2,3i-1} cos\theta + a_{3i-1,3i-1} sin\theta) U_{3i-1} + \cdots \\ + (a_{3i-2,N} cos\theta + a_{3i-1,N} sin\theta) U_N = F_{3i-2} cos\theta + F_{3i-1} sin\theta. \end{aligned} \tag{3.99}$$

Equation (3.99) takes the place of equation # $(3i-2)$ in the global assembled system. We use the following algorithm

$$\begin{aligned} a_{3i-2,j}^{new} &= a_{3i-2,j} cos\theta + a_{3i-1,j} sin\theta, \quad j = 1, 2, \cdots, N \\ F_{3i-2}^{new} &= F_{3i-2} cos\theta + F_{3i-1} sin\theta. \end{aligned} \tag{3.100}$$

After the completion of this algorithm, we replace equation # $(3i-1)$ in the global assembled system by Equation (3.93) using the following algorithm

$$\begin{aligned} a_{3i-1,j}^{new} &= 0, \quad j = 1, 2, \cdots, N \\ a_{3i-1,3i-2}^{new} &= -sin\theta \\ a_{3i-1,3i-1}^{new} &= cos\theta \\ F_{3i-1}^{new} &= 0. \end{aligned} \tag{3.101}$$

Example 3.3.

We developed a MATLAB code for plane frames such as the frame shown in Figure 3.19. We want to determine the reactions at the fixed ends and displacements and rotations of all joints. Given properties for all members: Young's modulus $E = 200$ GPa, moment of inertia $I = 1.091 \times 10^{-5}$ m^4, and area $A = 3.59 \times 10^{-3}$ m^2. Coordinates of the 8 global nodes are: (0,0), (0,8), (10,8), (20,8), (20,0), (0,17), (10,25), and (20,13) m.

Solution

We divide the frame into 8 elements using 8 global nodes. The span carrying uniformly distributed load is divided into 2 elements. The connectivity matrix is

$$[CM] = \begin{bmatrix} 1 & 2 \\ 2 & 3 \\ 3 & 4 \\ 5 & 4 \\ 4 & 8 \\ 8 & 7 \\ 6 & 7 \\ 2 & 6 \end{bmatrix}. \qquad (3.102)$$

There are two local nodes per element; either node can be taken as node # 1. The first node of an element is the origin of the local coordinate system $(\bar{x}, \bar{y})$, and the positive $\bar{x}$-axis is from node 1 to node 2. The $\bar{y}$ is obtained by rotating the $\bar{x}$-axis 90 degrees about a z-axis through local node 1. The global coordinates (x, y) of all nodes are given and are used to compute the element length h_e and the angle α_e for each of the elements. Knowing the element's properties: flexural rigidity $(EI)_e$, cross-section area A_e, length h_e, and angle α_e, we compute the element stiffness matrix and assemble it into a global stiffness matrix $[K]$. The forcing vectors due to distributed axial and transverse loadings are zero for all elements except for elements 2 and 3 that carry the transverse load $f_e^t = -1$ kN/m. Before the application of boundary conditions, the global system is

$$[K]\{U\} = \{F\} + \{R\}. \qquad (3.103)$$

This is a system of 24 equations [number of global nodes (8) $\times$ number of degrees of freedom per node (3)]. The boundary conditions at the fixed nodes 1 and 5 specify 6 of the 24 primary variables $\{U\}$. For $i = 1$ and 5, we set $U_{3i-2} = 0$, $U_{3i-1} = 0$, and $U_{3i} = 0$. The remaining 18 primary variables are unknown. The array of global secondary variables (the assembled secondary variables) $\{R\}$ is zero unless there are externally applied concentrated forces and moments. At the fixed nodes 1 and 5 the reactions are unknown. Thus, the six components R_{3i-2}, R_{3i-1}, and R_{3i} for $i = 1$ and 5 are unknown. The remaining 18 components are known as follows. There are no concentrated forces or moments at global nodes $i = 2$, 3, and 4. Hence, $R_{3i-2} = 0$, $R_{3i-1} = 0$, and $R_{3i} = 0$ for these three nodes. At node $i = 6$, we have $R_{3i-2} = 0$,

Table 3.1
Example 3.3. Solution to the Frame Shown in Figure 3.19

Node	u_x	u_y	θ_z	R_x	R_y	M_z
1	0	0	0	2.697	11.98	−6.20
2	14.57	−0.134	−0.964	0	0	0
3	14.56	−269.3	0.034	0	0	0
4	14.54	−0.134	0.827	0	0	0
5	0	0	0	−3.697	12.02	10.85
6	−8.41	−0.159	0.741	0	0	8
7	−10.04	1.809	−0.216	0	−4	0
8	−12.47	−0.1474	0.060	1	0	0

Displacements u_x and u_y in mm, rotation θ_z in degrees, externally applied concentrated forces R_x and R_y in kN, and moments M_z in kN.m.

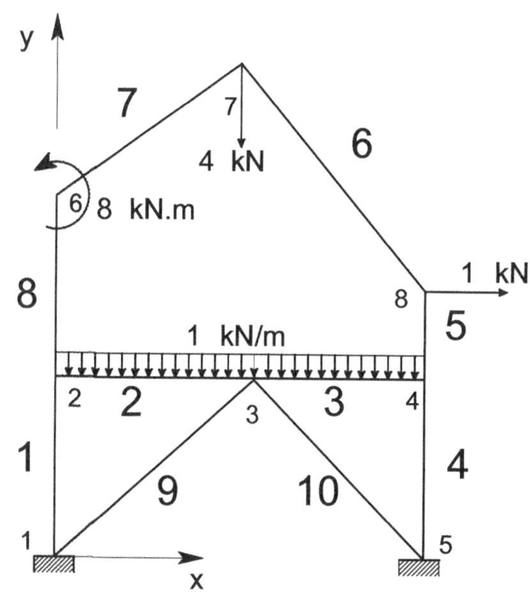

Figure 3.27 A modified frame, 8 global nodes and 10 members.

$R_{3i-1} = 0$, $R_{3i} = 8$ kN.m. At node $i = 7$, we have $R_{3i-2} = 0$, $R_{3i-1} = -4$ kN, and $R_{3i} = 0$. Finally, at node $i = 8$, we have $R_{3i-2} = 1$ kN, $R_{3i-1} = 0$, and $R_{3i} = 0$. The system Equation (3.103) is solved for the unknown 18 primary and 6 secondary variables. The displacements, rotations, and externally applied forces at the global nodes are given in Table 3.1. We note that node 7 is displaced upward by 1.809 mm. This is mainly due to the rotations at nodes 2 and 4. The vertical displacement at node 3 is −269.3 mm; which is too large. Let's add two members with the same properties from node 3 to nodes 1 and 5, and solve the problem again. Obtained results show that the rotations at nodes 2 and 4 are significantly reduced and now node 7 is displaced downward 35.19 mm. The vertical displacement at node 3 becomes −0.2356 mm.

3.5 PLANE TRUSSES

A truss is an idealized structure of straight members joined by pins at their ends. It is assumed that the forces exerted by the pins are collinear with the members. The global nodes of a truss are the joints (or pins) that can be displaced in the plane of the truss. The displacement vector of a node is projected onto global axes (x,y). A truss member is shown in Figure 3.28a; its two local nodes are labeled 1 and 2. At node 1, the global displacement components are denoted by U_1^e and U_2^e, and at node 2 they are denoted by U_3^e and U_4^e. A truss member offers resistance (stiffness) only to changes in length. Therefore, we resolve the displacement vectors at each node into components parallel to the member and normal to it. The parallel components are shown in Figure 3.28b. At node # 1 the parallel component is denoted by $\bar{U}_1$ and at node # 2 it is denoted by $\bar{U}_2$. The normal components are not shown in the figure but they are not necessarily zero. However, the normal components represent rigid body translation and rotation (assumed to be very small) with no change in length. The element equations in local coordinates are

$$\frac{(EA)_e}{h_e}\begin{bmatrix} 1 & -1 \\ -1 & 1 \end{bmatrix}\begin{Bmatrix} \bar{U}_1^e \\ \bar{U}_2^e \end{Bmatrix} = \frac{f_e^a h_e}{2}\begin{Bmatrix} 1 \\ 1 \end{Bmatrix} + (EA)_e\,\alpha_e^t \Delta T_e \begin{Bmatrix} -1 \\ 1 \end{Bmatrix} + \begin{Bmatrix} \bar{R}_1^e \\ \bar{R}_2^e \end{Bmatrix}. \quad (3.104)$$

The second array on the right side is thermal loading where ΔT_e is the change in temperature of the element relative to the state of zero stress, and α_e^t is the coefficient of thermal expansion, see Equation (2.313) of Section (2.7). The bar element equations depend only on the displacement components parallel to the bar. In order to write the equations in terms of global degrees of freedom, we first write those components in terms of the global displacement components. The transformation is

$$\begin{Bmatrix} \bar{U}_1^e \\ \bar{U}_2^e \end{Bmatrix} = \begin{bmatrix} C & S & 0 & 0 \\ 0 & 0 & C & S \end{bmatrix}\begin{Bmatrix} U_1^e \\ U_2^e \\ U_3^e \\ U_4^e \end{Bmatrix} \quad (3.105)$$

where $S = sin\alpha_e$ and $C = cos\alpha_e$. The pins exert only axial forces on the member as shown in Figure 3.28c. We resolve those forces into global coordinates as shown in Figure 3.28d.

$$\begin{Bmatrix} R_1^e \\ R_2^e \\ R_3^e \\ R_4^e \end{Bmatrix} = \begin{bmatrix} C & 0 \\ S & 0 \\ 0 & C \\ 0 & S \end{bmatrix}\begin{Bmatrix} \bar{R}_1^e \\ \bar{R}_2^e \end{Bmatrix}. \quad (3.106)$$

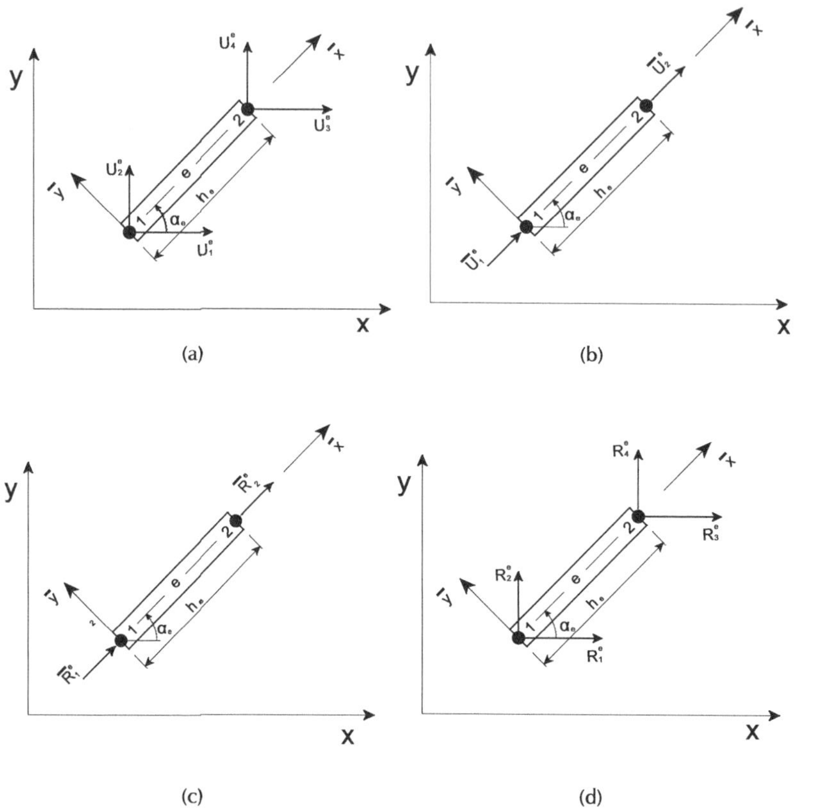

Figure 3.28 Local and global variables for a truss member. (a) Global displacement components, (b) displacement components parallel to the member; normal components are not shown, (c) axial forces exerted by pins on the member, and (d) global components of axial forces exerted by pins on member.

We resolve the forcing vector and thermal loading vector into global components,

$$\left\{\begin{array}{c} F_1^e \\ F_2^e \\ F_3^e \\ F_4^e \end{array}\right\} = \left[\begin{array}{cc} C & 0 \\ S & 0 \\ 0 & C \\ 0 & S \end{array}\right] \left\{\begin{array}{c} \bar{F}_1^e \\ \bar{F}_2^e \end{array}\right\}. \tag{3.107}$$

We define the transformation matrix

$$[T^e] = \left[\begin{array}{cccc} C & S & 0 & 0 \\ 0 & 0 & C & S \end{array}\right] \tag{3.108}$$

and write Equations (3.105) and (3.107) as

$$\{\bar{U}^e\} = [T^e]\{U^e\} \tag{3.109}$$

and

$$\{F^e\} = [T^e]^T \{\bar{F}^e\}. \tag{3.110}$$

Similarly, we have

$$\{R^e\} = [T^e]^T \{\bar{R}^e\} \tag{3.111}$$

Multiplying Equation (3.104) by $[T^e]^T$ and substituting Equation (3.109), we obtain the truss element equation in the global coordinate system,

$$[K^e]\{U^e\} = \{F^e\} + \{F_T^e\} + \{R^e\}. \tag{3.112}$$

The element stiffness matrix is

$$[K^e] = \frac{(EA)_e}{h_e}\begin{bmatrix} C & 0 \\ S & 0 \\ 0 & C \\ 0 & S \end{bmatrix}\begin{bmatrix} 1 & -1 \\ -1 & 1 \end{bmatrix}\begin{bmatrix} C & S & 0 & 0 \\ 0 & 0 & C & S \end{bmatrix} \tag{3.113}$$

the forcing vector is

$$\{F^e\} = \frac{f_e^a h_e}{2}\begin{bmatrix} C & 0 \\ S & 0 \\ 0 & C \\ 0 & S \end{bmatrix}\begin{bmatrix} 1 \\ 1 \end{bmatrix} \tag{3.114}$$

and the thermal loading is

$$\{F_T^e\} = (EA)_e \alpha_e^t \Delta T_e \begin{bmatrix} C & 0 \\ S & 0 \\ 0 & C \\ 0 & S \end{bmatrix}\begin{bmatrix} -1 \\ 1 \end{bmatrix}. \tag{3.115}$$

Knowing the element properties (geometric, mechanical, and physical), we compute the matrices defined by Equations (3.113)–(3.115), and assemble them into global stiffness matrix and forcing vector. The assembled system takes the general form

$$[K]\{U\} = \{F\} + \{R\}. \tag{3.116}$$

Remarks:

(1) For a plane truss, each node has two degrees of freedom (primary variables); two displacement components in the x- and y-directions.

(2) At global node # j, the primary variables are x-displacement U_{2j-1}, and y-displacement U_{2j}. These are shown in Figure 3.29.

(3) Distributed axial force (f^a) is prescribed as force per unit length in the element's local coordinates. f^a is positive along the $\bar{x}$-axis; but it is usually zero.

(4) At a joint that connects more than one element, the sums of secondary variables at that node are equal to the externally applied concentrated forces. Externally

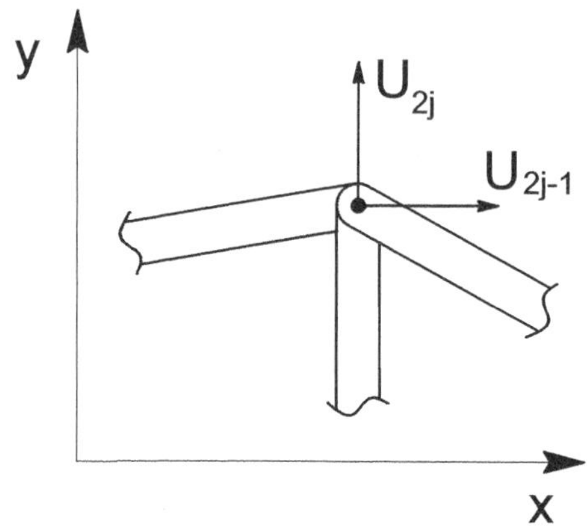

Figure 3.29 Primary variables at node number j of a plane truss, U_{2j-1} is the horizontal displacement, and U_{2j} is the vertical displacement.

applied concentrated forces at the global joints (nodes) are specified by the x and y force components.

For a truss of N joints, we have $2N$ equations. The coefficient matrix is of the dimensions $(2N, 2N)$, and the arrays of primary and secondary variables are of the dimensions $(2N, 1)$. The procedure for applying boundary conditions and solving for the unknown primary and secondary variables is similar to the case of frames, and will be demonstrated by examples. If global joint # i in a truss is supported by a roller on an inclined plane, the relevant conditions can be applied by using Equations (3.93)–(3.101) after we replace the subscripts $3i-2$ and $3i-1$ in those equations by $2i-1$ and $2i$, respectively.

Example 3.4. Statically Indeterminate Truss

The aluminum truss shown in Figure 3.30 is composed of eight members and five global nodes. The coordinates of the fives nodes are (0,0), (1.2,0), (2.4,0), (0,1), and (1.2,1) m. All members have the same cross-sectional area of 625×10^{-6} m^2. For aluminum $E = 70$ GPa, and $\alpha^t = 23.6 \times 10^{-6}$°C^{-1}. The truss is loaded by a vertical force $P_v = 4$ kN at joint 2, and a horizontal force $P_h = 2$ kN at joint 5. We use the finite element method to determine (a) the reactions at rollers 1 and 3 and pin 4, (b) the displacements of all nodes, and (c) the internal forces in the members.

Solution

We developed a MATLAB code for solving plane trusses, and used it for this problem. The connectivity matrix is

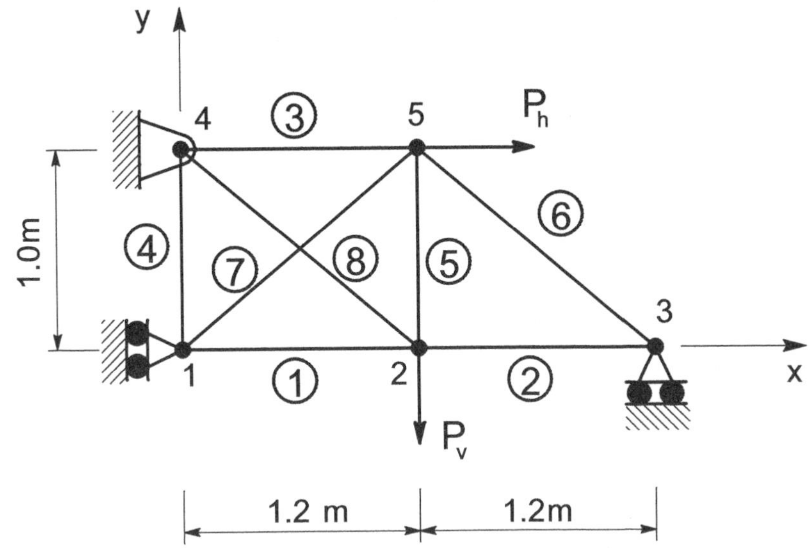

Figure 3.30 Statically indeterminate truss.

$$[CM] = \begin{bmatrix} 1 & 2 \\ 2 & 3 \\ 4 & 5 \\ 1 & 4 \\ 2 & 5 \\ 3 & 5 \\ 1 & 5 \\ 2 & 4 \end{bmatrix}. \tag{3.117}$$

The global system is 10 equations [number of global nodes (5) × number of degrees of freedom per node (2)]. At joint number i, the x- and y-displacements are denoted by U_{2i-1} and U_{2i}, respectively. The boundary conditions at the rollers 1 and 3 give $U_1 = 0$, and $U_6 = 0$. The boundary conditions at the pin joint at node 4 give $U_7 = 0$ and $U_8 = 0$. The array of global secondary variables (the assembled secondary variables) $\{R\}$ is zero unless there are externally applied concentrated forces. At the roller 1 the horizontal reaction R_1 is unknown, and at roller 3 the vertical reaction R_6 is unknown. At pin 4 both components of reaction R_7 and R_8 are unknown. The remaining six components are known as follows. At node 1 the vertical force is $R_2 = 0$. At node 2 the horizontal force is $R_3 = 0$, and the vertical force is $R_4 = -4$ kN. At node 3 the horizontal force is $R_5 = 0$. At node 5 the horizontal force is $R_9 = 2$ kN and the vertical force is $R_{10} = 0$. There is no axially distributed load or thermal loading, hence $\{F\} = 0$.

The assembled global system Equation (3.116) takes the form

$$10^7 [K]\{U\} = 10^3 \{R\} \tag{3.118}$$

where the matrix $[K]$ is

$$
\begin{bmatrix}
5.2988 & 1.3774 & -3.6458 & 0 & 0 & 0 & 0 & 0 & -1.6529 & -1.3774 \\
1.3774 & 5.5229 & 0 & 0 & 0 & 0 & 0 & -4.3750 & -1.3774 & -1.1479 \\
-3.6458 & 0 & 8.9446 & -1.3774 & -3.6458 & 0 & -1.6529 & 1.3774 & 0 & 0 \\
0 & 0 & -1.3774 & 5.5229 & 0 & 0 & 1.3774 & -1.1479 & 0 & -4.3750 \\
0 & 0 & -3.6458 & 0 & 5.2988 & -1.3774 & 0 & 0 & -1.6529 & 1.3774 \\
0 & 0 & 0 & 0 & -1.3774 & 1.1479 & 0 & 0 & 1.3774 & -1.1479 \\
0 & 0 & -1.6529 & 1.3774 & 0 & 0 & 5.2988 & -1.3774 & -3.6458 & 0 \\
0 & -4.3750 & 1.3774 & -1.1479 & 0 & 0 & -1.3774 & 5.5229 & 0 & 0 \\
-1.6529 & -1.3774 & 0 & 0 & -1.6529 & 1.3774 & -3.6458 & 0 & 6.9517 & 0 \\
-1.3774 & -1.1479 & 0 & -4.3750 & 1.3774 & -1.1479 & 0 & 0 & 0 & 6.6707
\end{bmatrix}. \tag{3.119}
$$

Table 3.2
Example 3.4. Solution to the Truss Shown in Figure 3.30

Node	u_x	u_y	R_x	R_y
1	0	-0.0175	1.3940	0
2	-0.0131	-0.1739	0	-4
3	0.0336	0	0	1.4192
4	0	0	-3.3940	2.5808
5	0.0333	-0.1240	2	0

Displacements u_x and u_y in mm, and external concentrated forces R_x and R_y in kN.

and the vectors are

$$\{U\} = (0\, U_2\, U_3\, U_4\, U_5\, 0\, 0\, 0\, U_9\, U_{10})^T$$

$$\{R\} = (R_1\, 0\, 0\, \text{-}4\, 0\, R_6\, R_7\, R_8\, 2\, 0)^T.$$

The system Equation (3.118) is solved for the unknown six primary and four secondary variables. The displacements and externally applied forces and reactions at the global nodes are given in Table 3.2. Now that we know the displacements at all nodes, we can compute the internal forces as follows. For each member (e), we extract the horizontal and vertical displacements at its nodes $(U_1^e, U_2^e, U_3^e, U_4^e)$. We then use Equation (3.105) to compute the local axial displacements $(\bar{U}_1^e, \bar{U}_2^e)$. Finally, we use Equation (3.104) to determine the forces at the member's nodes $(\bar{R}_1^e, \bar{R}_2^e)$. In most applications, we have $f_e^a = 0$, and hence $\bar{R}_1^e = -\bar{R}_2^e$, that is to say the two forces are equal and opposite as we expect for a truss member. The internal forces in the members 1 through 8 are $(-0.4768, 1.7030, 1.2143, 0.7644, 2.1835, -2.2168, -1.1940, 2.8374)$ kN; where a negative sign indicates that the member is under compression.

Example 3.5. Thermoelastic Forces in Trusses

For statically indeterminate trusses, differential heating and cooling of the truss members can induce internal forces and external reactions. We consider the aluminum truss of Example (3.4) shown in Figure 3.30 and remove the externally applied loads at nodes 2 and 5. All members are assembled at the same temperature without initial stresses in the members. Afterward, members 3 and 6 are heated and their temperatures increase by 20°C, while member 1 is cooled and its temperature decreases by 10°C. We wish to determine the nodal displacements and internal forces and external reactions.

Solution

The assembled system is

$$10^7\,[K]\,\{U\} = 10^4\,\{F\} + \{R\} \tag{3.120}$$

Table 3.3
Example 3.5. Thermoelastic Deformation and Forces in the Truss Shown in Figure 3.30

Node	u_x	u_y	R_x	R_y
1	0	0.1113	-17.9059	0
2	0.0477	0.2832	0	0
3	0.2933	0	0	7.4608
4	0	0	17.9059	-7.4608
5	0.1606	0.3425	0	0

Displacements u_x and u_y in mm, and external reactions R_x and R_y in kN.

the matrix $[K]$ is given by Equation (3.119), and the vectors are

$$\{U\} = (0\, U_2\, U_3\, U_4\, U_5\, 0\,0\,0\, U_9\, U_{10})^T$$

$$\{F\} = (1.0325 \;\; 0 \;\; -1.0325 \;\; 0 \;\; 1.5864 \;\; -1.3220 \;\; -2.0650 \;\; 0 \;\; 0.4786 \;\; 1.3220)^T.$$

$$\{R\} = (R_1\, 0\,0\,0\,0\, R_6\, R_7\, R_8\, 0\,0)^T$$

The system Equation (3.120) is solved for the unknown six primary and four secondary variables. The displacements and reactions are given in Table 3.3. The internal forces in the members 1 through 8 are

$$(12.0653, 8.9530, -14.7936, -4.8672, 2.5936, -11.6541, 7.6028, -4.0513)\ \text{kN};$$

where a negative sign indicates that the member is under compression. We conclude that thermal stresses can be significant in statically indeterminate trusses and must be accounted for in the design of such trusses.

3.6 PRINCIPLE OF VIRTUAL DISPLACEMENTS (WORK)

The principle of virtual displacements (or work) is frequently used in solid mechanics to develop finite-element models. Its derivation from the equations of equilibrium follows the same steps as the derivation of the weak form (e.g., [7] and [124]). The principle is mathematically equivalent to the weak form if the virtual displacement field is considered to be the test (weight) function of the weak form. We want to determine the displacement field within a continuum when acted upon by externally applied loads. Cartesian coordinates of a point in the undeformed medium are denoted by x_i and the components of the displacement vector by u_i for $i = 1, 2,$ and 3. The displacements and strains are assumed to be small enough to allow linear relations between strain and displacements. The equations of equilibrium of an infinitesimally small element are

$$\frac{\partial \sigma_{ji}}{\partial x_j} + f_i = 0 \tag{3.121}$$

where σ_{ij} is the stress tensor that exists in the continuum at equilibrium, and f_i is the body force (force per unit volume). In component form, we have three equations of equilibrium,

$$\begin{aligned}
\frac{\partial \sigma_{xx}}{\partial x} + \frac{\partial \sigma_{yx}}{\partial y} + \frac{\partial \sigma_{zx}}{\partial z} + f_x &= 0 \\
\frac{\partial \sigma_{xy}}{\partial x} + \frac{\partial \sigma_{yy}}{\partial y} + \frac{\partial \sigma_{zy}}{\partial z} + f_y &= 0 \\
\frac{\partial \sigma_{xz}}{\partial x} + \frac{\partial \sigma_{yz}}{\partial y} + \frac{\partial \sigma_{zz}}{\partial z} + f_z &= 0.
\end{aligned} \tag{3.122}$$

We want to solve the equations of equilibrium for the displacement field u_i that takes place at equilibrium. The solution must satisfy certain boundary conditions. Let's divide the bounding surface of the body into three contiguous but non-overlapping segments, so the body surface is $S = \Gamma_E + \Gamma_N + \Gamma_M$. Different boundary conditions are prescribed for each of the segments.

$$\begin{aligned}
u_i = \hat{u}_i \quad &: \quad \text{on } \Gamma_E \\
t_i = \hat{t}_i \quad &: \quad \text{on } \Gamma_N \\
t_i + \beta u_i = \gamma_i \quad &: \quad \text{on } \Gamma_M.
\end{aligned} \tag{3.123}$$

At points on Γ_E the displacement vector u_i is specified by the function $\hat{u}_i$, and at points on Γ_N the surface traction (stress vector) t_i is given by the function $\hat{t}_i$. At points on Γ_M a linear relation between the displacement vector u_i and the surface traction (stress vector) t_i is known, where β and γ_i are prescribed functions of position on the surface.

To derive the principle of virtual displacements, we first define a virtual displacement field $\bar{u}_i$ (the usual notation used for virtual displacement is δu_i, but we reserve that notation for variations of the actual displacement field). We recall that we want to determine the actual displacement field u_i that takes place when the continuum is in equilibrium. A virtual displacement field $\bar{u}_i$ is an *arbitrary* field that we imagine to take place in the deformed body. We specify it arbitrarily, and as such it is a known displacement field. There are few conditions that we impose on $\bar{u}_i$. It has to be kinematically consistent with the actual displacement field, so it must be infinitesimally small, continuous, and its gradient (first partial derivatives with respect to space coordinates) exists. One more condition will be specified later. We multiply the equation of equilibrium Equation (3.121) by $\bar{u}_i$, invoke the summation convention, and integrate the resulting equation over the body volume. This is equivalent to multiplying each of the three equations of equilibrium Equation (3.122) by the corresponding virtual displacement component ($\bar{u}$, $\bar{v}$, $\bar{w}$) and summing the three equations.

We obtain

$$\int_\Omega \bar{u}_i \frac{\partial \sigma_{ji}}{\partial x_j} d\Omega + \int_\Omega \bar{u}_i f_i d\Omega = 0 \tag{3.124}$$

We consider the identity

$$\begin{aligned} \bar{u}_i \frac{\partial \sigma_{ji}}{\partial x_j} &= \frac{\partial(\bar{u}_i \sigma_{ji})}{\partial x_j} - \sigma_{ji} \frac{\partial \bar{u}_i}{\partial x_j} \\ &= \frac{\partial(\bar{u}_i \sigma_{ji})}{\partial x_j} - \sigma_{ji} \left[\frac{1}{2}\left(\frac{\partial \bar{u}_i}{\partial x_j} + \frac{\partial \bar{u}_j}{\partial x_i} \right) + \frac{1}{2}\left(\frac{\partial \bar{u}_i}{\partial x_j} - \frac{\partial \bar{u}_j}{\partial x_i} \right) \right] \\ &= \frac{\partial(\bar{u}_i \sigma_{ji})}{\partial x_j} - \sigma_{ji}(\bar{\varepsilon}_{ij} + \bar{\Theta}_{ij}) \end{aligned} \tag{3.125}$$

where $\bar{\varepsilon}_{ij}$ is the virtual linear strain tensor,

$$\bar{\varepsilon}_{ij} = \frac{1}{2}\left(\frac{\partial \bar{u}_i}{\partial x_j} + \frac{\partial \bar{u}_j}{\partial x_i} \right) \tag{3.126}$$

and $\bar{\Theta}_{ij}$ is the virtual rotation tensor,

$$\bar{\Theta}_{ij} = \frac{1}{2}\left(\frac{\partial \bar{u}_i}{\partial x_j} - \frac{\partial \bar{u}_j}{\partial x_i} \right). \tag{3.127}$$

The product $\sigma_{ji}\bar{\Theta}_{ij}$ is identically zero because σ_{ji} is a symmetric tensor and $\bar{\Theta}_{ij}$ is a skew symmetric tensor. Using identity Equation (3.125), we rewrite Equation (3.124) as

$$\int_\Omega \frac{\partial(\bar{u}_i \sigma_{ji})}{\partial x_j} d\Omega - \int_\Omega \sigma_{ji}\bar{\varepsilon}_{ij} d\Omega + \int_\Omega \bar{u}_i f_i d\Omega = 0. \tag{3.128}$$

Using the divergence theorem, we replace the first volume integral by a surface integral

$$\int_\Omega \frac{\partial(\bar{u}_i \sigma_{ji})}{\partial x_j} d\Omega = \oint_S \bar{u}_i \sigma_{ji} n_j dS. \tag{3.129}$$

Writing the stress vector in terms of the stress tensor $t_i = \sigma_{ji} n_j$, and rearranging terms in Equation (3.128), we obtain

$$\int_\Omega \sigma_{ji}\bar{\varepsilon}_{ij} d\Omega = \oint_S \bar{u}_i t_i dS + \int_\Omega \bar{u}_i f_i d\Omega. \tag{3.130}$$

The surface integral on the right side is the boundary term, which we write as integrals over the three segments making up the surface,

$$\int_\Omega \sigma_{ji}\bar{\varepsilon}_{ij} d\Omega = \int_{\Gamma_E} \bar{u}_i t_i dS + \int_{\Gamma_N} \bar{u}_i t_i dS + \int_{\Gamma_M} \bar{u}_i t_i dS + \int_\Omega \bar{u}_i f_i d\Omega. \tag{3.131}$$

On Γ_E, the actual displacement vector u_i is prescribed while the surface traction t_i is unknown. The integral on Γ_E will be zero if we impose the condition $\bar{u}_i = 0$ at points on Γ_E where the actual displacement is known. The surface traction is known on Γ_N, thus the second surface integral can be evaluated. On Γ_M both the actual displacement and traction vectors are unknown. We use the boundary condition to write t_i in terms of u_i, ($t_i = \gamma_i - \beta u_i$) and substitute into Equation (3.131), and obtain

$$\int_\Omega \sigma_{ji}\bar{\varepsilon}_{ij} d\Omega + \int_{\Gamma_M} \beta \bar{u}_i u_i dS = \int_{\Gamma_N} \bar{u}_i \hat{t}_i dS + \int_{\Gamma_M} \bar{u}_i \gamma_i dS + \int_\Omega \bar{u}_i f_i d\Omega. \tag{3.132}$$

Equation (3.132) is the principle of virtual displacements. It states that for a body in equilibrium the virtual strain energy, which is the work done by the actual stresses (assumed constant) through virtual deformation (virtual change in shape and volume), is equal to the virtual work of externally applied surface and body loads. The displacement u_i and stress σ_{ij} fields that exist in a body at equilibrium satisfy this principle for any virtual displacement field $\bar{u}_i$ that we arbitrarily select provided that it is zero on the part of the body surface where the actual displacement is prescribed. The spatial gradient of the selected virtual displacement field must also exist throughout the body. We note that in deriving the principle, the relationship between stresses and strains (constitutive equation) is not specified, and hence the principle is valid for elastic and non-elastic deformation [124].

The principle of virtual displacements Equation (3.132) is the weak form of the equations of equilibrium if we take the virtual displacement $\bar{u}_i$ to be the test (weight) function w_i, which must have three components; one component for each of the three equations of equilibrium. The boundary term is the surface integral in the Equation (3.130). Hence, the primary variable is the displacement field u_i and the secondary variable is the stress vector t_i. The first condition in Equation (3.123) is an essential (or Dirichlet) boundary condition, hence the test function must be zero on Γ_E as it is for the virtual displacement. The second condition on Γ_N is a natural (or Neumann) boundary condition as it specifies the secondary variable, and the third condition on Γ_M is a mixed (or Robin) boundary condition since it gives a relationship between the primary and secondary variables but both of them are unknown.

Equivalent Nodal Forces and Moments

Interpretation of the weak form of the beam element Equation (3.22) as a statement of virtual displacements illuminates the physical meaning of the weak form and provides a clear understanding of the equivalent nodal loads. The left-hand side of Equation (3.22) is the total virtual work of internal bending moment $M = Dd^2v/dx^2$ acting on an element of length dx through the change in slope $(d^2w/dx^2)dx$ of virtual deflection $w(x)$. The virtual work of shear stresses is neglected in the Euler–Bernoulli beam model. The first term on the right is the virtual work of distributed load $f(x)$, and the remaining terms are the virtual work of concentrated forces and moments at the nodes of the finite element.

Understanding the term

$$\int_{x_1^e}^{x_2^e} w(x)f(x)dx$$

as the virtual work of distributed load $f(x)$ through virtual displacement $w(x)$ provides a mean for replacing concentrated forces and moments applied within the element domain by equivalent nodal force vector $\{F^e\}$. First, the force vector $\{F^e\}$

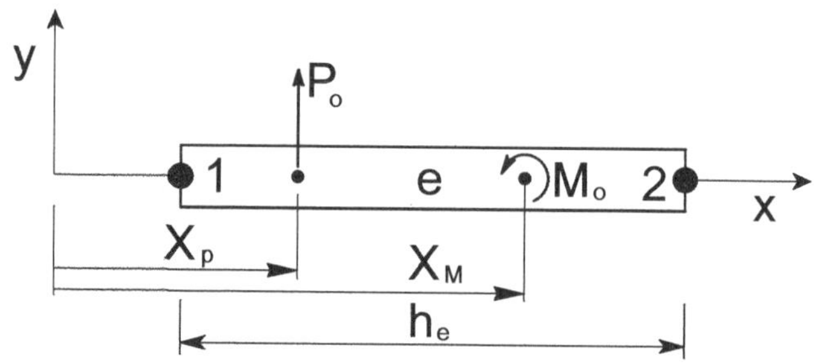

Figure 3.31 A beam element with concentrated force and moment.

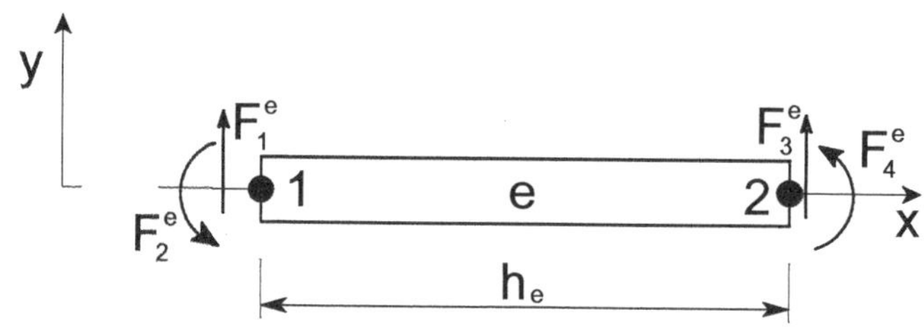

Figure 3.32 Equivalent nodal forces and moments.

given by Equation (3.54) is the equivalent nodal forces and moments for uniformly distributed load. It can be shown that if the beam element is given any virtual displacement $w(x)$ that is a linear combination of the element shape functions, then the work done by the distributed load is equal to the work done by the equivalent force vector (this statement is an interpretation of the equivalent nodal forces; not to be confused with the principle of virtual displacements).

Now, we consider an element subjected to concentrated force P_o at x_P and concentrated moment M_o at x_M as shown in Figure 3.31. We want to find the force vector $\{F^e\}$ that is equivalent to P_o and M_o. The equivalent forces and moments are depicted in Figure 3.32. To that end, we give the element a virtual displacement $w(x) = \psi_1(x)$, and equate the sum of the virtual work of all components of the equivalent force vector $\{F^e\}$ to the sum of the virtual work of P_o and M_o.

$$F_1^e \psi_1(x_1^e) + F_2^e \left(\frac{d\psi_1}{dx}\right)_{x=x_1^e} + F_3^e \psi_1(x_2^e) + F_4^e \left(\frac{d\psi_1}{dx}\right)_{x=x_2^e} = P_o \psi_1(x_P) + M_o \left(\frac{d\psi_1}{dx}\right)_{x=x_M}. \tag{3.133}$$

We note that F_1^e and F_3^e are forces, so they do work through vertical displacement. Whereas F_2^e and F_4^e are moments, so they do work through angular rotation (slope).

Using the properties of the Hermite cubic shape functions, we get

$$F_1^e = P_o \psi_1(x_P) + M_o \left(\frac{d\psi_1}{dx} \right)_{x=x_M} . \tag{3.134}$$

Next, we give the element a virtual displacement $w(x) = \psi_2(x)$, and obtain

$$F_2^e = P_o \psi_2(x_P) + M_o \left(\frac{d\psi_2}{dx} \right)_{x=x_M} . \tag{3.135}$$

Then we use $w(x) = \psi_3(x)$ and $w(x) = \psi_4(x)$, and obtain

$$F_3^e = P_o \psi_3(x_P) + M_o \left(\frac{d\psi_3}{dx} \right)_{x=x_M} \tag{3.136}$$

$$F_4^e = P_o \psi_4(x_P) + M_o \left(\frac{d\psi_4}{dx} \right)_{x=x_M} . \tag{3.137}$$

The nodal force vector that is equivalent to P_o and M_o is

$$\{F^e\} = P_o \left\{ \begin{array}{c} \psi_1 \\ \psi_2 \\ \psi_3 \\ \psi_4 \end{array} \right\}_{x=x_P} + M_o \left\{ \begin{array}{c} d\psi_1/dx \\ d\psi_2/dx \\ d\psi_3/dx \\ d\psi_4/dx \end{array} \right\}_{x=x_M} . \tag{3.138}$$

The nodal force vector F^e given by Equation (3.138) is equivalent to P_o and M_o in the sense that the two systems have the same virtual work if the beam element is subjected to any virtual displacement constructed by a linear superposition of the element shape functions. With the equivalent nodal forces given by Equation (3.138), we can use beam elements with concentrated forces and moments applied anywhere in the element span. However, we recall that the shear force diagram of a beam is discontinuous at the section of concentrated force, and the bending moment diagram is also discontinuous at a concentrated moment. These discontinuities indicate legitimate discontinuities of the third and second derivatives of the deflection curve $v(x)$, respectively. Therefore, the use of the Hermite cubic shape functions, which have continuous second and third derivatives, for an element with concentrated forces or moments will smooth out the discontinuities. It is recommended that beam elements are defined to make concentrated forces and moments act at the common nodes between elements rather than in the span of the element.

3.7 PRINCIPLE OF MINIMUM TOTAL POTENTIAL ENERGY

The principle of minimum total potential energy is usually used for developing finite element models for problems involving elastic deformations. Sokolnikoff [110], Argyris and Kelsey [7], and Washizu [124], among others, derive the principle of minimum total potential energy from the principle of virtual displacements. The total potential energy is the sum of the elastic strain energy and the potential energy of externally applied loads.

Strain Energy

The strain energy density (energy per unit volume), also called strain energy function, for a linear elastic material is given by

$$\begin{aligned} U_o &= \frac{1}{2}\{\sigma\}^T\{\varepsilon\} \\ &= \frac{1}{2}(\sigma_{xx}\varepsilon_{xx}+\sigma_{yy}\varepsilon_{yy}+\sigma_{zz}\varepsilon_{zz}+\sigma_{xy}\gamma_{xy}+\sigma_{xz}\gamma_{xz}+\sigma_{yz}\gamma_{yz}) \end{aligned} \tag{3.139}$$

where

$$\{\varepsilon\} = \begin{Bmatrix} \varepsilon_{xx} \\ \varepsilon_{yy} \\ \varepsilon_{zz} \\ \gamma_{xy} \\ \gamma_{xz} \\ \gamma_{yz} \end{Bmatrix} = \begin{Bmatrix} \partial u/\partial x \\ \partial v/\partial y \\ \partial w/\partial z \\ \partial u/\partial y+\partial v/\partial x \\ \partial u/\partial z+\partial w/\partial x \\ \partial w/\partial y+\partial v/\partial z \end{Bmatrix} \tag{3.140}$$

and

$$\{\sigma\} = \begin{Bmatrix} \sigma_{xx} \\ \sigma_{yy} \\ \sigma_{zz} \\ \sigma_{xy} \\ \sigma_{xz} \\ \sigma_{yz} \end{Bmatrix}. \tag{3.141}$$

The strain energy U of a body of volume Ω is

$$\begin{aligned} U &= \int_\Omega U_o d\Omega \\ &= \int_\Omega \frac{1}{2}\{\sigma\}^T\{\varepsilon\}d\Omega. \end{aligned} \tag{3.142}$$

For a linear elastic isotropic material, the constitutive equation is

$$\sigma_{ij} = \lambda\varepsilon_{kk}\delta_{ij}+2\mu\varepsilon_{ij} \tag{3.143}$$

where ε_{ij} is the linear strain tensor,

$$\varepsilon_{ij} = \frac{1}{2}\left(\frac{\partial u_i}{\partial x_j}+\frac{\partial u_j}{\partial x_i}\right). \tag{3.144}$$

Potential Energy of External Load

The externally applied loads are body force f_i (force per unit volume) and surface traction (stress vector) $\hat{t}_i$ prescribed on part of the body surface Γ_N. The displacement is prescribed on the remainder of the surface Γ_E. The potential energy of the body forces and externally applied surface forces is defined by

$$V = -\left(\int_\Omega f_i u_i d\Omega + \int_{\Gamma_N} \hat{t}_i u_i dS\right) \tag{3.145}$$

Thus, the potential energy is the negative of the work done by the body force and surface traction, assumed to be constant in magnitude and direction, during the deformation of the body from the initial state to the state of displacement u_i.

Total Potential Energy Π

The total potential energy is the sum of the strain energy and potential energy,

$$\begin{aligned}\Pi(u_i) &= U+V \\ &= \int_{\Omega}\frac{1}{2}\{\sigma\}^T\{\varepsilon\}d\Omega-\left(\int_{\Omega}f_iu_id\Omega+\int_{\Gamma_N}\hat{t}_iu_idS\right).\end{aligned} \tag{3.146}$$

A kinematically admissible displacement field u_i is continuous, differentiable, and takes the values prescribed on Γ_E. If we *arbitrarily* choose a kinematically admissible displacement field u_i, then compute ε_{ij} and σ_{ij} from Equations (3.140) and (3.143), respectively, we can compute the total potential energy Π from Equation (3.146) for the given f_i, and $\hat{t}_i$. There are unlimited choices for u_i, and each choice provides us with a stress field within the body σ_{ij} and a value for the total potential energy Π. However, there is no guarantee that any of the computed states of stress satisfies the equations of equilibrium, Equation (3.122).

Statement of the Principle of Minimum Total Potential Energy

The principle asserts that among all kinematically admissible displacement fields, the field that satisfies the equations of equilibrium makes the total potential energy an absolute minimum [124]. Let the actual displacement field at equilibrium be u_i, and assume that u_i^* be another kinematically admissible field in the neighborhood of u_i such that $u_i^* = u_i + \delta u_i$, where δu_i is an arbitrary variation, which of course must be zero on Γ_E. The total potential energy for u_i^* is

$$\begin{aligned}\Pi(u_i^*) &= \Pi(u_i+\delta u_i) \\ &= \Pi(u_i)+\delta\Pi+\text{ higher order variations.}\end{aligned} \tag{3.147}$$

Since $\Pi(u_i)$ is minimum (stationary), it is necessary that the first variation must be zero.

$$\delta\Pi = 0. \tag{3.148}$$

3.7.1 RAYLEIGH–RITZ METHOD

Equation (3.148) is the basis of the Rayleigh–Ritz method for finding approximate solutions and building finite-element models in elasticity and other problems in mechanics. We present two examples.

Example 3.6. Rayleigh–Ritz Solution for a Bar Problem

Uniform bar of length L and cross-sectional area A is subjected to uniformly distributed load $f(x) = f_o$. The bar is fixed at one end and supported by a linear spring

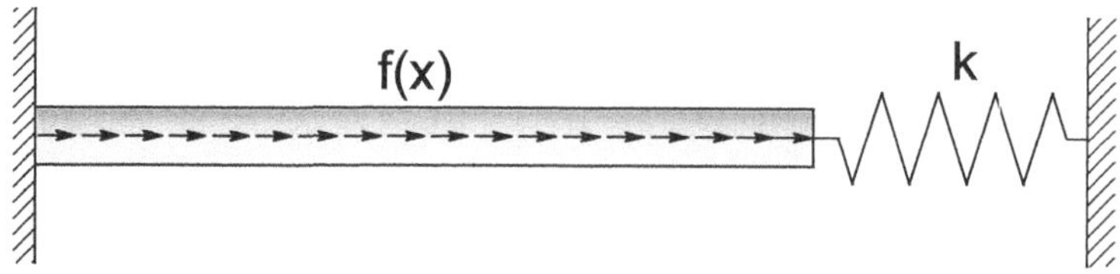

Figure 3.33 Axially loaded bar.

at the other end as shown in Figure 3.33. The spring is undeformed before application of the axial load. We wish to determine the axial displacement $u(x)$ by the Rayleigh–Ritz method.

Solution

The strain energy of the bar is due to normal strain $\varepsilon = du/dx$ and stress $\sigma = E\varepsilon$,

$$\begin{aligned} U(u) &= \int_0^L \frac{1}{2}\sigma\varepsilon A dx \\ &= \int_0^L \frac{1}{2}E\varepsilon^2 A dx \\ &= \int_0^L \frac{1}{2}EA\left(\frac{du}{dx}\right)^2 dx. \end{aligned} \tag{3.149}$$

We take the bar and spring as one assembly, and add the strain energy of the spring $ku^2(L)/2$ to that of the bar. Thus, the strain energy is

$$U(u) = \int_0^L \frac{1}{2}EA\left(\frac{du}{dx}\right)^2 dx + \frac{1}{2}ku^2(L). \tag{3.150}$$

The potential energy of the distributed load is

$$V(u) = -\int_0^L f_o u dx. \tag{3.151}$$

The total potential energy is

$$\Pi(u) = \int_0^L \frac{1}{2}EA\left(\frac{du}{dx}\right)^2 dx + \frac{1}{2}ku^2(L) - \int_0^L f_o u dx. \tag{3.152}$$

The simplest kinematically admissible displacement is $u = C_1 x$ as it satisfies the geometric boundary condition $u(0) = 0$. The total potential energy becomes

$$\Pi(C_1) = \frac{1}{2}EALC_1^2 + \frac{1}{2}kL^2C_1^2 - \frac{1}{2}f_oL^2C_1. \tag{3.153}$$

The necessary condition for minimum Π is $\partial\Pi/\partial C_1 = 0$, which gives $C_1 = f_oL/2(EA+kL)$, and hence the one-term Rayleigh–Ritz approximation is

$$u(x) = \frac{f_oL}{2(EA+kL)}x. \tag{3.154}$$

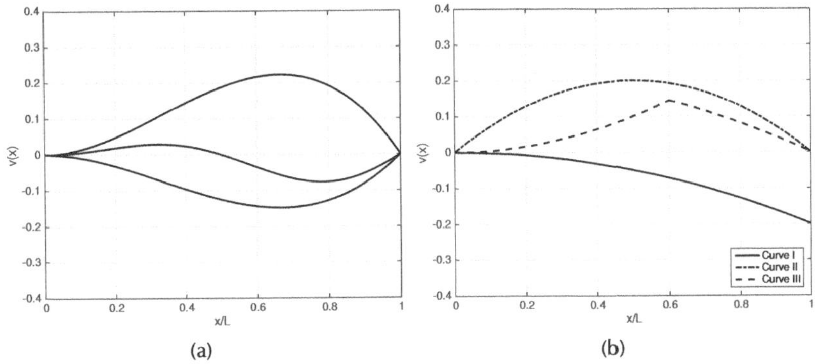

Figure 3.34 Deflection curves $v(x)$ for a beam fixed at $x = 0$ and simply supported at $x = L$. (a) Kinematically admissible curves, deflection and slope are zero at $x = 0$, and deflection is zero at $x = L$. (b) Kinematically inadmissible deflection curves. Curve I violates deflection condition at $x = L$, curve II violates slope condition at $x = 0$, and slope of curve III is discontinuous.

If we try a two-term solution $u = C_1x + C_2x^2$, and determine the coefficients C_1 and C_2 by applying the two necessary conditions $\partial\Pi/\partial C_1 = 0$ and $\partial\Pi/\partial C_2 = 0$, we obtain the exact solution

$$u(x) = \frac{f_o L^2}{2EA}\left[\frac{2EA + kL}{EA + kL}\frac{x}{L} - \frac{x^2}{L^2}\right]. \tag{3.155}$$

Example 3.7. Rayleigh–Ritz Solution for a Beam Problem

Let's consider the beam of Example 3.1 and shown in Figure 3.7. The strain energy of the beam is

$$U(v) = \int_0^L \frac{1}{2}EI\left(\frac{d^2v}{dx^2}\right)^2 dx \tag{3.156}$$

and the potential energy of the distributed load ($f(x) = -f_o$) is

$$V(v) = -\left[\int_0^L (-f_o)v\,dx\right]. \tag{3.157}$$

The total potential energy is

$$\Pi(v) = \int_0^L \frac{1}{2}EI\left(\frac{d^2v}{dx^2}\right)^2 dx + \int_0^L f_o v\,dx. \tag{3.158}$$

The deflection and slope at $x = 0$ are zero, and the deflection at $x = L$ is zero. Admissible deflection curves must satisfy these geometric conditions (essential boundary conditions). Examples of admissible and inadmissible curves are shown in Figures 3.34a and 3.34b, respectively. The trial solution of Example 3.1 satisfies these conditions, and hence we assume

$$v = C_1 x^2 (L - x) \tag{3.159}$$

where C_1 is a constant, we compute Π,

$$\begin{aligned} \Pi(C_1) &= \int_0^L \frac{1}{2} EIC_1^2 (2L - 6x)^2 dx + \int_0^L f_o C_1 x^2 (L - x) dx \\ &= 2EIL^3 C_1^2 - \frac{1}{12} f_o L^4 C_1. \end{aligned} \tag{3.160}$$

The necessary condition for minimum Π is

$$\begin{aligned} \frac{\partial \Pi}{\partial C_1} &= 0, \\ 4EIL^3 C_1 + \frac{1}{12} f_o L^4 &= 0 \end{aligned} \tag{3.161}$$

hence $C_1 = -f_o L/48$, and the approximate solution is

$$v = -\frac{f_o L^4}{48EI} (1 - \xi) \xi^2 \tag{3.162}$$

where $\xi = x/L$, this is the same solution Equation (3.18) we obtained by the principle of virtual displacements in Example 3.1. The exact solution is given by Equation (3.19). Using a one-term Rayleigh–Ritz solution, we find the error in deflection at $x = L/2$ to be 50%. Let's try a two-term solution. The trial solution is

$$v(x) = C_1 \phi_1(x) + C_2 \phi_2(x) \tag{3.163}$$

where

$$\begin{aligned} \phi_1(x) &= x^2 (L - x) \\ \phi_2(x) &= x^3 (L - x) \end{aligned} \tag{3.164}$$

where C_1 and C_2 are constants to be determined. We compute Π

$$\begin{aligned} \Pi(C_1, C_2) = &\int_0^L \frac{1}{2} EI \left(C_1 \frac{d^2 \phi_1}{dx^2} + C_2 \frac{d^2 \phi_2}{dx^2} \right)^2 dx \\ &- \int_0^L (-f_o)(C_1 \phi_1(x) + C_2 \phi_2(x)) dx \\ = &\frac{1}{2} a_{11} C_1^2 + a_{12} C_1 C_2 + \frac{1}{2} a_{22} C_2^2 - b_1 C_1 - b_2 C_2 \end{aligned} \tag{3.165}$$

where

$$\begin{aligned} a_{ij} &= \int_0^L EI \frac{d^2 \phi_i}{dx^2} \frac{d^2 \phi_j}{dx^2} dx \\ b_i &= \int_0^L (-f_o) \phi_i dx. \end{aligned} \tag{3.166}$$

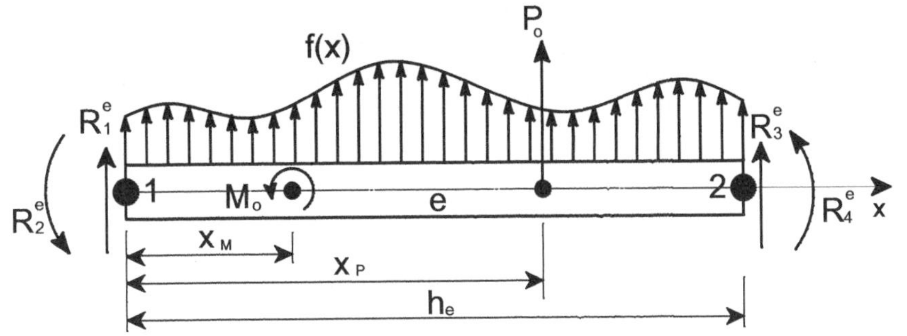

Figure 3.35 Free-body diagram of a beam element, all forces and moments are shown in their positive directions.

The necessary conditions for minimum Π are

$$\begin{aligned}\frac{\partial \Pi}{\partial C_1} &= 0\\ \frac{\partial \Pi}{\partial C_2} &= 0\end{aligned} \tag{3.167}$$

hence we obtain two linear equations that we solve for C_1 and C_2

$$\begin{aligned}a_{11}C_1 + a_{12}C_2 &= b_1\\ a_{12}C_1 + a_{22}C_2 &= b_2.\end{aligned} \tag{3.168}$$

In this example, the two-term Rayleigh–Ritz solution is identical to the exact solution.

Finite-Element Model using Potential Energy Approach

The Rayleigh–Ritz method can be used to derive beam element equations. We consider the free-body diagram of a beam element loaded by distributed load $f(x)$ and concentrated forces and moments as shown in Figure 3.35. The total potential energy of the element is

$$\begin{aligned}\Pi^e(v) &= \int_{x_1^e}^{x_2^e} \frac{1}{2}(EI)_e \left(\frac{d^2v}{dx^2}\right)^2 dx - \int_0^L f v dx - P_o v(x)|_{x=x_P} - M_o \frac{dv}{dx}|_{x=x_M}\\ &\quad -(R_1^e U_1^e + R_2^e U_2^e + R_3^e U_3^e + R_4^e U_4^e)\end{aligned} \tag{3.169}$$

where (U_1^e, U_2^e) and (U_3^e, U_4^e) are deflection and slope at nodes 1 and 2, respectively; they are the primary variables of the beam element as shown in Figure 3.11. And where (R_1^e, R_2^e) and (R_3^e, R_4^e) are shear force and bending moment at nodes 1 and 2, respectively, they are the secondary variables of the beam element as shown in Figure 3.35. The first term is the strain energy, and the remaining terms are the potential energy of externally applied loads. The element is in equilibrium under the action of the shown forces. We write the trial solution in terms of the Hermite cubic polynomials Equation (3.39) so that we can enforce continuity of slope at inter-elements nodes. In matrix form, Equation (3.38) becomes

$$V^e(x) = \{\psi\}\{U^e\} \tag{3.170}$$

and hence

$$\begin{aligned}\frac{d^2V^e(x)}{dx^2} &= \left[\frac{d^2\psi}{dx^2}\right]\{U^e] \\ &= [B^e]\{U^e\}\end{aligned} \tag{3.171}$$

where

$$[B^e] = \left[\frac{d^2\psi_1}{dx^2}\ \frac{d^2\psi_2}{dx^2}\ \frac{d^2\psi_3}{dx^2}\ \frac{d^2\psi_4}{dx^2}\right]. \tag{3.172}$$

We write

$$\begin{aligned}\left(\frac{d^2V^e(x)}{dx^2}\right)^2 &= \frac{d^2V^e(x)}{dx^2}\frac{d^2V^e(x)}{dx^2} \\ &= \{U^e\}^T[B^e]^T[B^e]\{U^e\}.\end{aligned} \tag{3.173}$$

Substituting Equations (3.170) and (3.173) into Equation (3.169), we get

$$\Pi^e(U^e) = \frac{1}{2}\{U^e\}^T[K^e]\{U^e\} - \{U^e\}^T(\{F^e\} + \{R^e\}) \tag{3.174}$$

where the stiffness matrix is

$$[K^e] = \int_{x_1^e}^{x_2^e}[B^e]^T[D^e][B^e]\,dx \tag{3.175}$$

where $[D^e] = [(EI)_e]$. The forcing vector including concentrated force P_o and moment M_o is

$$[F^e] = \int_{x_1^e}^{x_2^e}\{\psi\}^T f(x)dx + P_o\{\psi\}|_{x=x_P} + M_o\left\{\frac{d\psi}{dx}\right\}|_{x=x_M}. \tag{3.176}$$

The necessary conditions for minimum total potential energy for the element are

$$\frac{\partial\Pi^e(U^e)}{\partial U_i^e} = 0 \qquad i = 1,2,3,4. \tag{3.177}$$

We note here that only the elements of $\{U^e\}$ are subjected to variations while $\{F^e\}$ and $\{R^e\}$ are held constant since these are the loads that prevail at equilibrium. For a square matrix $[C]$ and vector x, we define a quadratic function G by

$$G = \frac{1}{2}\{x\}^T[C]\{x\}. \tag{3.178}$$

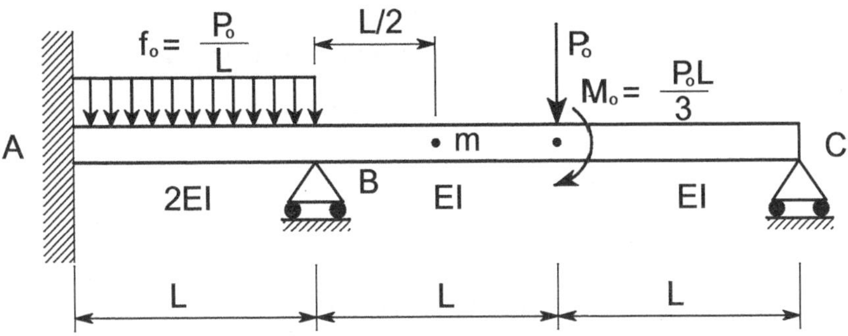

Figure 3.36 Continuous beam.

It can be shown that

$$\frac{\partial G}{\partial x_i} = \frac{1}{2}(c_{ij} + c_{ji})x_j \tag{3.179}$$

and if the matrix $[C]$ is symmetric, we get

$$\frac{\partial G}{\partial x_i} = c_{ij}x_j. \tag{3.180}$$

Noting that the stiffness matrix is symmetric, and applying the four necessary conditions Equation (3.177) for minimum Π given by Equation (3.174), we get the beam element equations

$$[K^e]\{U^e\} = \{F^e\} + \{R^e\}. \tag{3.181}$$

These are the same equations we obtained using the weak form or the principle of virtual displacements.

3.8 PROBLEMS

Beams

(1) We wish to determine an approximate solution to the deflection problem for the beam shown in Figure 3.36. Use Euler–Bernoulli beam theory and a mesh of three (cubic Hermitian) elements, and determine:

(a) The assembled stiffness matrix $[K]$ and force vector $\{F\}$ before applying boundary conditions. Give the answers in terms of P_o, EI, and L.
For $P_o = 2400$ N, $L = 3$ m, and $EI = 20 \times 10^7$ N.m^2, provide:

(b) The reactions at the three supports A, B, and C.

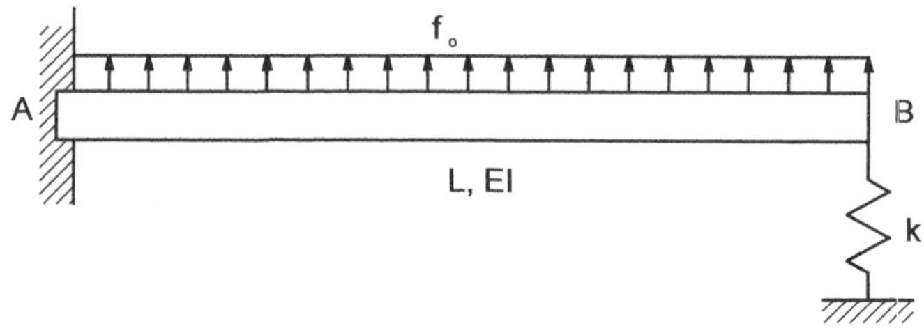

Figure 3.37 Cantilever supported by a linear spring at the tip.

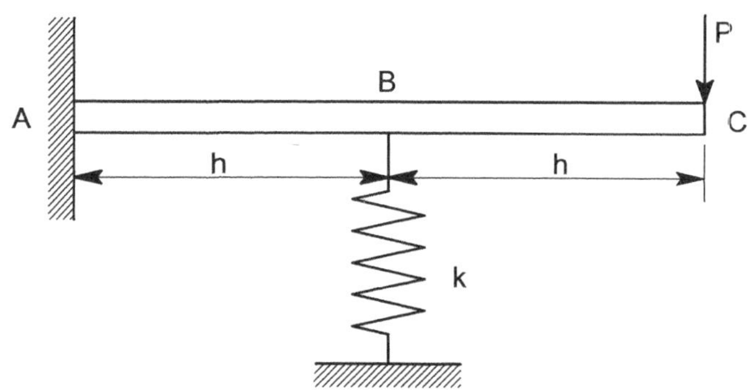

Figure 3.38 Loaded beam with linear spring at mid-span.

   (c) The bending moment diagram over part AB only.

   (d) The deflection and slope at section m.

(2) The beam AB shown in Figure 3.37 (length L and uniform flexural rigidity EI) carries uniformly distributed load f_o. It is fixed at A and attached at B to a linear spring of stiffness $k = EI/L^3$; the spring is undeformed before application of the load. Use Euler–Bernoulli beam theory and a mesh of two (cubic Hermitian) elements, and provide:

   (a) The assembled stiffness matrix $[K]$ and force vector $\{F\}$ before applying boundary conditions.

   (b) The reaction at A.

   (c) The deflection and slope at a distance $3L/4$ from end A.

(3) The beam ABC shown in Figure 3.38 (length $2h$ and flexural rigidity EI) is fixed at A and is attached to a linear spring of stiffness k at midpoint B. A vertical force P is applied to end C. Model the beam by two Euler–Bernoulli elements.

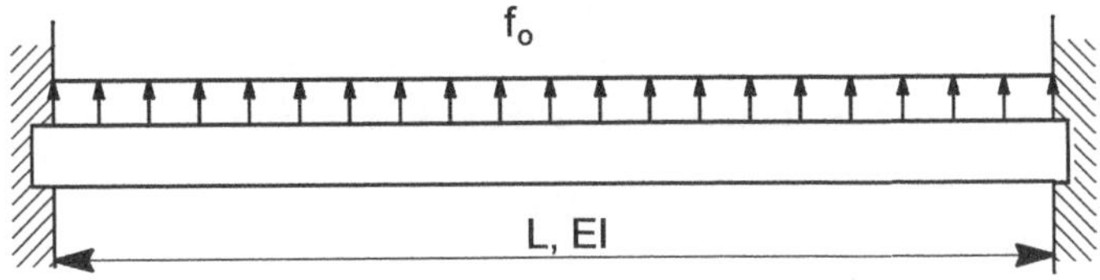

Figure 3.39 Fixed-fixed beam.

(a) Obtain the global (assembled) system of equations before applying the boundary conditions.

(b) Given EI, h, k, and P, apply the boundary conditions and obtain a system of equations for the unknown primary variables.

(c) Take $k = 6EI/h^3$ and determine the reactions.

(4) A high-order Euler–Bernoulli beam element is made of three nodes with two degrees of freedom at each node (deflection and slope) as shown in Figure 3.18. The approximation of deflection over the element is given by Equation (3.71). Derive the expressions of the shape functions Equation (3.72), stiffness matrix Equation (3.73), and forcing vector Equation (3.74).

(5) The beam shown in Figure 3.39 (length L and flexural rigidity EI) carries uniformly distributed load f_o, and is fixed at both ends. Determine the deflection at the midpoint and (reaction) moments at the fixed ends. Show that one high-order Euler–Bernoulli element will give the exact solution. You may also use one (cubic Hermitian) element, and explore the symmetry of the problem. (Hint: symmetry about the mid-section indicates that the shear force in the beam is zero at this section.)

(6) The beam ABC shown in Figure 3.40 (length $2h$ and flexural rigidity EI) is fixed at A and simply supported at C. It is also attached to a linear spring of stiffness k at midpoint B. The beam carries uniformly distributed load f_o. The spring is undeformed before application of the load. Model the beam by two Euler–Bernoulli elements.

(a) Obtain the global (assembled) system of equations before applying the boundary conditions.

(b) Given EI, h, $k = 6EI/h^3$, and f_o, determine the deflection at B.

(7) The beam AB shown in Figure 3.41 (length h and uniform flexural rigidity EI) is fixed at A and simply supported at B. It carries uniformly distributed load f_o. Determine the maximum deflection by (a) using one Hermitian cubic element, and (b) using one Hermitian fifth-order (quintic) element; see Equations (3.71)–(3.74).

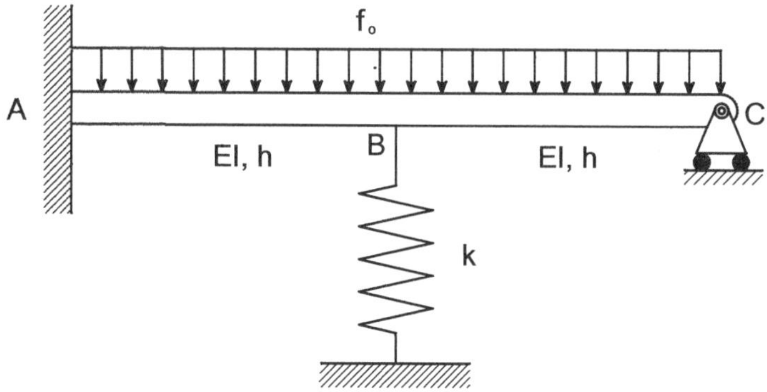

Figure 3.40 Loaded beam with spring at the midpoint.

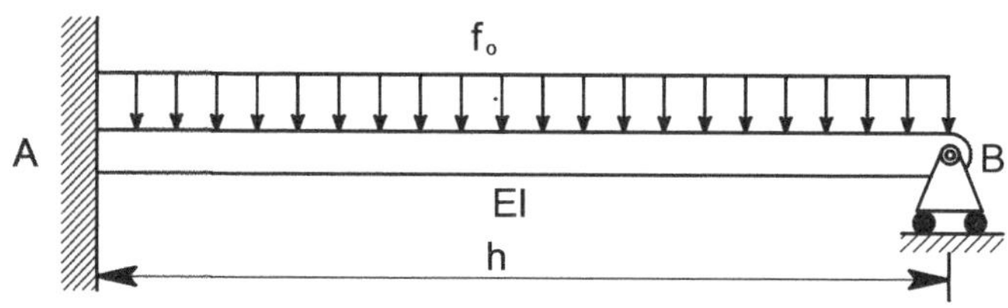

Figure 3.41 Beam fixed at A and simply supported at B.

(8) The beam ABC shown in Figure 3.42 (length $2h$ and flexural rigidity EI) is fixed at A and simply supported at C. It is also attached to a linear spring of stiffness $k = 2EI/h^3$ at the midpoint B. A vertical force P is then applied at point B. Model the beam by two Euler–Bernoulli elements.

(a) Obtain the global (assembled) system of equations before applying the boundary conditions.

(b) Given EI, h, and P, apply the boundary conditions and solve for the unknown primary variables.

(c) Determine the reaction at C.

(9) The continuous beam ABC shown in Figure 3.43 (length $2h$ and uniform flexural rigidity EI) carries uniformly distributed load f_o as shown. It is fixed at A and C, and it is attached to a linear spring of stiffness $k = EI/h^3$ at the midpoint B. The spring is undeformed before application of the load. Use Euler–Bernoulli beam theory and a mesh of two (cubic Hermitian) elements, and determine:

(a) The deflection at the midpoint B.

(b) The reactions at fixed end A.

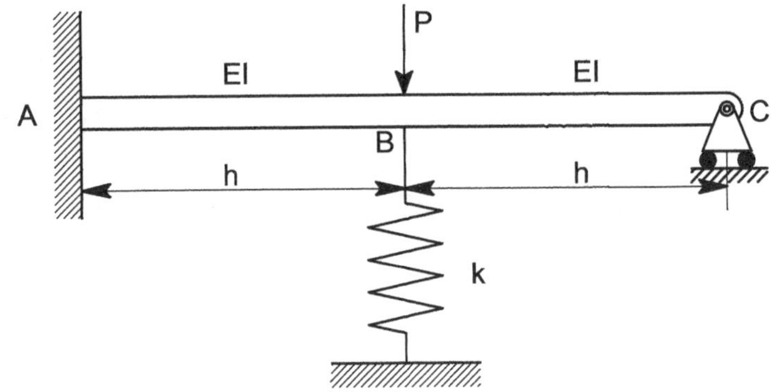

Figure 3.42 Beam fixed at *A* and simply supported at *C*.

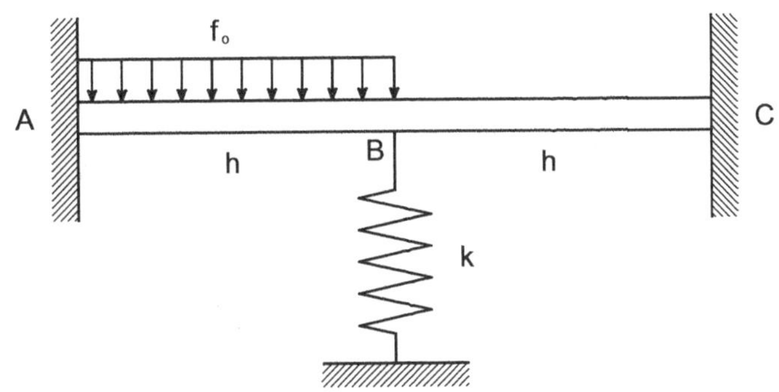

Figure 3.43 Fixed-fixed beam attached to a linear spring at midpoint *B*.

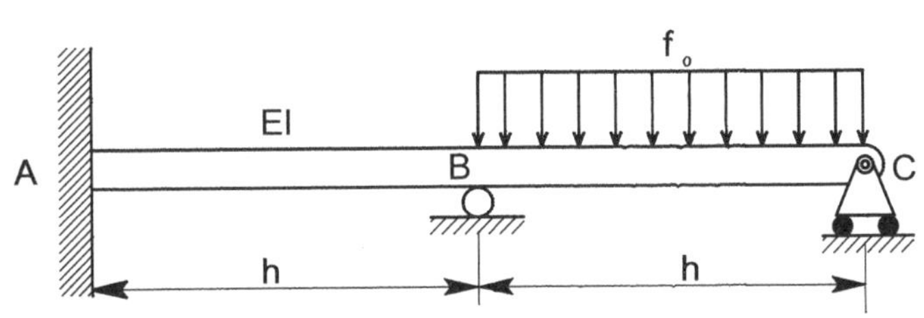

Figure 3.44 Continuous beam.

(10) The continuous beam *ABC* shown in Figure 3.44 (length $2h$ and uniform flexural rigidity EI) carries uniformly distributed load f_o as shown. It is fixed at *A* and simply supported at *C*, and it runs over a roller at the midpoint *B*. Use Euler–Bernoulli beam theory and a mesh of two (cubic Hermitian) elements, and provide:

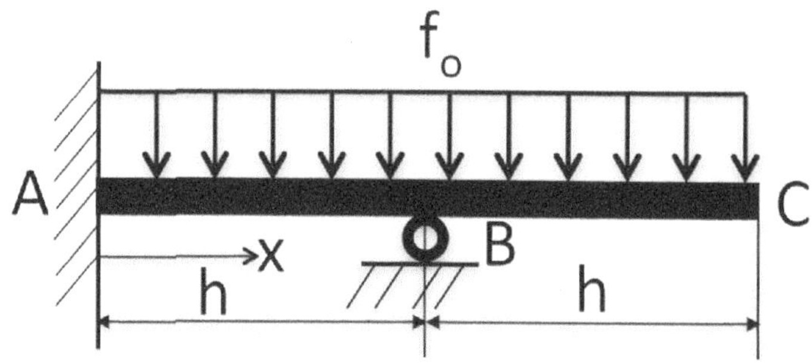

Figure 3.45 Uniform beam, fixed at A and runs over a roller at B.

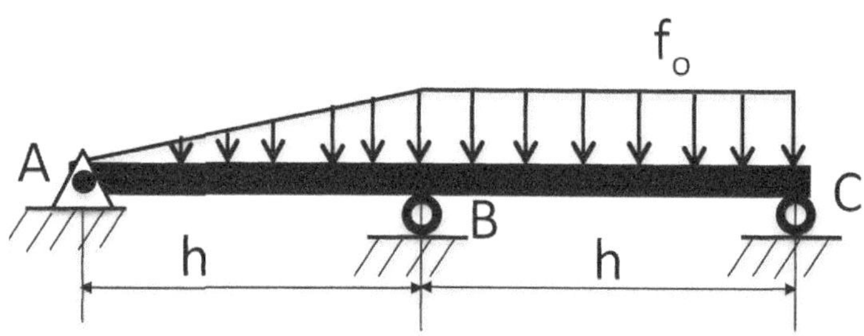

Figure 3.46 Uniform beam, hinged at A and runs over rollers at B and C.

(a) The assembled stiffness matrix $[K]$ and forcing vector $\{F\}$ before applying boundary conditions.

(b) The reaction at the roller B.

(11) The continuous beam ABC shown in Figure 3.45 (length $2h$ and uniform flexural rigidity EI) carries uniformly distributed load f_o as shown. It is fixed at A and runs over a roller at the midpoint B. Using Euler–Bernoulli beam theory and a mesh of two (cubic Hermitian) elements, determine analytic expressions for the deflection $v(x)$.

(12) The continuous beam ABC (length $2h$ and uniform flexural rigidity EI) carries a distributed load as shown in Figure 3.46. It is hinged at A and supported by rollers at the midpoint B and end C. Using Euler–Bernoulli beam theory, determine the reactions at the supports and draw the shear and bending moment diagrams. First, solve the problem on a mesh of four Hermite cubic elements. Second, solve the problem on a mesh of two Hermite quintic elements, and compare with the first solution.

(13) The continuous beam ABC shown in Figure 3.47 (length $2h$ and uniform flexural rigidity EI) carries uniformly distributed load f_o. It is fixed at A and simply

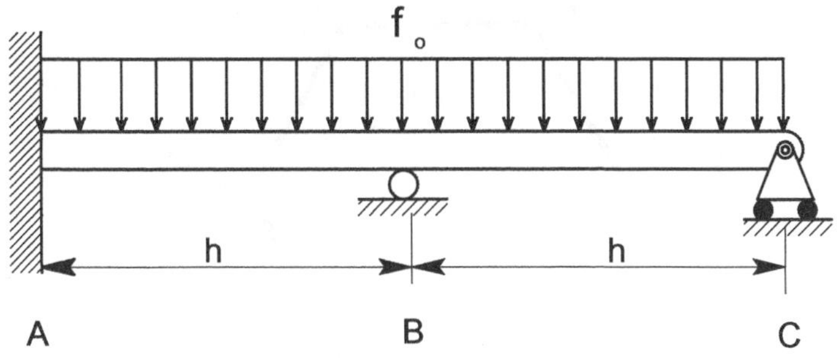

Figure 3.47 Continuous beam with uniformly distributed load.

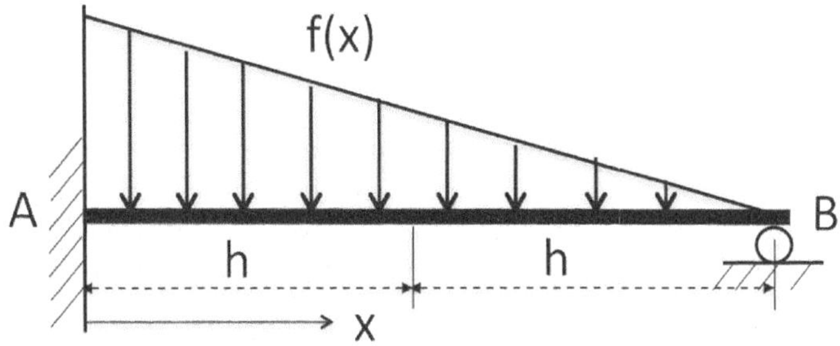

Figure 3.48 Beam with linearly varying distributed load.

supported at C, and it runs over a roller at the midpoint B. Use Euler–Bernoulli beam theory and a mesh of two (cubic Hermitian) elements, and provide:

(a) The assembled stiffness matrix $[K]$ and force vector $\{F\}$ before applying boundary conditions.

(b) The reaction at the roller B.

(c) The deflection at a distance $h/2$ from end C.

(14) The beam AB (length $2h$ and uniform flexural rigidity EI) is fixed at A and supported by a roller at B. It carries a distributed load $f(x) = f_o[1 - x/(2h)]$. Use Euler–Bernoulli beam theory and a mesh of two (cubic Hermitian) elements; each element of length h.

(a) Provide the assembled system before application of boundary conditions.

(b) Find the reactions at A and B.

(c) Draw the bending moment diagram of the beam.

(15) The beam of Example (3.2) was solved using two Hermite cubic elements. Solve this example again by using a cubic element for element # 1 and a 5th-order (quintic) element for element # 2. Note that there will be 4 global nodes. Plot the deflection, shear, and bending moment diagrams of the finite element method and compare with the exact solution as in Figure 3.17.

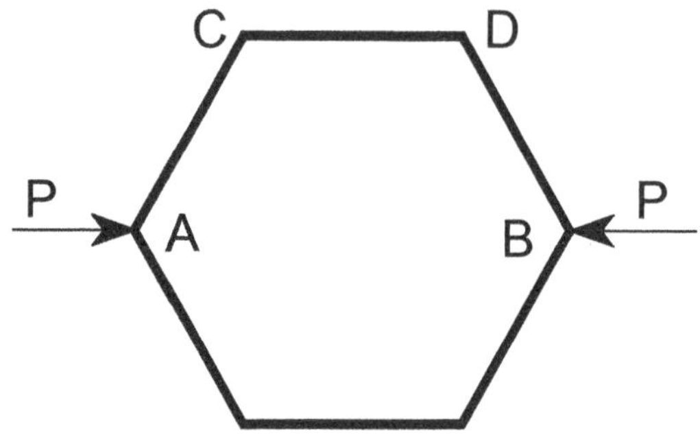

Figure 3.49 Hexagonal frame.

Frames

(16) A hexagonal frame of side h is made of aluminum ($E = 73$ GPa) rods of circular cross-section of diameter d and is acted upon by two equal and opposite forces on the diagonal AB as shown in Figure 3.49. Given $d = 8$ mm, and $h = 300$ mm, determine the force P needed to change the length of diagonal AB by 4 mm, and determine the change of length of side CD.

(17) The frame shown in Figure 3.19 is made of rolled-steel wide-flange section W130×28.1 ($E = 200$ GPa, $A = 3590 \times 10^{-6}$ m^2, $I = 10.91 \times 10^{-6}$ m^4). It is divided into eight elements using eight global nodes. Nodes 1 and 5 are fixed ends. The frame is loaded by a 4-kN vertical force at node 7, a 1-kN horizontal force at node 8, a 8-kN.m moment at node 6, and a 1-kN/m distributed load on span 234. Neglect the weight of the frame, compute and plot the bending moment diagrams for elements 2 and 6. The diagrams should show the bending moment against the local coordinate $\bar{x}$ for each element in the domain $0 \leq \bar{x} \leq h_e$, where h_e is the element length.

(18) Remove all externally applied loads on the frame of Figure 3.19 and determine the reactions at the fixed ends due to the weight of the members (the mass per unit length is 28.1 kg/m). Also determine the displacement at node 3.

(19) Before application of the force P, the vertical gap between the tip D of the Z-frame $ABCD$ and the tip E of cantilever EF is δ, see Figure 3.50. The frame and beam are made of copper ($E = 120$ GPa), and the cross-section is a square with side a. Given: $h = 30$ mm, $\alpha = 30$ degrees, $a = 2$ mm, $\delta = 0.3$ mm, and $P = 30$ N, determine: (a) The displacement of point D, (b) reactions at A and F, and (c) the contact force at D.

(20) The VT-frame shown in Figure 3.51 is made of pure gold, and is fixed at A and hinged at B. The connections C, D, and F are rigid. The frame is loaded by vertical forces V and T as shown in the figure. For all members the length is h and the cross-section is a circle of diameter d. For the numerical values of

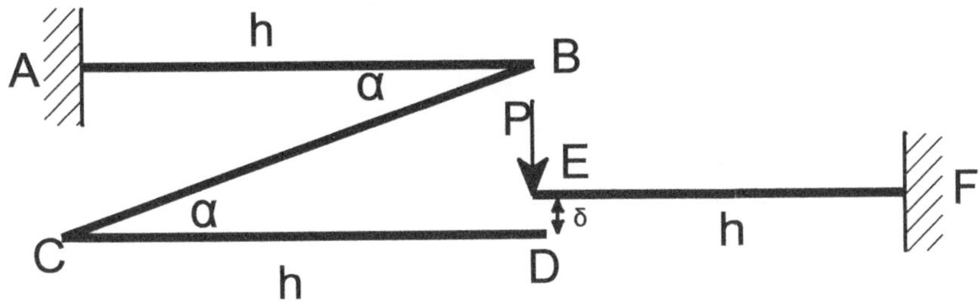

Figure 3.50 Z-frame and cantilever.

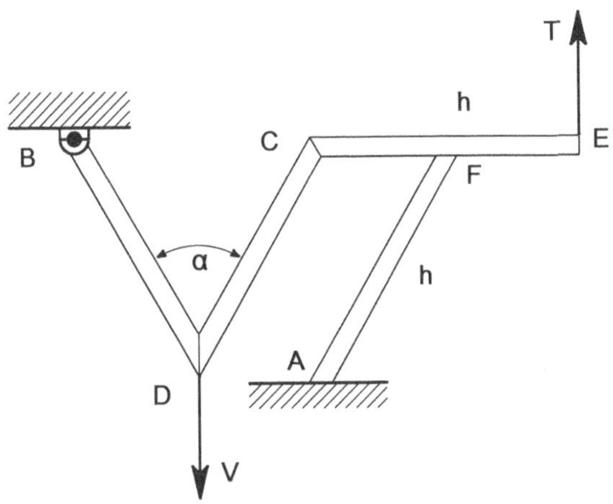

Figure 3.51 *VT*-frame.

$E = 79$ GPa, $h = 600$ mm, $d = 20$ mm, $\alpha = 50$ degrees, $T = 200$ N, and $V = 200$ N, determine: (a) the reactions at A and B, (b) displacements and rotations at the connections, and (c) draw the bending moment diagram of CFE.

(21) The square frame shown in Figure 3.52 with side $a = 500$ mm, thickness $t = 25$ mm, and depth $b = 30$ mm is made of steel ($E = 200$ GPa). The frame is in equilibrium under action of a concentrated horizontal force $P_o = 2$ kN on one side and uniformly distributed load $f_o = 4$ kN/m on the opposite side. A horizontal side may be represented by one frame element, and the vertical side carrying the concentrated force by two elements. Use at least two elements for the side carrying the distributed load. Determine the vertical displacement at the middle section of the horizontal sides. Also, draw the bending moment diagrams of the two vertical sides.

(22) The traffic lights frame is made of steel tapered tubes. The outside diameter of the vertical column varies linearly from d_1 at the base to d_2 at the tip, and its

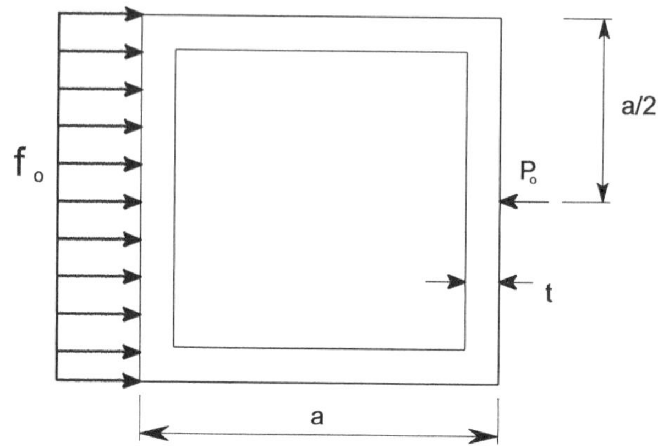

Figure 3.52 Square frame.

thickness, t_v, is uniform. The diameter of the horizontal arm also varies linearly from d_3 at the base to d_4 at its tip, and its thickness, t_h, is uniform. The horizontal arm carries three signals of masses m_1, m_2, and m_3. The finite element method is used to analyze the displacement and rotation of the light signals due to gravity. The weight of the tubular structure is treated as distributed axial or transverse loads, whereas the weights of the three light signals are replaced by concentrated vertical forces. For the data: $v_1 = 5$ m, $v_2 = 2$ m, $h_1 = 5$ m, $h_2 = 1.8$ m, $h_3 = 1.8$ m, $h_4 = 0.4$ m, $d_1 = 400$ mm, $d_2 = 150$ mm, $d_3 = 210$ mm, $d_4 = 70$ mm, $t_v = 25$ mm, $t_h = 10$ mm, $m_1 = 15$ kg, $m_2 = 15$ kg, and $m_3 = 25$ kg:

(a) Determine the displacements and rotations at the light signals.

(b) Draw the bending moment diagram for the horizontal arm.

Properties of steel: $\rho = 7830$ kg/m^3, $E = 200$ GPa.

(23) Horizontal member AB (length h and flexural rigidity EI) is fixed at A and hinged at B to a vertical member BC (length h and axial stiffness EA), which is hinged at C as shown in Figure 3.54. Member AB carries uniformly distributed load f_o. Model AB by a Euler–Bernoulli beam element and BC by a bar element.

(a) Assemble the two elements and provide the global system before application of boundary conditions.

(b) Apply proper boundary conditions and determine the deflection and slope at B.

(c) Calculate the deflection at the midpoint of AB.

(24) In the structure shown in Figure 3.55, the link AB is pin connected at A and B, and the two members BC and CD are rigidly connected at C and pin connected at D. All members have the properties: $E = 72$ GPa, $A = 2.304 \times 10^{-3}$ m^2, and

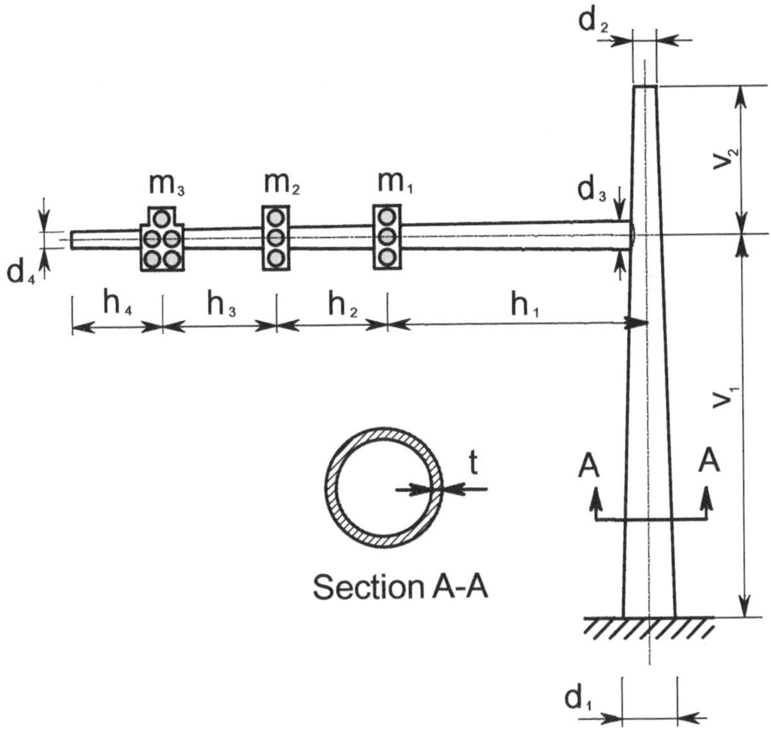

Figure 3.53 Traffic lights frame.

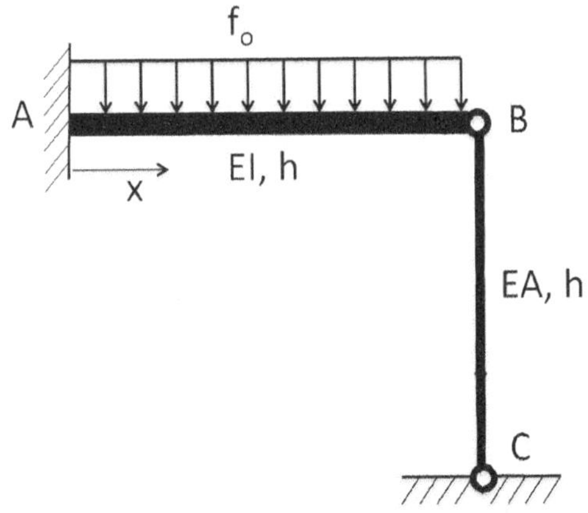

Figure 3.54 Beam *AB*, fixed at *A* and supported by link *BC*.

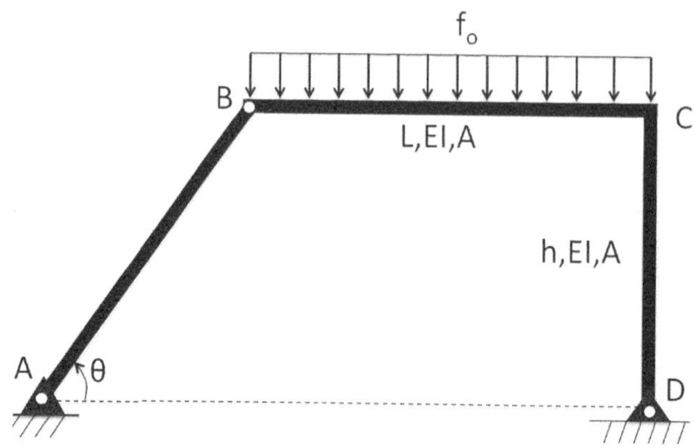

Figure 3.55 Frame *BCD*, hinged at *D* and supported by link *AB*.

$I = 4.42368 \times 10^{-7}$ m^4. Given $L = 1.2$ m, $h = 0.8$ m, $\theta = 60$ degrees, and $f_o = 2$ kN/m. Use the finite element method:

(a) Determine displacement and rotation at *C*.

(b) Determine the maximum deflection (vertical displacement) in *BC*.

(c) Draw the bending moment diagram for *BC*.

Hints

Model *AB* as a truss element, and model *BC* and *CD* as frame elements. Use the following notation:

Horizontal and vertical displacement at *A*: U_1 and U_2.

Horizontal and vertical displacement at *B*: U_3 and U_4, and rotation U_5.

Horizontal and vertical displacement at *C*: U_6 and U_7, and rotation U_8.

Horizontal and vertical displacement at *D*: U_9 and U_{10}, and rotation U_{11}.

Assemble the three elements into a global system, which will be 11 equations, then apply the boundary conditions.

(25) Frame *ABCD* is fixed at *A* and supported by a roller on an incline at *D* as shown in Figure 3.56. Given $L = 2$ m, $E = 200$ GPa, $A = 14.4 \times 10^{-3}$ m^2, $I = 17.28 \times 10^{-6}$ m^4, $P = 2$ kN, $f_o = 2$ kN/m, and $\theta = 40$ degrees.

(a) Determine the displacement (in mm) of the roller.

(b) Draw the shear force and bending moment diagrams of member *BC*.

Trusses

(26) Three members of equal length of 300 mm and cross-sectional area of 100 mm^2 form the truss shown in Figure 3.57. Member 1 is gold $E_g = 79$ GPa, member 2

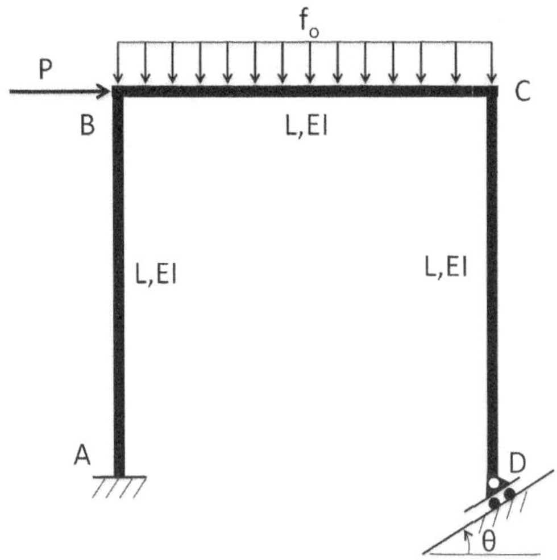

Figure 3.56 Frame, fixed at A and supported by inclined roller.

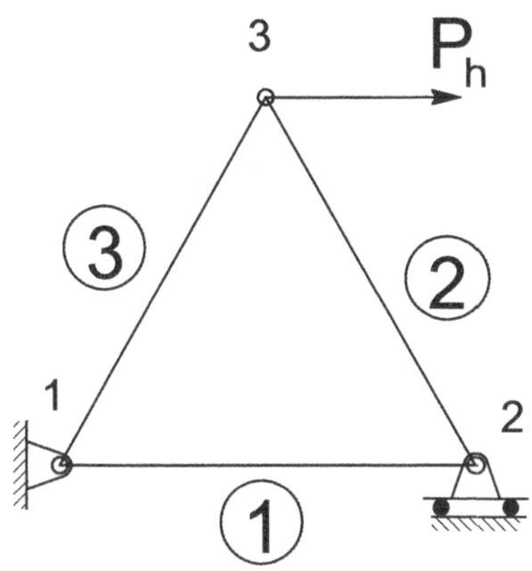

Figure 3.57 Gold-silver-bronze truss.

is silver $E_s = 83$ GPa, and member 3 is bronze $E_b = 105$. The truss is loaded by a horizontal force $P_h = 400$ N. Assemble the element matrices and provide the global stiffness matrix of the truss. Determine the displacements at the joints.

(27) The five-member truss shown in Figure 3.58 is made of titanium ($E = 115$ GPa, $\alpha^t = 9.5 \times 10^{-6} {}^\circ\mathrm{C}^{-1}$) and is loaded by a horizontal force $P_h = 2$ kN. The coordinates (x,y) of the joints 1, 2, 3, and 4 are (0,0), (1.5, 0), (0, 1.5), (1.5,1.5) m, respectively. The cross-sectional area for all members is 400 mm^2. Determine the reactions and internal forces in all members.

(28) We reconsider the five-member truss of Problem (27). The force P_h is removed. All members were assembled at the same temperature with no internal stresses.

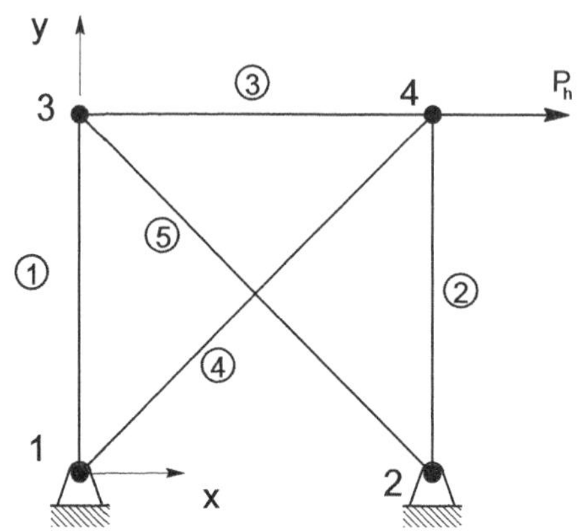

Figure 3.58 Statically indeterminate truss.

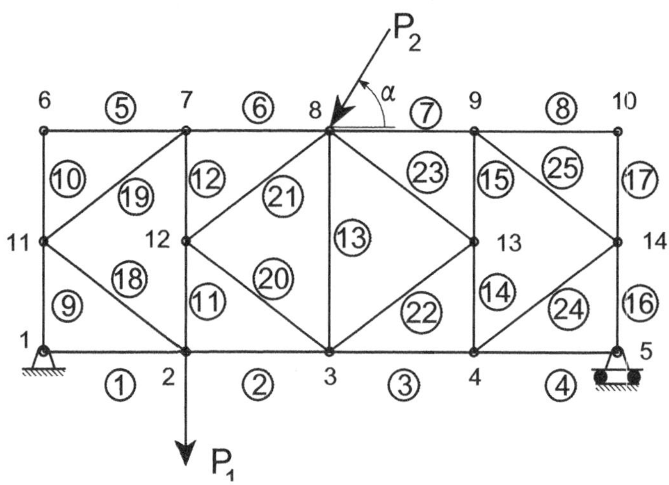

Figure 3.59 K-truss.

The temperature of member 3 drops by 25°C below its initial temperature. Determine the reactions and internal forces in all members.

(29) The steel *K*-truss shown in Figure 3.59 is composed of 25 members and 14 global nodes. The length of horizontal members is 2 m, and that for short vertical members is 1.5 m. All members have the same cross-sectional area of 400×10^{-6} m^2. For steel $E = 200$ GPa. The truss is loaded by a downward vertical force = 4 kN at node 2, and a force = 2 kN at angle $\alpha = 60$ degrees with the horizontal at node 8. Determine the displacements of all nodes and the internal forces in all members.

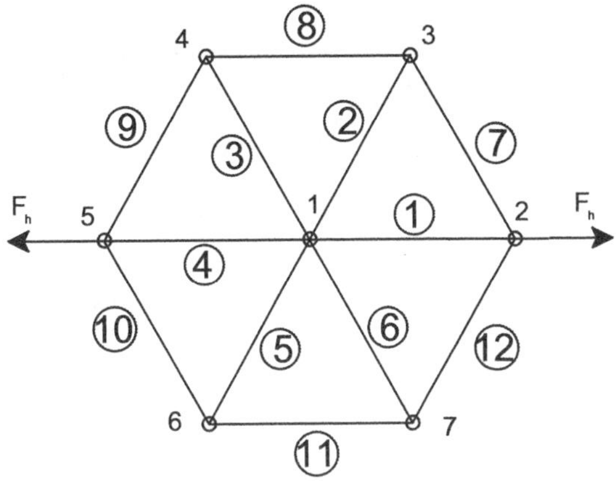

Figure 3.60 Hexagonal-truss, members of equal length of 1 m.

(30) In Problem (29) the structure shown in Figure 3.59 was treated as a truss where all joints are pin connections. In this problem, we want to analyze the same structure shown in Figure 3.59 as a frame where all joints are rigid connections. Use the same data in Problem 3.29, and assume that the members cross-section to be a square.

(a) Determine the displacement of node 8, and compare with the truss answer.

(b) Determine the axial force in members 13 and 21 and compare with the truss answer.

(c) Plot the bending moment diagram for members 13 and 21.

(31) Twelve aluminum bars are pin-connected to form a hexagonal truss as shown in Figure 3.60. The bar length is 1 m, and the cross-sectional area is 625×10^{-6} m^2. For aluminum $E = 70$ GPa, and $\alpha^t = 23.6 \times 10^{-6}{}^\circ\mathrm{C}^{-1}$. The truss is in equilibrium under the action of two equal and opposite forces at joints 2 and 5. The force magnitude is $F_h = 2$ kN. Determine the internal forces in all members of the truss.

(32) Six members are pin-connected to form the truss shown in Figure 3.61. It is supported by a roller on an incline at joint 1, and hinged at joint 2. Given $E = 70$ GPa, and $A = 25 \times 10^{-6}$ m^2 for all members, $P = 3$ kN, and $\theta = 30$ degrees.

(a) Determine the displacement (in mm) of the roller.

(b) Determine the internal forces (in kN) in all members.

(33) Remove the externally applied forces acting on the hexagonal truss of Problem (31). Let's assume that the temperature of member 8 is increased by 20°C. Determine the internal forces in all members of the truss.

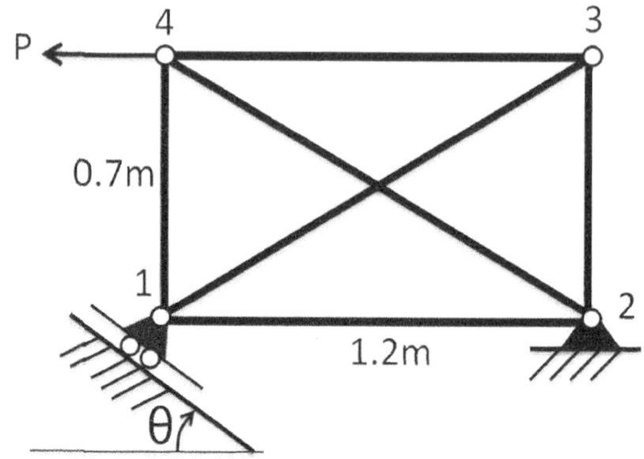

Figure 3.61 Truss supported by inclined roller and hinge.

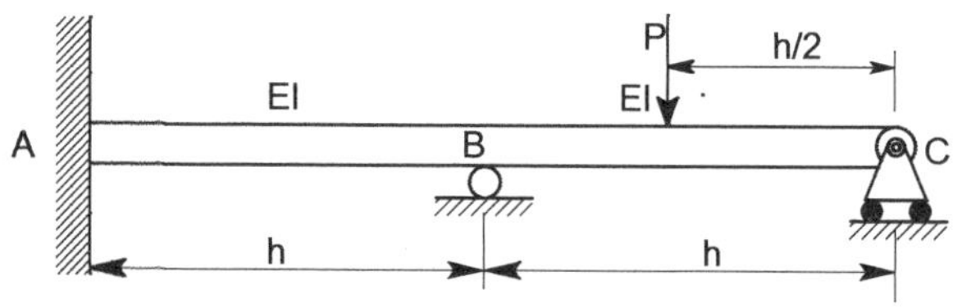

Figure 3.62 A continuous beam loaded by a concentrated force.

Virtual Displacements and Minimum Total Potential Energy

(34) The continuous beam *ABC* shown in Figure 3.62 is fixed at *A*, runs over a roller at *B*, and is simply supported at *C*. The beam is loaded by a concentrated force *P* at the midpoint of *BC*. The beam is discretized into two Euler–Bernoulli elements; one element for *AB* and another for *BC*. Determine the deflection at the midpoint of element *BC*.

(35) Cantilever beam *AB* (length h, flexural rigidity EI) is loaded by two concentrated forces at $x = h/2$ and $x = h$ as shown in Figure (3.63). We want to determine the beam deflection $v(x)$.

   (a) Use one Hermite cubic element, and determine the deflection at *B*.

   (b) Write down an expression for the total potential energy of the beam. Use the principle of minimum total potential energy to determine the deflection at *B*.

   (c) Compare the answers in (a) and (b) and explain any discrepancy, if any.

(36) The continuous beam *ABC* (length $2h$, flexural rigidity EI) is simply supported at *A*, runs over a roller at the midpoint *B*, and is fixed at *C* as shown in Figure (3.64). It is loaded by a concentrated force P_o at the midpoint of *AB* and a

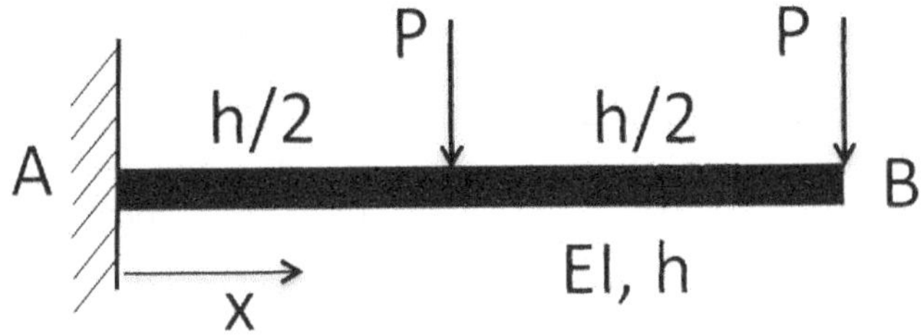

Figure 3.63 Cantilever *AB* loaded by two concentrated forces.

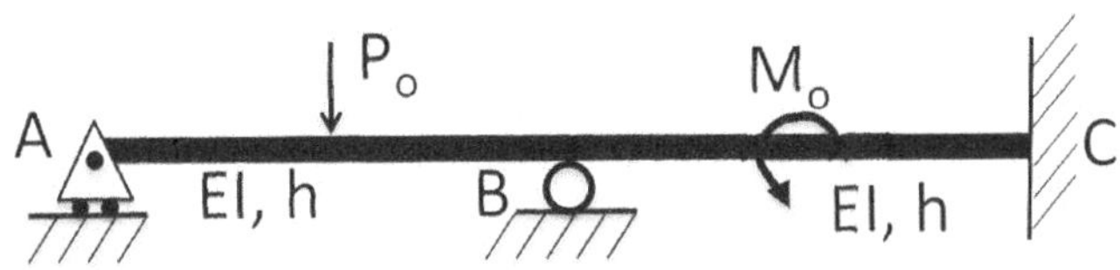

Figure 3.64 Continuous beam *ABC* loaded by concentrated force and moment.

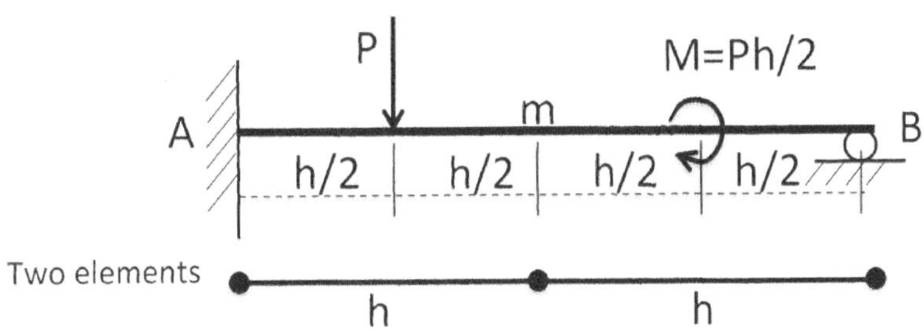

Figure 3.65 Beam of length $2h$ and flexural rigidity EI.

concentrated moment $M_o = hP_o/2$ at the midpoint of BC. Model the beam by two Euler–Bernoulli elements. Determine the reaction at B, and slope at the midpoint of BC.

(37) The uniform beam AB shown in Figure 3.65 is fixed at A and simply supported at B. Its length is $2h$ and flexural rigidity EI. The beam is loaded by a concentrated force P and moment $M = Ph/2$ as shown in the figure. Use the Euler–Bernoulli beam theory and the finite element method, determine the deflection at the midsection m and the reactions at A and B. Solve the problem by using two elements only, each of length h.

(38) Use two elements only to model the beam given in Problem (1) and find the reactions at the supports. Treat AB as one element and BC as a second element. Compare reactions to those obtained by using three elements in Problem (1).

(39) The beam ABC shown in Figure 3.66 (uniform EI, and length $2h$) is fixed at A and C. Section AB carries uniformly distributed load f_o, while a concentrated

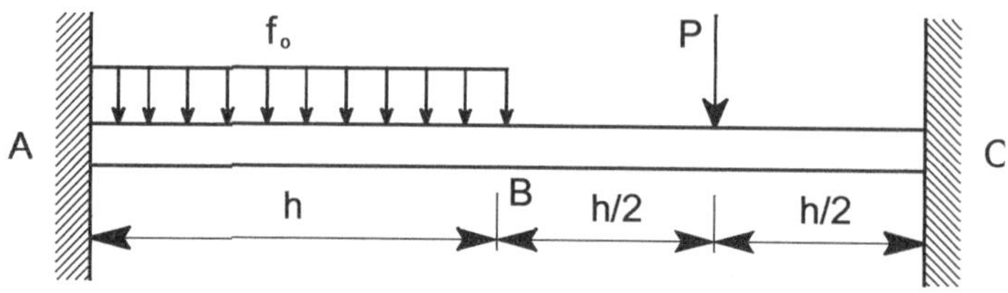

Figure 3.66 Fixed-fixed beam loaded by distributed load and a concentrated force.

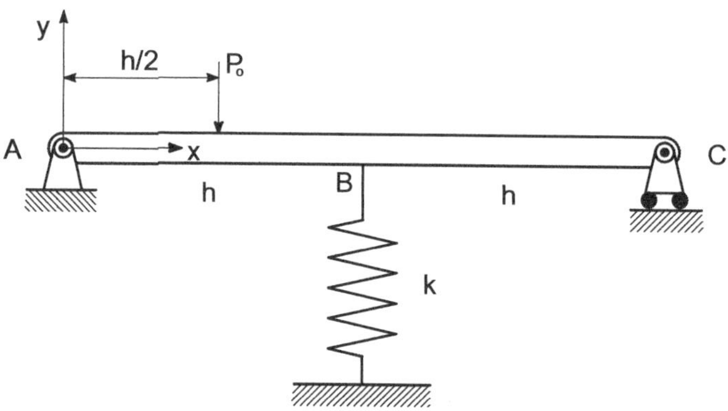

Figure 3.67 Simply supported beam.

force $P = 2f_oh$ acts at the midpoint of section BC. Given h, f_o, and EI, use the finite element method to find the reactions at A. Use one element for AB and one element only for BC.

(40) A simply supported beam AB shown in Figure 3.67 (length $2h$ and uniform flexural rigidity EI) is loaded by a concentrated force at $x = h/2$ and is restrained in the vertical direction by a linear spring of stiffness k. We wish to determine the deflection at the midpoint by the principle of minimum total potential energy.

(a) Write down an expression for the total potential energy of the system.

(b) Consider the following list of functions:

(i) $v(x) = Csin(\pi x/2h)$.

(ii) $v(x) = Ccos(\pi x/2h)$.

(iii) $v(x) = Cx(x - 2h)$.

(iv) $v(x) = Cx^2(x - h)$.
Identify all functions that are kinematically admissible deflection curves.

(c) Use one of the kinematically admissible deflection curves, determine the constant C and the deflection at the midpoint of the beam.

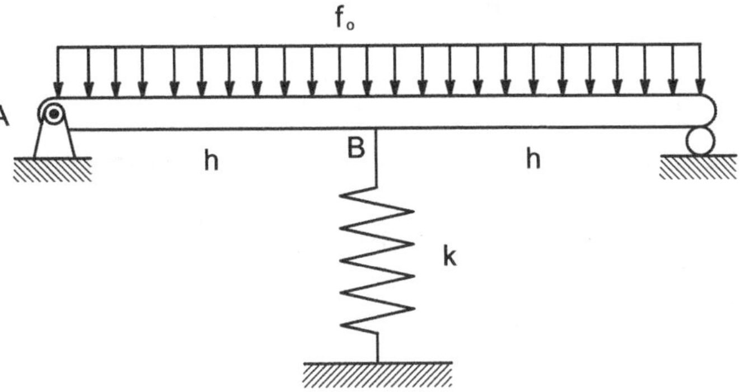

Figure 3.68 Simply supported beam AC is attached to a spring at midpoint B.

(41) A simply supported beam shown in Figure 3.68 (flexural rigidity EI and length $L = 2h$) is attached at mid-span ($x = 0.5L$) to a linear elastic spring of stiffness k. The spring is not deformed before application of the uniformly distributed load f_o.

(a) Write down an expression for the total potential energy, Π.

(b) If we assume the deflection to be

$$v(x) = C(A + Bx + x^2)$$

determine the constants A and B for the assumed deflection to be kinematically admissible.

(c) Use the principle of minimum total potential energy to determine the constant C.

(42) Write down an expression for the total potential energy of the structure given in Problem (1).

(43) Write down an expression for the total potential energy of the beam given in Problem (2). Assume a kinematically admissible deflection curve and use the Rayleigh–Ritz method to determine the deflection at end B. Compare your answer to the finite-element solution and explain the difference, if any.

(44) Consider the structure shown in Figure 3.54. We want to determine the displacement and slope at end B of beam AB by using the principle of minimum total potential energy.

(a) Treat AB as Euler–Bernoulli beam in which the shear strain energy is neglected and treat BC as a bar in axial deformation. Write down an expression for the total potential energy of the structure composed of the beam and bar.

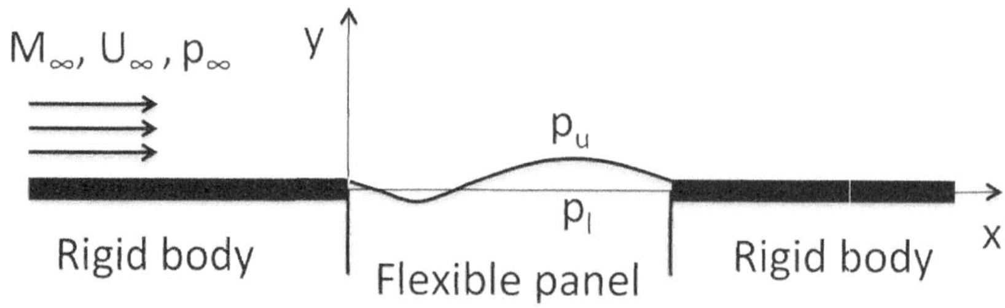

Figure 3.69 One-dimensional flexible panel.

(b) The transverse displacement of AB is assumed to be
$v = c_o + c_1 x + c_2 x^2 + c_3 x^3$
What are the restrictions on the coefficients c_o, c_1, c_2, and c_3?
Apply the principle and calculate the displacement and slope at B.

(45) **Fluid-Structure Interaction Problem: Prelude to Panel Flutter**
This problem is introduced here as an example of static fluid-structure interaction problem related to panel flutter. Thin elastic panels are usually used on surfaces of supersonic airplanes and rockets. Because of the high dynamic pressure ($\rho_\infty U_\infty^2/2$) on one side of the panel (see Figure 3.69), a small transverse disturbance in the panel position may trigger sustained vibrations of the panel known as panel flutter, which can lead to fatigue and eventual failure of the panel. Panel divergence (buckling) and flutter can also happen at subsonic speeds and incompressible flow. Such panels may be used on hulls of light fast ships, glass windows/curtains on high rise buildings, casings of submersible vehicles, and wings/bodies of unmanned and micro aerial vehicles (UAVs and MAVs). Panel flutter will be studied in Chapters 9 and 10. Deformation of an elastic structure due to fluid pressure is a complex problem with applications in aerospace, mechanical, civil, and biomedical engineering. We need to recognize that the fluid load is a function of the structure deformation, which in turn is a function of the applied fluid load. Thus, the two problems of finding the fluid load and the structural deformation are coupled and must be solved simultaneously. As a simple model of the problem, we consider a thin elastic panel surrounded by a rigid surface. The pressure on the panel lower surface is uniform p_ℓ and the pressure on the upper surface p_u is the static pressure of the fluid flow. The incoming flow far upstream of the deflected panel is uniform: velocity U_∞, pressure p_∞, density ρ_∞, and Mach number M_∞. The flexible plate properties are: elastic modulus E, Poisson's ratio ν, thickness h, and streamwise length L. The span (length in the z direction) is infinite, and flow pressure is assumed to be independent of z. The equation of static equilibrium of a small element of the plate is

$$\frac{d^2}{dx^2}\left(D\frac{d^2 v}{dx^2}\right) - f(x) = 0 \tag{3.182}$$

where f is the distributed load, which is the pressure difference $f(x) = p_\ell - p_u$, and D is the flexural rigidity of the plate

$$D = \frac{Eh^3}{12(1-\nu^2)} \tag{3.183}$$

Note that Equation (3.182) is the Euler–Bernoulli's beam equation if we take the beam cross-section to be a rectangle of height h and unit width (moment of inertia $I = bh^3/12$). Thus, the structural model is the Euler–Bernoulli beam theory (or the classical plate theory as given in Chapter 9). The pressure on the upper side of the plate is unknown as it depends on the shape of the deflected plate. The simplest fluid model is an inviscid potential flow model, which we use here. Furthermore, we consider the case of supersonic flow $M_\infty > 1$. Assuming small disturbances, we learn in aerodynamics that the pressure coefficient for linearized supersonic flow is

$$C_p(x) = \frac{2\theta(x)}{\sqrt{M_\infty^2 - 1}} \tag{3.184}$$

where $\theta = dv/dx$ is the local slope of the deflected panel. Thus, the pressure on the upper surface is

$$\begin{aligned} p_u(x) &= \frac{\rho_\infty U_\infty^2}{\sqrt{M_\infty^2 - 1}} \frac{dv}{dx} + p_\infty \\ &= \frac{\gamma p_\infty M_\infty^2}{\sqrt{M_\infty^2 - 1}} \frac{dv}{dx} + p_\infty \end{aligned} \tag{3.185}$$

where γ is the specific heat ratio.

(a) Introduce non-dimensional variables

$$x^* = \frac{x}{L}, \qquad v^* = \frac{v}{h}, \qquad f_o^* = \frac{(p_\ell - p_\infty)L^4}{hD} \tag{3.186}$$

and show that Equation (3.182) becomes

$$\frac{d^4 v^*}{dx^{*4}} + \Lambda \frac{dv^*}{dx} - f_o^* = 0 \tag{3.187}$$

where

$$\Lambda = \frac{M_\infty^2}{\sqrt{M_\infty^2 - 1}} \frac{\gamma p_\infty L^3}{D} \tag{3.188}$$

For simply supported panels, the boundary conditions are

$$\begin{aligned} x^* &= 0: \quad v^* = 0 \quad \text{and} \quad \frac{d^2 v^*}{dx^{*2}} = 0 \\ x^* &= 1: \quad v^* = 0 \quad \text{and} \quad \frac{d^2 v^*}{dx^{*2}} = 0 \end{aligned} \tag{3.189}$$

(b) Derive the weak form of Equation (3.187) for an element.

(c) For a two-node cubic Hermitian element (length h_e), use the Galerkin method to derive the element equations in the form:

$$\left[[K^e]+\Lambda\,[H_2^e]\right]\{U^e\}=\{F^e\}+\{R^e\} \tag{3.190}$$

and provide analytic expressions for the matrices $[K^e]$ and $[H_2^e]$. These matrices are functions of the element length only.

(d) Now, we consider a numerical example. Take $\gamma=1.4$, $M_\infty=2$, $p_\infty=10^5$ Pa, $p_\ell-p_\infty=300$ Pa, $h=3$ mm, $L=1$ m, $E=70$ GPa, and $\nu=0.35$. Model the panel by two elements, and plot the deflection v (mm) and pressure difference $p_\ell-p_u$ (Pa) as a function of x (m).

(e) Determine the exact analytic solution to Equation (3.187) with boundary conditions Equation (3.189), and compare finite element solution to the exact solution.

$$v^*=B_1exp(r_1x^*)+B_2exp(r_2x^*)+B_3exp(r_3x^*)+B_4+\frac{f_o^*}{\Lambda}x^* \tag{3.191}$$

where

$$\begin{aligned}
r_1 &= \left(\frac{1}{2}+i\frac{\sqrt{3}}{2}\right)\Lambda^{1/3}\\
r_2 &= \left(\frac{1}{2}-i\frac{\sqrt{3}}{2}\right)\Lambda^{1/3}\\
r_3 &= -\Lambda^{1/3}
\end{aligned} \tag{3.192}$$

Determine the coefficients $(B_1,\cdots B_4)$ by applying the boundary conditions.

4 Elliptic Equations: Equilibrium in Two Dimensions

In this chapter, we formulate the finite-element method to solve for a function u that depends on two space dimensions (x,y). The equation for the unknown function $u(x,y)$ is an elliptic second-order linear partial differential equation. Such a problem arises in many fields of engineering and science including torsion of prismatic bars of non-circular sections, steady fully developed laminar flow in ducts, steady heat conduction in solids, incompressible potential flow around airfoils, seepage flow, electrostatics, etc.

The problem domain is an area in a two-dimensional space that could be $x-y$ Cartesian plane, $r-\theta$ or $r-z$ plane in cylindrical coordinates. We first consider problems with simple boundaries such that the domain can be easily divided into triangles and/or rectangles whose sides are parallel to the coordinate axes. Furthermore, certain conditions are imposed on the problem parameters so that the element matrices can be evaluated analytically. We start with very coarse meshes so that assembly of the element matrices, applications of boundary conditions, and solving for the unknown primary and secondary variables become amenable to hand calculations or with the help of a calculator. We then develop MATLAB codes for mesh refinement and error assessment. We introduce isoparametric elements and solve problems over domains with curved boundaries and complex problem parameters. We develop MATLAB codes for such problems including mesh generation, numerical evaluation of element matrices, assembly of elements, applications of boundary conditions, solving for primary variables, and post-processing to obtain quantities of interest.

4.1 MODEL PROBLEM

The differential equation considered here is of the general form

$$-\left[\frac{\partial}{\partial x}\left(p_{11}\frac{\partial u}{\partial x}+p_{12}\frac{\partial u}{\partial y}\right)+\frac{\partial}{\partial y}\left(p_{21}\frac{\partial u}{\partial x}+p_{22}\frac{\partial u}{\partial y}\right)\right]+qu-f=0 \quad \text{on} \quad \Omega \tag{4.1}$$

where p_{ij}, q, and f are given functions of x and y. The matrix

$$[D]=\begin{bmatrix} p_{11} & p_{12} \\ p_{21} & p_{22} \end{bmatrix} \tag{4.2}$$

DOI: 10.1201/9781003323150-4

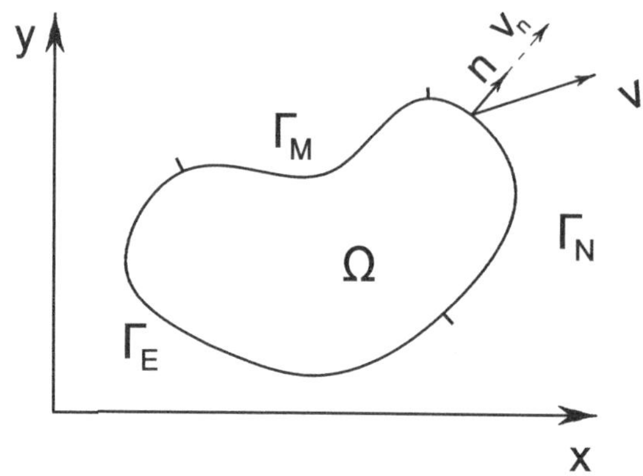

Figure 4.1 Domain of an elliptic problem; the domain is the area bounded by the closed curve, and **n** is outward unit normal. The domain boundary is divided into three non-overlapping segments where different boundary conditions are applied on each segment.

is assumed to be symmetric ($p_{ij} = p_{ji}$) positive definite (all eigenvalues are positive). These conditions ensure ellipticity of the problem. We define a vector $\vec{v} = v_x\hat{i} + v_y\hat{j}$ by

$$\begin{aligned} v_x &= p_{11}\frac{\partial u}{\partial x} + p_{12}\frac{\partial u}{\partial y} \\ v_y &= p_{21}\frac{\partial u}{\partial x} + p_{22}\frac{\partial u}{\partial y} \end{aligned} \tag{4.3}$$

and rewrite the differential equation as

$$-\left[\frac{\partial v_x}{\partial x} + \frac{\partial v_y}{\partial y}\right] + qu - f = 0 \qquad \text{on} \quad \Omega \tag{4.4}$$

or in vector form

$$-\nabla\cdot\vec{v} + qu - f = 0 \qquad \text{on} \quad \Omega. \tag{4.5}$$

The problem domain is sketched in Figure 4.1. On the domain boundary, the outward unit normal is denoted by $\hat{n}$. The component of the vector $\vec{v}$ normal to the boundary is $v_n = \hat{n}.\vec{v}$. To determine a unique solution $u(x,y)$ to Equation (4.1) boundary conditions must be specified on the entire boundary of the domain Ω. A variety of boundary conditions can be enforced on the solution. We consider a case where the domain boundary is divided into three non-overlapping segments $\partial\Omega = \Gamma_E \sqcup \Gamma_N \sqcup \Gamma_M$, and apply the boundary conditions

$$\begin{aligned} u &= \hat{u} & on\ \Gamma_E \quad &\text{(Dirichlet)} \\ v_n &= \hat{v}_n & on\ \Gamma_N \quad &\text{(Neumann)} \\ v_n + \lambda u &= \alpha & on\ \Gamma_M \quad &\text{(Robin)} \end{aligned} \tag{4.6}$$

where $\hat{u}$, $\hat{v}_n$, λ, and α are known functions of position on the respective boundary segment. At points of segment Γ_E, the function u must equal a given distribution $\hat{u}$; this is called a Dirichlet boundary condition. At points of segment Γ_N, the normal component of the vector $\vec{v}$ must equal a prescribed distribution $\hat{v}_n$; this is known as Neumann boundary condition. At points of segment Γ_M, both the function u and the normal component v_n of the vector $\vec{v}$ are unknown, but v_n is related to u as shown by the third line of Equation (4.6); this is called a Robin-type boundary condition. Depending on the physical problem at hand, only one or two of the three types of boundary conditions suffice to specify the conditions on the whole boundary.

4.2 WEAK FORM

To derive the weak form of Equation (4.1), we multiply Equation (4.5) by a test (weight) function $w(x,y)$ that we select, and integrate over the problem domain Ω.

$$\int_\Omega w[-\nabla\cdot\vec{v}+qu-f]\,d\Omega=0. \tag{4.7}$$

Using the vector identity

$$-w\nabla\cdot\vec{v}=-\nabla\cdot(w\vec{v})+\vec{v}\cdot\nabla w \tag{4.8}$$

we rewrite Equation (4.7) as

$$\int_\Omega -\nabla\cdot(w\vec{v})d\Omega+\int_\Omega(\vec{v}\cdot\nabla w+wqu)\,d\Omega-\int_\Omega wf d\Omega=0. \tag{4.9}$$

Using the divergence theorem

$$\int_\Omega \nabla\cdot\vec{B}d\Omega=\oint_{\partial\Omega}\vec{B}\cdot\vec{n}dS \tag{4.10}$$

and letting $\vec{B}=w\vec{v}$, we replace the first volume integral in Equation (4.9) by a boundary integral and obtain

$$\int_\Omega(\nabla w\cdot\vec{v}+wqu)\,d\Omega=\int_\Omega wf d\Omega+\oint_{\partial\Omega}wv_n ds \tag{4.11}$$

where $v_n=\hat{n}\cdot\vec{v}$ is the projection of the vector $\vec{v}$ onto the direction of the outward unit normal $\hat{n}$. Let $\hat{n}=n_x\hat{i}+n_y\hat{j}$, we find

$$\begin{aligned} v_n &= n_xv_x+n_yv_y \\ &= n_x\left(p_{11}\frac{\partial u}{\partial x}+p_{12}\frac{\partial u}{\partial y}\right)+n_y\left(p_{21}\frac{\partial u}{\partial x}+p_{22}\frac{\partial u}{\partial y}\right). \end{aligned} \tag{4.12}$$

Looking at the boundary term (boundary integral) in Equation (4.11), we define u as the primary variable and v_n as the secondary variable.

Boundary conditions are classified as essential boundary condition (EBC) (also called Dirichlet) when the primary variable $u=\hat{u}$ is specified, and natural boundary condition (NBC) (also called Neumann) when the secondary variable $v_n = \hat{v}_n$ is specified. We can also specify a relationship between the primary variable u and secondary variable v_n on the boundary, in this case we call it a mixed boundary condition (MBC) (also called Robin). We construct the test (weight) function w so that it is zero at the points of an EBC. Therefore, the boundary term vanishes over the EBC part of the boundary, and becomes a known integral over the NBC part. In the case of a mixed boundary condition, both of the primary and secondary variables are unknown. At an MBC, we use the boundary condition to express the secondary variable in terms of the primary variable and then substitute the result into the boundary term in the weak form. Thus, only the unknown primary variable appears in the weak form.

We consider a case where the domain boundary is divided into three non-overlapping segments $\partial\Omega = \Gamma_E \sqcup \Gamma_N \sqcup \Gamma_M$, and apply the boundary conditions

$$\begin{aligned} u &= \hat{u} && on\ \Gamma_E \quad (EBC) \\ v_n &= \hat{v}_n && on\ \Gamma_N \quad (NBC) \\ v_n + \lambda u &= \alpha && on\ \Gamma_M \quad (MBC) \end{aligned} \tag{4.13}$$

where $\hat{u}$, $\hat{v}_n$, λ, and α are known functions of position on the respective boundary segment.

$$\int_\Omega (\nabla w \cdot \vec{v} + wqu)\, d\Omega = \int_\Omega wf d\Omega + \int_{\Gamma_E} wv_n ds + \int_{\Gamma_N} wv_n ds + \int_{\Gamma_M} wv_n ds. \tag{4.14}$$

The boundary integral is divided into three integrals over the three segments. Since the primary variable u is prescribed on Γ_E, we construct the test function such that $w = 0$ on that part of the boundary, and hence the corresponding integral is zero. On the segment Γ_N, the secondary variable is specified and hence the integral can be evaluated because v_n and w are known functions. On the third segment Γ_M, we have a mixed boundary condition where both the secondary variable v_n and primary variable u are unknown. On this boundary, we express the secondary variable in terms of the primary variable $v_n = -\lambda u + \alpha$, and substitute it into the integral on Γ_M. This integral is still unknown.

$$\int_\Omega (\nabla w \cdot \vec{v} + wqu)\, d\Omega + \int_{\Gamma_M} w\lambda u ds = \int_\Omega wf d\Omega + \int_{\Gamma_N} w\hat{v}_n ds + \int_{\Gamma_M} w\alpha ds. \tag{4.15}$$

This is the weak form of Equation (4.1) with boundary conditions Equation (4.13). A solution to the weak form is a function $u(x,y)$ that satisfies Equation (4.15) for all test (weight) functions $w(x,y)$ that have first derivatives as indicated by the gradient ∇w and is zero on Γ_E. The function $u(x,y)$ must also possess first derivatives (as they appear in the vector $\vec{v}$) and satisfy the essential boundary conditions ($u = \hat{u}$ on Γ_E). Such a solution is called a weak solution to the strong statement given by the differential Equation (4.1) and boundary conditions Equation (4.13).

The finite element method facilitates the finding of an approximate solution ($u(x,y) \approx U(x,y)$) to the weak form by dividing the problem domain Ω into non-overlapping small elements (**Mesh generation step**). Each element is defined by a finite number of nodes, and the collection of the nodes of all elements define a global set of nodes. We establish an algebraic system of equations that we solve for the unknown values of the function $U(x,y)$ at the global nodes. To that end, we first use the weak form over the typical element domain to derive algebraic equations for the element (**Derive element equation step**). Derivation of the element equations requires expressing $U(x,y)$ over the typical element domain in terms of the values of U at the element nodes. Therefore, we have to derive interpolation functions (or element shape functions) in two space dimensions (**Element shape functions step**). No boundary conditions are specified for the typical element, and as such the element equations contain both unknown primary and secondary variables. We then assemble all the element equations into a global system to which we apply the boundary conditions and balance of (weighted) secondary variables at interior nodes (**Assemble elements into a global system and apply boundary condition step**). We solve the global system for the unknown primary variables (values of $U(x,y)$ at the global nodes), and determine unknown (weighted) secondary variables if any. Knowing the values of U at the global nodes and the element shape functions, we have an analytic expression for $U(x,y)$. It is important to note that the finite-element method does not only give values of U at a discrete set of nodes, but it gives an analytic expression for U that can be evaluated at any point (x,y) or used to compute any quantities of interest (**Post-processing step.**)

4.3 MESH GENERATION AND CONNECTIVITY MATRIX

Dividing the problem domain into elements is an important step as the geometric properties of the element (severely distorted elements) can impair convergence to the correct solutions. An example of a mesh of triangular elements is shown in Figures 4.2a and 4.2b for a semi-circle. The mesh consists of 286 elements and 169 global nodes. The global nodes are numbered 1 through 169 in no particular order. Each element is described by three local nodes numbered 1, 2, and 3 in the counterclockwise sense. Any of the three nodes can be numbered 1. Examples of a mesh of quadrilateral elements are shown in Figures 4.3a and 4.3b. In Figure 4.3a an L-shaped domain is divided into 14 rectangular elements using 24 global nodes. Each element is defined by four local nodes. Starting at any of the four nodes of the element, we number them 1, 2, 3, and 4 in the counterclockwise direction. Figure 4.3b shows a square domain with 1/4 circle removed from the lower left corner. The domain is divided into 100 quadrilateral elements using 121 global nodes. The elements and global nodes can be numbered in any order.

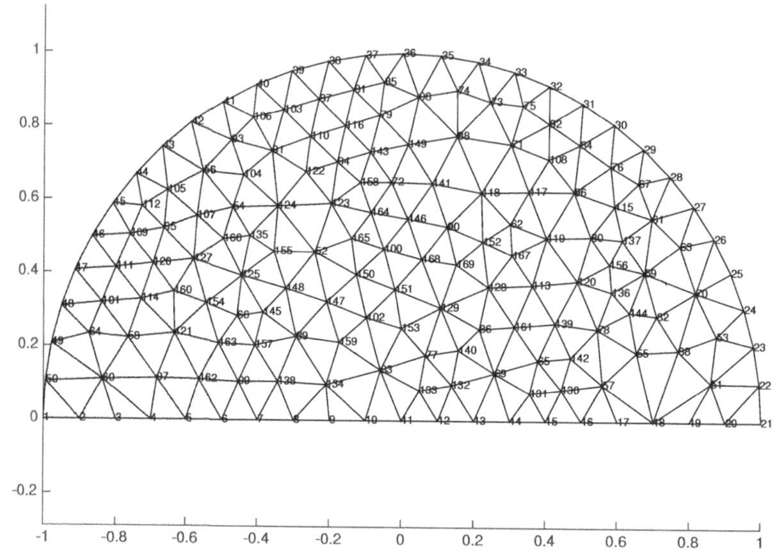

(a) Global node numbers

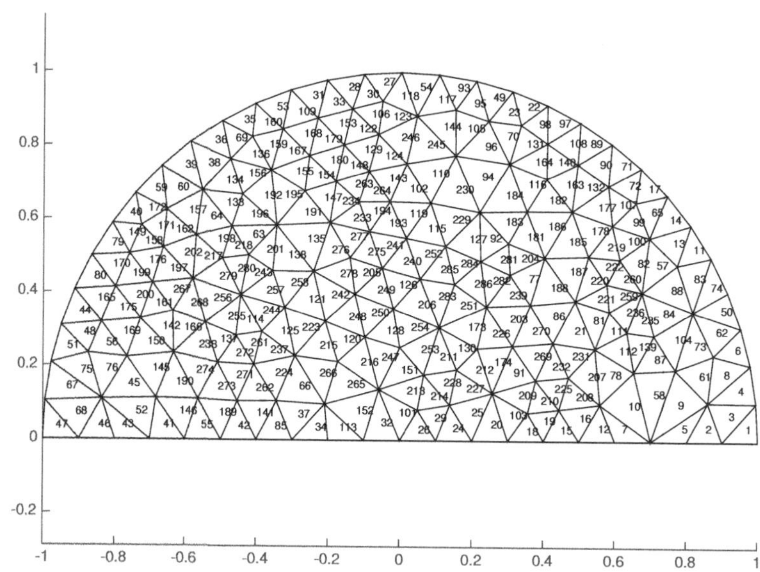

(b) Elements numbers

Figure 4.2 A semi-circle divided into triangular elements. The mesh is composed of 169 global nodes and 286 elements.

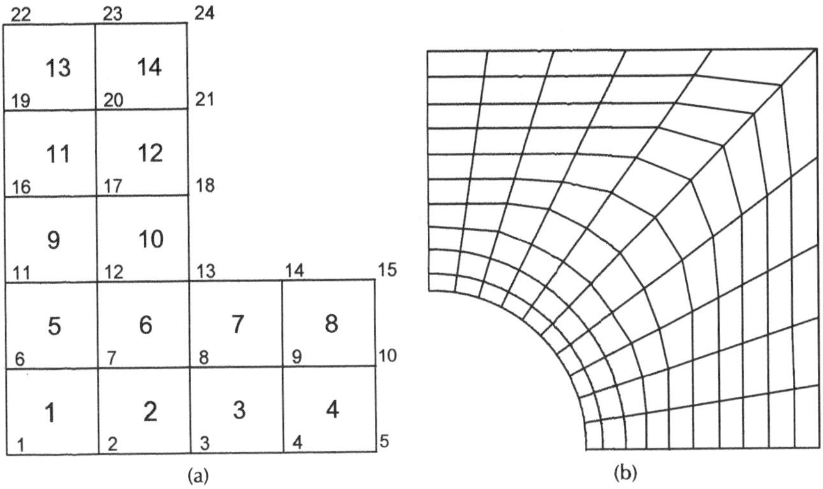

Figure 4.3 Examples of quadrilateral elements. (a) Rectangular elements on an L-shaped domain, and (b) quadrilateral elements on square domain with 1/4 circle removed.

Connectivity Matrix

As we shall see, the element equations will be obtained in terms of variables defined at the local nodes of the element. To be able to assemble the elements equations into a global system, we need to associate the local numbers of the nodes of each element with the corresponding global node numbers. Each local node of an element corresponds to a global node. The connectivity matrix (also called the identification matrix) gives the global nodes that corresponds to the element local nodes. The i-th row in the connectivity matrix gives the global nodes of the i-th element. Let's consider a mesh of n_e elements and all elements are 3-node triangles. For such a mesh, the connectivity matrix has n_e rows and three columns or dimensions (n_e, n) where $n = 3$. For the mesh shown in Figures 4.2a and 4.2b, element number 1 is given by global nodes 20, 21, and 22, and element number 2 is identified by global nodes 19, 20, and 51, and element number 3 has global nodes 20, 22, and 51, and so on. Therefore, the connectivity matrix for this mesh is

$$[CM] = \begin{bmatrix} 20 & 21 & 22 \\ 19 & 20 & 51 \\ 20 & 22 & 51 \\ . & . & . \\ . & . & . \\ 169 & 128 & 152 \end{bmatrix}. \tag{4.16}$$

The first row gives the global nodes of element #1, and the last row gives the global nodes of element # 286. The matrix is of the dimensions (286,3). For the mesh shown in Figure 4.3a, the number of elements is $n_e = 14$ and the number of nodes per element is $n = 4$. Hence, the connectivity matrix is of the dimensions 14 rows and 4 columns. Let's number the local nodes of each element such that node 1 is the lower left corner of the element, and the other nodes must follow in the counterclockwise direction. The connectivity matrix for this mesh is

$$[CM] = \begin{bmatrix} 1 & 2 & 7 & 6 \\ 2 & 3 & 8 & 7 \\ 3 & 4 & 9 & 8 \\ . & . & . & \\ . & . & . & \\ 20 & 21 & 24 & 23 \end{bmatrix}. \tag{4.17}$$

The first row gives the global nodes of element #1, and the last row gives the global nodes of element # 14. The matrix is of the dimensions (14,4). In the step of mesh generation, we also compute the global coordinates (x,y) of all nodes in the mesh, and store them in an array $(x_j, y_j, j = 1,2,...,N)$, where N is the number of global nodes.

4.4 APPROXIMATIONS AND ELEMENT SHAPE FUNCTIONS

In this section, we develop approximations of the primary variable u over an element. The primary variable $u(x,y)$ over the element domain is approximated by polynomials in x and y.

$$u(x,y) \approx U^e(x,y) = \sum_{j=1}^{n} U_j^e \psi_j(x,y) \tag{4.18}$$

where n is the number of nodes per element. The shape functions satisfy the condition $\psi_j(x_i, y_i) = \delta_{ij}$. To write $U^e(x,y)$ in matrix form, we define the arrays

$$\{\psi(x,y)\} = \{\psi_1(x,y)\ \psi_2(x,y) \ldots \psi_j(x,y) \ldots \psi_n(x,y)\} \tag{4.19}$$

$$\{U^e\} = \begin{Bmatrix} U_1^e \\ U_2^e \\ . \\ . \\ . \\ U_j^e \\ . \\ . \\ . \\ U_n^e \end{Bmatrix}. \tag{4.20}$$

In matrix form, the trial solution over element # e is

$$U^e(x,y) = \{\psi(x,y)\}\{U^e\}. \tag{4.21}$$

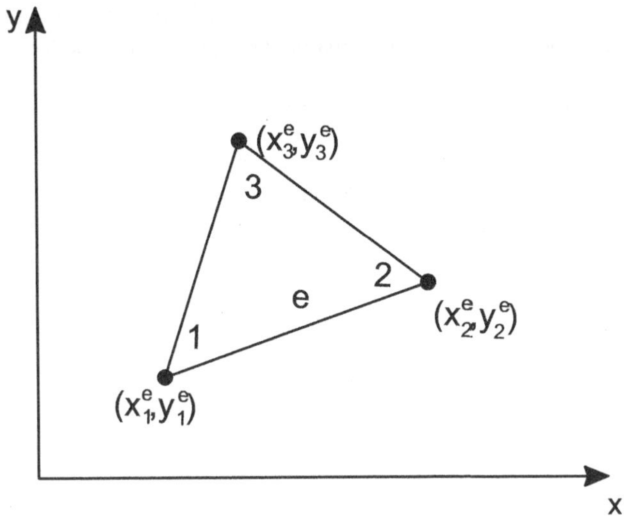

Figure 4.4 Bilinear triangle element, x and y are global coordinates. 1, 2, and 3 are the element local nodes.

We note that the dependence of the trial solution on x and y is given by the array $\{\psi(x,y)\}$, whereas the array $\{U^e\}$ on the right are values of the primary variable at the element nodes.

In this section, we derive shape functions $\psi_j(x,y)$ for two types of elements. The first element is a triangular element defined by three nodes ($n = 3$), and the second element is a rectangular element defined by four nodes ($n = 4$).

4.4.1 BILINEAR TRIANGULAR ELEMENT

A typical triangle element is shown in Figure 4.4. The global coordinates of the three nodes are known. We want to find an expression for $U^e(x,y)$ at any point within the element domain in terms of U_1^e, U_2^e, and U_3^e, which are the values of $U^e(x,y)$ at the three nodes. These are called the element's degrees of freedom. The lowest order polynomial is linear in x and y, so this element is called a bilinear triangular element. We assume

$$u(x,y) \approx U^e(x,y) = c_1 + c_2 x + c_3 y \tag{4.22}$$

that must satisfy the following conditions at the element nodes:

$$\begin{aligned} c_1 + c_2 x_1^e + c_3 y_1^e &= U_1^e \\ c_1 + c_2 x_2^e + c_3 y_2^e &= U_2^e \\ c_1 + c_2 x_3^e + c_3 y_3^e &= U_3^e. \end{aligned} \tag{4.23}$$

We solve for c_1, c_2, and c_3 in terms of U_1^e, U_2^e, and U_3^e, and substitute the result into Equation (4.22). Then we rearrange terms in that equation into the form:

$$U^e(x,y) = \psi_1(x,y)U_1^e + \psi_2(x,y)U_2^e + \psi_3(x,y)U_3^e \tag{4.24}$$

that is to say the collection of all terms multiplying U_1^e is the function ψ_1, the collection of all terms multiplying U_2^e is ψ_2, and the collection of all terms multiplying U_3^e is ψ_3. The obtained expressions for the three shape functions for the bilinear triangular element can be written as [97]:

$$\psi_i(x,y) = \frac{1}{2A_e}(\alpha_i + \beta_i x + \gamma_i y), \qquad i = 1,2,3 \tag{4.25}$$

$$\begin{aligned} \alpha_i &= x_j^e y_k^e - x_k^e y_j^e \\ \beta_i &= y_j^e - y_k^e \\ \gamma_i &= -x_j^e + x_k^e \end{aligned} \tag{4.26}$$

where (x_j^e, y_j^e) are the global coordinates of node j of element # e, and A_e is the element area. The indices i, j, and k take the following sequences:

If $i = 1$, then $j = 2$ and $k = 3$.

If $i = 2$, then $j = 3$ and $k = 1$.

If $i = 3$, then $j = 1$ and $k = 2$.

For example $\psi_2(x,y) = (\alpha_2 + \beta_2 x + \gamma_2 y)/(2A_e)$, where $\alpha_2 = x_3^e y_1^e - x_1^e y_3^e$, $\beta_2 = y_3^e - y_1^e$, and $\gamma_2 = -x_3^e + x_1^e$. The gradient of the shape functions is

$$\begin{aligned} \frac{\partial \psi_i(x,y)}{\partial x} &= \frac{\beta_i}{2A_e} \\ \frac{\partial \psi_i(x,y)}{\partial y} &= \frac{\gamma_i}{2A_e} \end{aligned} \tag{4.27}$$

which is constant for an element, but varies from one element to another. The bilinear triangular element is also called the constant strain element when applied to elasticity problems. The three shape functions are shown in Figures 4.5–4.7. Each function is a plane (flat surface) when the magnitude is drawn normal to the element. For later use, we summarize certain properties of these functions. The function $\psi_1(x,y)$ is equal 1 at node 1, and zero at nodes 2 and 3. It drops linearly from 1 to zero on edges 1-2 and 1-3, and is zero on edge 2-3. The function $\psi_2(x,y)$ is equal 1 at node 2, and zero at nodes 1 and 3. It drops linearly from 1 to zero on edges 2-1 and 2-3, and is zero on edge 1-3. The function $\psi_3(x,y)$ is equal 1 at node 3, and zero at nodes 1 and 2. It drops linearly from 1 to zero on edges 3-1 and 3-2, and is zero on edge 1-2. Since the starting point of the approximation is a linear function in x and y, the approximation over the element domain is a plane as shown in Figure 4.8. The linear distribution of $U^e(x,y)$ on any of the three edges is uniquely determined by

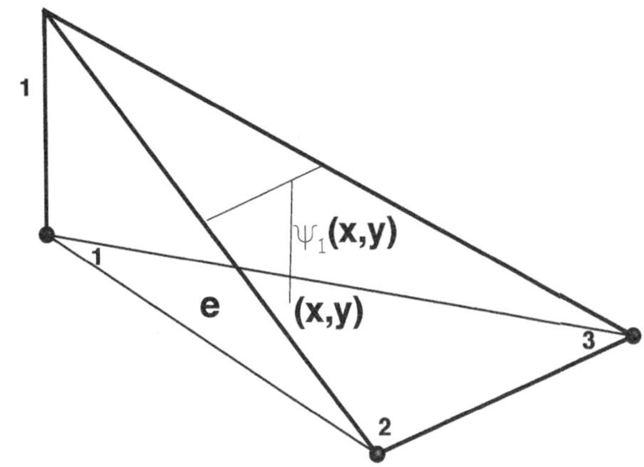

Figure 4.5 Shape function $\psi_1(x,y)$ for bilinear triangle element, x and y are global coordinates. ψ_1 is equal 1 at node 1, and zero at nodes 2 and 3. It varies linearly on edges 1-2 and 1-3, and is zero on edge 2-3.

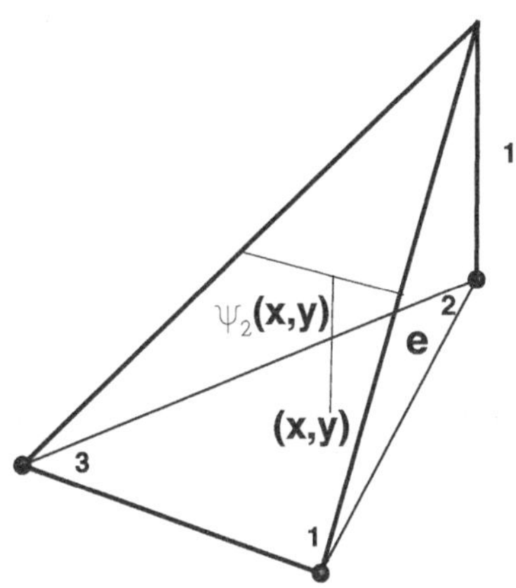

Figure 4.6 Shape function $\psi_2(x,y)$ for bilinear triangle element, x and y are global coordinates. ψ_2 is equal 1 at node 2, and zero at nodes 1 and 3. It varies linearly on edges 2-1 and 2-3, and is zero on edge 1-3.

the values of $U^e(x,y)$ at the two nodes on that edge; and the value at the third node has no effects on that linear distribution. Therefore, the distribution of $U(x,y)$ on an edge that is common to two adjacent elements will be the same when viewed from either element provided that the values of $U(x,y)$ are equal at the common nodes defining that edge. That is to say, $U(x,y)$ is continuous as the common edge is crossed from one element to the adjacent element. However, the space derivative of $U(x,y)$ normal to the common edge is not necessarily continuous. Such a discontinuity is

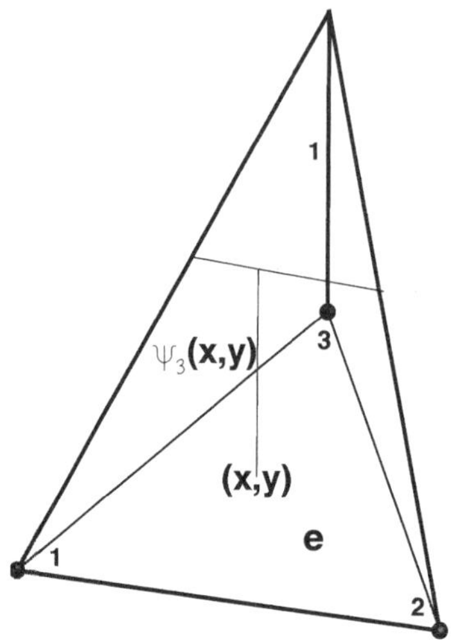

Figure 4.7 Shape function $\psi_3(x,y)$ for bilinear triangle element, x and y are global coordinates. ψ_3 is equal 1 at node 3, and zero at nodes 1 and 2. It varies linearly on edges 3-1 and 3-2, and is zero on edge 1-2.

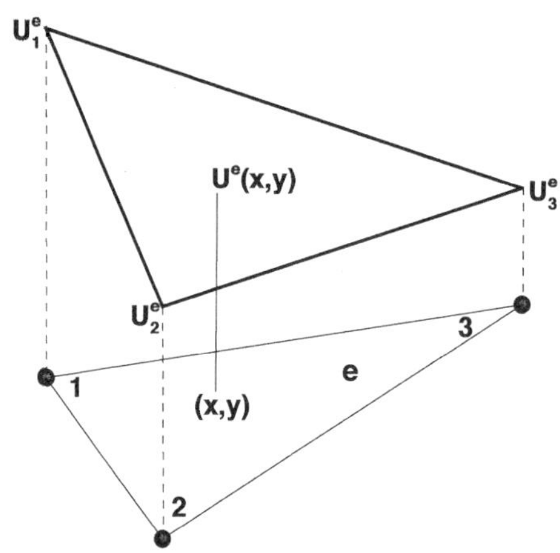

Figure 4.8 A function $u(x,y)$ is approximated in terms of the shape functions of the bilinear triangle element and values of u at the element nodes, $U^e(x,y) = \psi_1(x,y)U_1^e + \psi_2(x,y)U_2^e + \psi_3(x,y)U_3^e$. The approximating surface is a plane. The approximation varies linearly on the element edges.

allowed by the weak form because only the first derivatives appear in that equation, and therefore the first derivatives of the trial solution need not be continuous. For the model problem in this chapter, when the distribution on a common edge between two elements is identical for both elements, it is said that the element satisfies the

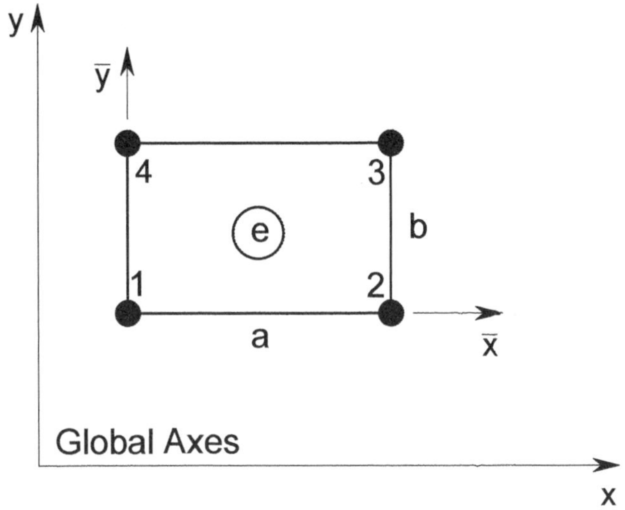

Figure 4.9 Bilinear rectangular element, $\bar{x}$ and $\bar{y}$ are local coordinates and x and y are global coordinates. 1, 2, 3, and 4 are the element local nodes.

compatibility condition. In that respect, the bilinear triangle element satisfies the compatibility condition.

4.4.2 BILINEAR RECTANGULAR ELEMENT

A typical rectangular element is shown in Figure 4.9. In this section, we consider a special case where, for all the elements in the mesh, the rectangle sides are parallel to a global Cartesian system (x,y). The approximation over the typical element and shape functions are derived for a local coordinate system $(\bar{x},\bar{y})$, and hence the global coordinates are not used. The more general case of any quadrilateral element is considered in Section 4.8. The rectangle sides are a and b, and the local coordinates of the four nodes are $(0,0)$, $(a,0)$, (a,b), and $(0,b)$ for nodes 1, 2, 3, and 4, respectively. We want to find an expression for $U^e(\bar{x},\bar{y})$ at any point within the element domain in terms of U_1^e, U_2^e, U_3^e, and U_4^e, which are the values of $U^e(\bar{x},\bar{y})$ at the four local nodes. These are called the element's degrees of freedom. The lowest order polynomial is linear in $\bar{x}$ and $\bar{y}$, but we need a fourth term. We choose the fourth term to be $\bar{x}\bar{y}$, and assume

$$u(x,y) \approx U^e(\bar{x},\bar{y}) = c_1 + c_2\bar{x} + c_3\bar{y} + c_4\bar{x}\bar{y} \tag{4.28}$$

that must satisfy the following conditions at the element nodes:

$$\begin{aligned} c_1 &= U_1^e \\ c_1 + c_2 a &= U_2^e \\ c_1 + c_2 a + c_3 b + c_4 ab &= U_3^e \\ c_1 + c_3 b &= U_4^e. \end{aligned} \tag{4.29}$$

Solving for c_1, c_2, c_3, and c_4, substituting the results into Equation (4.28), and rearranging terms, we rewrite it in the form

$$U^e(\bar{x},\bar{y}) = \psi_1(\bar{x},\bar{y})U_1^e + \psi_2(\bar{x},\bar{y})U_2^e + \psi_3(\bar{x},\bar{y})U_3^e + \psi_4(\bar{x},\bar{y})U_4^e. \tag{4.30}$$

This equation defines the shape functions for the bilinear rectangular element by

$$\begin{aligned}
\psi_1(\bar{x},\bar{y}) &= \left(1-\frac{\bar{x}}{a}\right)\left(1-\frac{\bar{y}}{b}\right)\\
\psi_2(\bar{x},\bar{y}) &= \frac{\bar{x}}{a}\left(1-\frac{\bar{y}}{b}\right)\\
\psi_3(\bar{x},\bar{y}) &= \frac{\bar{x}}{a}\frac{\bar{y}}{b}\\
\psi_4(\bar{x},\bar{y}) &= \left(1-\frac{\bar{x}}{a}\right)\frac{\bar{y}}{b}.
\end{aligned} \tag{4.31}$$

Consider the assumed approximation Equation (4.28). If we put $\bar{x}$ =constant, the approximation is a linear function of $\bar{y}$. And if we put $\bar{y}$ =constant, the approximation is a linear function of $\bar{x}$. Because the approximation is linear in $\bar{x}$ and $\bar{y}$, we call this element a bilinear rectangular element. Since on the element edges either $\bar{x}$ or $\bar{y}$ is constant, the trial solution varies linearly on the rectangle edges. Such a linear distribution is uniquely determined by the values of $U^e(\bar{x},\bar{y})$ at the two nodes on that edge; the values at the other two nodes have no effects on that linear distribution. Therefore, the distribution of $U^e(\bar{x},\bar{y})$ on an edge that is common to two adjacent elements will be the same when viewed from either element provided that the values of $U^e(\bar{x},\bar{y})$ are equal at the common nodes defining that edge. That is to say, $U^e(\bar{x},\bar{y})$ is continuous as the common edge is crossed from one element to the adjacent element. However, the space derivative of $U^e(\bar{x},\bar{y})$ normal to the common edge is not necessarily continuous. Such a discontinuity is allowed by the weak form because only the first derivatives appear in that equation, and therefore the trial solution need not have continuous first derivatives. The bilinear rectangular element satisfies the compatibility condition. Choosing the fourth term to be $\bar{x}\bar{y}$ is essential for element compatibility. Had we selected any other fourth term such as $\bar{x}^2$ or $\bar{y}^2$, the assumed solution on the element edges would be quadratic in $\bar{x}$ or $\bar{y}$. Since a quadratic function is not uniquely determined by two points, the values at the other two nodes participate in defining the quadratic distribution, and the element would be incompatible.

Three of the four shape functions are shown in Figures 4.10–4.12. The magnitude of the function is drawn normal to the element domain; surfaces are not planes. The lines drawn on the surface are lines of constant magnitude. For later use, we emphasize certain properties of these functions. The first property is $\psi_j(\bar{x}_i,\bar{y}_i) = \delta_{ij}$. Each function is 1 at one of the nodes and zero at the other three nodes. For example, $\psi_1(\bar{x},\bar{y})$ is equal 1 at node 1, and zero at nodes 2, 3, and 4. The second property is that each function varies linearly on two sides emanating from one node and zero on the other two sides emanating from the opposite node. For example, $\psi_3(\bar{x},\bar{y})$ varies linearly on edges 3-2 and 3-4, and is zero on edges 1-2 and 1-4. The third property

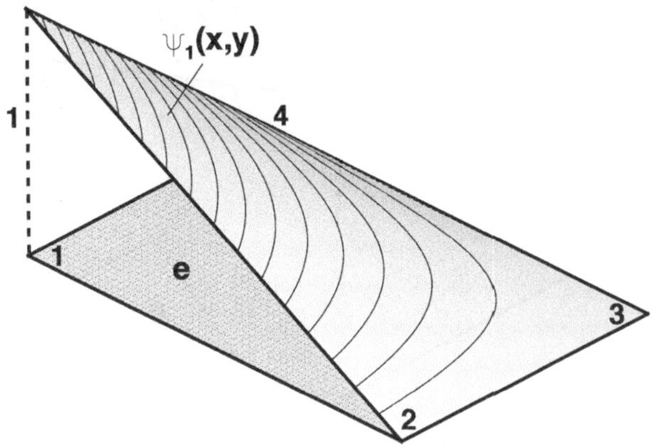

Figure 4.10 Shape function $\psi_1(\bar{x},\bar{y})$ for bilinear rectangle element. ψ_1 is equal 1 at node 1, and zero at nodes 2, 3, and 4. It varies linearly on edges 1-2 and 1-4, and is zero on edges 2-3 and 3-4.

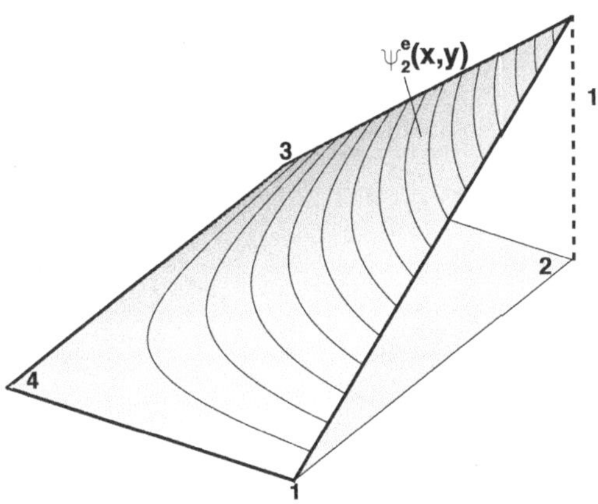

Figure 4.11 Shape function $\psi_2(\bar{x},\bar{y})$ for bilinear rectangle element. ψ_2 is equal 1 at node 2, and zero at nodes 1, 3, and 4. It varies linearly on edges 2-1 and 2-3, and is zero on edges 3-4 and 4-1.

is $\psi_1(\bar{x},\bar{y})+\psi_2(\bar{x},\bar{y})+\psi_3(\bar{x},\bar{y})+\psi_4(\bar{x},\bar{y})=1$ at any point. The surface representing the trial solution over a rectangular element is shown in Figure (4.13); it is not a flat surface, but varies linearly on all four sides of the rectangle.

4.5 ELEMENT EQUATIONS AND MATRICES

The analysis presented in this section applies to elements of arbitrary shape not necessarily a triangle or rectangle. To derive the element equations and matrices, we first write the weak form for an element, which is exactly the same as the weak form for

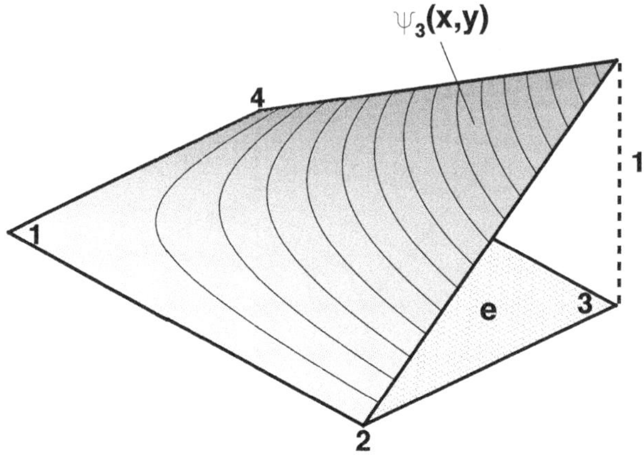

Figure 4.12 Shape function $\psi_3(\bar{x},\bar{y})$ for bilinear rectangle element. ψ_3 is equal 1 at node 3, and zero at nodes 1, 2, and 4. It varies linearly on edges 3-2 and 3-4, and is zero on edges 1-2 and 1-4.

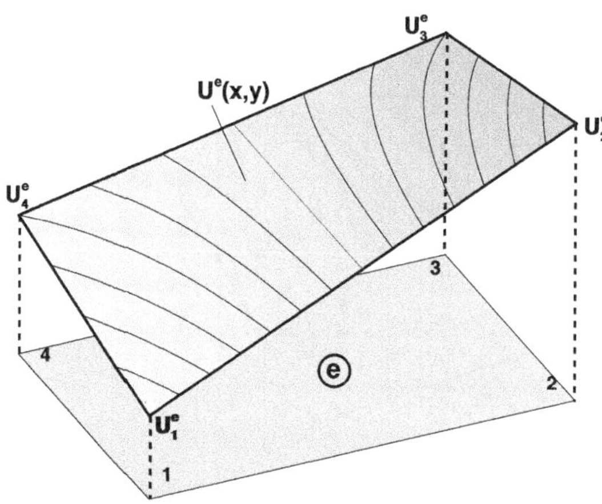

Figure 4.13 A function $u(x,y)$ is approximated in terms of the shape functions of the bilinear rectangular element and values of u at the element nodes, $U^e(\bar{x},\bar{y}) = \psi_1(\bar{x},\bar{y})U_1^e + \psi_2(\bar{x},\bar{y})U_2^e + \psi_3(\bar{x},\bar{y})U_3^e + \psi_4(\bar{x},\bar{y})U_4^e$. The approximating surface is not necessarily flat. The approximation varies linearly on the element edges.

the full domain except that the boundary integral remains unknown because there are no boundary conditions to be applied to individual elements.

$$\int_{\Omega^e} (\nabla w \cdot \vec{v} + wqu)\, d\Omega = \int_{\Omega^e} wf d\Omega + \oint_{\Gamma^e} wv_n ds \tag{4.32}$$

where Ω^e is the element domain and Γ^e its boundary. On the element boundary, v_n is unknown. The trial solution over the element is given in matrix form by Equation (4.21), and is repeated here.

$$U^e(x,y) = \{\psi_1(x,y)\ \psi_2(x,y) \ldots \psi_n(x,y)\} \begin{Bmatrix} U_1^e \\ U_2^e \\ \cdot \\ \cdot \\ \cdot \\ U_n^e \end{Bmatrix}, \quad U^e(x,y) = \{\psi\}\{U^e\}. \tag{4.33}$$

To determine the vector $\vec{v}$, we first find the gradient of the trial solution.

$$\frac{\partial U^e(x,y)}{\partial x} = \left\{ \frac{\partial \psi_1(x,y)}{\partial x}\ \frac{\partial \psi_2(x,y)}{\partial x} \cdots \frac{\partial \psi_n(x,y)}{\partial x} \right\} \begin{Bmatrix} U_1^e \\ U_2^e \\ \cdot \\ \cdot \\ \cdot \\ U_n^e \end{Bmatrix} \tag{4.34}$$

$$\frac{\partial U^e(x,y)}{\partial y} = \left\{ \frac{\partial \psi_1(x,y)}{\partial y}\ \frac{\partial \psi_2(x,y)}{\partial y} \cdots \frac{\partial \psi_n(x,y)}{\partial y} \right\} \begin{Bmatrix} U_1^e \\ U_2^e \\ \cdot \\ \cdot \\ \cdot \\ U_n^e \end{Bmatrix} \tag{4.35}$$

$$\begin{Bmatrix} \frac{\partial U^e(x,y)}{\partial x} \\ \frac{\partial U^e(x,y)}{\partial y} \end{Bmatrix} = \begin{bmatrix} \frac{\partial \psi_1(x,y)}{\partial x} & \frac{\partial \psi_2(x,y)}{\partial x} & \cdots & \frac{\partial \psi_n(x,y)}{\partial x} \\ \frac{\partial \psi_1(x,y)}{\partial y} & \frac{\partial \psi_2(x,y)}{\partial y} & \cdots & \frac{\partial \psi_n(x,y)}{\partial y} \end{bmatrix} \begin{Bmatrix} U_1^e \\ U_2^e \\ \cdot \\ \cdot \\ \cdot \\ U_n^e \end{Bmatrix}. \tag{4.36}$$

We define the gradient matrix $[B]$ by

$$[B] = \begin{bmatrix} \frac{\partial \psi_1(x,y)}{\partial x} & \frac{\partial \psi_2(x,y)}{\partial x} & \cdots & \frac{\partial \psi_n(x,y)}{\partial x} \\ \frac{\partial \psi_1(x,y)}{\partial y} & \frac{\partial \psi_2(x,y)}{\partial y} & \cdots & \frac{\partial \psi_n(x,y)}{\partial y} \end{bmatrix} \tag{4.37}$$

and write the approximation to $\vec{v}$ as

$$\begin{aligned}
\begin{Bmatrix} V_x^e \\ V_y^e \end{Bmatrix} &= \begin{Bmatrix} p_{11}\frac{\partial U^e(x,y)}{\partial x} + p_{12}\frac{\partial U^e(x,y)}{\partial y} \\ p_{21}\frac{\partial U^e(x,y)}{\partial x} + p_{22}\frac{\partial U^e(x,y)}{\partial y} \end{Bmatrix} \\
&= \begin{bmatrix} p_{11} & p_{12} \\ p_{21} & p_{22} \end{bmatrix} \begin{Bmatrix} \frac{\partial U^e(x,y)}{\partial x} \\ \frac{\partial U^e(x,y)}{\partial y} \end{Bmatrix} \\
&= \begin{bmatrix} p_{11} & p_{12} \\ p_{21} & p_{22} \end{bmatrix} \begin{bmatrix} \frac{\partial \psi_1(x,y)}{\partial x} & \dots & \frac{\partial \psi_n(x,y)}{\partial x} \\ \frac{\partial \psi_1(x,y)}{\partial y} & \dots & \frac{\partial \psi_n(x,y)}{\partial y} \end{bmatrix} \begin{Bmatrix} U_1^e \\ \cdot \\ \cdot \\ \cdot \\ U_n^e \end{Bmatrix} \qquad (4.38)
\end{aligned}$$

$$\begin{Bmatrix} V_x^e \\ V_y^e \end{Bmatrix} = [D][B]\{U^e\}. \qquad (4.39)$$

Galerkin

In the Galerkin method, the test (weight) function $w(x,y)$ is a linear combination of the basis functions. Thus for the typical element #e, we have the approximation for w,

$$W^e(x,y) = \{\psi_1(x,y)\ \psi_2(x,y) \dots \psi_n(x,y)\} \begin{Bmatrix} d_1 \\ d_2 \\ \cdot \\ \cdot \\ \cdot \\ d_n \end{Bmatrix}, \ W^e(x,y) = \{\psi\}\{d\}. \qquad (4.40)$$

Similar to the gradient of $U^e(x,y)$, we write the gradient of $W^e(x,y)$ using the matrix $[B]$,

$$\begin{Bmatrix} \frac{\partial W^e(x,y)}{\partial x} \\ \frac{\partial W^e(x,y)}{\partial y} \end{Bmatrix} = [B]\{d\}. \qquad (4.41)$$

Now, we can write the first term of the integrand in the weak form in matrix form,

$$\begin{aligned}
\nabla W^e \cdot \vec{V}^e &= \left(\frac{\partial W^e(x,y)}{\partial x}\hat{i} + \frac{\partial W^e(x,y)}{\partial y}\hat{j} \right) \cdot (V_x^e\hat{i} + V_y^e\hat{j}) \\
&= \frac{\partial W^e(x,y)}{\partial x}V_x^e + \frac{\partial W^e(x,y)}{\partial y}V_y^e \\
&= \begin{Bmatrix} \frac{\partial W^e(x,y)}{\partial x} & \frac{\partial W^e(x,y)}{\partial y} \end{Bmatrix} \begin{Bmatrix} V_x^e \\ V_y^e \end{Bmatrix} \\
&= \{d\}^T[B]^T[D][B]\{U^e\}. \qquad (4.42)
\end{aligned}$$

Next, we write the remaining terms in the weak form using matrix notation,

$$\begin{aligned} wqu &\approx W^e q U^e \\ &= \{d\}^T \{\psi\}^T q \{\psi\} \{U^e\} \end{aligned} \tag{4.43}$$

$$\begin{aligned} wf &\approx W^e f \\ &= \{d\}^T f \{\psi\}^T \end{aligned} \tag{4.44}$$

$$v_n w = \{d\}^T v_n \{\psi\}^T. \tag{4.45}$$

Substituting Equations (4.42)–(4.45) into the weak form for the element Equation (4.32), and factoring out the constant terms $\{d\}^T$ and $\{U^e\}$, we get

$$\begin{aligned} &\{d\}^T \{ \left[\int_{\Omega^e} \left([B]^T [D][B] + q\{\psi\}^T \{\psi\} \right) d\Omega \right] \{U^e\} \\ &- \int_{\Omega^e} f\{\psi\}^T d\Omega - \oint_{\Gamma^e} v_n \{\psi\}^T ds \} = 0. \end{aligned} \tag{4.46}$$

Next, we define the element matrices:

the coefficient matrix

$$\begin{aligned} [A^e] &= \int_{\Omega^e} \left([B]^T [D][B] + q\{\psi\}^T \{\psi\} \right) d\Omega \\ &= [K^e] + [M^e] \end{aligned} \tag{4.47}$$

the stiffness matrix

$$[K^e] = \int_{\Omega^e} [B]^T [D][B] d\Omega \tag{4.48}$$

the mass matrix

$$[M^e] = \int_{\Omega^e} q\{\psi\}^T \{\psi\} d\Omega \tag{4.49}$$

the forcing vector

$$\{F^e\} = \int_{\Omega^e} f\{\psi\}^T d\Omega \tag{4.50}$$

and vector of weighted secondary variables

$$\{R^e\} = \oint_{\Gamma^e} v_n \{\psi\}^T ds. \tag{4.51}$$

In terms of these matrices, Equation (4.46) becomes

$$\{d\}^T \{ [A^e] \{U^e\} - \{F^e\} - \{R^e\} \} = 0. \tag{4.52}$$

Since $\{d\}$ is an arbitrary array, the multiplying array must be zero. We obtain the element equations

$$[A^e] \{U^e\} = \{F^e\} + \{R^e\}. \tag{4.53}$$

Using index notation, we write the entries of the element matrices

$$k_{ij}^e = \int_{A_e} \left[\frac{\partial \psi_i}{\partial x} \left(p_{11} \frac{\partial \psi_j}{\partial x} + p_{12} \frac{\partial \psi_j}{\partial y} \right) + \frac{\partial \psi_i}{\partial y} \left(p_{21} \frac{\partial \psi_j}{\partial x} + p_{22} \frac{\partial \psi_j}{\partial y} \right) \right] dA \quad (4.54)$$

$$m_{ij}^e = \int_{A_e} q \psi_i \psi_j dA \quad (4.55)$$

$$F_i^e = \int_{A_e} f \psi_i dA \quad (4.56)$$

$$R_i^e = \oint_{\Gamma_e} v_n^e \psi_i ds. \quad (4.57)$$

4.5.1 MATRICES FOR BILINEAR TRIANGULAR ELEMENT

The gradient matrix for bilinear triangular elements is

$$\begin{aligned} [B] &= \begin{bmatrix} \frac{\partial \psi_1(x,y)}{\partial x} & \frac{\partial \psi_2(x,y)}{\partial x} & \frac{\partial \psi_3(x,y)}{\partial x} \\ \frac{\partial \psi_1(x,y)}{\partial y} & \frac{\partial \psi_2(x,y)}{\partial y} & \frac{\partial \psi_3(x,y)}{\partial y} \end{bmatrix} \\ &= \frac{1}{2A_e} \begin{bmatrix} \beta_1 & \beta_2 & \beta_3 \\ \gamma_1 & \gamma_2 & \gamma_3 \end{bmatrix}. \end{aligned} \quad (4.58)$$

This is a constant matrix. Let's assume that the matrix $[D] = [D^e]$ to be element-wise constant (it may be different for different elements). The integrals can be evaluated analytically, and we obtain for the stiffness matrix

$$[K^e] = A_e [B]^T [D^e][B] \quad (4.59)$$

And if $q = q_e$ is element-wise constant, the mass matrix

$$m_{ij} = \int_{A_e} q \psi_i \psi_j dA \quad (4.60)$$

becomes

$$\begin{aligned} m_{ij}^e = \tfrac{q_e}{4A_e^2} [&\alpha_i \alpha_j I_{00} + \beta_i \beta_j I_{20} + \gamma_i \gamma_j I_{02} + (\alpha_i \beta_j + \alpha_j \beta_i) I_{10} + \\ &(\alpha_i \gamma_j + \alpha_j \gamma_i) I_{01} + (\beta_i \gamma_j + \beta_j \gamma_i) I_{11}]. \end{aligned} \quad (4.61)$$

The I_{mn} are integrals over the triangle area defined by

$$I_{mn} = \int_{A_e} x^m y^n dxdy \quad (4.62)$$

We developed a MATLAB code using symbolic variables to evaluate the needed integrals, and obtained

$$\begin{aligned} I_{00} &= \int_{A_e} dxdy \\ &= A_e \\ &= \frac{1}{2} [(x_1^e y_2^e + x_2^e y_3^e + x_3^e y_1^e) - (y_1^e x_2^e + y_2^e x_3^e + y_3^e x_1^e)] \end{aligned} \quad (4.63)$$

$$\begin{aligned} I_{10} &= \int_{A_e} x dx dy \\ &= A_e x_c^e \\ I_{01} &= \int_{A_e} y dx dy \\ &= A_e y_c^e \end{aligned} \tag{4.64}$$

where (x_c^e, y_c^e) is the triangle centroid,

$$\begin{aligned} x_c^e &= \frac{1}{3}(x_1^e + x_2^e + x_3^e) \\ y_c^e &= \frac{1}{3}(y_1^e + y_2^e + y_3^e) \end{aligned} \tag{4.65}$$

$$\begin{aligned} I_{11} &= \int_{A_e} xy dx dy \\ &= \frac{A_e}{6}\left[x_1^e y_1^e + x_2^e y_2^e + x_3^e y_3^e + \frac{1}{2}(x_1^e y_2^e + x_2^e y_1^e + x_1^e y_3^e + x_3^e y_1^e + x_2^e y_3^e + x_3^e y_2^e)\right] \\ I_{20} &= \int_{A_e} x^2 dx dy \\ &= \frac{A_e}{6}(x_1^e x_1^e + x_2^e x_2^e + x_3^e x_3^e + x_1^e x_2^e + x_2^e x_3^e + x_3^e x_1^e) \\ I_{02} &= \int_{A_e} y^2 dx dy \\ &= \frac{A_e}{6}(y_1^e y_1^e + y_2^e y_2^e + y_3^e y_3^e + y_1^e y_2^e + y_2^e y_3^e + y_3^e y_1^e). \end{aligned} \tag{4.66}$$

If $f = f_e$ is also uniform, the forcing vector will be

$$\begin{aligned} F_i^e &= \int_{A_e} \frac{1}{2A_e}(\alpha_i + \beta_i x + \gamma_i y) f dA \\ &= \frac{f_e}{2}(\alpha_i + \beta_i x_c^e + \gamma_i y_c^e) \\ &= \frac{1}{3} A_e f_e \end{aligned} \tag{4.67}$$

$$\{F^e\} = \frac{A_e f_e}{3}\begin{Bmatrix} 1 \\ 1 \\ 1 \end{Bmatrix}. \tag{4.68}$$

Special Cases

For the Poisson's equation $-k\nabla^2 u = f$, the stiffness matrix for a bilinear right-angle triangle with local nodes labeled as shown in Figure (4.14) is

$$[K^e] = \frac{k}{2ab}\begin{bmatrix} c^2 & -b^2 & -a^2 \\ -b^2 & b^2 & 0 \\ -a^2 & 0 & a^2 \end{bmatrix} \tag{4.69}$$

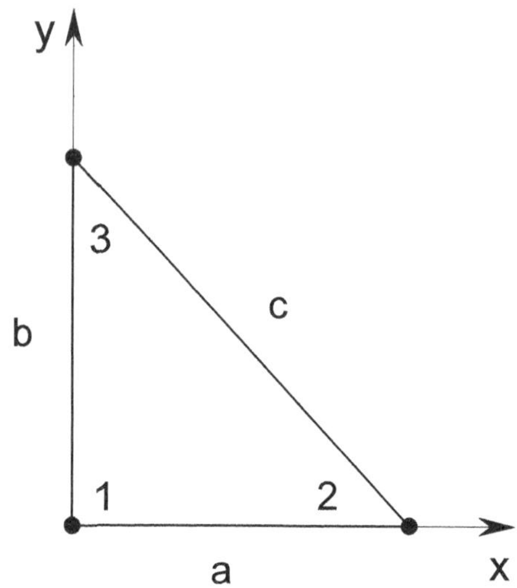

Figure 4.14 Right-angle triangular element.

where $c^2 = a^2 + b^2$, and the vertex at node 1 is 90 degrees. This matrix applies regardless of the element's orientation. For an equilateral triangular element, the stiffness matrix is

$$[K^e] = \frac{\sqrt{3}k}{6}\begin{bmatrix} 2 & -1 & -1 \\ -1 & 2 & -1 \\ -1 & -1 & 2 \end{bmatrix}. \tag{4.70}$$

4.5.2 MATRICES FOR BILINEAR RECTANGULAR ELEMENT

The shape functions for the bilinear rectangular element are written in terms of local coordinates $(\bar{x}, \bar{y})$ and are given by Equations (4.31). The element sides are $a = x_2^e - x_1^e$ and $b = y_4^e - y_1^e$. The global coordinates (x, y) for any point in the element are related to its local coordinates by

$$\begin{aligned} x &= x_1^e + \bar{x} \\ y &= y_1^e + \bar{y} \end{aligned} \tag{4.71}$$

hence

$$\begin{aligned} dx &= d\bar{x} \\ dy &= d\bar{y}. \end{aligned} \tag{4.72}$$

Knowing the shape functions, we find the gradient matrix

$$\begin{aligned}[B] &= \begin{bmatrix} \frac{\partial\psi_1(x,y)}{\partial x} & \frac{\partial\psi_2(x,y)}{\partial x} & \frac{\partial\psi_3(x,y)}{\partial x} & \frac{\partial\psi_4(x,y)}{\partial x} \\ \frac{\partial\psi_1(x,y)}{\partial y} & \frac{\partial\psi_2(x,y)}{\partial y} & \frac{\partial\psi_3(x,y)}{\partial y} & \frac{\partial\psi_4(x,y)}{\partial y} \end{bmatrix} \\ &= \frac{1}{ab}\begin{bmatrix} -b+\bar{y} & b-\bar{y} & \bar{y} & -\bar{y} \\ -a+\bar{x} & -\bar{x} & \bar{x} & a-\bar{x} \end{bmatrix}. \end{aligned} \tag{4.73}$$

If the coefficients $p_{ij} = p^e_{ij}$ are element-wise constants, the stiffness matrix becomes

$$\begin{aligned} k^e_{ij} = p^e_{11} \int_{A_e} \frac{\partial\psi_i}{\partial x}\frac{\partial\psi_j}{\partial x} dA + p^e_{12} \int_{A_e} \frac{\partial\psi_i}{\partial x}\frac{\partial\psi_j}{\partial y} dA + \\ p^e_{21} \int_{A_e} \frac{\partial\psi_i}{\partial y}\frac{\partial\psi_j}{\partial x} dA + p^e_{22} \int_{A_e} \frac{\partial\psi_i}{\partial y}\frac{\partial\psi_j}{\partial y} dA \\ = p^e_{11} l^{11}_{ij} + p^e_{12} l^{12}_{ij} + p^e_{21} l^{21}_{ij} + p^e_{22} l^{22}_{ij} \end{aligned} \tag{4.74}$$

where

$$\left[L^{11}\right] = \frac{b}{6a}\begin{bmatrix} 2 & -2 & -1 & 1 \\ -2 & 2 & 1 & -1 \\ -1 & 1 & 2 & -2 \\ 1 & -1 & -2 & 2 \end{bmatrix} \tag{4.75}$$

$$\left[L^{12}\right] = \frac{1}{4}\begin{bmatrix} 1 & 1 & -1 & -1 \\ -1 & -1 & 1 & 1 \\ -1 & -1 & 1 & 1 \\ 1 & 1 & -1 & -1 \end{bmatrix} \tag{4.76}$$

$$\left[L^{21}\right] = \left[L^{12}\right]^T \tag{4.77}$$

$$\left[L^{22}\right] = \frac{a}{6b}\begin{bmatrix} 2 & 1 & -1 & -2 \\ 1 & 2 & -2 & -1 \\ -1 & -2 & 2 & 1 \\ -2 & -1 & 1 & 2 \end{bmatrix}. \tag{4.78}$$

For element-wise constant $q = q_e$, the mass matrix is

$$\begin{aligned} m^e_{ij} &= \int_{A_e} q\psi_i\psi_j dA \\ &= q_e \int_{A_e} \psi_i\psi_j dA \\ &= q_e l^{00}_{ij} \end{aligned} \tag{4.79}$$

$$[M^e] = q_e\left[L^{00}\right] \tag{4.80}$$

$$\left[L^{00}\right] = \frac{ab}{36}\begin{bmatrix} 4 & 2 & 1 & 2 \\ 2 & 4 & 2 & 1 \\ 1 & 2 & 4 & 2 \\ 2 & 1 & 2 & 4 \end{bmatrix}. \tag{4.81}$$

If $f = f_e$ is also uniform, the forcing vector will be

$$\begin{aligned} F_i^e &= \int_{A_e} \psi_i f dA \\ &= f_e \int_{A_e} \psi_i dA \end{aligned} \tag{4.82}$$

$$\{F^e\} = \frac{A_e f_e}{4} \begin{Bmatrix} 1 \\ 1 \\ 1 \\ 1 \end{Bmatrix}. \tag{4.83}$$

4.5.3 MATRICES FOR NATURAL AND MIXED BOUNDARY CONDITIONS

The weighted secondary variables for an element are defined by the integral

$$R_i^e = \oint_{\Gamma^e} v_n \psi_i ds. \tag{4.84}$$

At a natural boundary condition, the secondary variable distribution v_n is prescribed on the boundary. If an edge of an element falls on a natural boundary condition, then v_n is known on that edge. The total amount (integral) of v_n on that edge is replaced by equivalent weighted secondary variables R_i^e at the element nodes according to the integral Equation (4.84). On an edge of a bilinear triangular or a rectangular element, any of the shape functions ψ_i, if not zero, varies linearly with the arc length s. Let's assume linear distribution of v_n on the edge connecting local nodes 1 and 2 of a bilinear triangular or rectangular element as shown in Figure 4.15,

$$v_n = v_{n1} + (v_{n2} - v_{n1})\frac{s}{h_{12}}, \qquad 0 \le s \le h_{12} \tag{4.85}$$

where v_{n1} and v_{n2} are known values at the nodes, and h_{12} is the edge length. On that edge, the shape functions are

$$\begin{aligned} \psi_1(s) &= 1 - \frac{s}{h_{12}} \\ \psi_2(s) &= \frac{s}{h_{12}}. \end{aligned} \tag{4.86}$$

The distributed secondary variable is now replaced by equivalent nodal values at nodes 1 and 2 by the integrals

$$\begin{aligned} R^e_{1(1-2)} &= \int_0^{h_{12}} \left(1 - \frac{s}{h_{12}}\right) \left[v_{n1} + (v_{n2} - v_{n1})\frac{s}{h_{12}}\right] ds \\ &= \frac{2}{3}\left(\frac{1}{2}h_{12}v_{n1}\right) + \frac{1}{3}\left(\frac{1}{2}h_{12}v_{n2}\right) \\ R^e_{2(1-2)} &= \int_0^{h_{12}} \left(\frac{s}{h_{12}}\right) \left[v_{n1} + (v_{n2} - v_{n1})\frac{s}{h_{12}}\right] ds \\ &= \frac{1}{3}\left(\frac{1}{2}h_{12}v_{n1}\right) + \frac{2}{3}\left(\frac{1}{2}h_{12}v_{n2}\right). \end{aligned} \tag{4.87}$$

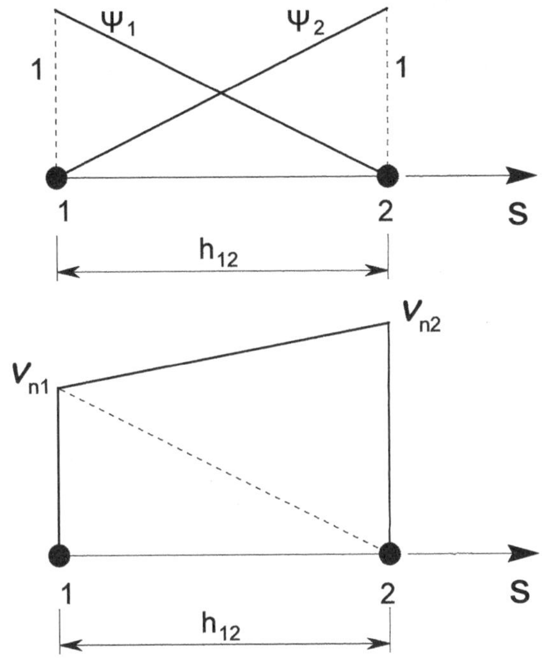

Figure 4.15 Shape functions and prescribed linearly varying secondary variable on edge 1-2, s is arc length along the edge.

We note that the total amount of distributed secondary variable v_n [the area $h_{12}(v_{n2}+v_{n1})/2$] is allocated to the element nodes such that $R^e_{1(1-2)}$ is the share of node # 1 and $R^e_{2(1-2)}$ is the share of node # 2. The share of the other nodes (not on edge 1-2) is zero. The result in Equation (4.87) is easy to recall. The linear distribution of v_n is divided into two triangles as shown in Figure (4.15). Node # 1 receives 2/3 of the area of the lower triangle and 1/3 of the area of the upper triangle. And node # 2 receives 1/3 of the area of lower triangle and 2/3 of the area of the upper triangle. This simple rule is useful for hand calculations. However, if the distributed v_n is more complex than linear, we have to perform the integral Equation (4.84). We also note that a node that is common to two edges receives contributions from both edges if they fall on a natural boundary condition.

Mixed Boundary Conditions

On a mixed boundary condition, both of the primary variable u and secondary variable v_n are unknown, but a relation is given

$$v_n + \lambda u = \alpha \qquad \text{on } \Gamma^e \tag{4.88}$$

that we use to eliminate the secondary variable from the weak form in favor of the primary variable, $v_n = \alpha - \lambda u$. First we consider a case where the entire boundary of the element is subjected to mixed boundary conditions. The array of weighted secondary variables at the element nodes can be written in terms of the array of primary variable as follows:

$$\begin{aligned} \{R^e\} &= \oint_{\Gamma^e} v_n \{\psi\}^T ds \\ &= \oint_{\Gamma^e} (\alpha - \lambda u) \{\psi\}^T ds \\ &= \oint_{\Gamma^e} \alpha \{\psi\}^T ds - \oint_{\Gamma^e} \lambda \{\psi\}^T \{\psi\} \{U^e\} ds \\ &= \{c^e\} - [G^e] \{U^e\} \end{aligned} \tag{4.89}$$

where

$$[G^e] = \oint_{\Gamma^e} \lambda \{\psi\}^T \{\psi\} ds \tag{4.90}$$

and

$$\{c^e\} = \oint_{\Gamma^e} \alpha \{\psi\}^T ds. \tag{4.91}$$

The equations for an element with mixed boundary conditions now reads

$$\begin{aligned} [A^e] \{U^e\} &= \{F^e\} + \{R^e\} \\ &= \{F^e\} + \{c^e\} - [G^e] \{U^e\}. \end{aligned} \tag{4.92}$$

The array of unknown primary variables must be moved to the left-hand side,

$$[[A^e] + [G^e]] \{U^e\} = \{F^e\} + \{c^e\}. \tag{4.93}$$

Assuming constant α and λ, we evaluate the integrals. For a bilinear triangular element we obtain:

$$[G^e] = \frac{\lambda_{12}^e h_{12}}{6} \begin{bmatrix} 2 & 1 & 0 \\ 1 & 2 & 0 \\ 0 & 0 & 0 \end{bmatrix} + \frac{\lambda_{23}^e h_{23}}{6} \begin{bmatrix} 0 & 0 & 0 \\ 0 & 2 & 1 \\ 0 & 1 & 2 \end{bmatrix} + \frac{\lambda_{31}^e h_{31}}{6} \begin{bmatrix} 2 & 0 & 1 \\ 0 & 0 & 0 \\ 1 & 0 & 2 \end{bmatrix} \tag{4.94}$$

$$\{c^e\} = \frac{\alpha_{12}^e h_{12}}{2} \left\{ \begin{matrix} 1 \\ 1 \\ 0 \end{matrix} \right\} + \frac{\alpha_{23}^e h_{23}}{2} \left\{ \begin{matrix} 0 \\ 1 \\ 1 \end{matrix} \right\} + \frac{\alpha_{31}^e h_{31}}{2} \left\{ \begin{matrix} 1 \\ 0 \\ 1 \end{matrix} \right\}. \tag{4.95}$$

The case of the entire boundary of the element is subjected to a mixed boundary condition is not practical. In practice only one or two edges of an element fall on a mixed boundary conditions and the remaining edges are either embedded within the domain or subjected to natural or essential boundary conditions. Thus in general, the element matrices are written as

$$[[A^e] + [G^e]_{MBC}] \{U^e\} = \{F^e\} + \{c^e\}_{MBC} + \{R^e\} \tag{4.96}$$

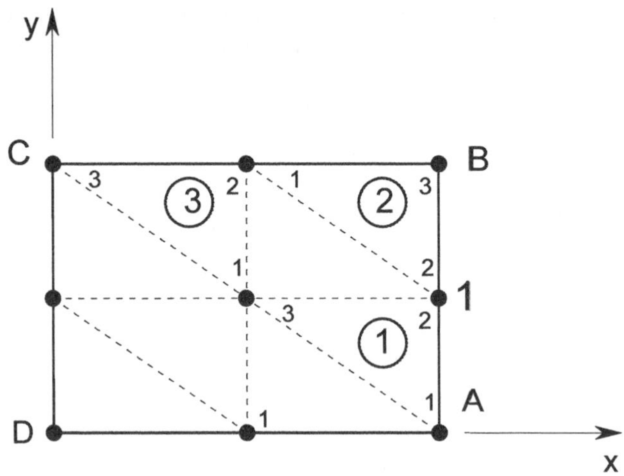

Figure 4.16 Sides AB and BC of the domain are subjected to mixed boundary conditions. The domain is divided into bilinear triangular elements.

where the subscript MBC indicates that the line integrals in Equations (4.90) and (4.91) extend over the edges subjected to mixed boundary conditions only. The array $\{R^e\}$ contains known contributions from natural boundary conditions and/or unknown contributions from essential boundary conditions and embedded edges. Figure 4.16 shows a rectangular domain $ABCD$ that is divided into bilinear triangular elements. Let's assume that the boundary conditions on sides AB and BC are of the mixed type. The matrices $[G^e]_{MBC}$ and $\{c^e\}_{MBC}$ are zero for all elements except for the elements labeled 1, 2, and 3. For element #1, only edge 1-2 is on the mixed boundary condition whereas the other two edges are embedded in the domain. Thus, for this element the matrices are

$$[G^e] = \frac{\lambda_{12}^e h_{12}}{6} \begin{bmatrix} 2 & 1 & 0 \\ 1 & 2 & 0 \\ 0 & 0 & 0 \end{bmatrix} \tag{4.97}$$

$$\{c^e\} = \frac{\alpha_{12}^e h_{12}}{2} \begin{Bmatrix} 1 \\ 1 \\ 0 \end{Bmatrix}. \tag{4.98}$$

For element #2, only edges 2-3 and 3-1 are on the mixed boundary condition whereas the third edge is embedded in the domain. Thus, for this element the matrices are

$$[G^e] = \frac{\lambda_{23}^e h_{23}}{6} \begin{bmatrix} 0 & 0 & 0 \\ 0 & 2 & 1 \\ 0 & 1 & 2 \end{bmatrix} + \frac{\lambda_{31}^e h_{31}}{6} \begin{bmatrix} 2 & 0 & 1 \\ 0 & 0 & 0 \\ 1 & 0 & 2 \end{bmatrix} \tag{4.99}$$

$$\{c^e\} = \frac{\alpha_{23}^e h_{23}}{2} \begin{Bmatrix} 0 \\ 1 \\ 1 \end{Bmatrix} + \frac{\alpha_{31}^e h_{31}}{2} \begin{Bmatrix} 1 \\ 0 \\ 1 \end{Bmatrix}. \tag{4.100}$$

For element #3, only edge 2-3 is on the mixed boundary condition whereas the other two edges are embedded in the domain. Thus, for this element the matrices are

$$[G^e] = \frac{\lambda_{23}^e h_{23}}{6} \begin{bmatrix} 0 & 0 & 0 \\ 0 & 2 & 1 \\ 0 & 1 & 2 \end{bmatrix} \tag{4.101}$$

$$\{c^e\} = \frac{\alpha_{23}^e h_{23}}{2} \begin{Bmatrix} 0 \\ 1 \\ 1 \end{Bmatrix}. \tag{4.102}$$

The element matrices for a bilinear rectangular element are given by Equation (4.96). If the four sides of the element are subjected to mixed boundary conditions, we get

$$\begin{aligned}[G^e] = {} & \frac{\lambda_{12}^e h_{12}}{6} \begin{bmatrix} 2 & 1 & 0 & 0 \\ 1 & 2 & 0 & 0 \\ 0 & 0 & 0 & 0 \\ 0 & 0 & 0 & 0 \end{bmatrix} + \frac{\lambda_{23}^e h_{23}}{6} \begin{bmatrix} 0 & 0 & 0 & 0 \\ 0 & 2 & 1 & 0 \\ 0 & 1 & 2 & 0 \\ 0 & 0 & 0 & 0 \end{bmatrix} \\ & + \frac{\lambda_{34}^e h_{34}}{6} \begin{bmatrix} 0 & 0 & 0 & 0 \\ 0 & 0 & 0 & 0 \\ 0 & 0 & 2 & 1 \\ 0 & 0 & 1 & 2 \end{bmatrix} + \frac{\lambda_{41}^e h_{41}}{6} \begin{bmatrix} 2 & 0 & 0 & 1 \\ 0 & 0 & 0 & 0 \\ 0 & 0 & 0 & 0 \\ 1 & 0 & 0 & 2 \end{bmatrix} \end{aligned} \tag{4.103}$$

$$\begin{aligned}\{c^e\} = {} & \frac{\alpha_{12}^e h_{12}}{2} \begin{Bmatrix} 1 \\ 1 \\ 0 \\ 0 \end{Bmatrix} + \frac{\alpha_{23}^e h_{23}}{2} \begin{Bmatrix} 0 \\ 1 \\ 1 \\ 0 \end{Bmatrix} \\ & + \frac{\alpha_{34}^e h_{34}}{2} \begin{Bmatrix} 0 \\ 0 \\ 1 \\ 1 \end{Bmatrix} + \frac{\alpha_{41}^e h_{41}}{2} \begin{Bmatrix} 1 \\ 0 \\ 0 \\ 1 \end{Bmatrix}. \end{aligned} \tag{4.104}$$

However, we must drop from Equations (4.103) and (4.104) the terms that correspond to edges not subjected to mixed boundary conditions. For example, Figure 4.17 shows the rectangular domain *ABCD* again but now it is divided into bilinear rectangular elements. As before, the boundary conditions on sides *AB* and *BC* are of the mixed type. The matrices $[G^e]_{MBC}$ and $\{c^e\}_{MBC}$ are zero for element #1 because none of its edges is on the mixed boundary condition. For element #2, only edge 2-3 is on the mixed boundary condition whereas the other three edges are embedded in the domain. Thus, for this element the matrices are

$$[G^e] = \frac{\lambda_{23}^e h_{23}}{6} \begin{bmatrix} 0 & 0 & 0 & 0 \\ 0 & 2 & 1 & 0 \\ 0 & 1 & 2 & 0 \\ 0 & 0 & 0 & 0 \end{bmatrix} \tag{4.105}$$

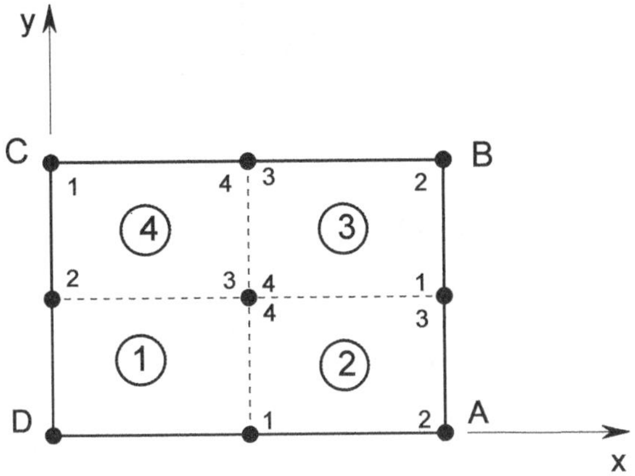

Figure 4.17 Sides AB and BC of the domain are subjected to mixed boundary conditions. The domain is divided into bilinear rectangular elements.

$$\{c^e\} = \frac{\alpha^e_{23} h_{23}}{2} \begin{Bmatrix} 0 \\ 1 \\ 1 \\ 0 \end{Bmatrix}. \tag{4.106}$$

For element #3, only edges 1-2 and 2-3 are on the mixed boundary condition whereas the other two edges are embedded in the domain. Thus, for this element the matrices are

$$[G^e] = \frac{\lambda^e_{12} h_{12}}{6} \begin{bmatrix} 2 & 1 & 0 & 0 \\ 1 & 2 & 0 & 0 \\ 0 & 0 & 0 & 0 \\ 0 & 0 & 0 & 0 \end{bmatrix} + \frac{\lambda^e_{23} h_{23}}{6} \begin{bmatrix} 0 & 0 & 0 & 0 \\ 0 & 2 & 1 & 0 \\ 0 & 1 & 2 & 0 \\ 0 & 0 & 0 & 0 \end{bmatrix} \tag{4.107}$$

$$\{c^e\} = \frac{\alpha^e_{12} h_{12}}{2} \begin{Bmatrix} 1 \\ 1 \\ 0 \\ 0 \end{Bmatrix} + \frac{\alpha^e_{23} h_{23}}{2} \begin{Bmatrix} 0 \\ 1 \\ 1 \\ 0 \end{Bmatrix}. \tag{4.108}$$

For element #4, only edge 1-4 is on the mixed boundary condition. Thus, for this element the matrices are

$$[G^e] = \frac{\lambda^e_{41} h_{41}}{6} \begin{bmatrix} 2 & 0 & 0 & 1 \\ 0 & 0 & 0 & 0 \\ 0 & 0 & 0 & 0 \\ 1 & 0 & 0 & 2 \end{bmatrix} \tag{4.109}$$

$$\{c^e\} = \frac{\alpha_{41}^e h_{41}}{2} \begin{Bmatrix} 1 \\ 0 \\ 0 \\ 1 \end{Bmatrix}. \tag{4.110}$$

4.6 ELEMENTS ASSEMBLY AND GLOBAL SYSTEM

No boundary conditions are applied when deriving the element matrices and equations. Boundary conditions are specified on the outer boundary of the entire domain. Therefore, the elements equations are assembled into a global system of equations that contains all primary and weighted secondary variables at all nodes. Let N be the number of global nodes, and n be the number of nodes per element. For the model problem considered in this chapter, there is only one degree of freedom for each node; that is, the primary variable at the node. Thus, the global system before application of the boundary conditions is N equations of the general form

$$[A]\{U\} = \{F\} + \{R\} \tag{4.111}$$

where the coefficient matrix $[A]$ is of dimensions (N,N), and the forcing vector $\{F\}$ and the vector of weighted secondary variables $\{R\}$ are of dimensions $(N,1)$. The entries of the coefficient matrix $a_{i_g j_g}$ are obtained by accumulating contributions from the coefficient matrices of all the elements whose local nodes include global nodes i_g and j_g. The connectivity (or identity) matrix is very helpful in the assembly of elements. Once the element matrices are computed for an element, they are assembled into the global system and the system is updated. Let a_{ij}^m denote an entry in the element coefficient matrix of element number m. The connectivity matrix identifies the corresponding position in the global system where this element entry is to be accumulated as $i_g = CM(m,i)$ and $j_g = CM(m,j)$, where CM is the connectivity matrix. Assembly of the elements can be accomplished by using this algorithm:

Initialize global matrices: $[A] = 0$ and $\{F\} = 0$

(1) For element m: compute element matrices $[A^e]$ and $\{F^e\}$.

(2) Assemble element matrices into the global system:

$$\begin{aligned} &\text{for j=1:n} \\ &\quad j_g = CM(m,j) \\ &\quad f_{j_g} = f_{j_g} + f_j^m \\ &\text{for i=1:n} \\ &\quad i_g = CM(m,i) \\ &\quad a_{i_g j_g} = a_{i_g j_g} + a_{ij}^m \\ &\text{end} \\ &\text{end} \end{aligned} \tag{4.112}$$

(3) Repeat steps (1) and (2) for all elements.

We remark that the array of weighted secondary variables $\{R\}$ is not assembled because it is unknown for the typical element. It will be computed directly for the global system rather than element-by-element. The entries of $\{R\}$ are zero for all interior nodes unless there is a concentrated source not accounted for by the distributed source $f(x,y)$. For boundary nodes where a natural boundary condition is prescribed, the entries of the weighted secondary variables are computed and added to the global forcing vector. For boundary nodes where essential boundary conditions are prescribed, the entries of $\{R\}$ are unknown whereas the corresponding entries of $\{U\}$ are known. We can either remove the corresponding equations that contain the unknown weighted secondary variables thereby obtaining a reduced system for the unknown primary variables, or replace those equations with equations that assign the known values of primary variables. For boundary nodes where mixed boundary conditions are prescribed, the element equations are given by Equation (4.96). Thus, we add $[G^e]_{MBC}$ to $[A^e]$, then assemble the augmented coefficient matrix into the global matrix. Also, we add $\{c^e\}_{MBC}$ to $\{F^e\}$, then assemble the augmented source vector into the global forcing vector. The use of augmented element coefficient matrix and forcing vector replaces the unknown secondary variable at the mixed boundary condition in terms of the primary variable. Thus, all the unknown entries of $\{R\}$ are either removed by applying EBC or replaced in terms of primary variables at MBC. After applying the boundary conditions, the modified (or reduced) system takes the form

$$[\bar{A}]\{U\} = \{\bar{F}\} \tag{4.113}$$

which is solved for the array of primary variables at all global nodes $\{U\}$. Afterward, we compute the unknown secondary variables and other quantities of interest. If we save the coefficient matrix and forcing vector before applications of boundary conditions, the array of secondary variables (known and unknown) can be found from Equation (4.111) rewritten here as

$$\{R\} = [A]\{U\} - \{F\}. \tag{4.114}$$

To reduce computer memory requirements, we need only save the coefficients of the equations for the unknown secondary variables. We remark that the construction of the global system requires only the element matrices $[A^e]$, $\{F^e\}$, $[G^e]_{MBC}$, and $\{c^e\}_{MBC}$. Weighted secondary variables at the element nodes disappear upon assembly except for the nodes on the outer domain boundary that can be decided by the boundary conditions. This is an expected result because the assembled global system could have been obtained by the use of global basis functions and the weak form for the entire domain without discretization into elements.

4.7 APPLICATIONS

4.7.1 HEAT CONDUCTION IN SOLIDS

We consider heat transfer by conduction in a solid. The objective is to determine the temperature distribution in the solid. Conservation of energy (or the first law of thermodynamics) is the basic law that governs the internal energy or temperature distribution within the solid, it can be written as

$$\rho C_p \frac{\partial T}{\partial t} + \nabla \cdot \vec{\tilde{q}} = f \tag{4.115}$$

where ρ (kg m^{-3}) is mass density, C_p is the heat capacity (J kg^{-1} °C^{-1}), $\vec{\tilde{q}}$ is the heat flux vector (W m^{-2}), and f (W m^{-3}) is the rate of internal energy added per unit volume. Fourier's law of heat conduction is used as the constitutive equation,

$$\vec{\tilde{q}} = -k\nabla T \tag{4.116}$$

where k (W m^{-1} °C^{-1}) is the material thermal conductivity. For steady conditions ($\partial T/\partial t = 0$), the temperature distribution in a two-dimensional solid $T(x,y)$ is governed by

$$-\left[\frac{\partial}{\partial x}\left(k\frac{\partial T}{\partial x}\right) + \frac{\partial}{\partial y}\left(k\frac{\partial T}{\partial y}\right)\right] - f = 0. \tag{4.117}$$

Comparison with the model problem shows that for the heat conduction problem we have $p_{11} = p_{22} = k$, $p_{12} = p_{21} = 0$, and $q = 0$. We also note that $\vec{v} = -\vec{\tilde{q}}$. Three different boundary conditions may be applied on the bounding surface of the solid:

(a) We may prescribe the temperature on the entire boundary or on a part of the boundary

$$T = T_s \qquad \text{Dirichlet boundary condition} \tag{4.118}$$

(b) Or we may prescribe the rate of heat flowing out of solid by conduction per unit area, which is the component of the heat flux vector normal to the boundary

$$\vec{n}.\vec{\tilde{q}} = (\tilde{q}_n)_s \qquad \text{Neumann boundary condition} \tag{4.119}$$

where $\vec{n}$ is the outward unit normal (pointing out of the solid domain). For an *insulated* boundary or wall, we have $\tilde{q}_n = 0$.

(c) Newton's law of cooling may be used to prescribe a relationship between the temperature and heat flux vector

$$\vec{n}.\vec{\tilde{q}} = \tilde{h}(T - T_\infty) \qquad \text{Robin boundary condition} \tag{4.120}$$

where $\tilde{h}$ (W m^{-2} °C^{-1}) is the coefficient of convection (forced or natural), and T_∞ is the temperature of the cooling fluid (gas or liquid) away from the surface (outside the boundary layer). Also, the body may exchange heat by radiation with the surrounding environment. The radiation boundary condition is nonlinear in the unknown surface temperature, and an iterative method can be used to solve problems with such a condition. During the iteration, the boundary condition is quasilinearized into a Robin type boundary condition as given by Equation (4.120). We may specify different types of boundary conditions on different (non-overlapping) segments of the boundary. However, only one of the three types of those conditions can be prescribed at a point on the boundary.

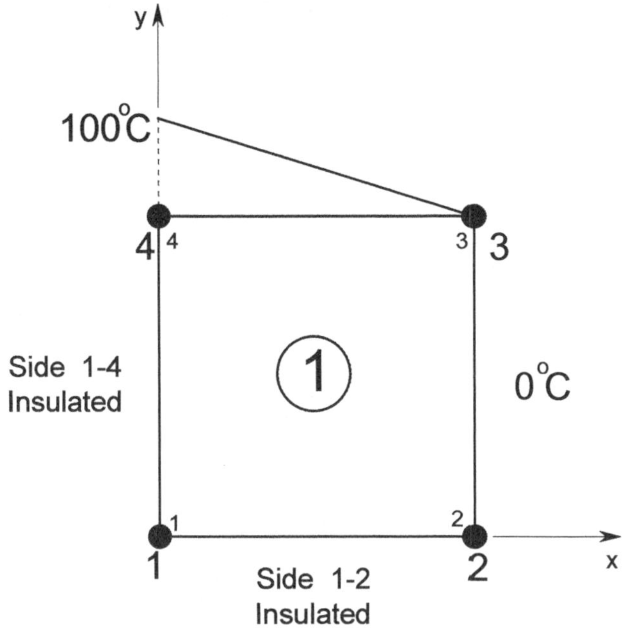

Figure 4.18 One element model of a square section, the element four nodes are also the global nodes.

Example 4.1. Heat Conduction using Rectangular Elements

Consider steady heat conduction in a square rod of side $d = 0.4$ m and thermal conductivity $k = 150$ W m^{-1}°C^{-1} as shown in Figure 4.18. There is no energy source within the domain ($f = 0$). The energy equation is

$$-k\nabla^2 T = 0 \tag{4.121}$$

The boundary conditions are

$$\begin{aligned} x &= 0 & \frac{\partial T}{\partial x} &= 0 & 0 \leq y \leq d \\ x &= d & T &= 0 & 0 \leq y \leq d \\ y &= 0 & \frac{\partial T}{\partial y} &= 0 & 0 \leq x \leq d \\ y &= d & T &= 100\left(1 - \frac{x}{d}\right) \text{°C} & 0 \leq x \leq d. \end{aligned} \tag{4.122}$$

Determine the temperature at the corner $(x,y) = (0,0)$, and check conservation of energy. (a) Solve this problem using one element, and (b) using four elements.

Solution using one square element ($b = a$)
Referring to the model problem, we conclude that $p_{11} = k$, $p_{22} = k$, $p_{12} = p_{21} = 0$,

and $q = 0$; hence, the mass matrix is zero and the stiffness matrix Equation (4.74) is

$$\begin{aligned} [K^e] &= k\left[L^{11}\right] + k\left[L^{22}\right] \\ &= \frac{k}{6}\begin{bmatrix} 4 & -1 & -2 & -1 \\ -1 & 4 & -1 & -2 \\ -2 & -1 & 4 & -1 \\ -1 & -2 & -1 & 4 \end{bmatrix} \end{aligned} \tag{4.123}$$

and because $f = 0$, the forcing vector is zero

$$\{F^e\} = \begin{Bmatrix} 0 \\ 0 \\ 0 \\ 0 \end{Bmatrix}. \tag{4.124}$$

For one element, the global system is

$$\frac{k}{6}\begin{bmatrix} 4 & -1 & -2 & -1 \\ -1 & 4 & -1 & -2 \\ -2 & -1 & 4 & -1 \\ -1 & -2 & -1 & 4 \end{bmatrix}\begin{Bmatrix} T_1 \\ T_2 \\ T_3 \\ T_4 \end{Bmatrix} = \begin{Bmatrix} R_1 \\ R_2 \\ R_3 \\ R_4 \end{Bmatrix} \tag{4.125}$$

$$R_i^e = \oint_{\Gamma_e} v_n^e \psi_i ds. \tag{4.126}$$

We recall the definition of v_n Equation (4.12),

$$\begin{aligned} v_n &= n_x v_x + n_y v_y \\ &= n_x\left(p_{11}\frac{\partial u}{\partial x} + p_{12}\frac{\partial u}{\partial y}\right) + n_y\left(p_{21}\frac{\partial u}{\partial x} + p_{22}\frac{\partial u}{\partial y}\right) \\ &= n_x\left(k\frac{\partial T}{\partial x}\right) + n_y\left(k\frac{\partial T}{\partial y}\right) \\ &= -\vec{n}\cdot\vec{q}. \end{aligned} \tag{4.127}$$

Hence v_n is the (negative) of the heat flux distribution on the element boundary. It is important to remember the negative sign when we interpret the physical meaning of the secondary variable in the problem of heat conduction as formulated here. If the secondary variable is negative (v_n is negative), then heat is flowing out of the domain. The two sides $x = 0$ and $y = 0$ are insulated, so $v_n = 0$ on these sides

$$\begin{aligned} R_1^1 &= \oint_{\Gamma_e} v_n^e \psi_1 ds \\ &= \int_{1-2} v_n^e \psi_1 ds + \int_{2-3} v_n^e \psi_1 ds + \int_{3-4} v_n^e \psi_1 ds + \int_{4-1} v_n^e \psi_1 ds \\ &= 0. \end{aligned} \tag{4.128}$$

The integrals on sides 1-2 and 4-1 are zero because $v_n = 0$ on these sides. The integrals on sides 2-3 and 3-4 are also zero because ψ_1 is zero on these sides whereas v_n is unknown.

$$\begin{aligned} R_2^1 &= \oint_{\Gamma_e} v_n^e \psi_2 ds \\ &= \int_{1-2} v_n^e \psi_2 ds + \int_{2-3} v_n^e \psi_2 ds + \int_{3-4} v_n^e \psi_2 ds + \int_{4-1} v_n^e \psi_2 ds \\ &= R_{2(2-3)}^1. \end{aligned} \tag{4.129}$$

The integrals on side 3-4 and 4-1 are zero because ψ_2 is zero on those sides. The integral on side 2-3 remains unknown because v_n is unknown and ψ_2 is not zero. Hence R_2 is unknown.

$$\begin{aligned} R_3^1 &= \oint_{\Gamma_e} v_n^e \psi_3 ds \\ &= \int_{1-2} v_n^e \psi_3 ds + \int_{2-3} v_n^e \psi_3 ds + \int_{3-4} v_n^e \psi_3 ds + \int_{4-1} v_n^e \psi_3 ds \\ &= R_{3(2-3)}^1 + R_{3(3-4)}^1. \end{aligned} \tag{4.130}$$

The integrals on sides 1-2 and 4-1 are zero because ψ_3 is zero on those sides. The integrals on sides 2-3 and 3-4 remain unknown because v_n is unknown and ψ_3 is not zero on those sides. Hence R_3 is unknown.

$$\begin{aligned} R_4^1 &= \oint_{\Gamma_e} v_n^e \psi_4 ds \\ &= \int_{1-2} v_n^e \psi_4 ds + \int_{2-3} v_n^e \psi_4 ds + \int_{3-4} v_n^e \psi_4 ds + \int_{4-1} v_n^e \psi_4 ds \\ &= R_{4(3-4)}^1. \end{aligned} \tag{4.131}$$

The integrals on sides 1-2 and 2-3 are zero because ψ_4 is zero on those sides. The integral on side 4-1 is zero because that side is insulated $v_n = 0$. The integral on side 3-4 remains unknown because v_n is unknown and ψ_4 is not zero on that side. Hence R_4 is unknown. The system before application of the boundary conditions is

$$\frac{k}{6}\begin{bmatrix} 4 & -1 & -2 & -1 \\ -1 & 4 & -1 & -2 \\ -2 & -1 & 4 & -1 \\ -1 & -2 & -1 & 4 \end{bmatrix} \begin{Bmatrix} T_1 \\ T_2 \\ T_3 \\ T_4 \end{Bmatrix} = \begin{Bmatrix} 0 \\ R_2 \\ R_3 \\ R_4 \end{Bmatrix}. \tag{4.132}$$

From the given boundary conditions, we know that $T_2 = 0$, $T_3 = 0$, and $T_4 = 100$°C. The first equation gives

$$\frac{k}{6}(4T_1 - T_2 - 2T_3 - T_4) = 0 \tag{4.133}$$

hence,

$$T_1 = 25°C. \tag{4.134}$$

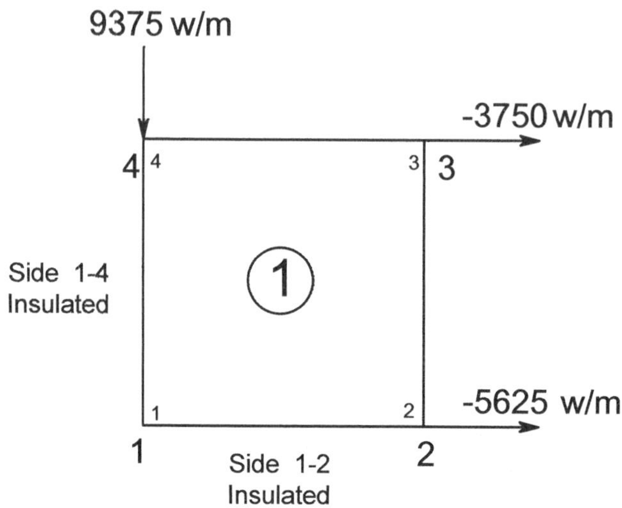

Figure 4.19 Heat flow on the domain boundary appears allocated to the nodes, positive when heat flows into the domain.

The unknown weighted secondary variables are found from the remaining three equations

$$
\begin{aligned}
R_2 &= \frac{k}{6}(-T_1 + 4T_2 - T_3 - 2T_4) \\
&= -5,625\mathrm{W}/m \\
R_3 &= \frac{k}{6}(-2T_1 - T_2 + 4T_3 - T_4) \\
&= -3,750\mathrm{W}/m \\
R_4 &= \frac{k}{6}(-T_1 - 2T_2 - T_3 + 4T_4) \\
&= 9,375\mathrm{W}/m.
\end{aligned} \tag{4.135}
$$

Although the finite-element solution provides heat flow at the nodes, it is important to recall that heat conduction takes place over an area made of the element edge and unit depth along the rod length. The heat flow at the nodes is equivalent to the heat transferred along the edges. The share of each node from the heat transferred along the edges meeting at the node is determined by the shape functions along those edges. For example, node #3 receives heat from edge 2-3 determined by

$$R^1_{3(2-3)} = \int_{2-3} v_n \psi_3 ds \tag{4.136}$$

and it receives heat from edge 3-4 determined by

$$R^1_{3(3-4)} = \int_{3-4} v_n \psi_3 ds. \tag{4.137}$$

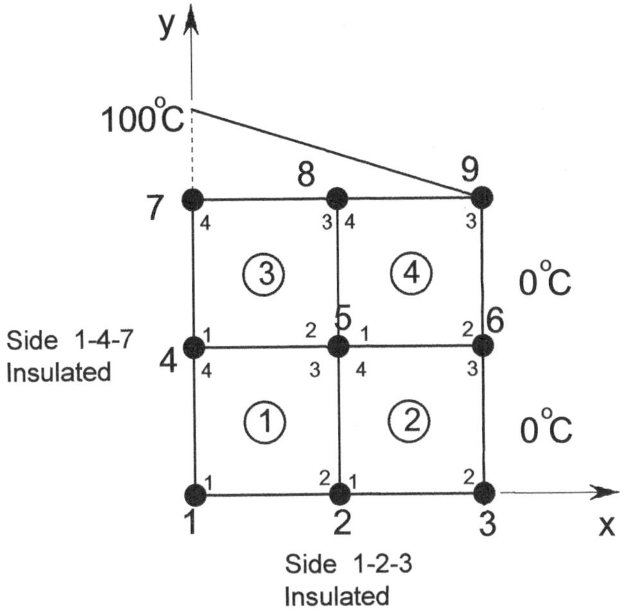

Figure 4.20 Four-element model of a square section, there are nine global nodes for the entire domain and four local nodes for each element.

The sum of these two contributions is $R_3 = -3{,}750\text{W}/m$, and the negative sign means heat is flowing out of the domain. The net heat flow over the boundary must equal the rate of energy generated within the domain. In this example, there is no energy generated within the domain, hence the net heat flow out of the boundary must be zero, as can be verified $R_1 + R_2 + R_3 + R_4 = 0$.

Solution using four square elements

We divide the square domain of side 0.4 m into 4 square elements each of side $a =$ 0.2 m. There are nine global nodes as shown in Figure 4.20. Since all elements are squares and have equal k, the element stiffness matrix is the same for all elements.

$$\begin{aligned}[K^e] &= k\left[L^{11}\right] + k\left[L^{22}\right] \\ &= \frac{k}{6}\begin{bmatrix} 4 & -1 & -2 & -1 \\ -1 & 4 & -1 & -2 \\ -2 & -1 & 4 & -1 \\ -1 & -2 & -1 & 4 \end{bmatrix}. \end{aligned} \tag{4.138}$$

Because each node has one degree of freedom, that is the temperature, the global coefficient matrix is of dimensions (9×9). This example is intended for hand calculations rather than computer implementation. Thus, we assemble the elements into a global system by writing an equation for each global node one-at-a-time.

In the global (assembled elements) system, the equation for global node #1 reads

$$k_{11}T_1 + k_{12}T_2 + \cdots + k_{18}T_8 + k_{19}T_9 = F_1 + R_1 \tag{4.139}$$

where

$$\begin{aligned}
k_{11} = k^1_{11} &= 4k/6 \\
k_{12} = k^1_{12} &= -k/6 \\
k_{13} &= 0 \\
k_{14} = k^1_{14} &= -k/6 \\
k_{15} = k^1_{13} &= -2k/6 \\
k_{16} &= 0 \\
k_{17} &= 0 \\
k_{18} &= 0 \\
k_{19} &= 0 \\
F_1 = F^1_1 &= 0 \\
R_1 &= R^1_1.
\end{aligned} \tag{4.140}$$

Looking at Figure 4.20, we realize that global node 1 is local node 1 of element 1, and it is not shared by any other element, hence $k_{11} = k^1_{11}$. We realize that global node 1 is connected to global node 5 only by the diagonal 1-3 of element 1, hence $k_{15} = k^1_{13}$. There is no connection between global node 1 and global nodes 3, 6, 7, 8, or 9; hence, the corresponding elements in the stiffness matrix are zero. The equation for global node 2 is

$$k_{21}T_1 + k_{22}T_2 + \cdots + k_{28}T_8 + k_{29}T_9 = F_2 + R_2 \tag{4.141}$$

where

$$\begin{aligned}
k_{21} = k^1_{21} &= -k/6 \\
k_{22} = k^1_{22} + k^2_{11} &= 4k/6 + 4k/6 = 8k/6 \\
k_{23} = k^2_{12} &= -k/6 \\
k_{24} = k^1_{24} &= -2k/6 \\
k_{25} = k^1_{23} + k^2_{14} &= -k/6 - k/6 = -2k/6 \\
k_{26} = k^2_{13} &= -2k/6 \\
k_{27} &= 0 \\
k_{28} &= 0 \\
k_{29} &= 0 \\
F_2 &= F^1_2 + F^2_1 = 0 \\
R_2 &= R^1_2 + R^2_1.
\end{aligned} \tag{4.142}$$

We realize that global node 2 is common to elements 1 and 2. It is local node 2 of element 1 and local node 1 of element 2, hence $k_{22} = k^1_{22} + k^2_{11}$. We note that global node 2 is connected to global node 5 by edge 2-3 of element 1 and edge 1-4 of element 2; hence, $k_{25} = k_{23}^1 + k_{14}^2$. For global node #5,

$$
\begin{aligned}
k_{51} = k_{31}^1 &= -2k/6 \\
k_{52} = k_{32}^1 + k_{41}^2 &= -k/6 - k/6 = -2k/6 \\
k_{53} = k_{42}^2 &= -2k/6 \\
k_{54} = k_{34}^1 + k_{21}^3 &= -k/6 - k/6 = -2k/6 \\
k_{55} = k_{33}^1 + k_{44}^2 + k_{22}^3 + k_{11}^4 &= 4k/6 + 4k/6 + 4k/6 + 4k/6 = 16k/6 \\
k_{56} = k_{43}^2 + k_{12}^4 &= -2k/6 \\
k_{57} = k_{24}^3 &= -2k/6 \\
k_{58} = k_{23}^3 + k_{14}^4 &= -k/6 - k/6 = -2k/6 \\
k_{59} = k_{13}^4 &= -2k/6 \\
F_5 &= F_3^1 + F_4^2 + F_2^3 + F_1^4 = 0 \\
R_5 &= R_3^1 + R_4^2 + R_2^3 + R_1^4.
\end{aligned}
\tag{4.143}
$$

Global node 5 is common to four elements. It is local node 3, 4, 2, and 1 of elements 1, 2, 3, and 4, respectively. Hence $k_{55} = k_{33}^1 + k_{44}^2 + k_{22}^3 + k_{11}^4$. Writing an equation for each of the nine global nodes, we put the global (assembled) system before application of the boundary conditions in matrix form

$$
\frac{k}{6}\begin{bmatrix}
4 & -1 & 0 & -1 & -2 & 0 & 0 & 0 & 0 \\
-1 & 8 & -1 & -2 & -2 & -2 & 0 & 0 & 0 \\
0 & -1 & 4 & 0 & -2 & -1 & 0 & 0 & 0 \\
-1 & -2 & 0 & 8 & -2 & 0 & -1 & -2 & 0 \\
-2 & -2 & -2 & -2 & 16 & -2 & -2 & -2 & -2 \\
0 & -2 & -1 & 0 & -2 & 8 & 0 & -2 & -1 \\
0 & 0 & 0 & -1 & -2 & 0 & 4 & -1 & 0 \\
0 & 0 & 0 & -2 & -2 & -2 & -1 & 8 & -1 \\
0 & 0 & 0 & 0 & -2 & -1 & 0 & -1 & 4
\end{bmatrix}
\begin{Bmatrix} T_1 \\ T_2 \\ T_3 \\ T_4 \\ T_5 \\ T_6 \\ T_7 \\ T_8 \\ T_9 \end{Bmatrix} =
$$

$$
\begin{Bmatrix}
F_1^1 \\ F_2^1 + F_1^2 \\ F_2^2 \\ F_4^1 + F_1^3 \\ F_3^1 + F_4^2 + F_2^3 + F_1^4 \\ F_3^2 + F_2^4 \\ F_4^3 \\ F_3^3 + F_4^4 \\ F_3^4
\end{Bmatrix} +
\begin{Bmatrix}
R_1^1 \\ R_2^1 + R_1^2 \\ R_2^2 \\ R_4^1 + R_1^3 \\ R_3^1 + R_4^2 + R_2^3 + R_1^4 \\ R_3^2 + R_2^4 \\ R_4^3 \\ R_3^3 + R_4^4 \\ R_3^4
\end{Bmatrix}.
\tag{4.144}
$$

In this example, the distributed source of internal energy f is zero. Therefore, all the components of the forcing vector are zero $F_i^e = 0$. The weighted secondary variables at internal nodes (not on the boundary) R_i is zero, unless there is a concentrated

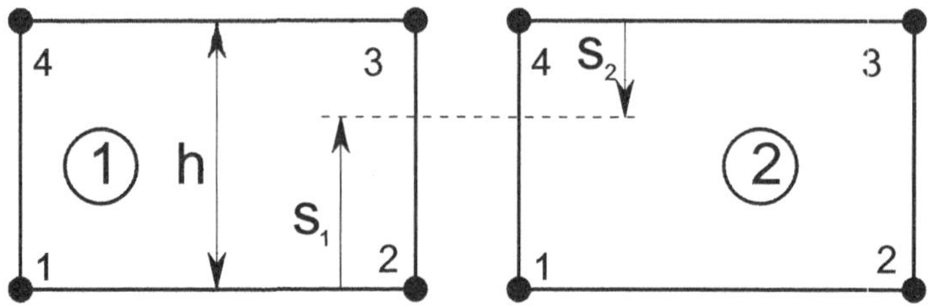

Figure 4.21 Distances measured along the common edge between elements 1 and 2, $s_1 + s_2 = h$.

source of energy not accounted for by the distributed source f. Let's consider R_1^1,

$$\begin{aligned} R_1^1 &= \oint_{\Gamma_e} v_n^1 \psi_1 ds \\ &= \int_{1-2} v_n^1 \psi_1 ds + \int_{2-3} v_n^1 \psi_1 ds + \int_{3-4} v_n^1 \psi_1 ds + \int_{4-1} v_n^1 \psi_1 ds. \\ &= 0 \end{aligned} \tag{4.145}$$

The integrals on sides 1-2 and 4-1 are zero because theses sides are insulated $v_n^1 = 0$. The integrals on sides 2-3 and 3-4 are also zero because ψ_1 is zero on these sides whereas v_n^1 is unknown. Next we consider R_2^1 and R_1^2,

$$\begin{aligned} R_2^1 &= \oint_{\Gamma^e} v_n^1 \psi_2 ds \\ &= \int_{1-2} v_n^1 \psi_2 ds + \int_{2-3} v_n^1 \psi_2 ds + \int_{3-4} v_n^1 \psi_2 ds + \int_{4-1} v_n^1 \psi_2 ds \\ &= \int_{2-3} v_n^1 \psi_2 ds. \end{aligned} \tag{4.146}$$

The integral on edge 1-2 is zero because $v_n^1 = 0$, and the integrals on edges 3-4 and 4-1 are zero because $\psi_2 = 0$ on these edges. The integral on edge 2-3 is unknown

$$\begin{aligned} R_1^2 &= \oint_{\Gamma^e} v_n^2 \psi_1 ds \\ &= \int_{1-2} v_n^2 \psi_1 ds + \int_{2-3} v_n^2 \psi_1 ds + \int_{3-4} v_n^2 \psi_1 ds + \int_{4-1} v_n^2 \psi_1 ds \\ &= \int_{4-1} v_n^2 \psi_1 ds. \end{aligned} \tag{4.147}$$

The integral on edge 1-2 is zero because it is insulated $v_n^2 = 0$, and the integrals on edges 2-3 and 3-4 are zero because $\psi_1 = 0$ on these edges. The integral on edge 4-1 is unknown. With the definitions of h, s_1 and $s_2 = h - s_1$ as shown in Figure 4.21, we show that the sum $R_2^1 + R_1^2$ is zero,

$$
\begin{aligned}
R_2^1 + R_1^2 &= \int_{2-3} v_n^1(s_1)\psi_2(s_1)ds_1 + \int_{4-1} v_n^2(s_2)\psi_1(s_2)ds_2 \\
&= \int_0^h v_n^1(s_1)(1-\frac{s_1}{h})ds_1 + \int_0^h v_n^2(s_2)\frac{s_2}{h}ds_2 \\
&= \int_0^h v_n^1(s_1)(1-\frac{s_1}{h})ds_1 - \int_h^0 v_n^2(s_2)(1-\frac{s_1}{h})ds_1 \\
&= \int_0^h \left[v_n^1(s_1)+v_n^2(s_2)\right](1-\frac{s_1}{h})ds_1 \\
&= 0 \qquad (4.148)
\end{aligned}
$$

because $\left[v_n^1(s_1) = -v_n^2(s_2)\right]$. The physical interpretation of this result is that the heat flowing out of element 1 along edge 2-3 must equal the heat flowing into element 2 along edge 1-4. It is left as an exercise to show that at global node 5, $R_5 = R_3^1 + R_4^2 + R_2^3 + R_1^4 = 0$. The vanishing of global weighted secondary variables at internal nodes such as node 5 is not surprising. This is an expected result because the assembled global system could have been obtained by the use of global basis functions and the weak form for the entire domain without discretization into elements. In such a formulation, the boundary term is an integral over the boundary of the entire domain, and weighted secondary variables (known or unknown) appear only at the nodes on the domain boundary. In summary, we have for the current example

$$
\begin{aligned}
R_1 &= R_1^1 = 0 \\
R_2 &= R_2^1 + R_1^2 = 0 \\
R_3 &= R_2^2 \qquad \text{unknown, but } T_3 = 0 \\
R_4 &= R_4^1 + R_1^3 = 0 \\
R_5 &= R_3^1 + R_4^2 + R_2^3 + R_1^4 = 0 \\
R_6 &= R_3^2 + R_2^4 \qquad \text{unknown, but } T_6 = 0 \\
R_7 &= R_4^3 \qquad \text{unknown, but } T_7 = 100°C \\
R_8 &= R_3^3 + R_4^4 \qquad \text{unknown, but } T_8 = 50°C \\
R_9 &= R_3^4 \qquad \text{unknown, but } T_9 = 0. \qquad (4.149)
\end{aligned}
$$

We note that whenever the primary variable (temperature) is known at a node the corresponding global weighted secondary variable is unknown, and whenever the global weighted secondary variable is known, the primary variable is unknown. The unknown temperatures are T_1, T_2, T_4, and T_5. The reduced system for the unknown temperatures is given by the first, second, fourth, and fifth equations of the global system Equation (4.144),

$$
\begin{bmatrix} 4 & -1 & -1 & -2 \\ -1 & 8 & -2 & -2 \\ -1 & -2 & 8 & -2 \\ -2 & -2 & -2 & 16 \end{bmatrix}
\begin{Bmatrix} T_1 \\ T_2 \\ T_4 \\ T_5 \end{Bmatrix} = \begin{Bmatrix} 0 \\ 0 \\ 200 \\ 300 \end{Bmatrix}. \qquad (4.150)
$$

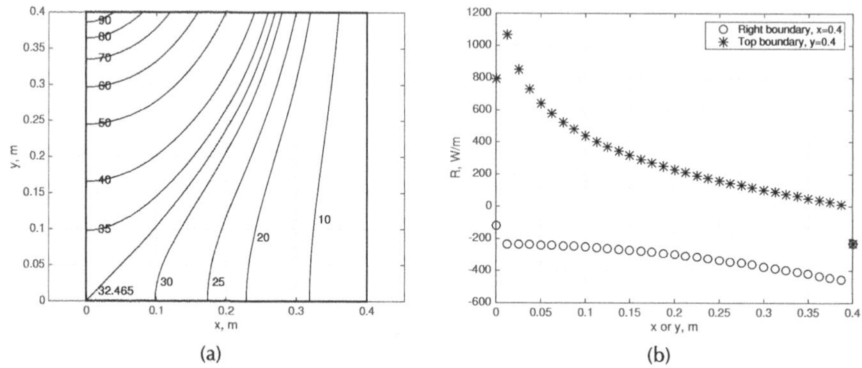

Figure 4.22 Steady heat conduction on a square domain with insulated walls at $x = 0$ and $y = 0$, and prescribed temperature on $x = 0.4$ m and $y = 0.4$ m. Mesh $32 \times 32 = 1024$ bilinear rectangular elements. (a) Temperature contours, and (b) heat flow at global nodes on the right and top boundaries.

The nonzero terms on the right are produced by rearranging the known temperatures. We solve the system for the unknown temperatures. Once the temperature is known at all global nodes, we determine the unknown weighted secondary variables from the global system before application of the boundary conditions Equation (4.144),

$$\{R\} = [K]\{T\} - \{F\}. \tag{4.151}$$

It is left as an exercise to compute the unknown temperatures and weighted secondary variables, and to check global conservation of energy.

Solutions using one or four bilinear elements do not provide sufficient accuracy for engineering applications. Such solutions are presented here only to demonstrate basic steps of the finite element method without the need of a computer. Computations on finer meshes require the use of a computer. To that end, we developed a MATLAB code and computed the solution to this example using a uniform mesh of $32 \times 32 = 1024$ bilinear rectangular elements.

Temperature contours are presented in Figure 4.22a. The contours appear perpendicular to the insulated sides $x = 0$ and $y = 0$, as expected, and appear nearly parallel to the right boundary $x = 0.4$ m of uniform temperature. The temperature at the corner $(x, y) = (0, 0)$ is 32.465°C. The weighted global secondary variables at the nodes on the right boundary $x = 0.4$ m, and top boundary $y = 0.4$ m are depicted in Figure 4.22b. There are 33 points on each boundary. The sum of the weighted secondary variables on each boundary give the rate of heat flow over that boundary. The rate of heat flow is positive on the top boundary indicating heat flowing into the domain, and is negative on the right boundary indicating equal heat flowing out of the domain. The other two sides are insulated and no energy is generated within the domain, thus global energy conservation is satisfied.

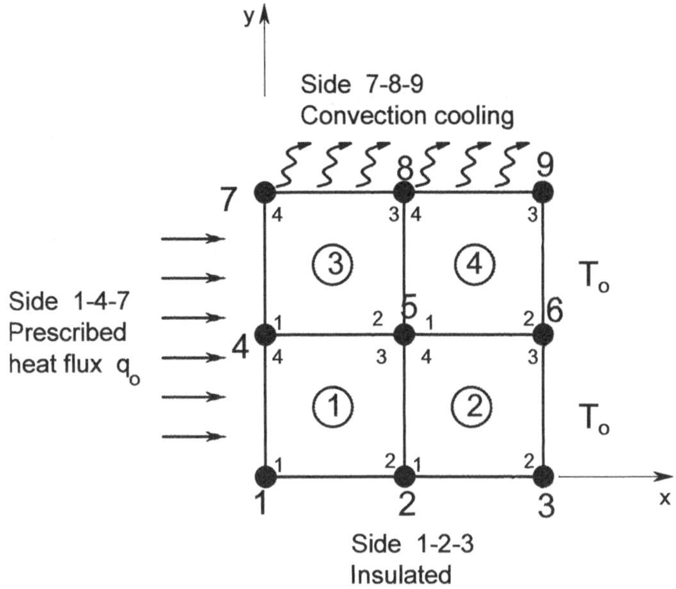

Figure 4.23 Four-element model of a square section, there are nine global nodes for the entire domain and four local nodes for each element.

Example 4.2. Heat Conduction with Convection Boundary Conditions

Consider steady heat conduction in a square rod of side $d = 0.2$ m and thermal conductivity $k = 90$ W m^{-1}°C^{-1}. Energy is generated within the domain at uniform rate of $f_o = 1\text{MW}\ m^{-3}$. The energy equation is

$$-k\nabla^2 T = f. \tag{4.152}$$

The boundary conditions are

$$\begin{aligned}
x &= 0 \qquad -k\frac{\partial T}{\partial x} = q_o \qquad 0 \le y \le d \\
x &= d \qquad T = T_o \qquad 0 \le y \le d \\
y &= 0 \qquad \frac{\partial T}{\partial y} = 0 \qquad 0 \le x \le d \\
y &= d \qquad -k\frac{\partial T}{\partial y} = \tilde{h}(T - T_\infty) \qquad 0 \le x \le d
\end{aligned} \tag{4.153}$$

where $q_o = 25$ kW m^{-2}, $T_o = 300$°C, $\tilde{h} = 180$ W m^{-2}°C^{-1} is the coefficient of convection, and $T_\infty = 10$°C is the cooling fluid temperature away from the surface. Determine: (a) Unknown nodal temperature, (b) heat flow at global nodes, and (c) check energy balance. Solve this problem using four rectangular elements.

Solution

The element side is a=d/2=0.1 m. For elements # 1 and 2, $[G^e] = 0$ and $\{c^e\} = 0$ because the boundaries of these elements are not subjected to convection. Their stiffness matrix is

$$[K^e] = \frac{k}{6}\begin{bmatrix} 4 & -1 & -2 & -1 \\ -1 & 4 & -1 & -2 \\ -2 & -1 & 4 & -1 \\ -1 & -2 & -1 & 4 \end{bmatrix} = \begin{bmatrix} 60 & -15 & -30 & -15 \\ -15 & 60 & -15 & -30 \\ -30 & -15 & 60 & -15 \\ -15 & -30 & -15 & 60 \end{bmatrix}. \tag{4.154}$$

For elements # 3 and 4, there is convection on side 3-4 of each of the elements, thus their stiffness matrices must be augmented with matrix $[G^e]$ for side 3-4 only since the other three sides of the elements are not subjected to convection.

$$[G^e] = \frac{\tilde{h}^e_{34} h_{34}}{6}\begin{bmatrix} 0 & 0 & 0 & 0 \\ 0 & 0 & 0 & 0 \\ 0 & 0 & 2 & 1 \\ 0 & 0 & 1 & 2 \end{bmatrix} = \begin{bmatrix} 0 & 0 & 0 & 0 \\ 0 & 0 & 0 & 0 \\ 0 & 0 & 6 & 3 \\ 0 & 0 & 3 & 6 \end{bmatrix} \tag{4.155}$$

$$[K^e]+[G^e] = \begin{bmatrix} 60 & -15 & -30 & -15 \\ -15 & 60 & -15 & -30 \\ -30 & -15 & 66 & -12 \\ -15 & -30 & -12 & 66 \end{bmatrix} \tag{4.156}$$

$$\{c^e\} = \frac{\alpha^e_{34} h_{34}}{2}\begin{Bmatrix} 0 \\ 0 \\ 1 \\ 1 \end{Bmatrix} = \begin{Bmatrix} 0 \\ 0 \\ 90 \\ 90 \end{Bmatrix} \tag{4.157}$$

because $\alpha = \tilde{h}T_\infty = 1800$. The forcing vector due to energy generation within the domain is the same for all four elements,

$$
\begin{aligned}
\{F^e\} &= \frac{f_e A_e}{4}\begin{Bmatrix}1\\1\\1\\1\end{Bmatrix}\\
&= \begin{Bmatrix}2500\\2500\\2500\\2500\end{Bmatrix} W/m.
\end{aligned}
\tag{4.158}
$$

Heat flow at global nodes

$$
\begin{aligned}
R_1 &= \frac{1}{2}q_o a = 1250\,W/m\\
R_2 &= 0\\
R_4 &= 2500\,W/m\\
R_5 &= 0.
\end{aligned}
\tag{4.159}
$$

The temperatures at nodes 3, 6, and 9 are given; $T_3 = T_6 = T_9 = 300°\text{C}$. The corresponding weighted secondary variables R_3, R_6, R_9 are unknown. At node # 8, both temperature T_8 and heat flow R_8 are unknown (convection boundary condition), but we have already expressed R_8 in terms of T_8 and modified the coefficient matrix as shown by Equation (4.156). Similarly, the temperature at node # 7 is unknown. However, the heat flow at node 7 comes from two edges of element # 3. On edge 4-1, the heat flux into the domain is given, and node # 7 receives a known quantity of 1250 W/m, but it also receives an unknown heat flow by convection on edge 3-4. The unknown heat flow by convection is expressed in terms of T_7 and is accounted for by the matrix $[G^e]$ and the vector $\{c^e\}$.

$$
15\begin{bmatrix}
4 & -1 & 0 & -1 & -2 & 0 & 0 & 0 & 0\\
-1 & 8 & -1 & -2 & -2 & -2 & 0 & 0 & 0\\
0 & -1 & 4 & 0 & -2 & -1 & 0 & 0 & 0\\
-1 & -2 & 0 & 8 & -2 & 0 & -1 & -2 & 0\\
-2 & -2 & -2 & -2 & 16 & -2 & -2 & -2 & -2\\
0 & -2 & -1 & 0 & -2 & 8 & 0 & -2 & -1\\
0 & 0 & 0 & -1 & -2 & 0 & 4.4 & -0.8 & 0\\
0 & 0 & 0 & -2 & -2 & -2 & -0.8 & 8.8 & -0.8\\
0 & 0 & 0 & 0 & -2 & -1 & 0 & -0.8 & 4.4
\end{bmatrix}
\begin{Bmatrix}T_1\\T_2\\T_3\\T_4\\T_5\\T_6\\T_7\\T_8\\T_9\end{Bmatrix} =
\begin{Bmatrix}2500\\5000\\2500\\5000\\10000\\5000\\2500\\5000\\2500\end{Bmatrix}
+\begin{Bmatrix}1250\\0\\R_3\\2500\\0\\R_6\\1250\\0\\R_9\end{Bmatrix}
+\begin{Bmatrix}0\\0\\0\\0\\0\\0\\90\\180\\90\end{Bmatrix}.
\tag{4.160}
$$

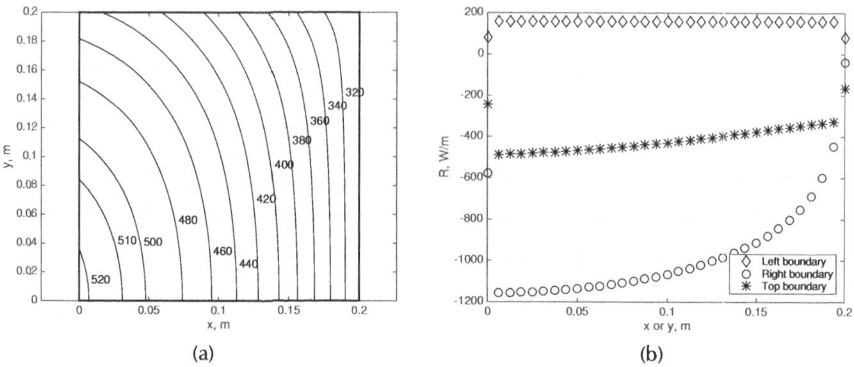

Figure 4.24 Steady heat conduction on a square domain with energy generated within the domain. Boundary conditions: insulated wall on $y = 0$ (bottom), prescribed heat flux on $x = 0$ (left), convection heat transfer on $y = 0.2$ m (top), and prescribed temperature on $x = 0.2$ m (right). Mesh $32 \times 32 = 1024$ bilinear rectangular elements. (a) Temperature contours, and (b) heat flow at global nodes on the left (conduction), right (conduction), and top (convection) boundaries.

We apply boundary conditions $T_3 = T_6 = T_9 = 300°$C, and solve for the unknown temperatures and weighted secondary variables. The solution is: $T_1 = 527.4, T_2 = 458.8, T_3 = 300, T_4 = 508.7, T_5 = 446.1, T_6 = 300, T_7 = 447.9, T_8 = 392.4, T_9 = 300°$C. The weighted secondary variables give heat flow by conduction at the boundary nodes (positive when entering domain): $R_1 = 1{,}250, R_2 = 0, R_3 = -9{,}264, R_4 = 2{,}500, R_5 = 0, R_6 = -16{,}919, R_7 = 1{,}250, R_8 = 0, R_9 = -5{,}382$ W/m. Heat flow by conduction at nodes 1, 4, and 7 is prescribed; it is positive indicating heat flow into the domain. At nodes 3, 6, and 9, the temperature is prescribed and computed heat flow by conduction is negative as heat is flowing out of the domain on the right boundary. Heat flow by convection per element is obtained from Equation (4.89),

$$\{R^e_{conv}\} = -[G^e]\{T^e\} + \{c^e\} \tag{4.161}$$

which gives: $R_7 = -3{,}775, R_8 = -6{,}773, R_9 = -2{,}887$ W/m. The negative heat flow at nodes 7, 8, and 9 is due to cooling of the top surface. The net heat flow out of the domain (conduction plus convection on all sides of the square) = 40,000 W/m, which is precisely the rate of energy generated within the domain. Thus, global energy is conserved.

The MATLAB code developed for the previous example is modified for convection boundary conditions and is used to solve this example on a mesh of $32 \times 32 = 1024$ bilinear rectangular elements. Temperature contours are depicted in Figure 4.24a. The contours appear perpendicular to the insulated side $y = 0$, as expected, and nearly parallel to the right boundary where uniform temperature is prescribed. The temperature at the corner $(x, y) = (0, 0)$ is 522.23°C. The weighted global secondary variables at the nodes on the left ($x = 0$), right ($x = 0.2$ m), and top ($y = 0.2$ m) boundaries are depicted in Figure 4.22b. On the left side heat is supplied by conduction at the rate of 5 kW/m, on the right side heat is leaving by conduction at the rate of 31.539 kW/m, and on the top heat is removed by convection at the rate of 13.461 kW/m. Thus, the net heat flowing out of the domain is 40 kW/m, which is equal to the rate of energy generated within the domain. Thus, global energy conservation is satisfied.

4.7.2 FULLY DEVELOPED LAMINAR FLOW IN NONCIRCULAR DUCTS

We consider incompressible flow and assume Newtonian fluid with constant density ρ and dynamic viscosity μ. In rectangular Cartesian coordinates (x, y, z), the velocity vector is projected into components (u, v, w). Conservation of mass and the principle of linear momentum become:

Continuity Equation

$$\frac{\partial u}{\partial x} + \frac{\partial v}{\partial y} + \frac{\partial w}{\partial z} = 0 \tag{4.162}$$

Navier–Stokes Equations

$$\rho\left(\frac{\partial u}{\partial t} + u\frac{\partial u}{\partial x} + v\frac{\partial u}{\partial y} + w\frac{\partial u}{\partial z}\right) = -\frac{\partial p}{\partial x} + \mu\left[\frac{\partial^2 u}{\partial x^2} + \frac{\partial^2 u}{\partial y^2} + \frac{\partial^2 u}{\partial z^2}\right] \tag{4.163}$$

$$\rho\left(\frac{\partial v}{\partial t} + u\frac{\partial v}{\partial x} + v\frac{\partial v}{\partial y} + w\frac{\partial v}{\partial z}\right) = -\frac{\partial p}{\partial y} + \mu\left[\frac{\partial^2 v}{\partial x^2} + \frac{\partial^2 v}{\partial y^2} + \frac{\partial^2 v}{\partial z^2}\right] \tag{4.164}$$

$$\rho\left(\frac{\partial w}{\partial t} + u\frac{\partial w}{\partial x} + v\frac{\partial w}{\partial y} + w\frac{\partial w}{\partial z}\right) = -\frac{\partial p}{\partial z} + \mu\left[\frac{\partial^2 w}{\partial x^2} + \frac{\partial^2 w}{\partial y^2} + \frac{\partial^2 w}{\partial z^2}\right]. \tag{4.165}$$

Conservative body force is assumed, and its potential is included with the pressure in the gradient terms.

Now, we simplify these equations for the case of a long straight duct with arbitrary cross-section. For low enough Reynolds number, we may assume laminar flow moving in straight lines parallel to the duct axis, which is taken as the z axis. Hence, $u = v = 0$ and the only velocity component is w. With these assumptions, the continuity equation gives $\partial w/\partial z = 0$, and thus w is a function of x, y, and t only. We find from the momentum Equations (4.163) and (4.164), that $\partial p/\partial x = 0$, and $\partial p/\partial y = 0$. We conclude that the pressure is a function of z and t only. However, since w is not a function of z, it follows from Equation (4.165) that $\partial p/\partial z$ is a function of time only or a constant. The equation for w simplifies to

$$\rho\frac{\partial w}{\partial t} = -\frac{\partial p}{\partial z} + \mu\nabla^2 w. \tag{4.166}$$

Further, we assume steady flow driven by a constant (negative) pressure gradient dp/dz. The momentum equation becomes

$$-\mu\nabla^2 w = -\frac{dp}{dz} \tag{4.167}$$

or

$$-\mu\left(\frac{\partial^2 w}{\partial x^2}+\frac{\partial^2 w}{\partial y^2}\right) = -\frac{dp}{dz}. \tag{4.168}$$

Comparison with the model problem shows that for the viscous flow problem we have $p_{11} = p_{22} = \mu$, and $p_{12} = p_{21} = 0$, $q = 0$, and $f = -dp/dz$. The boundary condition is the no-slip condition on the duct wall; that the velocity is zero at the wall

$$w = 0 \qquad \text{on} \qquad \partial\Omega. \tag{4.169}$$

The boundary condition is a Dirichlet-type condition; or an essential boundary condition.

A quantity of interest is the volume flow rate, which is given by

$$Q = \int_A w dA \tag{4.170}$$

where A is area of the duct cross-section. The nonzero components of the viscous stress tensor are also of interest,

$$\tau_{zx} = \tau_{xz} = \mu\frac{\partial w}{\partial x} \tag{4.171}$$

$$\tau_{yz} = \tau_{zy} = \mu\frac{\partial w}{\partial y}. \tag{4.172}$$

A word about the notation is in order. In this problem w is axial velocity in the duct; it is the primary variable, and should not be confused with the weight or test function. We recall the definition of the secondary variable v_n Equation (4.12),

$$\begin{aligned} v_n &= n_x v_x + n_y v_y \\ &= n_x(p_{11}\frac{\partial u}{\partial x}+p_{12}\frac{\partial u}{\partial y}) + n_y(p_{21}\frac{\partial u}{\partial x}+p_{22}\frac{\partial u}{\partial y}) \\ &= n_x(\mu\frac{\partial w}{\partial x}) + n_y(\mu\frac{\partial w}{\partial y}) \\ &= n_x\tau_{xz} + n_y\tau_{yz} \\ &= \tau_{nz} \end{aligned} \tag{4.173}$$

the secondary variable is the z-component of the shear stress on a surface defined by the unit normal $\hat{n}$. It is the total shear stress on such a surface. On the duct wall, the primary variable is prescribed by the no-slip condition, and the distribution of

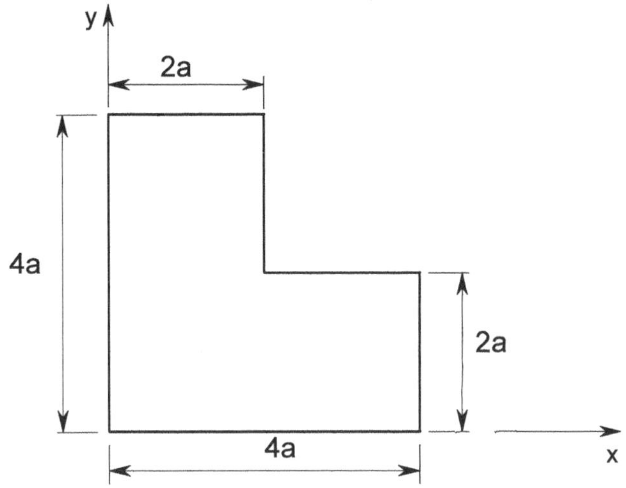

Figure 4.25 Cross-section of a straight duct carrying viscous flow.

secondary variable (wall shear stress) is unknown. Integrating Equation (4.168) over the duct cross-section and using the divergence theorem, we get

$$-\oint_{\Gamma_E} v_n ds = -\frac{dp}{dz}A \tag{4.174}$$

Over a segment of the duct between axial locations z_2 and z_1, Equation (4.174) shows that the total shear force on the duct wall is in equilibrium with the pressure force $(p_1 - p_2)A$.

Example 4.3. Laminar Flow in a Straight Duct with an Angle-Shaped Section

We wish to use the finite-element method to determine the velocity $w(x,y)$, and volume flow rate in a duct with the angle-shaped cross-section shown in Figure (4.25). Since μ is constant, we rewrite Equation (4.168) as

$$-\left(\frac{\partial^2 w}{\partial x^2} + \frac{\partial^2 w}{\partial y^2}\right) = -\frac{1}{\mu}\frac{dp}{dz}. \tag{4.175}$$

Solution Using Bilinear Square Elements

We divide the section into 12 bilinear square elements as shown in Figure 4.26. All elements are identical and the typical element is also shown in the figure; the element side is a. There are 21 global nodes, and from the no-slip boundary condition we conclude that the velocity is zero at all the nodes on the boundary; we need to determine the velocity at the interior nodes 7, 8, 9, 12, and 17 only. We assume the velocity distribution on the cross-section to be symmetric about the line connecting global nodes 1, 7, and 13; hence, $W_{12} = W_8$ and $W_{17} = W_9$. Thus, we need to write down and solve the three equations for global nodes 7, 8, and 9 only.

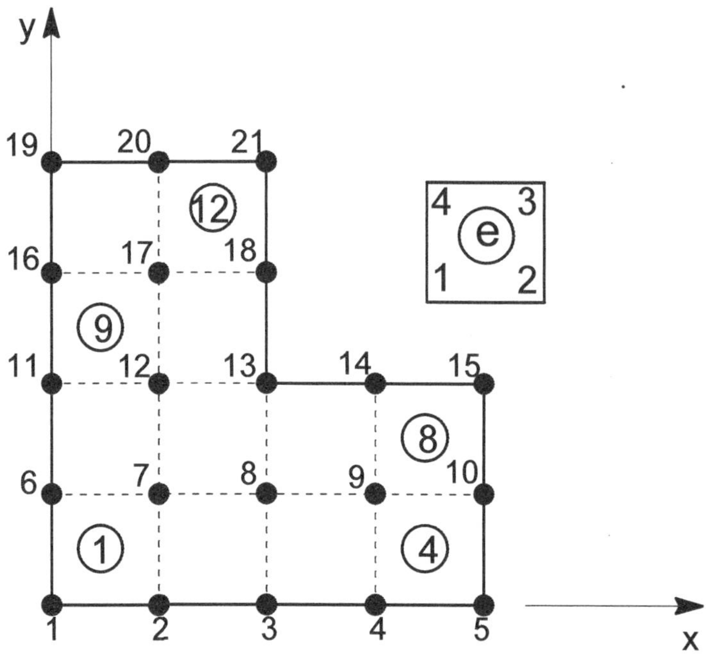

Figure 4.26 A mesh on the angle-shaped section, 12 square bilinear elements and 21 global nodes. The typical element (e) has four local nodes.

For Poisson's equation Equation (4.175), we have $p_{11}^e = 1$, $p_{22}^e = 1$, and $p_{12}^e = p_{21}^e = 0$, and the element stiffness matrix for a square element is

$$\begin{aligned}[K^e] &= p_{11}^e\left[L^{11}\right] + p_{22}^e\left[L^{22}\right] \\ &= \frac{1}{6}\begin{bmatrix} 4 & -1 & -2 & -1 \\ -1 & 4 & -1 & -2 \\ -2 & -1 & 4 & -1 \\ -1 & -2 & -1 & 4 \end{bmatrix}. \end{aligned} \tag{4.176}$$

The forcing vector is

$$\{F^e\} = \frac{f_o a^2}{4}\begin{Bmatrix} 1 \\ 1 \\ 1 \\ 1 \end{Bmatrix} \tag{4.177}$$

where $f_o = -(1/\mu)dp/dz$. The equation for global node #7 is

$$k_{7,1}W_1 + k_{7,2}W_2 + \cdots + k_{7,7}W_7 + \cdots + k_{7,21}W_{21} = F_7 + R_7.$$

Using the boundary conditions and the balance of weighted secondary variables (at an interior node $R = 0$), we obtain

$$k_{7,7}W_7 + k_{7,8}W_8 + k_{7,9}W_9 + k_{7,12}W_{12} + k_{7,17}W_{17} = F_7.$$

Because global nodes 7 and 9 do not belong to the same element, we get $k_{7,9} = 0$, similarly $k_{7,17} = 0$, and the equation for node # 7 simplifies to

$$k_{7,7}W_7 + k_{7,8}W_8 + k_{7,12}W_{12} = F_7.$$

Noting that global node #7 is common to four elements with the same matrices, we obtain $k_{7,7} = 4k^e_{11} = 4 \times (4/6) = 8/3$. Edge $7-8$ is common to two elements, hence $k_{7,8} = 2k^e_{12} = 2 \times (-1/6) = -1/3$. Similarly, we get $k_{7,12} = -1/3$. The forcing term F_7 is the sum of the forcing terms of the four elements that share global node # 7, hence $F_7 = 4(f_o a^2/4) = f_o a^2$. Using the symmetry condition, $W_{12} = W_8$, we write the equation at global node # 7 as

$$\frac{8}{3}W_7 - \frac{2}{3}W_8 = f_o a^2 \tag{4.178}$$

Keeping the non-zero terms, we write the equation for global node #8 as

$$k_{8,7}W_7 + k_{8,8}W_8 + k_{8,9}W_9 + k_{8,12}W_{12} = F_8$$

Because nodes 8 and 12 belong to the same element, we obtain $k_{8,12} = k^e_{24} = -2/6$. We evaluate the other coefficients as we have done for node #7, and obtain

$$-\frac{1}{3}W_7 + \frac{7}{3}W_8 - \frac{1}{3}W_9 = f_o a^2. \tag{4.179}$$

Following the same procedure, we obtain the equation for global node # 9,

$$-\frac{1}{3}W_8 + \frac{8}{3}W_9 = f_o a^2. \tag{4.180}$$

Solving the system of Equations (4.178)–(4.180), we get

$$\left\{ \begin{array}{c} W_7 \\ W_8 \\ W_9 \end{array} \right\} = \frac{3f_o a^2}{424} \left\{ \begin{array}{c} 73 \\ 80 \\ 63 \end{array} \right\}. \tag{4.181}$$

The volume flow rate through the duct is

$$Q = \int_A w(x,y)dA. \tag{4.182}$$

The integral over the typical element gives

$$\begin{aligned} Q^e &= \int_{A_e} W^e(x,y)dA \\ &= \int_{A_e} (W^e_1\psi_1 + W^e_2\psi_2 + W^e_3\psi_3 + W^e_4\psi_4)dA \\ &= \frac{A_e}{4}(W^e_1 + W^e_2 + W^e_3 + W^e_4). \end{aligned} \tag{4.183}$$

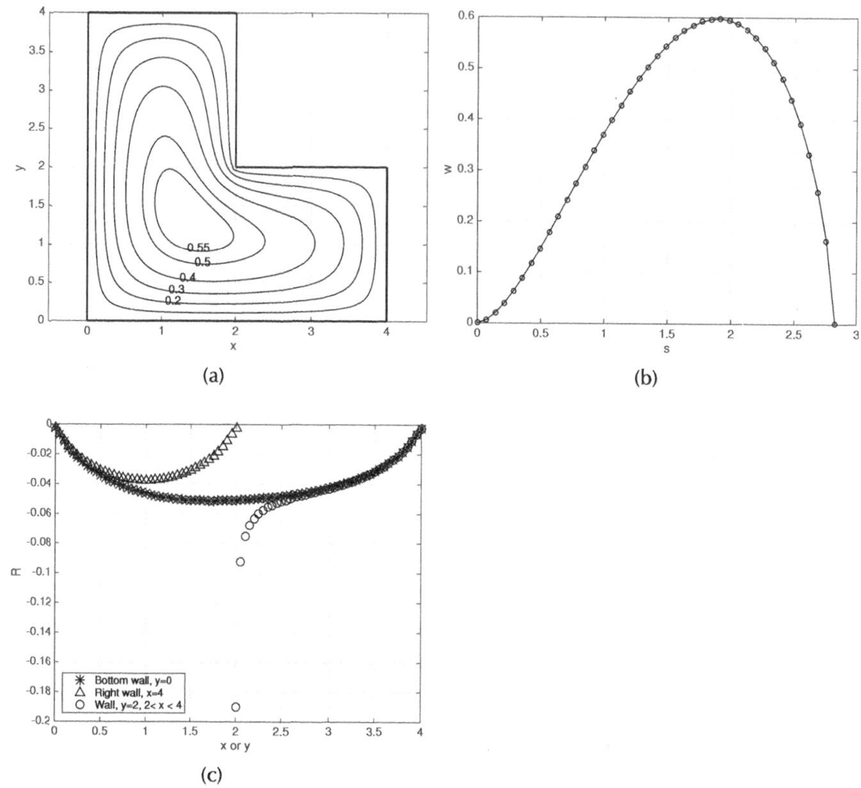

Figure 4.27 Fully developed laminar flow in a duct. (a) Velocity contours, (b) velocity distribution on diagonal line from (0,0) to (2,2), and (c) shear force (per unit duct-length) at global nodes on the duct boundaries.

Summing the contributions of all 12 elements, we get

$$\begin{aligned} Q &= \frac{a^2}{4}(4W_7 + 8W_8 + 8W_9) \\ &= \frac{1077}{424}a^4 f_o \\ &= 2.540a^4 f_o. \end{aligned} \tag{4.184}$$

An alternative approach for utilizing the symmetry, is to solve for half the domain on the right (or left) of diagonal 1-7-13. Essential boundary conditions are applied on the wall, but natural boundary condition $v_n = 0$ is applied on the diagonal 1-7-13. The problem is solved again on a finer mesh of 4800 bilinear rectangular elements. We divide the long side of the angle-shaped duct into 80 uniform segments, and the short side into 40 segments. We use a as a reference length and $f_o a^2$ as reference velocity. Velocity contours are shown in Figure 4.27a, and velocity distribution on the

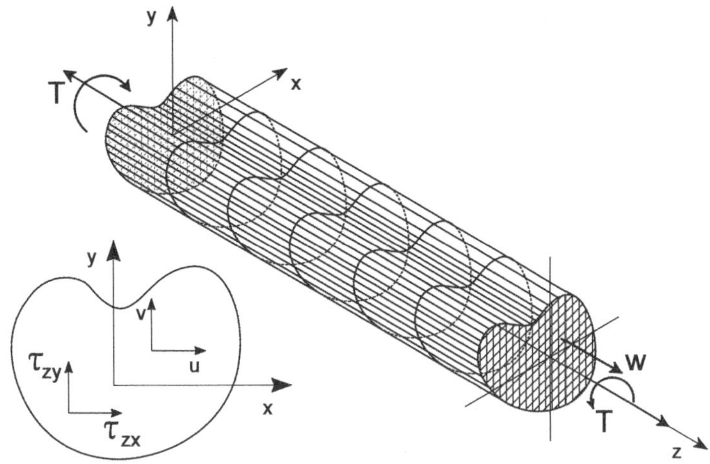

Figure 4.28 A non-circular prismatic bar twisted by two equal and opposite torques T. Displacements (u,v,w), and shear stresses (τ_{zx},τ_{zy}).

diagonal connecting the corners (0,0) to (2,2) is shown in Figure 4.27b. The maximum velocity of 0.5970 is on the diagonal. The volume flow rate computed by integrating w over the cross-section is $Q = 3.4209 f_o a^4$. The global weighted secondary variables R at the boundary nodes on the walls of half the duct are depicted in Figure 4.27c; values on the other half are symmetric. The values of R in the figure add up to 6, and for the total duct we get 12, which is the total nondimensional shear force per unit axial length. This result is true for any grid resolution. We note a relatively large value of R at the convex corner $(x,y) = (2,2)$. This is due to a singularity in the wall shear stress at the corner, but the integral of the shear stress over the wall is finite.

4.7.3 TORSION OF NONCIRCULAR SECTIONS

We want to determine the stresses in an elastic prismatic bar of an arbitrary cross-section twisted by two equal and opposite torques applied at its ends as shown in Figure 4.28. The twist angle per unit length is denoted by θ, and shear modulus of the material is G. The twist axis is an axis parallel to the prism generatrix, which is taken as the z-axis, and the cross-section is in the $x-y$ plane. The Cartesian components of displacements are denoted by (u,v,w). The displacement in the $x-y$ plane is a rigid body rotation about the z-axis, that is $u = -\theta zy$ and $v = \theta zx$. A non-circular section also warps, so there is a nonuniform displacement $w(x,y)$ in the z- direction. Boresi [14] presented approaches for solving the problem for solid sections (simply connected domains) and tubular sections (multiply connected domains); here we consider only solid sections. He also showed that if the twist axis is transferred to another axis parallel to the z-axis, the displacement field differs by a rigid body displacement; which has no effects on the stresses and torque. Theory of linear elasticity (e.g., Boresi [14]) reduces the problem to solving Poisson's equation for

Prandtl stress function, $\phi(x,y)$.

$$-\nabla^2\phi = 2G\theta \tag{4.185}$$

or

$$-\left(\frac{\partial^2\phi}{\partial x^2}+\frac{\partial^2\phi}{\partial y^2}\right) = 2G\theta \qquad on \qquad \Omega \tag{4.186}$$

subjected to the boundary condition

$$\phi = 0 \qquad on \qquad \partial\Omega \tag{4.187}$$

where the domain Ω is the cross-sectional area of the bar, A. The torque is given by

$$T = 2\int_A \phi dA \tag{4.188}$$

and the shear stresses are given by

$$\begin{aligned} \tau_{zx} &= \frac{\partial\phi}{\partial y} \\ \tau_{zy} &= -\frac{\partial\phi}{\partial x}. \end{aligned} \tag{4.189}$$

The other components of the stress tensor are zero. The warping function $\psi = w/\theta$ is related to the Prandtl stress function by

$$\begin{aligned} \frac{\partial\psi}{\partial x} &= y+\frac{1}{G\theta}\frac{\partial\phi}{\partial y} \\ \frac{\partial\psi}{\partial y} &= -\left(x+\frac{1}{G\theta}\frac{\partial\phi}{\partial x}\right). \end{aligned} \tag{4.190}$$

Comparison of Equation (4.186) with the model problem shows that for the torsion problem we have $p_{11} = p_{22} = 1$, $p_{12} = p_{21} = 0$, $q = 0$, and $f = 2G\theta$. The primary variable is Prandtl stress function ϕ, and the boundary condition is a Dirichlet condition; an essential boundary condition. Furthermore, comparison between Equation (4.186) subjected to boundary condition Equation (4.187) and Equation (4.168) subjected to boundary condition Equation (4.169) shows that Prandtl stress function ϕ (with proper nondimensionalization) for a section is mathematically identical to the axial velocity of a laminar flow in a duct with the same cross-section. Other problems, such as the transverse deflection of a membrane stretched over the section and subjected to a uniform pressure, are also governed by the same equation and boundary condition as the torsion and viscous flow problems.

Exact analytic solutions for the torsion or viscous flow problems are given in the literature. For example, White [125] gives solutions for viscous flow in nine noncircular sections including rectangle, triangle, and elliptic sections. We consider a rectangular section of dimensions $2a$ and $2b$. Taking the origin at the rectangle center,

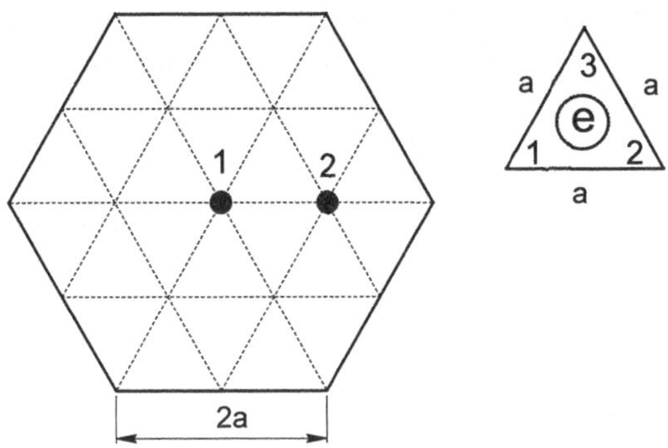

Figure 4.29 Torsion of a hexagonal section.

and defining nondimensional coordinates $\xi = x/a$ and $\eta = y/b$, we find the Prandtl stress function

$$\frac{\phi(\xi,\eta)}{G\theta a^2} = 1 - \xi^2 - 4\sum_{m=0}^{\infty} \frac{(-1)^m cosh\left(k_m \frac{b}{a}\eta\right) cos(k_m\xi)}{k_m^3 cosh(k_m \frac{b}{a})} \tag{4.191}$$

where $k_m = (2m+1)\pi/2$. The shear stresses are

$$\frac{\tau_{zx}(\xi,\eta)}{G\theta a} = -4\sum_{m=0}^{\infty} \frac{(-1)^m sinh\left(k_m \frac{b}{a}\eta\right) cos(k_m\xi)}{k_m^2 cosh(k_m \frac{b}{a})} \tag{4.192}$$

$$\frac{\tau_{zy}(\xi,\eta)}{G\theta a} = 2\xi - 4\sum_{m=0}^{\infty} \frac{(-1)^m cosh\left(k_m \frac{b}{a}\eta\right) sin(k_m\xi)}{k_m^2 cosh(k_m \frac{b}{a})}. \tag{4.193}$$

The warping function is

$$\frac{w(\xi,\eta)}{\theta a^2} = \frac{b}{a}\xi\eta - 4\sum_{m=0}^{\infty} \frac{(-1)^m sinh\left(k_m \frac{b}{a}\eta\right) sin(k_m\xi)}{k_m^3 cosh(k_m \frac{b}{a})} \tag{4.194}$$

and the torque T is given by

$$\frac{T}{G\theta a^4} = 16\left[\frac{b}{3a} - 2\sum_{m=0}^{\infty} \frac{tanh(k_m \frac{b}{a})}{k_m^5}\right]. \tag{4.195}$$

Example 4.4. Torsional Rigidity of a Hexagonal Section

We wish to use the finite element method to solve the torsion problem for a bar with a hexagonal section of side $2a$ as shown in Figure 4.29, and determine its torsional rigidity T/θ.

Solution using bilinear triangle elements

The section is divided into 24 bilinear equilateral triangular elements each of side a as shown in Figure 4.29. The stress function is assumed to be symmetric about all planes of geometric symmetry. The element stiffness matrix for a bilinear triangular element is given by Equation (4.59). The gradient matrix B given by Equation (4.58) depends only the differences in coordinates of the nodes. For the current problem, we have $p_{11}^e = 1$ and $p_{22}^e = 1$. Regardless of its position or orientation, the stiffness matrix for an equilateral triangular element is

$$[K^e] = \frac{\sqrt{3}}{6}\begin{bmatrix} 2 & -1 & -1 \\ -1 & 2 & -1 \\ -1 & -1 & 2 \end{bmatrix}. \tag{4.196}$$

The forcing vector is

$$\{F^e\} = \frac{f_o A_e}{3}\begin{Bmatrix} 1 \\ 1 \\ 1 \end{Bmatrix} \tag{4.197}$$

where $f_o = 2G\theta$. The element area is $A_e = \sqrt{3}a^2/4$.

The primary variable is the Prandtl stress function; we denote its values at global nodes 1 and 2 shown in Figure 4.29 by U_1 and U_2, respectively. On the boundary of the hexagon, Prandtl stress function is zero. Considering the symmetry of the elements arrangement, we find that the six nodes surrounding node 1 will have the same value U_2. Thus, we only need to write the two equations for nodes 1 and 2 in the assembled global system. The equation for global node #1 is

$$6k_{11}^e U_1 + 12k_{12}^e U_2 = 6F_1^e$$

and the equation for global node #2 is

$$2k_{21}^e U_1 + 6k_{22}^e U_2 + 4k_{21}^e U_2 = 6F_1^e$$

and the solution is: $U_1 = (5/6)f_o a^2$ and $U_2 = (7/12)f_o a^2$. For a linear triangular element, we have:

$$\int_{A_e} (U_1^e\psi_1 + U_2^e\psi_2 + U_3^e\psi_3)dA = \frac{A_e}{3}(U_1^e + U_2^e + U_3^e)$$

Thus, the torque is $T = (52\sqrt{3}/12)(2G\theta)a^4$, and the torsional rigidity $= T/\theta$. For the purpose of comparison with a circular section, we find $T/G\theta = (52\sqrt{3}/6)a^4 = 15.01a^4$. The radius of maximum inscribed circle in the hexagon is $r = \sqrt{3}a$, and its polar moment of inertia $J = \pi r^4/2$, and hence for such a circle $T/G\theta = (9\pi/2)a^4 = 14.137a^4$.

Example 4.5. Torsional Rigidity of a Semicircular Section

We consider the torsional problem of a semicircular section of radius a. An exact solution to the relevant Poisson's equation and boundary conditions can be found in

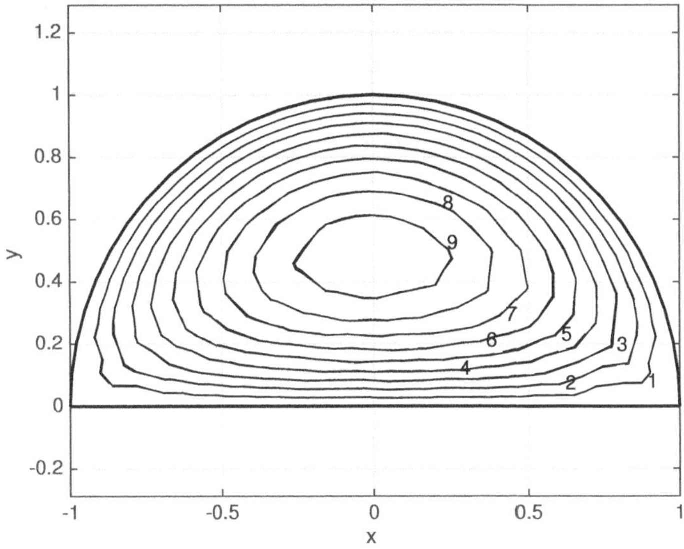

Figure 4.30 Contours of Prandtl stress function ($50\ \phi/(G\theta a^2)$). Comparison between finite element and exact analytic solution.

White [125] or Lebedev et al. [64]. Let (r, φ) denote the polar coordinates with origin at the circle center. The solution for Prandtl stress function is [64]

$$\begin{aligned}\frac{\phi(r,\varphi)}{G\theta a^2} = & -\frac{r^2 sin^2\varphi}{a^2} + \frac{1}{\pi}\left[\left(\frac{r}{a} - \frac{a}{r}\right) sin\varphi\right. \\ & -\frac{1}{2}\left(\left(\frac{r^2}{a^2} + \frac{a^2}{r^2}\right) cos2\varphi - 2\right) tan^{-1}\frac{2arsin\varphi}{a^2 - r^2} \\ & \left. +\frac{1}{4}\left(\frac{r^2}{a^2} - \frac{a^2}{r^2}\right) sin2\varphi\, ln\frac{a^2 + r^2 - 2arcos\varphi}{a^2 + r^2 + 2arcos\varphi}\right] \end{aligned} \tag{4.198}$$

and the torque is [125],

$$\frac{T}{G\theta a^4} = -\frac{\pi}{4} - \frac{32}{\pi} \sum_{m=1,3,5,\ldots}^{\infty} \frac{1}{m^2(m+2)^2(m-2)}. \tag{4.199}$$

We developed a MATLAB code for solving the torsion problem of a semicircle section. The mesh of bilinear triangular elements is shown in Figures 4.2a–4.2b. Contours of Prandtl stress function ($\phi/(G\theta a^2)$) by the finite-element method are compared with the exact analytic solution in Figure 4.30. The differences between the two solutions are negligible. Torque by the finite element method is $T/(G\theta a^4) = 0.2911$, whereas the exact value is 0.2976; the error is less than 2.5%.

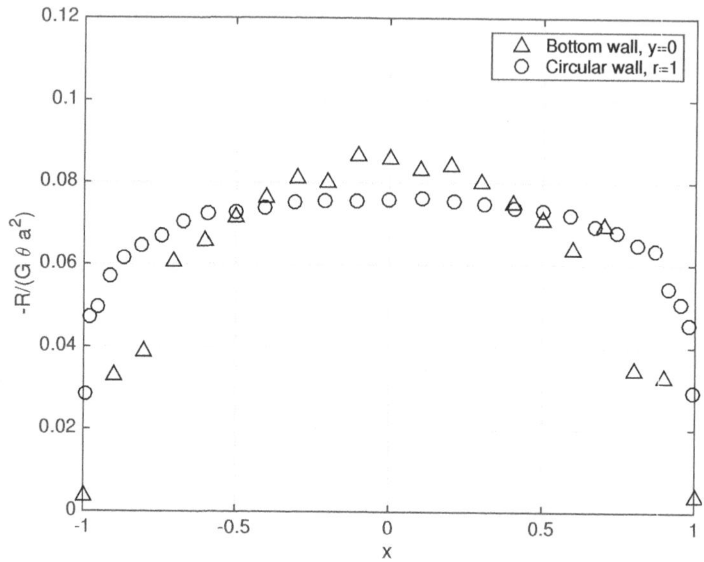

Figure 4.31 Weighted secondary variables at the global nodes on the boundary of the semicircle.

We recall the definition of the secondary variable v_n Equation (4.12),

$$\begin{aligned} v_n &= n_x v_x + n_y v_y \\ &= n_x(p_{11}\frac{\partial u}{\partial x} + p_{12}\frac{\partial u}{\partial y}) + n_y(p_{21}\frac{\partial u}{\partial x} + p_{22}\frac{\partial u}{\partial y}) \\ &= n_x\frac{\partial \phi}{\partial x} + n_y\frac{\partial \phi}{\partial y} \\ &= -n_x\tau_{zy} + n_y\tau_{zx} \\ &= \tau_{zt} \end{aligned} \tag{4.200}$$

the secondary variable is the component of the shear stress on the z-plane in the direction of unit vector $\hat{t} = \hat{n} \times \hat{k}$. Integrating Equation (4.186) over the bar cross-section and using the divergence theorem, we get

$$-\oint_{\Gamma_E} v_n ds = 2G\theta A \tag{4.201}$$

where Γ_E is the contour of the cross-section. For the semicircle ($A = \pi a^2/2$), the result is $\pi G\theta a^2$.

After we determine all the primary variables, we determine the weighted secondary variables $\{R\}$ at the global nodes by using Equation (4.114). The sum of the elements of $\{R\}$ for the global nodes on the boundary of the semicircle must equal the integral in Equation (4.201). Those elements are shown in Figure (4.31). There

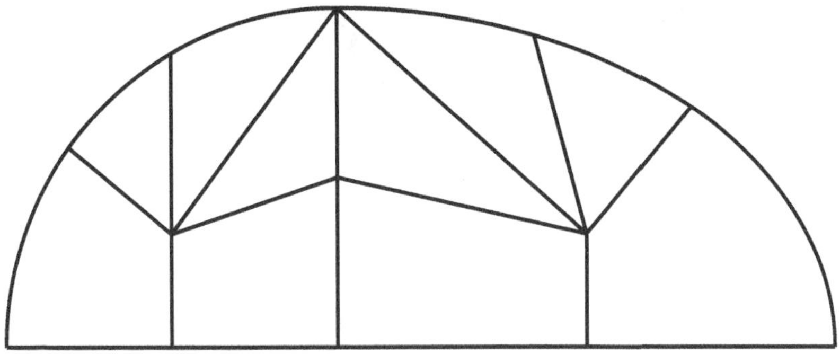

Figure 4.32 A domain with curved boundary is divided into triangular and quadrilateral elements.

are 21 nodes on the bottom wall and 29 nodes on the circular arc. The sum of $(-R)$ at the 50 nodes on the boundary is 3.1359 $G\theta a^2$, which is 0.18% less than the exact value $\pi G\theta a^2$. It is also interesting to report that the sum of the areas of all elements in the mesh is also 0.18% less than the exact area of the semicircle. There are no gaps or overlaps between the elements, but representation of the curved boundary of the semicircle by straight line segments introduces errors, albeit very small in the present example. Domains with curved boundary are better represented by elements with curved boundaries.

4.8 ISOPARAMETRIC ELEMENTS AND NUMERICAL INTEGRATION

The bilinear triangular and rectangular elements are suitable for the simple domains of the examples presented in this chapter so far. We used triangular elements for the semicircle domain, but we had to divide the circular arc into a large number of segments for accurate representation of its curvature. As such, elements with straight edges are inefficient for curved boundaries. The boundary curvature is better represented by elements with curved sides with more than two nodes on the side. Furthermore, the use of elements with high-order polynomials (quadratic, cubic, etc.) usually provides better resolution and is computationally more efficient than increasing the number of elements of low-order polynomials.

The domain shown in Figure 4.32 is divided into 10 elements; four quadrilateral elements and six trilateral elements. Elements with curved edges are used on the curved boundary of the domain. For the model problem under consideration in this chapter and for each of the elements shown in Figure 4.32, the element equations are still given by Equation (4.53) and the element matrices are given by Equations (4.48)–(4.51). However, it is not easy to construct interpolation functions (element shape functions) or numerically evaluate the element matrices in terms of the coordinates (x, y) normally used in Figure 4.32. These difficulties are satisfactorily resolved by mapping the element domain in the (x, y) into another element for which

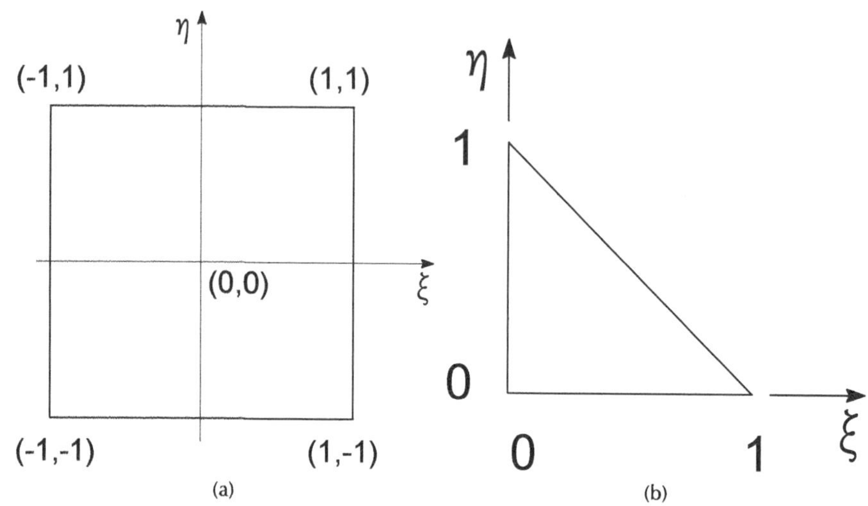

Figure 4.33 Canonical elements. (a) Square element, and (b) triangular element.

the construction of the interpolation functions (shape functions) is straightforward. For example, each of the quadrilateral elements in Figure 4.32 can be mapped into the square element shown in Figure 4.33a in the (ξ, η) plane. And each of the trilateral elements in Figure 4.32 can be mapped into the triangle element shown in Figure 4.33b in the (ξ, η) plane. We call the elements in the (ξ, η) plane canonical elements. For the mapping to be successful, each point in the canonical element maps to only one point in the element in the (x, y) domain, and vice versa; that is to say the mapping is one-to-one.

4.8.1 SHAPE FUNCTIONS OF CANONICAL ELEMENTS

The shape functions for the canonical elements are needed for both the mapping and approximation of the primary variable over the element.

Canonical Bilinear Quadrilateral Element

The canonical 4-node square element is shown in Figure 4.34. The coordinates of the four nodes in the (ξ, η) plane are (−1,−1), (1,−1), (1,1), and (−1,1) for the nodes 1, 2, 3, and 4, respectively. The construction of the 4 shape functions $\psi_j(\xi,\eta)$ is facilitated by recalling the shape functions of the linear element in one dimension. When written in terms of ξ, the 1D shape functions are $\psi_1^{(1D)}(\xi) = (1-\xi)/2$ and $\psi_2^{(1D)}(\xi) = (1+\xi)/2$ for $-1 \leq \xi \leq 1$, and when written in terms of η they are $\psi_1^{(1D)}(\eta) = (1-\eta)/2$ and $\psi_2^{(1D)}(\eta) = (1+\eta)/2$ for $-1 \leq \eta \leq 1$. The shape functions of the 2D element can be obtained as products of the 1D functions. Referring to Figure 4.34, we think of node #1 as the first node of a 1D linear element placed parallel to the ξ-axis. It is also the first node of a 1D linear element placed parallel

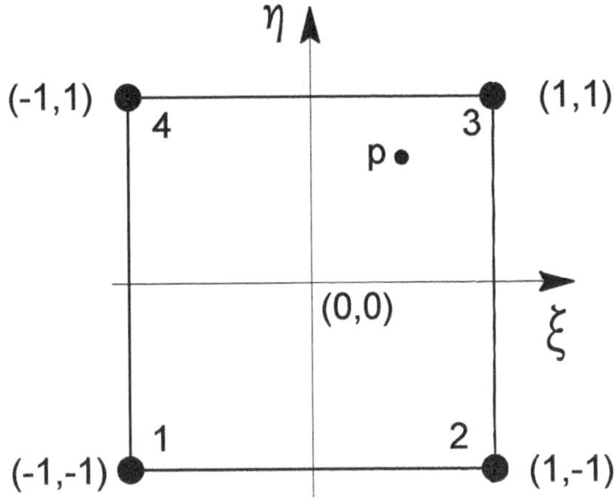

Figure 4.34 Canonical bilinear square element.

to the η-axis. Thus, its 2D shape function is the product of the relevant 1D shape functions. Hence, we obtain

$$\begin{aligned}\psi_1(\xi,\eta) &= \psi_1^{(1D)}(\xi)\psi_1^{(1D)}(\eta)\\ &= \frac{1}{4}(1-\xi)(1-\eta).\end{aligned}$$

Similarly, we think of node #2 as the second node of a 1D linear element placed parallel to the ξ-axis, and it is also the first node of a 1D linear element placed parallel to the η-axis. Thus, its 2D shape function is the product of the relevant 1D shape functions. Hence, we obtain

$$\begin{aligned}\psi_2(\xi,\eta) &= \psi_2^{(1D)}(\xi)\psi_1^{(1D)}(\eta)\\ &= \frac{1}{4}(1+\xi)(1-\eta).\end{aligned}$$

As for node # 3, it is the second node of a 1D linear element placed parallel to the ξ-axis, and the second node of a 1D linear element placed parallel to the η-axis. Hence, we obtain

$$\begin{aligned}\psi_3(\xi,\eta) &= \psi_2^{(1D)}(\xi)\psi_2^{(1D)}(\eta)\\ &= \frac{1}{4}(1+\xi)(1+\eta).\end{aligned}$$

Finally, for node #4, we have

$$\begin{aligned}\psi_4(\xi,\eta) &= \psi_1^{(1D)}(\xi)\psi_2^{(1D)}(\eta)\\ &= \frac{1}{4}(1-\xi)(1+\eta).\end{aligned}$$

In summary, the shape functions for the canonical bilinear square element are

$$\begin{aligned}\psi_1(\xi,\eta) &= \frac{1}{4}(1-\xi)(1-\eta)\\ \psi_2(\xi,\eta) &= \frac{1}{4}(1+\xi)(1-\eta)\\ \psi_3(\xi,\eta) &= \frac{1}{4}(1+\xi)(1+\eta)\\ \psi_4(\xi,\eta) &= \frac{1}{4}(1-\xi)(1+\eta).\end{aligned} \tag{4.202}$$

It is easy to verify that

$$\sum_{j=1}^{j=4} \psi_j(\xi,\eta) = 1 \quad \text{for } -1 \le \xi \le 1 \text{ and } -1 \le \eta \le 1. \tag{4.203}$$

Let us denote the coordinates of node # i by (ξ_i, η_i). One can verify that

$$\psi_j(\xi_i, \eta_i) = \delta_{ij}. \tag{4.204}$$

Canonical Biquadratic Quadrilateral Element

The canonical biquadratic square element is shown in Figure 4.35, which has 9 nodes labeled as shown in the figure. We construct its 9 shape functions by using the shape functions of the 1D quadratic element, rewritten here for $-1 \le \xi \le 1$,

$$\begin{aligned}\psi_1^{(1D)}(\xi) &= \frac{1}{2}(\xi^2-\xi)\\ \psi_2^{(1D)}(\xi) &= (1-\xi^2)\\ \psi_3^{(1D)}(\xi) &= \frac{1}{2}(\xi^2+\xi)\end{aligned} \tag{4.205}$$

where the end nodes # 1 and # 3 are at $\xi = -1$ and $\xi = 1$, whereas node # 2 is the middle node at $\xi = 0$. For convenience, we also repeat these 1D functions in terms of η,

$$\begin{aligned}\psi_1^{(1D)}(\eta) &= \frac{1}{2}(\eta^2-\eta)\\ \psi_2^{(1D)}(\eta) &= (1-\eta^2)\\ \psi_3^{(1D)}(\eta) &= \frac{1}{2}(\eta^2+\eta).\end{aligned} \tag{4.206}$$

Referring to Figure 4.35, we think of node #1 as the first node of a 1D quadratic element placed parallel to the ξ-axis. It is also the first node of a 1D quadratic element placed parallel to the η-axis. Thus, its 2D shape function is the product of the relevant 1D shape functions. Hence, we obtain

$$\begin{aligned}\psi_1(\xi,\eta) &= \psi_1^{(1D)}(\xi)\psi_1^{(1D)}(\eta)\\ &= \frac{1}{4}(\xi^2-\xi)(\eta^2-\eta).\end{aligned}$$

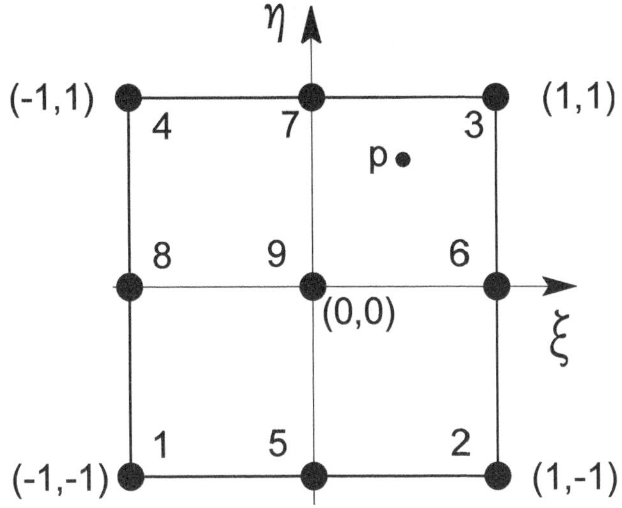

Figure 4.35 Canonical biquadratic square element.

Similarly, we think of node #2 as the third node of a 1D quadratic element placed parallel to the ξ-axis, and it is also the first node of a 1D quadratic element placed parallel to the η-axis. Thus, its 2D shape function is the product of the relevant 1D shape functions. Hence, we obtain

$$\begin{aligned}\psi_2(\xi,\eta) &= \psi_3^{(1D)}(\xi)\psi_1^{(1D)}(\eta)\\ &= \frac{1}{4}(\xi^2+\xi)(\eta^2-\eta).\end{aligned}$$

As for node # 5, it is the second node of a 1D quadratic element placed parallel to the ξ-axis, and the first node of a 1D quadratic element placed parallel to the η-axis. Hence, we obtain

$$\begin{aligned}\psi_5(\xi,\eta) &= \psi_2^{(1D)}(\xi)\psi_1^{(1D)}(\eta)\\ &= \frac{1}{2}(1-\xi^2)(\eta^2-\eta).\end{aligned}$$

As for node # 9, it is the second node of a 1D quadratic element placed parallel to the ξ-axis, and the second node of a 1D quadratic element placed parallel to the η-axis. Hence, we obtain

$$\begin{aligned}\psi_9(\xi,\eta) &= \psi_2^{(1D)}(\xi)\psi_2^{(1D)}(\eta)\\ &= (1-\xi^2)(1-\eta^2).\end{aligned}$$

This procedure of construction of the shape functions for multidimensional elements from the shape functions of the one-dimensional element is known as the tensor product procedure. We can continue this procedure to obtain the remaining shape functions of the biquadratic element. In summary they are

$$
\begin{aligned}
\psi_1(\xi,\eta) &= \frac{1}{4}(\xi^2-\xi)(\eta^2-\eta)\\
\psi_2(\xi,\eta) &= \frac{1}{4}(\xi^2+\xi)(\eta^2-\eta)\\
\psi_3(\xi,\eta) &= \frac{1}{4}(\xi^2+\xi)(\eta^2+\eta)\\
\psi_4(\xi,\eta) &= \frac{1}{4}(\xi^2-\xi)(\eta^2+\eta)\\
\psi_5(\xi,\eta) &= \frac{1}{2}(1-\xi^2)(\eta^2-\eta)\\
\psi_6(\xi,\eta) &= \frac{1}{2}(\xi^2+\xi)(1-\eta^2)\\
\psi_7(\xi,\eta) &= \frac{1}{2}(1-\xi^2)(\eta^2+\eta)\\
\psi_8(\xi,\eta) &= \frac{1}{2}(\xi^2-\xi)(1-\eta^2)\\
\psi_9(\xi,\eta) &= (1-\xi^2)(1-\eta^2).
\end{aligned}
\tag{4.207}
$$

It is easy to verify that

$$
\sum_{j=1}^{j=9} \psi_j(\xi,\eta) = 1 \quad \text{for } -1 \le \xi \le 1 \text{ and } -1 \le \eta \le 1 \tag{4.208}
$$

and that

$$
\psi_j(\xi_i,\eta_i) = \delta_{ij}. \tag{4.209}
$$

Canonical Bicubic Quadrilateral Element

The canonical bicubic square element is shown in Figure 4.36, it has 16 nodes labeled as shown in the figure. The nodes are equally spaced on each edge; for example, the coordinates (ξ,η) for nodes 2, 3, 7, and 11 are $(-1/3,-1)$, $(1/3,-1)$, $(1/3,-1/3)$, and $(1/3,1/3)$, respectively. We construct its 16 shape functions by using the shape functions of the 1D cubic element, written here for $-1 \le \xi \le 1$,

$$
\begin{aligned}
\psi_1^{(1D)}(\xi) &= -\frac{1}{16}(3\xi+1)(3\xi-1)(\xi-1)\\
\psi_2^{(1D)}(\xi) &= \frac{9}{16}(\xi+1)(3\xi-1)(\xi-1)\\
\psi_3^{(1D)}(\xi) &= -\frac{9}{16}(\xi+1)(3\xi+1)(\xi-1)\\
\psi_4^{(1D)}(\xi) &= \frac{1}{16}(\xi+1)(3\xi+1)(3\xi-1).
\end{aligned}
\tag{4.210}
$$

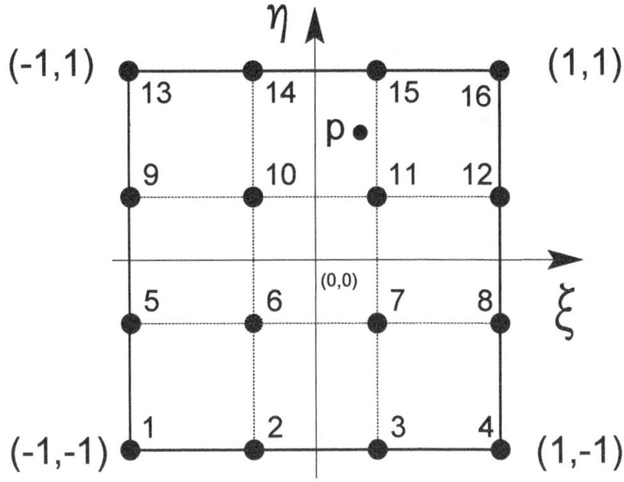

Figure 4.36 Canonical bicubic square element.

We can also write these functions in terms of η. The 16 shape functions for the 2D bicubic element are constructed by tensor product of the 1D cubic shape functions. Some of the functions are

$$\begin{aligned}\psi_1(\xi,\eta) &= \psi_1^{(1D)}(\xi)\psi_1^{(1D)}(\eta)\\ &= \frac{1}{256}(3\xi+1)(3\xi-1)(\xi-1)(3\eta+1)(3\eta-1)(\eta-1)\end{aligned} \quad (4.211)$$

$$\begin{aligned}\psi_7(\xi,\eta) &= \psi_3^{(1D)}(\xi)\psi_2^{(1D)}(\eta)\\ &= -\frac{81}{256}(\xi+1)(3\xi+1)(\xi-1)(\eta+1)(3\eta-1)(\eta-1)\end{aligned} \quad (4.212)$$

$$\begin{aligned}\psi_9(\xi,\eta) &= \psi_1^{(1D)}(\xi)\psi_3^{(1D)}(\eta)\\ &= \frac{9}{256}(3\xi+1)(3\xi-1)(\xi-1)(\eta+1)(3\eta+1)(\eta-1)\end{aligned} \quad (4.213)$$

$$\begin{aligned}\psi_{16}(\xi,\eta) &= \psi_4^{(1D)}(\xi)\psi_4^{(1D)}(\eta)\\ &= \frac{1}{256}(\xi+1)(3\xi+1)(3\xi-1)(\eta+1)(3\eta+1)(3\eta-1).\end{aligned} \quad (4.214)$$

These functions satisfy the conditions

$$\sum_{j=1}^{j=16} \psi_j(\xi,\eta) = 1 \quad \text{for } -1 \le \xi \le 1 \text{ and } -1 \le \eta \le 1 \quad (4.215)$$

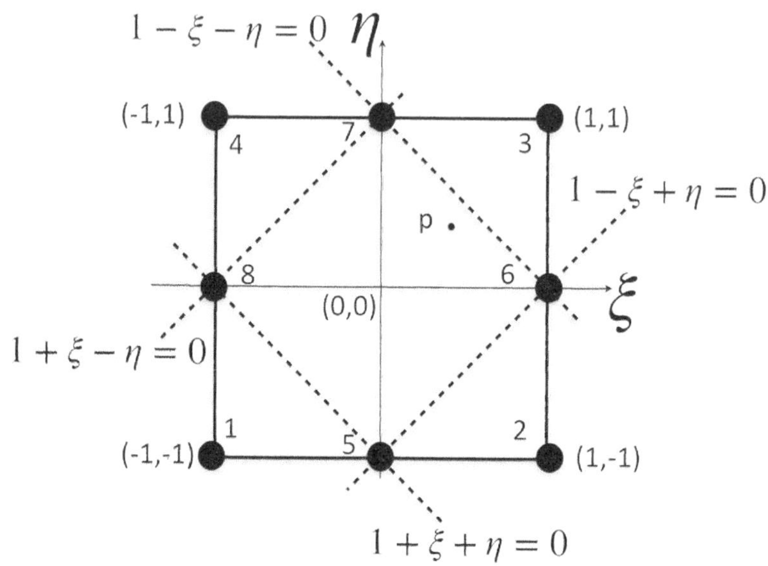

Figure 4.37 Canonical serendipity biquadratic square element.

and

$$\psi_j(\xi_i,\eta_i)=\delta_{ij}. \tag{4.216}$$

Canonical Serendipity Biquadratic Square Element

The canonical serendipity biquadratic square element is shown in Figure 4.37, which has eight nodes labeled as shown in the figure. Its 8 shape functions satisfy the property $\psi_i(\xi_j,\eta_j)=\delta_{ij}$. Thus, $\psi_1=0$ at all nodes except at node 1 we have $\psi_1=1$. Since the equations of the lines 2-6-3, 4-7-3, and 5-8 are $1-\xi=0$, $1-\eta=0$, and $1+\xi+\eta=0$, respectively, we try $\psi_1=c(1-\xi)(1-\eta)(1+\xi+\eta)$. The condition $\psi_1=1$ at $(\xi,\eta)=(-1,-1)$ gives $c=-1/4$, hence $\psi_1=-(1/4)(1-\xi)(1-\eta)(1+\xi+\eta)$. Similar procedure is used to construct ψ_2, ψ_3, and ψ_4. The function ψ_5 is zero at all nodes except at node 5 we have $\psi_5=1$. Since the equations of the lines 1-8-4, 2-6-3, and 4-7-3 are $1+\xi=0$, $1-\xi=0$, and $1-\eta=0$, respectively, we try $\psi_5=c(1+\xi)(1-\xi)(1-\eta)$. The condition $\psi_5=1$ at $(\xi,\eta)=(0,-1)$ gives $c=1/2$, hence $\psi_5=(1/2)(1+\xi)(1-\xi)(1-\eta)$. Similar procedure is used to construct ψ_6, ψ_7, and ψ_8. The shape functions of the serendipity biquadratic element are

$$\begin{aligned}
\psi_1(\xi,\eta) &= \frac{1}{4}(1-\xi)(1-\eta)(-1-\xi-\eta) \\
\psi_2(\xi,\eta) &= \frac{1}{4}(1+\xi)(1-\eta)(-1+\xi-\eta) \\
\psi_3(\xi,\eta) &= \frac{1}{4}(1+\xi)(1+\eta)(-1+\xi+\eta) \\
\psi_4(\xi,\eta) &= \frac{1}{4}(1-\xi)(1+\eta)(-1-\xi+\eta) \\
\psi_5(\xi,\eta) &= \frac{1}{2}(1-\xi^2)(1-\eta) \\
\psi_6(\xi,\eta) &= \frac{1}{2}(1+\xi)(1-\eta^2) \\
\psi_7(\xi,\eta) &= \frac{1}{2}(1-\xi^2)(1+\eta) \\
\psi_8(\xi,\eta) &= \frac{1}{2}(1-\xi)(1-\eta^2).
\end{aligned} \tag{4.217}$$

It is easy to verify that

$$\sum_{j=1}^{j=8} \psi_j(\xi,\eta) = 1 \quad \text{for } -1 \le \xi \le 1 \text{ and } -1 \le \eta \le 1 \tag{4.218}$$

and that

$$\psi_j(\xi_i,\eta_i) = \delta_{ij}. \tag{4.219}$$

Canonical Bilinear Triangular Element

The canonical bilinear triangular element is shown in Figure 4.38. The coordinates (ξ_j,η_j) of its 3 nodes are $(1,0)$, $(0,1)$, and $(0,0)$. For the purpose of deriving the shape functions, we define a variable $\zeta = 1-\xi-\eta$. The shape functions satisfy the property $\psi_i(\xi_j,\eta_j) = \delta_{ij}$. Thus, ψ_1 is zero at nodes 2 and 3. Since the equation of the line 2-3 is $\xi = 0$, we try $\psi_1 = c\xi$. The condition $\psi_1 = 1$ at node 1 gives $c = 1$. Thus, we have $\psi_1 = \xi$. Similarly, we get $\psi_2 = \eta$. The shape function ψ_3 is zero at nodes 1 and 2. Since the equation of the line 1-2 is $\zeta = 0$, we try $\psi_3 = c\zeta$, and we find $\psi_3 = \zeta$, or $\psi_3 = 1-\xi-\eta$. The three shape functions are

$$\psi_1(\xi,\eta) = \xi,\ \psi_2(\xi,\eta) = \eta,\ \psi_3(\xi,\eta) = \zeta \tag{4.220}$$

where $\zeta = 1-\xi-\eta$. One can verify that

$$\sum_{j=1}^{3} \psi_j(\xi,\eta) = 1. \tag{4.221}$$

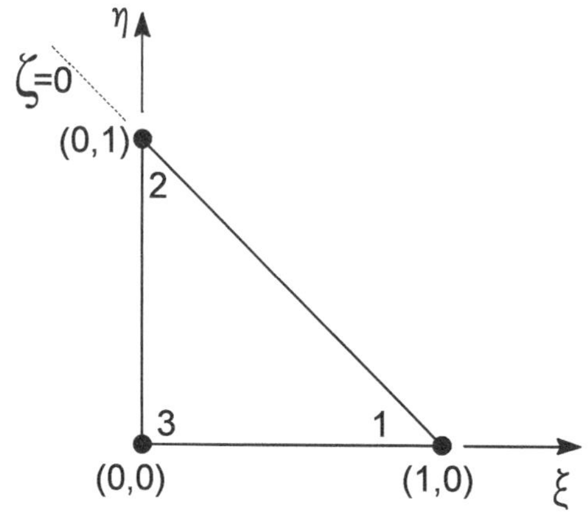

Figure 4.38 Canonical bilinear triangular element.

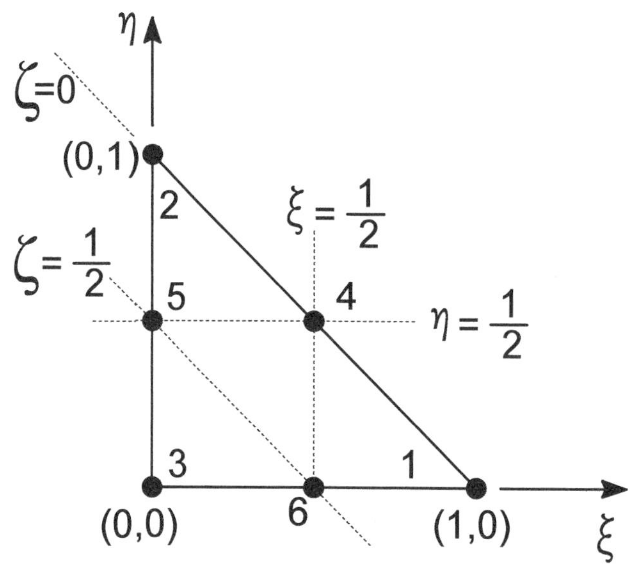

Figure 4.39 Canonical biquadratic triangular element.

Canonical Biquadratic Triangular Element

The canonical biquadratic triangular element is shown in Figure 4.39. The element has six nodes; three nodes at the vertexes and three nodes at the midpoints of the sides. The coordinates (ξ, η) of the middle nodes 4, 5, and 6 are $(1/2, 1/2)$, $(0, 1/2)$, and $(1/2, 0)$, respectively. The shape functions satisfy the property $\psi_i(\xi_j, \eta_j) = \delta_{ij}$. Thus, $\psi_1 = 0$ at all nodes except at node 1 we have $\psi_1 = 1$. Since the equation of the line 2-5-3 is $\xi = 0$, and the equation of the line 4-6 is $\xi - 1/2 = 0$, we try $\psi_1 =$

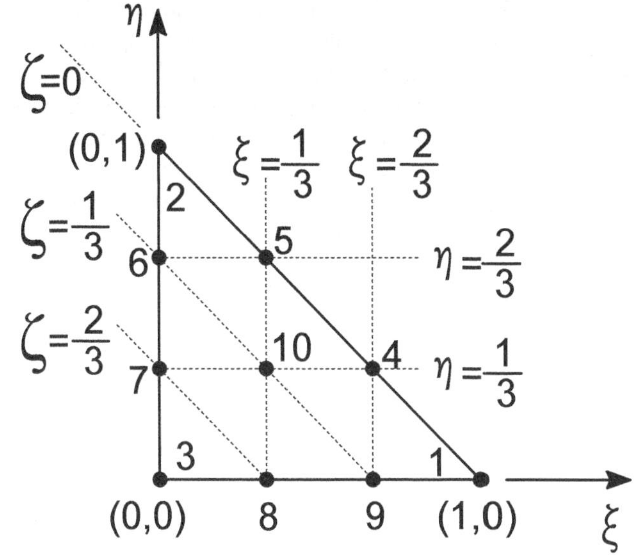

Figure 4.40 Canonical bicubic triangular element.

$c\xi(\xi-1/2)$. The condition $\psi_1=1$ at node 1 gives $c=2$. Thus, $\psi_1=2\xi(\xi-1/2)$. Similarly, we find $\psi_2=2\eta(\eta-1/2)$. The shape function ψ_3 is zero at all nodes except at node 3 we have $\psi_3=1$. Since the equation of the line 1-4-2 is $\zeta=0$ and the equation of line 5-6 is $\zeta-1/2=0$, we try $\psi_3=c\zeta(\zeta-1/2)$, and we find $\psi_3=2\zeta(\zeta-1/2)$. The equations for the lines 3-6-1 and 3-5-2 are $\eta=0$ and $\xi=0$, respectively. Hence, we try $\psi_4=c\xi\eta$. The condition $\psi_4=1$ at node 4 gives $c=4$. Hence, $\psi_4=4\xi\eta$. The six shape functions are

$$\begin{aligned}&\psi_1(\xi,\eta)=\xi(2\xi-1),\ \psi_2(\xi,\eta)=\eta(2\eta-1),\ \psi_3(\xi,\eta)=\zeta(2\zeta-1)\\&\psi_4(\xi,\eta)=4\xi\eta,\ \psi_5(\xi,\eta)=4\eta\zeta,\ \psi_6(\xi,\eta)=4\xi\zeta\end{aligned}\tag{4.222}$$

where $\zeta=1-\xi-\eta$. One can verify that

$$\sum_{j=1}^{6}\psi_j(\xi,\eta)=1.\tag{4.223}$$

Canonical Bicubic Triangular Element

The canonical bicubic triangular element is shown in Figure 4.40, it has 10 nodes labeled as shown in the figure. The nodes are equally spaced on each edge; for example, the coordinates (ξ,η) for nodes 7, 8, and 5 are $(0,1/3)$, $(1/3,0)$, and $(1/3,2/3)$, respectively. The equations for the lines 3-7-6-2, 8-10-5, and 9-4 are $\xi=0$, $\xi-1/3=0$, and $\xi-2/3=0$, respectively. Hence, the expression of ψ_1

should be $\psi_1 = c\xi(\xi - 1/3)(\xi - 2/3)$. The condition $\psi_1 = 1$ gives $c = 9/2$, and we find $\psi_1 = \xi(3\xi - 1)(3\xi - 2)/2$. Similarly, we get $\psi_2 = \eta(3\eta - 1)(3\eta - 2)/2$. The equations of the lines 1-4-5-2, 9-10-6, and 7-8, are $\zeta = 0$, $\zeta - 1/3 = 0$, and $\zeta - 2/3 = 0$, respectively. The expression of ψ_3 is constructed as the product of the factors on the left-hand side of the three equations, and requiring that its value to be 1 at node 3. We get $\psi_3 = \zeta(3\zeta - 1)(3\zeta - 2)/2$. The equations for the lines 3-7-6-2, 3-8-9-1, and 1-4-5-2 are $\xi = 0$, $\eta = 0$, and $\zeta = 0$. Thus, we have $\psi_{10} = 27\xi\eta\zeta$. In summary, the shape functions of the canonical bicubic element are

$$
\begin{aligned}
\psi_1(\xi,\eta) &= \frac{1}{2}\xi(3\xi - 1)(3\xi - 2) \\
\psi_2(\xi,\eta) &= \frac{1}{2}\eta(3\eta - 1)(3\eta - 2) \\
\psi_3(\xi,\eta) &= \frac{1}{2}\zeta(3\zeta - 1)(3\zeta - 2) \\
\psi_4(\xi,\eta) &= \frac{9}{2}\xi\eta(3\xi - 1) \\
\psi_5(\xi,\eta) &= \frac{9}{2}\xi\eta(3\eta - 1) \\
\psi_6(\xi,\eta) &= \frac{9}{2}\eta\zeta(3\eta - 1) \\
\psi_7(\xi,\eta) &= \frac{9}{2}\eta\zeta(3\zeta - 1) \\
\psi_8(\xi,\eta) &= \frac{9}{2}\xi\zeta(3\zeta - 1) \\
\psi_9(\xi,\eta) &= \frac{9}{2}\xi\zeta(3\xi - 1) \\
\psi_{10}(\xi,\eta) &= 27\xi\eta\zeta
\end{aligned}
\tag{4.224}
$$

where $\zeta = 1 - \xi - \eta$. One can verify that

$$
\sum_{j=1}^{10} \psi_j(\xi,\eta) = 1. \tag{4.225}
$$

The canonical element shape functions can be used to define (or interpolate) a function $G(\xi,\eta)$ in terms of its values at the element nodes. We denote the values of the function G at the n nodes of the element by the array

$$\{G^e\} = \begin{Bmatrix} G_1^e \\ G_2^e \\ \cdot \\ \cdot \\ \cdot \\ G_j^e \\ \cdot \\ \cdot \\ \cdot \\ G_n^e \end{Bmatrix}. \tag{4.226}$$

For quadrilateral elements we have $n = 4$, 9, and 16 for the bilinear, biquadratic, and bicubic elements, respectively. And for triangular elements we have $n = 3$, 6, and 10 for the bilinear, biquadratic, and bicubic elements, respectively. At any point in the element domain, we define the function $G^e(\xi,\eta)$ by

$$\begin{aligned} G^e(\xi,\eta) &= \psi_1(\xi,\eta)G_1^e + \cdots + \psi_j(\xi,\eta)G_j^e + \cdots + \psi_n(\xi,\eta)G_n^e \\ &= \sum_{j=1}^{n} \psi_j(\xi,\eta)G_j^e \end{aligned} \tag{4.227}$$

or in matrix form

$$G^e(\xi,\eta) = \{\psi\}\{G^e\} \tag{4.228}$$

where

$$\{\psi\} = \{\psi_1(\xi,\eta)\cdots\psi_j(\xi,\eta)\cdots\psi_n(\xi,\eta)\}. \tag{4.229}$$

The Equation (4.228) can be used to compute G^e at any point p in the canonical element. Because of the property Equation (4.216), the function defined by Equation (4.228) gives $G^e(\xi_i,\eta_i) = G_i^e$ as required. And the property Equation (4.215) also shows that if all the nodal values are equal to C, then $G^e(\xi,\eta) = C$.

4.8.2 MAPPING

Mapping an element from the physical domain in the $x-y$ plane to the canonical element in the (ξ,η) plane is explained for a biquadratic quadrilateral element. The analysis can be easily applied to any other element. A quadrilateral element is shown in Figure 4.41a. The only data known about the element geometry are the global coordinates (x,y) of the nine points labeled 1, 2, .., and 9. The coordinates are denoted by $(x_1^e,y_1^e),\cdots,(x_9^e,y_9^e)$. We want to define the boundary of the element and render a correspondence between points within its boundary and points in the biquadratic canonical element shown in Figure 4.41b. To this end, we define the arrays

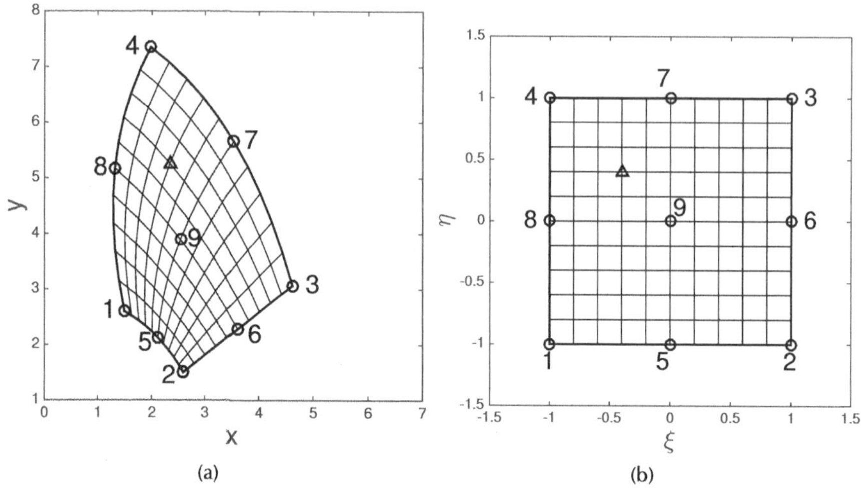

Figure 4.41 Mapping of a quadrilateral element. (a) A biquadratic element in the physical domain, (x,y) coordinates. (b) The corresponding canonical biquadratic square element, (ξ,η) coordinates.

$$\{x^e\} = \begin{Bmatrix} x_1^e \\ x_2^e \\ \cdot \\ \cdot \\ \cdot \\ \cdot \\ \cdot \\ x_9^e \end{Bmatrix}, \quad \{y^e\} = \begin{Bmatrix} y_1^e \\ y_2^e \\ \cdot \\ \cdot \\ \cdot \\ \cdot \\ \cdot \\ y_9^e \end{Bmatrix} \tag{4.230}$$

and identify the vector $\{G^e\}$ with $\{x^e\}$ and $\{y^e\}$. The mapping is defined by

$$\begin{aligned} x^e(\xi,\eta) &= \psi_1(\xi,\eta)x_1^e + \psi_2(\xi,\eta)x_2^e + \cdots + \psi_9(\xi,\eta)x_9^e \\ &= \sum_{j=1}^{9} \psi_j(\xi,\eta)x_j^e \end{aligned} \tag{4.231}$$

and

$$\begin{aligned} y^e(\xi,\eta) &= \psi_1(\xi,\eta)y_1^e + \psi_2(\xi,\eta)y_2^e + \cdots + \psi_9(\xi,\eta)y_9^e \\ &= \sum_{j=1}^{9} \psi_j(\xi,\eta)y_j^e \end{aligned} \tag{4.232}$$

or more concisely,

$$\{x^e(\xi,\eta)\ y^e(\xi,\eta)\} = \{\psi\}\{\{x^e\}\ \{y^e\}\}. \tag{4.233}$$

Choosing a point such as that marked by Δ in the canonical element as shown in Figure 4.41b, we use Equations (4.231) and (4.232) to find its coordinates in the (x,y) plane, which is also marked by Δ in Figure 4.41a. The same equations are used to map lines of constant ξ and constant η into lines in the element in the (x,y) plane. In particular, $\eta = -1$ on the edge $1-5-2$, and hence points on that edge are given by

$$\begin{aligned} x &= \psi_1(\xi,-1)x_1^e + \psi_2(\xi,-1)x_2^e + \cdots + \psi_9(\xi,-1)x_9^e \\ y &= \psi_1(\xi,-1)y_1^e + \psi_2(\xi,-1)y_2^e + \cdots + \psi_9(\xi,-1)y_9^e. \end{aligned}$$

Putting $\eta = -1$ in the shape functions as given by Equation (4.207), we find

$$x = \frac{1}{2}(\xi^2 - \xi)x_1^e + \frac{1}{2}(\xi^2 + \xi)x_2^e + (1 - \xi^2)x_5^e \tag{4.234}$$

$$y = \frac{1}{2}(\xi^2 - \xi)y_1^e + \frac{1}{2}(\xi^2 + \xi)y_2^e + (1 - \xi^2)y_5^e \tag{4.235}$$

these are the parametric equations of the curve passing through the nodes 1-5-2 in the (x,y)-plane. We can eliminate ξ between these two equations and obtain an equation $f(x,y) = 0$ as the equation for edge 1-5-2; however, such an equation is not necessary for the implementation of the finite element method. Each of the Equations (4.234) and (4.235) is quadratic in ξ; thus, it is uniquely defined by the three values of the coordinates of the points 1, 5, and 2 on that edge. The coordinates of the other six nodes of the element do not influence the equations of edge 1-5-2. This is important for geometric *compatibility* of the element. Two adjacent elements sharing nodes 1-5-2 will have identical edges connecting these nodes; there will be no gaps or overlaps between the two elements along that edge. Same conclusions apply to the other three edges of the element.

For the example element shown in Figure 4.41a, the given coordinates are such that the points 2, 6, and 3 are collinear; thus, edge 2-6-3 is a straight line segment. If all sides of a quadrilateral element are straight line segments, then the bilinear canonical element will be sufficient for mapping.

4.8.3 APPROXIMATIONS

The trial solution for the unknown primary variable $u(x,y)$ is approximated over the element domain in terms of its nodal values U_j^e, $(j = 1,2,\cdots,n)$ where n is the number of nodes per element. The approximation is written in terms of the element shape functions as in Equation (4.18), which is repeated here

$$u(x,y) \approx U^e(x,y) = \sum_{j=1}^{n} U_j^e \psi_j(x,y). \tag{4.236}$$

The shape functions satisfy the condition $\psi_j(x_i,y_i) = \delta_{ij}$. Explicit expressions in terms of x and y of the shape functions for elements such as those depicted in Figures 4.32 or 4.41a are difficult to derive. Mapping of the element into a canonical element in the $\xi - \eta$ plane has been done to avert the need for explicit expressions for $\psi_j(x,y)$. The nodal values of the primary variable U_j^e are assigned to the corresponding nodes of the canonical element. Using the shape functions of the canonical element, we write the approximation over the domain of the canonical element explicitly in terms of ξ and η

$$U^e(\xi,\eta) = \sum_{j=1}^{n} U_j^e \psi_j(\xi,\eta). \tag{4.237}$$

The value of the trial solution at a point (x,y) is taken to be its value at the corresponding point in the canonical element. The mapping functions provide the correspondence between points in the two elements

$$x(\xi,\eta) = \sum_{j=1}^{n} x_j^e \psi_j(\xi,\eta) \tag{4.238}$$

$$y(\xi,\eta) = \sum_{j=1}^{n} y_j^e \psi_j(\xi,\eta). \tag{4.239}$$

If we could invert the mapping by solving Equations (4.238) and (4.239) for ξ and η in terms of x and y

$$\xi = \xi(x,y) \tag{4.240}$$

$$\eta = \eta(x,y) \tag{4.241}$$

we would write the trial solution as

$$U^e(x,y) = \sum_{j=1}^{n} U_j^e \psi_j\left[\xi(x,y), \eta(x,y)\right]. \tag{4.242}$$

In writing Equations (4.237)–(4.239), we used the same canonical element (bilinear, biquadratic, bicubic, etc.) for both the trial solution and geometric mapping. In this case, the element is called isoparametric element. It is possible that geometric mapping is done with an element of order lower than that for the trial solution. For example, for a quadrilateral element whose sides are straight line segments, a bilinear square element may be used for mapping, but a biquadratic or bicubic element may be used for the approximation of the trial solution on that element. In such a case, the element is called subparametric. If the mapping is of higher order than the trial solution, the element is called superparametric [97]. In this book, we use only isoparametric elements.

4.8.4 ELEMENT MATRICES

The element equations are given by Equation (4.53) and the element matrices are given by Equations (4.48)–(4.51). The integrals will be transformed into integrals over the canonical element domain and evaluated by numerical quadrature. An important quantity that we need to compute before numerical integration is the gradient matrix $[B]$, which is defined by Equation (4.37) and is repeated here

$$[B] = \begin{bmatrix} \frac{\partial \psi_1(x,y)}{\partial x} & \frac{\partial \psi_2(x,y)}{\partial x} & \cdots & \frac{\partial \psi_n(x,y)}{\partial x} \\ \frac{\partial \psi_1(x,y)}{\partial y} & \frac{\partial \psi_2(x,y)}{\partial y} & \cdots & \frac{\partial \psi_n(x,y)}{\partial y} \end{bmatrix}. \tag{4.243}$$

To determine the required partial derivatives of the shape functions, we proceed as follows;

$$\begin{aligned} \frac{\partial \psi_i}{\partial \xi} &= \frac{\partial \psi_i}{\partial x}\frac{\partial x}{\partial \xi} + \frac{\partial \psi_i}{\partial y}\frac{\partial y}{\partial \xi} \\ \frac{\partial \psi_i}{\partial \eta} &= \frac{\partial \psi_i}{\partial x}\frac{\partial x}{\partial \eta} + \frac{\partial \psi_i}{\partial y}\frac{\partial y}{\partial \eta} \end{aligned} \tag{4.244}$$

or in matrix form

$$\left\{ \begin{matrix} \frac{\partial \psi_i}{\partial \xi} \\ \frac{\partial \psi_i}{\partial \eta} \end{matrix} \right\} = \begin{bmatrix} \frac{\partial x}{\partial \xi} & \frac{\partial y}{\partial \xi} \\ \frac{\partial x}{\partial \eta} & \frac{\partial y}{\partial \eta} \end{bmatrix} \left\{ \begin{matrix} \frac{\partial \psi_i}{\partial x} \\ \frac{\partial \psi_i}{\partial y} \end{matrix} \right\}. \tag{4.245}$$

The Jacobian Matrix

The matrix

$$[J] = \begin{bmatrix} \frac{\partial x}{\partial \xi} & \frac{\partial y}{\partial \xi} \\ \frac{\partial x}{\partial \eta} & \frac{\partial y}{\partial \eta} \end{bmatrix} \tag{4.246}$$

is called the Jacobian matrix. We denote its inverse by $[J]^{-1}$, and obtain from Equation (4.245)

$$\left\{ \begin{matrix} \frac{\partial \psi_i}{\partial x} \\ \frac{\partial \psi_i}{\partial y} \end{matrix} \right\} = [J]^{-1} \left\{ \begin{matrix} \frac{\partial \psi_i}{\partial \xi} \\ \frac{\partial \psi_i}{\partial \eta} \end{matrix} \right\}. \tag{4.247}$$

The Jacobian matrix can be evaluated at any point (ξ, η) in the canonical element. The mapping functions Equations (4.238) and (4.239) provide

$$\begin{aligned} \frac{\partial x}{\partial \xi} &= \sum_{j=1}^{n} x_j^e \frac{\partial \psi_j(\xi,\eta)}{\partial \xi} \\ \frac{\partial x}{\partial \eta} &= \sum_{j=1}^{n} x_j^e \frac{\partial \psi_j(\xi,\eta)}{\partial \eta} \end{aligned} \tag{4.248}$$

$$\begin{aligned} \frac{\partial y}{\partial \xi} &= \sum_{j=1}^{n} y_j^e \frac{\partial \psi_j(\xi,\eta)}{\partial \xi} \\ \frac{\partial y}{\partial \eta} &= \sum_{j=1}^{n} y_j^e \frac{\partial \psi_j(\xi,\eta)}{\partial \eta}. \end{aligned} \tag{4.249}$$

We define the matrix

$$\left[\tilde{B}\right] = \begin{bmatrix} \frac{\partial \psi_1(\xi,\eta)}{\partial \xi} & \frac{\partial \psi_2(\xi,\eta)}{\partial \xi} & \cdots & \frac{\partial \psi_n(\xi,\eta)}{\partial \xi} \\ \frac{\partial \psi_1(\xi,\eta)}{\partial \eta} & \frac{\partial \psi_2(\xi,\eta)}{\partial \eta} & \cdots & \frac{\partial \psi_n(\xi,\eta)}{\partial \eta} \end{bmatrix} \tag{4.250}$$

and obtain

$$\begin{aligned} \begin{Bmatrix} \frac{\partial x}{\partial \xi} \\ \frac{\partial x}{\partial \eta} \end{Bmatrix} &= \left[\tilde{B}\right] \{x^e\} \\ \begin{Bmatrix} \frac{\partial y}{\partial \xi} \\ \frac{\partial y}{\partial \eta} \end{Bmatrix} &= \left[\tilde{B}\right] \{y^e\}. \end{aligned} \tag{4.251}$$

The Jacobian matrix becomes

$$[J] = \left[\tilde{B}\right] \left[\ \{x^e\} \quad \{y^e\}\ \right] \tag{4.252}$$

For the model problem under consideration in this chapter, the gradient matrix defined by Equation (4.243) can be written simply as

$$[B] = [J]^{-1} \left[\tilde{B}\right]. \tag{4.253}$$

To compute the gradient matrix $[B]$ at a selected point (ξ, η) in the domain of the canonical element, we follow the steps: (1) compute $\left[\tilde{B}\right]$ from Equation (4.250). (2) Compute the Jacobian matrix from Equation (4.252); the coordinates of the element nodes $\{x^e\}$ and $\{y^e\}$ are given. (3) Compute the inverse of the Jacobian matrix, and (4) compute $[B]$ from Equation (4.253). The values of the shape functions $\psi_j(\xi, \eta)$ and the determinant of the Jacobian will also be computed at the selected point as these will be needed later for the numerical integration of the element matrices.

Transformation of Domain Integrals

The element matrices Equations (4.48)–(4.50) are integrals over the element area in the $x-y$ plane. We want to transform them to integrals over the area of the canonical element. A differential element of area in the canonical element is shown in Figure 4.42a, where $a_1 b_1 = d\xi$ and $a_1 c_1 = d\eta$. The four points a_1, b_1, c_1, and d_1 are mapped into the points a, b, c, and d in the $x-y$ plane as depicted in Figure 4.42b. The differential element in the $x-y$ plane is given by the latter four points. The vectors $\vec{ab}$ and $\vec{ac}$ can be written as

$$\begin{aligned} \vec{ab} &= \left(\frac{\partial x}{\partial \xi}\hat{i} + \frac{\partial y}{\partial \xi}\hat{j}\right) d\xi \\ \vec{ac} &= \left(\frac{\partial x}{\partial \eta}\hat{i} + \frac{\partial y}{\partial \eta}\hat{j}\right) d\eta. \end{aligned} \tag{4.254}$$

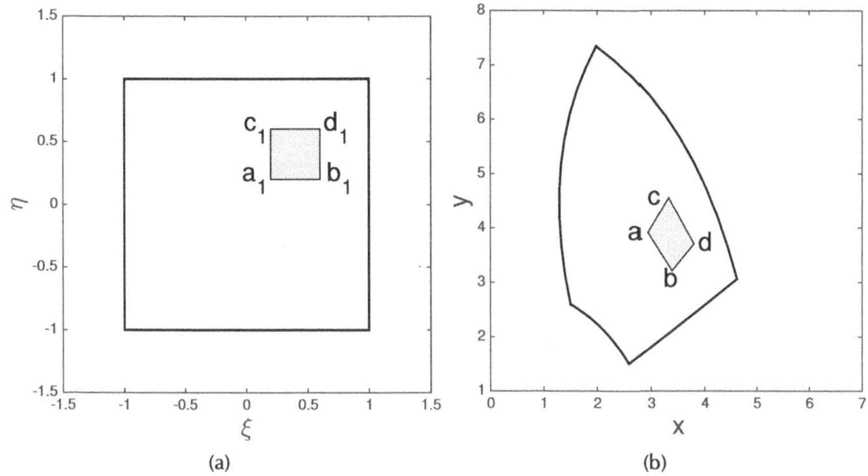

Figure 4.42 (a) A differential element in the canonical element, and (b) the mapped differential element in the physical problem domain.

The vector area of the differential element in the $x-y$ plane $dA\hat{k}$ is the cross-product of these two vectors

$$\begin{aligned} dA\hat{k} &= \vec{ab} \times \vec{ac} \\ &= \left(\frac{\partial x}{\partial \xi}\frac{\partial y}{\partial \eta} - \frac{\partial x}{\partial \eta}\frac{\partial y}{\partial \xi} \right) d\xi d\eta \hat{k} \end{aligned} \tag{4.255}$$

hence

$$dA = \|J\| d\xi d\eta \tag{4.256}$$

where $\|J\|$ is the Jacobian determinant; it is the ratio of the area of a differential element in the $x-y$ plane to its area in the canonical element and it must be positive. The Jacobian matrix and its determinant depend on the location of the point, that is to say, they are functions of ξ and η.

Using Equation (4.256), we write the element matrices Equations (4.48)–(4.50) for the isoparametric element as integrals over the canonical element.

The stiffness matrix,

$$[K^e] = \int_{-1}^{+1}\int_{-1}^{+1} [B]^T[D][B]\|J\| d\xi d\eta \tag{4.257}$$

the mass matrix,

$$[M^e] = \int_{-1}^{+1}\int_{-1}^{+1} q\{\psi\}^T\{\psi\}\|J\| d\xi d\eta \tag{4.258}$$

and the forcing vector,

$$\{F^e\} = \int_{-1}^{+1}\int_{-1}^{+1} f\{\psi\}^T\|J\| d\xi d\eta. \tag{4.259}$$

Transformation of Boundary Integrals

The vector of weighted secondary variables

$$\{R^e\} = \oint_{\Gamma^e} v_n \{\psi\}^T ds \tag{4.260}$$

is a contour integral around the element boundary. The arc length ds is

$$ds = \left[(dx)^2 + (dy)^2\right]^{1/2}. \tag{4.261}$$

On a line of constant η, the arc length can be written as

$$ds = \left[\left(\frac{\partial x}{\partial \xi}\right)^2 + \left(\frac{\partial y}{\partial \xi}\right)^2\right]^{1/2} d\xi, \qquad \eta = C \tag{4.262}$$

and on a line of constant ξ, the arc length can be written as

$$ds = \left[\left(\frac{\partial x}{\partial \eta}\right)^2 + \left(\frac{\partial y}{\partial \eta}\right)^2\right]^{1/2} d\eta, \qquad \xi = C. \tag{4.263}$$

We define

$$\begin{aligned} h_\xi &= \left[\left(\frac{\partial x}{\partial \xi}\right)^2 + \left(\frac{\partial y}{\partial \xi}\right)^2\right]^{1/2} \\ h_\eta &= \left[\left(\frac{\partial x}{\partial \eta}\right)^2 + \left(\frac{\partial y}{\partial \eta}\right)^2\right]^{1/2}. \end{aligned} \tag{4.264}$$

Going around the canonical element boundary in the counter-clockwise direction, we have

$$\eta = -1: \qquad ds = h_\xi d\xi, \tag{4.265}$$

$$\xi = 1: \qquad ds = h_\eta d\eta, \tag{4.266}$$

$$\eta = 1: \qquad ds = h_\xi(-d\xi), \tag{4.267}$$

$$\xi = -1: \qquad ds = h_\eta(-d\eta), \tag{4.268}$$

$$\begin{aligned} \{R^e\} &= \int_{-1}^{+1} \left[v_n\{\psi\}^T h_\xi\right]_{\eta=-1} d\xi + \int_{-1}^{+1} \left[v_n\{\psi\}^T h_\eta\right]_{\xi=1} d\eta \\ &+ \int_{+1}^{-1} \left[v_n\{\psi\}^T h_\xi\right]_{\eta=1} (-d\xi) + \int_{+1}^{-1} \left[v_n\{\psi\}^T h_\eta\right]_{\xi=-1} (-d\eta). \end{aligned} \tag{4.269}$$

The vector of weighted secondary variables now becomes

$$\begin{aligned} \{R^e\} &= \int_{-1}^{+1} \left[v_n\{\psi\}^T h_\xi\right]_{\eta=-1} d\xi + \int_{-1}^{+1} \left[v_n\{\psi\}^T h_\xi\right]_{\eta=1} d\xi \\ &+ \int_{-1}^{+1} \left[v_n\{\psi\}^T h_\eta\right]_{\xi=1} d\eta + \int_{-1}^{+1} \left[v_n\{\psi\}^T h_\eta\right]_{\xi=-1} d\eta. \end{aligned} \tag{4.270}$$

Table 4.1
Gauss-Legendre Quadrature Base Points (ζ_m) and Weight Factors (ω_m)

M	m	ζ_m	ω_m
1	1	0	2
2	1	$-1/\sqrt{3}$	1
	2	$+1/\sqrt{3}$	1
3	1	$-\sqrt{3/5}$	5/9
	2	0	8/9
	3	$+\sqrt{3/5}$	5/9
4	1	-0.86113 63115 94053	0.34785 48451 37454
	2	-0.33998 10435 84856	0.65214 51548 62546
	3	+0.33998 10435 84856	0.65214 51548 62546
	4	+0.86113 63115 94053	0.34785 48451 37454

Numerical Integration: Gauss–Legendre Quadrature Gauss–Legendre Rules for Quadrilateral Elements

The formulation of isoparametric elements calls for numerical methods for the computation of element matrices. The one-dimensional Gauss-Legendre quadrature method outlined in Section 2.8 can be extended to 2D (and 3D) for the computation of the double integrals in Equations (4.257)–(4.259). And of course, it can be used for the line integral in Equation (4.270). The double integrals are of the general form:

$$I = \int_{-1}^{+1} \int_{-1}^{+1} G(\xi, \eta) d\xi d\eta. \tag{4.271}$$

Gauss–Legendre quadrature formula in 1D is

$$\int_{-1}^{+1} G(\zeta) d\zeta = \sum_{m=1}^{M} \omega_m G(\zeta_m). \tag{4.272}$$

The base points ζ_m and corresponding weight factors ω_m ($m = 1, 2, ..., M$) are given in Table 4.1 for $M = 1$, 2, 3, and 4. Parameters for higher-order quadrature formulas are given by Abramowitz and Stegun [3]. The M-point formula is an exact integration of a polynomial of degree less than or equal to $(2M-1)$. The 3-point formula ($M = 3$) is

$$\int_{-1}^{+1} G(\zeta) d\zeta = \frac{5}{9} G\left(-\sqrt{\frac{3}{5}}\right) + \frac{8}{9} G(0) + \frac{5}{9} G\left(\sqrt{\frac{3}{5}}\right). \tag{4.273}$$

This formula is an exact value of the integral of the polynomial $G = a_o + a_1\zeta + a_2\zeta^2 + a_3\zeta^3 + a_4\zeta^4 + a_5\zeta^5$ for any combination of the six constants $(a_o, a_1, \cdots, a_5)$. Now, we consider the 2D integral. Applying the 1D Gauss–Legendre quadrature first

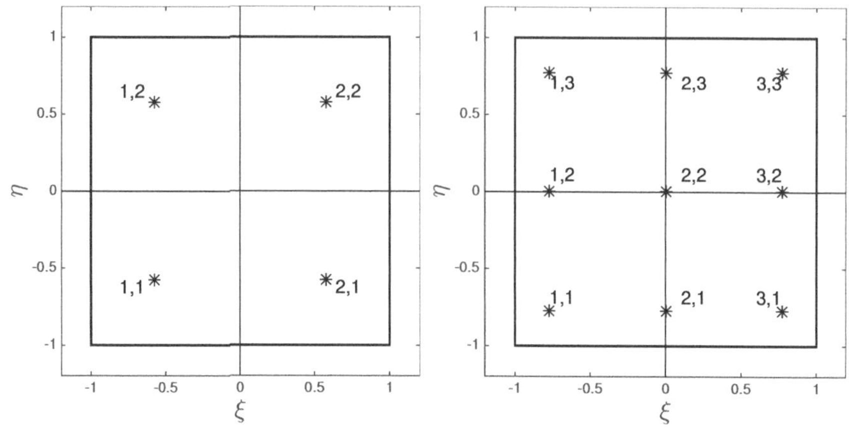

(a) (b)

Figure 4.43 Base points of Gauss–Legendre quadrature for double integrals. (a) Four points of the 2×2 rule, and (b) nine points of the 3×3 rule.

for integral with respect to ξ, we get

$$I = \int_{-1}^{+1} \sum_{m=1}^{M} \omega_m G(\xi_m, \eta) d\eta \tag{4.274}$$

then we apply it for the η integral, and obtain

$$I = \sum_{k=1}^{M} \sum_{m=1}^{M} \omega_k \omega_m G(\xi_m, \eta_k). \tag{4.275}$$

The $(M \times M)$-formula is an exact integral of a polynomial that is a linear combination of terms of the form $\xi^m \eta^n$ for $m \leq (2M-1)$ and $n \leq (2M-1)$.

Gaussian Rules for Triangular Elements

Element matrices for isoparametric triangular elements are usually evaluated numerically. Integrals over the area of the canonical element are of the general form:

$$I = \int_{A_e} G(\xi, \eta) d\xi d\eta \tag{4.276}$$

or

$$I = \int_0^1 \int_0^{1-\eta} G(\xi, \eta) d\xi d\eta. \tag{4.277}$$

Gaussian quadrature rules of different order of accuracy are given by Dunavant [33] and Hughes [53], among others. The result is

$$I = \frac{1}{2} \sum_{m=1}^{M} \omega_m G(\xi_m, \eta_m). \tag{4.278}$$

Table 4.2

Gaussian Quadrature Base Points (ξ_m, η_m) and Weight Factors (ω_m) for Integration on the Canonical Triangle, M= 4 and 7

M	m	ξ_m	η_m	ω_m
4	1	0.333333333333333	0.333333333333333	-0.562500000000000
	2	0.600000000000000	0.200000000000000	0.520833333333333
	3	0.200000000000000	0.600000000000000	0.520833333333333
	4	0.200000000000000	0.200000000000000	0.520833333333333
7	1	0.333333333333333	0.333333333333333	0.225000000000000
	2	0.797426985353087	0.101286507323456	0.125939180544827
	3	0.101286507323456	0.797426985353087	0.125939180544827
	4	0.101286507323456	0.101286507323456	0.125939180544827
	5	0.470142064105115	0.470142064105115	0.132394152788506
	6	0.470142064105115	0.059715871789770	0.132394152788506
	7	0.059715871789770	0.470142064105115	0.132394152788506

Table 4.3

Gaussian Quadrature Base Points (ξ_m, η_m) and Weight Factors (ω_m) for Integration on the Canonical Triangle, M= 13.

M	m	ξ_m	η_m	ω_m
13	1	0.333333333333333	0.333333333333333	-0.149570044467682
	2	0.479308067841920	0.260345966079040	0.175615257433208
	3	0.260345966079040	0.479308067841920	0.175615257433208
	4	0.260345966079040	0.260345966079040	0.175615257433208
	5	0.065130102902216	0.869739794195568	0.053347235608838
	6	0.869739794195568	0.065130102902216	0.053347235608838
	7	0.065130102902216	0.065130102902216	0.053347235608838
	8	0.048690315425316	0.312865496004874	0.077113760890257
	9	0.312865496004874	0.048690315425316	0.077113760890257
	10	0.638444188569810	0.048690315425316	0.077113760890257
	11	0.048690315425316	0.638444188569810	0.077113760890257
	12	0.638444188569810	0.312865496004874	0.077113760890257
	13	0.312865496004874	0.638444188569810	0.077113760890257

The base points (ξ_m, η_m) and weight factors ω_m are given Table 4.2 for $M = 4$ and 7, and in Table 4.3 for $M = 13$. These values of M give exact integration for complete polynomials of order 3, 5, and 7, respectively. Higher-order formulas can be found in [33].

4.9 APPLICATIONS

4.9.1 POTENTIAL FLOW AROUND 2D AIRFOILS, LIFT

A fundamental problem in aero-hydro-dynamics is the flow over a finite-span wing. The purpose of the wing is to produce lift, which is a force normal to the direction of flight. A simplified model is the flow around a two-dimensional airfoil, which is a wing section but the span is assumed to be infinite. The flow variables (velocity, pressure, etc.) are functions of two space dimensions. Introducing practical assumptions, we reduce the basic equations of fluid motion to Laplace's equation with Neumann or Dirichlet boundary conditions that we solve using the finite element method presented in this chapter. To that end, we define certain quantities and summarize the basic equations and assumptions leading to Laplace's equation.

Basic Equations

We consider incompressible homogeneous flow. In such a flow we neglect changes in fluid density due to pressure or temperature variations, hence the density ρ is constant. This assumption is valid if the maximum Mach number (ratio of fluid speed to local speed of sound) is much less than 1; in practice a maximum Mach number less than 0.3 may be acceptable. With the incompressible flow assumption, we have to exclude flows with strong heating or cooling at the boundaries because these effects can produce important density changes. The principle of conservation of mass is expressed by the continuity equation,

$$\nabla.\vec{V} = 0$$
$$\frac{\partial u}{\partial x} + \frac{\partial v}{\partial y} + \frac{\partial w}{\partial z} = 0 \tag{4.279}$$

where $\vec{V}(\vec{r},t) = u\hat{i} + v\hat{j} + w\hat{k}$ is the fluid velocity. For a Newtonian fluid, the principle of linear momentum (or the equation of motion) is expressed by the Navier–Stokes equation,

$$\rho\frac{D\vec{V}}{Dt} = -\nabla p + \mu\nabla^2\vec{V} + \rho\vec{f}_b \tag{4.280}$$

where p is the pressure, μ is the coefficient of dynamic viscosity assumed constant, and $\vec{f}_b$ is the body force per unit mass.

Circulation

Consider an arbitrary simple closed contour, C, in the fluid domain. The circulation is a scalar quantity defined as the line integral of the velocity component tangent to the contour,

$$\begin{aligned}\Gamma &= \oint_C \vec{V}.d\vec{r} \\ &= \oint_C v_t ds\end{aligned} \tag{4.281}$$

where v_t is the velocity component tangent to the contour C, and ds is the arc length. Circulation is an important quantity in aerodynamics as it determines the lift produced by a lifting surface.

Vorticity

Another important kinematic measure of fluid motion is vorticity, which is a vector quantity defined as the curl of the velocity,

$$\vec{\omega} = \nabla \times \vec{V}, \tag{4.282}$$

Consider a small fluid element surrounding an arbitrary point A. The velocity vector at point A describes the translational motion of the element. However, as the element translates it also deforms changing its shape, volume, and orientation. These kinematic quantities are characterized by the relative velocity between point A and other points in the element; and can be expressed in terms of the velocity gradient at the point. The relative motion can be decomposed into two motions; one part is due to rate of straining of the element and the other is due to rigid body spin. The vorticity vector is twice the angular velocity of the rigid body spin. The vorticity is a point function whereas circulation is a macro measure of the asymmetry of the velocity field. The two kinematic measures are related by Stokes theorem:

$$\oint_C \vec{V}.d\vec{r} \quad = \quad \int_S \vec{\omega} \cdot \hat{n} dS \tag{4.283}$$

We emphasize that the surface S must fall completely in the fluid domain and have contour C as its rim. If the vorticity flux on the surface S is zero, then the circulation around C is also zero. For singly connected domains we can always construct the surface S for any circuit C. However, for multiply connected domains, such as the domain outside a torus (a donut) or the domain outside an infinitely long cylinder, we cannot construct the surface S for certain circuits drawn in the domain. On such circuits the circulation may not be zero even if the vorticity is zero in the domain.

Potential (or Irrotational) Flow

A fluid motion is said to be irrotational if the vorticity is zero $\vec{\omega} = 0$. This is a strong kinematic constraint on the velocity field; that is, the curl of velocity is zero. The vorticity in a stagnant body of fluid or in uniform motion is obviously zero. The fluid moves in response to the forces acting on it. Thus, if the ensuing motion of such a body of fluid is to remain irrotational forces that impart angular momentum to a fluid element must be excluded. Derivation of Kelvin's circulation theorem elucidates certain mechanisms that can generate vorticity. A circuit composed of the same fluid elements at all times is called a material circuit. Such a circuit moves and deforms with the fluid. The circulation around a material circuit is in general a function of time. Using the momentum equations and certain vector identities, we find the rate of change of circulation around a material circuit,

$$\frac{d\Gamma_{mc}}{dt} = \oint -\frac{d\vec{r} \cdot \nabla p}{\rho} + \oint \frac{d\vec{r} \cdot \nabla \cdot \bar{\bar{\tau}}}{\rho} + \oint_{mc} \vec{f}_b . d\vec{r} \tag{4.284}$$

where $\bar{\bar{\tau}}$ is the viscous stress tensor. It can be shown that the first integral is zero for incompressible (constant density ρ) or barotropic flow $\rho = \rho(p)$. The second integral is also zero if we neglect viscous stresses and treat the flow as an inviscid flow. The third integral is zero when the body force field is conservative with single valued potential. Under the three conditions (barotropic, inviscid, and conservative body force), Kelvin's theorem states that the circulation around a material circuit remains constant with time. If the circulation is initially zero, it will remain zero at later times. Using Kelvin's circulation theorem and Stokes theorem, one can prove that if the motion of a body of fluid is initially irrotational, it will continue to be irrotational at later times provided that the aforementioned conditions are satisfied.

The velocity field of an irrotational flow can be written as the gradient of a scalar function

$$\begin{aligned} \vec{V} &= \nabla\phi \\ &= \frac{\partial\phi}{\partial x}\hat{i}+\frac{\partial\phi}{\partial y}\hat{j}+\frac{\partial\phi}{\partial z}\hat{k} \end{aligned} \tag{4.285}$$

ϕ is called the velocity potential, and thus an irrotational flow is also called a potential flow. The justification for Equation (4.285) follows from the vector identity $\nabla\times\vec{V} = \nabla\times\nabla\phi = 0$. Equation (4.285) applies whether the flow is compressible or incompressible, and steady or unsteady. Substituting Equation (4.285) into the continuity equation Equation (4.279), we get the equation for ϕ,

$$\begin{aligned} \nabla^2\phi &= 0 \\ \frac{\partial^2\phi}{\partial x^2}+\frac{\partial^2\phi}{\partial y^2}+\frac{\partial^2\phi}{\partial z^2} &= 0 \end{aligned} \tag{4.286}$$

which is Laplace's equation. Therefore, the major step in finding the velocity field $\vec{V}$ of an irrotational flow is to solve Laplace's equation subjected to proper boundary conditions. The pressure is found by integrating the momentum equation Equation (4.280). Assuming incompressible inviscid flow and conservative body force, the integrated momentum equation is Bernoulli's equation, which we write here for steady or unsteady potential flow,

$$\frac{\partial\phi}{\partial t}+\frac{1}{2}V^2+\frac{p}{\rho}+G=C \tag{4.287}$$

where $V = |\nabla\phi|$ is the velocity magnitude, G is the potential energy of the body force $\vec{f_b} = -\nabla G$, and C is constant but it may be a function of time, t.

In this section we use the finite-element method to solve the steady potential flow problem around a 2D airfoil of an arbitrary shape. The airfoil is placed at an angle of attack α in an otherwise uniform flow. We want to determine the velocity field, pressure distribution on the airfoil surface, and the lift. The relevant equations are

$$-\left(\frac{\partial^2\phi}{\partial x^2}+\frac{\partial^2\phi}{\partial y^2}\right)=0 \tag{4.288}$$

$$\vec{V} = \frac{\partial \phi}{\partial x}\hat{i} + \frac{\partial \phi}{\partial y}\hat{j}. \tag{4.289}$$

The airfoil profile in the $x-y$ plane is given by an equation

$$y = y_a(x). \tag{4.290}$$

The solution to Equation (4.288) must satisfy certain boundary conditions. Since the airfoil is stationary, the fluid velocity component normal to the airfoil surface must be zero while the tangential component (slip velocity) is unknown,

$$\begin{aligned} &\hat{n}.\vec{V} = 0 \\ \text{or} \quad &\hat{n}.\nabla\phi = 0 \\ \text{or} \quad &\frac{\partial \phi}{\partial n} = 0 \qquad \text{on} \qquad y = y_a. \end{aligned} \tag{4.291}$$

We note that the velocity potential ϕ is the primary variable. The flow domain is unbounded. Far from the airfoil we expect the disturbance velocity produced by airfoil to decay, and the flow returns to uniform flow. Taking the origin at the leading edge or at any nearby point, we have

$$\nabla\phi = U_\infty(\hat{i}cos\alpha + \hat{j}sin\alpha) \qquad \text{as} \qquad r \to \infty \tag{4.292}$$

where U_∞ is the magnitude of the free-stream velocity and α is the angle it makes with the positive x-axis. We place the airfoil chord on the x-axis, and thus α is the angle of attack. The boundary conditions Equations (4.291) and (4.292) are not sufficient to determine a unique solution to Equation (4.288). To elucidate this point, let's place a circular cylinder in uniform flow U_∞ parallel to the x-axis. The cylinder radius is b and its center is (0,0). The velocity potential is

$$\phi = U_\infty(r + \frac{b^2}{r})cos\theta - \frac{\Gamma}{2\pi}\theta. \tag{4.293}$$

where (r, θ) are polar coordinates, $(x = rcos\theta, y = rsin\theta)$. It can be shown that ϕ given by Equation (4.293) satisfies Equation (4.288) and the boundary conditions Equations (4.291) and (4.292) for any value of the parameter Γ. The tangency boundary condition (normal velocity is zero on the surface) and the farfield boundary condition do not render a unique solution to the potential flow over a 2D profile of any smooth profile. The parameter Γ is the circulation (in the clockwise sense) around any circuit that encloses the cylinder. An additional condition is required to make the solution unique. For example, we may specify the amount of circulation around the cylinder or we may specify a value for the tangential velocity at a point on the cylinder.

Kutta Condition

Unlike the case of smooth cylinders such as circular or elliptic cylinders, for which the amount of circulation can be arbitrarily prescribed, a definite value of the circulation around practical airfoil sections can be determined. A practical airfoil is constructed with sharp trailing edge to help fix the amount of circulation produced around the airfoil. Potential flow theory shows that for an airfoil section with a sharp trailing edge, the velocity at the trailing edge would be infinite if the circulation is arbitrarily prescribed. Requiring a finite velocity at a sharp trailing edge of an airfoil is known as the Kutta condition. Another statement of the condition is that the pressure is continuous across the trailing edge. That is to say, the pressure at the trailing edge has the same value whether we approach the trailing along the upper surface or along the lower surface of the airfoil. The angle between the tangents to the upper and lower surfaces at the trailing edge of the airfoil is called the trailing edge angle. If the trailing edge angle is greater than zero, then the sharp trailing edge is a stagnation point (zero velocity) since the tangential velocity at the trailing edge cannot have two different directions. However, if there is a common tangent to the upper and lower surface at the trailing edge, then the trailing edge angle is zero (cusp) and the velocity there need not be zero.

Generation of circulation around a 2D airfoil can be explained by analysis of fluid motion around an airfoil started from rest. Initially the vorticity and circulation around the airfoil are zero. As the airfoil moves, vorticity is generated on the airfoil surface because the fluid adjacent to the surface cannot slip relative to the surface; the no-slip condition cannot be enforced in an inviscid flow theory as viscous terms are neglected. Viscosity diffuses vorticity lateral to the surface at the same time it is convected. At high Reynolds number, the generated vorticity remains in the boundary layer, which is a thin layer next to the surface. Even though there is vorticity confined in the thin layer around the airfoil, the circulation around a circuit encompassing the airfoil and the vorticity region is still zero. This is because the generated vorticity is of mixed sign (positive and negative) such that the integral of vorticity over the enclosed area is zero. However, the resulting flow cannot negotiate the sharp trailing edge, and the boundary layer separates there. The separated vorticity layer rolls into a "starting" vortex downstream of the trailing edge. Vorticity continues to separate from the trailing edge and feeds into the starting vortex, consequently vorticity of opposite sign continues to build up in the boundary layer around the airfoil. When smooth flow is established at the trailing edge, the rate of vorticity shed from the upper surface will be equal to the rate of vorticity shed from the lower surface, and the amount of vorticity and circulation around the airfoil reach a steady level. The shed starting vortex is convected to the far field by the main stream; eventually it has no effects on the flow around the airfoil. For high Reynolds number flows, and as long as the boundary layer remains thin and attached to the airfoil surface, vorticity in the boundary layer can be modeled in an inviscid flow theory as an infinitesimally thin vortex sheet bound to the airfoil surface. The total strength of that vortex sheet is the circulation around the airfoil.

For the purpose of applying the finite-element method, the unbounded flow domain around the airfoil is truncated into a finite domain, and the farfield boundary condition is imposed on the outer boundary of the truncated domain. Unless the circulation is zero, the farfield boundary condition as given by Equation (4.292) produces incorrect results if applied to a finite domain because it neglects the disturbance on the boundary due to circulation. The circulation around the airfoil Γ is unknown but its contribution to the disturbance on the farfield boundary must be manifested in the boundary condition. If the outer boundary is far enough, say six chords away from the airfoil, the disturbance on the boundary may be approximated by the field of a line vortex located at the 1/4-chord point of the airfoil. Therefore, on the outer boundary of the computational domain, we specify the velocity potential by

$$\phi = U_\infty(xcos\alpha + ysin\alpha) - \frac{\Gamma}{2\pi}\theta \qquad \text{on} \qquad \partial\Omega \tag{4.294}$$

where Γ is the yet unknown circulation around the airfoil, and θ is the angle between the positive x-axis and the radial line from the 1/4-chord point to a point on the outer boundary. The circulation is considered positive when clockwise. In summary, the problem for ϕ is given by Laplace's Equation (4.288), and the boundary conditions Equation (4.291) and Equation (4.294). We must also apply a Kutta condition to fix the unknown circulation.

Gonzalez and Taha [41] (also, Taha and Gonzalez [112]), used Hertz' principle of least curvature and developed a novel variational formulation of Euler's equations for the dynamics of ideal fluids. They showed that the amount of circulation around a 2D airfoil in potential flow can be uniquely determined by minimization of the Appellian, which is the integral over the flow domain of squared acceleration magnitude. Their method applies to airfoils with or without a sharp trailing edge; and hence an alternative to the Kutta condition which is meaningful only for a sharp trailing edge.

C-Mesh

We developed a MATLAB code for finite-element solution of the 2D potential flow problem around an airfoil of arbitrary profile. It also includes a mesh generation function. The mesh topology is shown in Figure 4.44, which is known as a C-grid. Although the mesh is a structured mesh, the MATLAB code uses the connectivity matrix of the elements whether they are structured or unstructured. In Figure 4.44, the points B, C, D, E, and F are anchor points on the airfoil surface. Points B and F are one point, which is the airfoil trailing edge. Point B is the trailing edge as it is approached along the lower surface, and point F is the trailing edge as it is approached along the upper surface. The double definition of the trailing edge point facilitates the implementation of the Kutta condition. Point D is the airfoil leading edge; which is taken as the origin of the $x-y$ coordinate system. The trailing edge is on the x-axis at the point $(C,0)$, where C is the chord. We define the wake cut as any line (straight or curved) from the trailing edge to a point on the outer boundary. Lines $B-A$ and $F-G$ are identical lines, and we take them as the wake cut and place them on the x-axis. Nodes on the wake cut are also doubly defined; each mesh node on BA corresponds to a node on FG with the same global coordinates but different node number. The outer boundary is defined by the circuit $A-N-M-L-K-J-I-H-G$. With the introduction of the wake cut, the doubly connected domain around the airfoil is rendered singly connected domain enclosed by the boundary $A-B-C-D-E-F-G-H-I-J-K-L-M-N-A$. An elliptic grid generator is used, and

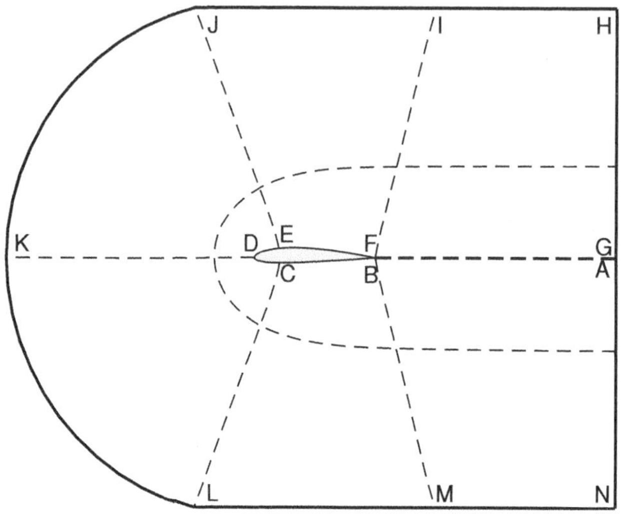

Figure 4.44 C-grid topology.

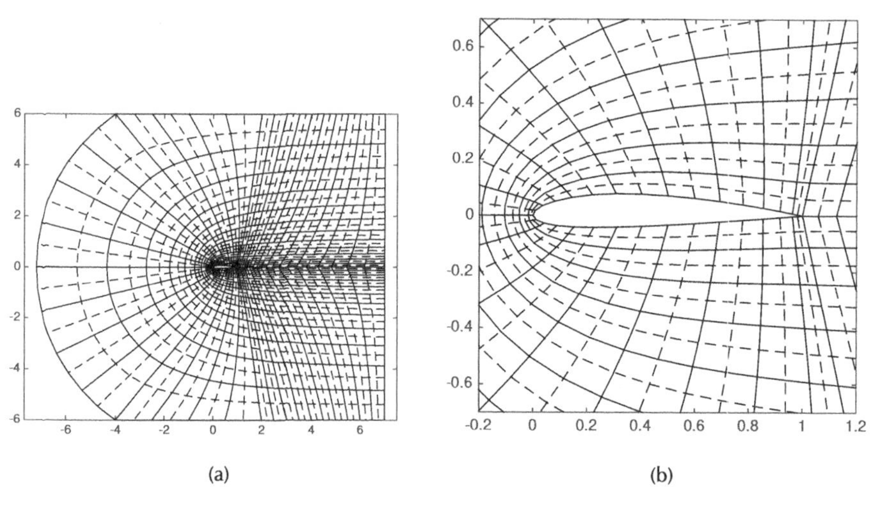

Figure 4.45 An example of a C-grid.

an example of a C-grid is shown in Figures 4.45a and 4.45b. The mesh is composed of quadrilateral elements that can be either isoparametric 4-node bilinear elements or 9-node biquadratic elements. They can be divided into triangular elements, but that option is not implemented in the developed code. The mesh generator function provides the global (x, y) coordinates of all nodes, and the connectivity matrix for the elements.

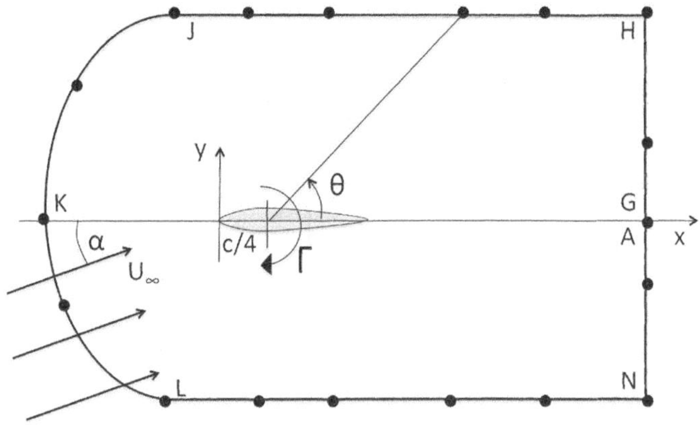

Figure 4.46 Flow parameters for farfield boundary condition.

The Assembled Global System

For Laplace's equation we have $p_{11} = p_{22} = 1$, $p_{12} = p_{21} = 0$, $q = 0$, and $f = 0$. The element stiffness matrix is Equation (4.257), and the integral is evaluated numerically by Gauss–Legendre quadrature. The mass matrix Equation (4.258) and forcing vector Equation (4.259) are zero. The assembled global system before application of the boundary conditions takes the form

$$[K]\{U\} = \{R\}. \tag{4.295}$$

There are N global nodes (we double count the nodes on the wake cut), and there are N unknown nodal values of the velocity potential ϕ, which are denoted by the array $\{U\}$. The global coefficient matrix is of the dimensions (N,N). The array of weighed secondary variables $\{R\}$ will be decided by the boundary conditions. The circulation Γ is also unknown. In reality we have $N+1$ unknowns. We augment the array $\{U\}$ with the unknown $U_{N+1} \equiv \Gamma$, and expand the coefficient matrix to a matrix with dimensions $(N+1, N+1)$; the additional column and row will be defined shortly.

Application of Farfield Boundary Condition

For global nodes on the farfield boundary as shown in Figure 4.46, the velocity potential is given by Equation (4.294). The effects of circulation are represented by a line vortex at the 1/4-chord point, and assumed positive in the clockwise direction. Let i denote the index of a global node on that boundary. We apply the boundary condition by simply replacing the equation corresponding to node i (row number i) in the global system Equation (4.295) by

$$U_i + \frac{\theta_i}{2\pi} U_{N+1} = U_\infty (x_i cos\alpha + y_i sin\alpha). \tag{4.296}$$

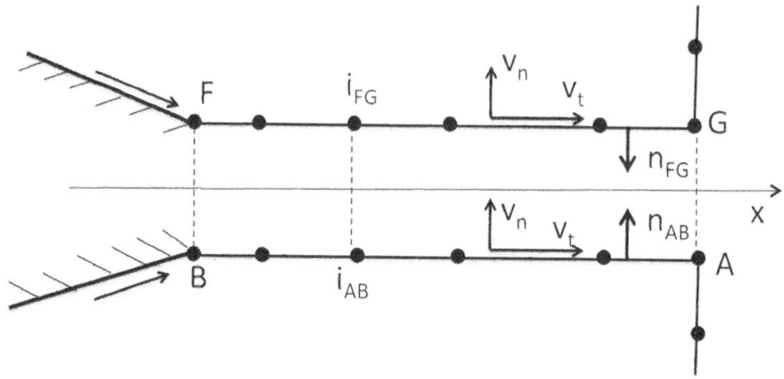

Figure 4.47 Conditions on the wake cut.

The angle θ is defined in Figure 4.46, and α is the angle of attack. Thus, $K_{ij} = 0$ for $j = 1,...,N$ except $K_{ii} = 1$, and $K_{i,N+1} = \theta_i/(2\pi)$. The right-hand side of Equation (4.296) becomes the forcing term for nodes on the farfield boundary. Equation (4.296) is used for all nodes on the farfield boundary including the nodes A and G. It is important that we take θ in the range $0 \leq \theta \leq 2\pi$, with $\theta = 0$ being point G, and $\theta = 2\pi$ being point A. The farfield condition may be classified as an essential boundary condition.

Application of Boundary Condition on Airfoil Surface

The secondary variable is $v_n = \partial\phi/\partial n$. The boundary condition Equation (4.291) on the airfoil surface prescribes $v_n = 0$, it is a Neumann or natural boundary condition. Since $v_n = 0$, we put the weighted secondary variables $R_i = 0$ for all the nodes on the airfoil surface, we exclude the trailing edge points B and F because we treat them as points on the wake cut.

Compatibility Conditions on the Wake Cut

The wake cut is introduced to render the doubly connected domain singly connected. Figure 4.47 shows the nodes on the wake cut. The node i_{AB} on side AB corresponds to node i_{FG} on side FG. The two nodes are the same point in space but they are assigned different node numbers. The wake cut is an interior line in the flow domain, and hence there are no boundary conditions prescribed on the cut. The velocity components (normal v_n and tangential v_t) are continuous as we cross the cut from side AB to side FG. However, the velocity potential ϕ is discontinuous across the wake cut. That is to say, on the wake cut: $\phi(x,0^+) \neq \phi(x,0^-)$; however, the jump in ϕ is not a function of x. To clarify this point, we consider the circulation around any circuit that surrounds the airfoil. Such a circuit intersects the wake cut at node i_{AB}, which is the same as node i_{FG}. Going around the circuit in the clockwise sense, we have

$$\begin{aligned} \Gamma &= \oint \vec{V}.d\vec{r} \\ &= \oint \nabla\phi.d\vec{r} \\ &= \oint d\phi \\ &= \int_{i_{AB}}^{i_{FG}} d\phi \\ &= \phi_{i_{FG}} - \phi_{i_{AB}}. \end{aligned} \tag{4.297}$$

Thus, the velocity potential at all the nodes on the wake cut (nodes on AB and FG) satisfy

$$\Gamma + \phi_{i_{AB}} - \phi_{i_{FG}} = 0. \tag{4.298}$$

The secondary variable is the velocity component normal to the boundary. On the wake cut the velocity is continuous; however, the outward unit normal on AB is opposite to that on FG. Thus, the weighted secondary variables on the wake cut satisfy

$$R_{i_{AB}} + R_{i_{FG}} = 0. \tag{4.299}$$

The conditions specified by Equations (4.298) and (4.299) are applied for all the nodes on wake cut (nodes on AB and FG) excluding nodes A and G that are treated as points on the farfield boundary. Before application of the wake cut conditions, the equations for the nodes i_{AB} and i_{FG} are

$$\sum_{j=1}^{N} K_{i_{AB},j} U_j = R_{i_{AB}} \tag{4.300}$$

$$\sum_{j=1}^{N} K_{i_{FG},j} U_j = R_{i_{FG}}. \tag{4.301}$$

In view of condition Equation (4.299), we add Equation (4.300) to Equation (4.301) and obtain

$$\sum_{j=1}^{N} \left(K_{i_{AB},j} + K_{i_{FG},j}\right) U_j = 0. \tag{4.302}$$

We rewrite Equation (4.298),

$$U_{i_{FG}} - U_{i_{AB}} - U_{N+1} = 0. \tag{4.303}$$

To apply the wake cut conditions, we replace the Equation (4.300) and Equation (4.301) in the original system by Equation (4.302) and Equation (4.303), respectively. The unknown weighted secondary variables are now eliminated or replaced by farfield condition Equation (4.296).

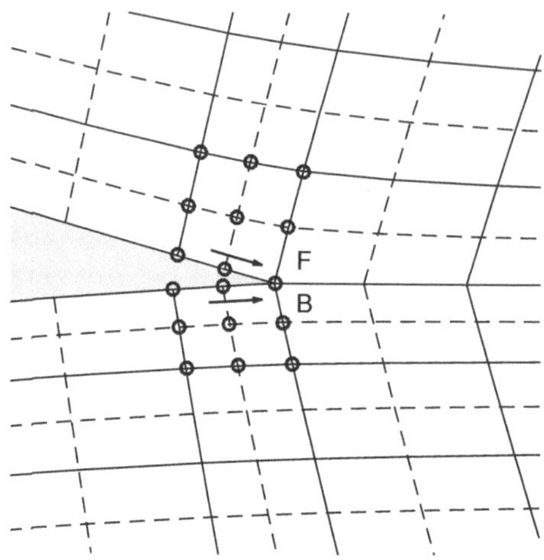

Figure 4.48 Global nodes involved in the Kutta condition at the trailing edge.

Application of Kutta Condition

Kutta condition provides the needed equation to close the system of $N+1$ unknowns; the unknowns being ϕ at N global nodes and circulation Γ. The trailing edge is doubly defined by the two points B on the lower surface and F on the upper surface. Bernoulli's equation shows that equal pressure at the two points implies equal velocity magnitude. Figure 4.48 shows the mesh near the trailing edge. We consider two points on the airfoil surface; one very near point B on the lower surface and the other point very near point F on the upper surface. At these two points, the velocity component normal to the surface is zero as required by the boundary condition. Thus, the tangential velocity at these two points must be equal. The Kutta condition, as implemented in the developed code, is an equation that equates the component of velocity tangent to the airfoil surface at points B and F.

$$[\hat{t}\cdot\nabla\phi]_B = [\hat{t}\cdot\nabla\phi]_F \tag{4.304}$$

where $\hat{t} = t_x\hat{i} + t_y\hat{j}$ is the unit tangent at the trailing edge. The gradient of the velocity potential is expressed in terms of the velocity potential at the neighboring nodes. To fix matters, we consider a mesh of biquadratic quadrilateral elements and show in Figure 4.48 the two elements on the airfoil surface above and below the trailing edge. The velocity gradient for each element is given by

$$\left\{ \begin{array}{c} \frac{\partial\phi^e(x,y)}{\partial x} \\ \frac{\partial\phi^e(x,y)}{\partial y} \end{array} \right\} = [B]\{\phi^e\} \tag{4.305}$$

Table 4.4
Geometric Parameters of Three Grids for an Airfoil Section Using Bi-quadratic Elements

	n_{AB}	n_{BF}	n_{AN}	$n_{elements}$	n_{nodes}
Coarse mesh (G_1)	10	16	12	432	1,825
Medium mesh (G_2)	16	32	20	1,280	5,289
Fine mesh (G_3)	20	64	32	3,328	13,585

For all grids: $x_A = 7, x_M = 2, x_L = -4, -y_N = y_H = 6, s_{AB} = 1.15, s_{AN} = 1.1.$

where the gradient matrix $[B]$ for the isoparametric element is given by Equation (4.253). The final form of the Kutta condition is

$$[\{\ t\ \}[B]\{U^e\}]_F - [\{\ t\ \}[B]\{U^e\}]_B = 0 \tag{4.306}$$

where $\{t\} = \{t_x, t_y\}$ is the unit tangent. This is equation number $N+1$ that closes the global system. The system of $N+1$ linear equations is solved only once, and the solution provides the velocity potential at the N global nodes and the circulation Γ as variable number $N+1$. We post-process the solution to obtain velocity, pressure, and streamlines. The lift per unit span is $\rho U_\infty \Gamma$.

Example: Flow around 2D Airfoils

Joukowski Airfoil
Exact analytic solutions for potential flow around 2D airfoils can be obtained by conformal mapping of the airfoil to a circle (See Milne-Thomson [75], for example). Joukowski transformation

$$z = \zeta + \frac{a^2}{\zeta} \tag{4.307}$$

where $a > 0$ is a real constant, maps a circle in the complex ζ-plane into a Joukowski airfoil in the complex z-plane. We place the circle center in the second quadrant $\zeta_c = \xi_c + i\eta_c$, and let the circle pass through the point $(a, 0)$ on the positive ξ-axis. A flat plate airfoil of chord $C = 4a$ is obtained if the circle center is at the origin $(\xi_c = 0,\ \eta_c = 0)$. And a circular arc airfoil of chord $C = 4a$ is obtained if the center is located on the imaginary axis $(\xi_c = 0)$. The trailing edge angle is zero for the Joukowski family of airfoils. We solve the flow over a Joukowski airfoil with finite thickness and camber by choosing $a = 1$, and $\zeta_c = -0.1 + i0.06$. The airfoil chord is defined by $C = x_{max} - x_{min}$, and is used as a reference length. To investigate the effects of mesh refinement on surface pressure distribution and lift, we generated three meshes using either biquadratic or bilinear quadrilateral elements. The parameters for the biquadratic mesh are given in Table 4.4. The number of elements on the wake

Table 4.5
Circulation $\Gamma/U_\infty C$

	Mesh (G_1)	Mesh (G_2)	Mesh (G_3)
Bi-linear element	0.7551	0.7729	0.7779
Bi-quadratic element	0.7716	0.7772	0.7791
Exact (conformal mapping)	0.7792	0.7792	0.7792

Joukowski airfoil ($\zeta_c = -0.1 + i0.06$, and $a = 1$) at an angle of attack $\alpha = 10$ degrees.

cut is n_{AB}. For the coarse mesh, there are 21 global nodes on the wake cut giving 10 biquadratic elements or 20 bilinear elements. The number of elements on the airfoil surface is n_{BF}. For the coarse mesh, there are 33 global nodes on the airfoil surface giving 16 biquadratic elements or 32 bilinear elements. The number of elements on the outflow boundary is $n_{AN} = n_{GH}$. For the coarse mesh, there are 25 global nodes on AN and GH giving 12 biquadratic elements or 24 bilinear elements. We use geometric stretching for distributing the global nodes on the wake cut and the outflow boundary, so that small elements are used near the trailing edge and normal to the wake cut. The stretching parameters are $s_{AB} = 1.15$ and $s_{AN} = 1.1$, which are used for all meshes. The number of elements for the three meshes are given in the table. At an angle of attack $\alpha = 10$ degrees, we present in Table 4.5 the computed circulation Γ and the exact value by conformal mapping. The error in circulation for the coarse mesh (G_1) of bilinear elements is 3%, and for the biquadratic element is 1%. The bilinear and biquadratic meshes have equal number of global nodes but different number of elements. For the fine mesh (G_3), the error for the bilinear element is 0.17%, and for biquadratic is 0.01%.

NACA-2412 Airfoil

A well known series of wing sections is the NACA four-digit wing sections [2]. A section is defined by a thickness function and a mean (or camber) line. The thickness distribution function is

$$\pm y_t = \frac{t}{0.2}\left(0.2969\sqrt{x} - 0.1260x - 0.3516x^2 + 0.2843x^3 - 0.1015x^4\right) \tag{4.308}$$

where t is the maximum thickness expressed as a fraction of the chord. The leading edge is $x = 0$, and the trailing edge is $x = 1$. At $x = 1$, the thickness given by Equation (4.308) is not exactly zero, hence the trailing edge has a finite thickness. For potential flow computation, we prefer to work with an airfoil of zero thickness at the trailing edge. The function Equation (4.308) is zero at $x = 1.0089304113$. Therefore, we still use Equation (4.308) to define the thickness function but we use it for $0 \leq x \leq 1.0089304113$, and define the chord to be $C = 1.0089304113$. The camber (or mean)

Table 4.6
Circulation $\Gamma/U_\infty C$

	Mesh (G_1)	Mesh (G_2)	Mesh (G_3)
Bi-linear element	0.7090	0.7219	0.7253
Bi-quadratic element	0.7206	0.7248	0.7262
Panel method	0.7223	0.7253	0.7261

NACA2412 airfoil at an angle of attack $\alpha = 10$ degrees.

line is defined by

$$
\begin{aligned}
y_c &= \frac{m}{p^2}\left(2px - x^2\right) && \text{for } 0 \leq x \leq p \\
y_c &= \frac{m}{(C-p)^2}\left[C^2 - 2pC + 2px - x^2\right] && \text{for } p \leq x \leq C.
\end{aligned} \tag{4.309}
$$

The surface coordinates are generated by combining the thickness function and mean line as described by Abbot and Doenhoff [2]. To generate the airfoil NACA-2412 we used $t = 0.12$, $p = 0.4$, and $m = 0.02$. Finite element solutions are obtained for three grids with the parameters given in Table 4.4. The computed circulations at $\alpha = 10$ degrees are compared with those obtained by a panel code (also developed by the authors) in Table 4.6. The panel code uses the same surface grid used in the finite element mesh. The results of finite element are in good agreement with the panel method. The results of biquadratic element are almost identical to the panel method. Pressure coefficient distributions $C_p(x) = 2(p - p_\infty)/\rho U_\infty^2$ on the airfoil surface are compared with results of conformal mapping for Joukowski airfoil in Figures 4.49a, 4.49c, and 4.49e for the three grids. Quadratic elements are used. The suction peak at the leading edge and the details of the pressure near the trailing edge are well predicted by the finite element computations. The trailing edge of the Joukowski airfoil is a cusp (zero angle), and hence the velocity is not zero there. The current implementation of the Kutta condition produces the correct behavior of the flow at the trailing edge.

Surface pressure coefficient distributions for NACA2412 airfoil are compared with results of the panel method (a boundary integral method) in Figures 4.49b, 4.49d, and 4.49f for the three grids. The finite element results are in excellent agreement with the panel method results. The suction peak at the leading edge is well predicted by the finite element method. The trailing edge angle for the NACA2412 airfoil is greater than zero. In an inviscid potential flow, the trailing edge is a stagnation point and the pressure coefficient is one. A very dense mesh is needed near the trailing to capture the severe pressure gradient at the stagnation point.

Karman–Trefftz Airfoil

We recall that the trailing edge of a Joukowski airfoil is a cusp (zero angle). An airfoil that is generated by conformal mapping with a non-zero angle at the trailing

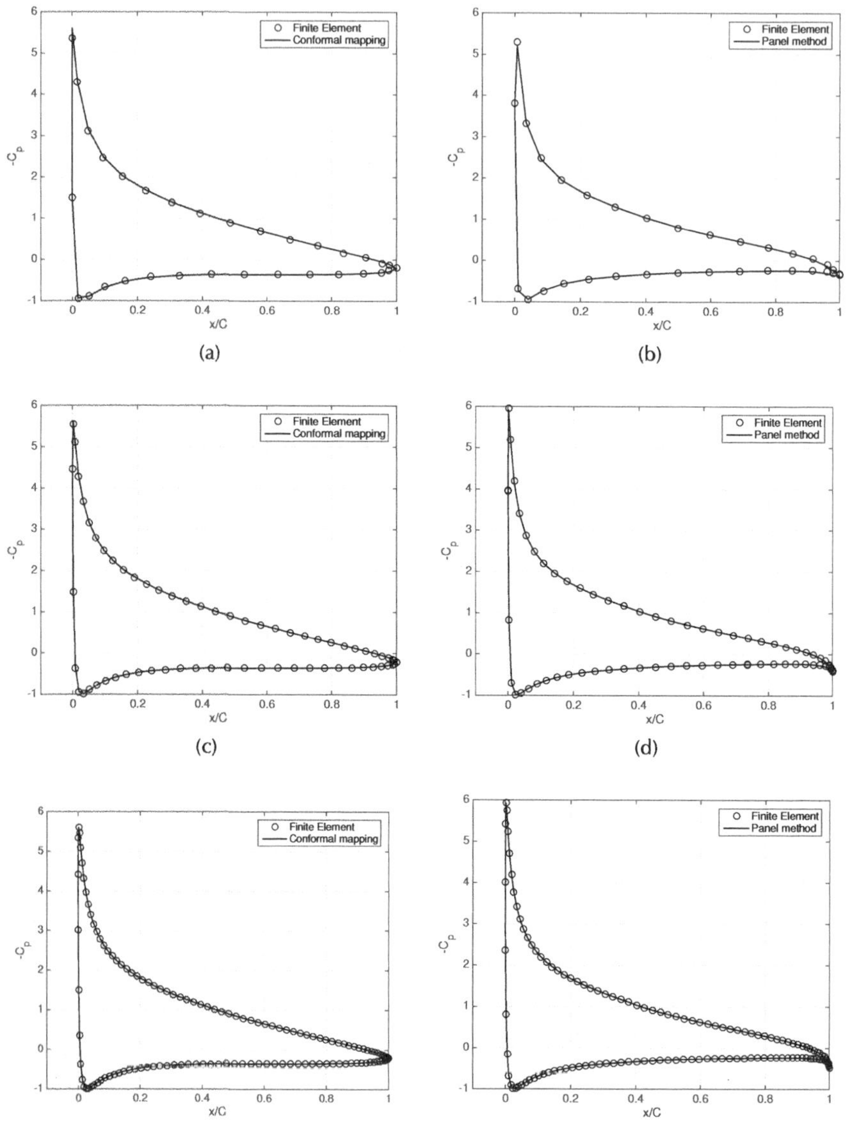

Figure 4.49 Pressure coefficient distribution on airfoil surface. Mesh G_1: (a) and (b), mesh G_2: (c) and (d), and mesh G_3: (e) and (f). Joukowski airfoil: (a), (c), and (e). NACA2412 airfoil: (b), (d), and (f).

edge is the Karman–Trefftz airfoil (see Milne-Thomson [75], for example) defined by the mapping

$$\begin{aligned} f &= \left(\frac{\zeta - a}{\zeta + a}\right)^n \\ z &= na\frac{1+f}{1-f}. \end{aligned} \tag{4.310}$$

A circle in the ζ-plane with center at ζ_c and passing through the point $\zeta = a$ is transformed into a Karman–Trefftz airfoil with trailing-edge angle equals $(2-n)\pi$. A Joukowski airfoil is a special case for $n = 2$. By conformal mapping, we can determine exact solutions for the pressure distribution and streamlines for this family of airfoils.

Streamlines

For two-dimensional incompressible flow, the velocity components can be defined in terms of a stream function $\psi(x,y)$ by

$$\begin{aligned} u &= \frac{\partial \psi}{\partial y} \\ v &= -\frac{\partial \psi}{\partial x}. \end{aligned} \tag{4.311}$$

The continuity equation ($\nabla \cdot \vec{V} = 0$) is identically satisfied by these velocity components, and for irrotational flow ($\nabla \times \vec{V} = 0$) we obtain Laplace's equation

$$-\left(\frac{\partial^2 \psi}{\partial x^2} + \frac{\partial^2 \psi}{\partial y^2}\right) = 0. \tag{4.312}$$

The stream function $\psi(x,y)$ is constant on a streamline. Therefore, streamlines around the airfoil are plotted as stream function contours. After we determine the velocity potential ϕ and Circulations Γ, we solve Equation (4.312) for ψ subjected to the boundary conditions,

$$\psi = 0 \qquad \text{on the airfoil surface} \qquad y = y_a(x). \tag{4.313}$$

This is a Dirichlet boundary condition (or *EBC*) in contrast to the Neumann condition (or *NBC*) used for the velocity potential ϕ. On the farfield boundary of the computational domain, we specify the component of velocity tangential to the boundary,

$$\frac{\partial \psi}{\partial n} = \left[U_\infty(\hat{i}cos\alpha + \hat{j}sin\alpha) + \frac{\Gamma}{2\pi r^2}(\hat{i}x - \hat{j}y)\right] \cdot \hat{t} \tag{4.314}$$

where $\hat{t}$ is the unit tangent to the outer boundary of the computational domain. We note that the velocity on the outer boundary includes the field due to circulation Γ, which is determined by the solution to the velocity potential problem. The farfield condition is a Neumann boundary condition (or *NBC*) in contrast to the Dirichlet condition (or *EBC*) used for the velocity potential ϕ. The weighted secondary variables are contour integrals around the element boundary. Thus, integration of the tangential component of the velocity can be written in terms of the velocity potential, and hence the weighted secondary variables at the farfield nodes are expressed in terms of the velocity potential there, which has been already determined by solving the ϕ problem. We also note that the stiffness matrix for the ϕ problem before applying the boundary conditions can be saved and used for the ψ problem since both functions are governed by Laplace's equation. The wake cut is not a streamline, but the velocity must be continuous across the cut. Continuity of the velocity normal to the cut is enforced by the equality of the stream function at corresponding nodes on the cut; that is to say, there is no jump in the stream function across the cut. Continuity of the tangential velocity component is enforced by the sum of weighted secondary variables which is zero at the corresponding nodes on the cut. We first solve the velocity potential problem and determine the circulation Γ that satisfies the Kutta condition, then we solve for ψ using the known Γ. As such, a Kutta condition is not enforced when solving for ψ. The streamlines should show smooth flow near the trailing edge if the correct Γ is used, which appears only in the farfield condition.

The streamlines around Joukowski and Karman–Trefftz airfoils at $\alpha = 30$ degrees are shown in Figures 4.50a and 4.50b, respectively. The finite element streamlines are in excellent agreement with the conformal mapping streamlines for both airfoils. The flow leaves smoothly at the trailing edge indicating satisfaction of the Kutta condition. The location of the stagnation streamline is well predicted.

Example: Potential Flow Past a Cascade of Airfoils

Flow between blades of turbomachines such as axial flow compressors (e.g., [26], [116]) may be simplified as flow through two-dimensional cascades. A 2D cascade is a set of infinite identical airfoils equally spaced on a line as shown in Figure 4.51. The airfoil chord is inclined at an angle β with the x-axis, it is called the stagger angle. The spacing between two consecutive airfoils measured parallel to the y-axis is the cascade pitch s. The objective of this example is to demonstrate implementation of periodic boundary conditions in finite-element analysis of potential flow in cascades. We assume steady, incompressible, potential flow. Far upstream of the cascade, the incident flow is uniform with velocity $\vec{W}_1$ inclined at an angle α_1 with the x-axis, and far down stream the flow returns to uniform flow with velocity $\vec{W}_2$ inclined at an angle α_2. We assume the flow to be periodic in the y-direction, hence, velocity components satisfy $u(x,y \pm ms) = u(x,y)$, $v(x,y \pm ms) = v(x,y)$, where m is an integer. It suffices to determine the flow around one airfoil in a domain (also called control volume) such as *PLNHJQP* shown Figure 4.51, where *PQ* and *NH* are parallel to the y-axis and of length s. We note that *PL* is equal and parallel to *QJ*, and *LN* is equal and parallel to *JH*, the flow is periodic on these parallel lines. *PQ* is inflow boundary and *NH* is outflow boundary of the control volume. We resolve the velocity vectors into Cartesian components,

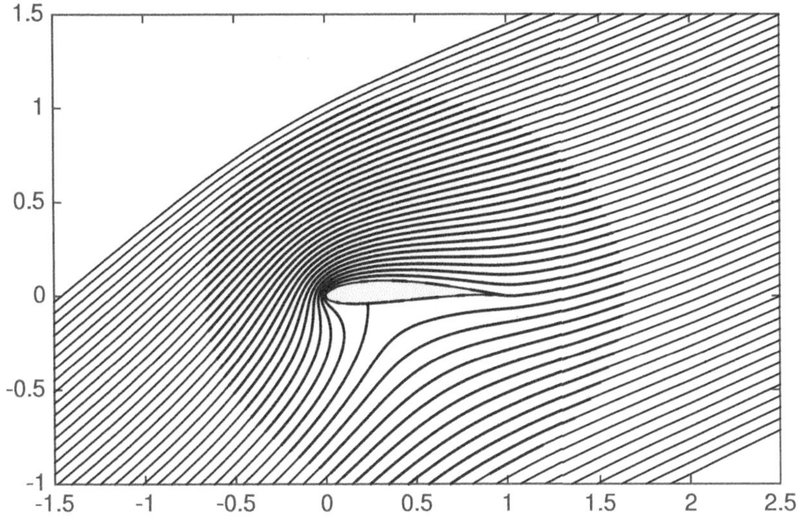

(a)

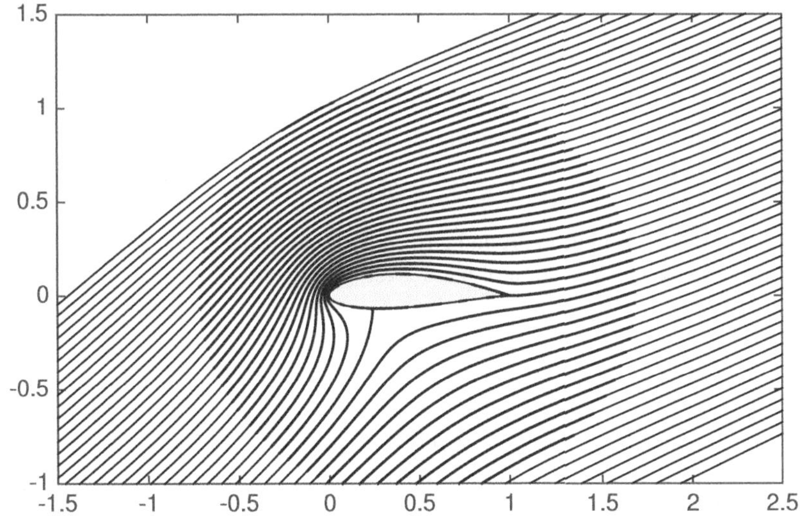

(b)

Figure 4.50 Streamlines at $\alpha = 30$ degrees, comparison between finite element and conformal mapping solutions. Conformal mapping streamlines are thicker lines within the circle. (a) Joukowski airfoil, and (b) Karman–Trefftz airfoil, $\zeta_c = -0.1 + i0.06$, and $n = 1.9$.

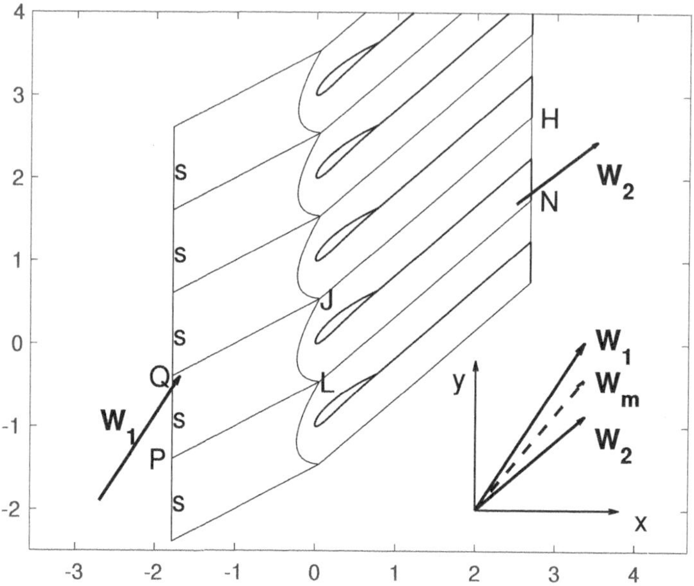

Figure 4.51 Cascade geometry.

$$\vec{W}_1 = U_1\hat{\imath} + V_1\hat{\jmath}, \qquad \vec{W}_2 = U_2\hat{\imath} + V_2\hat{\jmath}. \tag{4.315}$$

The velocity field is periodic on the lines *PLN* and *QJH*; thus, the circulation (measured clockwise) around the computational domain boundary is

$$\Gamma = (V_1 - V_2)s \tag{4.316}$$

and is equal to the circulation around any contour enclosing the airfoil because the flow is irrotational. Conservation of mass gives

$$U_1 = U_2 \equiv U \tag{4.317}$$

and Bernoulli's equation gives

$$p_2 - p_1 = \frac{1}{2}\rho\left(V_1^2 - V_2^2\right). \tag{4.318}$$

The fluid force (F_x, F_y) acting on the airfoil can be determined by the principle of linear momentum applied to the control volume *PLNHJQP* and enforcing periodicity of pressure and velocity on the lines *PLN* and *QJH*. We obtain,

$$F_x = -\rho V_m \Gamma, \qquad F_y = \rho U \Gamma \tag{4.319}$$

where $V_m = (V_1 + V_2)/2$, and ρ is fluid density. The force is

$$\vec{F} = \rho \vec{W}_m \times (-\Gamma\hat{k}), \qquad F = \rho W_m \Gamma \tag{4.320}$$

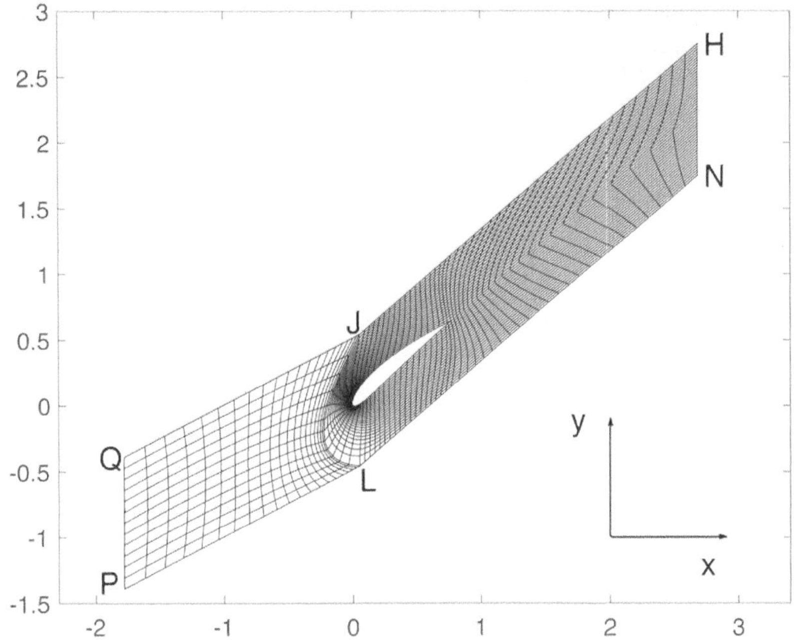

Figure 4.52 Mesh over a periodic segment containing one blade.

where W_m is the magnitude of the mean of the inlet and outlet velocities,

$$\vec{W}_m = \frac{1}{2}\left(\vec{W}_1 + \vec{W}_2\right). \tag{4.321}$$

Boundary conditions treatment in finite-element method

We generate a mesh over a periodic segment *PLNHJQP* as shown Figure 4.52. To facilitate application of the periodic boundary condition, there must be a one-to-one correspondence between global nodes on *PLN* and *QJH*. Each node on *PLN* must correspond to a node on *QJH*, and the two nodes must fall on a line parallel to the y-axis. The primary variable is the velocity potential ϕ. On the inflow boundary *PQ*, the velocity is prescribed $U_1 = W_1 cos\alpha_1$, and $V_1 = W_1 sin\alpha_1$. The velocity potential on that boundary is determined by integration of V_1 with respect to y. Hence, we apply Dirichlet (*EBC*) at the global nodes on *PQ*,

$$\phi = V_1(y - y_P), \qquad for \quad y_P \le y \le y_Q. \tag{4.322}$$

On the outflow boundary *NH*, only the normal velocity $U_2 = U = W_1 cos\alpha_1$ is known. We apply Neumann (*NBC*) on *NH*,

$$\frac{\partial \phi}{\partial n} = U \qquad for \quad y_N < y < y_H \tag{4.323}$$

Weighted secondary variables R_j are computed and used at the global nodes on NH. However, the nodes N and H are treated as points on the periodic boundary. Let a denote a point on PLN that corresponds to point b on QJH, periodicity of normal and tangential velocity components means

$$\left(\frac{\partial \phi}{\partial n}\right)_a = -\left(\frac{\partial \phi}{\partial n}\right)_b$$
$$\left(\frac{\partial \phi}{\partial s}\right)_a = \left(\frac{\partial \phi}{\partial s}\right)_b. \tag{4.324}$$

Let global node j on PLN corresponds to global node k on QJH. Integration of the first condition with respect to arc length on the boundary shows that the weighted secondary variables satisfy

$$R_j + R_k = 0. \tag{4.325}$$

Also, integration of the second condition with respect to arc length on the boundary gives

$$\phi_j - \phi_k = \phi_P - \phi_Q. \tag{4.326}$$

Note that $\phi_P - \phi_Q = V_1 s$, is known. We assemble the element equations into a global system. The equation at global node j is replaced by the sum of the two equations for global nodes j and k, we then enforce Equation (4.325), which eliminates the unknown weighted secondary variables at those nodes. The equation at global node k is replaced by Equation (4.326). Boundary conditions on the airfoil surface and wake cut, and Kutta condition are applied as in the case of an isolated airfoil.

Baskharone and Hamed [10] work is a good example of the analysis of potential flow over a cascade using the finite-element method. They presented a novel method for computing the circulation around the airfoil. They obtained excellent agreement with an exact solution for surface pressure distribution. Bhimarasetty and Govardhan [12] combined conformal mapping and vortex panel method to solve potential flow over cascades of arbitrary airfoil sections. The periodic mapping function [116]

$$\zeta = tanh\left(\frac{\pi z}{s}\right) \tag{4.327}$$

transforms the cascade in the z-plane onto a single profile in the ζ-plane. Uniform flow far upstream of the cascade becomes a source of strength ($Q = Us$) and a (counter-clockwise) vortex of strength ($\Gamma_1 = -V_1 s$) at $\zeta = -1$. Uniform flow far downstream becomes a sink of strength ($Q = Us$) and a (clockwise) vortex of strength ($\Gamma_2 = V_2 s$) at $\zeta = 1$. A panel method is used to determine the flow over the single body in the ζ-plane induced by the singularities at $\zeta = \pm 1$. We note that $\Gamma + \Gamma_1 + \Gamma_2 = 0$. A Kutta condition is applied at the trailing edge in the ζ-plane.

Comparison of finite-element and panel method solutions

We consider a cascade of NACA-4412 airfoils at a stagger angle $\beta = 40°$ and pitch-to-chord ratio $s/C = 1$. The incidence flow angle is $\alpha_1 = 56°$. A mesh of bilinear

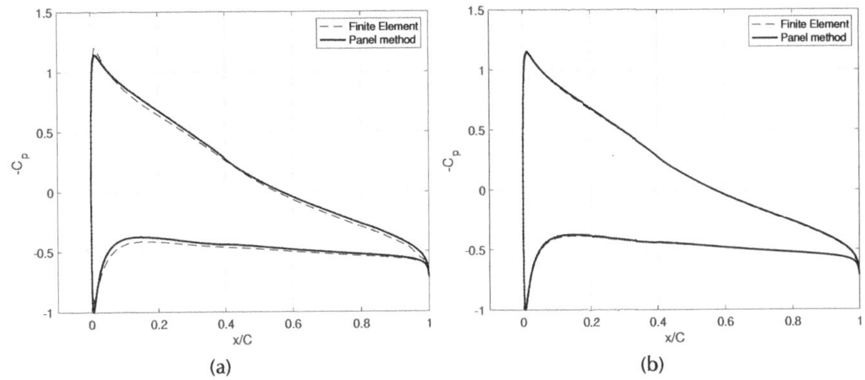

Figure 4.53 Cascade surface pressure coefficient, C_p. (a) Coarse mesh. (b) Fine mesh.

quadrilateral elements is presented in Figure 4.52. A C-mesh is used for the region *LNHJL*, and an H-mesh is used for *PLJQP*. The extended H-mesh upstream of the cascade uses 20 intervals on line *PL*. Results are obtained for two mesh resolutions.

Coarse mesh:
Number of segments on wake cut $n_{AB} = 24$, and on airfoil surface $n_{BF} = 48$. The number of segments on downstream exit is $n_{AN} = n_{GH} = 24$. The total number of elements is 2544, and global nodes 2685.

Fine mesh:
Number of segments on wake cut $n_{AB} = 48$, and on airfoil surface $n_{BF} = 96$. The number of segments on downstream exit is $n_{AN} = n_{GH} = 96$. The total number of elements is 18912, and global nodes 19221.
The panel method uses a fine surface mesh of 800 panels. Finite-element pressure coefficient distributions C_p on the airfoil surface are compared with panel method solutions in Figures 4.53a and 4.53b for the two mesh resolutions. The fine mesh C_p is nearly identical to that of the panel method. The circulation around the airfoil is Γ/W_1C =0.4034 and 0.3998 for coarse and fine meshes, respectively, and 0.3994 for the panel method.

Example: Flow over a Cylinder in a Channel
A circular cylinder is placed in a channel in an otherwise uniform stream of velocity U_∞ as shown in Figure 4.54. The cylinder radius is b and the channel half height is $2b$. The flow is assumed to be two-dimensional inviscid incompressible potential flow. We want to determine the pressure distribution on the cylinder using the finite-element method and compare the results with the exact analytic solution.

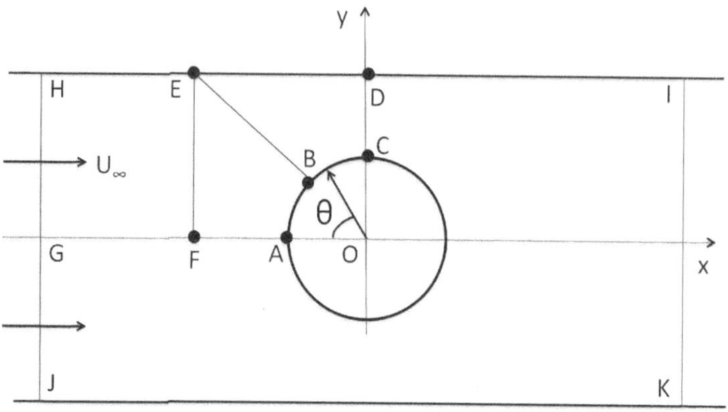

Figure 4.54 A circular cylinder in a channel.

Analytic Solution Using the Method of Images

The potential flow over a circular cylinder in an unbounded domain is given by the superposition of a dipole at the circle center and uniform flow. In terms of the complex variable $z = x + iy$, the complex potential $F(z) = \phi + i\psi$ is

$$F(z) = U_\infty z + \frac{\mu_o}{z} \tag{4.328}$$

where $\mu_o = U_\infty b^2$ is the dipole strength. The velocity potential and stream function are

$$\begin{aligned} \phi &= U_\infty \left(r + \frac{b^2}{r} \right) cos\theta_1 \\ \psi &= U_\infty \left(r - \frac{b^2}{r} \right) sin\theta_1 \end{aligned} \tag{4.329}$$

where $\theta_1 = \pi - \theta$; the angle θ is as shown in Figure 4.54. The complex velocity is

$$u - iv = U_\infty \left(1 - \frac{b^2}{z^2} \right) \tag{4.330}$$

from which we obtain the tangential velocity on the cylinder $r = b$ as $V = 2U_\infty sin\theta$. This solution, however, is not valid for flow over a cylinder in a channel because it violates the boundary conditions on the channel walls which is $v = 0$ at $y = \pm 2b$. The method of images in conjunction with the Milne–Thomson circle theorem [75] can be used to solve this problem. In addition to the dipole of strength μ_o at the circle center $(0,0)$ we add its images in the channel walls, which are dipoles of the same strength μ_o at $y_o = \pm 2kh$ for $k = 1,2,..$, where $h = 2b$ is half the channel height. We call these images the first generation of dipoles. Each member of the first generation has an image of smaller strength $\mu_1 = \mu_o(b/y_o)^2$ at the inverse point $y_c = b^2/y_o$

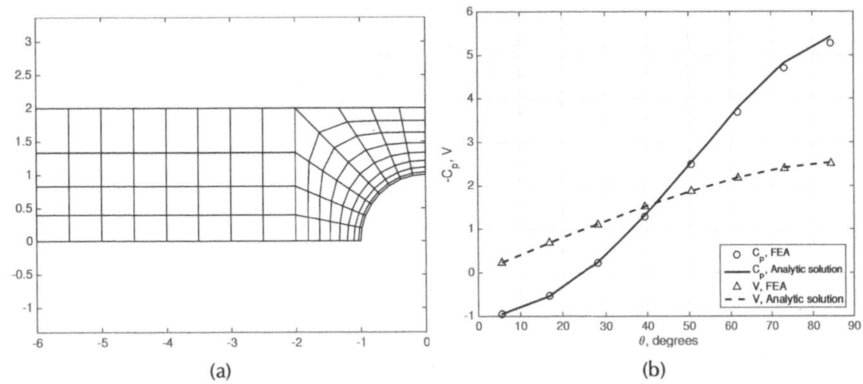

Figure 4.55 Comparison of finite element and analytic solution for a coarse mesh. (a) A mesh of bilinear quadrilateral elements, and (b) tangential velocity and pressure coefficient on the cylinder surface.

within the circle. The new dipole within the circle also creates images in the channel walls of strength μ_1 at the points $y_1 = 2h - y_c$. This is the second generation of dipoles. Again each member of the second generation has an image within the circle of smaller strength $\mu_2 = \mu_1 (b/y_1)^2$ at $y_2 = b^2/y_1$. The third generation of dipoles are the images of μ_2 in the channel walls. In our computations, we used 300, 100, and 40 pairs of the first, second, and third generations of dipoles.

Finite Element Solution and Comparison with Analytic Solution

Assuming symmetry about the x- and y- axes, we solve for the velocity potential ϕ in one quadrant; say the second quadrant as shown in Figure 4.54. The coordinates of points A, B, C, D, E, F, and G are $(-b,0)$, $(-b/\sqrt{2}, b/\sqrt{2})$, $(0,b)$, $(0,2b)$, $(-2b,2b)$, $(-2b,0)$, and $(-6b,0)$, respectively. The computational domain is bounded by the curve $ABCDEHGFA$. The boundary conditions are $\partial\phi/\partial n = 0$ on GFA, ABC, and HED, and $\partial\phi/\partial n = -1$ on GH. The y-velocity component on CD is $v = \partial\phi/\partial y = 0$; thus, ϕ is constant on CD, and hence we take $\phi = 0$ on CD.

The domain is divided into bilinear quadrilateral elements as shown in Figure 4.55a.

The element matrices are evaluated by the 2×2 Gauss–Legendre quadrature formula. We divided AC and FG into eight uniform segments, and BE into eight nonuniform segments with smaller elements near the cylinder surface. The velocity magnitude V and pressure coefficient at the center of the elements adjacent to the cylinder surface are shown in Figure 4.55b. The finite element results are in good agreement with the analytic solution determined by the method of images. The velocity vectors at the center of each element are shown in Figure 4.56. A fine mesh of 32 segments on AC and 16 segments on BE and FG is constructed and shown Figure 4.57a. The results for this mesh are shown in Figure 4.57b, which shows excellent agreement

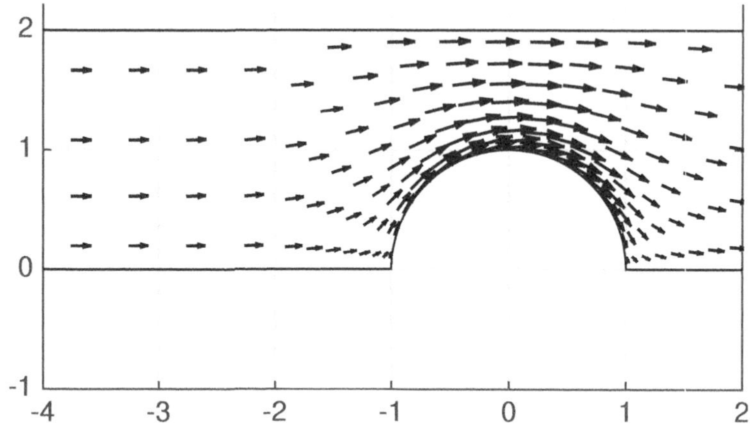

Figure 4.56 Velocity vectors at the center of each element.

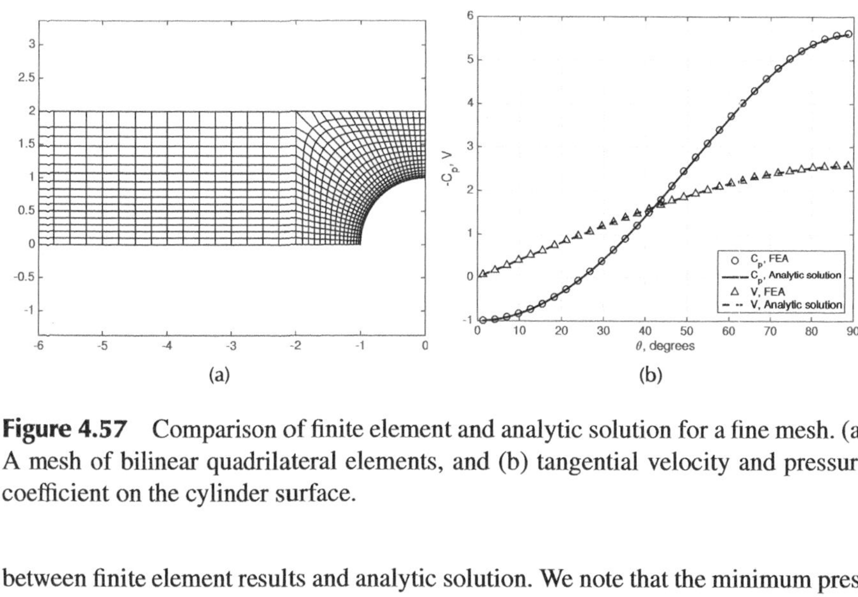

Figure 4.57 Comparison of finite element and analytic solution for a fine mesh. (a) A mesh of bilinear quadrilateral elements, and (b) tangential velocity and pressure coefficient on the cylinder surface.

between finite element results and analytic solution. We note that the minimum pressure coefficient for a cylinder in unbounded domain is -3, but for the current case it is -5.6104.

Example: Torsional Rigidity of a Semicircular Section Using Isoparametric Elements

In Example 4.5, we solved the torsional problem for a semicircular section of radius a using bilinear triangular elements. Here, we solve the same problem using high-order isoparametric elements. The cross-section is shown in Figure 4.58a. For this problem, Prandtl stress function u is symmetric about the y-axis. Thus, the secondary

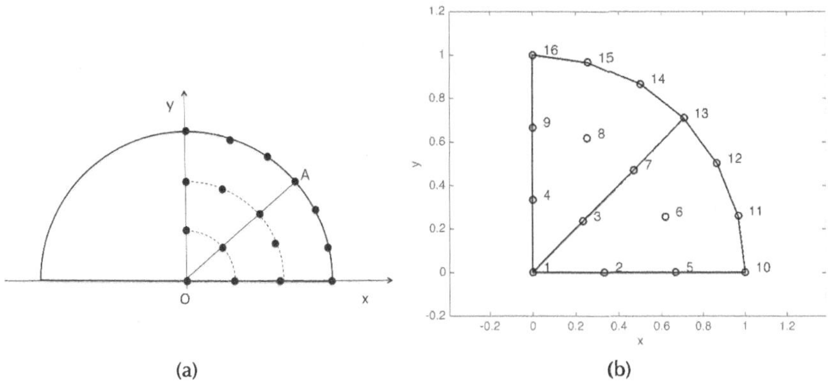

Figure 4.58 Torsion of a prismatic bar with semicircular cross-section. (a) Two bicubic triangular elements. (b) Global nodes.

variable v_n is zero on that axis. Accurate representation of the curved boundary is obtained by using two bicubic triangular elements as shown in Figure 4.58b. There are 16 global nodes. The nodes on the semicircular boundary are connected by straight lines but the actual boundaries of the isoparametric elements are curved. Nodes are equally spaced on the circular arcs of radii $a/3$, $2a/3$, and a. The function u is zero at all nodes except nodes 3, 4, 6, 7, 8, and 9. The element matrices are computed using a 7-point Gauss quadrature formula. Essential boundary conditions $u = 0$ are applied at nodes 1, 2, 5, 10, 11, 12, 13, 14, 15, and 16. And natural $v_n = 0$ condition is applied on side 1-4-9-16. Prandtl stress function at the global nodes by the finite-element method and the exact analytic solution are $(u_3, u_4, u_6, u_7, u_8, u_9) = (0.14146, 0.18057, 0.10166, 0.14254, 0.16337, 0.16632)$ and $(0.14014, 0.17802, 0.10193, 0.14486, 0.16356, 0.16894)$, respectively.

The computed torque by the FEM is 0.29654 as compared to the exact value of 0.29756. The excellent comparison between the finite-element method and exact solution demonstrate the efficiency of high-order isoparametric elements.

4.9.2 OXYGEN TRANSPORT AND CONSUMPTION IN KROGH CAPILLARY-TISSUE CYLINDER

A Krogh cylinder is a simplified mathematical model for the transport of oxygen by capillaries and its diffusion and consumption in the surrounding tissues. Such a cylinder is shown in Figures 4.59a and 4.59b. As blood flows through the capillary from the arterial side to the venous side, dissolved oxygen diffuses through the capillary wall into the surrounding tissue where it is consumed by the tissue cells. As the oxygen tension (partial pressure) of dissolved oxygen decreases, oxygen that is chemically combined with hemoglobin is also released into the blood stream providing a source of dissolved oxygen. Reneau et al. [101] presented a mathematical model for the transient transport and consumption of oxygen in a Krogh cylinder.

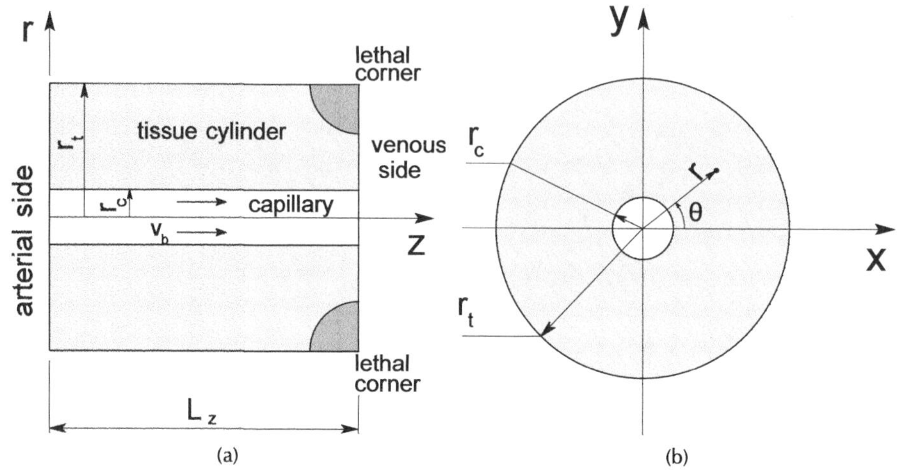

Figure 4.59 A sketch of Krogh capillary-tissue cylinder. (a) A meridional plane (θ = constant). (b) Cross-section of the tissue cylinder.

The model is given by two partial differential equations for the oxygen tension P_{O_2}; one equation for the capillary and another for the tissue. Here we combine the two equations into one equation, which is more appropriate for finite element treatment.

In this section, we will present two formulations of the problem. In the first formulation, the primary variable is the dissolved oxygen concentration C_{O_2}, and in the second formulation the primary variable is the oxygen tension (partial pressure) P_{O_2}. The two variables are linearly related by *Henry's law* $P_{O_2} = HC_{O_2}$ where H is Henry's constant; its value for blood at 37°C is 0.74 mmHg/μM [38]. As a side note, Henry's law relates oxygen partial pressure to concentration of dissolved oxygen; but it does not apply to oxygen concentration bound to hemoglobin. We can also write $C_{O_2} = \mu P_{O_2}$ where μ is oxygen solubility coefficient. We denote μ by S_b in the capillary region, and by S_t in the tissue region.

Equations in Terms of Dissolved Oxygen Concentration C_{O_2}

In cylindrical coordinates (r,θ,z), and for axisymmetric distribution of oxygen concentration ($\partial C_{O_2}/\partial\theta = 0$), the equation is

$$\frac{\partial C_{O_2}}{\partial t} - \left[\frac{1}{r}\frac{\partial}{\partial r}\left(D_r r\frac{\partial C_{O_2}}{\partial r}\right) + \frac{\partial}{\partial z}\left(D_z\frac{\partial C_{O_2}}{\partial z}\right)\right] + v_b\left(\frac{\partial C_{O_2}}{\partial z} + C_{Hbs}\frac{\partial \Upsilon}{\partial z}\right) + B(C_{O_2}) = 0 \qquad (4.331)$$

where D_r and D_z are oxygen diffusivity (m^2/s) in the radial and axial directions, respectively. In general, their values vary in space, and hence their values in the capillary are different from their values in the tissue. The blood velocity in the axial direction is v_b. In this model v_b is a prescribed constant in the capillary and is zero in the tissue. Reneau et al. [101] give the rate of oxygen release from hemoglobin by

$$v_b C_{Hbs} \frac{\partial \Upsilon}{\partial z}$$

where Υ is the percent saturation of erythrocytes with oxygen for which they used the Hill equation

$$\Upsilon(P_{O_2}) = \frac{(P_{O_2}/P_{50})^n}{1+(P_{O_2}/P_{50})^n} \tag{4.332}$$

and C_{Hbs} is the saturated concentration of oxygen bound to hemoglobin, which corresponds to $\Upsilon = 1$. Fournier [38] estimated $C_{Hbs} = 8800$ μM. Based on available experimental data, Fournier gives values of $P_{50} = 26$ mmHg and $n = 2.34$. Hence, the rate of oxygen release from hemoglobin becomes

$$v_b C_{Hbs} \frac{\partial \Upsilon}{\partial z} = v_b \phi(C_{O_2}) \frac{\partial C_{O_2}}{\partial z} \tag{4.333}$$

where

$$\begin{aligned} \phi(C_{O_2}) &= \frac{C_{Hbs}}{C_{50}} \frac{d}{d(C_{O_2}/C_{50})} \left[\frac{(C_{O_2}/C_{50})^n}{1+(C_{O_2}/C_{50})^n} \right] \\ &= \frac{C_{Hbs}}{C_{50}} \frac{n(C_{O_2}/C_{50})^{n-1}}{[1+(C_{O_2}/C_{50})^n]^2} \end{aligned} \tag{4.334}$$

where $C_{50} = P_{50}/H$. Substituting Equation (4.333) into Equation (4.331), we get

$$\begin{aligned} \frac{\partial C_{O_2}}{\partial t} &- \left[\frac{1}{r} \frac{\partial}{\partial r} \left(D_r r \frac{\partial C_{O_2}}{\partial r} \right) + \frac{\partial}{\partial z} \left(D_z \frac{\partial C_{O_2}}{\partial z} \right) \right] \\ &+ v_b [1+\phi(C_{O_2})] \frac{\partial C_{O_2}}{\partial z} + B(C_{O_2}) = 0. \end{aligned} \tag{4.335}$$

The rate of oxygen uptake $B(C_{O_2})$ is zero in the capillary, and in the tissue it is assumed to be given by a Michaelis–Menten reaction, as will be presented shortly.

Boundary Conditions

We consider a capillary-tissue cylinder of axial length L_z, capillary radius r_c, and tissue radius r_t. We assume that there is no oxygen flux on the bounding surface of the tissue, and hence the boundary conditions are

$$\begin{aligned} r = r_t: &\quad \frac{\partial C_{O_2}}{\partial r} = 0 \quad 0 \le z \le L_z \\ z = 0: &\quad \frac{\partial C_{O_2}}{\partial z} = 0 \quad r_c \le r \le r_t \\ z = L_z: &\quad \frac{\partial C_{O_2}}{\partial z} = 0 \quad r_c \le r \le r_t. \end{aligned} \tag{4.336}$$

For the capillary region, we use the conditions

$$
\begin{aligned}
r=0: &\quad \frac{\partial C_{O_2}}{\partial r}=0 \quad 0\leq z\leq L_z \\
z=0: &\quad C_{O_2}=\tilde{C}_{O_2} \quad 0\leq r\leq r_c \\
z=L_z: &\quad \frac{\partial C_{O_2}}{\partial z}=0 \quad 0\leq r\leq r_c
\end{aligned}
\tag{4.337}
$$

where $\tilde{C}_{O_2}$ is a prescribed oxygen concentration in the capillary at the arterial side. Reneau et al. [101] neglected the resistance to mass transfer at the interface between the capillary and tissue, and hence both oxygen concentration and flux are continuous at the interface.

$$
\begin{aligned}
r=r_c: (C_{O_2})_{capillary} &= (C_{O_2})_{tissue} \quad 0\leq z\leq L_z \\
\left(D_r\frac{\partial C_{O_2}}{\partial r}\right)_{capillary} &= \left(D_r\frac{\partial C_{O_2}}{\partial r}\right)_{tissue} \quad 0\leq z\leq L_z.
\end{aligned}
\tag{4.338}
$$

By the first condition of Equation (4.338) we enforce continuity of dissolved oxygen concentration at the interface between capillary and tissue, while Reneau et al. [101] enforce continuity of the tension. The two conditions would be equivalent if oxygen solubility coefficient in blood is approximately equal to that in tissue. Moreover, in the present finite-element model, the diffusion coefficient in capillary wall thickness can be different from that of the tissue. Reneau et al. use zero-order reaction for uptake of oxygen in the tissue. Here, the rate of oxygen uptake $B(C_{O_2})$ is zero in the capillary, and is given by a Michaelis–Menten reaction in the tissue,

$$
B(C_{O_2})=V_{max}\frac{C_{O_2}}{K_m+C_{O_2}} \tag{4.339}
$$

where V_{max} and K_m are the maximum reaction rate and Michaelis constant, respectively.

We choose reference length L_o, concentration C_o, and diffusivity D_o, and introduce nondimensional variables $t^*=D_o t/L_o^2$, $r^*=r/L_o$, $z^*=z/L_o$, $u^*=C_{O_2}(r,z)/C_o$, $p_{11}^*=D_r/D_o$, $p_{22}^*=D_z/D_o$, $v_b^*=v_b L_o/D_o$, and consumption rate $B^*=BL_o^2/C_oD_o$. We drop the star for simplicity of notation, and write the governing equation in nondimensional form as

$$
\frac{\partial u}{\partial t}-\left[\frac{1}{r}\frac{\partial}{\partial r}\left(p_{11}r\frac{\partial u}{\partial r}\right)+\frac{\partial}{\partial z}\left(p_{22}\frac{\partial u}{\partial z}\right)\right]+v_b\left[1+\phi(u)\right]\frac{\partial u}{\partial z}+B(u)=0. \tag{4.340}
$$

The nondimensional Michaelis–Menten rate of oxygen consumption is

$$
B(u)=\frac{\alpha u}{\beta+u} \tag{4.341}
$$

where $\alpha = V_{max}L_o^2/(D_oC_o)$ is nondimensional maximum reaction rate and $\beta = K_m/C_o$ is nondimensional Michaelis constant.

$$\phi(u) = \frac{\gamma n\,(u/u_{50})^{n-1}}{[1+(u/u_{50})^n]^2} \tag{4.342}$$

where $u_{50} = S_bP_{50}/C_o$, and $\gamma = C_{Hbs}/(S_bP_{50})$.

Let us use the tissue radius as the reference length $L_o = r_t$, and the prescribed concentration at the arterial end as a reference concentration $C_o = \tilde{C}_{O_2}$. The nondimensional boundary conditions become

$$\begin{aligned}
r=1: &\quad \frac{\partial u}{\partial r}=0 \quad 0\le z\le L_z\\
z=0: &\quad \frac{\partial u}{\partial z}=0 \quad r_c\le r\le 1\\
z=L_z: &\quad \frac{\partial u}{\partial z}=0 \quad r_c\le r\le 1
\end{aligned} \tag{4.343}$$

$$\begin{aligned}
r=0: &\quad \frac{\partial u}{\partial r}=0 \quad 0\le z\le L_z\\
z=0: &\quad u=1 \quad 0\le r\le r_c\\
z=L_z: &\quad \frac{\partial u}{\partial z}=0 \quad 0\le r\le r_c
\end{aligned} \tag{4.344}$$

$$\begin{aligned}
r=r_c: \quad (u)_{capillary} &= (u)_{tissue} \quad 0\le z\le L_z\\
\left(p_{11}\frac{\partial u}{\partial r}\right)_{capillary} &= \left(p_{11}\frac{\partial u}{\partial r}\right)_{tissue} \quad 0\le z\le L_z.
\end{aligned} \tag{4.345}$$

In this section, we solve the steady state problem, $\partial u/\partial t = 0$; in which case the differential equation is an elliptic equation. We note that Equation (4.340) is nonlinear in u because the convection $v_b(1+\phi)\partial u/\partial z$ and the uptake $B(u)$ terms are nonlinear in u. An iteration method is used. We assume a solution $u(r,z) = g(r,z)$ and update it by iteration. We first quasi-linearize $B(u)$ around a guessed solution $g(r,z)$ as we have done in Section 2.9.1.

$$B(u) = \frac{\alpha\beta u}{(\beta+g)^2} + \frac{\alpha g^2}{(\beta+g)^2} \tag{4.346}$$

and we also evaluate the oxygen release source $\phi(u)$ at the guessed solution g. Now the equation for the oxygen concentration takes the form

$$\begin{aligned}
-\left[\frac{1}{r}\frac{\partial}{\partial r}\left(p_{11}r\frac{\partial u}{\partial r}\right) + \frac{\partial}{\partial z}\left(p_{22}\frac{\partial u}{\partial z}\right)\right] + v_b[1+\phi(g)]\frac{\partial u}{\partial z}\\
+q(g)u - f(g) = 0
\end{aligned} \tag{4.347}$$

where

$$
\begin{aligned}
q(g) &= \frac{\theta\beta}{(\beta+g)^2} \\
f(g) &= -\frac{\theta g^2}{(\beta+g)^2}
\end{aligned}
\tag{4.348}
$$

Since g is a known function of r and z, Equation (4.347) is linear in the unknown function u. We denote the guessed solution for the kth iteration by g^k and the updated solution by u^k. For the next iteration, the guessed solution is

$$g^{k+1} = \omega u^k + (1-\omega)g^k \tag{4.349}$$

where $0 < \omega \leq 1$ is an under-relaxation factor. The differential equation is solved again to determine u^{k+1}. The absolute error

$$E(r,z) = |u(r,z)^{k+1} - u(r,z)^k| \tag{4.350}$$

is found, and its maximum over all grid points E_{max} is calculated. The iteration is terminated if $E_{max} < \varepsilon$, where ε is a prescribed tolerance.

In the present finite-element method, the capillary and tissue are treated as one domain, and one differential equation Equation (4.347) is solved in that domain. The distinction between the capillary region and tissue region is reflected only in the prescribed parameters. For elements in the capillary region we prescribe $\alpha = 0$ and $\beta = 0$, and for elements in the tissue region we prescribe $v_b = 0$. The continuity conditions of dissolved oxygen concentration and its flux at the interface between capillary and tissue requires no special treatment because those conditions are readily satisfied by the compatibility of the trial solution at the boundary between adjacent elements and balance of secondary variables. The weak formulation of the problem allows discontinuities of the diffusion coefficients, and seamlessly satisfies continuity of the primary (concentration) and secondary variables (flux).

The Weak Form and Element Matrices

The differential equation (4.347) is nearly the same as the model problem given by Equation (4.1) except for the convection term in the capillary domain. We define two vectors

$$
\begin{aligned}
\vec{v} &= p_{11}\frac{\partial u}{\partial r}\hat{e}_r + p_{22}\frac{\partial u}{\partial z}\hat{e}_z \\
\vec{b} &= b_r\hat{e}_r + b_z\hat{e}_z
\end{aligned}
\tag{4.351}
$$

where in the present application $b_z = v_b[1+\phi(g)]$ and $b_r = 0$, and rewrite Equation (4.347) as

$$-\nabla\cdot\vec{v} + \vec{b}\cdot\nabla u + q(g)u - f(g) = 0 \tag{4.352}$$

where

$$\begin{aligned} \nabla \cdot \vec{v} &= \frac{1}{r}\frac{\partial (rv_r)}{\partial r} + \frac{\partial v_z}{\partial z} \\ \nabla u &= \frac{\partial u}{\partial r}\hat{e_r} + \frac{\partial u}{\partial z}\hat{e_z}. \end{aligned} \tag{4.353}$$

Following the steps for deriving the weak form for the model problem, we obtain the weak form for Equation (4.352),

$$\int_{\Omega^e} \left(\nabla w \cdot \vec{v} + w\vec{b} \cdot \nabla u + wqu\right) d\Omega = \int_{\Omega^e} wf d\Omega + \oint_{\partial\Omega^e} wv_n dS \tag{4.354}$$

where Ω^e is the finite element domain and $\partial\Omega^e$ is its boundary. The weight (test) function is $w(r,z)$. To define the element of volume $d\Omega$ for axisymmetric problems, we consider a plane that passes through the axis of symmetry, that is a plane of constant θ, which we call the $r-z$ plane. Let dA denote an infinitesimal element of area in that plane at a radial distance r from the axis of symmetry. The infinitesimal volume is a ring obtained by rotating the element dA through an angle of 2π about the axis, and hence $d\Omega = 2\pi r dA$. Similarly, an infinitesimal surface element is $dS = 2\pi r ds$, where ds is an infinitesimal arc length. The weak form is written as integrals over the element area A_e and its contour Γ^e in the $r-z$ plane,

$$\int_{A_e} \left(\nabla w \cdot \vec{v} + w\vec{b} \cdot \nabla u + wqu\right) r dA = \int_{A_e} wf r dA + \oint_{\Gamma^e} wv_n r ds. \tag{4.355}$$

Element Matrices

The derivation of the element equations and matrices follows precisely the derivation presented in Section 4.5. It is repeated here with notations appropriate for axisymmetric problems. The trial solution over the element is given in matrix form

$$\begin{aligned} U^e(r,z) &= \{\psi_1(r,z)\ \psi_2(r,z) \ldots \psi_n(r,z)\} \begin{Bmatrix} U_1^e \\ U_2^e \\ \cdot \\ \cdot \\ \cdot \\ U_n^e \end{Bmatrix} \\ &= \{\psi\}\{U^e\}. \end{aligned} \tag{4.356}$$

To determine the vector $\vec{v}$, we first find the gradient of the trial solution.

$$\frac{\partial U^e(r,z)}{\partial r} = \left\{ \frac{\partial \psi_1(r,z)}{\partial r}\ \frac{\partial \psi_2(r,z)}{\partial r} \ldots \frac{\partial \psi_n(r,z)}{\partial r} \right\} \begin{Bmatrix} U_1^e \\ U_2^e \\ \cdot \\ \cdot \\ \cdot \\ U_n^e \end{Bmatrix} \tag{4.357}$$

$$\frac{\partial U^e(r,z)}{\partial z} = \left\{ \frac{\partial \psi_1(r,z)}{\partial z} \; \frac{\partial \psi_2(r,z)}{\partial z} \cdots \frac{\partial \psi_n(r,z)}{\partial z} \right\} \left\{ \begin{array}{c} U_1^e \\ U_2^e \\ \cdot \\ \cdot \\ \cdot \\ U_n^e \end{array} \right\} \tag{4.358}$$

$$\left\{ \begin{array}{c} \frac{\partial U^e(r,z)}{\partial r} \\ \frac{\partial U^e(r,z)}{\partial z} \end{array} \right\} = \left[\begin{array}{cccc} \frac{\partial \psi_1(r,z)}{\partial r} & \frac{\partial \psi_2(r,z)}{\partial r} & \cdots & \frac{\partial \psi_n(r,z)}{\partial r} \\ \frac{\partial \psi_1(r,z)}{\partial z} & \frac{\partial \psi_2(r,z)}{\partial z} & \cdots & \frac{\partial \psi_n(r,z)}{\partial z} \end{array} \right] \left\{ \begin{array}{c} U_1^e \\ U_2^e \\ \cdot \\ \cdot \\ \cdot \\ U_n^e \end{array} \right\}. \tag{4.359}$$

For the Krogh tissue cylinder problem, $p_{12} = p_{21} = 0$. We define the gradient matrix $[B]$ by

$$[B] = \left[\begin{array}{cccc} \frac{\partial \psi_1(r,z)}{\partial r} & \frac{\partial \psi_2(r,z)}{\partial r} & \cdots & \frac{\partial \psi_n(r,z)}{\partial r} \\ \frac{\partial \psi_1(r,z)}{\partial z} & \frac{\partial \psi_2(r,z)}{\partial z} & \cdots & \frac{\partial \psi_n(r,z)}{\partial z} \end{array} \right] \tag{4.360}$$

and write $\vec{v}$ as

$$\begin{aligned} \left\{ \begin{array}{c} V_r^e \\ V_z^e \end{array} \right\} &= \left\{ \begin{array}{c} p_{11}\frac{\partial U^e(r,z)}{\partial r} + p_{12}\frac{\partial U^e(r,z)}{\partial z} \\ p_{21}\frac{\partial U^e(r,z)}{\partial r} + p_{22}\frac{\partial U^e(r,z)}{\partial z} \end{array} \right\} \\ &= \left[\begin{array}{cc} p_{11} & p_{12} \\ p_{21} & p_{22} \end{array} \right] \left\{ \begin{array}{c} \frac{\partial U^e(r,z)}{\partial r} \\ \frac{\partial U^e(r,z)}{\partial z} \end{array} \right\} \\ &= \left[\begin{array}{cc} p_{11} & p_{12} \\ p_{21} & p_{22} \end{array} \right] \left[\begin{array}{ccc} \frac{\partial \psi_1(r,z)}{\partial r} & \cdots & \frac{\partial \psi_n(r,z)}{\partial r} \\ \frac{\partial \psi_1(r,z)}{\partial z} & \cdots & \frac{\partial \psi_n(r,z)}{\partial z} \end{array} \right] \left\{ \begin{array}{c} U_1^e \\ \cdot \\ \cdot \\ \cdot \\ U_n^e \end{array} \right\} \end{aligned} \tag{4.361}$$

$$\left\{ \begin{array}{c} V_x^e \\ V_y^e \end{array} \right\} = [D][B]\{U^e\}. \tag{4.362}$$

Galerkin

For small blood velocity, we expect the Galerkin finite-element method to work without difficulty. However, for convection dominated problems (high Peclet or Reynolds number), formulation other than the Galerkin method may be needed for the convection term. In the Galerkin method, the test (weight) function $w(r,z)$ is a linear combination of the basis functions. Thus, for element # e, we have the approximation for

w,

$$W^e(r,z) = \{\psi_1(r,z)\ \psi_2(r,z) \ldots \psi_n(r,z)\} \left\{ \begin{array}{c} d_1 \\ d_2 \\ \cdot \\ \cdot \\ \cdot \\ d_n \end{array} \right\}. \tag{4.363}$$

Similar to the gradient of $U^e(r,z)$, we write the gradient of $W^e(r,z)$ using the matrix $[B]$,

$$\left\{ \begin{array}{c} \frac{\partial W^e(r,z)}{\partial r} \\ \frac{\partial W^e(r,z)}{\partial z} \end{array} \right\} = [B]\{d\}. \tag{4.364}$$

Now, we can write the first term of the integrand in the weak form in matrix form,

$$\begin{aligned} \nabla W^e \cdot \vec{V}^e &= \left(\frac{\partial W^e(r,z)}{\partial r}\hat{e}_r + \frac{\partial W^e(r,z)}{\partial z}\hat{e}_z \right) \cdot \left(V_r^e \hat{e}_r + V_z^e \hat{e}_z \right) \\ &= \frac{\partial W^e(r,z)}{\partial r} V_r^e + \frac{\partial W^e(r,z)}{\partial z} V_z^e \\ &= \left\{ \begin{array}{cc} \frac{\partial W^e(r,z)}{\partial r} & \frac{\partial W^e(r,z)}{\partial z} \end{array} \right\} \left\{ \begin{array}{c} V_r^e \\ V_z^e \end{array} \right\} \\ &= \{d\}^T [B]^T [D][B]\{U^e\}. \end{aligned} \tag{4.365}$$

Next, we write the remaining terms in the weak form using matrix notation,

$$\begin{aligned} w\vec{b} \cdot \nabla u &\approx \{d\}^T \{\psi\}^T \{b^e\}^T \nabla U^e \\ &= \{d\}^T \{\psi\}^T \{b^e\}^T [B]\{U^e\} \end{aligned} \tag{4.366}$$

where $\{b^e\}^T = \{b_r\ b_z\}^T$.

$$\begin{aligned} wqu &\approx W^e q U^e \\ &= \{d\}^T \{\psi\}^T q \{\psi\}\{U^e\} \end{aligned} \tag{4.367}$$

$$\begin{aligned} wf &\approx W^e f \\ &= \{d\}^T f \{\psi\}^T \end{aligned} \tag{4.368}$$

$$v_n w = \{d\}^T v_n \{\psi\}^T. \tag{4.369}$$

Substituting Equations (4.365)–(4.369) into the weak form for the element Equation (4.355), and factoring out the constant terms $\{d\}^T$ and $\{U^e\}$, we get

$$\begin{aligned} &\{d\}^T \{ \left[\int_{A_e} \left([B]^T [D][B] + \{\psi\}^T \{b^e\}^T [B] + q\{\psi\}^T \{\psi\} \right) r dA \right] \{U^e\} \\ &\quad - \int_{A_e} f \{\psi\}^T r dA - \oint_{\Gamma^e} v_n \{\psi\}^T r ds \} = 0. \end{aligned} \tag{4.370}$$

Next, we define the element matrices; the coefficient matrix

$$\begin{aligned}[A^e] &= \int_{A_e} \left([B]^T[D][B] + \{\psi\}^T\{b^e\}^T[B] + q\{\psi\}^T\{\psi\}\right) rdA \\ &= [K^e] + [C^e] + [M^e] \end{aligned} \tag{4.371}$$

the stiffness matrix

$$[K^e] = \int_{A_e} [B]^T[D][B] rdA \tag{4.372}$$

the convection matrix

$$[C^e] = \int_{A_e} \{\psi\}^T\{b^e\}^T[B] rdA \tag{4.373}$$

the mass matrix

$$[M^e] = \int_{A_e} q\{\psi\}^T\{\psi\} rdA \tag{4.374}$$

the forcing vector

$$\{F^e\} = \int_{\Omega^e} f\{\psi\}^T rdA \tag{4.375}$$

and vector of weighted secondary variables

$$\{R^e\} = \oint_{\Gamma^e} v_n\{\psi\}^T rds. \tag{4.376}$$

In terms of these matrices, the Equation (4.370) becomes

$$\{d\}^T\left\{[A^e]\{U^e\} - \{F^e\} - \{R^e\}\right\} = 0. \tag{4.377}$$

Since $\{d\}$ is an arbitrary array, the multiplying array must be zero. We obtain the element equations

$$[A^e]\{U^e\} = \{F^e\} + \{R^e\}. \tag{4.378}$$

Using index notation, we write the entries of the element matrices.

$$k_{ij}^e = \int_{A_e} \left[\frac{\partial \psi_i}{\partial r}\left(p_{11}\frac{\partial \psi_j}{\partial r} + p_{12}\frac{\partial \psi_j}{\partial z}\right) + \frac{\partial \psi_i}{\partial z}\left(p_{21}\frac{\partial \psi_j}{\partial r} + p_{22}\frac{\partial \psi_j}{\partial z}\right)\right] rdA \tag{4.379}$$

$$C_{ij}^e = \int_{A_e} \psi_i \left(b_r\frac{\partial \psi_j}{\partial r} + b_z\frac{\partial \psi_j}{\partial z}\right) rdA \tag{4.380}$$

$$m_{ij}^e = \oint_{A_e} q\psi_i\psi_j rdA \tag{4.381}$$

$$F_i^e = \int_{\Gamma_e} f\psi_i rdA \tag{4.382}$$

$$R_i^e = \oint_{\Gamma_e} v_n^e \psi_i rds. \tag{4.383}$$

Equations in Terms of Oxygen Tension P_{O_2}

In cylindrical coordinates (r,θ,z), and for axisymmetric distribution of oxygen tension ($\partial P_{O_2}/\partial\theta = 0$), the governing equation is

$$\frac{\partial P_{O_2}}{\partial t} - \left[\frac{1}{r}\frac{\partial}{\partial r}\left(D_r r\frac{\partial P_{O_2}}{\partial r}\right) + \frac{\partial}{\partial z}\left(D_z\frac{\partial P_{O_2}}{\partial z}\right)\right] + v_b\left(\frac{\partial P_{O_2}}{\partial z} + \frac{C_{Hbs}}{S_b}\frac{\partial \Upsilon}{\partial z}\right) + \frac{B(P_{O_2})}{S_t} = 0. \tag{4.384}$$

Reneau et al. [101] give the rate of oxygen release from hemoglobin by

$$v_b\frac{C_{Hbs}}{S_b}\frac{\partial \Upsilon}{\partial z}$$

where Υ is the percent saturation of erythrocytes with oxygen for which they used the Hill equation

$$\Upsilon(P_{O_2}) = \frac{(P_{O_2}/P_{50})^n}{1+(P_{O_2}/P_{50})^n} \tag{4.385}$$

and C_{Hbs} is the saturated concentration of oxygen bound to hemoglobin, which corresponds to $\Upsilon = 1$. Hence, the rate of oxygen release from hemoglobin becomes

$$v_b\frac{C_{Hbs}}{S_b}\frac{\partial \Upsilon}{\partial z} = v_b\phi(P_{O_2})\frac{\partial P_{O_2}}{\partial z} \tag{4.386}$$

where

$$\begin{aligned}\phi(P_{O_2}) &= \frac{C_{Hbs}}{S_b P_{50}}\frac{d}{d(P_{O_2}/P_{50})}\left[\frac{(P_{O_2}/P_{50})^n}{1+(P_{O_2}/P_{50})^n}\right] \\ &= \frac{C_{Hbs}}{S_b P_{50}}\frac{n(P_{O_2}/P_{50})^{n-1}}{[1+(P_{O_2}/P_{50})^n]^2}.\end{aligned} \tag{4.387}$$

Substituting Equation (4.386) into Equation (4.384), we get

$$\frac{\partial P_{O_2}}{\partial t} - \left[\frac{1}{r}\frac{\partial}{\partial r}\left(D_r r\frac{\partial P_{O_2}}{\partial r}\right) + \frac{\partial}{\partial z}\left(D_z\frac{\partial P_{O_2}}{\partial z}\right)\right] + v_b[1+\phi(P_{O_2})]\frac{\partial P_{O_2}}{\partial z} + \frac{B(P_{O_2})}{S_t} = 0. \tag{4.388}$$

The rate of oxygen uptake $B(P_{O_2})$ is zero in the capillary, and in the tissue it is assumed to be given by a Michaelis–Menten reaction.

Boundary Conditions

We assume that there is no oxygen flux on the bounding surface of the tissue, and hence the boundary conditions are

$$
\begin{aligned}
r = r_t: &\quad \frac{\partial P_{O_2}}{\partial r} = 0 \quad 0 \le z \le L_z \\
z = 0: &\quad \frac{\partial P_{O_2}}{\partial z} = 0 \quad r_c \le r \le r_t \\
z = L_z: &\quad \frac{\partial P_{O_2}}{\partial z} = 0 \quad r_c \le r \le r_t.
\end{aligned} \tag{4.389}
$$

For the capillary region, we use the conditions

$$
\begin{aligned}
r = 0: &\quad \frac{\partial P_{O_2}}{\partial r} = 0 \quad 0 \le z \le L_z \\
z = 0: &\quad P_{O_2} = \tilde{P}_{O_2} \quad 0 \le r \le r_c \\
z = L_z: &\quad \frac{\partial P_{O_2}}{\partial z} = 0 \quad 0 \le r \le r_c
\end{aligned} \tag{4.390}
$$

where $\tilde{P}_{O_2}$ is a prescribed oxygen tension in the capillary at the arterial side. When solving for P_{O_2}, Reneau et al. [101] and Lagerlund and Low [63] enforced continuity of oxygen tension and flux at the interface between capillary and tissue,

$$
\begin{aligned}
r = r_c: (P_{O_2})_{capillary} &= (P_{O_2})_{tissue} \quad 0 \le z \le L_z \\
S_b \left(D_r \frac{\partial P_{O_2}}{\partial r} \right)_{capillary} &= S_t \left(D_r \frac{\partial P_{O_2}}{\partial r} \right)_{tissue} \quad 0 \le z \le L_z.
\end{aligned} \tag{4.391}
$$

The first condition of Equation (4.391), which enforces continuity of the primary variable, poses no problem to the finite-element formulation because compatibility of the element shape functions guarantees continuity of the primary variable on the common boundary between two adjacent elements. However, the second condition of Equation (4.391), which enforces continuity of the dissolved oxygen flux, calls for a special treatment when applying balance of secondary variables at nodes on the interface between capillary and tissue. The weak form of Equation (4.388) shows that the secondary variable on the interface between capillary and tissue is

$$
\begin{aligned}
v_n &= \vec{v}.\hat{n} \\
&= D_r \frac{\partial P_{O_2}}{\partial r} \quad \text{on} \quad r = r_c
\end{aligned} \tag{4.392}
$$

which is not the oxygen flux, and since in Equation (4.391) $S_b \neq S_t$, we conclude that continuity of the secondary variable for the pressure formulation does not guarantee continuity of the flux. This issue does not arise with the problem formulation in terms of concentration in which continuity of both concentration and flux are enforced on the boundary between capillary and tissue. To enforce continuity of flux in the pressure formulation, we will implement a simple modification to the element equations with nodes on the capillary-tissue boundary.

The rate of oxygen uptake $B(P_{O_2})$ is zero in the capillary, and is given by a Michaelis–Menten reaction in the tissue,

$$
B(P_{O_2}) = V_{max} \frac{P_{O_2}}{P_m + P_{O_2}} \tag{4.393}
$$

where V_{max} and P_m are the maximum reaction rate and Michaelis constant, respectively. We choose reference length L_o, tension P_o, and diffusivity D_o, and introduce nondimensional variables $t^* = D_o t/L_o^2$, $r^* = r/L_o$, $z^* = z/L_o$, $u^* = P_{O_2}(r,z)/P_o$, $p_{11}^* = D_r/D_o$, $p_{22}^* = D_z/D_o$, $v_b^* = v_b L_o/D_o$, and consumption rate $B^* = BL_o^2/(S_t P_o D_o)$. We drop the star for simplicity of notation, and write the governing equation in nondimensional form as

$$\frac{\partial u}{\partial t} - \left[\frac{1}{r}\frac{\partial}{\partial r}\left(p_{11} r \frac{\partial u}{\partial r}\right) + \frac{\partial}{\partial z}\left(p_{22}\frac{\partial u}{\partial z}\right)\right] + v_b\left[1+\phi(u)\right]\frac{\partial u}{\partial z} + B(u) = 0. \tag{4.394}$$

The nondimensional Michaelis–Menten rate of oxygen consumption is

$$B(u) = \frac{\alpha u}{\beta + u} \tag{4.395}$$

where $\alpha = V_{max}L_o^2/(S_t P_o D_o)$ is nondimensional maximum reaction rate and $\beta = P_m/P_o$ is nondimensional Michaelis constant.

$$\phi(u) = \frac{\gamma n \left(u/u_{50}\right)^{n-1}}{\left[1+\left(u/u_{50}\right)^n\right]^2} \tag{4.396}$$

where $u_{50} = P_{50}/P_o$, and $\gamma = C_{Hbs}/(S_b P_{50})$.

In this section, we solve the steady state problem, $\partial u/\partial t = 0$; in which case the differential equation is an elliptic equation. Let us use the tissue radius as the reference length $L_o = r_t$, and the prescribed tension at the arterial end as a reference tension $P_o = \tilde{P}_{O_2}$. The nondimensional equation is

$$-\left[\frac{1}{r}\frac{\partial}{\partial r}\left(p_{11} r \frac{\partial u}{\partial r}\right) + \frac{\partial}{\partial z}\left(p_{22}\frac{\partial u}{\partial z}\right)\right] + v_b\left[1+\phi(u)\right]\frac{\partial u}{\partial z} + B(u) = 0 \tag{4.397}$$

and boundary conditions are

$$\begin{aligned}
r = 1: &\quad \frac{\partial u}{\partial r} = 0 \quad 0 \le z \le L_z \\
z = 0: &\quad \frac{\partial u}{\partial z} = 0 \quad r_c \le r \le 1 \\
z = L_z: &\quad \frac{\partial u}{\partial z} = 0 \quad r_c \le r \le 1
\end{aligned} \tag{4.398}$$

$$\begin{aligned}
r = 0: &\quad \frac{\partial u}{\partial r} = 0 \quad 0 \le z \le L_z \\
z = 0: &\quad u = 1 \quad 0 \le r \le r_c \\
z = L_z: &\quad \frac{\partial u}{\partial z} = 0 \quad 0 \le r \le r_c
\end{aligned} \tag{4.399}$$

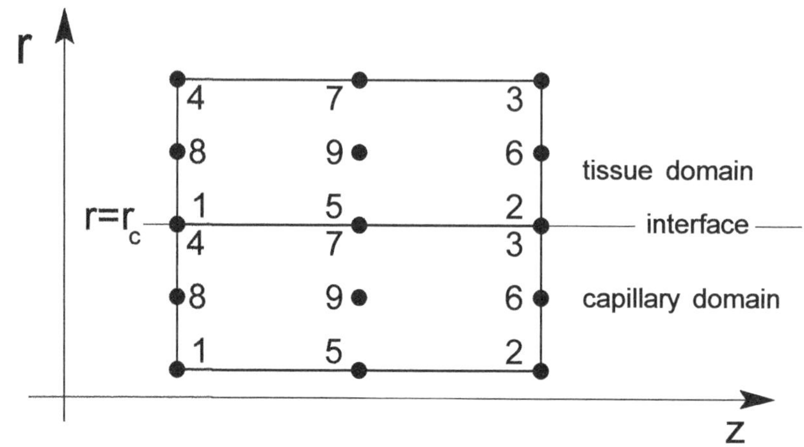

Figure 4.60 Compatibility conditions at the capillary-tissue interface.

$$
\begin{aligned}
r = r_c: \quad (u)_{capillary} &= (u)_{tissue} \qquad 0 \leq z \leq L_z \\
S_r \left(p_{11} \frac{\partial u}{\partial r} \right)_{capillary} &= \left(p_{11} \frac{\partial u}{\partial r} \right)_{tissue} \qquad 0 \leq z \leq L_z
\end{aligned} \tag{4.400}
$$

where $S_r = S_b/S_t$ is the ratio of oxygen solubility in blood to that in tissue. In nondimensional form, the differential equation for pressure Equation (4.394) is identical to Equation (4.340) for the concentration, and the boundary conditions for the pressure problem Equations (4.398)–(4.400) are also identical to the boundary conditions Equations (4.343)–(4.345) for the concentration formulation except for the factor S_r in the second condition in Equation (4.400). This factor has no effects on the derivation of the weak form for the typical element or the element matrices. Therefore, the element equations for the pressure formulation are given by Equation (4.378), and the element matrices are given by Equations (4.371)–(4.376). The factor S_r affects only the balance of secondary variables at the global nodes that fall on the interface between capillary and tissue. We demonstrate the method we developed in order to account for the S_r factor for the case of bi-quadratic elements; the same method can be applied to any other element. In Figure 4.60 we show two elements; the first element is on the capillary side and the other is on the tissue side. The common nodes on the capillary-tissue interface are labeled 3, 4, and 7 for the element on the capillary side and 1, 2, and 5 for the element on the tissue side. After we compute the element coefficient matrix $[A^e]$ and forcing vector $\{F^e\}$ for the element on the capillary side, we multiply the third, fourth, and seventh rows of the matrix and vector by S_r. Then we assemble that element into the global system. This modification is done only for the elements whose nodes 3, 4, and 7 fall on the capillary-tissue interface. No changes are implemented for the elements on the tissue side.

Table 4.7
Geometric and Physical Parameters for Rat Sciatic Nerve [63], with Permission of ElSevier

	Symbol		Units
Capillary radius	r_c	4.5	μm
Tissue radius	r_t	63	μm
Axial length	L_z	1000	μm
Oxygen solubility in blood	S_b	1.517	μM mmHg^{-1}
Oxygen solubility in tissue	S_t	1.34	μM mmHg^{-1}
Henry's constant for blood	$H_b = S_b^{-1}$	0.659	$mmHg\,\mu M^{-1}$
Henry's constant for tissue	$H_t = S_t^{-1}$	0.746	$mmHg\,\mu M^{-1}$
Oxygen Diffusivity in blood	D_b	1.1×10^3	$(\mu m)^2\ s^{-1}$
Oxygen Diffusivity in tissue	D_t	1.7×10^3	$(\mu m)^2\ s^{-1}$
Saturated oxyhemoglobin	C_{Hbs}	8930	μM
50% O_2 hemoglobin concentration	P_{50}	26	mmHg
Hill exponent	n	2.8	
Blood velocity in capillary	v_b	420	$\mu m\ s^{-1}$
Michaelis constant	K_m	4.02	μM
Michaelis constant	P_m	3	mmHg
Maximum O_2 consumption rate	V_{max}	7.59	$\mu M\ s^{-1}$
Arterial side oxygen tension	$\tilde{P}_{O_2}$	90	mmHg

$1\mu M = 10^{-6}$ moles/Liter, and $1\mu m = 10^{-6}$ meters.

Example: Oxygen Delivery in Rat Peripheral Nerve

Lagerlund and Low [62] conducted an extensive numerical study of the steady oxygen release, diffusion, and consumption in a Krogh capillary-tissue model of peripheral nerve. The consumption rate obeys Michaelis–Menten kinetics. In this section, we use the numerical values provided by Lagerlund and Low [63] for the various geometric and physical parameters of the model. These values are listed in Table 4.7. We converted the units of the parameters that involve oxygen concentration from $cm^3\ O_2/cm^3$ blood to the units of μM by taking the volume of 1 mole of gas at standard conditions (0°C and 1 atm) to be 22,400 cm^3. Lagerlund and Low refer to the values in Table 4.7 as the "normal" conditions. We developed a MATLAB code for solving the steady oxygen delivery and consumption in the Krogh cylinder model as described in this section. Because the problem is axisymmetric and the radii of capillary and tissue are constant, we used a Cartesian mesh in the $r-z$ plane. Both bilinear and biquadratic isoparametric elements can be used. Early trials with the code indicated high axial and radial gradient of oxygen tension near the arterial side. Therefore, a nonuniform grid is used in the axial and radial directions. The grid is geometrically stretched in the z-direction. In the radial direction, the grid points are uniformly spaced in the capillary ($0 \le r \le r_c$) and geometrically stretched in the tissue ($r_c \le r \le r_t$). Because the problem is nonlinear, an initial guess is needed for

Table 4.8
Lethal Corner Endoneurial Oxygen Tension (mmHg)

P_{50} mmHg	$\tilde{P}_{O_2}$ mmHg	v_b μm s^{-1}	V_{max} μM s^{-1}	Present FEM
3	90	420	7.59	18.15
6	90	420	7.59	20.70
3	235	420	7.59	21.07
3	1423	420	7.59	52.13
3	90	840	7.59	28.54
3	90	420	3.79	35.65
3	45	280	7.59	8.67
3	90	630	11.38	12.55
3	235	280	7.59	14.53
3	90	210	3.79	24.75
3	90	210	11.38	2.09

the oxygen tension $u(r,z)$. Anticipating decaying behavior of tension with axial and radial directions, we assumed an initial guess as

$$g(r,z) = \tilde{P}_{O_2} exp\left(-z/r_t\right)\left\{0.6+0.4exp[-(r/r_t)^2]\right\}; \tag{4.401}$$

that we found to work for all cases presented in this section. We also tried an initial guess that is a solution to a linear problem in which the blood release function $\phi(u)$ is constant with value equal to $\phi(1)$, which is its value at the arterial side. Also, the Michaelis–Menten reaction rate $B(u)$ is linearized to $B(u) = \alpha[0.05 + u/(\beta+1)]$. The linear problem is solved only once without iteration, and its solution is used as the guessed value $g(r,z)$ to start the iteration for the nonlinear problem. All results presented here are based on the initial guess Equation (4.401).

Lagerlund and Low [62] computed the lethal corner endoneurial oxygen tension for the normal conditions and many other cases by varying one or two of the parameters of the normal conditions. We compare the present finite element results with their results. We use a mesh of biquadratic elements (9 nodes per element). We used 80 elements in the axial direction (161 points). The cylinder length is $L_z = 1000\,\mu$m. The steps are geometrically stretched so that the ratio of two consecutive steps $\Delta z_{j+1}/\Delta z_j = 1.03$. The minimum step is at the arterial side $\Delta z = 0.2673\,\mu$m, and the largest step is at the venous side $\Delta z = 29.386\,\mu$m. The number of elements on the capillary radius is 6 (13 points), and the step size is uniform $\Delta r = 4.5/12 = 0.375\,\mu$m. The steps in the tissue are geometrically stretched, and the ratio of two consecutive steps is $\Delta r_{k+1}/\Delta r_k = 1.07$. The minimum step is next to the capillary $\Delta r = 0.3928$, and the largest is at maximum tissue radius $\Delta r = 4.1942\,\mu$m. For the iteration method indicated by Equation (4.349), we used an under-relaxation parameter $\omega = 0.3$ and error tolerance $\varepsilon = 10^{-10}$. Most cases converge in about 60 iterations.. Lethal corner endoneurial oxygen tensions computed by the present finite-element method are shown in Table 4.8. The first row in the table corresponds to the normal conditions,

Table 4.9
Grid Convergence Study of Lethal Corner Endoneurial Oxygen Tension for Hyperbaric Condition $\tilde{P}_{O_2}$=1423 mmHg

Element type	Elements capillary	Elements tissue	Elements axial	Total elements	Tension mmHg
Bilinear	6	18	80	1,920	54.4864
	6	18	160	3,840	53.4687
	12	36	160	7,680	52.3144
	12	36	240	11,520	52.2063
Biquadratic	3	9	40	480	53.5858
	3	9	80	960	52.9501
	6	18	80	1,920	52.1341
	6	18	120	2,880	52.0915

and values for parameters not shown in Table 4.8 are given in Table 4.7. The difference in pressure $|\Delta P_{O_2}|$ between the finite element results and Lagerlund and Low simulations results are less than 1 mmHg (or 1 Torr) except for the hyperbaric condition where the arterial side tension is $\tilde{P}_{O_2}$ = 1423 mmHg. The difference in this case is 11.67 mmHg. It is important to note that the values given by Largerlund and Low are the average tension over a volume element of radius 12 μm centered at the lethal corner point, whereas the finite element results are the tension at one point that is the corner point. Lagerlund and Low [62] reported high sensitivity to grid resolution for the simulations of the hyperbaric conditions. The present finite-element method shows small sensitivity to grid resolution. A grid convergence study is conducted for the hyperbaric condition. The lethal corner endoneurial oxygen tension is shown in Table 4.9 for different grid resolution and two elements of different order. The results in the table show little sensitivity of the tension to grid resolution. A grid of 1920 biquadratic or 3840 bilinear elements provides reliable results, and further grid refinement is not needed.

Contours of oxygen tension (partial pressure) P_{O_2} for normal conditions (parameters are given in Table 4.7) are depicted in Figure 4.61. Figure 4.62a shows the variation of tension with the axial distance at different radial locations, and Figure 4.62b shows the variation with the radial distance at four axial sections. At the arterial end, the tension is 90 mmHg, but it decays rapidly both in the radial and axial directions. The effects of diffusion are strongest near the arterial end. As expected, the minimum tension of 18.15 mmHg is at the lethal corner ($z = L_z$, $r = r_t$).

Contours of oxygen tension (partial pressure) P_{O_2} for a hyperbaric condition (parameters are given in Table 4.7 except the tension at the arterial side is 1423 mmHg) are depicted in Figure 4.63. Figure 4.64a shows the variation of tension with the axial distance at different radial locations, and Figure 4.64b shows the variation with the radial distance at four axial sections. The high tension at the arterial end induces strong radial diffusion in a tissue layer of axial thickness of approximately 20 μm

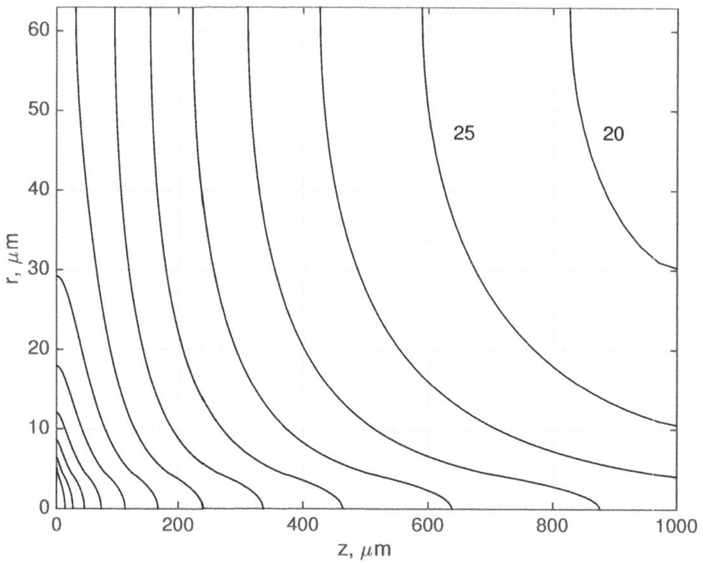

Figure 4.61 Oxygen tension for normal condition $\tilde{P}_{O_2} = 90$ mmHg. Mesh: 7680 bilinear quadrilateral elements. Tension contours; interval between contours is 5 mmHg.

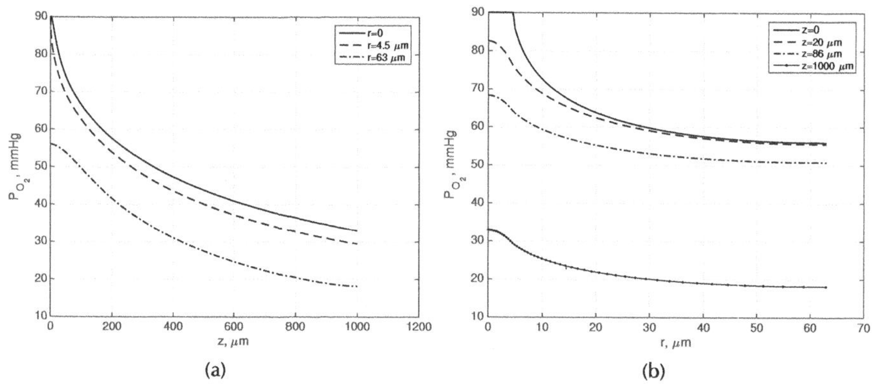

Figure 4.62 Oxygen tension for normal condition $\tilde{P}_{O_2} = 90$ mmHg. Mesh: 7680 bilinear quadrilateral elements. (a) Axial variation of tension at capillary centerline $r = 0$, capillary-tissue interface $r = r_c$, and tissue cylinder radius $r = r_t$, and (b) radial variation of tension at four axial sections.

resulting in high tension of about 700 mmHg at the extreme radial location ($r = r_t$). Axial diffusion now becomes significant in the tissue. As evident in Figure 4.64b, the radial distribution is nearly uniform for axial locations greater than 80 μm showing negligible radial diffusion in the tissue for farther locations. The minimum predicted tension of approximately 52 mmHg is at the lethal corner ($z = L_z$, $r = r_t$).

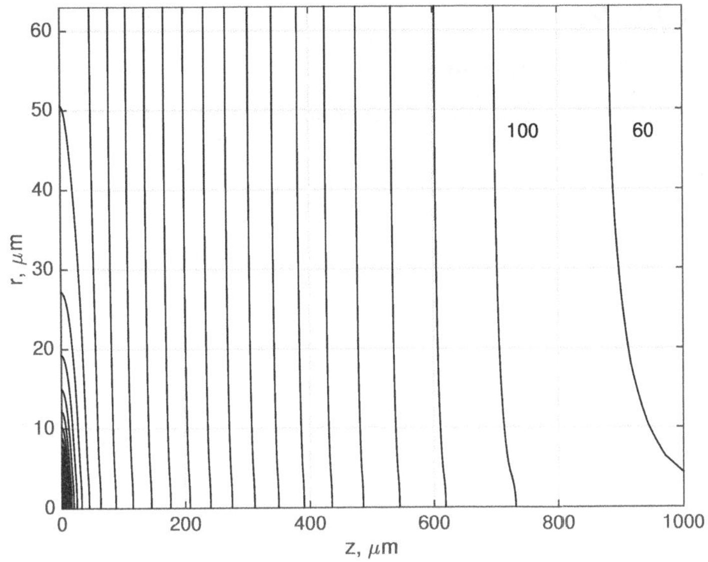

Figure 4.63 Oxygen tension for a hyperbaric condition $\tilde{P}_{O_2} = 1423$ mmHg. Mesh: 7680 bilinear quadrilateral elements. Tension contours; interval between contours is 40 mmHg.

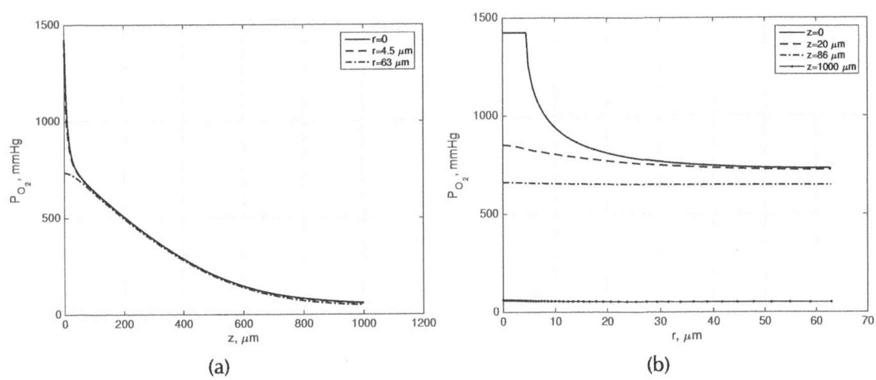

Figure 4.64 Oxygen tension for a hyperbaric condition $\tilde{P}_{O_2} = 1423$ mmHg. Mesh: 7680 bilinear elements. (a) Axial variation of tension at capillary centerline $r = 0$, capillary-tissue interface $r = r_c$, and tissue cylinder radius $r = r_t$, and (b) radial variation of tension at four axial sections.

4.10 PROBLEMS

Weak Form and Shape Functions

(1) Derive the weak form for the elliptic boundary-value problem

$$-[\frac{\partial}{\partial x}(k_x\frac{\partial u}{\partial x})+\frac{\partial}{\partial y}(k_y\frac{\partial u}{\partial y})]=f$$

on a unit square domain shown in Figure 4.65, with boundary conditions

$$x=0:\ k_x\frac{\partial u}{\partial x}=1 \quad \text{for} \quad 0<y<1$$

$$x=1:\ u=1 \quad \text{for} \quad 0<y<1$$

$$y=0:\ u=0 \quad \text{for} \quad 0<x<1$$

$$y=1:\ k_y\frac{\partial u}{\partial y}=sin(\pi x) \quad \text{for} \quad 0<x<1$$

and classify the boundary conditions as natural or essential. It is recommended that one multiplies the differential equation by a test function $w(x,y)$, integrate over the square, and proceed showing the details of the integration by parts and applications of boundary conditions.

(2) Derive the weak form for the elliptic boundary-value problem

$$-[\frac{\partial}{\partial x}(k_x\frac{\partial u}{\partial x})+\frac{\partial}{\partial y}(k_y\frac{\partial u}{\partial y})+\frac{\partial}{\partial z}(k_z\frac{\partial u}{\partial z})]=f$$

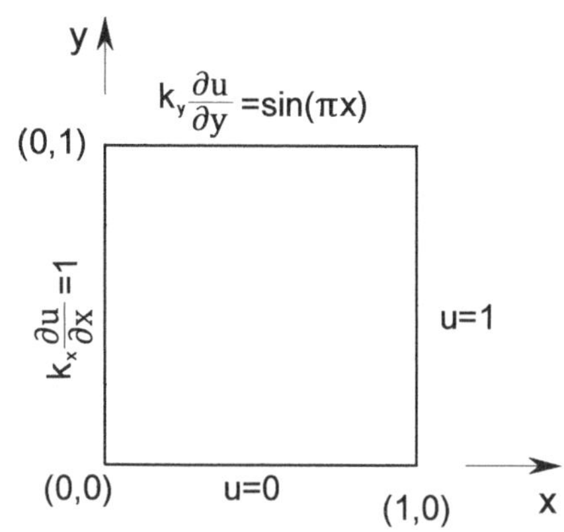

Figure 4.65 Problem 1.

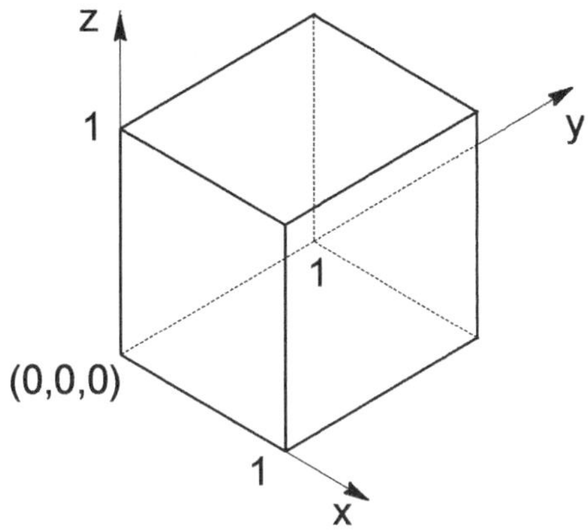

Figure 4.66 Problem 2.

on a domain given by the unit cube shown in Figure 4.66, with boundary conditions

$$\begin{aligned}
x &= 0: -k_x\frac{\partial u}{\partial x} + 3u = 1 \\
x &= 1: u = 1 \\
y &= 0: u = e^{-z}cos(\pi x) \\
y &= 1: k_y\frac{\partial u}{\partial y} = cos(\pi z)sin(\pi x) \\
z &= 0: u = 0 \\
z &= 1: k_z\frac{\partial u}{\partial z} = e^{-y}cos(\pi x)
\end{aligned}$$

and classify the boundary conditions as essential, natural, or mixed boundary conditions. It is recommended that one multiplies the differential equation by a test function $w(x,y,z)$, integrate over the cube, and proceed showing the details of the integration by parts and applications of boundary conditions.

(3) We wish to study steady heat conduction in a composite body made of two different materials as shown in Figure 4.67. In domain Ω_1 the thermal conductivity is k_1 and the rate of energy generation per unit volume is f_1. In domain Ω_2 the thermal conductivity is k_2 and there is no energy generation. The interface between the two domains is denoted by Γ_o. The rest of the boundary of Ω_1 is denoted by Γ_1 and that of Ω_2 by Γ_2. Boundary Γ_1 is insulated, and boundary Γ_2 is subjected to convection heat transfer with coefficient $\tilde{h}$ and ambient temperature T_∞. The temperature field is governed by:

$$-\nabla \cdot (k_1 \nabla T) - f = 0 \quad \text{in} \quad \Omega_1 \tag{4.402}$$

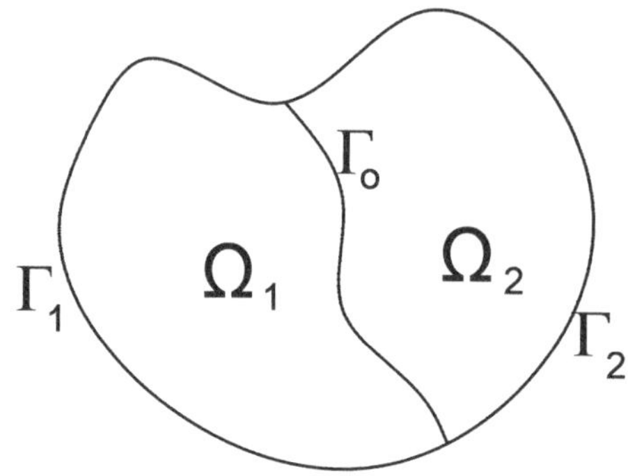

Figure 4.67 Problem 3.

$$-\nabla \cdot (k_2 \nabla T) = 0 \quad \text{in} \quad \Omega_2 \tag{4.403}$$

and boundary conditions

$$\text{on} \quad \Gamma_1 : \quad -k_1 \frac{\partial T}{\partial n_1} = 0 \tag{4.404}$$

$$\text{on} \quad \Gamma_2 : \quad -k_2 \frac{\partial T}{\partial n_2} = \tilde{h}(T - T_\infty) \tag{4.405}$$

$$\text{on} \quad \Gamma_o : \quad (T)_1 = (T)_2 \quad \text{and} \quad -k_1 \frac{\partial T}{\partial n_1} - k_2 \frac{\partial T}{\partial n_2} = 0 \tag{4.406}$$

where n_1 and n_2 are outward unit normal to the boundaries of Ω_1 and Ω_2, respectively. The conditions on Γ_o imply continuity of temperature and normal heat flux on the interface. Thus, there is a discontinuity in the temperature gradient across the interface because k_1 is not equal to k_2.
Derive a weak form that governs the temperature throughout the composite domain $\Omega_1 + \Omega_2$. This form should not include integrals over the interface Γ_o.

(4) The domain ($0 \leq x \leq 2$, $0 \leq y \leq 2$) is divided into four square bilinear elements using nine global nodes as shown in Figure 4.68. The function $u(x,y) = e^x cos(y/2)$ is evaluated at the global nodes. Using those values, we then approximate the function over each element by

$$u(x,y) \approx U^e(x,y) = \sum_{j=1}^{n} U_j^e \psi_j(x,y)$$

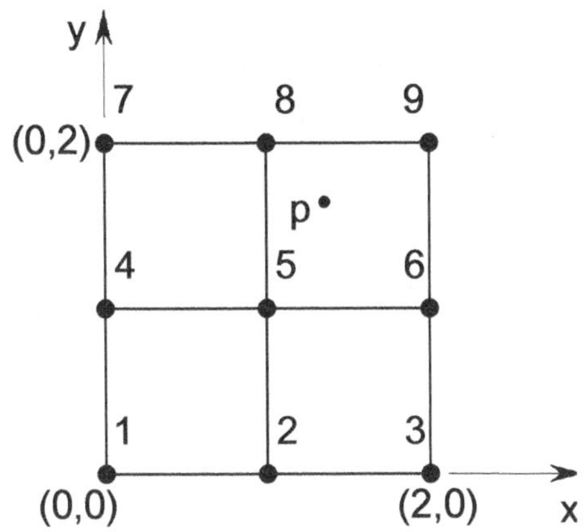

Figure 4.68 Problem 4.

where $\psi_j(x,y)$ are the shape functions of the bilinear rectangular element. Calculate:

(a) The percentage error in the function $u(x,y)$ and its gradient ($\partial u/\partial x$, $\partial u/\partial y$) at the point $(x,y) = (4/3, 5/3)$.

(b) The percentage error in the integral

$$I = \int_A u(x,y)dA$$

where A is the domain area.

(5) Repeat the error analysis of Problem (4) if the domain is divided into 16 square bilinear elements as shown in Figure 4.69.

(6) A finite element analysis of Poisson's equation $-\nabla^2 u = f$ provides approximate values for $u(x,y)$ at the nodes of two bilinear rectangular elements $\{U_1,U_2,U_3,U_4,U_5,U_6\}^T = \{5,3,0,1,0,0\}^T$; see Figure 4.70. Using the finite element solution, compute an approximate value of the integral

$I = \int_0^1 \int_0^2 u(x,y)dxdy$

(7) Given $(x_1^e, y_1^e) = (12,2)$, $(x_2^e, y_2^e) = (12,5)$, and $(x_3^e, y_3^e) = (8,5)$, write down the shape functions for the bilinear triangle element shown in Figure 4.71. If the nodal values for a function $U(x,y)$ are $U_1^e = 20$, $U_2^e = -10$ and $U_3^e = 15$, compute $U^e(x,y)$ at the centroid of the element. Also compute the gradient $\nabla U^e(x,y)$.

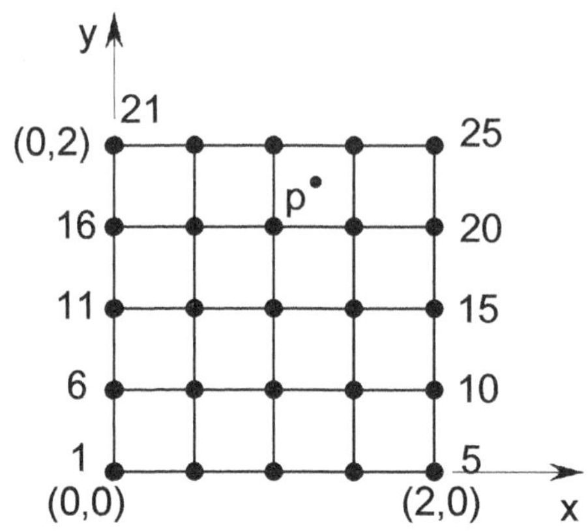

Figure 4.69 Problem 5.

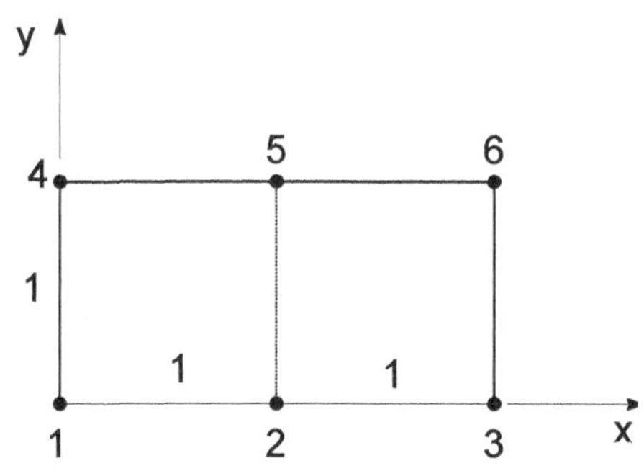

Figure 4.70 Problem 6.

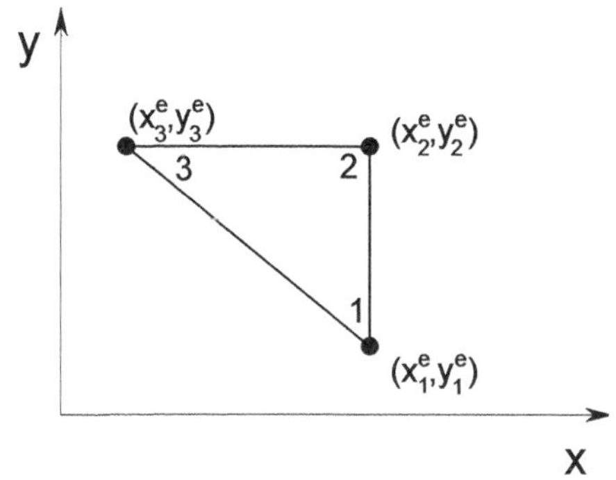

Figure 4.71 Problem 7.

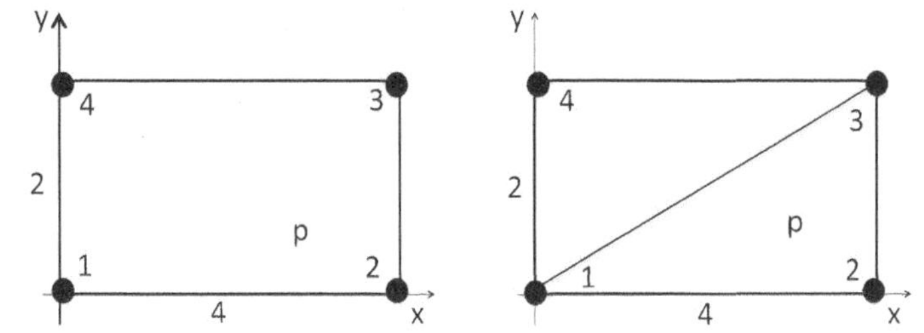

Figure 4.72 Problem 8.

(8) A function $u(x,y)$ has the values $u_1 = 8$, $u_2 = -4$, $u_3 = 0$, and $u_4 = 4$ at the four corners of the rectangular domain shown in Figure 4.72.

(a) The shape functions of the bilinear rectangular element are used to approximate the function $u(x,y)$ over the rectangular domain. Compute the integral $\int_A xudA$, where A is the area of the rectangle.

(b) Divide the rectangle into two triangles. Approximate the function over each triangle by a bilinear triangular element and estimate u at the point $p(x,y) = (3.5,0.5)$.

(9) (a) In two dimensions, the mass matrix for an element is defined by the integral

$$m_{ij}^e = \int_{A_e} q\psi_i\psi_j dA$$

where A_e is the element area and ψ_i are the element shape functions. Assuming that q is uniform over the element, verify Equation (4.61) for the bilinear triangular element.

(b) For Poisson's equation $-k\nabla^2 u = f$, obtain the stiffness matrix for a bilinear right-angle triangle with local nodes labeled as shown in Figure 4.73,

$$[K^e] = \frac{k}{2ab}\begin{bmatrix} c^2 & -b^2 & -a^2 \\ -b^2 & b^2 & 0 \\ -a^2 & 0 & a^2 \end{bmatrix} \quad (4.407)$$

where $c^2 = a^2 + b^2$, and the vertex at node 1 is 90 degrees. This matrix applies regardless of the element's orientation.

Applications in Heat Conduction, Fluid Mechanics, and Solid Mechanics

(10) Consider steady heat conduction in a square solid rod of side $a = 0.4$ m and unit depth, and thermal conductivity $k = 150\ \mathrm{Wm^{-1}\,{}^\circ C^{-1}}$, see Figure 4.74. No internal energy is generated within the rod ($f = 0$). The left and bottom sides ($x = 0$ and $y = 0$) are insulated. The temperature on the right side ($x = 0.4$ m)

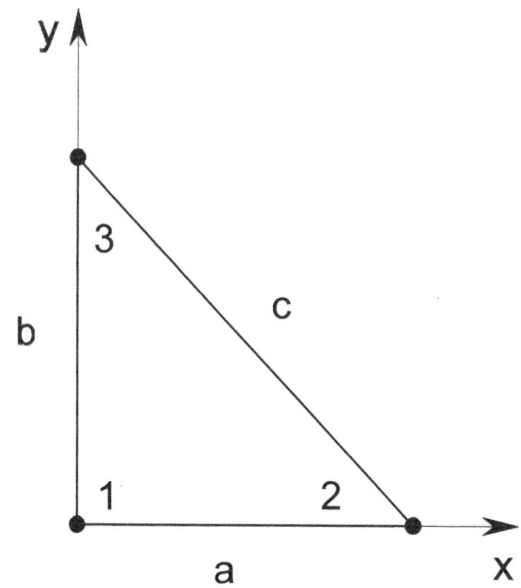

Figure 4.73 Triangular element.

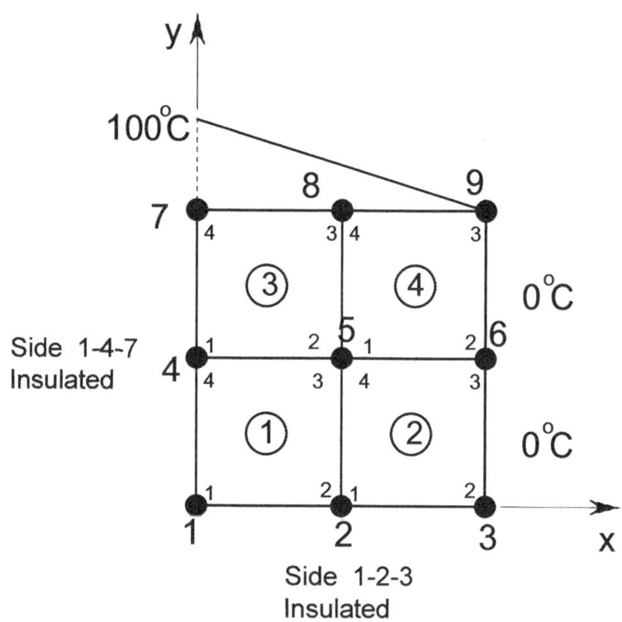

Figure 4.74 Heat conduction in a square rod using four elements.

is 0°C, and it varies linearly, $T = 100(1 - x/a)$°C, on the top side ($y = 0.4$ m). Use four bilinear rectangular elements and determine the temperature at the insulated corner and at the center of the square. Also determine the nodal values of the heat flow, and check conservation of energy (i.e., the net heat flow on the boundary must be zero).

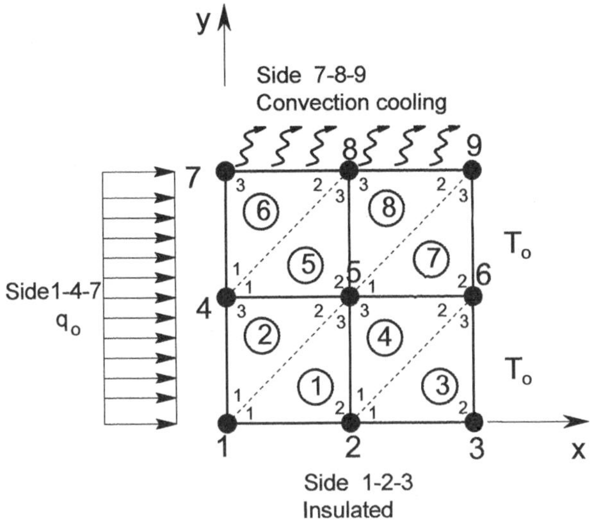

Figure 4.75 Eight bilinear triangular elements model of a square section, there are nine global nodes for the entire domain and three local nodes for each element.

(11) Consider steady heat conduction in a square rod of side $d = 0.2$ m and thermal conductivity $k = 90$ W m^{-1}°C^{-1}, see Figure 4.75. Energy is generated within the domain at the uniform rate of $f_o = 1$MW m^{-3}. The energy equation is

$$-k\nabla^2 T = f. \tag{4.408}$$

The boundary conditions are

$$\begin{aligned}
x &= 0 & -k\frac{\partial T}{\partial x} &= q_o & 0 \le y \le d \\
x &= d & T &= T_o & 0 \le y \le d \\
y &= 0 & \frac{\partial T}{\partial y} &= 0 & 0 \le x \le d \\
y &= d & -k\frac{\partial T}{\partial y} &= \tilde{h}(T - T_\infty) & 0 \le x \le d
\end{aligned} \tag{4.409}$$

where $q_o = 25$ kW m^{-2}, $T_o = 300$°C, $\tilde{h} = 180$ W m^{-2}°C^{-1} is the coefficient of convection, and $T_\infty = 10$°C is the cooling fluid temperature away from the surface. Determine: (a) Unknown nodal temperature, (b) heat flow at global nodes, and (c) check energy balance. Solve this problem using eight bilinear triangular elements as shown in Figure 4.75.

(12) Consider steady heat conduction in an L-shaped domain. The domain is divided into four square elements of side a, and the global nodes are marked as shown in Figure 4.76. No internal energy is generated within the domain ($f = 0$). The

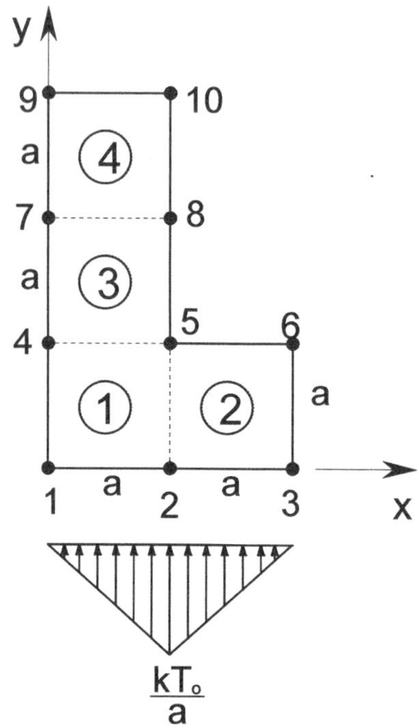

Figure 4.76 Problem 12.

long side on the left ($x = 0$) and the short side on the right ($x = 2a$) are insulated. The bottom side ($y = 0$) receives heat transfer by conduction as given by the distribution shown, where k is the thermal conductivity and T_o is a reference temperature. The rest of the L-shaped boundary is kept at a temperature of T_O. Take $k = 60\ \text{Wm}^{-1}{}^\circ\text{C}^{-1}$, $a = 0.1$ m, and $T_o = 10^\circ\text{C}$, and determine the maximum temperature in the domain. Also determine the nodal values of the heat flow, and check conservation of energy (i.e., the net heat flow on the boundary must be zero).

(13) Resolve Problem (12) on a finer mesh made of 16 elements obtained by dividing each of the four elements into four equal square elements. Determine the change in temperature at global nodes 1, 2, and 3 of Problem (12) as a result of refining the mesh.

(14) Consider steady heat conduction in an L-shaped domain as described in Problem (12). The domain is divided into four square elements of side a. No internal energy is generated within the domain ($f = 0$). The long side on the left ($x = 0$) and the short side on the right ($x = 2a$) are insulated. The bottom side ($y = 0$) receives heat transfer by conduction as given by the distribution shown in Figure 4.76, where k is the thermal conductivity and T_o is a reference temperature. The rest of the L-shaped boundary is subjected to convection heat transfer with

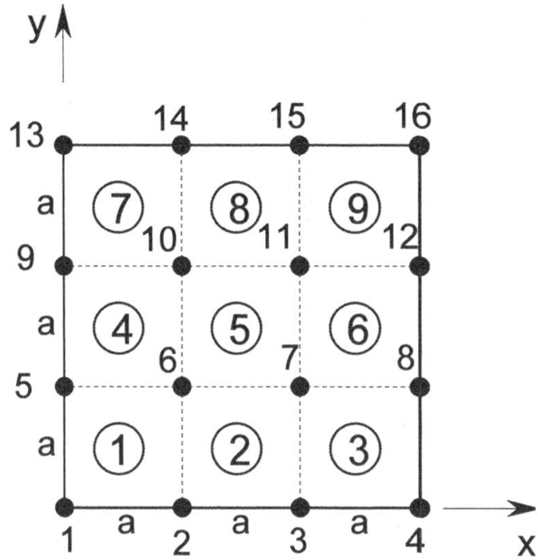

Figure 4.77 Problem 15.

parameters $T_\infty = T_o$ and $\tilde{h} = 90$ Wm^{-2}°C^{-1}. Take $k = 60$ Wm^{-1}°C^{-1}, $a = 0.1$ m, and $T_o = 10$°C, and determine the maximum temperature in the domain. Also determine the rate of heat flow at the nodes on the domain boundary, and check conservation of energy (i.e., the net heat flow on the boundary must be zero because no energy is generated within the domain).

(15) The axial velocity, $w(x,y)$, of a fully developed laminar flow in a straight duct is governed by Poisson's equation

$$-\nabla^2 w = -\frac{1}{\mu}\frac{dp}{dz}$$

where μ is the coefficient of viscosity and dp/dz is the pressure gradient. We wish to determine the velocity profile and volume flow rate in a square duct of side $3a$ using the finite-element method. We divide the section into nine bilinear rectangular elements as shown in Figure 4.77. Assuming symmetry about the vertical and horizontal lines through the duct center, we conclude that the velocities at the four interior nodes have equal values. Given a, μ, and dp/dz, determine the velocity at the interior nodes and the volume flow rate through the duct.

(16) An elastic prismatic bar with hexagonal cross-section of side a is twisted by two equal and opposite torques at its ends. Given the twist angle per unit length, θ, and shear modulus, G, determine the torque using the mesh shown in Figure 4.78.

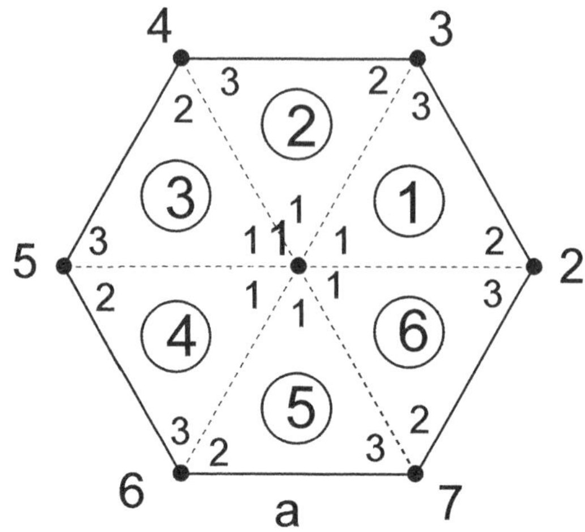

Figure 4.78 Problem 16.

Recall that in terms of Prandtl stress function, u, the torsion problem is governed by Poisson's equation

$$-\nabla^2 u = 2G\theta$$

subjected to the condition $u = 0$ on the boundary. The torque is given by

$$T = 2\int_A u dA$$

(17) Consider steady heat conduction in two dimensions governed by the equation

$$-k\nabla^2 T = f$$

where $k = 60$ W m^{-1}°C^{-1} and $f = 24$ kW m^{-3}, on the L-shaped domain shown in Figure 4.79. The domain is divided into eight bilinear triangular elements. All elements are 45-right triangles with side $a = 0.12$ m. The left side (1-4-7-9) and the right side (3-6) are insulated and the bottom side (1-2-3) is subjected to conduction heat transfer as depicted in the figure. The boundary (6-5-8-10-9) is subjected to convection heat transfer with parameters $T_\infty = 30$°C and $\tilde{h} = 150$ W m^{-2}°C^{-1}.

(a) The equations for element number 5 are written as

$$[[K^e] + [G^e]]\{T^e\} = \{F^e\} + \{R^e\} + \{c^e\}$$

Determine $[G^e]$.

(b) Determine a_{44}, a_{45}, and a_{48} of the assembled coefficient matrix.

(c) Determine R_2 at global node number 2.

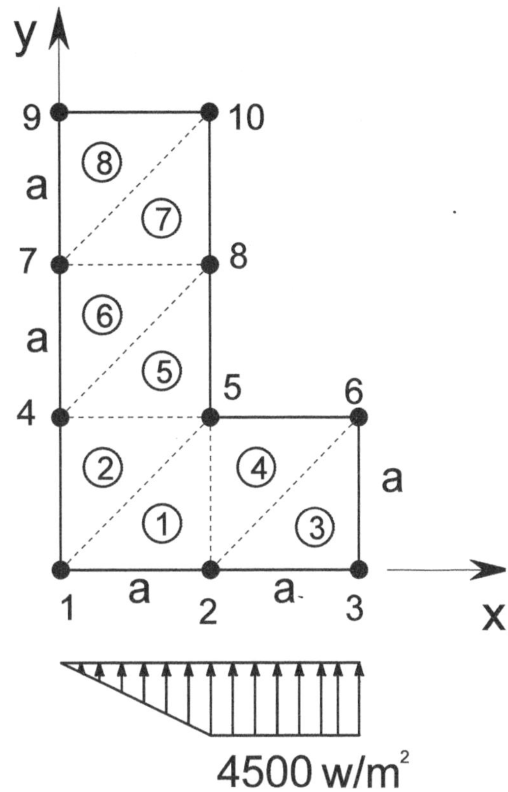

Figure 4.79 Problem 17.

(18) The two-dimensional heat equation

$$-k\nabla^2 T = f$$

where $k = 60$ W m^{-1}°C^{-1} and $f = 24$ kW m^{-3}, is to be solved on the domain shown in Figure 4.80. Sides (AC) and (BC) are insulated. The temperature distribution on side (AB) varies linearly with x: $T = 30x$°C, where x is in meters.

(a) Obtain the assembled system, and verify that it is given by

$$\begin{bmatrix} 40 & -10 & -10 & 0 & -20 & 0 \\ -10 & 100 & -20 & -30 & -40 & 0 \\ -10 & -20 & 100 & 0 & -40 & -30 \\ 0 & -30 & 0 & 30 & 0 & 0 \\ -20 & -40 & -40 & 0 & 100 & 0 \\ 0 & 0 & -30 & 0 & 0 & 30 \end{bmatrix} \begin{Bmatrix} T_1 \\ T_2 \\ T_3 \\ T_4 \\ T_5 \\ T_6 \end{Bmatrix} = \begin{Bmatrix} F_1 \\ F_2 \\ F_3 \\ F_4 \\ F_5 \\ F_6 \end{Bmatrix} + \begin{Bmatrix} R_1 \\ R_2 \\ R_3 \\ R_4 \\ R_5 \\ R_6 \end{Bmatrix} \tag{4.410}$$

(b) Compute the source vector $\{F\}$.

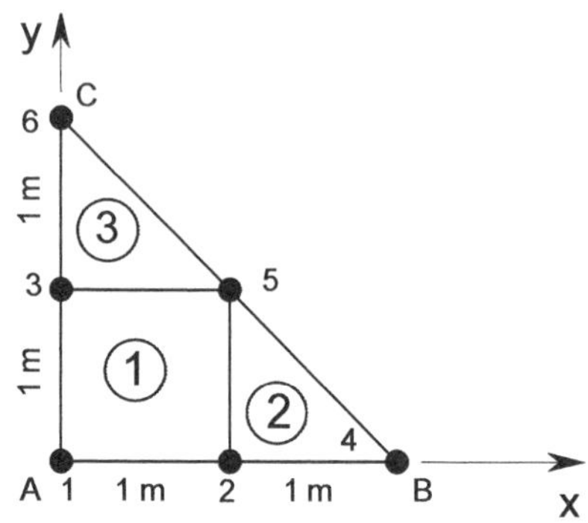

Figure 4.80 Problem 18.

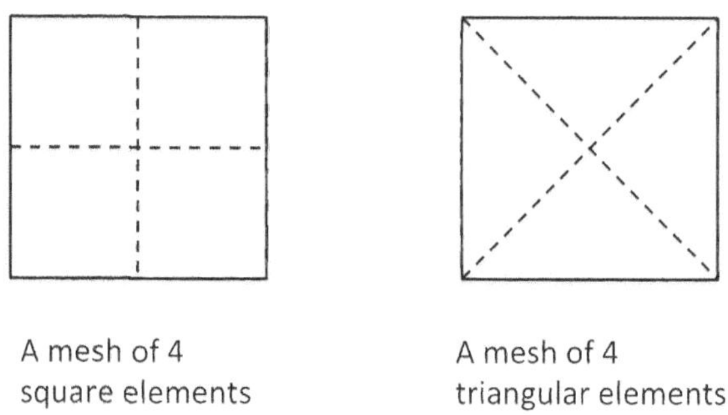

Figure 4.81 Torsion of a prismatic bar of square cross-section.

(c) Apply proper boundary conditions, and determine the unknown temperatures.

(19) Use the finite element method to compute the torsional rigidity of a prismatic bar of square cross-section of side a and shear modulus G. Obtain solutions for two different meshes:

(a) Divide the cross-section into four square elements.

(b) Divide the cross-section into four triangular elements.

(20) An elastic prismatic bar is twisted by two equal and opposite torques at its ends. The twist angle per unit length is denoted by θ, and shear modulus by G. For this problem, Prandtl stress function, u, is a solution to Poisson's equation

$$-\nabla^2 u = 2G\theta$$

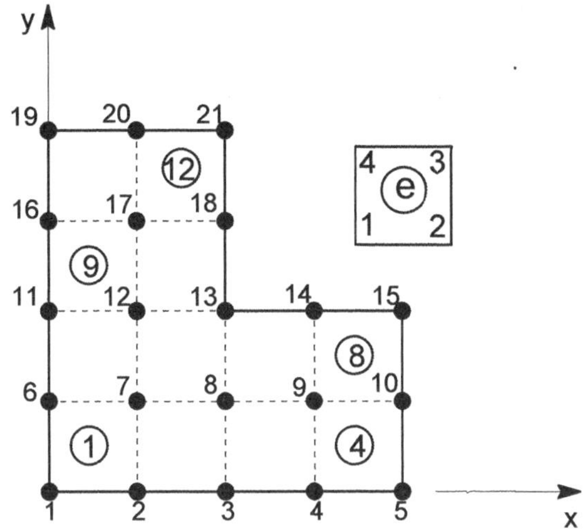

Figure 4.82 Problem 20.

subjected to the condition $u = 0$ on the boundary of the cross-section. The torque is given by

$$T = 2\int_A u dA$$

The finite element method is used to solve this problem for a bar with an angle-shaped section as shown in Figure 4.82. The section is divided into 12 square elements each of side a. Determine the torsional rigidity, T/θ, of the section. Also, determine the secondary variable (R_3) at global node # 3, and explain its physical meaning.

(21) Reconsider Problem (20) for the torsional rigidity of the angle-shaped section shown in Figure 4.82. Apply symmetry conditions $v_n = 0$ on the diagonal connecting the origin (0,0) to the corner $(2a, 2a)$, and solve the problem on half of the section. Use bilinear triangular elements or a combination of rectangular and triangular elements; you decide the type and number of elements.

(22) We wish to compute the torsional rigidity of a square section of side $4a$ and shear modulus G by the finite-element method.

(a) Utilize symmetry about the vertical and horizontal lines through the center, and solve the problem using the mesh 1 shown in Figure 4.83, which consists of four bilinear square elements each of side a and 90-45 triangular elements.

(b) Diagonals are also lines of symmetry. Solve the problem on (1/8) of the square using the mesh 2 shown in Figure 4.83, which consists of three triangles. Apply homogeneous natural boundary conditions on the diagonal and vertical side.

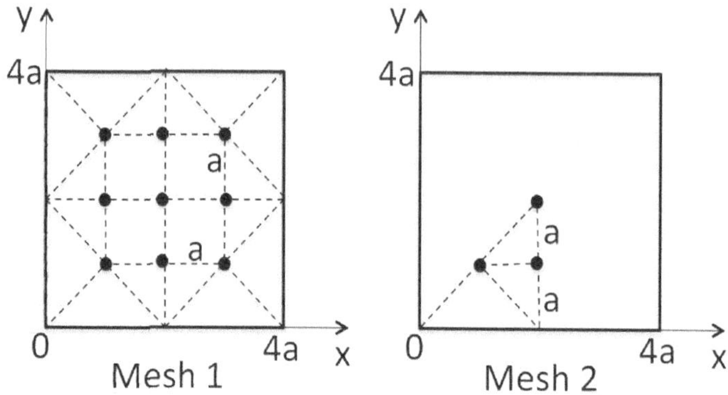

Figure 4.83 Torsion of a square section.

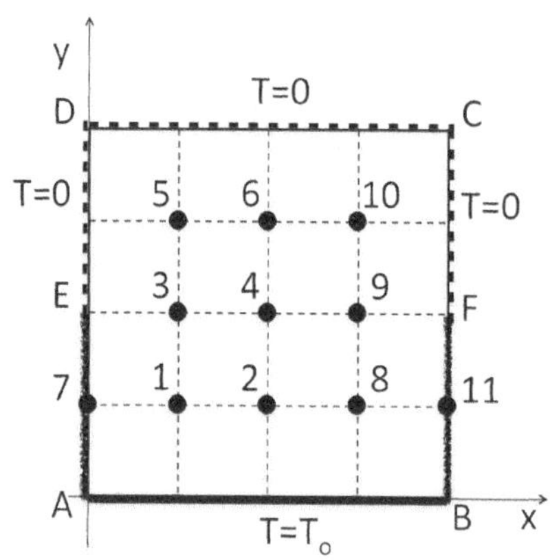

Figure 4.84 Problem 23.

(23) We want to determine the temperature in the square domain (side = a, thermal conductivity = k) shown in Figure 4.84 for steady heat conduction

$$\frac{\partial^2 T}{\partial x^2} + \frac{\partial^2 T}{\partial y^2} = 0$$

where the bottom surface AB is at temperature $T_o\,^\circ\mathrm{C}$, the segments AE and BF are insulated, and the rest of the boundary is at $0\,^\circ\mathrm{C}$. The domain is divided into 16 bilinear square elements each of side $a/4$. Note that the temperature field is symmetric about the vertical line 2-4-6. Hence, we can solve the problem on half the domain. Or we can take $T_8 = T_1$, $T_9 = T_3$, $T_{10} = T_5$, and $T_{11} = T_7$, and construct and solve seven equations for the unknown temperatures $T_1,\ldots,T_7$.

(a) Obtain the seven equations and write them in matrix form.

(b) Determine the unknown nodal temperatures.

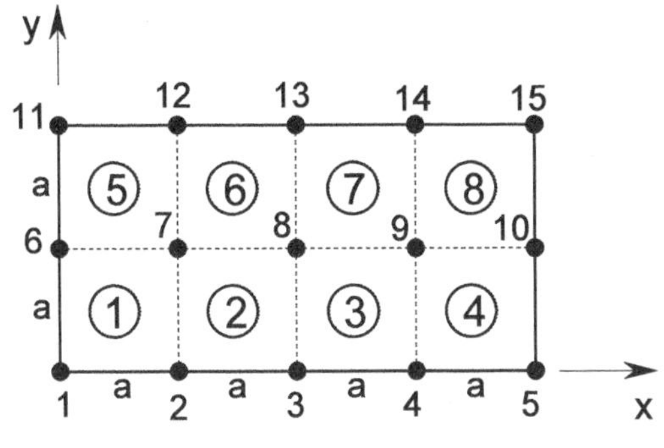

Figure 4.85 Finite element mesh for viscous flow in a rectangular duct.

(c) Determine the rate of heat flow on AB per unit length in the z direction.

(d) Determine the rate of heat flow on $EDCF$ per unit length in the z direction.

(24) The axial velocity, $w(x,y)$, of a fully developed laminar flow in a straight duct is governed by Poisson's equation

$$-\nabla^2 w = -\frac{1}{\mu}\frac{dp}{dz}$$

where μ is the coefficient of viscosity and dp/dz is the pressure gradient. We wish to determine the velocity profile and volume flow rate in a rectangular duct of width $4a$ and side $2a$ using the finite element method. We divide the section into eight bilinear rectangular elements as shown in Figure 4.85. Assuming symmetry about the vertical line through the duct center, we conclude that the velocity at node 7 is equal to that at node 9. Given a, μ and dp/dz, determine the velocity at the interior nodes and the volume flow rate through the duct.

(25) Reconsider the laminar flow in the rectangular duct of Problem (24). Symmetry condition implies that shear stress on the plane $x = 2a$ is zero. Solve the problem on half the rectangle by imposing natural boundary conditions $v_n = 0$ on the plane of symmetry. Compare the results with the solution of Problem (24).

(26) Consider steady heat conduction in an angle-shaped domain as shown in Figure 4.86. The domain is divided into 12 square elements each of side a. Internal energy is generated within the domain at the uniform rate of f_o. The long side on the left ($x = 0$) and the short side on the right ($x = 4a$) are insulated. The bottom side ($y = 0$) receives heat transfer by conduction as given by the distribution shown, where k is the thermal conductivity and T_o is a reference temperature. The rest of the angle-shaped boundary is subjected to convection heat transfer with parameters $T_\infty = T_o$ and $\tilde{h} = 120$ Wm^{-2}°C^{-1}. Take $k = 60$ Wm^{-1}°C^{-1}, $a = 0.1$ m, $T_o = 10$°C, and $f_o = 80$ kW m^{-3}.

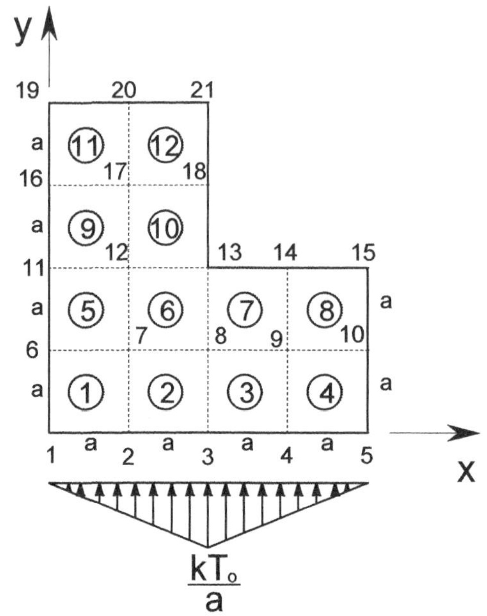

Figure 4.86 Problem 26.

All elements are assembled and boundary conditions are applied.

(a) The equation at global node number 4 is written as

$$K_{4,3}T_3 + K_{4,4}T_4 + K_{4,5}T_5 + K_{4,8}T_8 + K_{4,9}T_9 + K_{4,10}T_{10} = B_4.$$

Calculate $K_{4,8}$, $K_{4,9}$, and B_4. Note that B_4 includes the effects of heat conduction on the boundary and energy generated within the domain.

(b) The equation at global node number 21 is written as

$$K_{21,17}T_{17} + K_{21,18}T_{18} + K_{21,20}T_{20} + K_{21,21}T_{21} = B_{21}.$$

Calculate $K_{21,20}$ and B_{21}. Note that $K_{21,20}$ and B_{21} include the effects of heat transfer by convection and energy generated within the domain.

(27) The axial velocity, $w(x,y)$, of fully developed laminar flow in a straight duct is governed by Poisson's equation

$$-\nabla^2 w = -\frac{1}{\mu}\frac{dp}{dz}$$

where μ is the coefficient of viscosity and dp/dz is the pressure gradient. The finite-element method is used to determine the velocity profile and volume flow rate in a duct of the cross-section shown by Figure 4.87. ABC is an

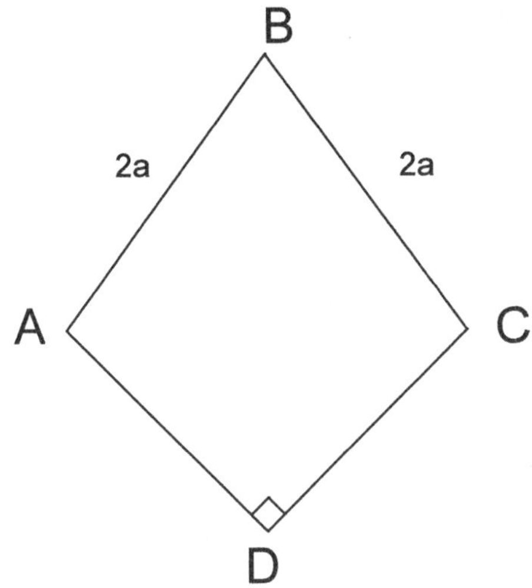

Figure 4.87 Problem 27.

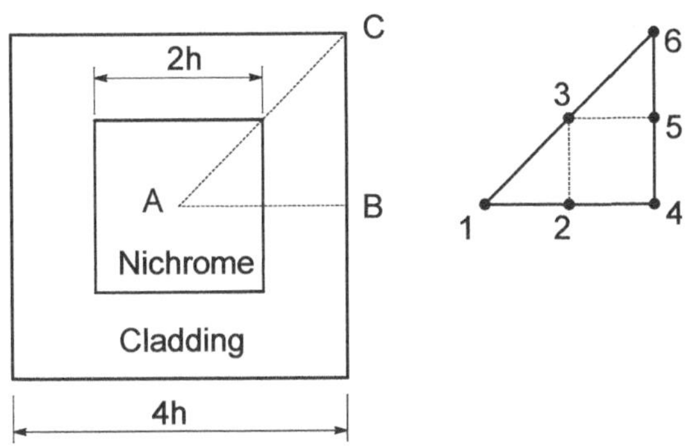

Figure 4.88 Nichrome square rod surrounded by square cladding material.

equilateral triangle of side $2a$ and ADC is 45° right-angle triangle. Divide the domain into bilinear triangular elements (the number of elements is your decision). Given a, μ and dp/dz, determine the rate of volume flow.

(28) Consider steady heat conduction in a body composed of a square rod of nichrome tightly surrounded by a square cladding material as shown in Figure 4.88. The thermal energy equation is

$$-\left[\frac{\partial}{\partial x}\left(k\frac{\partial T}{\partial x}\right)+\frac{\partial}{\partial y}\left(k\frac{\partial T}{\partial y}\right)\right]-f=0$$

where k is the thermal conductivity, and f is the rate of internal energy generated per unit volume. Heat is generated uniformly within the nichrome rod at the rate f_o, while no energy is generated within the cladding ($f = 0$). The outside

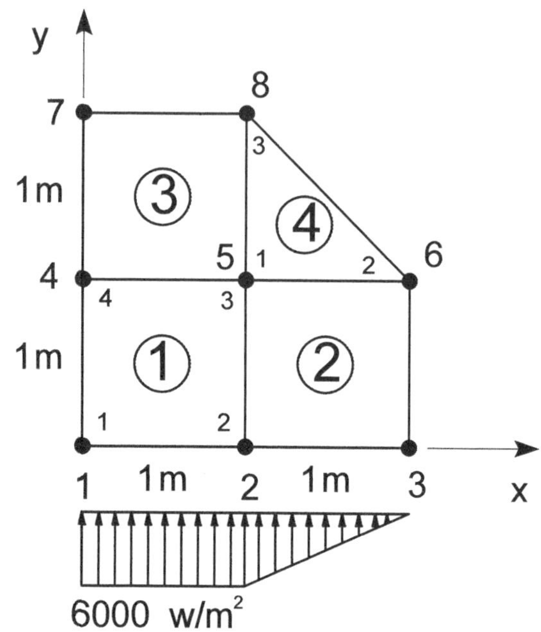

Figure 4.89 Problem 29.

temperature of cladding is kept at 0°C. The finite element method is used to determine the temperature distribution within the composite body. Because of symmetry, we consider a small domain *ABC*, where sides *AB* and *AC* are treated as insulated walls. The domain is divide into three elements using six global nodes. Given: $h = 2$ mm, $f_o = 48$ MW m^{-3}, thermal conductivity of nichrome $k_n = 12$ W m^{-1}°C^{-1}, and of cladding material $k_c = 4$ W m^{-1}°C^{-1}. Determine the unknown temperatures and rate of heat flow at node number 5.

(29) Consider the two-dimensional heat equation

$$-k\nabla^2 T = f$$

where $k = 60$ W m^{-1}°C^{-1} and $f = 24$ kW m^{-3}, on thc domain shown in Figure 4.89. The domain is divided into three square elements and one triangular element. The left side (1-4-7) is insulated and the bottom side (1-2-3) is subjected to conduction heat transfer as depicted in the figure. The rest of the domain boundary (3-6-8-7) is subjected to convection heat transfer with parameters $T_\infty = 30$°C and $\tilde{h} = 150$ W m^{-2}°C^{-1}.

(a) The equations for the triangular element are written as

$$[[K^e] + [G^e]]\{T^e\} = \{F^e\} + \{R^e\} + \{c^e\}.$$

Determine $[G^e]$.

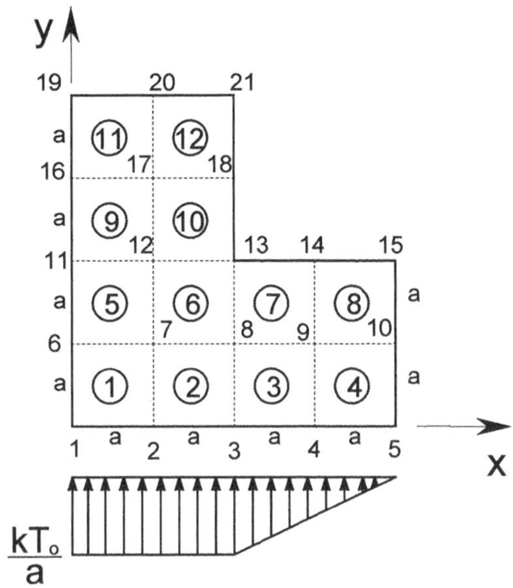

Figure 4.90 Problem 30.

(b) Determine a_{55} and a_{56} of the global assembled stiffness matrix.

(c) Determine R_2 at global node number 2.

(30) Consider steady heat conduction in an angled-shaped domain as shown in Figure 4.90. The domain is divided into 12 square elements each of side a. No internal energy is generated within the domain ($f = 0$). The long side on the left ($x = 0$) and the short side on the right ($x = 4a$) are insulated. The bottom side ($y = 0$) receives heat transfer by conduction as given by the distribution shown, where k is the thermal conductivity and T_o is a reference temperature. The rest of the angled-shaped boundary is subjected to convection heat transfer with parameters $T_\infty = T_o$ and $\tilde{h} = 90\ \text{Wm}^{-2\circ}\text{C}^{-1}$. Take $k = 60\ \text{Wm}^{-1\circ}\text{C}^{-1}$, $a = 0.1$ m and $T_o = 10°\text{C}$.

(a) The equations for element number 12 are written as

$$[[K^e]+[G^e]]\{T^e\} = \{F^e\}+\{R^e\}+\{c^e_\infty\}.$$

Determine $[G^e]$.

(b) Determine R_4 at global node number 4.

(31) Consider steady heat conduction in two dimensions governed by the equation

$$-k\nabla^2 T = f$$

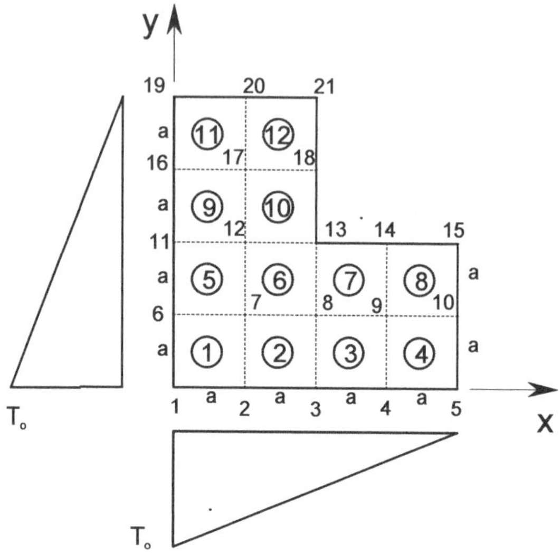

Figure 4.91 Problem 31.

where $k = 60$ W m^{-1}°C^{-1} and $f = 0$ kW m^{-3}. For the angle-shaped domain shown in Figure 4.91, the temperature is prescribed on the outer boundary. At corner A (node 1) it is $T_o = 40$°C, and it drops linearly to 0°C on sides AB (bottom side) and AC (left side). The temperature is 0°C on the rest of the boundary. The finite element method is used to solve for the temperature field within the domain. The domain is divided into 12 square elements each of side $a = 0.1$ m. Determine the temperature at all interior nodes.

Hint:
The temperature field is symmetric about the diagonal 1-7-13.

(32) The axial velocity, $w(x,y)$, of fully developed laminar flow in a straight duct is governed by Poisson's equation

$$-\nabla^2 w = -\frac{1}{\mu}\frac{dp}{dz}$$

where μ is the coefficient of viscosity and dp/dz is the pressure gradient. The finite element method is used to determine the velocity profile and volume flow rate in a hexagonal duct of side $2a$ as shown in Figure 4.92. The section is divided into 24 equilateral triangular elements each of side a, and bilinear elements are used. The flow is assumed to be symmetric about all planes of geometric similarity. Given a, μ and dp/dz, determine the velocity at the interior nodes, and the rate of volume flow.

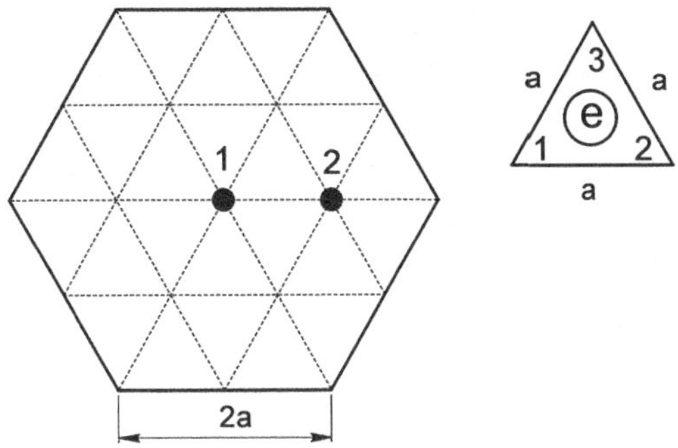

Figure 4.92 Problem 32.

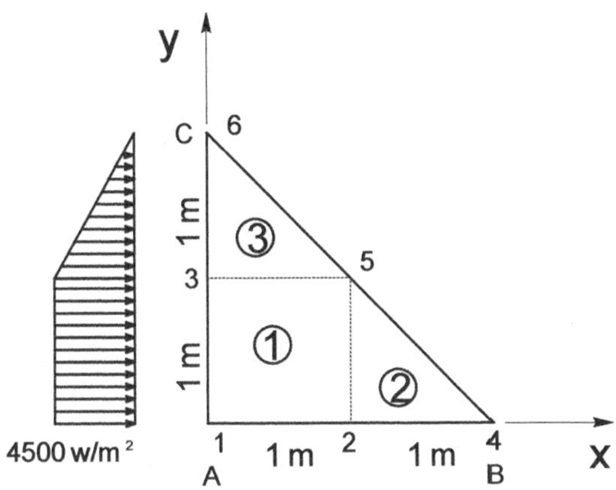

Figure 4.93 Problem 33.

(33) The two-dimensional heat equation

$$-k\nabla^2 T = f$$

where $k = 60$ W m^{-1}°C^{-1} and $f = 24$ kW m^{-3}, is to be solved on the domain shown in Figure 4.93. Side (AB) is insulated, and side (AC) is subjected to conduction heat transfer as depicted in the figure. Side (BC) is kept at a constant temperature of 10°C. The assembled global system before application of

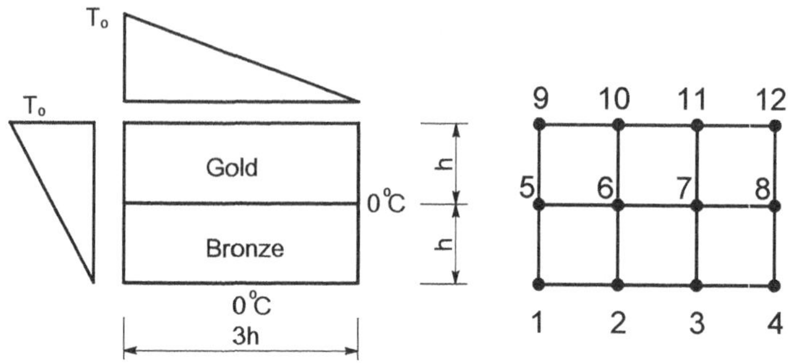

Figure 4.94 Problem 34.

boundary conditions is given by

$$\begin{bmatrix} 40 & -10 & -10 & 0 & -20 & 0 \\ -10 & 100 & -20 & -30 & -40 & 0 \\ -10 & -20 & 100 & 0 & -40 & -30 \\ 0 & -30 & 0 & 30 & 0 & 0 \\ -20 & -40 & -40 & 0 & 100 & 0 \\ 0 & 0 & -30 & 0 & 0 & 30 \end{bmatrix} \begin{Bmatrix} T_1 \\ T_2 \\ T_3 \\ T_4 \\ T_5 \\ T_6 \end{Bmatrix} = \begin{Bmatrix} F_1 \\ F_2 \\ F_3 \\ F_4 \\ F_5 \\ F_6 \end{Bmatrix} + \begin{Bmatrix} R_1 \\ R_2 \\ R_3 \\ R_4 \\ R_5 \\ R_6 \end{Bmatrix} \tag{4.411}$$

(a) Compute the source vector $\{F\}$.

(b) Apply proper boundary conditions, and obtain a reduced system for the unknown temperatures. Solve the reduced system.

(34) Consider steady heat conduction in a body composed of a block of gold clamped to a block of bronze as shown in Figure 4.94. The temperature is governed by the equation

$$-\left[\frac{\partial}{\partial x}\left(k\frac{\partial T}{\partial x}\right)+\frac{\partial}{\partial y}\left(k\frac{\partial T}{\partial y}\right)\right]=0$$

where k is the thermal conductivity.
The temperature on the boundary varies linearly on the upper and left surfaces, and is 0°C on the bottom and right surfaces. The finite element method is used to determine the temperature distribution within the composite body. Each block is divided into three square bilinear elements of side h. Given: $T_o = 60$°C, and thermal conductivity for gold $k_g = 317$ W m^{-1}°C^{-1}, and bronze $k_b = 52$ W m^{-1}°C^{-1}. Determine the unknown temperatures.

(35) Consider steady heat conduction in a square rod of side $d = 0.02$ m and thermal conductivity $k = 30$ W m^{-1}°C^{-1}. Energy is generated within the domain at the uniform rate of $f_o = 10$MW m^{-3}. The energy equation is

$$-k\nabla^2 T = f. \tag{4.412}$$

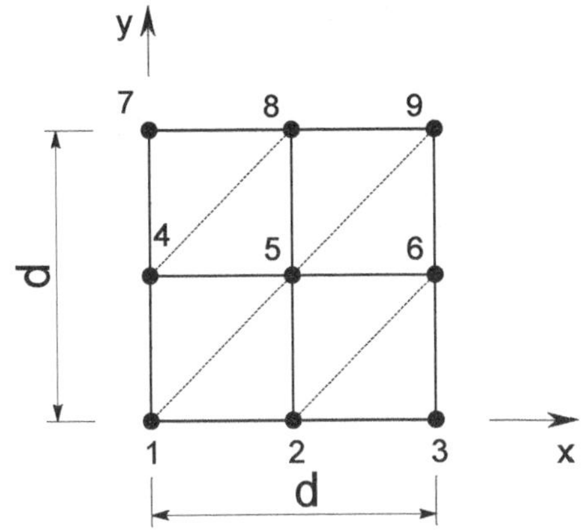

Figure 4.95 Problem 35.

The boundary conditions are

$$\begin{aligned} x &= 0 & -k\frac{\partial T}{\partial x} = q_o(1-y/d) && 0<y<d \\ x &= d & T = T_o && 0 \le y \le d \\ y &= 0 & \frac{\partial T}{\partial y} = 0 && 0<x<d \\ y &= d & T = T_o && 0 \le x \le d \end{aligned} \tag{4.413}$$

where $q_o = 0.2$ MW m^{-2} and $T_o = 10$°C. The domain is divided into eight bilinear triangular elements using nine global nodes as shown in Figure 4.95. The elements are assembled, and the equations at global nodes 1 and 4 are written as

$$\begin{aligned} K_{11}T_1 + K_{12}T_2 + K_{13}T_3 + K_{14}T_4 + K_{15}T_5 + \ldots &= F_1 + R_1 \\ K_{41}T_1 + K_{42}T_2 + K_{43}T_3 + K_{44}T_4 + K_{45}T_5 + \ldots &= F_4 + R_4. \end{aligned}$$

Determine: (a) K_{15}, F_1, and R_1, and (b) K_{44} and R_4

(36) An aluminum ($G = 28$ GPa) prismatic bar of square section of side $a = 40$ mm is twisted by two equal and opposite torques at its ends. The twist angle per unit length is $\theta = 0.8°$ m^{-1}. Using the minimum number of elements, determine the torque magnitude.
Recall that in terms of Prandtl stress function ϕ, the torsion problem is governed by Poisson's equation

$$-\nabla^2\phi = 2G\theta$$

where θ is in radians per meter, and $\phi = 0$ on the boundary of the cross-section. The torque is given by

$$T = 2\int_A \phi dA$$

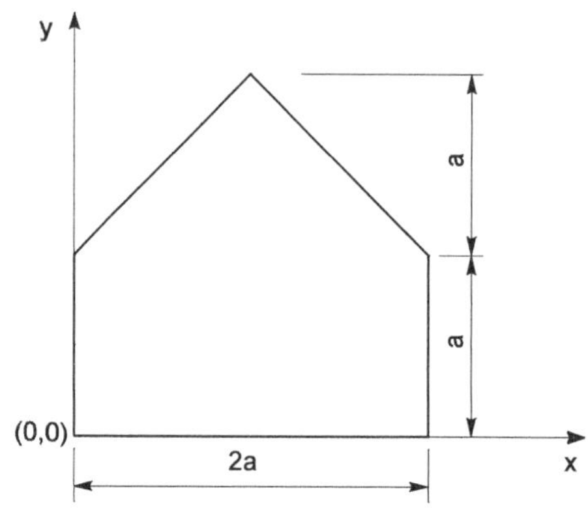

Figure 4.96 Cross-section of a prismatic bar.

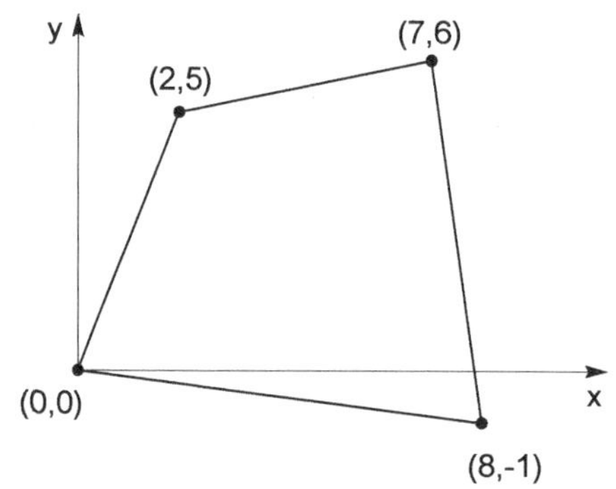

Figure 4.97 Problem 38.

(37) An elastic prismatic bar whose cross-section is shown in Figure 4.96 is twisted by two equal and opposite torques at its ends. The twist angle per unit length is denoted by θ, and shear modulus by G. Given a, θ, and G, determine the torque by the finite element method. You decide the number and type of elements.

(38) We wish to solve the 2D Poisson's equation

$$-\nabla^2 u = f$$

using the finite-element method. A mesh of 4-node bilinear quadrilateral elements is to be used. Compute the element matrices $[K^e]$ and $\{F^e\}$ for the element shown in Figure 4.97. The forcing term is given by $f(x,y) = ye^{-x/8}$. Use the 2×2 Gauss–Legendre quadrature rule.

(39) Consider the Poisson's equation of Problem (38), and compute the element matrices $[K^e]$ and $\{F^e\}$ for the 9-node biquadratic element shown in Figure 4.98.

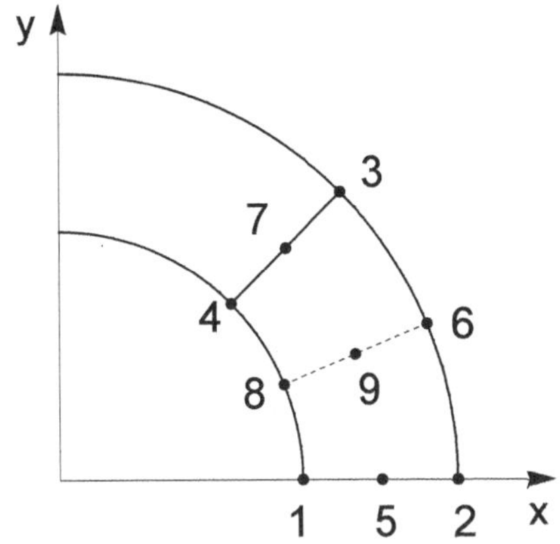

Figure 4.98 Problem 39.

Two of the sides are circular arcs with radii 1 and 2. The other two sides are radial lines at $\theta = 0$ and $\theta = 40°$. Nodes 5, 6, 7, and 8 are the mid-points of the respective side, and node 9 is the mid-point of 6-8. Use the 3×3 Gauss–Legendre quadrature rule.

(40) Consider the flow of oil in a straight duct whose cross-section is given by two circular arcs of radii $r_1 = 0.01$ m and $r_2 = 0.02$ m, and two radial lines subtending an angle of 45° at the center. The oil viscosity is $\mu = 0.1$ Ns/m^2, and the volume flow rate is 0.0002 m^3/s. Neglect body forces and assume fully developed laminar flow. The axial velocity $w(x,y)$ is governed by Poisson's equation:

$$-\nabla^2 w = -\frac{1}{\mu}\frac{dp}{dz}$$

where dp/dz is the axial pressure gradient; which is constant. The volume flow rate is given by

$$Q = \int_A w(x,y)dA.$$

Introduce nondimensional variables

$$x^* = x/r_2$$

$$y^* = y/r_2$$

$$w^* = w\mu/(-r_2^2 dp/dz)$$

and obtain the equation

$$-(\frac{\partial^2 w^*}{\partial x^{*2}} + \frac{\partial^2 w^*}{\partial y^{*2}}) = 1.$$

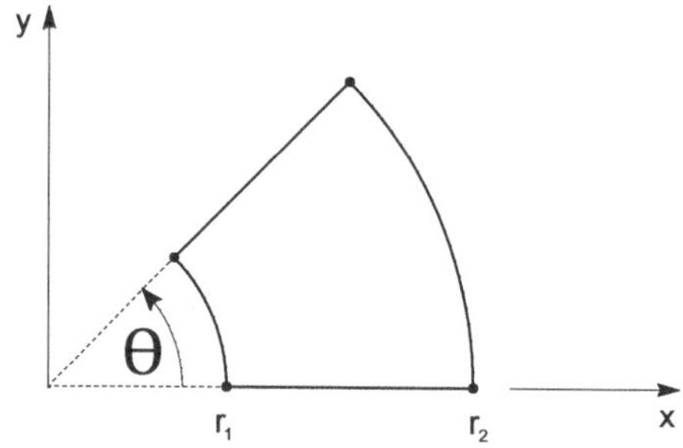

Figure 4.99 Problem 40.

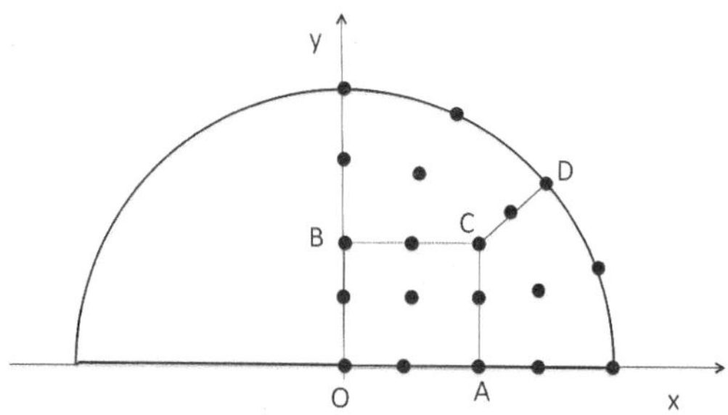

Figure 4.100 Problem 41.

The no-slip boundary condition is

$$w^* = 0.$$

Use the finite element method to solve Poisson's equation, and

(a) Determine the axial pressure gradient dp/dz.

(b) Determine the maximum axial velocity w_{max}.

(41) We wish to determine the torsional rigidity of a semicircular section. We utilize the symmetry of the section, and solve for the Prandtl stress function on half the domain using three biquadratic elements as shown in Figure 4.100. The coordinates of points A, B, C, and D are $(b/2, 0)$, $(0, b/2)$, $(b/2, b/2)$, and $(b/\sqrt{2}, b/\sqrt{2})$, respectively, where b is the circle radius. A node on an element's edge bisects the edge, whereas the coordinates of a central node are the averages of the corner nodes. Compare the finite element solution to the exact solution.

(42) Consider viscous laminar flow (viscosity μ) due to axial pressure gradient dp/dz in a duct with equilateral triangle section of side a. We solve Poisson's equation for the axial velocity $w(x,y)$ using one isoparametric bicubic triangular element with nodes as shown in Figure 4.101. The nodes are equally spaced

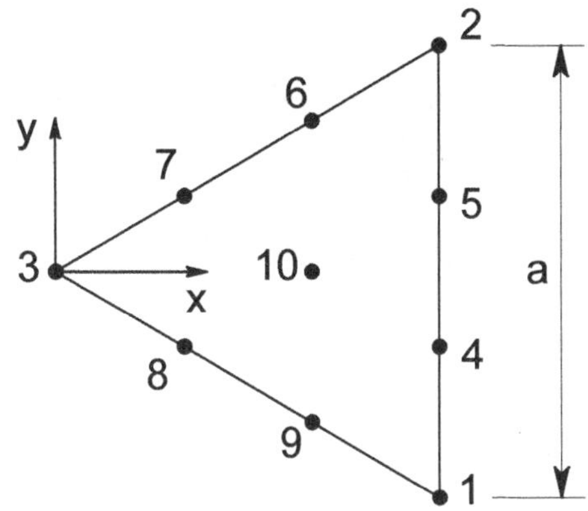

Figure 4.101 Problem 42.

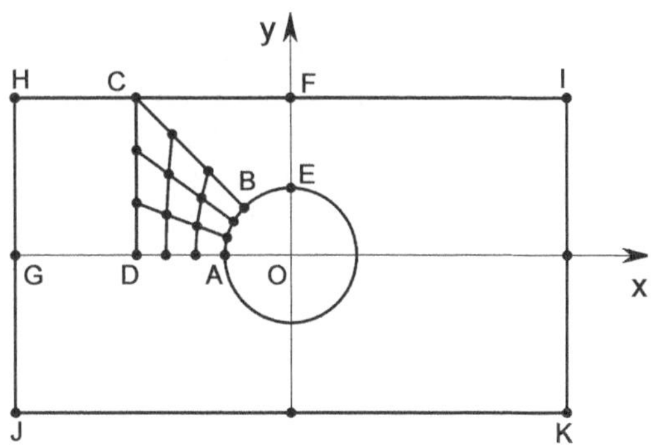

Figure 4.102 Problem 43.

on the sides and node 10 is at $x = \sqrt{3}a/3$. Compare the finite-element solution to the exact solution, which is given by

$$w(x,y) = \frac{\sqrt{3}}{2}\left(-\frac{1}{\mu}\frac{dp}{dz}a^2\right)\left(\frac{x}{a} - \frac{\sqrt{3}}{2}\right)\left(\frac{y^2}{a^2} - \frac{1}{3}\frac{x^2}{a^2}\right) \tag{4.414}$$

and the exact volume flow rate is

$$Q = \frac{\sqrt{3}}{320}\left(-\frac{1}{\mu}\frac{dp}{dz}\right)a^4. \tag{4.415}$$

(43) Hot liquid flows steadily in a circular conduit drilled in a long column of rectangular cross-section $JKIH$ as shown in Figure 4.102. The conduit radius is b and the rectangle height is $4b$ and length is $6b$. We assume that the temperature of the circular conduit surface to be uniform at 100°C and that the outer surface of

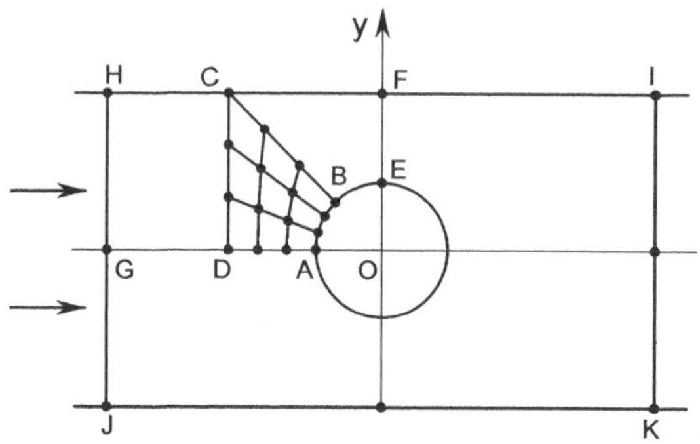

Figure 4.103 Problem 44.

the column also to be uniform at 0°C. We want to use the finite element method to analyze the temperature field in the column. Assuming symmetry about the x- and y- axes, we solve for the temperature $T(x,y)$ in one quadrant; say the second quadrant as shown in Figure 4.102. The domain is divided into three bicubic elements labeled *ABCD*, *BEFC*, and *DCHG*. Each element has 16 local nodes, and the total number of global nodes is 40. The coordinates of points *B*, *C*, *D*, *G*, and *H* are $(-b/\sqrt{2}, b/\sqrt{2})$, $(-2b, 2b)$, $(-2b, 0)$, $(-3b, 0)$, and $(-3b, 2b)$, respectively. The nodes of element *ABCD* are shown in the figure; the nodes are equally spaced on *AB* and on radial lines.
Take $b = 0.2$ m, and thermal conductivity $k = 204$ W m^{-1}°C^{-1}.

(a) Specify and classify (essential, natural, mixed) boundary conditions on *GDA*, *ABE* and *HCF*, *GH*, and *EF*.

(b) Prove that the rate of heat flow to the environment per unit length of the column is equal to the sum of weighted secondary variables on the outer boundary of the rectangle. Compute that rate. Ignore the change in the liquid temperature along conduit length.

(c) Report the values of temperature at the nodes of element *BEFC*.

(44) A circular cylinder of radius $b = 1$ is placed midway between two parallel plates in an otherwise uniform stream of velocity $U_\infty = 1$. The flow is assumed to be two-dimensional inviscid incompressible potential flow. Assuming symmetry about the x- and y- axes, we solve for the velocity potential ϕ in one quadrant; say the second quadrant as shown in Figure 4.103. The domain is divided into three bicubic elements labeled *ABCD*, *BEFC*, and *DCHG*. Each element has 16 local nodes, and the total number of global nodes is 40. The coordinates of points *B*, *C*, *D*, *G*, and *H* are $(-b/\sqrt{2}, b/\sqrt{2})$, $(-2b, 2b)$, $(-2b, 0)$, $(-6b, 0)$, and $(-6b, 2b)$, respectively. The nodes of element *ABCD* are shown in the figure; the nodes are equally spaced on *AB* and on radial lines.

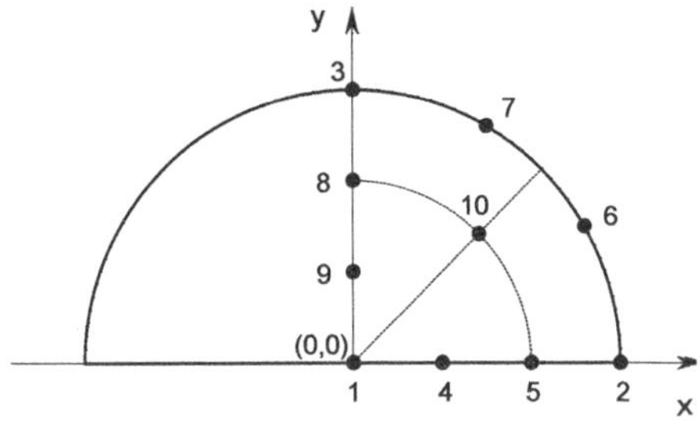

Figure 4.104 Problem 45.

(a) Justify and apply the following boundary conditions: $\partial\phi/\partial n = 0$ on GDA, ABE and HCF, $\partial\phi/\partial n = -1$ on GH, and $\phi = 0$ on EF.

(b) Evaluate the element matrices by the 4×4 Gauss–Legendre quadrature formula.

(c) Report the values of global weighted secondary variables R at the 4 nodes on GH and EF.

(d) Report the values of the fluid velocity at the base points of the (3×3) Gauss-Legendre rule for element $ABCD$.

(e) Compute and plot the pressure coefficient on the cylinder surface.

(45) Determine the torsional rigidity of the semicircular section using one isoparametric bicubic triangular element shown in Figure 4.104, and compare with the exact solution.

(46) An elastic prismatic bar whose cross-section is shown in Figure 4.105a is twisted by two equal and opposite torques at its ends. The height v of the cross-section is used as reference length ($(y_A - y_B)/v = 1$), and $AM = CD = EF = GH = CJ = HI = v/3$. The angle $\measuredangle ABC = \measuredangle MLK = 56$ degrees. A MATLAB code (*Mesh_VT.m*) for mesh generation is provided. It generates a mesh of bilinear quadrilateral elements such as that shown in Figure 4.105b, and provides the connectivity matrix. The number of elements on edges AB, BC, CJ, ED, AM, CD, and DG are 10, 8, 4, 8, 6, 4, and 5, respectively. This mesh is considered a coarse mesh. Use the finite-element method to solve Poisson's equation for Prandtl stress function using isoparametric bilinear quadrilateral elements. First, for the coarse mesh provide:

(a) Torque $T/\theta G v^4$, where θ is twist angle per unit length, and G is the shear modulus.

(b) Nondimensional $(\sigma/\theta G v)$ shear stress magnitude at the center $(\xi = \eta = 0)$ of an element nearest corner L. The element number is identified by the mesh generation code.

(c) Repeat item (b) for corners C, D, G, and H.

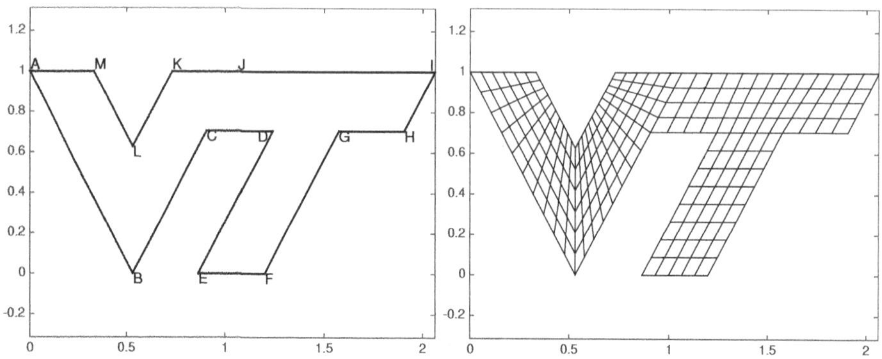

(a) (b)

Figure 4.105 (a) VT prismatic bar. (b) A coarse mesh of bilinear quadrilateral elements.

Second, refine the mesh by doubling the number of elements on each edge, and provide the answers to items (a), (b), and (c).

(47) An eccentric cylinder of radius $r_1 = 0.5$ m is drilled through a larger cylinder of radius $r_2 = 1$ m. The distance between centers is 0.3 m. The inner surface is at uniform temperature of 30 °C, while the temperature of the outer surface is given by $T = 80cos(\theta/2)$ °C, where θ is the angle a radial line from the center of the larger circle makes with line CD, so that $\theta = 0$ at point D and π at point B. The material thermal conductivity is $k = 204$ W $\text{m}^{-1}{}^{\circ}C^{-1}$. A MATLAB code (*eccentric_cylinders_Mesh.m*) for mesh generation is provided. It generates a mesh of biquadratic quadrilateral elements such as that shown in Figure 4.106, and provides the connectivity matrix. Assume AB and CD to be lines of symmetry of the temperature field.

(a) Specify and classify (essential, natural, mixed) boundary conditions on AB, AC, CD, and BD.

(b) Determine the net rate of heat transfer to the inner surface per unit axial length of the cylinder.

(c) Provide a plot for the weighted secondary variables at the global nodes on the outer boundary BD as a function of θ.

(48) Consider a 2D co-axial capacitor of two eccentric long cylinders. The radii of the inner and outer conductors are r_1 and r_2, respectively, and the distance between centers is d, see Figure 4.106. Permittivity of the dielectric material is ε. The electrostatic potential V in the dielectric is a solution to Laplace's equation

$$\frac{\partial^2 V}{\partial x^2} + \frac{\partial^2 V}{\partial y^2} = 0.$$

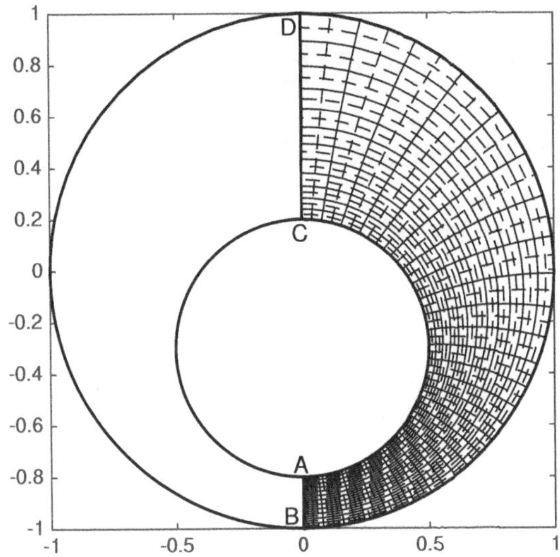

Figure 4.106 A mesh of biquadratic elements between eccentric cylinders.

The inner conductor AC is at potential $V = V_o$, the outer conductor BD is at potential $V = 0$. A MATLAB code (*eccentric_cylinders_Mesh.m*) for mesh generation is provided. It generates a mesh of biquadratic quadrilateral elements such as that shown in Figure 4.106, and provides the connectivity matrix. Assume AB and CD to be lines of symmetry of the electric potential field.

(a) The electric field (intensity) is $\vec{E} = -\nabla V$, hence, it is normal to the conductors surfaces. Plot E on the surface of the inner conductor.

(b) The electric charge on the conductor's surface is given by

$$Q = \varepsilon \int_A \vec{E} \cdot d\vec{A}. \tag{4.416}$$

Show that $Q/\varepsilon\ell$ is given by the sum of the weighted secondary variables at the nodes on the inner conductor, (which is also equal to that on the outer conductor but of opposite sign)). ℓ is the cylinder length in the z-direction. Compute the capacitance $C = Q/V_o$, and compare with the exact analytic solution

$$\frac{C}{\varepsilon\ell} = \frac{2\pi}{cosh^{-1}\left(\frac{r_1^2+r_2^2-d^2}{2r_1r_2}\right)}. \tag{4.417}$$

5 Parabolic Equations: Time-Dependent Diffusion Problems

The rate of heat transfer by a cooling fin varies with time in response to changes in the operating conditions of the attached devices. A solute in a region of high concentration will diffuse to the neighboring regions as time progress. Similarly, vorticity of fluid motion also diffuses from layers of high intensity to the surrounding fluid due to molecular viscosity. Dissolved oxygen concentration (or tension) diffuses in tissues and cells where it is consumed. The basic principles of continuum mechanics provide mathematical models for the space-time distributions of temperature in fins, concentration of solutes, vorticity of fluids, and oxygen tension in tissues. Simplified mathematical models for the aforementioned phenomena and other applications in transient heat and mass diffusion give rise to parabolic equations.

In this chapter, we formulate finite element models for partial differential equations of the parabolic type. The independent variable u is usually a function of space coordinate x and time t. The differential operator is first-order in time and second-order in space. Solutions need to satisfy prescribed boundary conditions in space and take certain initial distributions in time. The finite-element method will be used to replace the differential operator in space by an algebraic operator. This produces a system of coupled ordinary differential equations in time. Then, the finite-difference method will be used to approximate the time derivative thereby providing an algebraic system to be solved at each time step. The models are developed in one space dimension (x or r). Extension to two space dimensions (x, y) is straightforward, and the relevant matrices can be found in Chapter 4.

5.1 MODEL PROBLEM

A model parabolic problem is given by the partial differential equation (PDE)

$$\gamma(x)\frac{\partial u}{\partial t} - \frac{\partial}{\partial x}[p(x)\frac{\partial u}{\partial x}] + q(x)u - f(x,t) = 0 \qquad a < x < b \qquad (5.1)$$

and boundary conditions

$$x = a: \qquad -p\frac{\partial u}{\partial x} + \beta_1 u = \alpha_1$$

$$x = b: \qquad p\frac{\partial u}{\partial x} + \beta_2 u = \alpha_2 \qquad (5.2)$$

DOI: 10.1201/9781003323150-5

where γ, p, and q, are given functions of x, and f is a given function of x and t. α_i are given functions of t, and β_i are given constants. We want to determine $u(x,t)$ for $t > 0$ subject to initial condition

$$t = 0: \qquad u = u_o(x) \qquad a < x < b. \tag{5.3}$$

This initial boundary-value problem arises in many applications in engineering science and mechanics. Momentum, heat, and mass diffusion are common examples. In this chapter we use the finite-element method in space x to replace the PDE with a system of ODEs in time. Starting with the initial condition, we use time advancement methods such as the Crank–Nicolson method to obtain the solution at a future time.

5.2 THE WEAK FORM

Multiply Equation (5.1) by a test (or weight) function $w(x)$ and integrate over the problem domain,

$$\int_a^b w\gamma\frac{\partial u}{\partial t}dx + \int_a^b w\left[-\frac{\partial}{\partial x}\left(p\frac{\partial u}{\partial x}\right) + qu - f\right]dx = 0. \tag{5.4}$$

Next, integrate the second term by parts

$$\int_a^b w\gamma\frac{\partial u}{\partial t}dx + \left(-wp\frac{\partial u}{\partial x}\right)_{x=a}^{x=b} + \int_a^b \left[\frac{\partial w}{\partial x}p\frac{\partial u}{\partial x} + wqu\right]dx - \int_a^b wf dx = 0 \tag{5.5}$$

Definition of Primary and Secondary Variables

The secondary variable is the coefficient of the test function in the boundary term,

$$R = p\frac{\partial u}{\partial x}. \tag{5.6}$$

The primary variable is u. Introducing the definitions

$$R_a = -\left(p\frac{\partial u}{\partial x}\right)_{x=a} \tag{5.7}$$

and

$$R_b = \left(p\frac{\partial u}{\partial x}\right)_{x=b} \tag{5.8}$$

into Equation (5.5), we get

$$\int_a^b w\gamma\frac{\partial u}{\partial t}dx + \int_a^b \left[\frac{\partial w}{\partial x}p\frac{\partial u}{\partial x} + wqu\right]dx - \int_a^b wf dx - w(b)R_b - w(a)R_a = 0. \tag{5.9}$$

Classification of Boundary Conditions

To complete the derivation of the weak form, we need to apply the boundary conditions. First we classify the given boundary conditions. A boundary condition may be classified as one of three types: essential, natural, or mixed. If a boundary condition specifies a primary variable, it is called an essential boundary condition (EBC). If a boundary condition specifies a secondary variable, it is called a natural boundary condition (NBC). A boundary condition that gives a relationship between the primary and secondary variables on the boundary is called a mixed boundary condition (MBC). Thus, essential boundary condition means u is given on the boundary, and natural boundary condition means R is given on the boundary. Mixed boundary condition means that neither u nor R is given on the boundary but we have an equation that can be used to express R in terms of u on the boundary.

Now, we are in a position to make decisions regarding the boundary terms in Equation (5.9). At a natural boundary condition, the specified value of the secondary variable R is substituted into the corresponding boundary term in Equation (5.9). At a mixed boundary condition, we use the boundary condition to express the secondary variable R in terms of the primary variable u and the result is substituted into the corresponding boundary term in Equation (5.9). Finally, at an EBC we construct the test (weight) function such that it is zero at an EBC, and hence the corresponding boundary term in Equation (5.9) is zero. For example, if the boundary condition at $x = a$ is an EBC, then we choose $w(x)$ such that $w(a) = 0$, and if the boundary condition at $x = b$ is an EBC, then $w(b) = 0$. Therefore, the weak form depends on the type of boundary conditions given by the problem statement.

For the boundary conditions given by Equation (5.2), the weak form becomes,

$$\int_a^b w\gamma\frac{\partial u}{\partial t}dx + \int_a^b \left[\frac{\partial w}{\partial x}p\frac{\partial u}{\partial x} + wqu\right]dx + \beta_2 w(b)u(b,t) + \beta_1 w(a)u(a,t) = \int_a^b wf dx + w(b)\alpha_2 + w(a)\alpha_1. \tag{5.10}$$

We note that $u(a,t)$ and $u(b,t)$ are unknown.

5.3 THE WEAK FORM FOR AN ELEMENT

No boundary conditions are prescribed for the typical element; thus, the weak form is given by Equation (5.9) except that the integrals are evaluated over the element domain and the secondary variables are evaluated at the boundary nodes of the element. For an element with n local nodes denoted by $x_1^e, x_2^e, \cdots, x_n^e$, we have

$$\int_{x_1^e}^{x_n^e} w\gamma\frac{\partial u}{\partial t}dx + \int_{x_1^e}^{x_n^e} \left[\frac{\partial w}{\partial x}p\frac{\partial u}{\partial x} + wqu\right]dx - \int_{x_1^e}^{x_n^e} wf dx - w(x_n^e)R_n^e - w(x_1^e)R_1^e = 0 \tag{5.11}$$

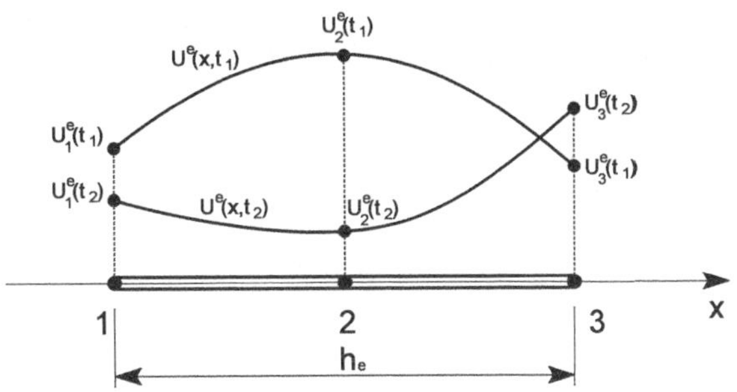

Figure 5.1 Approximation of $u(x,t)$ at two different times t_1 and t_2. For a 3-node element, the quadratic shape functions are used to interpolate the time-dependent nodal values.

$$\begin{aligned} R_1^e &= -\left(p\frac{\partial u}{\partial x}\right)_{x=x_1^e} \\ R_n^e &= \left(p\frac{\partial u}{\partial x}\right)_{x=x_n^e}. \end{aligned} \tag{5.12}$$

5.4 APPROXIMATIONS AND ELEMENT MATRICES

First we consider a 3-node element as shown in Figure 5.1. At certain instant of time t_1, the unknown function $u(x,t)$ is a function of x that can be approximated over the element by

$$u(x,t_1) \approx U^e(x,t_1) = U_1^e(t_1)\psi_1(x) + U_2^e(t_1)\psi_2(x) + U_3^e(t_1)\psi_3(x) \tag{5.13}$$

where $U_1^e(t_1)$, $U_2^e(t_1)$, and $U_3^e(t_1)$ are the nodal values of the primary variable at time t_1, and ψ_i are the quadratic element shape functions presented in Chapter 2. At another instant of time t_2, the approximation is

$$u(x,t_2) \approx U^e(x,t_2) = U_1^e(t_2)\psi_1(x) + U_2^e(t_2)\psi_2(x) + U_3^e(t_2)\psi_3(x). \tag{5.14}$$

We changed the nodal values of primary variables but used the same shape functions.

Therefore, for an element with n nodes and at any instant of time the approximation is

$$u(x,t) \approx U^e(x,t) = \sum_{j=1}^{n} U_j^e(t)\psi_j(x) \tag{5.15}$$

where n is the number of nodes per element. The shape functions are the Lagrange interpolation polynomials; they satisfy the condition $\psi_j(x_i) = \delta_{ij}$. To write $U^e(x,t)$

in matrix form, we define the arrays

$$\{\psi(x)\} = \{\psi_1(x)\ \psi_2(x) \ldots \psi_j(x) \ldots \psi_n(x)\} \tag{5.16}$$

$$\{U^e(t)\} = \begin{Bmatrix} U_1^e(t) \\ U_2^e(t) \\ \cdot \\ \cdot \\ \cdot \\ U_j^e(t) \\ \cdot \\ \cdot \\ \cdot \\ U_n^e(t) \end{Bmatrix}. \tag{5.17}$$

In matrix form, the trial solution over element # e is

$$U^e(x,t) = \{\psi(x)\}\{U^e(t)\}. \tag{5.18}$$

We note that the dependence of the trial solution on x is given by the array $\{\psi(x)\}$, whereas variation with time is given by the array $\{U^e(t)\}$ on the right-hand side, which is values of the primary variable at the element nodes.

The time derivative of the trial solution is

$$\frac{\partial U^e(x,t)}{\partial t} = \{\psi(x)\}\{\dot{U}^e(t)\} \tag{5.19}$$

where

$$\{\dot{U}^e(t)\} \equiv \frac{d}{dt}\{U^e(t)\} \tag{5.20}$$

and we also define

$$\dot{U}_j^e(t) \equiv \frac{dU_j^e(t)}{dt}. \tag{5.21}$$

The gradient of the trial solution is

$$\frac{\partial U^e(x,t)}{\partial x} = \left\{ \frac{d\psi_1(x)}{dx}\ \frac{d\psi_2(x)}{dx} \ldots \frac{d\psi_n(x)}{dx} \right\} \begin{Bmatrix} U_1^e(t) \\ U_2^e(t) \\ \cdot \\ \cdot \\ \cdot \\ U_n^e(t) \end{Bmatrix}. \tag{5.22}$$

We define the gradient matrix $[B]$ by

$$[B] = \left[\frac{d\psi_1(x)}{dx}\ \frac{d\psi_2(x)}{dx} \ldots \frac{d\psi_n(x)}{dx} \right] \tag{5.23}$$

and write the gradient of the trial solution in matrix form,

$$\frac{\partial U^e(x,t)}{\partial x} = [B]\{U^e(t)\}. \tag{5.24}$$

Galerkin

In the Galerkin method, the test (weight) function $w(x)$ is a linear combination of the basis functions $\psi_i(x)$. Thus for element # e, we have the approximation for w,

$$W^e(x) = \{\psi_1(x)\ \psi_2(x) \ldots \psi_n(x)\} \begin{Bmatrix} d_1 \\ d_2 \\ \cdot \\ \cdot \\ \cdot \\ d_n \end{Bmatrix} \tag{5.25}$$

$$W^e(x) = \{d\}^T\{\psi(x)\}^T. \tag{5.26}$$

Similar to the gradient of $U^e(x,t)$, we write the gradient of $W^e(x)$ using the matrix $[B]$,

$$\frac{dW^e(x)}{dx} = \{d\}^T[B]^T \tag{5.27}$$

Substituting Equations (5.18), (5.19), (5.24), (5.26), and (5.27) into the weak form for the element Equation (5.11), and factoring out the terms $\{d\}^T$, $\{U^e\}$, and $\{\dot{U}^e\}$ we get

$$\begin{aligned}\{d\}^T\left[\int_{x_1^e}^{x_n^e} \gamma\{\psi\}^T\{\psi\}dx\right]\{\dot{U}^e\} + \{d\}^T\left[\int_{x_1^e}^{x_n^e} \left([B]^T p[B] + q\{\psi\}^T\{\psi\}\right)dx\right]\{U^e\} \\ + \{d\}^T\left[-\int_{x_1^e}^{x_n^e} f\{\psi\}^T dx - \{\psi(x_n^e)\}^T R_n^e - \{\psi(x_1^e)\}^T R_1^e\right] = 0.\end{aligned} \tag{5.28}$$

We define the element matrices. The mass matrix corresponding to the time derivative term is

$$[M^e] = \int_{x_1^e}^{x_n^e} \gamma\{\psi\}^T\{\psi\}dx \tag{5.29}$$

the coefficient matrix is

$$\begin{aligned}[A^e] &= \int_{x_1^e}^{x_n^e} \left([B]^T p[B] + q\{\psi\}^T\{\psi\}\right)dx \\ &= [K^e] + [M^e]_q\end{aligned} \tag{5.30}$$

where the stiffness matrix is

$$[K^e] = \int_{x_1^e}^{x_n^e} [B]^T p[B]dx \tag{5.31}$$

and the mass matrix for the q term is

$$[M^e]_q = \int_{x_1^e}^{x_n^e} q\{\psi\}^T\{\psi\}dx. \tag{5.32}$$

The forcing vector is

$$\{F^e\} = \int_{x_1^e}^{x_n^e} f\{\psi\}^T dx \tag{5.33}$$

and the vector of secondary variables is

$$\{R^e\} = \{\psi(x_n^e)\}^T R_n^e + \{\psi(x_1^e)\}^T R_1^e. \tag{5.34}$$

In terms of these matrices, Equation (5.28) becomes

$$\{d\}^T\left\{[M^e]\left\{\dot{U}^e\right\} + [A^e]\{U^e\} - \{F^e\} - \{R^e\}\right\} = 0. \tag{5.35}$$

Since $\{d\}$ is an arbitrary array, the multiplying array must be zero. We obtain the element equations

$$[M^e]\left\{\dot{U}^e\right\} + [A^e]\{U^e\} = \{F^e\} + \{R^e\}. \tag{5.36}$$

Using index notation, we write the entries of the element matrices,

$$m_{ij}^e = \int_{x_1^e}^{x_n^e} \gamma\psi_i\psi_j dx \tag{5.37}$$

$$a_{ij}^e = \int_{x_1^e}^{x_n^e}\left(p\frac{d\psi_i}{dx}\frac{d\psi_j}{dx} + q\psi_i\psi_j\right)dx \tag{5.38}$$

$$F_i^e = \int_{x_1^e}^{x_n^e} f\psi_i dx. \tag{5.39}$$

We recall that the shape functions satisfy the property $\psi_j(x_i) = \delta_{ij}$, hence

$$R_i^e = \delta_{in}R_n^e + \delta_{i1}R_1^e. \tag{5.40}$$

We note that F_i^e and R_i^e are, in general, functions of time.

The shape functions for the linear and quadratic elements are given in Chapter 2. The corresponding element matrices for element-wise uniform γ, p, q, and f are also given in that chapter. The mass matrix is repeated here, for the linear element

$$[M^e] = \frac{\gamma_e h_e}{6}\begin{bmatrix} 2 & 1 \\ 1 & 2 \end{bmatrix} \tag{5.41}$$

and for the quadratic element

$$[M^e] = \frac{\gamma_e h_e}{30}\begin{bmatrix} 4 & 2 & -1 \\ 2 & 16 & 2 \\ -1 & 2 & 4 \end{bmatrix}. \tag{5.42}$$

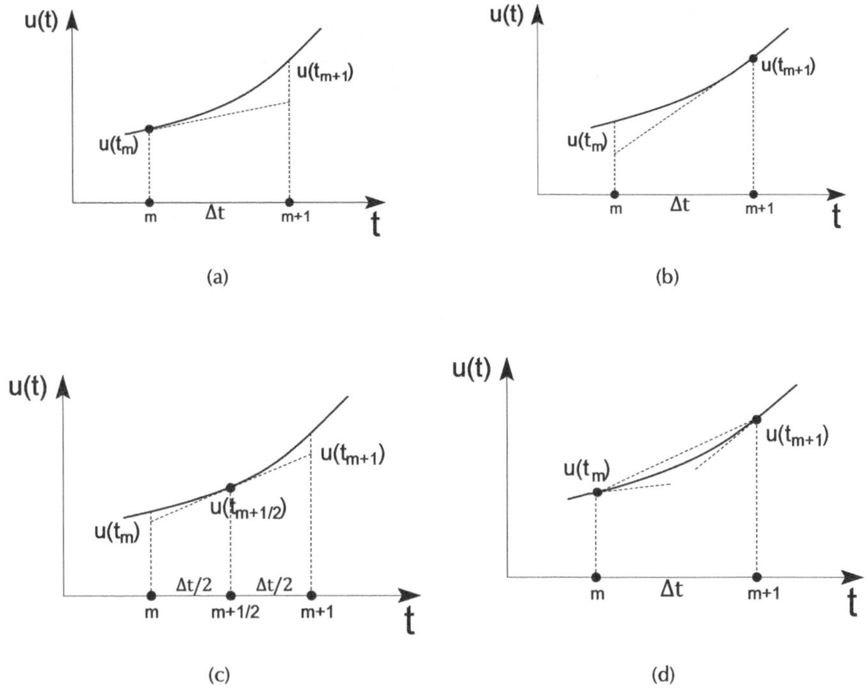

Figure 5.2 Geometric interpretation of time advancement methods. (a) Euler explicit, $u(t)$ is approximated by the tangent line at time step m, (b) Euler implicit, $u(t)$ is approximated by the tangent line at time step $m+1$, (c) Crank–Nicolson, $u(t)$ is approximated by the tangent line at mid-point $m+1/2$, and (d) the $\alpha-$method, the chord slope is approximated by a weighted average of the slopes at m and $m+1$.

5.5 TEMPORAL APPROXIMATION: TIME MARCHING

The element equations and hence the assembled global system is a system of coupled ordinary differential equations in time. The given initial condition at $t = 0$ provides the initial values of the primary variables at all global nodes. We need to solve the assembled system to determine the primary variables at future times. Let's consider a first-order differential equation for one variable $u(t)$.

$$\frac{du}{dt} + \lambda u = f(t) \tag{5.43}$$

with initial condition

$$t = 0, \qquad u = u_o. \tag{5.44}$$

We want to develop numerical methods to find $u(t)$, for $t > 0$. Geometric interpretations of the different time integration schemes are presented in Figures 5.2a–d.

Euler Explicit Method

A point on the time axis is denoted by t_m. The numerical solution at this instant of time is denoted by u^m. We want to find u^{m+1}, which is the numerical solution at time $t_{m+1} = t_m + \Delta t$, where Δt is a specified time step. Expand $u(t)$ in a Taylor series around t_m:

$$u(t_m + \Delta t) = u(t_m) + \left(\frac{du}{dt}\right)^m \Delta t + \ldots. \tag{5.45}$$

Denote $u(t_m) \equiv u^m$ and $u(t_m + \Delta t) \equiv u^{m+1}$, we have

$$u^{m+1} = u^m + \left(\frac{du}{dt}\right)^m \Delta t + \ldots. \tag{5.46}$$

The differential equation Equation (5.43) gives du/dt at any instant of time; we write it at step m,

$$\left(\frac{du}{dt}\right)^m = -\lambda u^m + f^m \tag{5.47}$$

where $f^m \equiv f(t_m)$. Substituting into Equation (5.46) and truncating the series after the second term, we get

$$u^{m+1} = u^m + (-\lambda u^m + f^m)\Delta t \tag{5.48}$$

$$u^{m+1} = (1 - \lambda\Delta t)u^m + f^m \Delta t. \tag{5.49}$$

Knowing u at time step m, we use Equation (5.49) to determine u^{m+1}. The initial condition gives u^o.

$$\begin{aligned} u^1 &= (1 - \lambda\Delta t)\, u^o + f^o \Delta t \\ u^2 &= (1 - \lambda\Delta t)\, u^1 + f^1 \Delta t \\ u^3 &= (1 - \lambda\Delta t)\, u^2 + f^2 \Delta t \end{aligned} \tag{5.50}$$

and so on. This method is first-order accurate (error is proportional to Δt), and is stable only if $0 < \lambda\Delta t < 2$. Unless Δt is smaller than a certain value, the error will accumulate with each time step and the calculations will diverge. This method is called explicit because the time derivative is evaluated using known quantities at step m.

Euler Implicit Method

We expand $u(t)$ into a Taylor series around the point t_{m+1} and use it to approximate $u(t_m)$,

$$u(t_m) = u(t_{m+1}) - \left(\frac{du}{dt}\right)^{m+1} \Delta t + \ldots. \tag{5.51}$$

or

$$u^m = u^{m+1} - \left(\frac{du}{dt}\right)^{m+1} \Delta t + \ldots. \tag{5.52}$$

The differential equation gives du/dt at any instant of time, thus

$$\left(\frac{du}{dt}\right)^{m+1} = -\lambda u^{m+1} + f^{m+1}. \tag{5.53}$$

Substituting into Equation (5.52) and keeping only two terms of the series, we get

$$\begin{aligned} u^m &= u^{m+1} - \left(-\lambda u^{m+1} + f^{m+1}\right)\Delta t \\ &= (1+\lambda\Delta t)u^{m+1} - f^{m+1}\Delta t. \end{aligned} \tag{5.54}$$

Rearranging terms,

$$(1+\lambda\Delta t)u^{m+1} = u^m + f^{m+1}\Delta t \tag{5.55}$$

and solving for u^{m+1}, we get

$$u^{m+1} = (1+\lambda\Delta t)^{-1}(u^m + f^{m+1}\Delta t). \tag{5.56}$$

Knowing u at time step m, we use Equation (5.56) to determine u^{m+1}

$$\begin{aligned} u^1 &= (1+\lambda\Delta t)^{-1}(u^o + f^1\Delta t) \\ u^2 &= (1+\lambda\Delta t)^{-1}(u^1 + f^2\Delta t) \\ u^3 &= (1+\lambda\Delta t)^{-1}(u^2 + f^3\Delta t). \end{aligned} \tag{5.57}$$

This method is also first-order accurate (truncation error is $O(\Delta t)$), and is stable for any Δt if $\lambda > 0$. The method converges but the results are not accurate unless Δt is small. This method is called implicit because the time derivative is expressed in terms of yet unknown quantities at step $m+1$.

Crank–Nicolson Method

The Euler explicit and implicit methods are first-order accurate in time. Second-order or higher methods are recommended for engineering computations. To achieve second-order accuracy in time, we expand $u(t)$ in a Taylor series around $t = t_m + \Delta t/2$, or time step $m+1/2$,

$$u^{m+1} = u^{m+\frac{1}{2}} + \left(\frac{du}{dt}\right)^{m+\frac{1}{2}}\frac{\Delta t}{2} + \frac{1}{2}\left(\frac{d^2u}{dt^2}\right)^{m+\frac{1}{2}}\left(\frac{\Delta t}{2}\right)^2 + \frac{1}{6}\left(\frac{d^3u}{dt^3}\right)^{m+\frac{1}{2}}\left(\frac{\Delta t}{2}\right)^3 + \ldots \tag{5.58}$$

$$u^{m} = u^{m+\frac{1}{2}} - \left(\frac{du}{dt}\right)^{m+\frac{1}{2}}\frac{\Delta t}{2} + \frac{1}{2}\left(\frac{d^2u}{dt^2}\right)^{m+\frac{1}{2}}\left(\frac{\Delta t}{2}\right)^2 - \frac{1}{6}\left(\frac{d^3u}{dt^3}\right)^{m+\frac{1}{2}}\left(\frac{\Delta t}{2}\right)^3 + \ldots \tag{5.59}$$

subtracting and dividing by Δt, we get

$$\frac{u^{m+1} - u^m}{\Delta t} = \left(\frac{du}{dt}\right)^{m+\frac{1}{2}} + O(\Delta t)^2. \tag{5.60}$$

The differential equation at $m+1/2$ gives

$$\begin{aligned}\left(\frac{du}{dt}\right)^{m+\frac{1}{2}} &= -\lambda u^{m+\frac{1}{2}} + f^{m+\frac{1}{2}} \\ &= -\frac{\lambda}{2}(u^{m+1}+u^m) + \frac{1}{2}(f^{m+1}+f^m). \end{aligned} \tag{5.61}$$

Note that the error in using $u^{m+1/2} = (u^m + u^{m+1})/2$ is $O(\Delta t)^2$. From Equations (5.60) and (5.61) we obtain

$$\frac{u^{m+1}-u^m}{\Delta t} = -\frac{\lambda}{2}(u^{m+1}+u^m) + \frac{1}{2}(f^{m+1}+f^m). \tag{5.62}$$

Rearranging

$$(1+\frac{1}{2}\lambda\Delta t)u^{m+1} = (1-\frac{1}{2}\lambda\Delta t)u^m + \frac{\Delta t}{2}(f^{m+1}+f^m) \tag{5.63}$$

and solving for u^{m+1}, we obtain

$$u^{m+1} = (1+\frac{1}{2}\lambda\Delta t)^{-1}\left[(1-\frac{1}{2}\lambda\Delta t)u^m + \frac{\Delta t}{2}(f^{m+1}+f^m)\right]. \tag{5.64}$$

Knowing u at time step m, we use Equation (5.64) to determine u^{m+1}. This method is second-order accurate (truncation error is $O(\Delta t)^2$), and is stable for any Δt if $\lambda > 0$. The Crank–Nicolson method is the preferred method for time integration of parabolic equations.

The α-Method

The Euler explicit method is based on the approximation

$$\frac{u^{m+1}-u^m}{\Delta t} = \left(\frac{du}{dt}\right)^m \tag{5.65}$$

and the Euler implicit method is based on the approximation

$$\frac{u^{m+1}-u^m}{\Delta t} = \left(\frac{du}{dt}\right)^{m+1} \tag{5.66}$$

whereas the the Crank–Nicolson method is based on the more accurate approximation

$$\frac{u^{m+1}-u^m}{\Delta t} = \frac{1}{2}\left(\frac{du}{dt}\right)^{m+1} + \frac{1}{2}\left(\frac{du}{dt}\right)^m. \tag{5.67}$$

In the α-method we take the slope of the chord $(u^{m+1}-u^m)/\Delta t$ to be a weighted average of the slopes at the end points of the time step,

$$\frac{u^{m+1}-u^m}{\Delta t} = \alpha\left(\frac{du}{dt}\right)^{m+1} + (1-\alpha)\left(\frac{du}{dt}\right)^m \tag{5.68}$$

where $0 \leq \alpha \leq 1$. Using the differential equation at time levels m and $m+1$,

$$\left(\frac{du}{dt}\right)^{m+1} = -\lambda u^{m+1} + f^{m+1} \tag{5.69}$$

$$\left(\frac{du}{dt}\right)^{m} = -\lambda u^{m} + f^{m} \tag{5.70}$$

we get

$$u^{m+1} = (1+\alpha\lambda\Delta t)^{-1}\left[\{1-(1-\alpha)\lambda\Delta t\}u^m + \Delta t\{\alpha f^{m+1} + (1-\alpha)f^m\}\right]. \tag{5.71}$$

The commonly used values of α are

$$\begin{aligned} \alpha &= 0 && \text{Euler explicit} \\ \alpha &= 1 && \text{Euler implicit} \\ \alpha &= \frac{1}{2} && \text{Crank–Nicolson.} \end{aligned} \tag{5.72}$$

The method is unconditionally stable for $0.5 \leq \alpha \leq 1$.

The Second-Order Backward-Difference Formula (BDF2) Method

Other methods such as the two-point backward difference formula (BDF2) is also recommended. In the BDF2 method, the time derivative at $m+1$ is approximated by

$$\left(\frac{du}{dt}\right)^{m+1} = \frac{3u^{m+1} - 4u^m + u^{m-1}}{2\Delta t}. \tag{5.73}$$

Using the differential equation at step $m+1$, we get

$$-\lambda u^{m+1} + f^{m+1} = \frac{3u^{m+1} - 4u^m + u^{m-1}}{2\Delta t}. \tag{5.74}$$

Solving for u^{m+1}, we get

$$u^{m+1} = (3+2\lambda\Delta t)^{-1}\left(4u^m - u^{m-1} + 2\Delta t f^{m+1}\right). \tag{5.75}$$

This method is second-order and unconditionally stable, but it requires storage of the solution at two previous steps. It is not self-starting because it cannot be used for the first time step. The α method or any other method can be used for the first step, and the BDF2 method can be used for the second step and thereafter.

Now, we apply the α-method to the system of equations (5.36). For a scalar equation, we write Equation (5.68) as

$$\frac{1}{\Delta t}\left(u^{m+1} - u^m\right) = \alpha\,(\dot{u})^{m+1} + (1-\alpha)\,(\dot{u})^m. \tag{5.76}$$

We apply this formula to every variable in the system,

$$\frac{1}{\Delta t}\left[\{U^e\}^{m+1} - \{U^e\}^m\right] = \alpha\left\{\dot{U}^e\right\}^{m+1} + (1-\alpha)\left\{\dot{U}^e\right\}^m. \tag{5.77}$$

We first multiply Equation (5.77) by $[M^e]$,

$$\frac{1}{\Delta t}\left[[M^e]\{U^e\}^{m+1} - [M^e]\{U^e\}^m\right] = \alpha[M^e]\{\dot{U}^e\}^{m+1} + (1-\alpha)[M^e]\{\dot{U}^e\}^m. \tag{5.78}$$

We use the element differential equations (5.36) to express $[M^e]\{\dot{U}^e\}$ at time levels m and $m+1$ as

$$[M^e]\{\dot{U}^e\}^m = -[A^e]\{U^e\}^m + \{F^e\}^m + \{R^e\}^m \tag{5.79}$$

$$[M^e]\{\dot{U}^e\}^{m+1} = -[A^e]\{U^e\}^{m+1} + \{F^e\}^{m+1} + \{R^e\}^{m+1} \tag{5.80}$$

then substitute Equations (5.79) and (5.80) into Equation (5.78),

$$\begin{aligned}&\frac{1}{\Delta t}\left[[M^e]\{U^e\}^{m+1} - [M^e]\{U^e\}^m\right] = \\ &\alpha\left[-[A^e]\{U^e\}^{m+1} + \{F^e\}^{m+1} + \{R^e\}^{m+1}\right] + \\ &(1-\alpha)[-[A^e]\{U^e\}^m + \{F^e\}^m + \{R^e\}^m].\end{aligned} \tag{5.81}$$

We assumed that the mass $[M^e]$ and coefficient $[A^e]$ matrices to be independent of time. Rearranging terms, we get

$$\begin{aligned}&\left[\frac{1}{\Delta t}[M^e] + \alpha[A^e]\right]\{U^e\}^{m+1} = \left[\frac{1}{\Delta t}[M^e] - (1-\alpha)[A^e]\right]\{U^e\}^m \\ &+\left[\alpha\{F^e\}^{m+1} + (1-\alpha)\{F^e\}^m\right] + \left[\alpha\{R^e\}^{m+1} + (1-\alpha)\{R^e\}^m\right].\end{aligned} \tag{5.82}$$

We define some auxiliary matrices

$$\begin{aligned}[AL^e] &= \frac{1}{\Delta t}[M^e] + \alpha[A^e] \\ [AR^e] &= \frac{1}{\Delta t}[M^e] - (1-\alpha)[A^e] \\ \{\tilde{F}^e\} &= \alpha\{F^e\}^{m+1} + (1-\alpha)\{F^e\}^m \\ \{\tilde{R}^e\} &= \alpha\{R^e\}^{m+1} + (1-\alpha)\{R^e\}^m\end{aligned} \tag{5.83}$$

and obtain algebraic element equations

$$[AL^e]\{U^e\}^{m+1} = [AR^e]\{U^e\}^m + \{\tilde{F}^e\} + \{\tilde{R}^e\}. \tag{5.84}$$

The element equations are assembled into a global system, and continuity of primary variables at common nodes between all elements, $U_n^e = U_1^{e+1}$ is enforced,

$$[AL]\{U\}^{m+1} = [AR]\{U\}^m + \{\tilde{F}\} + \{\tilde{R}\}. \tag{5.85}$$

Next, balance of secondary variables at interior nodes is enforced, $\tilde{R}_j = 0$, where j is a global interior node. For an EBC, the primary variable U_j, where j is a boundary node, is given. The secondary variable at that node is unknown and is to be determined from the corresponding global equation before application of the boundary conditions. At an NBC, the secondary variable $\tilde{R}_j$, where j is a boundary node, is given but the primary variable is unknown. At an MBC, the secondary variable $\tilde{R}_j$ is expressed in terms of the yet unknown primary variable and the result is used to eliminate the secondary variable from the corresponding global equation. After application of the boundary conditions, the modified global system becomes

$$\left[\overline{AL}\right]\{U\}^{m+1} = \left[\overline{AR}\right]\{U\}^m + \left\{\tilde{\overline{F}}\right\}. \tag{5.86}$$

Knowing $\{U\}^m$, we solve this system sequentially for $\{U\}^{m+1}$, $m = 0, 1, 2, \cdots$. The initial condition provides $\{U\}^0$.

5.6 TRANSIENT HEAT CONDUCTION

Example: Temperature Distribution in a Cooling Fin

A copper fin (diameter = 4 mm, length $L = 20$ mm, density $\rho = 8900$ Kg/m^3, specific heat $C = 375$ J/(Kg °C), thermal conductivity $k = 400$ W/(m °C)) is initially at temperature of 25°C. The fin surface is subjected to convection cooling with ambient temperature $T_\infty = 25$°C, and coefficient of convection $\tilde{h} = 150$ W/(m^2 °C). Suddenly, the temperature at the end $x = 0$ is increased to 85°C and maintained at that temperature thereafter. The end $x = L$ is assumed to be insulated. Use two linear elements, and the α-method with $\Delta t = 0.1$ s, and $\alpha = 2/3$, determine the temperature at all nodes and the heat flow at $x = 0$ at times $t = 0.1, 0.2$, and 0.3 s.

Solution

The energy equation (first law of thermodynamics) is

$$\rho CA\frac{\partial T}{\partial t} - \frac{\partial}{\partial x}[kA\frac{\partial T}{\partial x}] + \tilde{h}PT - (f_o + \tilde{h}PT_\infty) = 0 \qquad 0 < x < L \tag{5.87}$$

Comparison with the model equation (5.1) shows that

$$\begin{aligned}
\gamma &= \rho CA = (8900)(375) \times \frac{\pi}{4}\left(\frac{4}{1000}\right)^2 = 41.9404 \quad J/(m^\circ C) \\
p &= kA = (400) \times \frac{\pi}{4}\left(\frac{4}{1000}\right)^2 = 5.03 \times 10^{-3} \quad W\,m/^\circ C \\
q &= \tilde{h}P = (150) \times 2\pi\left(\frac{2}{1000}\right) = 1.885 \quad W/(m^\circ C) \\
f &= f_oA + \tilde{h}PT_\infty = \tilde{h}PT_\infty = 47.1239 \quad W/m \qquad (f_o = 0).
\end{aligned} \tag{5.88}$$

Element length $h_e = 0.01$ m. In this example all properties are independent of x, thus the two elements have the same matrices. The mass matrix for a linear element is

$$\begin{aligned}[M^e] &= \frac{\gamma_e h_e}{6}\begin{bmatrix} 2 & 1 \\ 1 & 2 \end{bmatrix} \\ &= \begin{bmatrix} 0.1398 & 0.0699 \\ 0.0699 & 0.1398 \end{bmatrix}.\end{aligned} \tag{5.89}$$

Element coefficient matrix is

$$\begin{aligned}[A^e] &= \frac{p_e}{h_e}\begin{bmatrix} 1 & -1 \\ -1 & 1 \end{bmatrix} + \frac{q_e h_e}{6}\begin{bmatrix} 2 & 1 \\ 1 & 2 \end{bmatrix} \\ &= \begin{bmatrix} 0.5089 & -0.4995 \\ -0.4995 & 0.5089 \end{bmatrix}.\end{aligned} \tag{5.90}$$

The forcing vector is

$$\begin{aligned}[F^e] &= \frac{f_e h_e}{2}\begin{bmatrix} 1 \\ 1 \end{bmatrix} \\ &= \begin{bmatrix} 0.2356 \\ 0.2356 \end{bmatrix}.\end{aligned} \tag{5.91}$$

We note that the forcing vector in this example is independent of time.

Use $\alpha = 2/3$ and $\Delta t = 0.1$ s, and obtain the auxiliary matrices as defined in Equation (5.83),

$$\begin{aligned}[AL^e] &= \frac{1}{\Delta t}[M^e] + \alpha [A^e] \\ &= \begin{bmatrix} 1.7373 & 0.3660 \\ 0.3660 & 1.7373 \end{bmatrix} \\ [AR^e] &= \frac{1}{\Delta t}[M^e] - (1-\alpha)[A^e] \\ &= \begin{bmatrix} 1.2284 & 0.8655 \\ 0.8655 & 1.2284 \end{bmatrix} \\ \{\tilde{F}^e\} &= \alpha \{F^e\}^{m+1} + (1-\alpha)\{F^e\}^m \\ &= \begin{Bmatrix} 0.2356 \\ 0.2356 \end{Bmatrix}.\end{aligned} \tag{5.92}$$

The element equations for element # 1 are

$$\begin{aligned}\begin{bmatrix} 1.7373 & 0.3660 \\ 0.3660 & 1.7373 \end{bmatrix}\begin{Bmatrix} T_1^1 \\ T_2^1 \end{Bmatrix}^{m+1} &= \begin{bmatrix} 1.2284 & 0.8655 \\ 0.8655 & 1.2284 \end{bmatrix}\begin{Bmatrix} T_1^1 \\ T_2^1 \end{Bmatrix}^{m} \\ &\quad + \begin{Bmatrix} 0.2356 \\ 0.2356 \end{Bmatrix} + \begin{Bmatrix} \tilde{R}_1^1 \\ \tilde{R}_2^1 \end{Bmatrix}\end{aligned} \tag{5.93}$$

and for element # 2 we have

$$\begin{bmatrix} 1.7373 & 0.3660 \\ 0.3660 & 1.7373 \end{bmatrix} \begin{Bmatrix} T_1^2 \\ T_2^2 \end{Bmatrix}^{m+1} = \begin{bmatrix} 1.2284 & 0.8655 \\ 0.8655 & 1.2284 \end{bmatrix} \begin{Bmatrix} T_1^2 \\ T_2^2 \end{Bmatrix}^{m} + \begin{Bmatrix} 0.2356 \\ 0.2356 \end{Bmatrix} + \begin{Bmatrix} \tilde{R}_1^2 \\ \tilde{R}_2^2 \end{Bmatrix}. \tag{5.94}$$

We assemble the element equations into a global system, and enforce continuity of temperature at common nodes between the two elements, $T_2^1 = T_1^2 = T_2$. The assembled system before application of the boundary conditions is

$$\begin{bmatrix} 1.7373 & 0.3660 & 0 \\ 0.3660 & 3.4746 & 0.3660 \\ 0 & 0.3660 & 1.7373 \end{bmatrix} \begin{Bmatrix} T_1 \\ T_2 \\ T_3 \end{Bmatrix}^{m+1} =$$

$$\begin{bmatrix} 1.2284 & 0.8655 & 0 \\ 0.8655 & 2.4568 & 0.8655 \\ 0 & 0.8655 & 1.2284 \end{bmatrix} \begin{Bmatrix} T_1 \\ T_2 \\ T_3 \end{Bmatrix}^{m} + \begin{Bmatrix} 0.2356 \\ 0.4712 \\ 0.2356 \end{Bmatrix} + \begin{Bmatrix} \tilde{R}_1 \\ \tilde{R}_2 \\ \tilde{R}_3 \end{Bmatrix}. \tag{5.95}$$

The boundary conditions are $T_1 = 85°\text{C}$ ($\tilde{R}_1$ is unknown.) At the interior node #2, $\tilde{R}_2 = 0$, and at the insulated end $\tilde{R}_3 = 0$. Before we apply the boundary conditions, we save the first equation

$$\tilde{R}_1 = 1.7373T_1^{m+1} + 0.3660T_2^{m+1} - (1.2284T_1^m + 0.8655T_2^m) - 0.2356 \tag{5.96}$$

that will be used to determine $\tilde{R}_1$, which is the rate of heat flow at node 1. Now, we replace the first equation of the system Equation (5.95) by the boundary condition $T_1^{m+1} = 85°\text{C}$. The modified system for the unknown temperatures becomes

$$\begin{bmatrix} 1 & 0 & 0 \\ 0.3660 & 3.4746 & 0.3660 \\ 0 & 0.3660 & 1.7373 \end{bmatrix} \begin{Bmatrix} T_1 \\ T_2 \\ T_3 \end{Bmatrix}^{m+1} =$$

$$\begin{bmatrix} 0 & 0 & 0 \\ 0.8655 & 2.4568 & 0.8655 \\ 0 & 0.8655 & 1.2284 \end{bmatrix} \begin{Bmatrix} T_1 \\ T_2 \\ T_3 \end{Bmatrix}^{m} + \begin{Bmatrix} 85 \\ 0.4712 \\ 0.2356 \end{Bmatrix}. \tag{5.97}$$

We write the modified system Equation (5.97) as

$$\left[\bar{AL}\right]\{T\}^{m+1} = \left[\bar{AR}\right]\{T\}^m + \{\bar{F}\}. \tag{5.98}$$

The inverse of $\left[\bar{AL}\right]$ is

$$\left[\bar{AL}\right]^{-1} = \begin{bmatrix} 1 & 0 & 0 \\ -0.1077 & 0.2943 & -0.0620 \\ 0.0227 & -0.0620 & 0.5887 \end{bmatrix}. \tag{5.99}$$

Multiply Equation (5.98) by the inverse of $\lfloor AL \rceil$, we get

$$\{T\}^{m+1} = \lfloor AL \rceil^{-1} \left[\lfloor AR \rceil \{T\}^m + \{\bar{F}\} \right]. \tag{5.100}$$

Knowing the initial temperature $\{T\}^0 = \{85\ 25\ 25\}^T$, we use Equation (5.100) to determine the temperature at future times, and then use Equation (5.96) to determine the rate of heat flow at node # 1.

$m = 0$ and $t = 0.1$ s, we get $\{T\}^1 = \{85\ 33.82\ 23.14\}^T$ °C, and $\tilde{R}_1 = 33.76$ W

$m = 1$ and $t = 0.2$ s, we get $\{T\}^2 = \{85\ 39.39\ 25.05\}^T$ °C, and $\tilde{R}_1 = 28.17$ W

$m = 2$ and $t = 0.3$ s, we get $\{T\}^3 = \{85\ 43.47\ 28.32\}^T$ °C, and $\tilde{R}_1 = 24.84$ W.

We note that the temperature at node # 3 drops below 25°C at $t = 0.1$ s; this is physically not possible, and it is due to the use of a coarse grid of two linear elements. A MATLAB code is developed for this problem that allows us to solve the problem over a much finer grid. We solve this example using 2, 4, and 8 linear elements. The time step for all meshes is $\Delta t = 0.02$ s and $\alpha = 2/3$. The temperature distributions at times $t = 0.3$ s and 6 s are shown in Figures 5.3a and 5.3b, respectively. Two linear elements cannot correctly represent the zero-gradient of temperature at the insulated fin tip. The temperature at the insulated end ($x = L$) is shown in Figure 5.3c; for 4 and 8 elements the temperature increases smoothly from the initial temperature of 25°C to about 80°C at steady state. However, with two linear elements the temperature drops below 25°C. The rate of heat flow into the fin (at $x = 0$) is shown in Figure 5.3d. Because of the sudden increase in temperature from the initial value of 25 to 85°C, we notice a singularity at $t = 0^+$. The singularity becomes more pronounced as the mesh is refined. If the increase in temperature at $x = 0$ has a finite rate, the rate of heat flow will also increase gradually.

Example: Transient Couette Flow

A layer of viscous fluid is initially at rest ($u = 0$) between two long parallel plates separated by a distance $H = 1$; see Figure 5.4a. The upper plate is given a velocity that varies with time as shown in Figure 5.4b. The fluid is assumed to move parallel to the plates with velocity $u(y,t)$ that is governed by the $x-$momentum equation in nondimensional terms

$$\frac{\partial u}{\partial t} = \frac{\partial^2 u}{\partial y^2} \tag{5.101}$$

The finite-element method in space and Crank–Nicolson method for time integration are used to determine the transient velocity. The thickness of the fluid layer is modeled by one quadratic element and the time step is $\Delta t = 0.2$.

At time $t = 0.8$, determine: (a) the velocity at $y = 0.5$, and (b) the shear stress at $y = 0$.

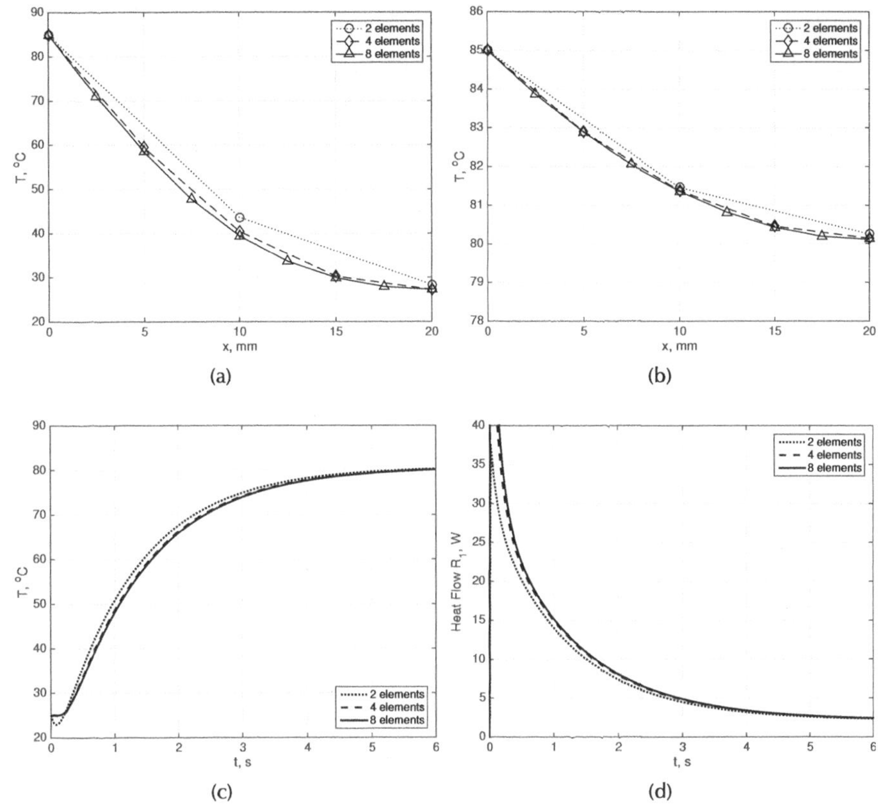

Figure 5.3 Effects of the number of linear elements on transient heat transfer in a cooling fin. (a) Temperature distribution at time $t = 0.3$ s, (b) temperature distribution at time $t = 6$ s, (c) temperature at the insulated end $x = L$, and (d) rate of heat flow into the fin at $x = 0$.

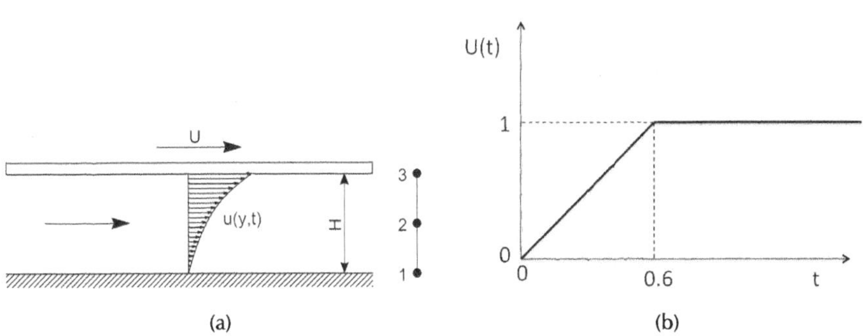

Figure 5.4 (a) Transient flow between parallel plates, and (b) prescribed velocity of the upper plate.

Solution

Comparison of Equation (5.101) to the model Equation (5.1) shows that $\gamma = 1, p = 1, q = 0$, and $f = 0$. The domain is modeled by one element, hence $h_e = 1$. The mass matrix is

$$[M^e] = \frac{\gamma_e h_e}{30}\begin{bmatrix} 4 & 2 & -1 \\ 2 & 16 & 2 \\ -1 & 2 & 4 \end{bmatrix} \tag{5.102}$$

where $\gamma_e = 1$, and $h_e = 1$. With $q_e = 0$, the coefficient matrix becomes

$$[A^e] = \frac{p_e}{3h_e}\begin{bmatrix} 7 & -8 & 1 \\ -8 & 16 & -8 \\ 1 & -8 & 7 \end{bmatrix} \tag{5.103}$$

where $p_e = 1$. For Crank–Nicolson method $\alpha = 1/2$, and using $\Delta t = 0.2$, we obtain the auxiliary matrices

$$[AL^e] = \frac{1}{\Delta t}[M^e] + \alpha [A^e],\ [AR^e] = \frac{1}{\Delta t}[M^e] - (1-\alpha)[A^e] \tag{5.104}$$

$$[AL] = \frac{1}{6}\begin{bmatrix} 11 & -6 & 0 \\ -6 & 32 & -6 \\ 0 & -6 & 11 \end{bmatrix},\ [AR] = \frac{1}{6}\begin{bmatrix} -3 & 10 & -2 \\ 10 & 0 & 10 \\ -2 & 10 & -3 \end{bmatrix}. \tag{5.105}$$

The system of equations for the nodal velocities before applications of the boundary conditions is

$$\frac{1}{6}\begin{bmatrix} 11 & -6 & 0 \\ -6 & 32 & -6 \\ 0 & -6 & 11 \end{bmatrix}\begin{Bmatrix} U_1 \\ U_2 \\ U_3 \end{Bmatrix}^{m+1} = \frac{1}{6}\begin{bmatrix} -3 & 10 & -2 \\ 10 & 0 & 10 \\ -2 & 10 & -3 \end{bmatrix}\begin{Bmatrix} U_1 \\ U_2 \\ U_3 \end{Bmatrix}^{m} + \begin{Bmatrix} \tilde{R}_1 \\ 0 \\ \tilde{R}_3 \end{Bmatrix} \tag{5.106}$$

(a) The boundary conditions give $U_1 = 0$ for all m, whereas U_3 is known but varies with time or m. We use the second equation of the system,

$$\frac{1}{6}\left(-6U_1^{m+1} + 32U_2^{m+1} - 6U_3^{m+1}\right) = \frac{1}{6}\left(10U_1^m + 10U_3^m\right) \tag{5.107}$$

hence

$$U_2^{m+1} = \frac{1}{32}\left(6U_3^{m+1} + 10U_3^m\right). \tag{5.108}$$

We use Figure 5.4b for U_3^m and obtain

$$\begin{aligned} m=0: t=0.2, U_2^1 &= \frac{1}{32}[6(1/3)+10(0)] = 1/16 \\ m=1: t=0.4, U_2^2 &= \frac{1}{32}[6(2/3)+10(1/3)] = 11/48 \\ m=2: t=0.6, U_2^3 &= \frac{1}{32}[6(1)+10(2/3)] = 19/48 \\ m=3: t=0.8, U_2^4 &= \frac{1}{32}[6(1)+10(1)] = 1/2. \end{aligned} \tag{5.109}$$

(b) The shear stress at the lower wall is $(\partial u/\partial y)_{y=0}$, which is given by $-\tilde{R}_1$. We use the first equation of the system Equation (5.106) and obtain

$$\begin{aligned} \tilde{R}_1 &= -U_2^{m+1} - (10/6)U_2^m + (2/6)U_3^m \\ &= -0.82639. \end{aligned} \tag{5.110}$$

Example: Pasteurization and Hard-Boiling of an Egg: A Spherical Model

Abbasnezhad et al. [1] presented an elaborate computational fluid dynamics model of thermal pasteurization of intact eggs. The purpose of pasteurization is to reduce the risk of microbial, such as *Salmonella enteritidis* and *Escherichia coli*, contamination. One method of pasteurization is to place the egg in a hot water bath at a certain temperature for a certain period of time. Microorganism destruction is assessed by a criterion that depends on the transient temperature within the egg and the time of heating. Abbasnezhad et al. developed a three-dimensional model and investigated the effects of air cavities in the egg on the pasteurization time.

In this example, we consider heating and hard-boiling of a simplified model of an egg. We want to determine the transient temperature in a 3-layer spherical domain with thermophysical properties as those of egg materials. The geometric model is shown in Figure 5.5a. Assuming that the temperature depends on radial position and time only, we write the equation of heat conduction in spherical coordinates

$$\rho C\frac{\partial T}{\partial t} - \frac{1}{r^2}\frac{\partial}{\partial r}\left[kr^2\frac{\partial T}{\partial r}\right] + qT - f = 0 \tag{5.111}$$

where ρ, C, and k are mass density, heat capacity, and thermal conductivity, respectively. The finite-element model for spherical and cylindrical coordinates is developed in Section 5.7.2. The element equations are

$$[M^e]\{\dot{U}^e\} + [A^e]\{U^e\} = \{F^e\} + \{R^e\}. \tag{5.112}$$

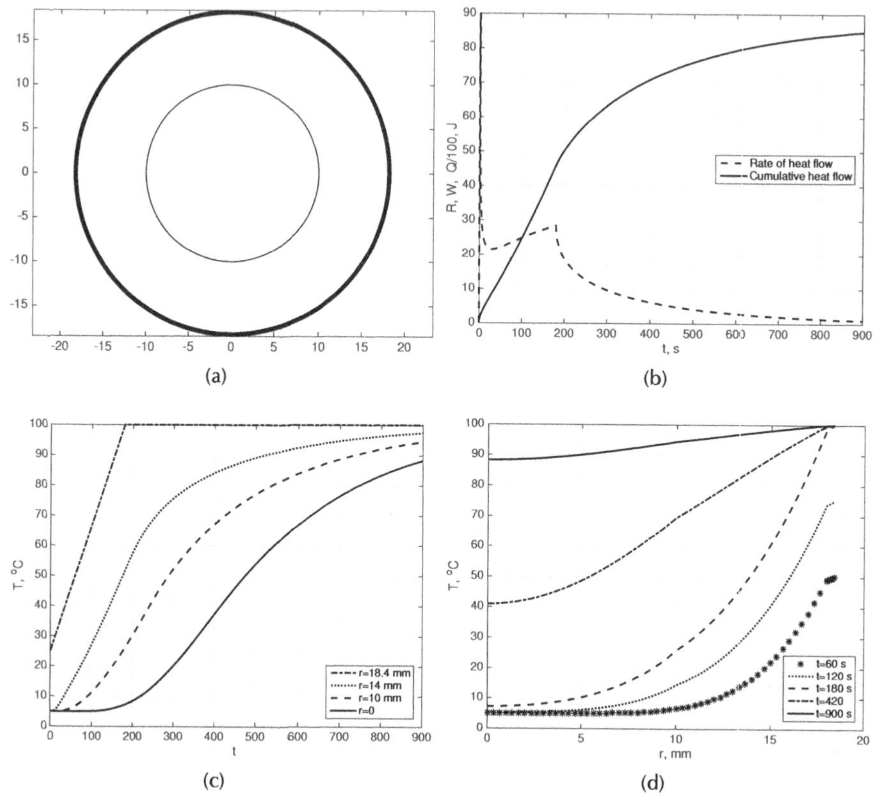

Figure 5.5 Transient temperature in a three-layer spherical domain. (a) A simplified egg model; yolk radius = 10 mm, white radial thickness = 8 mm, and shell thickness = 0.4 mm. (b) Rate of heat flow R and cumulative heat flow Q. (c) Temperature variation with time at different radial locations. (d) Temperature distributions at different times.

We consider a quadratic element (three local nodes per element). The radii of the element nodes are denoted by r_1^e, r_2^e, and r_3^e. The second node is the midpoint $r_2^e = (r_1^e + r_3^e)/2$, and the element length is $h_e = r_3^e - r_1^e$. The element matrices in spherical coordinates for element-wise uniform properties $\gamma = \gamma_e$, $p = p_e$, $q = q_e$, and $f = f_e$ are evaluated analytically.

$$[M^e] = \frac{\gamma_e h_e^3}{15} \begin{bmatrix} 2r_1^2 + r_1/2 + 1/14 & r_1^2 - 1/7 & -r_1^2/2 - r_1/2 - 5/28 \\ r_1^2 - 1/7 & 8r_1^2 + 8r_1 + 16/7 & r_1^2 + 2r_1 + 6/7 \\ -r_1^2/2 - r_1/2 - 5/28 & r_1^2 + 2r_1 + 6/7 & 2r_1^2 + 7r_1/2 + 11/7 \end{bmatrix} \tag{5.113}$$

Table 5.1
Properties of a Three-layer Spherical Egg Model

	Yolk	White	Shell
Radial thickness (Δr) mm	10	8	0.4
Density (ρ), kg/m^3	1035	1048	800
Heat capacity (C), J/kg °C	3560	3560	1000
Thermal conductivity (k), W/m °C	0.34	0.45	2.25

$$[A^e] = \frac{p_e h_e}{15}\begin{bmatrix} 35r_1^2+15r_1+3 & -20r_1-40r_1^2-6 & 5r_1^2+5r_1+3 \\ -20r_1-40r_1^2-6 & 80r_1^2+80r_1+32 & -60r_1-40r_1^2-26 \\ 5r_1^2+5r_1+3 & -60r_1-40r_1^2-26 & 35r_1^2+55r_1+23 \end{bmatrix} + \frac{q_e h_e^3}{15}\begin{bmatrix} 2r_1^2+r_1/2+1/14 & r_1^2-1/7 & -r_1^2/2-r_1/2-5/28 \\ r_1^2-1/7 & 8r_1^2+8r_1+16/7 & r_1^2+2r_1+6/7 \\ -r_1^2/2-r_1/2-5/28 & r_1^2+2r_1+6/7 & 2r_1^2+7r_1/2+11/7 \end{bmatrix} \tag{5.114}$$

$$\{F^e\} = \frac{f_e h_e^3}{60}\left\{\begin{matrix} 10r_1^2-1 \\ 40r_1^2+40r_1+12 \\ 10r_1^2+20r_1+9 \end{matrix}\right\} \tag{5.115}$$

where $r_1 = r_1^e/h_e$. In this example $p = k$, $q = 0$, and $f = 0$.

The boundary conditions are

$$\begin{aligned} r &= 0: \quad -pr^2\frac{\partial T}{\partial r} = 0 \qquad t > 0 \\ r &= r_s: \quad T = T_w(t) \qquad t > 0 \end{aligned} \tag{5.116}$$

where $r_s = 18.4$ mm is the spherical egg radius, and $T_w(t)$ is the water bath temperature assumed to be

$$\begin{aligned} T_w &= 25 + 75(t/180)\,^\circ C \qquad 0 \le t \le 180\text{s} \\ &= 100^\circ C \qquad t > 180\text{s}. \end{aligned} \tag{5.117}$$

We also specify an initial distribution of the temperature $T(r,0) = 5°$C. Properties of egg materials are given in Table 5.1. The Euler implicit method ($\alpha = 1$) is used for time integration and a mesh of quadratic elements is used. The number of elements are 15, 12, and 3 for the yolk, white, and shell regions, respectively.

The instantaneous rate of heat flow $R(t)$ and cumulative heat flow $Q(t)$ into the egg

$$\begin{aligned} R(t) &= 4\pi r_s^2 k\left(\frac{\partial T}{\partial r}\right)_{r=r_s} \\ Q(t) &= \int_0^t R(\tau)d\tau \end{aligned} \tag{5.118}$$

are shown in Figure 5.5b. The initial high rate of R is because of the difference in the initial temperature of the shell (5°C) and the initial temperature of water (25°C). As the water temperature increases, the rate R also increases and reaches its maximum of approximately 28 W when the water boils at time 180 s. The rate then continuously decreases as the water temperature remains at 100°C and the interior of the egg heats up. The cumulative heat flow increases nearly linearly with time for 180 s, then it levels off. After 300 s from the beginning of the heating, the heat supplied to the egg is approximately 6300 J. The evolution of temperature with time at different radial positions is shown in Figure 5.5c. The temperature at the shell surface is equal to the water temperature as prescribed by the boundary condition. As one anticipates, the slowest heating point is the center of the yolk. It takes 630 s for the center to reach 70°C. The temperature distributions at different times are depicted in Figure 5.5d. We recall that the egg initial temperature is 5°C. Even after 420 s, the entire yolk region of 10 mm radius is less that 70°C. Because of the difference in thermal conductivity of the shell and white, there is a clear discontinuity in the temperature gradient at the interface between them.

The results presented in this example should be taken with "a grain of salt" because of the various simplifying assumptions used in the model including neglecting phase changes, ignoring convection and air region, and constant thermal properties. A more realistic model can be found in [1].

5.7 BIOMEDICAL ENGINEERING APPLICATIONS

5.7.1 BIOHEAT: PENNES' HEAT CONDUCTION MODEL

Pennes [83] developed a heat transfer model for the temperature distribution in skin tissues taking into consideration blood perfusion and metabolic heat generation. In three-dimensions, the model is

$$\rho C\frac{\partial T}{\partial t} - \nabla \bullet (k\nabla T) + \rho_b C_b \omega_b (T - T_b) - f_m = 0 \tag{5.119}$$

where ρ, C, and k are the mass density, specific heat, and thermal conductivity of tissue, respectively, and ρ_b, and C_b are the density and specific heat of blood. Blood perfusion rate (rate of volume flow per unit volume) is given by ω_b, and T_b is arterial blood temperature. f_m is the rate of metabolic energy production (energy per unit volume per unit time). For one-dimensional space, Pennes' model becomes

$$\rho C\frac{\partial T}{\partial t} - \frac{\partial}{\partial x}\left(k\frac{\partial T}{\partial x}\right) + \rho_b C_b \omega_b (T - T_b) - f_m = 0. \tag{5.120}$$

This is the model problem given by Equation (5.1) with $\gamma = \rho C$, $p = k$, $q = \rho_b C_b \omega_b$, and $f = \rho_b C_b \omega_b T_b + f_m$. In this section, we neglected the metabolic energy production, $f_m = 0$.

Pennes' thermal model has many applications in bioheat. Torvi and Dale [120] used the 1D model to investigate skin thermal injuries of first-, second-, and third-degree burns due to exposure to flash fire. Henriques' thermal injury model [50], which is an Arrhenius-type chemical reaction model, is used to determine the degree of damage. Torvi and Dale modeled the skin tissue by a three-layer model of different thermophysical properties for each layer: the epidermis, dermis, and the subcutaneous region. Galerkin's weighted-residual (strong form) finite-element method is used in space and Crank–Nicolson is used for time integration. Johnson et al. [57] also investigated thermal injuries of a skin exposed to hot water. They modeled the skin by a four-layer tissue: the epidermis, dermis, the subcutaneous layer, and muscle. They used a second-order finite-difference method in space and an explicit first-order method in time.

Cetingul and Herman [20] used Pennes' model in 3D to study the perturbations in the skin surface temperature caused by near surface benign or malignant lesions. The skin surface is subjected to a cooling stress by specifying a low temperature at the surface (4°C) for a finite period of time (typically 120 s) followed by a recovery process in which the skin returns to its original temperature by natural convection at the surface. The transient surface temperature distribution during the stress test and recovery phase is influenced by the local variations in the thermophysical properties and blood perfusion of the lesion. They advocated monitoring such variations as an aid in the detection of malignant lesions near the skin surface.

Chang and Nguyen [22] also used Pennes' model to study hepatic ablation by a radio-frequency probe. The electromagnetic energy produced by the probe is converted into heat due to the tissue resistance. Thus, in the Pennes' equation the internal energy source is modified to include the effects of the electric current, hence $f = \rho_b C_b \omega_b T_b + f_m + \sigma|\nabla V|^2$, where σ is tissue conductivity (Siemens/meter) and V is the electric potential (Volts). The electric potential is determined by solving Laplace's equation. The tissue is modeled as a cylinder of finite height with a probe inserted on its axis, and axisymmetric conditions (temperature and voltage are functions of r, z, and t) are assumed. Proper boundary conditions are prescribed at the probe surface and the bounding tissue surface. Chang and Nguyen used Henriques–Moritz' thermal injury model to determine the shape and size of the lesion for different ablation time and probe voltage.

Example: Comparison Between Finite-Element and Analytic Solution for a Single-Layer Skin

We consider a simple example of a single-layer skin tissue. The objective is to compare the results of the finite-element method with the exact analytic solution. The thermophysical properties of the tissue are: Thickness $L = 18$ mm, density $\rho = 1200$ kg/m^3, heat capacity $C = 3200$ J/kg °C, and thermal conductivity $k = 0.4$ W/m °C, and blood perfusion and metabolic heat are neglected. The skin surface ($x = 0$) is subjected to heat flux $\tilde{q} = 4.183$ W/m^2 for an exposure time $t_{expo} = 30$ s, and then the surface is insulated afterward. The temperature of the core is assumed to be $T_c = 37$°C and is applied to the right side of the single-layer tissue at $x = L$. Thus the boundary conditions are

$$
\begin{aligned}
x = 0: \quad & -k\frac{\partial T}{\partial x} = \tilde{q} && 0 \le t \le t_{expo} \\
& -k\frac{\partial T}{\partial x} = 0 && t > t_{expo} \\
x = L: \quad & T = T_c && t > 0. \qquad (5.121)
\end{aligned}
$$

The initial ($t = 0$) temperature of the model is assumed to be uniform at core temperature $T(x,0) = T_c$. Crank–Nicolson method ($\alpha = 0.5$) is used for time integration with $\Delta t = 0.06$ s. Simulations are obtained for time $t = 60$ s; that is, 30 s after exposure to heat flux.

Buettner [19] provided the exact analytic solution for a semi-infinite layer ($0 \le x < \infty$) subjected to a constant heat flux at $x = 0$. Using Buettner's solution, Hardee and Lee [46] constructed the solution for heat flux of finite duration followed by insulated wall condition. The solution is

$$T(x,t) = T_c + \frac{2\tilde{q}}{\sqrt{k\rho C}}\left[ierfc(\xi)\sqrt{t} - ierfc(\xi_1)\sqrt{t_1}S(t_1)\right] \qquad (5.122)$$

where $\xi = x/2\sqrt{Dt}$, $t_1 = t - t_{expo}$, $\xi_1 = x/2\sqrt{Dt_1}$, $D = k/\rho C$ is the thermal diffusivity of tissue, $S(t_1)$ is the Heaviside step function, and

$$ierfc(\xi) = \frac{1}{\sqrt{\pi}}e^{-\xi^2} - \xi\, erfc(\xi) \qquad (5.123)$$

where $erfc(\xi)$ is the complementary error function.

Comparison between the finite-element and exact solutions is shown in Figures 5.6a and 5.6b for a mesh of nine linear elements, and in Figures 5.6c and 5.6d for a mesh of nine cubic elements. Figures 5.6a and 5.6c show the surface temperature against time, and Figures 5.6b and 5.6d show the temperature distribution within the tissue layer at time $t = 60$ s. The results for nine cubic elements are in excellent agreement with the exact solution. For the time of comparison (60 s), the solution for the semi-infinite domain is a very good approximation for the finite-thickness single layer of 18 mm.

Example: Skin Thermal Injury due to Direct Contact with Hot Water

The four-layer skin tissue model of Johnson et al. [57] is used in this example to determine thermal injury as a result of contact with hot water. A sketch of the model is shown in Figure 5.7, and the thermophysical properties also reported by Johnson et al. are given in Table 5.2. The skin surface ($x = 0$) is in direct contact with hot water at temperature $T_w = 70°$C for an exposure time of $t_{expo} = 20$ s. Thereafter, the skin is cooled by radiation/convection with an environment at a temperature $T_\infty = 20°$C and an effective convection/radiation coefficient of $\tilde{h} = 10$ W/m^2°C. The temperature of the core is assumed to be $T_c = 37°$C and is applied to the right side of the muscle layer ($x = L$), where L is the total thickness of the four layers. Thus, the boundary

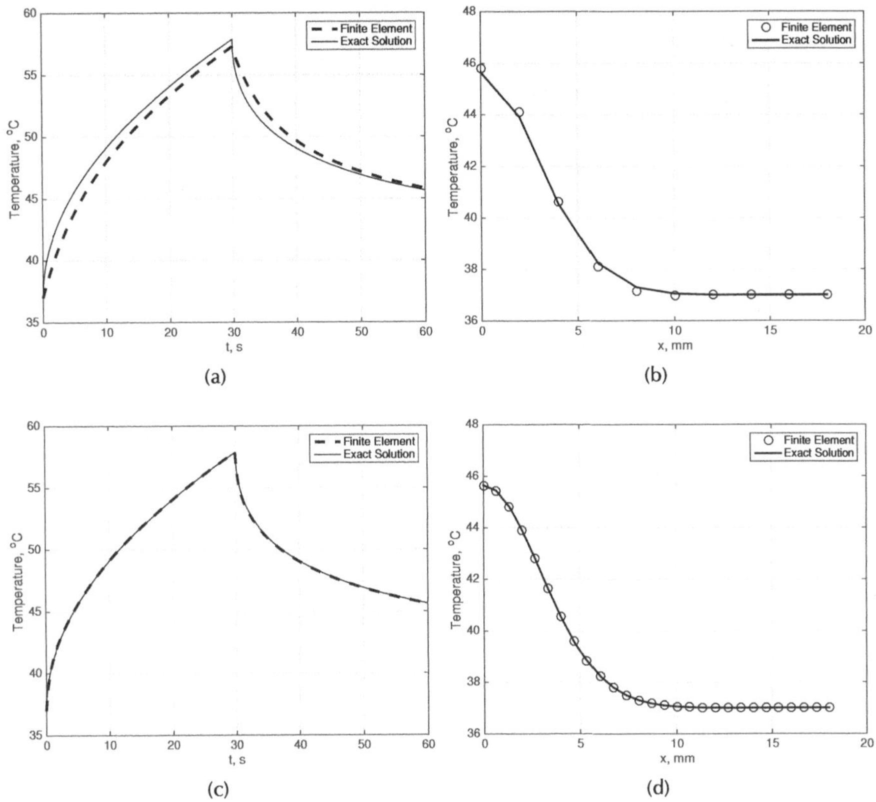

Figure 5.6 Comparison between finite-element solution and exact solution for a single-layer skin tissue. (a) and (b): uniform mesh of nine linear elements, and (c) and (d): uniform mesh of nine cubic elements. $x=0$ for (a) and (c), and $t=60$ s for (b) and (d).

conditions are

$$
\begin{aligned}
x=0: \quad & T=T_w && 0\le t\le t_{expo}\\
& -k\frac{\partial T}{\partial x}+\tilde{h}T=\tilde{h}T_\infty && t>t_{expo}\\
x=L: \quad & T=T_c && t>0. \qquad (5.124)
\end{aligned}
$$

The initial ($t=0$) temperature of the model is assumed to be uniform at core temperature $T(x,0)=T_c$. The blood properties are $\rho_b=1060$ kg/m^3, and heat capacity $C_b=3770$ J/kg°C. The degree of thermal injury is classified according to the value attained by the Arrhenius integral:

$$
\Omega(x,t)=P_f\int_{t_b}^{t}e^{-[\Delta E/RT(x,\tau)]}d\tau \qquad (5.125)
$$

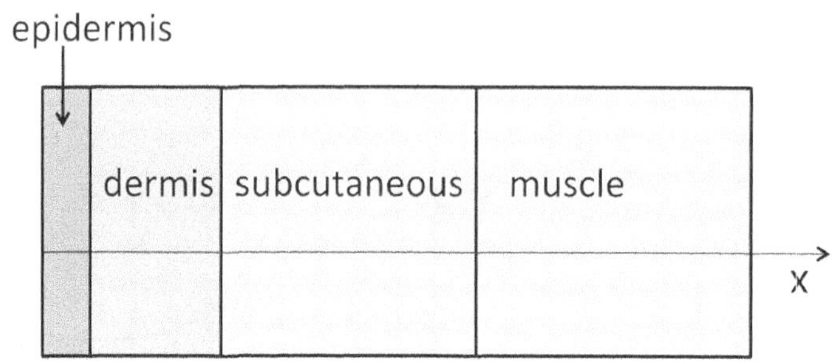

Figure 5.7 Sketch (not-to-scale) of a four-layer skin tissue model.

Table 5.2
Properties of a Four-layer Skin Tissue Model, with Permission of ASME

	Epidermis	Dermis	Subcutaneous	Muscle
Thickness (t) mm	0.08	2	10	30
Density (ρ), kg/m^3	1200	1200	1000	1000
Heat capacity (C), J/kg °C	3600	3600	2500	3800
Conductivity (k), W/m °C	0.22	0.4	0.2	0.45
Blood perfusion (ω_b), 1/s	0	0.00125	0.00125	0.00125

where $\Delta E = 6.28 \times 10^8$ (J/kmol K) is the activation energy, $R = 8314$ (J/kmol) is the universal gas constant, and $P_f = 3.1 \times 10^{98}$ s^{-1} is called the "frequency" factor. t_b is the time at, which the local temperature reaches 44°C. The temperature in Equation (5.125) must be in Kelvin, $T = T\,°\text{C} + 273.15$. Thermal injuries are defined as first-degree burns if Ω at the basal layer (the interface between the epidermis and dermis) reaches 0.53, and as second-degree burns if it reaches 1 there. Third-degree burns result if Ω reaches 1 at the interface between the dermis and subcutaneous layers. Torvi and Dale [120] used those thermal injuries definitions. Johnson et al. [57] adopted different classifications of burns. Burns arc classified as superficial (S), superficial partial-thickness (SP), deep partial-thickness (DP), and full thickness (FT) burns. The severity of burns increases from S to FT as the depth where Ω reaches 1 increases. Classes of burns S, SP, DP, and FT correspond to depths of $\Omega = 1$ in the epidermis layer, the first-half of the dermis layer, the second-half of the dermis layer, and at or deeper than the interface between the dermis and subcutaneous layers, respectively. Knowledge of the circumstances leading to thermal injuries and their classifications has implications for the required clinical treatments.

The finite-element mesh is composed of cubic elements where the number of elements are 1, 16, 30, and 10 for the epidermis, dermis, subcutaneous, and muscle layers, respectively. Crank–Nicolson method ($\alpha = 0.5$) is used for time integration

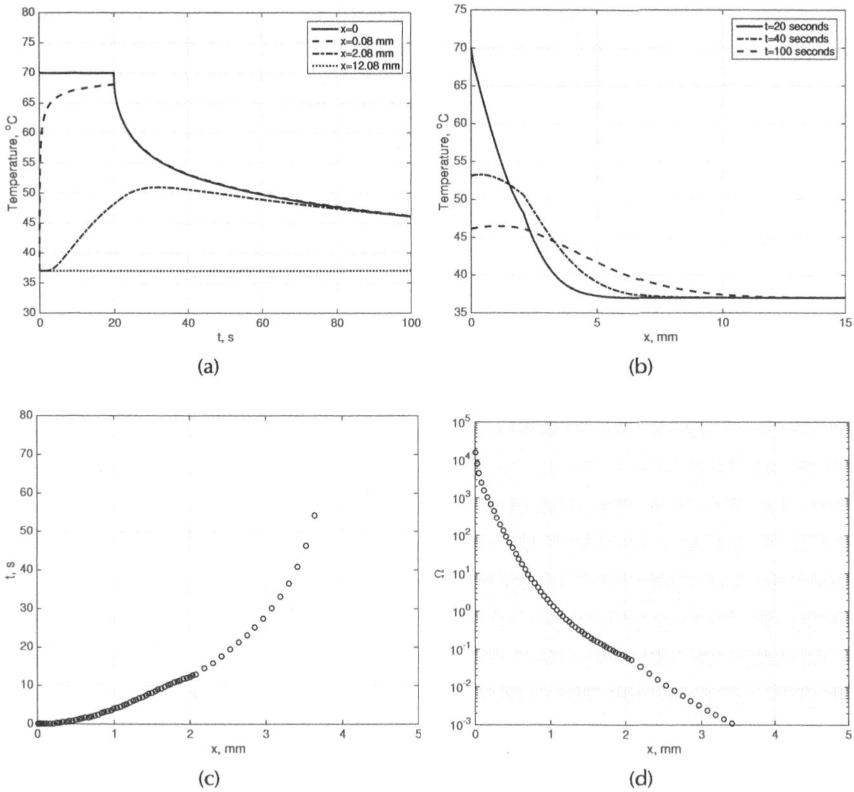

Figure 5.8 Thermal injury of skin tissue in contact with hot water at 70°C for exposure time of 20 s. (a): Temperature variation with time at the surface ($x = 0$), the interface between epidermis and dermis layers ($x = 0.08$ mm), the interface between dermis and subcutaneous layers ($x = 2.08$ mm), and the interface between subcutaneous and muscle layers ($x = 12.08$ mm). (b) Temperature distribution within tissue at different times, (c) The time at, which the local temperature reaches 44°C. (d) Distribution of thermal injury parameter Ω within the tissue at time 100 s.

with $\Delta t = 0.01$ s. Simulations are obtained for time $t = 100$ s; that is, 80 s after exposure to hot water. The temperature variation with time at different locations in the tissue is shown in Figure 5.8a. At the surface ($x = 0$), the temperature remains at 70° C for the exposure time of 20 s, and then it cools by natural convection and radiation. The temperature is still above 45° C after 80 s into the recovery phase. The temperature at the interface between the epidermis and dermis layers increases very rapidly from the initial temperature of 37°C to 68°C during exposure time. The temperature at the interface between the dermis and subcutaneous layers continues to increase even after the exposure time, and it reaches its maximum of 51°C at approximately 30 s. The increase in temperature of the muscle layer is insignificant. The temperature distribution in the tissue is shown in Figure 5.8b at the end of exposure time of 20 s, 40 s, and at the end of simulation time of 100 s. Even in the recovery phase, as the skin surface temperature cools down, the deeper tissue temperature (for example, at $x = 5$ mm) continues to heat up. During the exposure time internal energy accumulates in the surface layers and then diffuses to the deeper tissue at a later time. The temperature gradient $\partial T/\partial x$ is discontinuous at the interfaces between the layers because of the discontinuities in the thermal conductivity whereas the heat flux vector is continuous.

Figure 5.8c depicts the time when the local temperature reaches 44°C. For example, at $x = 3$ mm, that time is 28 s, which is greater than the exposure time. The distribution of the thermal injury indicator Ω at time 100 s is shown in Figure 5.8d. It reaches a value of 1 at a depth approximately 1.1 mm, that is in the second-half of the dermis layer. Thus, the burn due to contact with water at 70°C for 20 s would be classified as deep partial-thickness burn. It would also be classified as a second-degree burn.

Example: Skin Thermal Injury due to Radiation Heat Exchange with a Nearby Hot Surface

In this example, the source of heating is a hot flat plate placed parallel to the skin with a small gap between them; thus, heating is due to thermal radiation. We neglect convection in the gap between the skin and hot plate, and assume radiation shape factor of 1. Emissivities of skin ε_s and hot plate ε_p are assumed to be 0.9 and 1, respectively. The four-layer tissue model and properties of the previous example are also used here. The only difference is in the heating phase where the boundary condition is

$$x = 0: \qquad -k\frac{\partial T}{\partial x} + \sigma\varepsilon_s T^4 = \sigma\varepsilon_h T_h^4 \qquad 0 \le t \le t_{expo} \tag{5.126}$$

where $\sigma = 5.67 \times 10^{-8}$ W/m^2 K^4 is the Stefan–Boltzmann constant, and temperatures of skin surface T and hot plate T_h must be in Kelvin. Because the heat flux at the skin surface and its temperature are both unknown, this boundary condition is a mixed boundary condition, but it is nonlinear. Let T_k be a provisional skin surface temperature. We quasi-linearize the boundary condition about that temperature by replacing T^4 by

$$T^4 \approx 4T_k^3(T - T_k) + T_k^4 \tag{5.127}$$

and rewrite the radiation boundary condition as

$$x = 0: \qquad -k\frac{\partial T}{\partial x} + aT = b \qquad 0 \le t \le t_{expo} \tag{5.128}$$

where $a = 4\sigma\varepsilon_s T_k^3$ and $b = 3\sigma\varepsilon_s T_k^4 + \sigma\varepsilon_h T_h^4$. Now the boundary condition is linear and the finite-element method and the α-method for time integration as formulated in this chapter can be applied. In the finite-element mesh, the skin surface is global node # 1, and the temperature and heat flux are denoted by T_1^{m+1}, and $\tilde{R}_1 = \alpha R_1^{m+1} + (1-\alpha)R_1^m$, respectively. The radiation boundary condition becomes

$$\tilde{R}_1 + a_1 T_1^{m+1} = b_1 \qquad 0 \leq t \leq t_{expo} \tag{5.129}$$

where $a_1 = 4\alpha\sigma\varepsilon_s T_k^3$, and $b_1 = 3\alpha\sigma\varepsilon_s T_k^4 - (1-\alpha)\sigma\varepsilon_s (T_1^m)^4 + \sigma\varepsilon_h T_h^4$. Before we proceed to the next time step, we have to update the provisional skin surface temperature T_k until the change in that temperature drops below a specified tolerance. When solving for $\{T\}^{m+1}$, we update T_k by

$$T_k^{\nu+1} = \left(T_1^{m+1}\right)^{\nu} \tag{5.130}$$

where ν is a counter of this inner iteration. For $\nu = 1$, we set $T_k^1 = T_1^m$. With each update, the mixed boundary condition must be applied again to the global system. Depending on the magnitude of Δt, numerical examples show that a few (2 or 3) iterations are usually sufficient for convergence.

We consider thermal injury by a hot plate of $T_h = 160°C$. The exposure time is 5 minutes followed by a recovery phase of 25 minutes, and hence the total simulation time is 30 minutes. The time step is 0.1 s. The temperature variation with time at different locations in the tissue is shown in Figure 5.9a. The skin surface ($x = 0$) temperature starts at 37°C and increases to a maximum of 61°C during the exposure to thermal radiation for 300 s, then it cools by natural convection and radiation. The temperature drops below 35°C at the end of the recovery phase because the cooling environment is at 20°C. The temperature at the interface between the epidermis and dermis layers follows very closely the skin surface temperature. The temperature at the interface between the dermis and subcutaneous layers reaches its maximum of 55°C nearly at the end of exposure time. There is about 1.5°C increase in temperature of the muscle layer. The temperature distribution in the tissue is shown in Figure 5.9b at the end of exposure time of 300 s, 600 s, and at the end of simulation time of 1800 s. As expected, at the end of exposure of 300 s, the maximum temperature is the skin surface temperature. After 300 s into the recovery phase, the maximum temperature of approximately 41°C is in the subcutaneous layer. The temperature gradient $\partial T/\partial x$ is discontinuous at the interfaces between the layers because of the discontinuities in the thermal conductivity whereas the heat flux is continuous.

The net heat flux incident on the skin is shown in Figure 5.9c where positive flux indicates heat into the skin (heating) and negative means flux out of the skin (cooling). The maximum flux is about 1500 W/m^2, and it drops during the exposure time because of the heat emitted by the skin increases as its temperature increases. Once the hot plate is replaced by the cooling environment, the heat flux becomes negative (heat transfer from skin to the environment) and remains negative during the recovery phase. In this example, the temperature at the right boundary of muscle is prescribed to be the core temperature of 37°C. The heat flux into the core is shown in Figure 5.9d. It is negative (heat following out of the muscle into the core) but is very small in comparison to the flux incident on the skin surface.

Figure 5.9e depicts the time when the local temperature reaches 44°C. For example at $x = 3$ mm, that time is 125 s. The distribution of the thermal injury indicator Ω

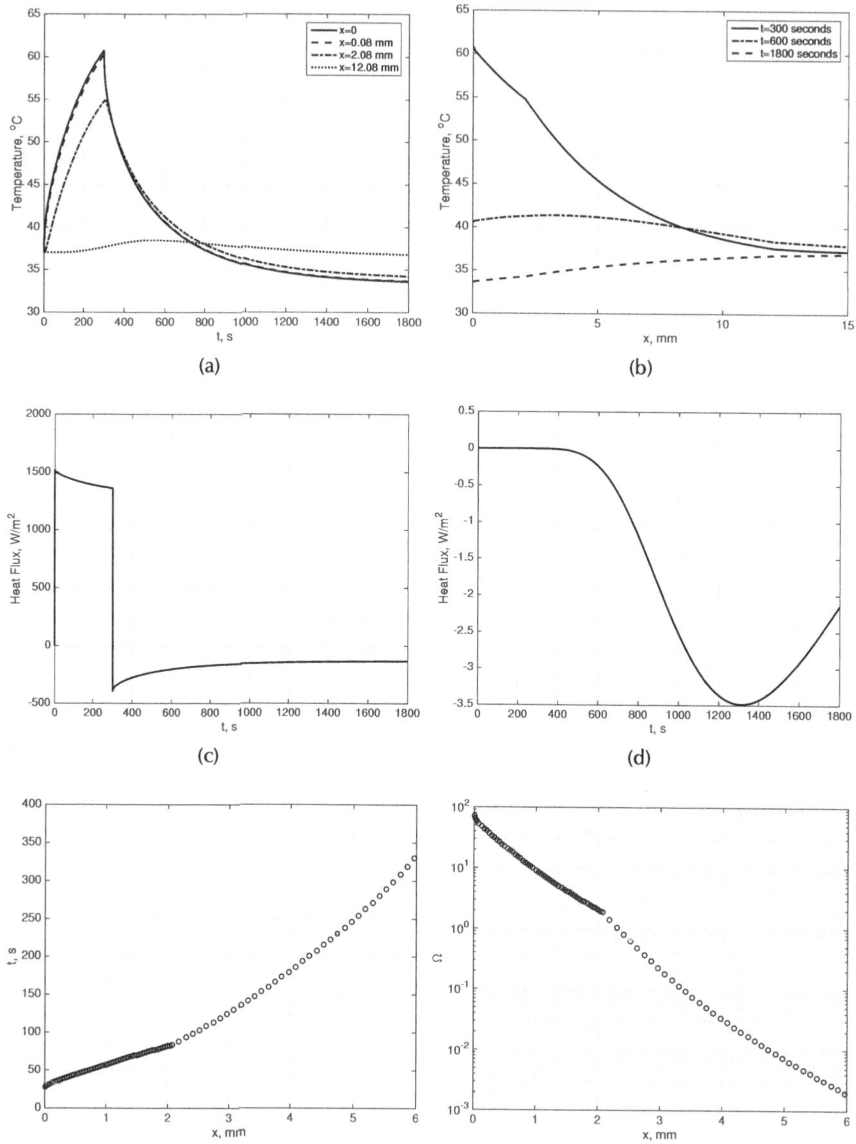

Figure 5.9 Thermal injury of skin tissue receiving heat by radiation from a hot plate at 160°C for an exposure time of 5 minuets. (a) Temperature variation with time at the surface ($x = 0$), the interface between epidermis and dermis layers ($x = 0.08$ mm), the interface between dermis and subcutaneous layers ($x = 2.08$ mm), and the interface between subcutaneous and muscle layers ($x = 12.08$ mm). (b) Temperature distribution at different times. (c) Heat flux from hot plate to epidermis surface. (d) Internal heat flux out of muscle layer, (e) The time when the local temperature reaches 44°C. (f) Distribution of thermal injury parameter Ω at time 30 minuets.

at time 1800 s is shown in Figure 5.9f. It reaches a value of 1 at a depth approximately 2.3 mm, that is in the subcutaneous layer. Thus, the burn in this example would be classified as full-thickness burn. It would also be classified as a third-degree burn.

5.7.2 UNSTEADY OXYGEN CONSUMPTION IN SPHERICAL AND CYLINDRICAL DOMAINS

In this section, we consider the transient oxygen consumption in a cylindrical or a spherical domain. In three-dimensional space, mass flux vector is given by Fick's law:

$$\vec{J} = -D_T \nabla C_{O_2} \tag{5.131}$$

where C_{O_2} is oxygen concentration (mass/volume) and D_T is diffusion coefficient (area/time) of oxygen in tissue or cell. In the absence of convection, mass balance in an infinitesimally small volume gives the governing equation for the concentration,

$$\frac{\partial C_{O_2}}{\partial t} - \nabla \bullet (D_T \nabla C_{O_2}) + B(C_{O_2}) = 0 \tag{5.132}$$

where $B(C_{O_2})$ is the rate of oxygen uptake (mass per unit time per unit volume). We consider a cylindrical or spherical domain and assume that the concentration varies only in the radial direction and time. The diffusion-reaction equation simplifies to

$$\frac{\partial C_{O_2}}{\partial t} - \frac{1}{r^k}\frac{\partial}{\partial r}\left(D_T r^k \frac{\partial C_{O_2}}{\partial r}\right) + B(C_{O_2}) = 0 \tag{5.133}$$

where $k = 1$ for cylindrical domains, and $k = 2$ for spherical domains. The rate of oxygen consumption is assumed to be a Michaelis–Menten reaction

$$B(C_{O_2}) = \frac{V_{max} C_{O_2}}{K_m + C_{O_2}} \tag{5.134}$$

where V_{max} is the maximum reaction rate [(mass/volume)/time] and K_m is the Michaelis constant (mass/volume).

We introduce nondimensional variables $t^* = tD_r/L_r^2$, $r^* = r/L_r$, $u(r,t) = C_{O_2}(r,t)/C_r$, $p = D_T/D_r$, $B^* = BL_r^2/C_r D_r$, where L_r, C_r, and D_r are reference length, concentration, and diffusivity, respectively. We drop the star for simplicity of notation, and write the governing equation in nondimensional form as

$$\frac{\partial u}{\partial t} - \frac{1}{r^k}\frac{\partial}{\partial r}\left(p r^k \frac{\partial u}{\partial r}\right) + B(u) = 0. \tag{5.135}$$

The nondimensional Michaelis–Menten rate of oxygen consumption is

$$B(u) = \frac{\theta u}{\beta + u} \tag{5.136}$$

where $\theta = V_{max}L_r^2/(D_T C_r)$ is nondimensional maximum reaction rate and $\beta = K_m/C_r$ is nondimensional Michaelis constant. Oxygen tension (partial pressure) is linearly proportional to concentration. Using the proper reference quantities, we find the nondimensional equations for oxygen tension are identical to the nondimensional equations for the oxygen concentration.

To derive a finite-element model, we first quasi-linearize $B(u)$ around a guessed solution $v(r,t)$ as we have done in Section 2.9.1.

$$B(u) = \frac{\theta \beta u}{(\beta + v)^2} + \frac{\theta v^2}{(\beta + v)^2}. \tag{5.137}$$

Now the equation for the transient oxygen concentration takes the form

$$\frac{\partial u}{\partial t} - \frac{1}{r^k}\frac{\partial}{\partial r}\left[pr^k \frac{\partial u}{\partial r}\right] + qu - f = 0 \tag{5.138}$$

where

$$\begin{aligned} q(v) &= \frac{\theta \beta}{(\beta + v)^2} \\ f(v) &= -\frac{\theta v^2}{(\beta + v)^2}. \end{aligned} \tag{5.139}$$

Let us consider a cylindrical annular domain or a spherical shell of inner radius a and outer radius b (a and b are nondimensional), and specify the boundary conditions:

$$\begin{aligned} r &= a: \quad -pr^k \frac{\partial u}{\partial r} + \gamma_a u = \alpha_a \qquad t > 0 \\ r &= b: \quad pr^k \frac{\partial u}{\partial r} + \gamma_b u = \alpha_b \qquad t > 0. \end{aligned} \tag{5.140}$$

We must also specify an initial distribution of the concentration $u(r,0) = C_{O_2}(r,0)/C_r$

$$u(r,0) = u_o(r) \qquad a \le r \le b. \tag{5.141}$$

To derive the weak form for the problem Equations (5.138)–(5.140), we multiply Equation (5.138) by a test (weight) function $w(r)$, and integrate over the domain volume. The differential element of volume is a cylindrical or a spherical shell of radial thickness dr. For a cylindrical shell of length L_z the volume is $2\pi r dr L_z$, and for a spherical shell it is $4\pi r^2 dr$. We assume a unit length in the z-direction for the cylindrical shell $L_z = 1$, and write the element volume as $2^k \pi r^k dr$ for both shells. Furthermore, for simplicity, we drop the common constant factor $2^k \pi$.

$$\int_a^b w\frac{\partial u}{\partial t} r^k dr + \int_a^b -w\frac{\partial}{\partial r}\left[pr^k \frac{\partial u}{\partial r}\right] dr + \int_a^b wqur^k dr - \int_a^b wfr^k dr = 0. \tag{5.142}$$

Integrating the second term by parts, we obtain

$$\int_a^b w\frac{\partial u}{\partial t}r^k dr + \int_a^b \left[\frac{\partial w}{\partial r}pr^k\frac{\partial u}{\partial r} + wqur^k\right] dr = \int_a^b wfr^k dr + w(a)R_a + w(b)R_b \tag{5.143}$$

where

$$\begin{aligned} R_a &= -\left(pr^k\frac{\partial u}{\partial r}\right)_{r=a} \\ R_b &= \left(pr^k\frac{\partial u}{\partial r}\right)_{r=b} \end{aligned} \tag{5.144}$$

are the secondary variables at the boundary points. For the boundary conditions Equation (5.140), the weak form becomes

$$\int_a^b w\frac{\partial u}{\partial t}r^k dr + \int_a^b \left[\frac{\partial w}{\partial r}pr^k\frac{\partial u}{\partial r} + wqur^k\right] dr$$

$$+\gamma_a w(a)u(a,t) + \gamma_b w(b)u(b,t) = \int_a^b wfr^k dr + \alpha_a w(a) + \alpha_b w(b). \tag{5.145}$$

To develop the finite-element model, we divide the radial distance $a \leq r \leq b$ into elements, each element is defined by n local nodes. The coordinate of the first node of a typical element (e) is r_1^e and its nth node is at r_n^e. The weak form for such an element follows from Equation (5.143),

$$\int_{r_1^e}^{r_n^e} w(r)\frac{\partial u}{\partial t}r^k dr + \int_{r_1^e}^{r_n^e} \left[\frac{\partial w}{\partial r}pr^k\frac{\partial u}{\partial r} + wqur^k\right] dr = \int_{r_1^e}^{r_n^e} wfr^k dr + w(r_1^e)R_1^e + w(r_n^e)R_n^e \tag{5.146}$$

where

$$\begin{aligned} R_1^e &= -\left(pr^k\frac{\partial u}{\partial r}\right)_{r=r_1^e} \\ R_n^e &= \left(pr^k\frac{\partial u}{\partial r}\right)_{r=r_n^e} \end{aligned} \tag{5.147}$$

are the secondary variables at the end nodes of the element; they are unknown since no boundary conditions are imposed for the typical element.

Approximation

An approximation of the primary variable over the element domain is written in terms of the element shape functions $\psi_i(r)$,

$$u(r,t) \approx U^e(r,t) = \sum_{j=1}^{n} U_j^e(t)\psi_j(r) \tag{5.148}$$

To write $U^e(r,t)$ in matrix form, we define the arrays

$$\{\psi(r)\} = \{\psi_1(r)\ \psi_2(r) \ldots \psi_j(r) \ldots \psi_n(r)\} \tag{5.149}$$

$$\{U^e(t)\} = \begin{Bmatrix} U_1^e(t) \\ U_2^e(t) \\ \cdot \\ \cdot \\ \cdot \\ U_j^e(t) \\ \cdot \\ \cdot \\ \cdot \\ U_n^e(t) \end{Bmatrix}. \tag{5.150}$$

In matrix form, the trial solution over element # e is

$$U^e(r,t) = \{\psi(r)\}\{U^e(t)\}. \tag{5.151}$$

We note that the dependence of the trial solution on r is given by the array $\{\psi(r)\}$, whereas the array $\{U^e(t)\}$ depends on t only; the later array are values of the primary variables at the element nodes. The time derivative of the trial solution is

$$\frac{\partial U^e(r,t)}{\partial t} = \{\psi(r)\}\{\dot{U}^e(t)\}. \tag{5.152}$$

Galerkin

In the Galerkin method, the test (or weight) function is taken as a linear superposition of the element shape functions.

$$W^e(r) = \sum_{i=1}^{n} d_i \psi_i(r) \tag{5.153}$$

where d_i ($i = 1, 2, \ldots n$) is an array of (arbitrary) constants that are the values of the test function at the element nodes. We define the array

$$\{d\} = \begin{Bmatrix} d_1 \\ d_2 \\ \cdot \\ \cdot \\ \cdot \\ d_i \\ \cdot \\ \cdot \\ \cdot \\ d_n \end{Bmatrix} \tag{5.154}$$

and write $W^e(r)$ in matrix form

$$W^e(r) = \{\psi(r)\}\{d\} \tag{5.155}$$

or

$$W^e(r) = \{d\}^T\{\psi(r)\}^T. \tag{5.156}$$

Note that the choice $W^e(r) = \psi_1(r)$ is obtained if $d_1 = 1$ and $d_i = 0$ if $i \neq 1$. Similarly, the choice $W^e(r) = \psi_2(x)$ is obtained if $d_2 = 1$ and $d_i = 0$ if $i \neq 2$, and so on.

Once we decided on the trial solution and test function, we use the weak form to obtain a system of ordinary differential equations in time for the array of unknown primary variables at the element nodes and the unknown secondary variables at the element end nodes. To that end, we substitute the trial solution Equation (5.151) and the test function Equation (5.156) into the weak form Equation (5.146).

$$\begin{aligned}
&\int_{r_1^e}^{r_n^e} \{d\}^T\{\psi(r)\}^T\{\psi(r)\}\{\dot{U}^e\}r^k dr + \\
&\int_{r_1^e}^{r_n^e} \left[\{d\}^T\frac{d\{\psi(r)\}^T}{dr}p\frac{d\{\psi(r)\}}{dr}\{U^e\} + \{d\}^T\{\psi(r)\}^T q\{\psi(r)\}\{U^e\}\right] r^k dr \\
&- \int_{r_1^e}^{r_n^e} \{d\}^T\{\psi(r)\}^T f r^k dr - \{d\}^T\{\psi(r_n^e)\}^T R_n^e - \{d\}^T\{\psi(r_1^e)\}^T R_1^e = 0.
\end{aligned} \tag{5.157}$$

Recall that $\{d\}$, $\{U^e\}$, and $\{\dot{U}^e\}$ are independent of r, we factor them out of the integration operation

$$\begin{aligned}
&\{d\}^T\Bigg\{\left(\int_{r_1^e}^{r_n^e} \{\psi(r)\}^T\{\psi(r)\}r^k dr\right)\{\dot{U}^e\} + \\
&\left(\int_{r_1^e}^{r_n^e} \left[\frac{d\{\psi(r)\}^T}{dr}p\frac{d\{\psi(r)\}}{dr} + \{\psi(r)\}^T q\{\psi(r)\}\right] r^k dr\right)\{U^e\} \\
&- \int_{r_1^e}^{r_n^e} \{\psi(r)\}^T f r^k dr - \{\psi(r_n^e)\}^T R_n^e - \{\psi(r_1^e)\}^T R_1^e \Bigg\} = 0.
\end{aligned} \tag{5.158}$$

We define the element coefficient matrix

$$[A^e] = \int_{r_1^e}^{r_n^e} \left[\frac{d\{\psi(r)\}^T}{dr}p(r)\frac{d\{\psi(r)\}}{dr} + \{\psi(r)\}^T q(v)\{\psi(r)\}\right] r^k dr \tag{5.159}$$

and the element consistent mass matrix

$$[M^e] = \int_{r_1^e}^{r_n^e} \{\psi(r)\}^T\{\psi(r)\}r^k dr. \tag{5.160}$$

The vector

$$\begin{aligned}
\{\tilde{b}^e\} &= \int_{r_1^e}^{r_n^e} \{\psi(r)\}^T f(v) r^k dr + \{\psi(r_n^e)\}^T R_n^e + \{\psi(r_1^e)\}^T R_1^e \\
&= \{F^e\} + \{R^e\}
\end{aligned} \tag{5.161}$$

is the sum of the element forcing vector

$$\{F^e\} = \int_{r_1^e}^{r_n^e} \{\psi(r)\}^T f(v) r^k dr \tag{5.162}$$

and the array of the element secondary variables. Recalling the properties of the element shape functions ($\psi_j(r_i) = \delta_{ij}$), we have

$$\{R^e\} = \begin{Bmatrix} R_1^e \\ 0 \\ \cdot \\ \cdot \\ \cdot \\ 0 \\ \cdot \\ \cdot \\ 0 \\ R_n^e \end{Bmatrix}. \tag{5.163}$$

We write Equation (5.158) as

$$\{d\}^T \left[[M^e] \{\dot{U}^e\} + [A^e] \{U^e\} - \{\tilde{b}^e\} \right] = 0. \tag{5.164}$$

Because $\{d\}$ is arbitrary, the bracket must be zero, and we obtain the desired element equations

$$[M^e] \{\dot{U}^e\} + [A^e] \{U^e\} = \{\tilde{b}^e\} \tag{5.165}$$

or

$$[M^e] \{\dot{U}^e\} + [A^e] \{U^e\} = \{F^e\} + \{R^e\}. \tag{5.166}$$

In index notation, the entries of the coefficient matrix are

$$a_{ij}^e = \int_{r_1^e}^{r_n^e} \left[\frac{d\psi_i(r)}{dr} p(r) \frac{d\psi_j(r)}{dr} + \psi_i(r) q(v) \psi_j(r) \right] r^k dr \tag{5.167}$$

and the mass matrix entries are

$$m_{ij}^e = \int_{r_1^e}^{r_n^e} \psi_i(r) \psi_j(r) r^k dr. \tag{5.168}$$

The entries of the right-hand vector are

$$\tilde{b}_i^e = \int_{r_1^e}^{r_n^e} \psi_i(r) f(v) r^k dr \qquad \text{if} \quad 2 \le i \le n-1 \tag{5.169}$$

$$\tilde{b}_1^e = \int_{r_1^e}^{r_n^e} \psi_1(r) f(v) r^k dr + R_1^e \tag{5.170}$$

$$\tilde{b}_n^e = \int_{r_1^e}^{r_n^e} \psi_n(r) f(v) r^k dr + R_n^e. \tag{5.171}$$

Time Integration

We apply the time-marching scheme of Section 5.5 to the element equations Equation (5.166) and obtain the algebraic system for element # e

$$[AL^e]\{U^e\}^{m+1} = [AR^e]\{U^e\}^m + \{\tilde{F}^e\} + \{\tilde{R}^e\} \tag{5.172}$$

where

$$\begin{aligned} [AL^e] &= \frac{1}{\Delta t}[M^e] + \alpha [A^e] \\ [AR^e] &= \frac{1}{\Delta t}[M^e] - (1-\alpha)[A^e] \\ \{\tilde{F}^e\} &= \alpha \{F^e\}^{m+1} + (1-\alpha)\{F^e\}^m \\ \{\tilde{R}^e\} &= \alpha \{R^e\}^{m+1} + (1-\alpha)\{R^e\}^m \end{aligned} \tag{5.173}$$

where m is the time step. In deriving the time marching scheme of Section 5.5, we assumed that the coefficient matrix $[A^e]$ to be independent of time. In the present case, the function $v(r,t)$ makes the coefficient matrix dependent on time. We will account for this dependence by computing the coefficient matrix using a value of v that is a weighted average of the solutions at time steps (m) and $(m+1)$. Since $v(r,t)$ is a guessed concentration, it has to be updated by iteration; as we shall show later.

The algebraic element equations are assembled into a global system, and boundary conditions at the global boundary nodes are applied. The modified global system takes the form

$$\left[\overline{AL}\right]\{U\}^{m+1} = \left[\overline{AR}\right]\{U\}^m + \{\overline{F}\}. \tag{5.174}$$

Knowing the solution array at time step m, $\{U\}^m$, we solve the system Equation (5.174) for $\{U\}^{m+1}$. The initial condition at $t=0$ (corresponds to $m=0$) provides the global solution array $\{U\}^0$. If the original equation were linear in the concentration u, it would have been sufficient to solve the system only once per time step. This is the case for the zero-order and first-order reaction rates. However, for the Michaelis–Menten reaction rate under consideration here, the oxygen consumption rate is nonlinear. The element matrices given by Equations (5.159) and (5.162) contain the functions $q(v)$ and $f(v)$ defined by Equation (5.139) that depend on a guessed concentration $v(r;t)$. Therefore, before we proceed to the next time step, we have to update the guessed concentration $v(r;t)$ until the changes in the solution $\{U\}^{m+1}$ drop below a specified small tolerance. When solving for $\{U\}^{m+1}$, we update the nodal values of $v(r;t)$ by

$$\{v\}^{\nu+1} = \alpha \{U^\nu\}^{m+1} + (1-\alpha)\{U\}^m \tag{5.175}$$

where ν is a counter of this inner iteration, and $\{U^\nu\}^{m+1}$ is the solution obtained with guessed solution $v^\nu(r;t)$. For $\nu = 1$, we set $\{v\}^1 = \{U\}^m$. With each update, new element matrices are computed and assembled; and a new solution $\{U^\nu\}^{m+1}$ is obtained. Depending on the magnitude of Δt, numerical examples show that a few (2 or 3) iterations are usually sufficient for convergence. We note that if an explicit method is used for time marching ($\alpha = 0$), then there is no need for iterating on $v(r;t)$; the nonlinear reaction rate is evaluated at the known solution $\{U\}^m$.

An Alternative Procedure For Deriving the Finite Element Equations

In the above derivation of the element equations, we quasi-linearized the Michaelis–Menten reaction rate before applying the finite-element in space and time discretizations. Alternatively, we first apply the time marching scheme (the α-method), which converts the PDE into ODE is space, then we apply the finite-element method in space to obtain the algebraic element equations. The α-method is given by Equation (5.68), and is repeated here

$$\frac{u^{m+1}-u^m}{\Delta t} = \alpha\left(\frac{\partial u}{\partial t}\right)^{m+1} + (1-\alpha)\left(\frac{\partial u}{\partial t}\right)^m \tag{5.176}$$

where $0 \le \alpha \le 1$. The concentration Equation (5.135) gives the time derivatives in terms of space derivatives, and we rewrite Equation (5.176) as

$$\begin{aligned}\frac{u^{m+1}-u^m}{\Delta t} &+ \alpha\left[-\frac{1}{r^k}\frac{\partial}{\partial r}\left(pr^k\frac{\partial u}{\partial r}\right) + B(u)\right]^{m+1} \\ &+(1-\alpha)\left[-\frac{1}{r^k}\frac{\partial}{\partial r}\left(pr^k\frac{\partial u}{\partial r}\right) + B(u)\right]^m = 0.\end{aligned} \tag{5.177}$$

Variables at time step m are known, whereas those at step $(m+1)$ are unknown. Thus, we quasi-linearize the reaction rate $B(u^{m+1})$ only

$$B(u^{m+1}) = q(v)u^{m+1} + f(v) \tag{5.178}$$

where

$$\begin{aligned} q(v) &= \frac{\theta\beta}{(\beta+v)^2} \\ f(v) &= -\frac{\theta v^2}{(\beta+v)^2}. \end{aligned} \tag{5.179}$$

Here $v(r)$ is a guessed value for $u^{m+1}(r)$. We rewrite Equation (5.177)

$$\begin{aligned}\frac{u^{m+1}-u^m}{\Delta t} &+ \alpha\left[-\frac{1}{r^k}\frac{\partial}{\partial r}\left(pr^k\frac{\partial u}{\partial r}\right)\right]^{m+1} + \alpha q(v)u^{m+1} + \alpha f(v) \\ &+(1-\alpha)\left[-\frac{1}{r^k}\frac{\partial}{\partial r}\left(pr^k\frac{\partial u}{\partial r}\right)\right]^m + (1-\alpha)B(u^m) = 0.\end{aligned} \tag{5.180}$$

This is a linear ordinary differential equation for $u^{m+1}(r)$. We derive its weak form by multiplying it by a test (weight) function $w(r)$ and integrate over the element domain. We then integrate the terms in brackets by parts and obtain

$$
\begin{aligned}
&\frac{1}{\Delta t}\int_{r_1^e}^{r_n^e} w(r)u^{m+1}r^k dr + \alpha\int_{r_1^e}^{r_n^e}\left[\frac{\partial w}{\partial r}p\frac{\partial u}{\partial r} + wq(v)u\right]^{m+1} r^k dr = \\
&\frac{1}{\Delta t}\int_{r_1^e}^{r_n^e} w(r)u^m r^k dr - (1-\alpha)\int_{r_1^e}^{r_n^e}\left[\frac{\partial w}{\partial r}p\frac{\partial u}{\partial r} + wq(u^m)u\right]^{m} r^k dr \\
&\qquad - \int_{r_1^e}^{r_n^e} w\left[\alpha f(v) + (1-\alpha)f(u^m)\right] r^k dr \\
&\qquad + w(r_1^e)\left[\alpha\left(R_1^e\right)^{m+1} + (1-\alpha)\left(R_1^e\right)^m\right] \\
&\qquad + w(r_n^e)\left[\alpha\left(R_n^e\right)^{m+1} + (1-\alpha)\left(R_n^e\right)^m\right].
\end{aligned} \tag{5.181}
$$

In deriving the above equation, we wrote, without any approximation, $B(u^m) = q(u^m)u^m + f(u^m)$. Substituting the trial solution Equation (5.151) and test function Equation (5.156) into the element weak form Equation (5.181), we get the element algebraic equations,

$$
[AL^e]\{U^e\}^{m+1} = [AR^e]\{U^e\}^m + \{\tilde{F}^e\} + \{\tilde{R}^e\} \tag{5.182}
$$

where

$$
\begin{aligned}
[AL^e] &= \frac{1}{\Delta t}[M^e] + \alpha[A^e] \\
[AR^e] &= \frac{1}{\Delta t}[M^e] - (1-\alpha)[A_*^e] \\
\{\tilde{F}^e\} &= \alpha\{F^e\}^{m+1} + (1-\alpha)\{F_*^e\}^m \\
\{\tilde{R}^e\} &= \alpha\{R^e\}^{m+1} + (1-\alpha)\{R^e\}^m
\end{aligned} \tag{5.183}
$$

where

$$
[M^e] = \int_{r_1^e}^{r_n^e}\{\psi(r)\}^T\{\psi(r)\}r^k dr \tag{5.184}
$$

$$
[A^e] = \int_{r_1^e}^{r_n^e}\left[\frac{d\{\psi(r)\}^T}{dr}p(r)\frac{d\{\psi(r)\}}{dr} + \{\psi(r)\}^T q(v)\{\psi(r)\}\right] r^k dr \tag{5.185}
$$

$$
[A_*^e] = \int_{r_1^e}^{r_n^e}\left[\frac{d\{\psi(r)\}^T}{dr}p(r)\frac{d\{\psi(r)\}}{dr} + \{\psi(r)\}^T q(u^m)\{\psi(r)\}\right] r^k dr \tag{5.186}
$$

$$
\{F^e\} = \int_{r_1^e}^{r_n^e}\{\psi(r)\}^T f(v) r^k dr \tag{5.187}
$$

$$\{F_*^e\} = \int_{r_1^e}^{r_n^e} \{\psi(r)\}^T f(u^m) r^k dr. \tag{5.188}$$

During the inner iteration, only the matrix $[A^e]$ and the vector $\{F^e\}$ need be updated, and because $v(r;t)$ is a guessed solution, we update its values at the global nodes by

$$\{v\}^{\nu+1} = \{U^\nu\}^{m+1}. \tag{5.189}$$

In this section, the element matrices are evaluated by Gauss–Legendre quadrature formula as presented in Section 2.8. For linear and quadratic elements we use two-point and three-point formulas, respectively.

Example: Comparison Between Finite-Element Solution and Exact Analytic Solution for a Spherical Cell

It is important to verify the finite-element solution by comparison with exact analytic solutions. Such solutions are not known for the nonlinear Michaelis–Menten reaction rate. For first- and zero-order consumption rates, the exact solutions are derived in the Appendix of this chapter.

We consider the oxygen uptake in a spherical cell with the following nondimensional parameters: $a = 0$, $b = 1$, and $p = 1$. The boundary conditions are

$$\begin{aligned} r &= 0: \quad \frac{\partial u}{\partial r} = 0 \qquad t > 0 \\ r &= 1: \quad \frac{\partial u}{\partial r} + Hu = H \qquad t > 0 \end{aligned} \tag{5.190}$$

and $H = 5$. The initial concentration is zero ($u_o(r) = 0$). A uniform mesh of 10 quadratic elements is used. Crank–Nicolson method ($\alpha = 0.5$) with a constant time step of $\Delta t = 0.001$ is used for time integration. The exact solution is an infinite series for, which the sum of the first 100 terms is sufficient for convergence. Comparisons between finite-element solutions and exact analytic solutions are shown in Figures 5.10a and 5.10b for first-order consumption rate $B(u) = 5u$, and in Figures 5.11a and 5.11b for zero-order reaction rate $B(u) = 1$. For both the reaction rates, the finite-element solutions are in excellent agreement with the exact analytic solutions. For first-order reaction rate, the time evolution of oxygen concentration at different radial locations is depicted in Figure 5.10a. The concentration starts at the initial value of zero and rises to a steady state value after time approximately $t = 0.8$. The radial distribution of oxygen concentration at different times is depicted in Figure 5.10b. The concentration never drops below the initial value of zero at any time or radius. The situation with the zero-order reaction is significantly different from the first-order reaction rate at earlier times. Figure 5.11a shows that the concentration at cell center $r = 0$ and at $r = 0.5$ drops below the initial concentration of zero and becomes negative until diffusion of oxygen from the environment remedies this nonphysical situation. Figure 5.11a shows that at $t = 0.05$ the concentration is negative (or below the initial zero concentration) for $0 \le r \le 0.39$; this is not a numerical error. One is tempted to declare this region as anoxic (zero oxygen concentration) and artificially

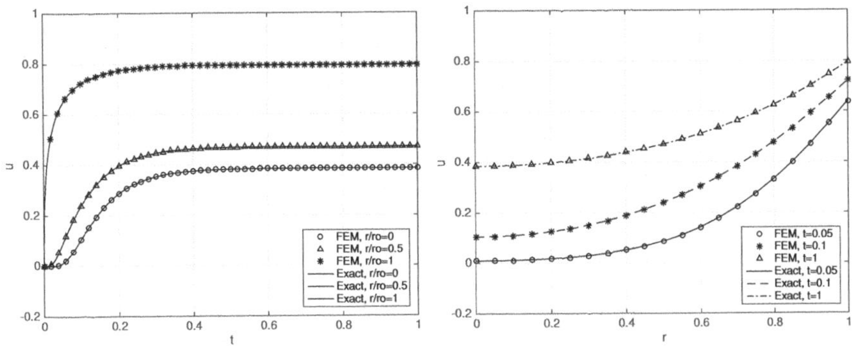

(a) (b)

Figure 5.10 Comparison between finite element solution and exact solution for oxygen consumption in a spherical cell with a first-order consumption rate. Initial concentration is zero. (a) Variation of concentration with time at different radial positions, and (b) radial distribution of concentration at different times.

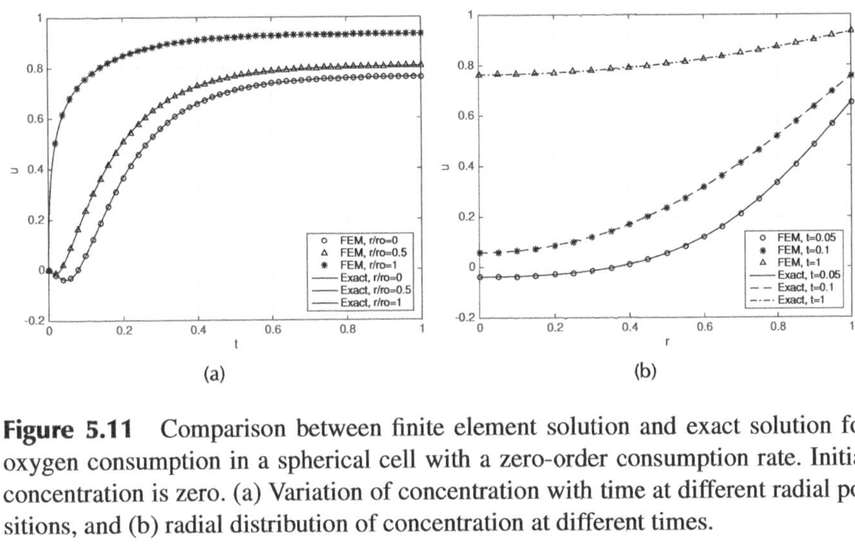

(a) (b)

Figure 5.11 Comparison between finite element solution and exact solution for oxygen consumption in a spherical cell with a zero-order consumption rate. Initial concentration is zero. (a) Variation of concentration with time at different radial positions, and (b) radial distribution of concentration at different times.

replace the negative concentration with zero, but that would be the wrong conclusion. The negative concentration is a result of the zero-order consumption rate, which is not valid at low concentration. Only first-order or Michaelis–Menten kinetics can predict reliable hypoxic or anoxic conditions in cells and tissues.

Example: Transient Oxygen Consumption in a Spherical Cell

Lin [65] studied the transient oxygen diffusion in a spherical cell with the uptake kinetics given by the Michaelis–Menten reaction rate. He solved the reaction-diffusion equation using a finite-difference method in space and Crank–Nicolson in time.

McElwain [73] solved the steady-state problem ($\partial u/\partial t = 0$) using a fourth-order Runge–Kutta method and a shooting technique (iteration) in order to satisfy the outer boundary condition. Lin and McElwain solved for the oxygen tension P_{O_2} (partial pressure); it is linearly proportional to the concentration; thus, the nondimensional equation for oxygen concentration is identical to the equation for the tension with adjustments in the reaction rate constants. Lin's and McElwain's data are:

Spherical cell radius, $r_o = 5\times 10^{-3}$ cm,

Diffusion coefficient in cell, $D_T = 2\times 10^{-5}$ cm^2 s^{-1},

Saturated oxygen tension in the environment outside the cell, $P_o = 155$ mm Hg,

Maximum reaction rate, $V_{max} = 47.2$ mmHg s^{-1},

Michaelis constant, $K_m = 4.834$ mmHg,

Permeability of the cell membrane, $h = 2\times 10^{-2}$ cm s^{-1}.

We note that 1 mmHg = 133.322365 Pa.

Nondimensional variables are defined by using r_o, D_T, and P_o as reference length, diffusivity, and tension, respectively. Thus, in the differential equation for the tension we have $u(r,t) = P_{O_2}(r,t)/P_o$, $\theta = V_{max}r_o^2/(D_T P_o)$, $\beta = K_m/P_o$, $p = 1$, and $k = 2$. The boundary conditions are

$$\begin{aligned} r &= 0: \quad \frac{\partial u}{\partial r} = 0 \qquad t > 0 \\ r &= 1: \quad \frac{\partial u}{\partial r} + Hu = H \qquad t > 0 \end{aligned} \tag{5.191}$$

where $H = hr_o/D_T$. The corresponding numerical values are: $\theta = 0.3806$, $\beta = 0.0312$, and $H = 5$. The initial tension in the spherical cell ($t = 0$) is zero. A mesh of 30 quadratic elements is used, and the Crank–Nicolson scheme ($\alpha = 0.5$) is used for time integration with $\Delta t = 0.001$. The finite element method results are depicted in Figures 5.12a and 5.12b. The evolution of oxygen tension with time at different radial locations is shown in Figure 5.12a. The tension increases very rapidly for regions near the outer boundary ($r = 1$) where oxygen diffuses from the environment. The oxygen near the cell center evolves slower as it awaits the diffusion of oxygen. The radial distribution of oxygen tension at different times is plotted in Figure 5.12b. After time $t = 1$, the tension reaches a steady state with radial distribution that varies slightly with radius.

Next we compare the results of Michaelis–Menten kinetics to the first- and zero-order kinetics. The finite-element method is used for the three reaction models. For the Michaelis–Menten kinetics the reaction rate is $B(u) = \theta u/(\beta + u)$. For the first-order kinetics, the rate is assumed to be $B(u) = \theta u$, and for zero-order it is $B(u) = \theta$. All parameters correspond to Lin's data. The evolution of oxygen tension with time at cell center is shown in Figure 5.13a. For the Michaelis–Menten and first-order reactions, the tension increases smoothly from the initial value of zero, whereas for

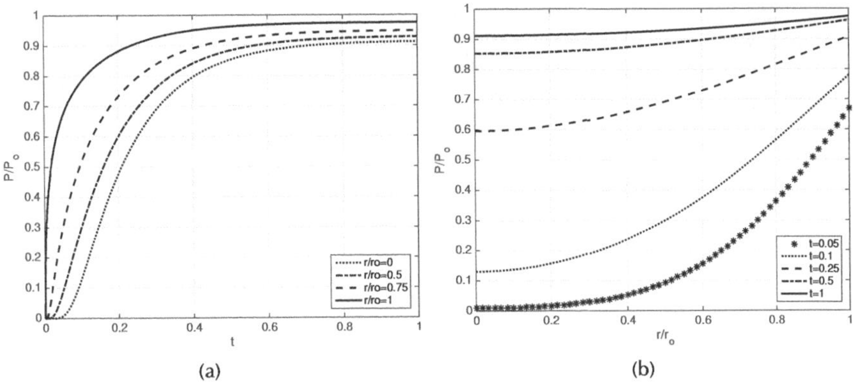

Figure 5.12 Finite-element solution for transient oxygen tension in a spherical cell with Michaelis–Menten kinetics. Initial tension is zero. (a) Variation of tension with time at different radial positions, and (b) radial distribution of tension at different times.

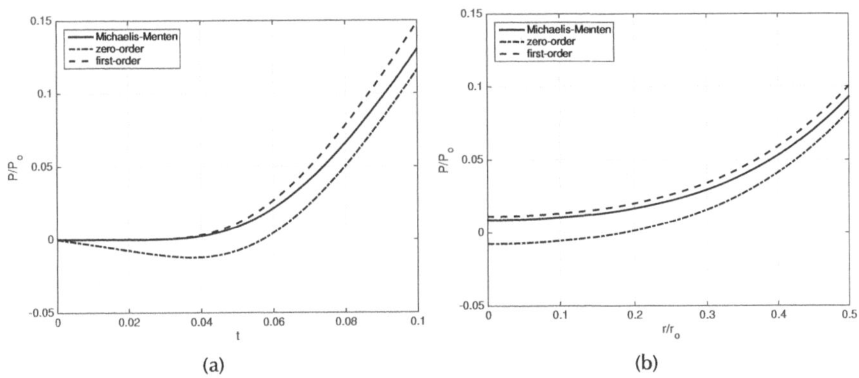

Figure 5.13 Finite-element solution for transient oxygen tension in a spherical cell. Initial tension is zero. Comparison of Michaelis–Menten kinetics with first- and zero-order kinetics, cell radius $r_o = 50\,\mu$m. (a) Variation of tension with time at cell center, and (b) radial distribution of tension at $t = 0.05$.

the zero-order rate the tension becomes negative (below initial tension) before it increases again as oxygen diffusion takes effect. The radial distribution of oxygen tension at time $t = 0.05$ is plotted in Figure 5.13b. Because the initial oxygen tension is zero, the results of first-order reaction, which is valid for low tension (or concentration), are closer to the more realistic Michaelis–Menten reaction. The results of the zero-order rate are not realistic because the model is not valid for tensions lower than the Michaelis constant. For a larger spherical cell of radius $r_o = 80\,\mu$m we have $\theta = 0.9745$, and $H = 8$. The discrepancies in the transient response among the different models increase for larger cells as shown in Figures 5.14a and 5.14b.

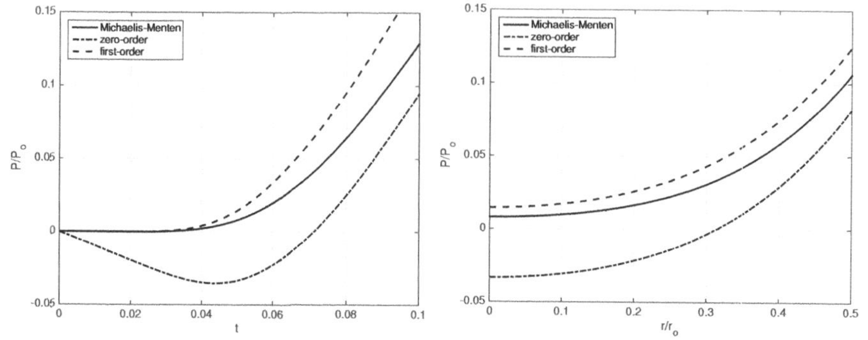

(a) (b)

Figure 5.14 Finite-element solution for transient oxygen tension in a spherical cell. Comparison of Michaelis–Menten kinetics with first- and zero-order kinetics, $r_o = 80\,\mu$m. (a) Variation of tension with time at cell center, and (b) radial distribution of tension at time $t = 0.05$. Initial tension is zero.

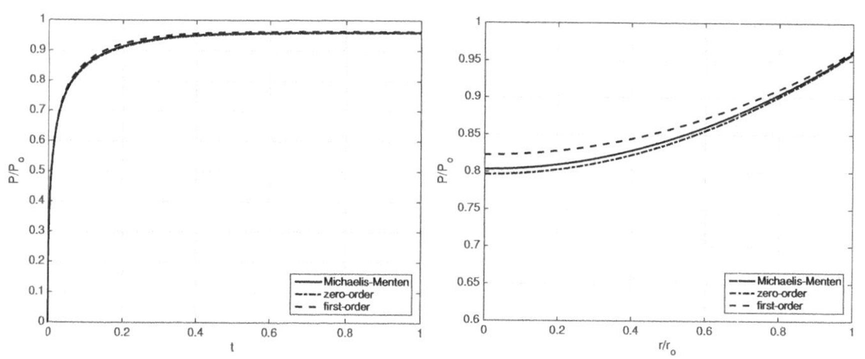

(a) (b)

Figure 5.15 Finite-element solution for transient oxygen tension in a spherical cell. Initial tension is zero. Comparison of Michaelis–Menten kinetics with first- and zero-order kinetics, $r_o = 80\,\mu$m. (a) Variation of tension with time at cell boundary $r/r_o = 1$, and (b) radial distribution of tension at time $t = 1$.

However, as the oxygen tension increases with time above the Michaelis constant, the zero-order reaction produces realistic results. The time evolution of the oxygen tension at the cell boundary $r = 1$, where the oxygen tension is high, is shown in Figure 5.15a and the radial distribution at time $t = 1$ is shown in Figure 5.15b. It is clear that the zero-order reaction is a good approximation to the Michaelis–Menten reaction in regions or at times of high tension or concentration.

5.7.3 TRANSIENT OXYGEN UPTAKE IN A KROGH CYLINDER TISSUE

Example: Comparison Between Finite-Element Solution and Exact Analytic Solution for a Krogh Cylinder Tissue

We consider a Krogh cylinder tissue and assume that the oxygen concentration to vary only with the radial distance and time. The capillary and tissue radii are $r_c = 4\,\mu$m and $r_t = 40\,\mu$m, respectively. Oxygen diffusivity in tissue is $D_T = 2\times10^{-9}$ m^2/s. The concentration at the capillary surface is a prescribed constant ($C_{O_2c} = 0.04$ mole m^{-3}), and zero flux is prescribed on the outer surface of the cylinder tissue. The concentration is made nondimensional by using C_{O_2c} as a reference concentration, and the cylinder tissue radius r_t is used as a reference length. Thus, the boundary conditions are

$$
\begin{aligned}
r &= r_c: \quad u = 1 \qquad t > 0 \\
r &= 1: \quad \frac{\partial u}{\partial r} = 0 \qquad t > 0
\end{aligned} \tag{5.192}
$$

where r_c is the nondimensional capillary radius. The parameters in the Michaelis–Menten kinetics are taken as $K_m = 0.002$ mole m^{-3}, and $V_{max} = 0.05$ mole m^{-3} s^{-1}. The corresponding nondimensional parameters are $\theta = V_{max}r_t^2/(D_T C_{O_2c}) = 1$, and $\beta = K_m/C_{O_2c} = 0.05$. In the differential equation we use $p = 1$, and $k = 1$. The uptake rate for the first-order reaction is assumed to be $B(u) = \theta u$, and for the zero-order uptake it is $B(u) = \theta$. The initial ($t = 0$) oxygen concentration in the tissue is zero ($u(r,0) = 0$).

In this example, the transient oxygen in the tissue for first- or zero-order rate is solved by the finite-element method and results are verified by comparison with the exact analytic solution, which is derived in the Appendix of this chapter. A uniform mesh of 20 quadratic elements is used in the domain $r_c \le r \le 1$, and the Crank–Nicolson method ($\alpha = 0.5$) is used for time integration with fixed time step $\Delta t = 0.001$. The exact solution is an infinite series; we summed the first 100 terms.

Comparisons between finite-element solutions and exact analytic solutions are shown in Figures 5.16a and5.16b for first-order consumption rate, and in Figures 5.17a and 5.17b for zero-order reaction rate. For both the reaction rates, the finite-element solutions are in excellent agreement with the exact analytic solutions. For first-order reaction rate, the time evolution of oxygen concentration at different radial locations is depicted in Figure 5.16a. The concentration starts at the initial value of zero and rises to a steady state value after time approximately $t = 2$. The radial distribution of oxygen concentration at different times is depicted in Figure 5.16b. The concentration never drops below the initial value of zero at any time or radius. As with the case of the spherical cell, the predictions with the zero-order uptake rate are significantly different from the first-order reaction rate at earlier times. Figure 5.17a shows that the concentration at the outer boundary of the cylinder tissue ($r = 1$) drops below the initial concentration of zero and becomes negative until diffusion of oxygen from the capillary remedies this nonphysical situation. Figure 5.17b shows that at $t = 0.1$ the concentration is negative (or below the initial zero concentration) for $0.65 \le r \le 1$; again this is not a numerical error. Declaring this region as anoxic

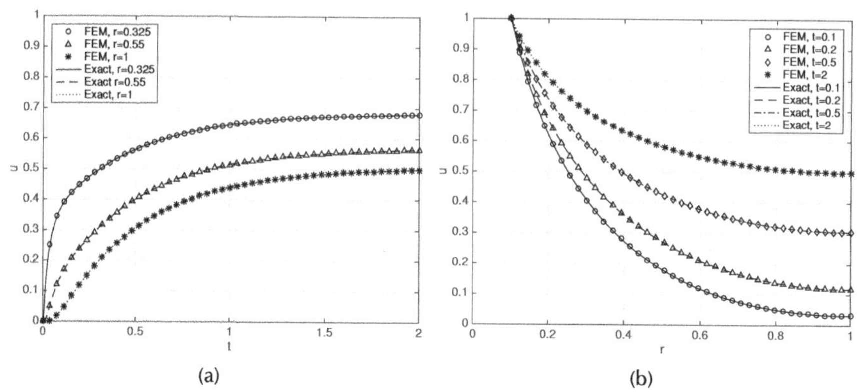

Figure 5.16 Comparison between finite element solution and exact solution for oxygen consumption in a Krogh cylinder tissue with a fist-order consumption rate. Initial concentration is zero. (a) Variation of concentration with time at different radial positions, and (b) radial distribution of concentration at different times.

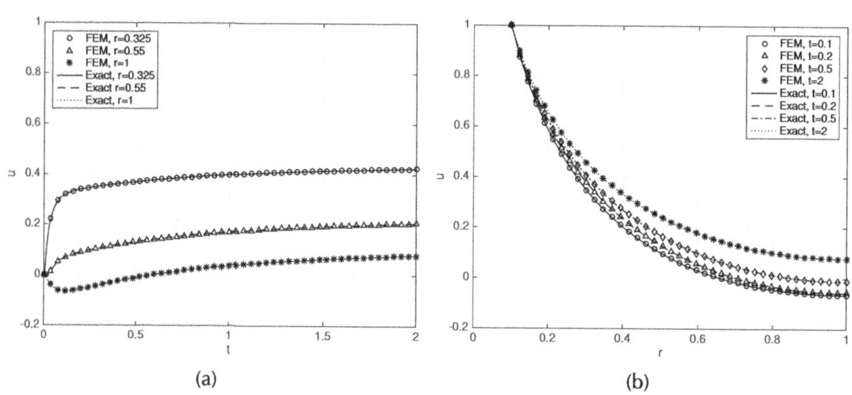

Figure 5.17 Comparison between finite element solution and exact solution for oxygen consumption in a Krogh cylinder tissue with a zero-order consumption rate. (a) Variation of concentration with time at different radial positions, and (b) radial distribution of concentration at different times. Initial concentration is zero.

(zero oxygen concentration) and artificially replacing the negative concentration with zero is not appropriate. The negative concentration is a result of the use of zero-order consumption rate, which is not valid at low concentration. Only first-order or Michaelis–Menten kinetics can predict reliable hypoxic or anoxic conditions in a Krogh cylinder tissue.

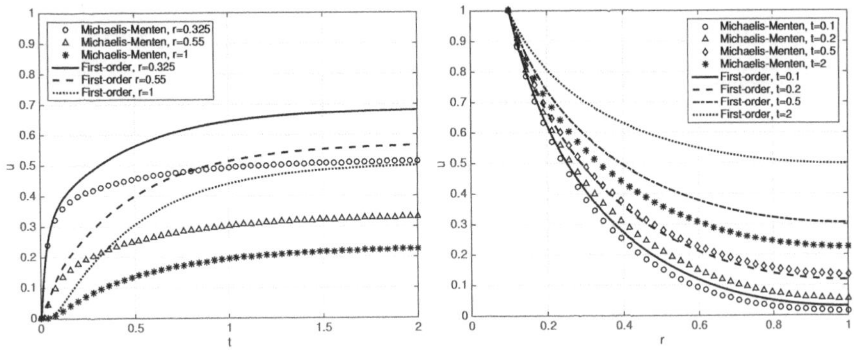

Figure 5.18 Comparison between Michaelis–Menten and first-order rate kinetics for oxygen consumption in a Krogh cylinder tissue. Initial concentration is zero. (a) Variation of concentration with time at different radial positions, and (b) radial distribution of concentration at different times.

Example: Comparison Between Different Kinetics Models for a Krogh Cylinder Tissue

In this example, we compare the results of Michaelis–Menten kinetics to the first- and zero-order kinetics. The finite-element method is used for the Michaelis–Menten kinetics whereas the exact solution is used for first- and zero-order kinetics; which has been proven to agree with the finite-element solution in the previous example. For the Michaelis–Menten kinetics the reaction rate is $B(u) = \theta u/(\beta + u)$. For the first-order kinetics, the rate is assumed to be $B(u) = \theta u$, and for zero-order it is $B(u) = \theta$. All parameters and conditions of the previous example are used here. The time evolution of oxygen concentration at different radial positions is shown in Figure 5.18a for the Michaelis–Menten and first-order reactions. The concentration increases smoothly from the initial value of zero, but saturates at different levels for the two models at steady state; the levels are higher for first-order rate. We note that if the first-order rate is defined by $B(u) = (\theta/\beta)u$ instead of $B(u) = \theta u$, the steady state concentration levels for the first-order rate would have been lower than Michaelis–Menten kinetics levels. The radial distributions of oxygen tension at different times are plotted in Figure 5.18b. The agreement between Michaelis–Menten and first-order kinetics is good only for $t < 0.1$. The time evolution of oxygen concentration at different radial positions is shown in Figure 5.18b for the Michaelis–Menten and zero-order reactions. Near the outer boundary of cylinder tissue, the zero-order rate predicts negative concentration for $t < 0.6$ before it increases again as the oxygen diffusion from the capillary takes effect. At steady state, the concentration saturates at lower levels than those of the Michaelis–Menten kinetics. The radial distributions of oxygen concentration at different times are plotted in Figure 5.19a. The results of the zero-order rate are not realistic because the model is not valid for concentrations lower than the Michaelis constant. This example shows that for Krogh cylinder

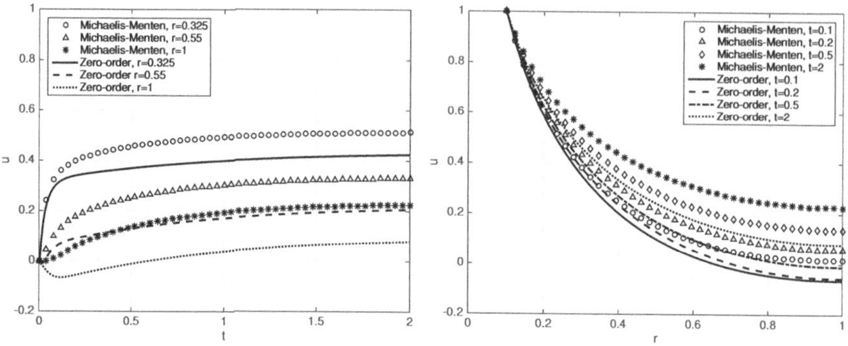

(a) (b)

Figure 5.19 Comparison between Michaelis–Menten and zero-order rate kinetics for oxygen consumption in a Krogh cylinder tissue. Initial concentration is zero. (a) Variation of concentration with time at different radial positions, and (b) radial distribution of concentration at different times.

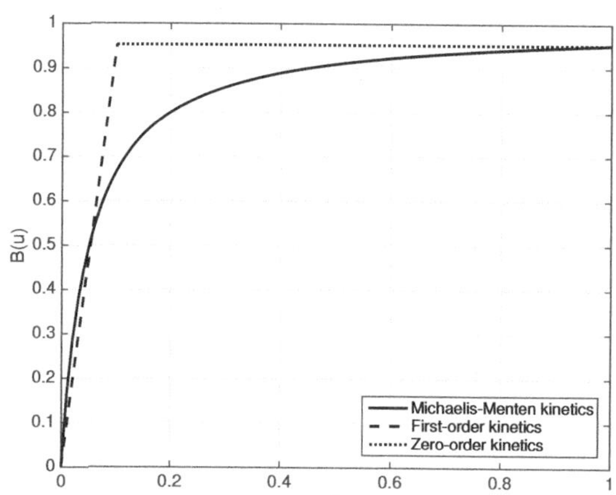

Figure 5.20 Approximation of the Michaelis–Menten kinetics $B(u) = \theta u/(\beta + u)$ by piece-wise first- and zero-order rates. $\theta = 1$ and $\beta = 0.05$.

tissues the Michaelis–Menten kinetics should be used, and that zero- and first-order rates may be only qualitatively correct.

In the interest of using linear uptake rates, Massaro and Fatt [71] approximated the nonlinear Michaelis–Menten kinetics by a first-order rate if the local oxygen concentration (or tension) falls below a certain value, and by a zero-order rate for concentration (or tension) higher than that value. Using the numerical data of the present example ($\theta = 1$ and $\beta = 0.05$), we plot in Figure 5.20 the Michaelis–Menten rate and the approximations by first- and zero-order rates. The first-order is $B(u) = (s/u_s)u$, and the zero-order rate is $B(u) = s$, where $s = \theta/(\beta + 1)$. The switch between the two linear rates happens at $u_s = 0.1$. Comparison between the Michaelis–Menten and the piece-wise linear rates for transient oxygen uptakes will be left as an exercise.

5.8 APPENDIX

Exact Analytic Solutions For Reaction-Diffusion in Spherical and Cylindrical Domains

We derive exact analytic solutions to unsteady oxygen consumption in a spherical cell or a Krogh cylinder. The oxygen concentration or tension is assumed to depend on radial distance and time only. Solutions are presented for two cases of reaction rate: (1) first-order $B(u) = qu$ and (2) zero-order $B(u) = f$, where q and f are constants. In these two cases the differential equation for the oxygen concentration is linear, and the exact solution for a reaction rate of the form $B(u) = qu + f$ can be found by superposition.

Spherical Cell

Case 1: First-order reaction rate

We first consider the case of first-order reaction rate. The equation for oxygen concentration or tension is

$$\frac{\partial u}{\partial t} - \frac{1}{r^2}\frac{\partial}{\partial r}\left[r^2\frac{\partial u}{\partial r}\right] + qu = 0 \tag{5.193}$$

where $q > 0$ is a constant. The boundary conditions are

$$\begin{aligned} r &= 0: \quad \frac{\partial u}{\partial r} = 0 \qquad t > 0 \\ r &= 1: \quad \frac{\partial u}{\partial r} + Hu = H \qquad t > 0. \end{aligned} \tag{5.194}$$

The initial distribution of the concentration is given

$$u(r,0) = u_o(r) \qquad 0 \le r \le 1. \tag{5.195}$$

We let

$$u(r,t) = v(r) + w(r,t) \tag{5.196}$$

and construct $v(r)$ to satisfy

$$-\frac{1}{r^2}\frac{\partial}{\partial r}\left[r^2\frac{\partial v}{\partial r}\right] + qv = 0 \tag{5.197}$$

and the boundary conditions

$$\begin{aligned} r &= 0: \quad \frac{\partial v}{\partial r} = 0 \\ r &= 1: \quad \frac{\partial v}{\partial r} + Hv = H. \end{aligned} \tag{5.198}$$

We also construct $w(r,t)$ to satisfy

$$\frac{\partial w}{\partial t} - \frac{1}{r^2}\frac{\partial}{\partial r}\left[r^2\frac{\partial w}{\partial r}\right] + qw = 0 \tag{5.199}$$

and the boundary conditions

$$\begin{aligned} r &= 0: \quad \frac{\partial w}{\partial r} = 0 \qquad t > 0 \\ r &= 1: \quad \frac{\partial w}{\partial r} + Hw = 0 \qquad t > 0. \end{aligned} \tag{5.200}$$

The solution to v can be written in terms of the modified spherical Bessel functions. The differential equation

$$z^2\frac{d^2Q(z)}{dz^2} + 2z\frac{dQ(z)}{dz} - [z^2 + n(n+1)]Q(z) = 0 \tag{5.201}$$

for $n = 0, \pm 1, \pm 2, \cdots$ is known as the modified spherical Bessel equation (see [3]). Its general solution is

$$Q(z) = \sqrt{\frac{\pi}{2z}}\left[A_o I_{n+\frac{1}{2}}(z) + B_o I_{-n-\frac{1}{2}}(z)\right] \tag{5.202}$$

where $I_{n+\frac{1}{2}}(z)$ and $I_{-n-\frac{1}{2}}(z)$ are modified Bessel functions of fractional order. Let $\eta = \sqrt{q}$, and $z = \eta r$, the equation for $v(r)$ Equation (5.197) can be written as

$$z^2\frac{d^2v(z)}{dz^2} + 2z\frac{dv(z)}{dz} - z^2 v(z) = 0. \tag{5.203}$$

Thus, its solution is

$$v(z) = \sqrt{\frac{\pi}{2z}}\left[A_o I_{\frac{1}{2}}(z) + B_o I_{-\frac{1}{2}}(z)\right] \tag{5.204}$$

which can also be written in terms of the hyperbolic functions

$$v(r) = A_o\frac{sinh(\eta r)}{\eta r} + B_o\frac{cosh(\eta r)}{\eta r}. \tag{5.205}$$

The boundary condition at $r = 0$ demands that $B_o = 0$, and the boundary condition at $r = 1$ gives

$$A_o = \frac{H\eta}{(H-1)sinh(\eta) + \eta cosh(\eta)} \tag{5.206}$$

and hence

$$v(r) = A_o\frac{sinh(\eta r)}{\eta r}. \tag{5.207}$$

We solve the equation for $w(r,t)$ by separation of variables. We assume

$$w(r,t) = T(t)R(r) \tag{5.208}$$

and substitute into Equation (5.199) to obtain

$$\frac{dT(t)}{dt}R(r) - T(t)\frac{1}{r^2}\frac{d}{dr}\left[r^2\frac{dR(r)}{dr}\right] + qT(t)R(r) = 0. \tag{5.209}$$

Using $q = \eta^2$, we get

$$\frac{1}{T(t)}\frac{dT(t)}{dt} = \frac{1}{r^2R(r)}\frac{d}{dr}\left[r^2\frac{dR(r)}{dr}\right] - \eta^2. \tag{5.210}$$

We set

$$\frac{1}{T(t)}\frac{dT(t)}{dt} = -\lambda^2 \tag{5.211}$$

and hence

$$\frac{1}{r^2R(r)}\frac{d}{dr}\left[r^2\frac{dR(r)}{dr}\right] - \eta^2 = -\lambda^2 \tag{5.212}$$

where λ is a constant. Let $\zeta^2 = \lambda^2 - \eta^2$ and $z = \zeta r$, we obtain

$$z^2\frac{d^2R(z)}{dz^2} + 2z\frac{dR(z)}{dz} + z^2R(z) = 0. \tag{5.213}$$

The solution to $R(r)$ can be written in terms of the spherical Bessel functions j_n and y_n (see [3]). Consider the differential equation

$$z^2\frac{d^2Q(z)}{dz^2} + 2z\frac{dQ(z)}{dz} + [z^2 - n(n+1)]Q(z) = 0. \tag{5.214}$$

The general solution to this equation is

$$Q(z) = Aj_n(z) + By_n(z). \tag{5.215}$$

Thus, the solution for $R(r)$ is

$$R(r) = Aj_o(\zeta r) + By_o(\zeta r). \tag{5.216}$$

The following properties are useful in the present analysis,

$$\begin{aligned}
j_o(z) &= \frac{sin(z)}{z} \\
j_1(z) &= \frac{sin(z)}{z^2} - \frac{cos(z)}{z} \\
y_o(z) &= -\frac{cos(z)}{z} \\
y_1(z) &= -\frac{cos(z)}{z^2} - \frac{sin(z)}{z} \\
j_{-1}(z) &= -y_o(z).
\end{aligned} \tag{5.217}$$

The solution for $T(t)$ is

$$T(t) = Ce^{-\lambda^2 t}. \tag{5.218}$$

The boundary condition at $r = 0$ demands that $B = 0$, and the boundary condition at $r = 1$ provides the equation for eigenvalue ζ,

$$\zeta cos(\zeta) + (H - 1)sin(\zeta) = 0 \tag{5.219}$$

the positive roots are denoted by ζ_k, $k = 1, 2, \cdots$, and hence $\lambda_k^2 = \zeta_k^2 + \eta^2$. The solution for $w(r,t)$ is

$$w(r,t) = \sum_{k=1}^{\infty} A_k e^{-\lambda_k^2 t} j_o(\zeta_k r). \tag{5.220}$$

The coefficients A_k are determined by imposing the initial condition $v(r) + w(r,0) = u_o(r)$,

$$\sum_{k=1}^{\infty} A_k j_o(\zeta_k r) = u_o(r) - v(r). \tag{5.221}$$

The functions $j_o(\zeta_k r)$ satisfy the orthogonality condition,

$$\int_0^1 j_o(\zeta_k r) j_o(\zeta_m r) r^2 dr = 0 \qquad \text{if } k \neq m \tag{5.222}$$

and if $k = m$, the result is

$$\int_0^1 [j_o(\zeta_k r)]^2 r^2 dr = \frac{1}{2\zeta_k^2}\left[1 - \frac{sin(2\zeta_k)}{2\zeta_k}\right]. \tag{5.223}$$

Multiplying Equation (5.221) by $r^2 j_o(\zeta_m)$, integrating over the domain, and using the orthogonality condition, we get

$$A_k = \frac{\int_0^1 j_o(\zeta_k r)\,[u_o(r) - v(r)]\, r^2 dr}{\int_0^1 [j_o(\zeta_k r)]^2\, r^2 dr}. \tag{5.224}$$

The exact solution for the oxygen consumption in a spherical cell with first-order reaction rate is

$$u(r,t) = A_o \frac{sinh(\eta r)}{\eta r} + \sum_{k=1}^{\infty} A_k e^{-\lambda_k^2 t} j_o(\zeta_k r) \tag{5.225}$$

where A_o is given by Equation (5.206) and A_k are given by Equation (5.224). We note that $v(r)$ is the steady state solution ($t \to \infty$).

Case 2: Zero-order reaction rate

The equation for oxygen concentration or tension with zero-order uptake rate is

$$\frac{\partial u}{\partial t} - \frac{1}{r^2}\frac{\partial}{\partial r}\left[r^2 \frac{\partial u}{\partial r}\right] + f = 0 \tag{5.226}$$

where $f > 0$ is a constant reaction rate. The boundary conditions are

$$\begin{aligned} r &= 0: \quad \frac{\partial u}{\partial r} = 0 \qquad t > 0 \\ r &= 1: \quad \frac{\partial u}{\partial r} + Hu = H \qquad t > 0. \end{aligned} \tag{5.227}$$

The initial distribution of the concentration is given

$$u(r,0) = u_o(r) \qquad 0 \le r \le 1. \tag{5.228}$$

We let

$$u(r,t) = v(r) + w(r,t) \tag{5.229}$$

and construct $v(r)$ to satisfy

$$-\frac{1}{r^2}\frac{\partial}{\partial r}\left[r^2\frac{\partial v}{\partial r}\right] + f = 0 \tag{5.230}$$

and the boundary conditions

$$\begin{aligned} r &= 0: \quad \frac{\partial v}{\partial r} = 0 \\ r &= 1: \quad \frac{\partial v}{\partial r} + Hv = H. \end{aligned} \tag{5.231}$$

We also construct $w(r,t)$ to satisfy

$$\frac{\partial w}{\partial t} - \frac{1}{r^2}\frac{\partial}{\partial r}\left[r^2\frac{\partial w}{\partial r}\right] = 0 \tag{5.232}$$

and the boundary conditions

$$\begin{aligned} r &= 0: \quad \frac{\partial w}{\partial r} = 0 \qquad t > 0 \\ r &= 1: \quad \frac{\partial w}{\partial r} + Hw = 0 \qquad t > 0. \end{aligned} \tag{5.233}$$

The solution to $v(r)$ is

$$v(r) = 1 + \frac{f}{6}r^2 - \frac{f}{6}\left(1 + \frac{2}{H}\right). \tag{5.234}$$

Using separation of variables we find $w(r,t)$ to be the same as $w(r,t)$ of the first-order reaction rate except that $\lambda_k^2 = \zeta_k^2$, hence

$$w(r,t) = \sum_{k=1}^{\infty} A_k e^{-\lambda_k^2 t} j_o(\zeta_k r) \tag{5.235}$$

where the eigenvalues ζ_k are the positive roots of Equation (5.219). The exact solution for the oxygen consumption in a spherical cell with zero-order reaction rate is

$$u(r,t) = 1 + \frac{f}{6}r^2 - \frac{f}{6}\left(1 + \frac{2}{H}\right) + \sum_{k=1}^{\infty} A_k e^{-\lambda_k^2 t} j_o(\zeta_k r) \tag{5.236}$$

where A_k are given by Equation (5.224), and $\lambda_k^2 = \zeta_k^2$. We note that $v(r)$ is the steady state solution ($t \to \infty$).

Krogh Cylinder

Case 1: First-order reaction rate

For first-order reaction rate, the equation for oxygen concentration or tension in nondimensional variables is

$$\frac{\partial u}{\partial t} - \frac{1}{r}\frac{\partial}{\partial r}\left[r\frac{\partial u}{\partial r}\right] + qu = 0 \tag{5.237}$$

where $q > 0$ is a constant. The boundary conditions are

$$\begin{aligned} r &= r_c: \quad u = 1 \qquad t > 0 \\ r &= 1: \quad \frac{\partial u}{\partial r} = 0 \qquad t > 0. \end{aligned} \tag{5.238}$$

The cylindrical tissue radius is used as a reference length, and r_c is the capillary radius. The oxygen concentration (or tension) on the capillary surface is used as a reference concentration. The initial radial distribution of the concentration is given

$$u(r,0) = u_o(r) \qquad r_c \leq r \leq 1. \tag{5.239}$$

We let

$$u(r,t) = v(r) + w(r,t) \tag{5.240}$$

and construct $v(r)$ to satisfy

$$-\frac{1}{r}\frac{\partial}{\partial r}\left[r\frac{\partial v}{\partial r}\right] + qv = 0 \tag{5.241}$$

and the boundary conditions

$$\begin{aligned} r &= r_c: \quad v = 1 \\ r &= 1: \quad \frac{\partial v}{\partial r} = 0. \end{aligned} \tag{5.242}$$

We also construct $w(r,t)$ to satisfy

$$\frac{\partial w}{\partial t} - \frac{1}{r}\frac{\partial}{\partial r}\left[r\frac{\partial w}{\partial r}\right] + qw = 0 \tag{5.243}$$

and the boundary conditions

$$\begin{aligned} r &= r_c: \quad w = 0 \qquad t > 0 \\ r &= 1: \quad \frac{\partial w}{\partial r} = 0 \qquad t > 0. \end{aligned} \tag{5.244}$$

The solution to v can be written in terms of the modified Bessel functions. The differential equation

$$z^2\frac{d^2Q(z)}{dz^2} + z\frac{dQ(z)}{dz} - [z^2 + \nu^2]Q(z) = 0 \tag{5.245}$$

is known as the modified Bessel equation (see [3]). Its general solution is

$$Q(z) = A_o I_\nu(z) + B_o K_\nu(z) \tag{5.246}$$

where $I_\nu(z)$ and $K_\nu(z)$ are modified Bessel functions of the first and second kind, respectively. Let $\eta = \sqrt{q}$, and $z = \eta r$, the equation for $v(r)$ can be written as

$$z^2\frac{d^2v(z)}{dz^2} + z\frac{dv(z)}{dz} - z^2 v(z) = 0. \tag{5.247}$$

Thus, its solution is

$$v(z) = A_o I_o(z) + B_o K_o(z). \tag{5.248}$$

The boundary conditions at $r = r_c$ and $r = 1$ give

$$A_o I_o(z_c) + B_o K_o(z_c) = 1 \tag{5.249}$$

$$A_o I_o^{'}(z_1) + B_o K_o^{'}(z_1) = 0 \tag{5.250}$$

where $z_c = \eta r_c$ and $z_1 = \eta$. Using $I_o^{'}(z) = I_1(z)$ and $K_o^{'}(z) = -K_1(z)$, we get

$$\begin{aligned} A_o &= \frac{K_1(z_1)}{I_o(z_c)K_1(z_1) + I_1(z_1)K_o(z_c)} \\ B_o &= \frac{I_1(z_1)}{I_o(z_c)K_1(z_1) + I_1(z_1)K_o(z_c)} \\ v(z) &= A_o I_o(z) + B_o K_o(z). \end{aligned} \tag{5.251}$$

We solve the equation for $w(r,t)$ by separation of variables. We assume

$$w(r,t) = T(t)R(r) \tag{5.252}$$

and substitute into Equation (5.243) to obtain

$$\frac{dT(t)}{dt}R(r) - T(t)\frac{1}{r}\frac{d}{dr}\left[r\frac{dR(r)}{dr}\right] + qT(t)R(r) = 0. \tag{5.253}$$

Using $q = \eta^2$, we get

$$\frac{1}{T(t)}\frac{dT(t)}{dt} = \frac{1}{rR(r)}\frac{d}{dr}\left[r\frac{dR(r)}{dr}\right] - \eta^2. \tag{5.254}$$

We set

$$\frac{1}{T(t)}\frac{dT(t)}{dt} = -\lambda^2 \tag{5.255}$$

and hence

$$\frac{1}{rR(r)}\frac{d}{dr}\left[r\frac{dR(r)}{dr}\right] - \eta^2 = -\lambda^2 \tag{5.256}$$

where λ is a constant. In Equation (5.256), we substitute $\zeta^2 = \lambda^2 - \eta^2$ and $z = \zeta r$, and get

$$z^2\frac{d^2R(z)}{dz^2} + z\frac{dR(z)}{dz} + z^2R(z) = 0. \tag{5.257}$$

The solution to $R(r)$ can be written in terms of the Bessel functions J_n and Y_n (see [3]). Consider the differential equation

$$z^2\frac{d^2Q(z)}{dz^2} + z\frac{dQ(z)}{dz} + [z^2 - \nu^2]Q(z) = 0. \tag{5.258}$$

The general solution to this equation is

$$Q(z) = AJ_\nu(z) + BY_\nu(z). \tag{5.259}$$

Thus, the solution for $R(r)$ is

$$R(r) = AJ_o(\zeta r) + BY_o(\zeta r). \tag{5.260}$$

The solution for $T(t)$ is

$$T(t) = Ce^{-\lambda^2 t}. \tag{5.261}$$

The boundary conditions at $r = r_c$ and $r = 1$ demand

$$\begin{aligned} AJ_o(z_c) + BY_o(z_c) &= 0 \\ AJ_o'(z_1) + BY_o'(z_1) &= 0. \end{aligned} \tag{5.262}$$

where $z_c = \zeta r_c$ and $z_1 = \zeta$. Using $J_o'(z) = -J_1(z)$ and $Y_o'(z) = -Y_1(z)$, we get

$$\begin{aligned} AJ_o(z_c) + BY_o(z_c) &= 0 \\ AJ_1(z_1) + BY_1(z_1) &= 0. \end{aligned} \tag{5.263}$$

For a non-trivial solution, we set the determinant to zero, and obtain

$$J_o(z_c)Y_1(z_1) - J_1(z_1)Y_o(z_c) = 0. \tag{5.264}$$

The consecutive positive roots are the eigenvalues ζ_k, $k = 1, 2, \cdots$, and hence $\lambda_k^2 = \zeta_k^2 + \eta^2$. The coefficient B can be written in terms of the coefficient A, and after some minor redefinition of A, we can write the solution to $w(r,t)$ as

$$w(r,t) = \sum_{k=1}^{\infty} A_k e^{-\lambda_k^2 t} G(\zeta_k r) \tag{5.265}$$

where the eigenfunction is

$$G(z) = \frac{J_o(z)}{J_o(z_c)} - \frac{Y_o(z)}{Y_o(z_c)}. \tag{5.266}$$

The coefficients A_k are determined by imposing the initial condition $v(r) + w(r,0) = u_o(r)$,

$$\sum_{k=1}^{\infty} A_k G(\zeta_k r) = u_o(r) - v(r). \tag{5.267}$$

The eigenfunctions $G(\zeta_k r)$ satisfy the orthogonality condition,

$$\int_{r_c}^{1} G(\zeta_k r) G(\zeta_m r) r dr = 0 \qquad \text{if } k \neq m. \tag{5.268}$$

Multiplying Equation (5.267) by $rG(\zeta_m)$, integrating over the domain, and using the orthogonality condition, we get

$$A_k = \frac{\int_{r_c}^{1} G(\zeta_k r)\,[u_o(r) - v(r)]\, r dr.}{\int_{r_c}^{1} [G(\zeta_k r)]^2\, r dr} \tag{5.269}$$

The integral

$$\int_{r_c}^{1} [G(\zeta_k r)]^2\, r dr$$

can be evaluated exactly as it can be written in terms of the following integrals:

$$\begin{aligned} \int J_o^2(z) z dz &= \frac{z^2}{2}\left[J_o^2(z) + J_1^2(z)\right] \\ \int Y_o^2(z) z dz &= \frac{z^2}{2}\left[Y_o^2(z) + Y_1^2(z)\right] \\ \int J_o(z) Y_o(z) z dz &= \frac{z^2}{2}\left[J_o(z) Y_o(z) + J_1(z) Y_1(z)\right]. \end{aligned} \tag{5.270}$$

The exact solution for the transient oxygen uptake in a Krogh cylinder tissue with first-order reaction rate is

$$u(r,t) = A_o I_o(\eta r) + B_o K_o(\eta r) + \sum_{k=1}^{\infty} A_k e^{-\lambda_k^2 t} G(\zeta_k r) \tag{5.271}$$

where A_o and B_o are given by Equation (5.251) and A_k are given by Equation (5.269). We note that $v(r)$ is the steady state solution ($t \to \infty$).

Case 2: Zero-order reaction rate

For zero-order reaction rate, the equation for oxygen concentration or tension in nondimensional variables is

$$\frac{\partial u}{\partial t} - \frac{1}{r}\frac{\partial}{\partial r}\left[r\frac{\partial u}{\partial r}\right] + f = 0 \tag{5.272}$$

where $f > 0$ is a constant reaction rate. The boundary conditions are

$$\begin{aligned} r &= r_c: \quad u = 1 \qquad t > 0 \\ r &= 1: \quad \frac{\partial u}{\partial r} = 0 \qquad t > 0. \end{aligned} \tag{5.273}$$

The initial radial distribution of the concentration is given

$$u(r,0) = u_o(r) \qquad r_c \leq r \leq 1. \tag{5.274}$$

We let

$$u(r,t) = v(r) + w(r,t) \tag{5.275}$$

and construct $v(r)$ to satisfy

$$-\frac{1}{r}\frac{\partial}{\partial r}\left[r\frac{\partial v}{\partial r}\right] + f = 0 \tag{5.276}$$

and the boundary conditions

$$\begin{aligned} r &= r_c: \quad v = 1 \\ r &= 1: \quad \frac{\partial v}{\partial r} = 0. \end{aligned} \tag{5.277}$$

We also construct $w(r,t)$ to satisfy

$$\frac{\partial w}{\partial t} - \frac{1}{r}\frac{\partial}{\partial r}\left[r\frac{\partial w}{\partial r}\right] = 0 \tag{5.278}$$

and the boundary conditions

$$\begin{aligned} r &= r_c: \quad w = 0 \qquad t > 0 \\ r &= 1: \quad \frac{\partial w}{\partial r} = 0 \qquad t > 0. \end{aligned} \tag{5.279}$$

The solutions for $v(r)$ is

$$v(r) = 1 + \frac{f}{4}\left[r^2 - r_c^2 - 2ln(r/r_c)\right] \tag{5.280}$$

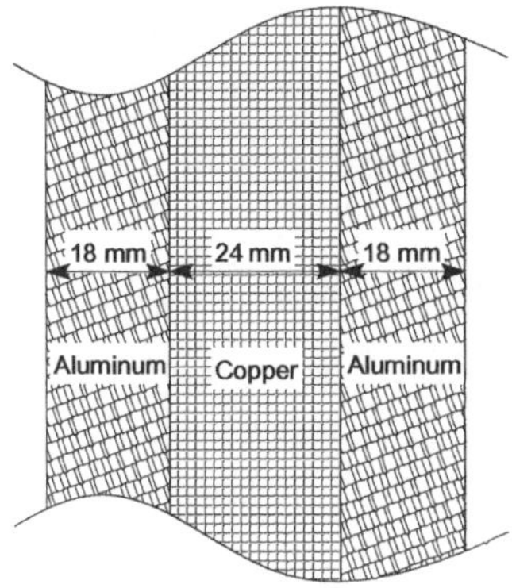

Figure 5.21 A composite wall of three slabs.

The solution for $w(r,t)$ is the same as $w(r,t)$ for the first-order reaction rate except that $\lambda_k^2 = \zeta_k^2$. The exact solution for the transient oxygen uptake in a Krogh cylinder tissue with zero-order reaction rate is

$$u(r,t) = 1 + \frac{f}{4}\left[r^2 - r_c^2 - 2ln(r/r_c)\right] + \sum_{k=1}^{\infty} A_k e^{-\lambda_k^2 t} G(\zeta_k r) \tag{5.281}$$

where A_k are given by Equation (5.269), $\lambda_k^2 = \zeta_k^2$, and ζ_k are the consecutive positive roots of Equation (5.264). We note that $v(r)$ is the steady state solution ($t \to \infty$).

5.9 PROBLEMS

(1) A wall is composed of three homogeneous layers, as shown in Figure 5.21. Properties for aluminum are: $\rho = 2700$ kg m^{-3}, $C = 963$ J/kg.K, and $k = 204$ W m^{-1}°C^{-1}, and for copper: $\rho = 8900$ kg m^{-3}, $C = 375$ J/kg.K, and $k = 400$ W m^{-1}°C^{-1}. To analyze transient temperature distribution in the composite wall, three quadratic elements are used; one for each layer. Compute the consistent mass matrix for each element, and the assembled mass matrix for the composite wall.

(2) A wall is composed of two homogeneous layers, as shown in Figure 5.22. Properties for aluminum are: $\rho = 2700$ kg m^{-3}, $C = 963$ J/kg.K, and $k = 204$ W m^{-1}°C^{-1}, and for copper: $\rho = 8900$ kg m^{-3}, $C = 375$ J/kg.K, and $k = 400$ W m^{-1}°C^{-1}. The initial temperature (time $t = 0$) of the composite wall is $T_o = 300$°C. Suddenly, the temperature on both sides of the wall ($x = 0$ and $x = 14$ mm) is reduced to $T_f = 0$°C and held at this temperature for subsequent times. The transient temperature is analyzed using the finite element method in space and the α-method in time. Use one linear element for each layer, and $\alpha = 1$ and $\Delta t = 0.1$ s, determine temperature at the interface between aluminum and copper at time $t = 0.3$ s.

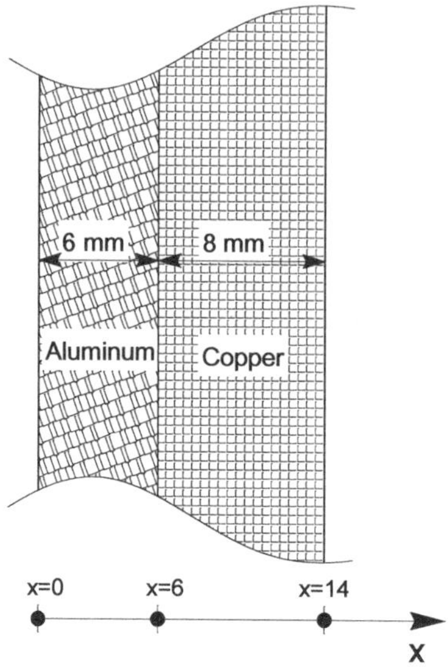

Figure 5.22 A composite wall of two slabs.

(3) A slab of pure gold of thickness of 16 mm is initially ($t = 0$) at temperature of $T_o = 300$°C. Suddenly, the temperature of one of the sides of the slab is dropped to 0°C while the other side is kept at 300°C. The material properties for gold are: $\rho_g = 19300$ kg m^{-3}, $C_g = 129$ J/kg.K, and $k_g = 317$ W m^{-1}°C^{-1}. Transient temperature distribution in the slab is governed by the heat equation

$$\rho C \frac{\partial T}{\partial t} = k \frac{\partial^2 T}{\partial x^2}.$$

The finite-element method in space and the α-method in time are to be used to solve for the temperature $T(x,t)$. Use one quadratic element, and take $\alpha = 1$ and $\Delta t = 0.2$ s. Determine the temperature at the middle node, T_2, at time $t = 0.8$ s.

(4) At time $t = 0$, a slab of pure gold of thickness of 32 mm at a temperature of $T_g = 300$°C is clamped between two slabs of commercial bronze (90% Cu, 10% Al) each of thickness of 16 mm at a temperature of $T_b = 10$°C. Properties for gold are: $\rho_g = 19300$ kg m^{-3}, $C_g = 129$ J/kg.K, and $k_g = 317$ W m^{-1} °C^{-1}, and for bronze: $\rho_b = 8800$ kg m^{-3}, $C_b = 420$ J/kg.K, and $k_b = 52$ W m^{-1} °C^{-1}. The outside surfaces of the bronze slabs are assumed to be insulated. Using the finite-element method, compute and plot the temperature distribution ($T - x$) curve in the composite wall at times $t = 1, 2$, and 3 s. Advance the solution in time by the first-order Euler implicit method ($\alpha = 1$), and use $\Delta t = 0.1$ s. Explain the clear discontinuity in $\partial T / \partial x$ at the interface between gold and bronze. Would it be safe to touch the gold slab after 3 seconds? If not, would it be safe after 1 hour?

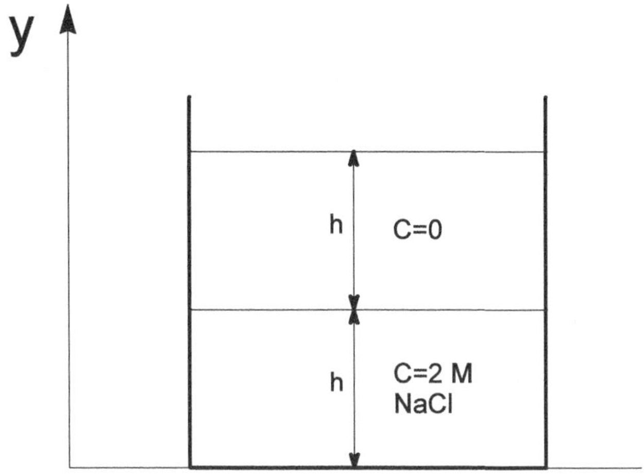

Figure 5.23 Stratified layers of fresh and salt-water.

Hints: Utilize the symmetry about the midplane of the golden slab, and use at least two quadratic elements; one gold element and one bronze element. The discontinuous initial ($t = 0$) temperature at the interface may be replaced by:

$$T_o = \frac{\beta_g T_g + \beta_b T_b}{\beta_g + \beta_b}$$

where

$$\beta_g = \sqrt{\rho_g C_g k_g}$$
$$\beta_b = \sqrt{\rho_b C_b k_b}.$$

There is a singularity at the interface at time $t = 0$, which is difficult to resolve without introducing coordinate transformation. It is possible to alleviate its effects by using the analytic solution for the semi-infinite interface problem to specify smooth initial conditions at a small time t_o. Such a solution can be obtained by Laplace transform or by seeking a self similar solution in the variable

$$\eta = x/\sqrt{\nu t}$$

where $\nu = k/\rho C$, and can be written in terms of error function (also called probability integral).

(5) Two stationary layers of fresh water and salt-water solution are stratified as shown in Figure 5.23. The concentration in the bottom layer is $C = 2M$ NaCl, and in the upper layer $C = 0$. The temperature is 25°C. Each layer is of thickness of $h = 30$ mm. Mass diffusion is assumed to be in the y-direction only. The concentration is governed by the diffusion equation:

$$\frac{\partial C}{\partial t} = D\frac{\partial^2 C}{\partial y^2}$$

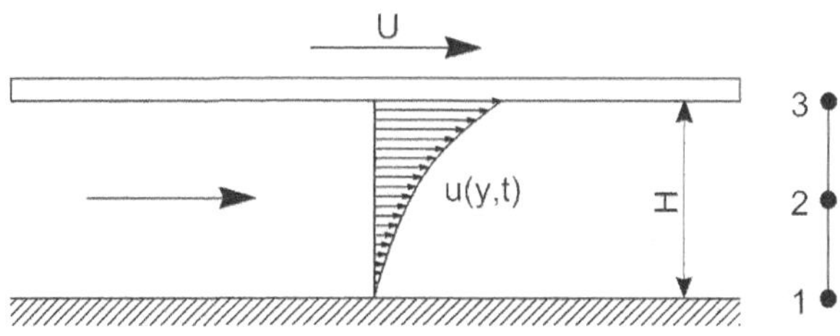

Figure 5.24 Transient Couette flow.

where $D = 1.6 \times 10^{-9}$ m^2/s is the diffusion coefficient at the given temperature. Use the finite-element method to determine the concentration $C(y,t)$ for $0 \leq y \leq 2h$ and $t > 0$. Use at least 5 global nodes, and make choices for: the type of element (linear, quadratic, cubic, etc.), the number of elements, and the time marching scheme (Euler explicit, Euler implicit, Crank–Nicolson, etc.). Also apply appropriate boundary conditions at $y = 0$ and $y = 2h$.
Determine the time when the concentration at $y = 2h$ reaches a value of $C = 0.5M$ NaCl, and plot the concentration distribution (C vs. y) at that time.

(6) An aluminum pin fin of diameter $D = 15$ mm, length $L = 60$ mm, thermal conductivity $k = 204$ W m^{-1}°C^{-1}, density $\rho = 2700$ kg/m^3 and specific heat $C = 963$ J/kg.K is exposed to ambient air at $T_\infty = 20$°C with a convection heat transfer coefficient $\tilde{h} = 300$ W m^{-2}°C^{-1}. The initial temperature of the fin is 20°C. The right end ($x = 60$ mm) of the fin is insulated. The temperature at the left end ($x = 0$) is suddenly increased to $T_o = 220$°C. We wish to solve for the transient temperature distribution in the fin by the finite-element method. Use a mesh of two quadratic elements and determine the assembled (global) mass matrix.

(7) We wish to solve for the transient temperature distribution in the cooling fin of Problem (6). Use a mesh of one cubic element, and the α-method for time advancement. Use appropriate values of α and Δt, determine: The temperature at the insulated end ($x = 60$ mm) and the rate of heat flow into the fin ($x = 0$) at times $t = 1, 2,$ and 3 s.

(8) A layer of viscous fluid is initially at rest ($u = 0$) between two long parallel plates separated by a distance $H = 1$; see Figure 5.24. Suddenly, the upper plate is given a horizontal velocity $U = 1$. The fluid is assumed to move parallel to the plates with velocity $u(y,t)$ that is governed by the x-momentum equation,

$$\frac{\partial u}{\partial t} = \frac{\partial^2 u}{\partial y^2}.$$

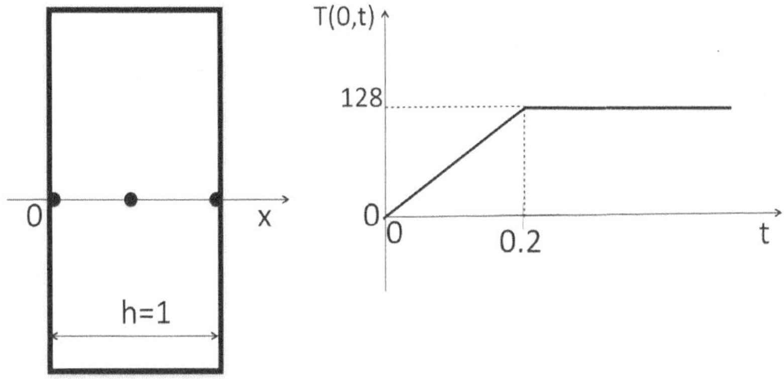

Figure 5.25 Wall of thickness $h=1$, and variation with time of the temperature of side $x=0$.

The finite-element method in space and the α-method in time are used to obtain the algebraic system:

$$[AL]\{U\}^{m+1}=[AR]\{U\}^{m}+\{\bar{F}\}+\{\bar{R}\}.$$

The thickness of the fluid layer is modeled by two linear elements. Use $\alpha=1$ and $\Delta t=0.2$. (a) Determine $[AL]$ and $[AR]$ before application of boundary conditions. (b) Determine the velocity at the middle node, U_2, at time $t=0.4$.

(9) At time $t=0$, a slab of pure gold of thickness of 16 mm at a temperature of $T_g=300°\text{C}$ is clamped between two slabs of silver and bronze each of thickness of 16 mm. Initially, the silver slab is at a temperature of $T_s=100°\text{C}$, and the bronze slab at $T_b=30°\text{C}$. The material properties are: Gold: $\rho_g=19300$ kg m^{-3}, $C_g=129$ J/kg.K, and $k_g=317$ W m^{-1}°C^{-1}. Silver: $\rho_s=10524$ kg m^{-3}, $C_s=234$ J/kg.K, and $k_s=419$ W m^{-1}°C^{-1}. Bronze: $\rho_b=8800$ kg m^{-3}, $C_b=420$ J/kg.K, and $k_b=52$ W m^{-1}°C^{-1}. Transient temperature distribution in the composite wall is to be analyzed by the finite element method. The composite wall is modeled by three quadratic elements; one element for each of the slabs.

(a) Compute consistent mass matrix for each element, and the assembled mass matrix for the composite wall.

(b) Estimate the temperature at the interfaces between the slabs immediately after clamping (averaging the initial temperatures is inaccurate, see the hint for Problem 4).

(10) A wall is initially at a temperature of 0°C, see Figure 5.25. The left surface ($x=0$) is heated such that its temperature increases with time as shown in the figure, whereas the right surface ($x=1$) is insulated for subsequent times. The

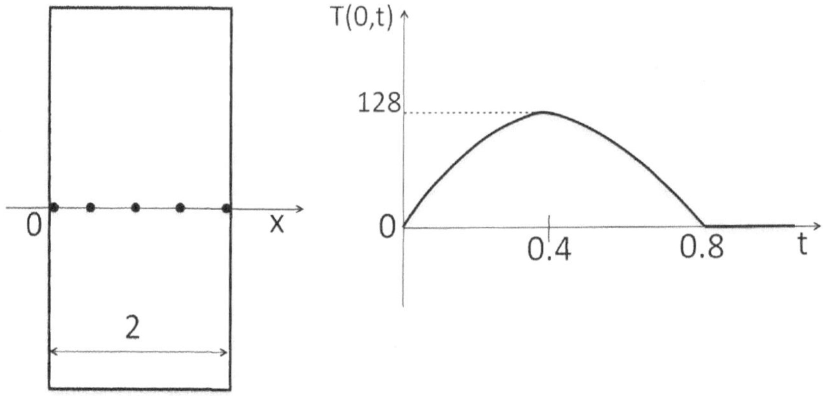

Figure 5.26 Wall of thickness 2 units, and variation with time of the temperature of side $x = 0$.

energy equation is

$$\frac{\partial T}{\partial t} = \frac{\partial^2 T}{\partial x^2} \tag{5.282}$$

where t and x are nondimensional time and space coordinate, respectively. The wall thickness is one unit.

The transient wall temperature is analyzed using the finite element method in space and the α-method in time. Use one quadratic element, and $\alpha = 1$ and $\Delta t = 0.1$. At time $t = 0.3$, determine: (a) the temperature at $x = 1$, and (b) the heat flux at $x = 0$.

(11) Wall temperature $T(x,t)$ in transient heat conduction is governed by the energy equation

$$\frac{\partial T}{\partial t} = \frac{\partial^2 T}{\partial x^2} \tag{5.283}$$

where t and x are nondimensional time and space coordinate, respectively, see Figure 5.26. The wall thickness is two units. The wall initial temperature is zero. The surface ($x = 2$) is insulated at all times. The surface ($x = 0$) is heated such that its temperature changes with time

$$T(0,t) = 128 sin\left(\frac{5\pi}{4}t\right) \qquad \text{for } 0 \leq t \leq 0.8. \tag{5.284}$$

Model the wall by two quadratic elements, and use the α-method in time with $\alpha = 1$ and $\Delta t = 0.1$. At time $t = 0.8$: (a) Find the temperature at all nodes, and (b) the heat flux at $x = 0$.

(12) A layer of viscous fluid is initially at rest ($u = 0$) between two long parallel plates separated by a distance $H = 1$; see Figure 5.24. Suddenly, the upper plate is given a horizontal velocity $U = 1$. The fluid is assumed to move parallel to the plates with velocity $u(y,t)$ that is governed by the $x-$momentum equation,

$$\frac{\partial u}{\partial t} = \frac{\partial^2 u}{\partial y^2}.$$

The finite element method in space and the $\alpha-$method in time are used to obtain the algebraic system:

$$[AL]\{U\}^{m+1} = [AR]\{U\}^m + \{\bar{F}\} + \{\bar{R}\}.$$

The thickness of the fluid layer is modeled by one quadratic element. Use $\alpha = 1$ and $\Delta t = \frac{1}{12}$. (a) Determine $[AL]$ and $[AR]$ before application of boundary conditions. (b) Determine the velocity at the middle node, U_2, at time $t = \frac{1}{4}$.

(13) Transient temperature for heat conduction in a solid is governed by

$$\rho C \frac{\partial T}{\partial t} = k(\frac{\partial^2 T}{\partial x^2} + \frac{\partial^2 T}{\partial y^2})$$

Consider heat transfer in a rectangular domain with boundary conditions:

$$\begin{aligned} x &= 0: \; k\frac{\partial T}{\partial x} = q_o \\ x &= a: \; T = 0 \\ y &= 0: \; T = T_o sin(\pi t) sin(\pi x) \\ y &= b: \; k\frac{\partial T}{\partial y} = 0. \end{aligned}$$

(a) Derive the weak form and classify the boundary conditions as natural or essential.
(b) A finite-element model of this problem is written as

$$[M^e]\{\dot{T}^e\} + [K^e]\{T^e\} = \{F^e\} + \{R^e\}$$

where the consistent mass matrix is

$$m_{ij}^e = \int_{A^e} \rho C \psi_i \psi_j dA.$$

Assume that ρC is element-wise uniform, write down the consistent mass matrix $[M^e]$ for a square element of side h.

(14) Transient heat conduction in a two-dimensional domain is governed by the energy equation:

$$\rho C \frac{\partial T}{\partial t} - \left[\frac{\partial}{\partial x}\left(k\frac{\partial T}{\partial x}\right) + \frac{\partial}{\partial y}\left(k\frac{\partial T}{\partial y}\right)\right] + \beta T - f = 0$$

where k is the thermal conductivity, f is the rate of energy generated per unit volume, and β is a constant coefficient. The domain is divided into rectangular elements and the temperature over each element is approximated by

$$T^e(x,y,t) = T_1^e \psi_1(x,y) + T_2^e \psi_2(x,y) + T_3^e \psi_3(x,y) + T_4^e \psi_4(x,y)$$

where ψ_i are the shape functions of the bilinear rectangular element. Using this approximation and the Galerkin method, we obtain the element equations:

$$[M^e]\{\dot{T}^e\} + [A^e]\{T^e\} = \{F^e\} + \{R^e\}.$$

Take $\rho C = 12$, $k = 36$, $\beta = 6$, and $f = 3$, write down the matrices $[M^e]$ and $[A^e]$ for an element of sides $a = 2$ and $b = 3$.

6 Hyperbolic Equations: Waves and Vibrations Problems

In this chapter we consider wave phenomena and vibrations of continuous media; problems that are described by hyperbolic equations. Such equations admit discontinuous solutions in space and time. The discontinuities may be present in the initial conditions or may develop in time for nonlinear problems or if the speed of wave propagation is a function of space. The Galerkin finite-element method can be used for hyperbolic equations as long as the solution remains smooth, but the method develops spurious oscillations at discontinuities. Therefore, the current treatment deals with simple problems whose solutions remain smooth in space and time, and thus we can use the Galerkin method. At the risk of obtaining less accurate solutions, artificial damping can be used to smooth discontinuities.

Problems in this chapter cover transverse, longitudinal, and torsional vibrations of one-dimensional structures such as strings, beams, and bars. We solve for the displacement field, and hence the acceleration is represented by a second-order time derivative. For strings and bars the space derivative is second order, whereas for Euler–Bernoulli's beam equation the space derivative is fourth order. Lagrange interpolation functions can be used for the first class of problems, but we need Hermite cubic interpolation functions for beams. Consequently, we consider two model problems: the first is second-order differential in space and the second is fourth order. Application of the the finite-element method to the space operator, whether it is second or fourth order, produces a coupled system of second-order ordinary differential equations in time. Starting from prescribed initial conditions, we use the finite-difference method to integrate the system in time. For all problems presented here, we use Newmark method. In this chapter we are interested in the response of the aforementioned structures in the time domain. The study of free vibrations of structures assuming harmonic response gives rise to an eigenvalue problem and will be presented in the next chapter.

6.1 MODEL PROBLEMS

We consider two model problems, the first problem is second-order differential equation in time and space,

$$\mu\frac{\partial^2 u}{\partial t^2}+\nu\frac{\partial u}{\partial t}-\frac{\partial}{\partial x}[p\frac{\partial u}{\partial x}]+qu-f=0 \tag{6.1}$$

DOI: 10.1201/9781003323150-6

where μ, ν, p, and q are given functions of space coordinate x, and f is a given function of space x and time t. This equation describes waves and vibrations phenomena including transverse waves on taut elastic strings, longitudinal elastic waves on slender rods (bars), torsional vibrations of shafts, the damped wave model for heat conduction, among other wave phenomena.

The **equation of motion for transverse vibrations of a taut string** is

$$\rho_l \frac{\partial^2 v}{\partial t^2} - T\frac{\partial^2 v}{\partial x^2} - f(x,t) = 0 \tag{6.2}$$

where v is the transverse displacement, ρ_l is mass density per unit length (kg/m), T (N) is the tension, and $f(x,t)$ is distributed transverse force per unit length (N/m). The wave propagates in the x-direction with wave speed of $\sqrt{T/\rho_l}$ whereas the particle velocity is normal to the string.

The **equation of motion for longitudinal vibrations of a bar** is

$$\rho A\frac{\partial^2 u}{\partial t^2} - \frac{\partial}{\partial x}[EA\frac{\partial u}{\partial x}] - f(x,t) = 0 \tag{6.3}$$

where u is the axial (longitudinal) displacement, ρ is mass density (kg/m^3), A is cross-sectional area, E is Young's modulus, and $f(x,t)$ is distributed axial force per unit length. The wave propagates in the x-direction with wave speed of $\sqrt{E/\rho}$ and the particle velocity is also along the bar.

The second model problem is second-order differential equation in time and fourth-order in space. The typical problem is the **equation of motion for transverse vibrations of a beam** according to Euler–Bernoulli's beam theory, that is

$$\rho A\frac{\partial^2 v}{\partial t^2} + \frac{\partial^2}{\partial x^2}\left[D\frac{\partial^2 v}{\partial x^2}\right] - \frac{\partial}{\partial x}\left(\rho I\frac{\partial^3 v}{\partial t^2 \partial x}\right) + qv - f(x,t) = 0 \qquad 0 < x < L \tag{6.4}$$

where $D = EI$ is the flexural rigidity, which may vary with x, ρ is the mass density (kg/m^3), A is the beam cross-sectional area, and I is section moment of inertia about the bending axis. The third term is the rotary inertia [119], and the fourth term (qv) represents the reaction due to an elastic foundation, where q is the stiffness coefficient of the foundation (N/m^2).

6.2 THE WEAK FORMS AND FINITE ELEMENT MODELS

6.2.1 FIRST MODEL PROBLEM: SECOND-ORDER IN SPACE AND TIME

The derivation of the weak form for model problem Equation (6.1) follows the same steps for the parabolic equation of Section 5.1. For an element with n local nodes denoted by $x_1^e, x_2^e, \cdots, x_n^e$, the weak form is

$$\int_{x_1^e}^{x_n^e} w\mu \frac{\partial^2 u}{\partial t^2} dx + \int_{x_1^e}^{x_n^e} w\nu \frac{\partial u}{\partial t} dx + \int_{x_1^e}^{x_n^e} \left[\frac{\partial w}{\partial x} p \frac{\partial u}{\partial x} + wqu \right] dx$$

$$- \int_{x_1^e}^{x_n^e} wf dx - w(x_n^e) R_n^e - w(x_1^e) R_1^e = 0 \tag{6.5}$$

where

$$\begin{aligned} R_1^e &= -\left(p \frac{\partial u}{\partial x} \right)_{x=x_1^e} \\ R_n^e &= \left(p \frac{\partial u}{\partial x} \right)_{x=x_n^e} . \end{aligned} \tag{6.6}$$

Using Lagrange interpolation polynomials for approximating the solution in space and the Galerkin method, we obtain the following element equations:

$$[M^e] \{\ddot{U}^e\} + [C^e] \{\dot{U}^e\} + [A^e] \{U^e\} = \{F^e\} + \{R^e\}. \tag{6.7}$$

The mass matrix corresponding to the inertia term is

$$[M^e] = \int_{x_1^e}^{x_n^e} \mu \{\psi\}^T \{\psi\} dx. \tag{6.8}$$

The damping coefficient matrix has the same structure as the mass matrix and is given by

$$[C^e] = \int_{x_1^e}^{x_n^e} \nu \{\psi\}^T \{\psi\} dx, \tag{6.9}$$

the coefficient matrix is

$$\begin{aligned} [A^e] &= \int_{x_1^e}^{x_n^e} \left([B]^T p [B] + q \{\psi\}^T \{\psi\} \right) dx \\ &= [K^e] + [M^e]_q \end{aligned} \tag{6.10}$$

where the stiffness matrix is

$$[K^e] = \int_{x_1^e}^{x_n^e} [B]^T p [B] dx. \tag{6.11}$$

The gradient matrix $[B]$ is defined by Equation (5.23). The mass matrix for the q term is

$$[M^e]_q = \int_{x_1^e}^{x_n^e} q \{\psi\}^T \{\psi\} dx. \tag{6.12}$$

The forcing vector is

$$\{F^e\} = \int_{x_1^e}^{x_n^e} f \{\psi\}^T dx. \tag{6.13}$$

and the vector of secondary variables is

$$\{R^e\} = \{\psi(x_n^e)\}^T R_n^e + \{\psi(x_1^e)\}^T R_1^e. \tag{6.14}$$

The shape functions for the Lagrange elements (linear, quadratic, cubic, and higher order) are given in Chapter 2. The corresponding element matrices for element-wise uniform μ, p, q, and f are also given in that chapter. The mass matrix is repeated here, for the linear element

$$[M^e] = \frac{\mu_e h_e}{6}\begin{bmatrix} 2 & 1 \\ 1 & 2 \end{bmatrix} \tag{6.15}$$

and for the quadratic element

$$[M^e] = \frac{\mu_e h_e}{30}\begin{bmatrix} 4 & 2 & -1 \\ 2 & 16 & 2 \\ -1 & 2 & 4 \end{bmatrix}. \tag{6.16}$$

We obtain the damping matrix for the ν and mass matrix for q terms by replacing μ by ν and q in Equations (6.15) and (6.16), respectively.

6.2.2 SECOND MODEL PROBLEM: TRANSVERSE VIBRATIONS OF BEAMS

Now, we consider the beam equation,

$$\rho A\frac{\partial^2 v}{\partial t^2} + \frac{\partial^2}{\partial x^2}\left[D\frac{\partial^2 v}{\partial x^2}\right] - \frac{\partial}{\partial x}\left(\rho I\frac{\partial^3 v}{\partial t^2 \partial x}\right) + qv - f(x,t) = 0 \qquad 0 < x < L \tag{6.17}$$

where $D = EI$ is the flexural rigidity, which may vary with x, ρ is the mass density (kg/m^3), and A is the beam cross-sectional area. In this section, we neglect the rotary inertia given by the third term of Equation (6.17). We consider an element that is defined by two nodes with global coordinates x_1^e and x_2^e. The weak form for element (e) is written as

$$\int_{x_1^e}^{x_2^e} \rho A w\frac{\partial^2 v}{\partial t^2}dx + \int_{x_1^e}^{x_2^e} \frac{\partial^2 w}{\partial x^2}D\frac{\partial^2 v}{\partial x^2}dx + \int_{x_1^e}^{x_2^e} wqv dx = \int_{x_1^e}^{x_2^e} wf dx$$
$$+ w(x_1^e)R_1^e + w(x_2^e)R_3^e + (\frac{dw}{dx}|_{x_1^e})R_2^e + (\frac{dw}{dx}|_{x_2^e})R_4^e \tag{6.18}$$

where $w(x)$ is a test (weight) function, and

$$\begin{aligned} R_1^e &= \left[\frac{\partial}{\partial x}(D\frac{\partial^2 v}{\partial x^2})\right]_{x_1^e} \\ R_2^e &= -\left(D\frac{\partial^2 v}{\partial x^2}\right)_{x_1^e} \\ R_3^e &= -\left[\frac{\partial}{\partial x}\left(D\frac{\partial^2 v}{\partial x^2}\right)\right]_{x_2^e} \\ R_4^e &= \left(D\frac{\partial^2 v}{\partial x^2}\right)_{x_2^e} \end{aligned} \tag{6.19}$$

are the secondary variables at the element's nodes. Note that they are defined in terms of the derivatives of the unknown deflection $v(x,t)$; thus, we have to consider them as unknowns. We note that R_1^e and $-R_2^e$ are the shear force (Q) and bending moment (M) at the first node of the element, while $-R_3^e$ and R_4^e are Q and M at the second node of the element. There are no boundary conditions to be imposed at the element level. Boundary conditions are only known at certain nodes. These conditions will be applied to the global system that we will obtain by assembling all elements. At a node that is common to two adjacent elements, say elements $e-1$ and e, the secondary variables satisfy

$$\begin{aligned} R_3^{e-1} + R_1^e &= F_{ext} \\ R_4^{e-1} + R_2^e &= M_{ext} \end{aligned} \tag{6.20}$$

where F_{ext} and M_{ext} are externally applied concentrated force and moment at the common node; they can be functions of time. F_{ext} is positive when upward, and M_{ext} is positive when counterclockwise. These are externally applied loads not accounted for by the distributed load $f(x,t)$. Secondary variables and externally applied loads are, in general, functions of time.

Approximation and Hermite Cubic Shape Functions

For longitudinal waves on rods or transverse vibrations of a taut string, the differential equations are second-order in space and time. The weak form contains only the first derivative $\partial u/\partial x$ of the unknown function, and hence it requires continuity of the variable $u(x,t)$ only. Lagrange family of shape functions (linear, quadratic, cubic, etc.) allows us to satisfy the continuity requirement. For transverse vibrations of beams, we have a fourth-order differential equation in space x. The weak form contains the second derivative $\partial^2 v/\partial x^2$, and hence it requires continuity of both the function $v(x,t)$ and its first derivative $\partial v/\partial x$. Lagrange interpolation (shape) functions are not admissible in this case because they do not allow us to enforce continuity of the slope at the inter-element nodes even if we use quadratic or cubic elements. This is because the approximation over the element domain is expressed in terms of the nodal values of the unknown function but not its derivative. To be able to enforce continuity of both the function and its slope between elements, we have to express the approximation over the element's domain in terms of the nodal values of the function v and its first derivative $\partial v/\partial x$. There are two degrees of freedom at each node (deflection and slope), and hence for a two-node element, there are four degrees of freedom. The approximation $V^e(x,t)$ over the element's domain must be cubic in x.

We introduce the definitions,

$$\begin{aligned} x &= x_1^e: & V^e &= U_1^e \\ x &= x_1^e: & \frac{\partial V^e}{\partial x} &= U_2^e \\ x &= x_2^e: & V^e &= U_3^e \\ x &= x_2^e: & \frac{\partial V^e}{\partial x} &= U_4^e. \end{aligned} \tag{6.21}$$

We note here that U_1^e and U_3^e are the deflections at nodes one and two of element e, respectively, and U_2^e and U_4^e are the slopes at those nodes. The trial solution over an element is

$$v(x,t) \approx V^e(x,t) = \sum_{j=1}^{4} U_j^e(t)\psi_j(x). \tag{6.22}$$

The shape functions $\psi_j(x)$ are the Hermite cubic polynomials given by Equation (3.39). To write $V^e(x,t)$ in matrix form, we define the arrays

$$\{\psi(x)\} = \{\psi_1(x)\ \psi_2(x)\ \psi_3(x)\ \psi_4(x)\} \tag{6.23}$$

$$\{U^e(t)\} = \begin{Bmatrix} U_1^e(t) \\ U_2^e(t) \\ U_3^e(t) \\ U_4^e(t) \end{Bmatrix}. \tag{6.24}$$

In matrix form, the trial solution over element # e is

$$V^e(x,t) = \{\psi(x)\}\{U^e(t)\}. \tag{6.25}$$

We note that the dependence of the trial solution on x is given by the array $\{\psi(x)\}$, whereas variation with time is given by the array $\{U^e(t)\}$, which are the values of the primary variable at the element nodes.

Transverse acceleration (the second-derivative with respect to time) of the trial solution is

$$\frac{\partial^2 V^e(x,t)}{\partial t^2} = \{\psi(x)\}\{\ddot{U}^e(t)\} \tag{6.26}$$

where

$$\{\ddot{U}^e(t)\} \equiv \frac{d^2}{dt^2}\{U^e(t)\} \tag{6.27}$$

and the acceleration at the elements nodes are

$$\ddot{U}_j^e(t) \equiv \frac{d^2 U_j^e(t)}{dt^2}. \tag{6.28}$$

The second-derivative with respect to x of the trial solution is

$$\frac{\partial^2 V^e(x,t)}{\partial x^2} = \left[\frac{d^2\psi_1(x)}{dx^2}\ \frac{d^2\psi_2(x)}{dx^2}\ \frac{d^2\psi_3(x)}{dx^2}\ \frac{d^2\psi_4(x)}{dx^2}\right]\begin{Bmatrix} U_1^e(t) \\ U_2^e(t) \\ U_3^e(t) \\ U_4^e(t) \end{Bmatrix}. \tag{6.29}$$

We define the curvature matrix $[B]$ by

$$[B] = \left[\frac{d^2\psi_1(x)}{dx^2}\ \frac{d^2\psi_2(x)}{dx^2}\ \frac{d^2\psi_3(x)}{dx^2}\ \frac{d^2\psi_4(x)}{dx^2}\right] \tag{6.30}$$

and obtain

$$\frac{\partial^2 V^e(x,t)}{\partial x^2} = [B]\{U^e(t)\}. \tag{6.31}$$

Galerkin

In the Galerkin method, the test (weight) function $w(x)$ is a linear combination of the basis functions. Thus for element # e, we have the approximation for $w(x)$,

$$W^e(x) = \{\psi_1(x)\ \psi_2(x)\ \psi_3(x)\ \psi_4(x)\}\left\{\begin{array}{c} d_1 \\ d_2 \\ d_3 \\ d_4 \end{array}\right\} \tag{6.32}$$

$$W^e(x) = \{d\}^T\{\psi(x)\}^T. \tag{6.33}$$

Similar to the space second-derivative of $V^e(x,t)$, we write the second-derivative of $W^e(x)$ using the matrix $[B]$,

$$\frac{d^2W^e(x)}{dx^2} = \{d\}^T[B]^T. \tag{6.34}$$

Element Matrices: Galerkin Finite Element Model

To obtain the element matrices, we choose four linearly independent weight (test) functions and apply the weak form to the approximate solution $V^e(x,t)$. In the Galerkin method, the weight (test) functions are the same as the shape functions, $W_i^e = \psi_i$ for $i = 1,...,4$. Substituting the approximate solution into the weak form, we get

$$\begin{aligned}
&\int_{x_1^e}^{x_2^e} \rho A\psi_i \sum_{j=1}^{4} \ddot{U}_j^e(t)\psi_j(x)dx + \int_{x_1^e}^{x_2^e} \frac{d^2\psi_i}{dx^2} D \sum_{j=1}^{4} U_j^e(t)\frac{d^2\psi_j}{dx^2}dx \\
&+ \int_{x_1^e}^{x_2^e} q\psi_i \sum_{j=1}^{4} U_j^e(t)\psi_j(x)dx = \int_{x_1^e}^{x_2^e} \psi_i f dx \\
&+ \psi_i(x_1^e)R_1^e + \psi_i(x_2^e)R_3^e + (\frac{d\psi_i}{dx}|_{x_1^e})R_2^e + (\frac{d\psi_i}{dx}|_{x_2^e})R_4^e
\end{aligned} \tag{6.35}$$

$$\begin{aligned}
&\sum_{j=1}^{4}\left(\int_{x_1^e}^{x_2^e} \rho A\psi_i\psi_j dx\right)\ddot{U}_j^e(t) + \sum_{j=1}^{4}\left(\int_{x_1^e}^{x_2^e}\left[\frac{d^2\psi_i}{dx^2}D\frac{d^2\psi_j}{dx^2} + q\psi_i\psi_j\right]dx\right)U_j^e(t) = \\
&\int_{x_1^e}^{x_2^e} \psi_i f dx + \psi_i(x_1^e)R_1^e + \psi_i(x_2^e)R_3^e + (\frac{d\psi_i}{dx}|_{x_1^e})R_2^e + (\frac{d\psi_i}{dx}|_{x_2^e})R_4^e.
\end{aligned} \tag{6.36}$$

In index notation, the element equations are

$$\sum_{j=1}^{4} m_{ij}^e\ddot{U}_j^e(t) + \sum_{j=1}^{4} a_{ij}^e U_j^e(t) = F_i^e + R_i^e \qquad i = 1,...,4. \tag{6.37}$$

The mass matrix:

$$m_{ij}^e = \int_{x_1^e}^{x_2^e} \rho A \psi_i \psi_j dx. \tag{6.38}$$

The coefficient matrix:

$$a_{ij}^e = \int_{x_1^e}^{x_2^e} \left(\frac{d^2\psi_i}{dx^2} D \frac{d^2\psi_j}{dx^2} + q\psi_i\psi_j \right) dx. \tag{6.39}$$

Forcing vector:

$$F_i^e = \int_{x_1^e}^{x_2^e} \psi_i f dx. \tag{6.40}$$

Secondary variables (shear force and bending moment at the nodes):

$$R_i^e = \psi_i(x_1^e)R_1^e + \psi_i(x_2^e)R_3^e + (\frac{d\psi_i}{dx}|_{x_1^e})R_2^e + (\frac{d\psi_i}{dx}|_{x_2^e})R_4^e. \tag{6.41}$$

Recalling the properties of the Hermite cubic interpolation polynomials:

$$\begin{aligned} \psi_i(x_1^e) &= \delta_{i1} \\ \psi_i(x_2^e) &= \delta_{i3} \\ \frac{d\psi_i}{dx}|_{x_1^e} &= \delta_{i2} \\ \frac{d\psi_i}{dx}|_{x_2^e} &= \delta_{i4} \end{aligned} \tag{6.42}$$

we obtain the finite-element model in matrix form

$$[M^e]\{\ddot{U}^e\} + [A^e]\{U^e\} = \{F^e\} + \{R^e\} \tag{6.43}$$

where the mass matrix is

$$[M^e] = \int_{x_1^e}^{x_2^e} \rho A \{\psi\}^T \{\psi\} dx \tag{6.44}$$

and the coefficient matrix is

$$\begin{aligned} [A^e] &= \int_{x_1^e}^{x_2^e} \left([B]^T D [B] + q\{\psi\}^T \{\psi\} \right) dx \\ &= [K^e] + [M^e]_q \end{aligned} \tag{6.45}$$

which is composed of the stiffness matrix

$$[K^e] = \int_{x_1^e}^{x_2^e} [B]^T D [B] dx \tag{6.46}$$

and the mass matrix due to the elastic foundation

$$[M^e]_q = \int_{x_1^e}^{x_2^e} q\{\psi\}^T \{\psi\} dx. \tag{6.47}$$

The forcing vector is

$$\{F^e(t)\} = \int_{x_1^e}^{x_2^e} f(x,t)\{\psi\}^T dx \tag{6.48}$$

and the vector of secondary variables is

$$\{R^e\} = \begin{Bmatrix} R_1^e \\ R_2^e \\ R_3^e \\ R_4^e \end{Bmatrix}. \tag{6.49}$$

To evaluate these matrices, we need to specify the functions ρA, $D(x) = EI$, and $f(x,t)$. For simplicity, we consider a special case where these functions are element-wise constant $\rho A = (\rho A)_e$, $D = (EI)_e$, $q = q_e$, and $f = f_e(t)$.

$$[M^e] = \frac{(\rho A)_e h_e}{420} \begin{bmatrix} 156 & 22h_e & 54 & -13h_e \\ 22h_e & 4h_e^2 & 13h_e & -3h_e^2 \\ 54 & 13h_e & 156 & -22h_e \\ -13h_e & -3h_e^2 & -22h_e & 4h_e^2 \end{bmatrix} \tag{6.50}$$

$$[K^e] = \frac{2(EI)_e}{h_e^3} \begin{bmatrix} 6 & 3h_e & -6 & 3h_e \\ 3h_e & 2h_e^2 & -3h_e & h_e^2 \\ -6 & -3h_e & 6 & -3h_e \\ 3h_e & h_e^2 & -3h_e & 2h_e^2 \end{bmatrix} \tag{6.51}$$

$$\{F^e(t)\} = \frac{f_e(t)h_e}{12} \begin{Bmatrix} 6 \\ h_e \\ 6 \\ -h_e \end{Bmatrix}. \tag{6.52}$$

The matrix for q is obtained from Equation (6.50) by using q_e in place of $(\rho A)_e$. To account for structural damping we add a damping term proportional to the velocity, and rewrite the element equations Equation (6.43),

$$[M^e]\{\ddot{U}^e\} + [C^e]\{\dot{U}^e\} + [A^e]\{U^e\} = \{F^e\} + \{R^e\}. \tag{6.53}$$

It is evident that the finite element model of the second model problem given by Equation (6.53) is identical to that of the first model problem given by Equation (6.7). After assembly of all elements, we use the same time integration scheme for both model problems. Newmark method [79] is used for time integration and is presented next.

6.3 TIME ADVANCEMENT SCHEME: NEWMARK METHOD

We assemble all elements into a global system and apply boundary conditions. The system will be a coupled system of second-order differential equations in time of the form

$$[M]\{\ddot{U}\} + [C]\{\dot{U}\} + [A]\{U\} = \{Q\}. \tag{6.54}$$

The matrix $[C]$ is added to account for structural damping, which is assumed to be proportional to the velocity. Integration of Equation (6.54) in time requires initial conditions for the generalized nodal displacements $\{U\}$ and velocity $\{\dot{U}\}$. The problem statement must provide initial values for the deflection and velocity. At time $t = 0$, we have

$$v = v_o(x) \tag{6.55}$$

and

$$\dot{v} = \dot{v}_o(x). \tag{6.56}$$

To obtain the initial slope and its time derivative, we simply differentiate Equations (6.55) and (6.56) with respect to x, hence at $t = 0$ we have

$$\frac{dv}{dx} = \frac{dv_o(x)}{dx} \tag{6.57}$$

$$\frac{d\dot{v}}{dx} = \frac{d\dot{v}_o(x)}{dx}. \tag{6.58}$$

Evaluation of Equations (6.55)–(6.58) at the global nodes provides the initial values $\{U\}^0$ and $\{\dot{U}\}^0$. The initial acceleration is obtained from the system Equation (6.54),

$$\{\ddot{U}\}^0 = [M]^{-1}\left[\{Q\}^0 - [C]\{\dot{U}\}^0 - [K]\{U\}^0\right]. \tag{6.59}$$

Now we have initial values for the generalized displacement, velocity, and acceleration. We want to find the displacement and velocity at future time.

Newmark Method

Newmark method [79] is a numerical scheme for time advancement of system Equation (6.54) for given initial values of $\{U\}$ and $\{\dot{U}\}$. Consider the Taylor series expansion of a function of time $f(t)$ about time level t_m,

$$f^{m+1} = f^m + \left(\frac{df}{dt}\right)^m \Delta t + \frac{1}{2}\left(\frac{d^2f}{dt^2}\right)^m (\Delta t)^2 + \frac{1}{6}\left(\frac{d^3f}{dt^3}\right)^m (\Delta t)^3 + \dots \tag{6.60}$$

where $f^m = f(t_m)$ and $f^{m+1} = f(t_m + \Delta t)$. Introducing the notation

$$\dot{f}^m = \left(\frac{df}{dt}\right)^m \tag{6.61}$$

$$\ddot{f}^m = \left(\frac{d^2f}{dt^2}\right)^m \tag{6.62}$$

we have the approximation,

$$f^{m+1} \approx f^m + \dot{f}^m \Delta t + \frac{1}{2}\ddot{f}^{m+\gamma}(\Delta t)^2 \tag{6.63}$$

where

$$\ddot{f}^{m+\gamma} = (1-\gamma)\ddot{f}^m + \gamma\ddot{f}^{m+1} \tag{6.64}$$

where γ is a parameter to be specified later. Equation (6.63) is the displacement formula for constant acceleration except that the acceleration is a weighted average of the acceleration at the two time levels m and $m+1$.

Applying formula Equation (6.63) to the array of generalized displacements, we obtain

$$\{U\}^{m+1} = \{U\}^m + \{\dot{U}\}^m \Delta t + \frac{1}{2}\{\ddot{U}\}^{m+\gamma}(\Delta t)^2 \tag{6.65}$$

where

$$\{\ddot{U}\}^{m+\gamma} = (1-\gamma)\{\ddot{U}\}^m + \gamma\{\ddot{U}\}^{m+1}. \tag{6.66}$$

Displacement Update

Substituting Equation (6.66) into Equation (6.65), we get

$$\{U\}^{m+1} = \{U\}^m + \{\dot{U}\}^m \Delta t + \frac{1}{2}(1-\gamma)(\Delta t)^2\{\ddot{U}\}^m + \frac{1}{2}\gamma(\Delta t)^2\{\ddot{U}\}^{m+1}. \tag{6.67}$$

Similarly, for the velocity approximation we use

$$\{\dot{U}\}^{m+1} = \{\dot{U}\}^m + \{\ddot{U}\}^{m+\alpha}\Delta t \tag{6.68}$$

where

$$\{\ddot{U}\}^{m+\alpha} = (1-\alpha)\{\ddot{U}\}^m + \alpha\{\ddot{U}\}^{m+1} \tag{6.69}$$

where α is a parameter; not necessarily equal to γ.

Velocity Update

Substituting Equation (6.69) into Equation (6.68), we get

$$\{\dot{U}\}^{m+1} = \{\dot{U}\}^m + (1-\alpha)\Delta t\{\ddot{U}\}^m + \alpha\Delta t\{\ddot{U}\}^{m+1}. \tag{6.70}$$

We assume the matrices $[M]$, $[C]$, and $[A]$ to be constant in time, and write the system Equation (6.54) at time $m+1$,

$$[M]\{\ddot{U}\}^{m+1} = \{Q\}^{m+1} - [C]\{\dot{U}\}^{m+1} - [A]\{U\}^{m+1}. \tag{6.71}$$

Using the velocity update formula Equation (6.70), we obtain

$$\begin{aligned} [M]\{\ddot{U}\}^{m+1} &= \{Q\}^{m+1} \\ -[C]\left[\{\dot{U}\}^m + (1-\alpha)\Delta t\{\ddot{U}\}^m + \alpha\Delta t\{\ddot{U}\}^{m+1}\right] &- [A]\{U\}^{m+1}. \end{aligned} \tag{6.72}$$

Rearranging terms, we get

$$\begin{aligned} ([M] + \alpha\Delta t[C])\{\ddot{U}\}^{m+1} &= \{Q\}^{m+1} \\ -[C]\left[\{\dot{U}\}^m + (1-\alpha)\Delta t\{\ddot{U}\}^m\right] &- [A]\{U\}^{m+1}. \end{aligned} \tag{6.73}$$

Solving for the acceleration at time step $m+1$, we have

$$\{\ddot{U}\}^{m+1} = ([M] + \alpha\Delta t\,[C])^{-1} \times \left[\{Q\}^{m+1} - [C]\left(\{\dot{U}\}^{m} + (1-\alpha)\Delta t\,\{\ddot{U}\}^{m}\right) - [A]\{U\}^{m+1}\right]. \tag{6.74}$$

Substituting into the displacement update formula Equation (6.67), and rearranging terms, we obtain

$$\left[[M] + \alpha\Delta t\,[C] + \frac{1}{2}\gamma(\Delta t)^2\,[A]\right]\{U\}^{m+1} = ([M] + \alpha\Delta t\,[C])\left[\{U\}^{m} + \{\dot{U}\}^{m}\Delta t + \frac{1}{2}(1-\gamma)(\Delta t)^2\{\ddot{U}\}^{m}\right] + \frac{1}{2}\gamma(\Delta t)^2\left[\{Q\}^{m+1} - [C]\left(\{\dot{U}\}^{m} + (1-\alpha)\Delta t\,\{\ddot{U}\}^{m}\right)\right]. \tag{6.75}$$

Knowing $\{U\}^m$, $\{\dot{U}\}^m$, and $\{\ddot{U}\}^m$, we determine the solution at time level $m+1$ as follows:

(1) Solve the system Equation (6.75) for $\{U\}^{m+1}$. Note that $\{Q\}$ is a known function of time, and thus can be evaluated at any m.

(2) Solve the system Equation (6.73) for $\{\ddot{U}\}^{m+1}$.

(3) Use Equation (6.70) to compute $\{\dot{U}\}^{m+1}$.

The parameters α and γ have significant effects on the stability and accuracy of the Newmark method. Hughes [53] presented in-depth analysis of the stability and accuracy of the method. For undamped systems ($[C]=0$), the method is unconditionally stable (no restriction on the magnitude of Δt) if $\gamma \geq \alpha \geq 1/2$. The method is second-order accurate if and only if $\alpha = 1/2$. Therefore, the values of $\alpha = 1/2$ and $\gamma = 1/2$ are recommended. Numerical examples presented in this chapter demonstrate that these values provide the best accuracy and stability of the Newmark method if the strain field remains smooth in space and time. However, if discontinuities in strain develops, spurious (numerical) high frequencies develop. Damping of high frequencies can be achieved by using $\alpha > 1/2$, and for that purpose Hughes [53] recommended $\gamma = (\alpha + 1/2)^2/2$. (The parameters γ, and 2β in [53] are α, and γ of the present chapter, respectively).

6.4 APPLICATIONS: WAVES AND VIBRATIONS ON STRINGS, BARS, AND BEAMS

Example 6.1. Waves on a Taut String

We consider a taut string of length $8b$, density per unit length $\rho\prime$, and tension T. We let $C = \sqrt{T/\rho\prime}$ denote the speed of propagation of transverse waves on the string.

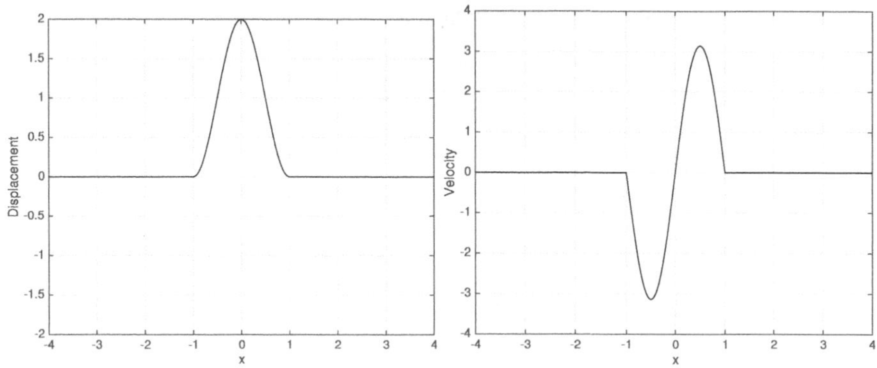

(a) (b)

Figure 6.1 Initial displacement and velocity of a taut string.

The initial conditions are

$$\begin{aligned} v(x,0) &= v_o[1+cos(\pi x/b)] & -b \leq x \leq b \\ &= 0 & |x| > b \end{aligned} \qquad (6.76)$$

$$\begin{aligned} \dot{v}(x,0) &= \pi C \frac{v_o}{b} sin(\pi x/b) & -b \leq x \leq b \\ &= 0 & |x| > b. \end{aligned} \qquad (6.77)$$

These conditions are constructed so that the initial motion is a right-running wave. The string is fixed at the end points $x = \pm 4b$, and hence the boundary conditions are $v = 0$ at $x = \pm 4b$. We introduce nondimensional variables $x^* = x/b$, $t^* = Ct/b$, and $v^* = v/v_o$. The nondimensional equation of motion takes the simple form

$$\frac{\partial^2 v^*}{\partial t^{*2}} - \frac{\partial^2 v^*}{\partial x^{*2}} = 0, \qquad (6.78)$$

and the initial conditions become

$$\begin{aligned} v^*(x^*,0) &= 1+cos(\pi x^*) & -1 \leq x^* \leq 1 \\ &= 0 & |x^*| > 1 \end{aligned} \qquad (6.79)$$

$$\begin{aligned} \dot{v^*}(x^*,0) &= \pi sin(\pi x^*) & -1 \leq x^* \leq 1 \\ &= 0 & |x^*| > 1. \end{aligned} \qquad (6.80)$$

The initial conditions are shown in Figures 6.1a and 6.1b. We use uniform mesh of 80 cubic elements giving a total of 241 global nodes. Newmark method with $\alpha = 1/2$, $\gamma = 1/2$, and fixed time step $\Delta t^* = 0.005$ is used for time integration.

For verification of the numerical solution, it is important to check conservation of energy. The total energy of the string, which is the sum of the kinetic energy

$$K = \frac{1}{2}\rho l \int_{-4b}^{4b} \left(\frac{\partial v}{\partial t}\right)^2 dx \tag{6.81}$$

and potential energy

$$V = \frac{1}{2} T \int_{-4b}^{4b} \left(\frac{\partial v}{\partial x}\right)^2 dx \tag{6.82}$$

is conserved at any instant of time since the flow of energy into the string is zero [36].

Comparison between the finite-element results and the exact analytic solution is shown in Figures 6.2a–6.2f for the displacement $v^*(x^*,t^*)$, and in Figures 6.3a–6.3f for the velocity $\dot{v}^*(x^*,t^*)$. The figures show that the pulse propagates to the right as expected and reflects from the fixed end $x^* = 4$. At time $t^* = 4$, the forward half of the pulse is reflected as an inverted pulse propagating to the left. Interference of the reflected forward half pulse with the incident rear half pulse flattens the string as shown in Figure 6.2c, and potential energy is momentarily zero. The instantaneous string velocity at this instant is shown in Figure 6.3c. The maximum velocity is double the initial maximum velocity, and total energy of the string appears as kinetic energy. At time $t^* = 5$, the pulse is completely reflected and propagates to the left. At time $t^* = 8$ the central point of the pulse arrives at the initial position $x^* = 0$, but the pulse is inverted. The pulse is reflected at the end $x^* = -4$, and then propagates to the right. The simulation is continued until $t^* = 16$ where the pulse returns to its initial position as shown in Figures 6.4a and 6.4b. The finite element results are in good agreement with the exact analytic solution except for small oscillations in the velocity distribution signifying dispersion errors.

Example 6.2. Longitudinal Free Vibrations of a Bar

We consider the free longitudinal vibrations of a bar of length L, uniform cross-sectional area A, mass density ρ, and Young's modulus E. We let $C = \sqrt{E/\rho}$ denote the speed of propagation of longitudinal infinitesimally small elastic waves on the rod. The bar is fixed at $x = 0$ and stretched by a force F_o that gives the end $x = L$ a displacement $u_o = F_oL/EA$. While the bar is at rest, the force is suddenly removed, and longitudinal oscillations start. The initial conditions are

$$u(x,0) = u_o \frac{x}{L} \qquad 0 \le x \le L \tag{6.83}$$

$$\dot{u}(x,0) = 0 \qquad 0 \le x \le L \tag{6.84}$$

and boundary conditions are

$$\begin{aligned} x = 0: &\qquad u(x,t) = 0, &\qquad t > 0 \\ x = L: &\qquad EA\frac{\partial u}{\partial x} = 0, &\qquad t > 0. \end{aligned} \tag{6.85}$$

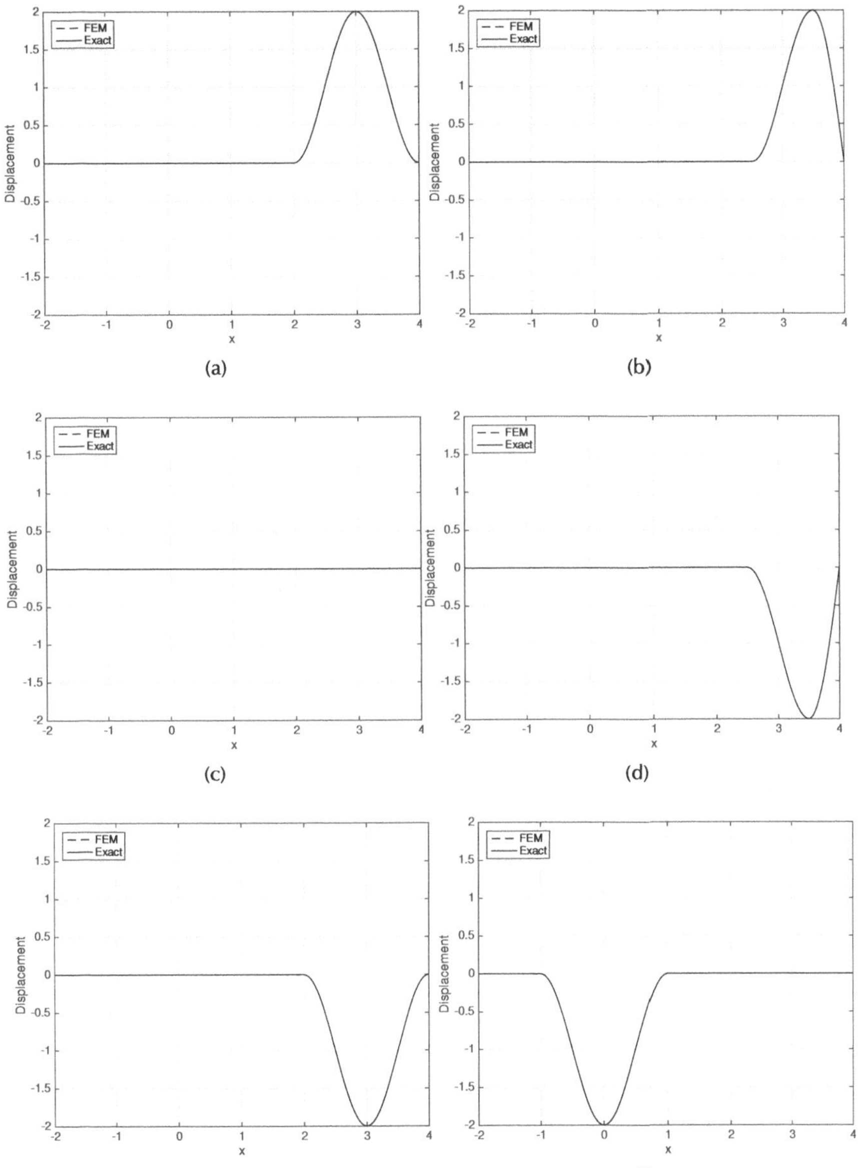

Figure 6.2 Comparison of finite-element and exact solution for the displacement of a propagating pulse on a taut string. (a) $t^* = 3$, the pulse center is at $x^* = 3$, (b) $t^* = 3.5$, forward 1/4 of pulse is reflected, (c) $t^* = 4$, forward 1/2 of pulse is reflected canceling the rear half of incident pulse, (d) $t^* = 4.5$, 3/4 of incident pulse is reflected, (e) $t^* = 5$, the pulse is completely reflected with inverted displacement, and (f) $t = 8$, the reflected pulse reaches initial position. A mesh of 80 cubic elements is used, and $\Delta t^* = 0.005$.

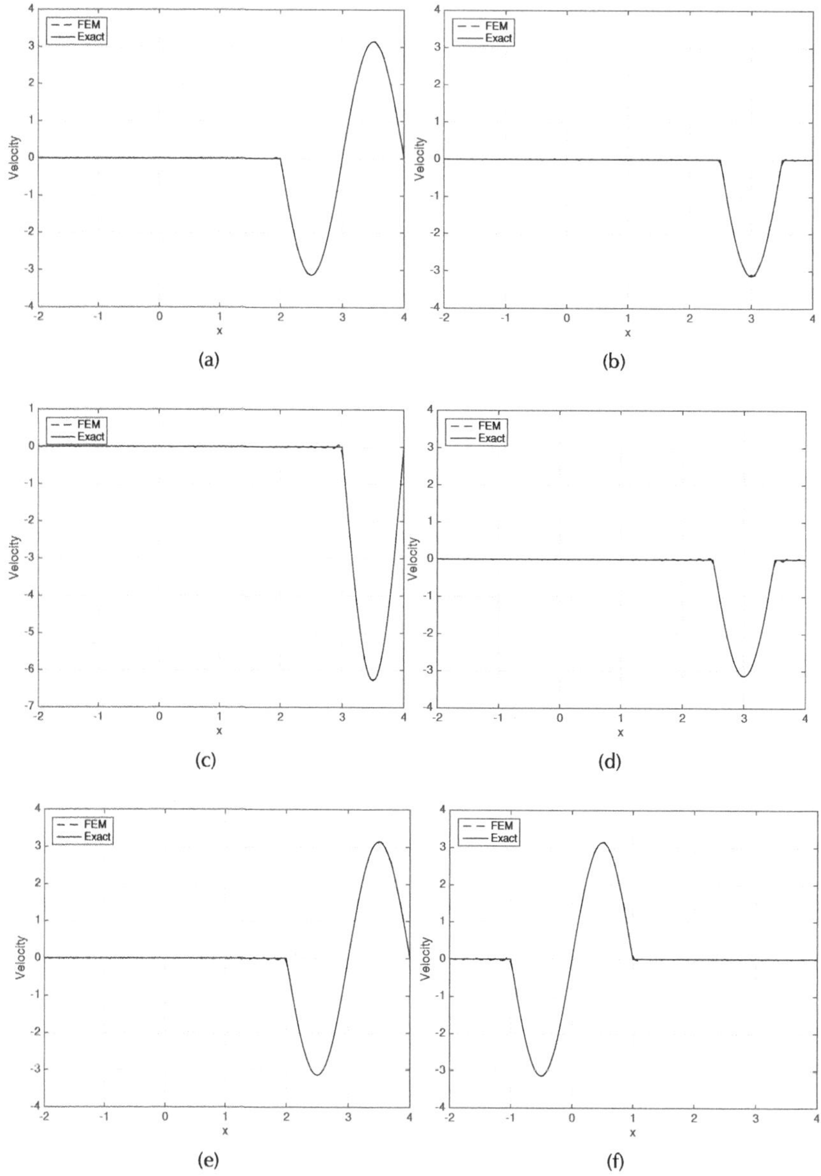

Figure 6.3 Comparison of finite-element and exact solutions for the velocity of a propagating pulse on a taut string. (a) $t = 3$, the pulse center is at $x = 3$, (b) $t = 3.5$, forward 1/4 of pulse is reflected, (c) $t = 4$, forward 1/2 of pulse is reflected doubling the velocity of the rear half of incident pulse, (d) $t = 4.5$, 3/4 of incident pulse is reflected, (e) $t = 5$, the reflected pulse propagates to the left; particle velocity is negative (downward) in the head and positive (upward) in the tail, and (f) $t = 8$, the reflected pulse reaches initial position. A mesh of 80 cubic elements is used, and $\Delta t = 0.005$.

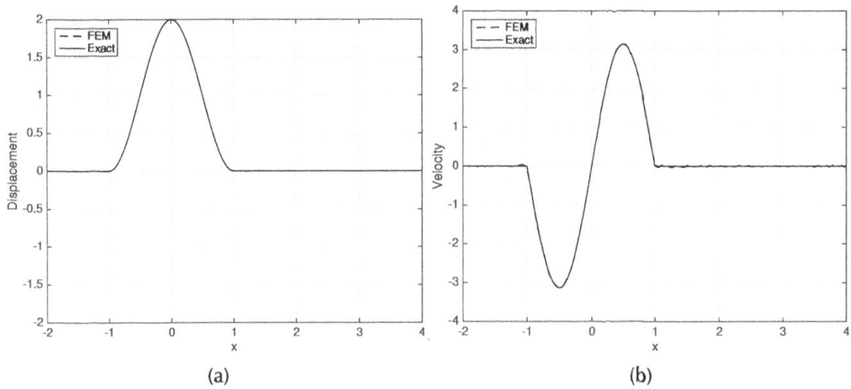

Figure 6.4 Displacement and velocity at $t^* = 16$. The pulse returns unchanged to initial position.

We introduce the nondimensional variables: $x^* = x/L$, $t^* = Ct/L$, and $u^* = u/u_o$, and write the nondimensional equation of motion,

$$\frac{\partial^2 u^*}{\partial t^{*2}} - \frac{\partial^2 u^*}{\partial x^{*2}} = 0 \tag{6.86}$$

and initial condition

$$u^*(x^*,0) = x^* \qquad 0 \le x^* \le 1 \tag{6.87}$$

$$\dot{u^*}(x^*,0) = 0 \qquad 0 \le x^* \le 1 \tag{6.88}$$

and the boundary conditions

$$\begin{aligned} x^* &= 0: & u^*(x^*,t^*) &= 0, & t^* &> 0 \\ x^* &= 1: & \frac{\partial u^*}{\partial x^*} &= 0, & t^* &> 0. \end{aligned} \tag{6.89}$$

The boundary condition at $x^* = 0$ is an EBC whereas that at $x^* = 1$ is an NBC. We divide the bar into 100 cubic elements using 301 global nodes. Newmark method is used for time integration with parameters $\alpha = 1/2$, $\gamma = 1/2$, and $\Delta t^* = 6.25 \times 10^{-4}$. The displacement at the tip $x^* = 1$ is compared with the exact analytic solution in Figures 6.5a and 6.5b. At time $t^* = 16$, the amplitude of the tip oscillation is slightly damped but distorted. The history of the reaction force at the fixed end, determined as the secondary variable at the boundary $x = 0$, is shown in Figures 6.5c and 6.5d. The exact force alternates between $+1$ and -1 with a period of 4. The finite-element results exhibit significant artificial spikes and high frequencies indicating dispersion errors. In order to damp high frequencies, we use $\alpha = 0.6$ and $\gamma = 0.606$, and repeat

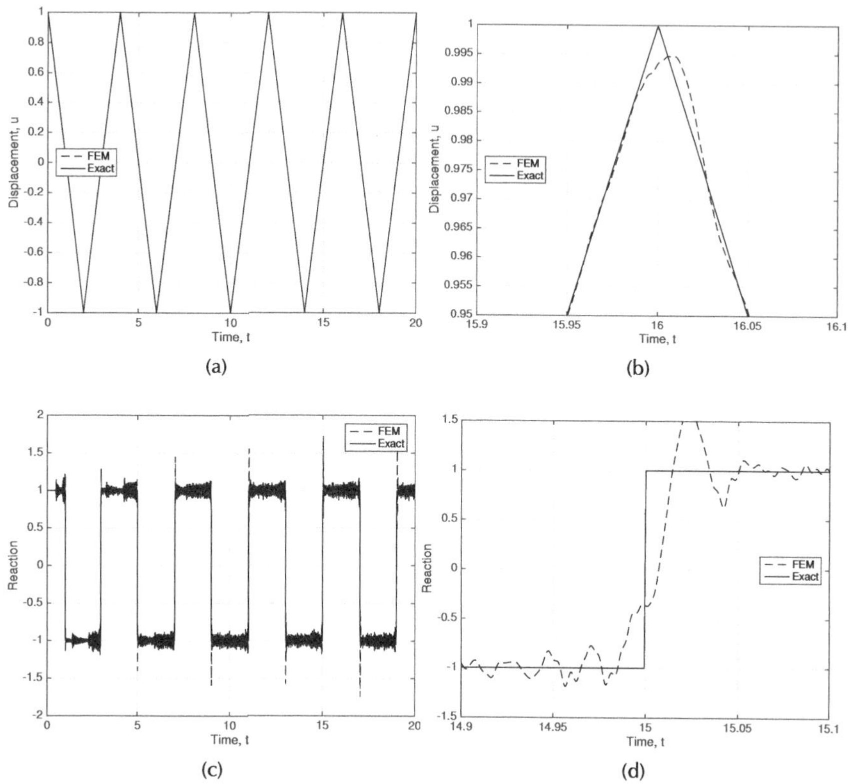

Figure 6.5 Comparison of finite-element and exact solutions for the longitudinal vibrations of a bar. (a) Tip displacement, $u^*(1, t^*)$, (b) closeup of tip displacement near $t^* = 16$, (c) force at the fixed end $x^* = 0$, (d) closeup of the force near time $t^* = 15$. A mesh of 100 cubic elements is used, and $\Delta t = 6.25 \times 10^{-4}$. Newmark method parameters $\alpha = 1/2$ and $\gamma = 1/2$.

the simulations. The new results are shown in Figures 6.6a–6.6d. The high frequencies are completely eliminated but the amplitude of the oscillations also damps in time as shown by the closeup of the tip displacement at time $t^* = 16$; the amplitude is 0.975 instead of 1. As the simulation continues in time the bar vibration will artificially subside. We also note that the discontinuous time response of the force at the fixed end is artificially smoothed. Increasing α to 0.8 (and $\gamma = 0.845$), we find the amplitude of the tip displacement at time $t^* = 16$ to be 0.955 instead of 1.

Example 6.3. Axial Impact of a Bar by a Moving Rigid Block

We want to study the longitudinal vibration and stress in a bar (slender rod), which is fixed at one end and impacted by a moving rigid block at the other end. The bar is uniform with length L, cross-sectional area A, mass density ρ, Young's modulus E,

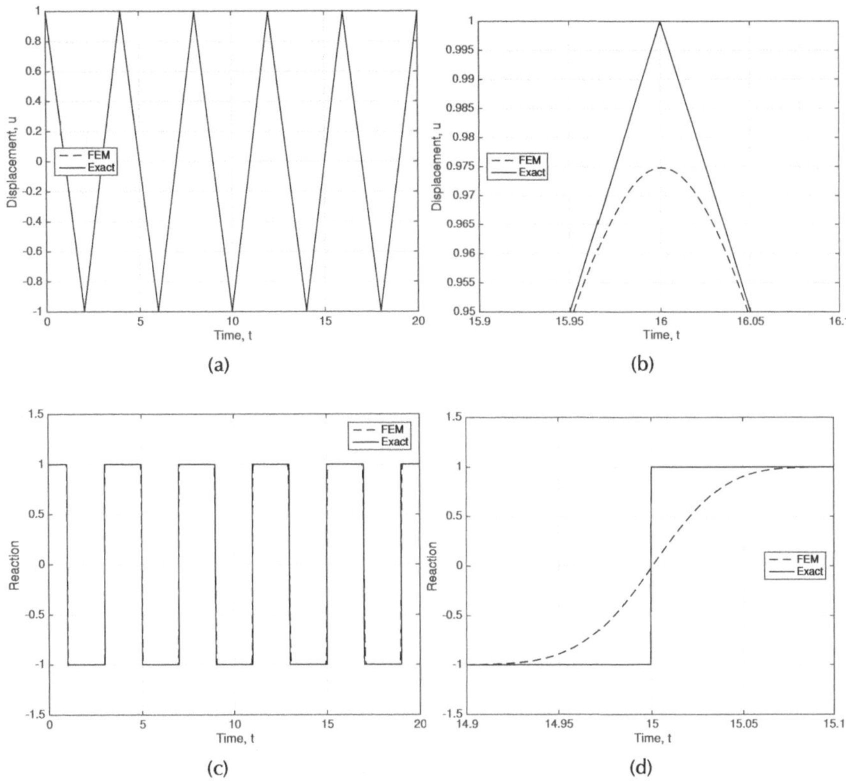

Figure 6.6 Comparison of finite-element and exact solutions for the longitudinal vibrations of a bar. (a) Tip displacement, $u^*(1,\ t^*)$, (b) closeup of tip displacement near $t^* = 16$, (c) force at the fixed end $x^* = 0$, (d) closeup of the force near time $t^* = 15$. A mesh of 100 cubic elements is used, and $\Delta t = 6.25 \times 10^{-4}$. Newmark method parameters $\alpha = 0.6$ and $\gamma = 0.6050$.

and is fixed at $x = 0$ and free at $x = L$. It is initially at rest without strain. A block of mass M_o moving in the negative x-axis with velocity V_o strikes the free end of the bar at time $t = 0$. The resulting motion after impact is assumed to be one-dimensional with longitudinal displacement $u(x,t)$, and the deformation is assumed to be in the linear elastic range. The block remains in contact with the bar as long as the contact force is compressive. We define the contact time t_c as the period of time during which the contact force between the block and the bar tip remains compressive. After that time the block is no longer in contact with the bar, and the bar performs free oscillations without the block at the tip. The equation of motion of the bar is

$$\rho A \frac{\partial^2 u}{\partial t^2} - EA \frac{\partial^2 u}{\partial x^2} = 0, \qquad 0 \leq x \leq L. \tag{6.90}$$

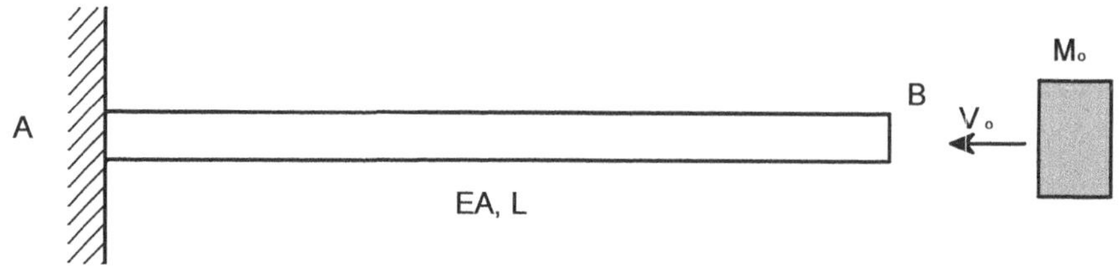

Figure 6.7 Impact of an elastic bar by a rigid block.

The boundary conditions are

$$\begin{aligned} x=0: &\quad u=0 && \text{for} \quad t>0 \\ x=L: &\quad EA\frac{\partial u}{\partial x}=-M_o\frac{\partial^2 u}{\partial t^2} && \text{for} \quad 0<t\leq t_c \\ &\quad EA\frac{\partial u}{\partial x}=0 && \text{for} \quad t>t_c \end{aligned} \tag{6.91}$$

The initial conditions at $t=0$ are

$$\begin{aligned} u(x,0) &= 0 && \text{for} \quad 0\leq x\leq L \\ \frac{\partial u}{\partial t} &= 0 && \text{for} \quad 0\leq x<L \\ \frac{\partial u}{\partial t} &= -V_o && \text{at} \quad x=L. \end{aligned} \tag{6.92}$$

We let $C=\sqrt{E/\rho}$ denote the speed of propagation of infinitesimal elastic pulse in the bar, and introduce nondimensional variables $x^*=x/L, t^*=Ct/L, u^*=uC/(V_oL)$, and $M_o^*=M_o/(\rho AL)$. The nondimensional equation of motion becomes

$$\frac{\partial^2 u^*}{\partial t^{*2}}-\frac{\partial^2 u^*}{\partial x^{*2}}=0, \qquad 0\leq x^*\leq 1. \tag{6.93}$$

The boundary conditions are

$$\begin{aligned} x^*=0: &\quad u^*=0 && \text{for} \quad t^*>0 \\ x^*=1: &\quad \frac{\partial u^*}{\partial x^*}=-M_o^*\frac{\partial^2 u^*}{\partial t^{*2}} && \text{for} \quad 0<t^*\leq t_c^* \\ &\quad \frac{\partial u^*}{\partial x^*}=0 && \text{for} \quad t^*>t_c^*. \end{aligned} \tag{6.94}$$

The initial conditions at $t^*=0$ are

$$\begin{aligned} u^*(x^*,0) &= 0 && \text{for} \quad 0\leq x^*\leq 1 \\ \frac{\partial u^*}{\partial t^*} &= 0 && \text{for} \quad 0\leq x^*<1 \\ \frac{\partial u^*}{\partial t^*} &= -1 && \text{at} \quad x^*=1. \end{aligned} \tag{6.95}$$

A quantity of interest in this example is the normal stress $\sigma = E\partial u/\partial x$. The nondimensional stress is $\sigma^* = \sigma/(\rho C V_o) = \partial u^*/\partial x^*$. Solution to the nondimensional problem (u^*, $\dot{u}^*$, σ^*) depends only on one parameter that is the ratio of the mass of the block to the mass of the rod M_o^*, for which we need to specify a numerical value.

An exact analytic solution to the problem defined by Equations (6.93)–6.95) can be found in Love [66] or Timoshenko [119]. The analytic solution is valid for $t \leq t_c$. We use the finite-element method in space and Newmark method for time integration. We divide the bar into 200 Lagrange cubic elements, which gives $N = 601$ global nodes. The first global node is the fixed end $x = 0$ and the impacted end is node $N = 601$. The array of displacement, velocity, and acceleration are denoted by $\{U\}$, $\{\dot{U}\}$, and $\{\ddot{U}\}$, respectively. The boundary conditions are

$$\begin{aligned} U_1 &= 0 && \text{for} \quad t^* > 0 \\ R_N &= -M_o^* \ddot{U}_N && \text{for} \quad 0 < t^* < t_c^* \\ R_N &= 0 && \text{for} \quad t^* > t_c^*. \end{aligned} \tag{6.96}$$

The first condition is an essential boundary condition. The second condition (may be classified as a mixed boundary condition) is used to remove the secondary variable at node N in terms of the acceleration at that node, and consequently the mass matrix is modified by replacing $m_{N,N}$ by $m_{N,N} + M_o^*$. This condition is valid as long as R_N is negative (compression force). During the simulation, we monitor the sign of R_N, and at the instant it becomes positive (tension) the block loses contact with the bar, we switch to the third condition $R_N = 0$ and use the unmodified mass matrix. The displacement, velocity, and acceleration of the block are the same as those of the bar tip as long as R_N is negative. The time step when R_N changes sign from negative to positive (tension) determines the contact time t_c^*. The initial conditions are

$$\begin{aligned} U_j &= 0 && \text{for} \quad j = 1, 2, ..., N \\ \dot{U}_j &= 0 && \text{for} \quad j = 1, 2, ..., N-1. \\ \dot{U}_N &= -1. \end{aligned} \tag{6.97}$$

The discontinuous initial velocity is problematic for the Newmark integration method if $\alpha = 0.5$; high frequency oscillations contaminate the time response. We used $\alpha = 0.55$, which significantly reduced the artificial oscillations but resulted in a small damping and smoothing of the discontinuities of the stress. For the given α, we compute the parameter $\gamma = (\alpha + 1/2)^2/2$. A fixed time step of $\Delta t^* = 2.5 \times 10^{-4}$ is used for this example.

The contact time t_c^*, the ratio of the block kinetic energy at the end of contact to its kinetic energy before impact, the maximum stress σ_{max}^* at the fixed end, and the tip displacement at the end of contact $u^*(1, t_c^*)$ are shown in Figures 6.8a–6.8d, respectively, as functions of M_o^* in the range $0.4 \leq M_o^* \leq 6.6$. The finite-element method results are in good agreement with the exact analytic solution. The discontinuities in some of these parameters are associated with the time when a reflected stress wave

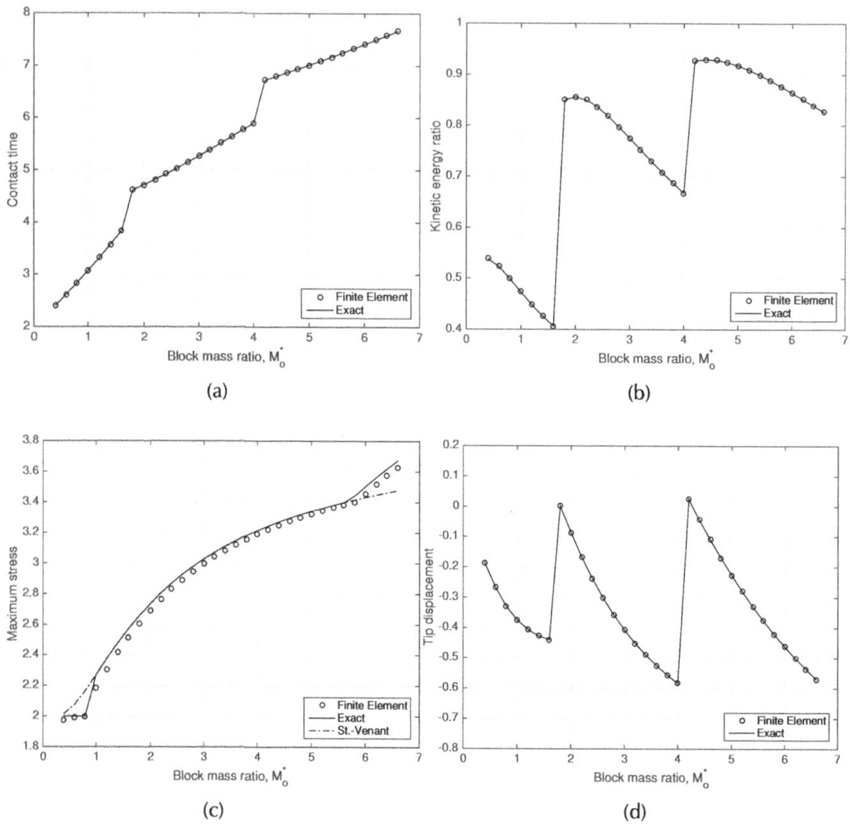

Figure 6.8 Comparison of finite-element and exact solutions for the impact of a bar with a rigid block, (a) contact time t_c^*, (b) ratio of block kinetic energy after impact to its kinetic energy before impact, (c) maximum compression stress at the fixed end σ_{max}^*, and (d) tip displacement at the end of contact between block and bar. A mesh of 200 cubic elements is used, and $\Delta t = 2.5 \times 10^{-4}$.

from the fixed end arrives at the free end. For some values of M_o^*, a stress wave reflected from the fixed end may arrive to the block just before the block loses contact with the bar or just after.

The contact time is determined by the condition $R_N > 0$; that is to say, when the secondary variable at node N changes sign from negative to positive. The contact force or stress R_N between the block and the bar tip also suffers discontinuous changes. If the discontinuity happens at a point of near zero value of R_N, the finite-element results may differ from the exact solution because of the artificial damping of the slightly positive peak. An example for this situation happens at $M_o^* = 4.15$; this point is not shown in Figure 6.8a. Figure 6.8c shows that the maximum stress at the fixed end predicted by the finite-element method is lower than the exact solution.

This is also due to artificial damping of the discontinuities of the stress waves. Timoshenko [119] attributed to St. Venant a formula for the maximum stress for $M_o^* < 5$,

$$\sigma_{max}^* = 2(1 + e^{-2/M_o^*}). \tag{6.98}$$

Predictions of this formula are plotted in Figure 6.8c; it is in remarkable agreement with the exact solution in the range $1 \leq M_o^* \leq 5.5$.

Next we consider the time response for $M_o^* = 2.4$. The contact time is $t^* = 4.9177$ by the finite element method, and 4.9175 by the exact solution. At this time, the displacement $u^*(x^*, t_c^*)$ and velocity $\dot{u}^*(x^*, t_c^*)$ in the bar are depicted in Figures 6.9a and 6.9b, respectively. The stress wave indicated by the discontinuity in the velocity is propagating toward the fixed end at the time when the block loses contact with the bar tip. We assume the block to move in the positive x-axis with constant velocity equal to the rebound velocity, which is found to be 0.9145 by the finite-element method and 0.9143 by the exact analytic solution. After time t_c^*, we modify the boundary condition so that $x^* = 1$ is free from stress ($R_N = 0$), and continue the simulation. We also continue the exact analytic solution using the method of characteristics. The initial conditions for the method of characteristics calculations are provided by the exact analytic solution at the contact time t_c^*. Thus, the finite-element and exact solutions from the beginning to the end of the simulation are independent of one another. The two solutions are compared in Figures 6.9c–6.9f for reaction at fixed end, the contact force between block and bar tip, the displacement of the tip and block, and velocity of the bar tip, respectively. Excellent agreement between the two solutions is evident. We note that the block trajectory depicted in Figure 6.9e shows that the bar tip does not strike the block after separation. The possibility of a secondary interaction between the bar tip and the block after separation at the contact time t_c^* is demonstrated for the case of $M_o^* = 3.6$. The contact time is $t^* = 5.6390$ by the finite-element method, and it is also 5.6390 by the exact solution. At this time, the displacement $u^*(x^*, t_c^*)$ and velocity $\dot{u}^*(x^*, t_c^*)$ in the bar are depicted in Figures 6.10a and 6.10b, respectively. The stress wave indicated by the discontinuity in the velocity is propagating toward the free end at the time when the block loses contact with the bar tip. We assume the block to move in the positive x-axis with constant velocity equal to the rebound velocity, which is found to be 0.8417 by the finite-element method and 0.8415 by the exact analytic solution. After time t_c^*, we modify the boundary condition so that $x^* = 1$ is free from stress ($R_N = 0$), and continue the simulation. We also continue the exact analytic solution using the method of characteristics. The initial conditions for the method of characteristics calculations are provided by the exact analytic solution at the contact time t_c^*. Thus, the finite-element and exact solutions from the beginning to the end of the simulation are independent of one another. The two solutions are compared in Figures 6.10c–6.10f for the reaction at fixed end, the contact force between block and bar tip, the displacement of the tip and block, and velocity of the bar tip, respectively. Excellent agreement between the two solutions is evident. The results shown in these figures assume that the block is removed from the field after time t_c^*. However, if the block continues to move along the positive

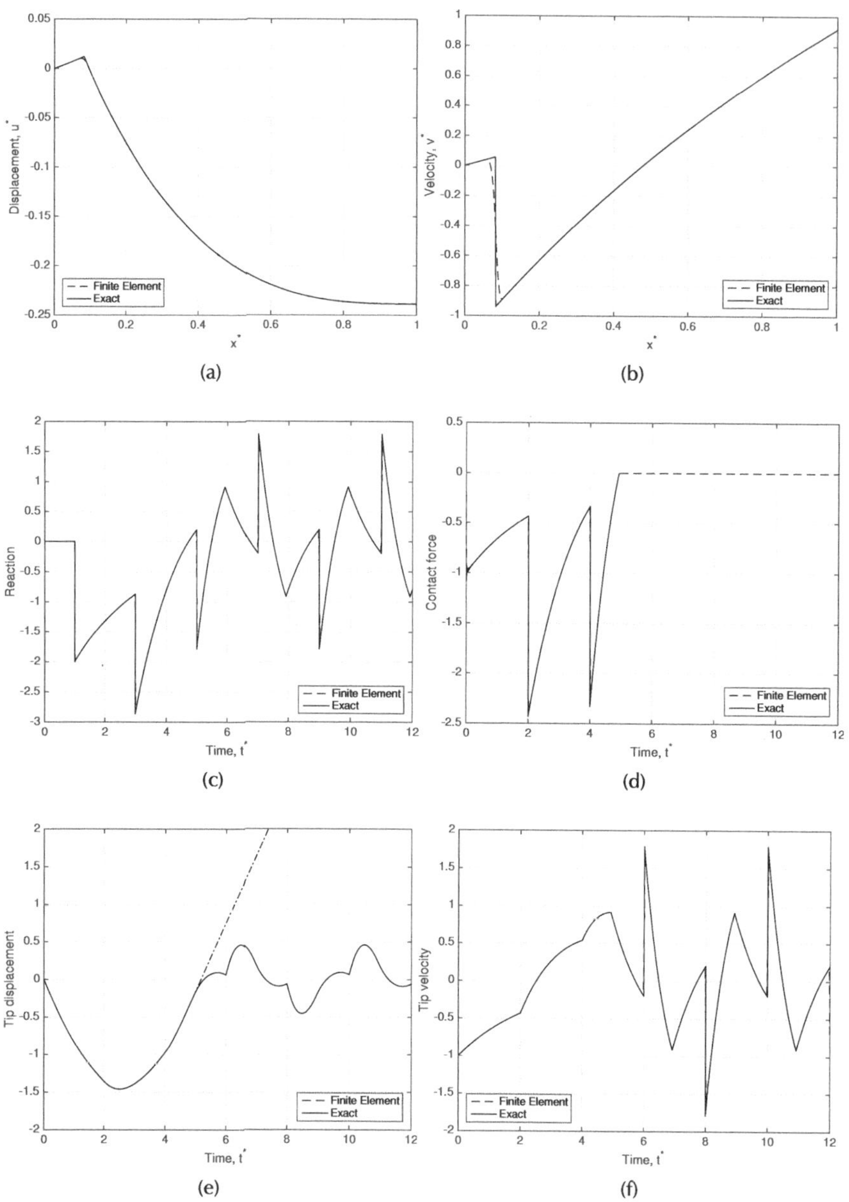

Figure 6.9 Comparison of finite-element and exact solutions for impact of a bar and a block of mass ratio $M_o^* = 2.4$. (a) Displacement $u^*(x^*, t_c^*)$ in the bar at end of contact, (b) velocity $\dot{u}^*(x^*, t_c^*)$ in the bar at end of contact, (c) reaction at the fixed end, (d) contact force between block and the bar tip, (e) displacement of bar tip and block; the dash-dot straight line is the block trajectory after rebound, and (f) velocity of bar tip during and after impact. A mesh of 200 cubic elements is used, and $\Delta t = 2.5 \times 10^{-4}$.

x-axis with the velocity of 0.8417, then the bar tip will strike the block at approximately $t^* = 6.0355$. As shown in Figure 6.10f the tip velocity decreases for a period of 0.36, but then increases suddenly at time $t^* = 6$ to a value approximately 2.5 due to the reflection of the compression wave from the free end as an expansion wave. The results shown in Figures 6.10c–6.10f do not account for this second interaction between the block and the bar.

Example 6.4. Free Vibrations of a Cantilever Beam

We want to evaluate the accuracy of the Newmark method for direct integration of the finite element equations. We consider free vibrations of a cantilever beam of length $L = 2h$ and uniform EI and ρA, and model the beam by two Euler–Bernoulli elements each of length h. There are three global nodes with two degrees of freedom (deflection and slope) at each node. The assembled system before application of the boundary conditions is

$$\frac{\rho Ah}{420}\begin{bmatrix} 156 & 22h & 54 & -13h & 0 & 0 \\ 22h & 4h^2 & 13h & -3h^2 & 0 & 0 \\ 54 & 13h & 312 & 0 & 54 & -13h \\ -13h & -3h^2 & 0 & 8h^2 & 13h & -3h^2 \\ 0 & 0 & 54 & 13h & 156 & -22h \\ 0 & 0 & -13h & -3h^2 & -22h & 4h^2 \end{bmatrix}\begin{Bmatrix} \ddot{U}_1 \\ \ddot{U}_2 \\ \ddot{U}_3 \\ \ddot{U}_4 \\ \ddot{U}_5 \\ \ddot{U}_6 \end{Bmatrix}$$

$$+\frac{2EI}{h^3}\begin{bmatrix} 6 & 3h & -6 & 3h & 0 & 0 \\ 3h & 2h^2 & -3h & h^2 & 0 & 0 \\ -6 & -3h & 12 & 0 & -6 & 3h \\ 3h & h^2 & 0 & 4h^2 & -3h & h^2 \\ 0 & 0 & -6 & -3h & 6 & -3h \\ 0 & 0 & 3h & h^2 & -3h & 2h^2 \end{bmatrix}\begin{Bmatrix} U_1 \\ U_2 \\ U_3 \\ U_4 \\ U_5 \\ U_6 \end{Bmatrix} = \begin{Bmatrix} R_1 \\ R_2 \\ R_3 \\ R_4 \\ R_5 \\ R_6 \end{Bmatrix}. \tag{6.99}$$

The boundary conditions are

At the fixed end ($x = 0$), we have $U_1 = 0$ and $U_2 = 0$. Because these conditions apply at any time, we also have $\dot{U}_1 = \ddot{U}_1 = 0$ and $\dot{U}_2 = \ddot{U}_2 = 0$.

At the free end ($x = L$), the shear force and bending moment are zero, hence we have $R_5 = 0$ and $R_6 = 0$.

At the inter-element node we have $R_3 = 0$ and $R_4 = 0$ because there are no externally applied force or moment at that node.

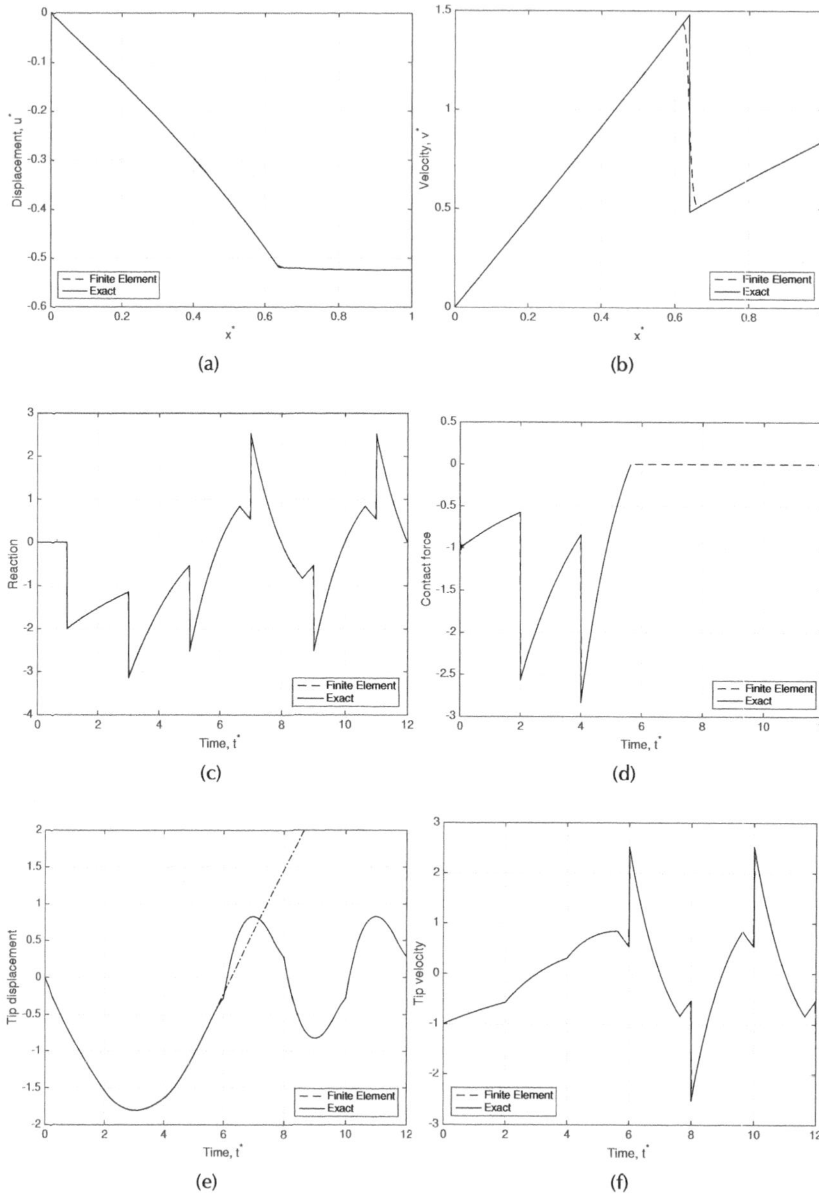

Figure 6.10 Comparison of finite-element and exact solutions for impact of a bar and a block of mass ratio $M_o^* = 3.6$. (a) Displacement $u^*(x^*, t_c^*)$ in the bar at end of contact, (b) velocity $\dot{u}^*(x^*, t_c^*)$ in the bar at end of contact, (c) reaction at the fixed end, (d) contact force between block and the bar tip, (e) displacement of bar tip and block; the dash-dot straight line is the block trajectory after rebound, and (f) velocity of bar tip during and after impact. A mesh of 200 cubic elements is used, and $\Delta t = 2.5 \times 10^{-4}$.

Before we apply the boundary conditions, we save the first two equations to be used for the computation of the unknown reactions R_1 and R_2; if needed.

$$\begin{Bmatrix} R_1 \\ R_2 \end{Bmatrix} = \frac{\rho A h}{420}\begin{bmatrix} 156 & 22h & 54 & -13h & 0 & 0 \\ 22h & 4h^2 & 13h & -3h^2 & 0 & 0 \end{bmatrix}\begin{Bmatrix} \ddot{U}_1 \\ \ddot{U}_2 \\ \ddot{U}_3 \\ \ddot{U}_4 \\ \ddot{U}_5 \\ \ddot{U}_6 \end{Bmatrix} + \frac{2EI}{h^3}\begin{bmatrix} 6 & 3h & -6 & 3h & 0 & 0 \\ 3h & 2h^2 & -3h & h^2 & 0 & 0 \end{bmatrix}\begin{Bmatrix} U_1 \\ U_2 \\ U_3 \\ U_4 \\ U_5 \\ U_6 \end{Bmatrix}. \tag{6.100}$$

Now, we apply the boundary conditions and obtain the modified system

$$\frac{\rho A h}{420}\begin{bmatrix} 1 & 0 & 0 & 0 & 0 & 0 \\ 0 & 1 & 0 & 0 & 0 & 0 \\ 54 & 13h & 312 & 0 & 54 & -13h \\ -13h & -3h^2 & 0 & 8h^2 & 13h & -3h^2 \\ 0 & 0 & 54 & 13h & 156 & -22h \\ 0 & 0 & -13h & -3h^2 & -22h & 4h^2 \end{bmatrix}\begin{Bmatrix} \ddot{U}_1 \\ \ddot{U}_2 \\ \ddot{U}_3 \\ \ddot{U}_4 \\ \ddot{U}_5 \\ \ddot{U}_6 \end{Bmatrix} + \frac{2EI}{h^3}\begin{bmatrix} 0 & 0 & 0 & 0 & 0 & 0 \\ 0 & 0 & 0 & 0 & 0 & 0 \\ -6 & -3h & 12 & 0 & -6 & 3h \\ 3h & h^2 & 0 & 4h^2 & -3h & h^2 \\ 0 & 0 & -6 & -3h & 6 & -3h \\ 0 & 0 & 3h & h^2 & -3h & 2h^2 \end{bmatrix}\begin{Bmatrix} U_1 \\ U_2 \\ U_3 \\ U_4 \\ U_5 \\ U_6 \end{Bmatrix} = \begin{Bmatrix} 0 \\ 0 \\ 0 \\ 0 \\ 0 \\ 0 \end{Bmatrix}. \tag{6.101}$$

We can also obtain a reduced system for the unknown degrees of freedom as

$$\frac{\rho A h}{420}\begin{bmatrix} 312 & 0 & 54 & -13h \\ 0 & 8h^2 & 13h & -3h^2 \\ 54 & 13h & 156 & -22h \\ -13h & -3h^2 & -22h & 4h^2 \end{bmatrix}\begin{Bmatrix} \ddot{U}_3 \\ \ddot{U}_4 \\ \ddot{U}_5 \\ \ddot{U}_6 \end{Bmatrix} + \frac{2EI}{h^3}\begin{bmatrix} 12 & 0 & -6 & 3h \\ 0 & 4h^2 & -3h & h^2 \\ -6 & -3h & 6 & -3h \\ 3h & h^2 & -3h & 2h^2 \end{bmatrix}\begin{Bmatrix} U_3 \\ U_4 \\ U_5 \\ U_6 \end{Bmatrix} = \begin{Bmatrix} 0 \\ 0 \\ 0 \\ 0 \end{Bmatrix}. \tag{6.102}$$

The problem is solved in terms of nondimensional parameters,

$$x^* = x/L \tag{6.103}$$

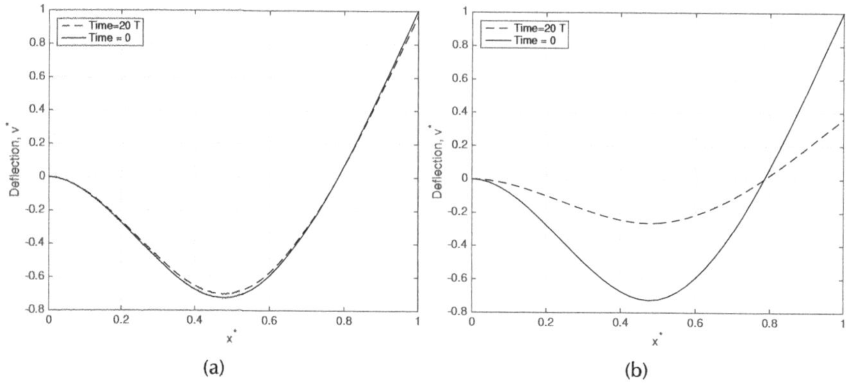

Figure 6.11 Effects of α on the free vibration of a cantilever beam. Initial deflection ($t^* = 0$) is the second mode of free vibration. (a) $\alpha = 0.5$, $\Delta t^* = T/40$, and (b) $\alpha = 0.6$, $\Delta t^* = T/40$.

$$t^* = t\sqrt{EI/\rho AL^4} \tag{6.104}$$

$$v^* = v/L \tag{6.105}$$

that reduce the equation of motion to

$$\frac{\partial^2 v^*}{\partial t^{*2}} + \frac{\partial^4 v^*}{\partial x^{*4}} = 0$$

Beam eigenvalue problems are presented in Chapter 7. In the present chapter, the beam response is determined in the time domain. The cantilever beam is given an initial deflection identical to the second-mode shape determined by solving the eigenvalue problem of the finite element model using two elements only. This is important for consistency with direct time integration of the two-element model of the dynamic response; the mode shape is shown in Figure 6.11a. Hence the errors are solely due to the time integration scheme. The beam is released from rest. The exact time response will be harmonic with frequency equal to the natural frequency of the second mode as obtained by solving the eigenvalue problem using two clements, which is $\omega_2^* = 22.22147447408264$ (nondimensional). The initial deflection and slope at the six global nodes are $\{U_1, U_2, U_3, U_4, U_5, U_6\} = \{0, 0, -0.7218, 0.4344, 1, 4.8145\}$; the tip deflection is scaled to be 1. The exact tip deflection will be $v^*(t) = cos(\omega_2 t)$ and its velocity will be $\dot{v}^*(t) = -\omega_2 sin(\omega_2 t)$.

The Newmark method is used to integrate Equation (6.102) in time for a final time of 20 T^* where $T^* = 2\pi/\omega_2^*$ is the period of the second mode. We used $\Delta t^* = T^*/40$. The effects of the parameter α on the beam deflection are shown in Figure 6.11a for $\alpha = 0.5$, and in Figure 6.11b for $\alpha = 0.6$, in both cases $\gamma = (\alpha + 1/2)^2/2$. For $\alpha = 0.6$, the vibration is significantly damped; whereas for $\alpha = 0.5$ the damping is negligible.

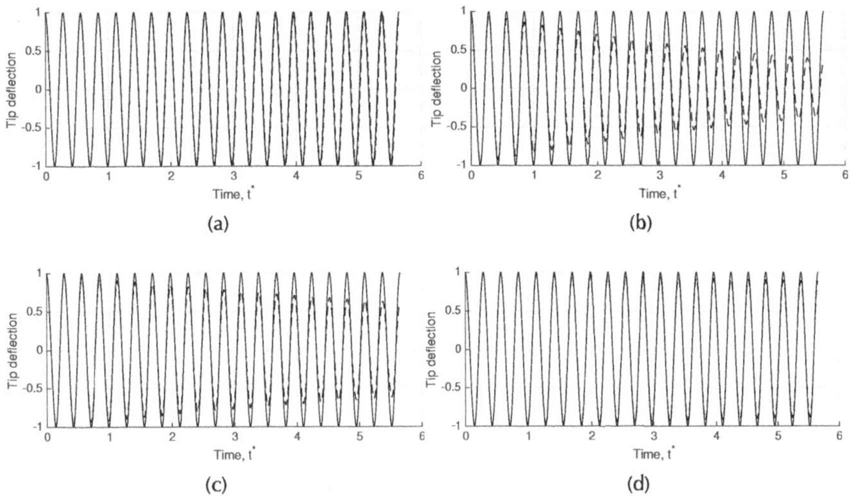

Figure 6.12 Effects of α and Δt^* on the free vibration of a cantilever beam. (a) $\alpha = 0.5$, $\Delta t^* = T/40$, (b) $\alpha = 0.6$, $\Delta t^* = T/40$, (c) $\alpha = 0.6$, $\Delta t^* = T/80$, (d) $\alpha = 0.6$, and $\Delta t^* = T/320$. Exact solution (solid line) and finite element method (dashed line).

The artificial damping can be reduced by using smaller time steps. Figure 6.12a shows the tip deflection as a function of time for $\alpha = 0.5$ and $\Delta t^* = T^*/40$; damping is negligible. For $\alpha = 0.6$, Figures 6.12b–6.12d show the tip deflection for $\Delta t^* = T^*/40$, $T^*/80$, and $T^*/320$, respectively. As the time step is refined, the artificial damping is reduced. For Newmark method, the values of α and time step should be selected carefully. An overall assessment of the order of accuracy for $\alpha = 1/2$ and $\gamma = 1/2$ is obtained by running simulations for various time steps. The errors in the tip deflection and its velocity after 20 periods of oscillations are shown by Figure 6.13 as a function of the time step Δt^*. On a log-log scale, the slope of the displacement error is 4. However, the slope for the velocity is 2, and hence the method is second-order.

Example 6.5. Forced Vibrations of a Beam

We consider a uniform beam (cross-sectional area A, mass density ρ, length L, flexural rigidity EI) that is simply supported at $x = 0$ and built-in at $x = L$ as shown in Figure 6.14. The beam is initially at rest when uniformly distributed downward load f_o is suddenly applied and maintained thereafter. We want to analyze the resulting beam vibration by direct time integration of the finite-element equations and verify the results by comparison with the exact analytic solution. To that end, we introduce nondimensional variables $x^* = x/L$, $t^* = \sqrt{EI/\rho AL^4}\,t$, $v^* = EIv/f_oL^4$, into the beam

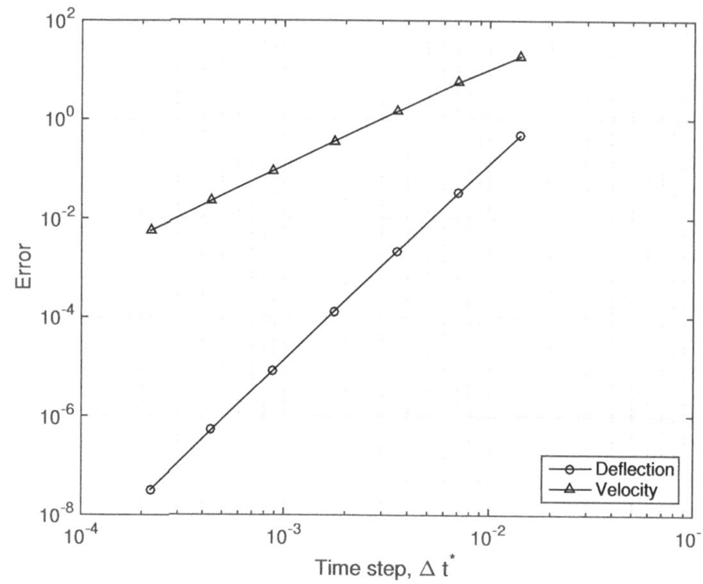

Figure 6.13 Errors in cantilever tip deflection and velocity after 20 periods. Newmark method, $\alpha = 0.5$.

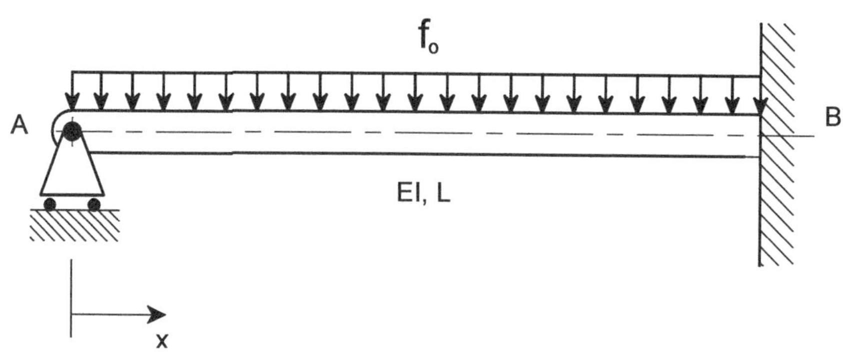

Figure 6.14 A beam forced by suddenly applied uniformly distributed load.

equation of motion to obtain

$$\frac{\partial^2 v^*}{\partial t^{*2}} + \frac{\partial^4 v^*}{\partial x^{*4}} + 1 = 0. \tag{6.106}$$

The boundary conditions are: at $x^* = 0$ we have deflection $v^* = 0$ and moment $\partial^2 v^*/\partial x^{*2} = 0$. At $x^* = 1$ we have deflection $v^* = 0$ and slope $\partial v^*/\partial x^* = 0$. The initial conditions at $t^* = 0$ are zero deflection and velocity, $v^*(x^*, 0) = 0$ and

$\partial v^*/\partial t^* = 0$.

First, the analytic solution to the static equilibrium problem is

$$v_s^*(x^*) = -\frac{1}{48}x^*(1-x^*)(1+x^*-2x^{*2}). \tag{6.107}$$

The maximum static deflection is $\delta_m^* = 5.41612 \times 10^{-3}$ at $x^* = 0.421535$. The forced response can be determined by separation of variables and is given by Lebedev et al. [64] as

$$v^*(x^*,t^*) = -\frac{1}{48}x^*(1-x^*)(1+x^*-2x^{*2}) - \sum_{m=1}^{\infty} B_m X_m(x^*) cos\left(\lambda_m^2 t^*\right) \tag{6.108}$$

where

$$\begin{aligned} B_m &= \frac{sinh(\lambda_m) - 2cosh(\lambda_m)sin(\lambda_m) + sin(\lambda_m)}{\lambda_m^5 sinh^2(\lambda_m)sin^2(\lambda_m)} \\ X_m(x^*) &= sinh(\lambda_m)sin(\lambda_m x^*) - sin(\lambda_m)sinh(\lambda_m x^*) \end{aligned} \tag{6.109}$$

and λ_m are consecutive positive roots of the characteristic equation $tan(\lambda_m) = tanh(\lambda_m)$. We note that the natural frequencies of the beam are $\omega_m = \lambda_m^2\sqrt{EI/\rho AL^4}$. The first 10 nondimensional frequencies are

```
1.541820572e+01 4.996486203e+01 1.042476965e+02 1.782697295e+02
2.720309713e+02 3.855314219e+02 5.187710813e+02 6.717499495e+02
8.444680266e+02 1.036925312e+03
```

We denote the lowest natural frequency by $\omega_1^* = 15.41820572$, and the corresponding period by $T_1^* = 2\pi/\omega_1^*$. The finite-element computation uses 10 Hermite cubic elements. For time integration, we use the Newmark method with parameters $\alpha = \gamma = 1/2$ and $\Delta t^* = T_1^*/100 = 4.075172833 \times 10^{-03}$. Simulations are conducted for $t^* = 9.5T_1^*$, and the deflection v^*/δ_m^* is compared with the exact solution in Figure 6.15a. The maximum dynamic deflection is approximately twice the static deflection at equilibrium. The deflection v^*/δ_m^* at $x^* = 0.4$ as a function of time is shown in Figure 6.16a; good agreement between the two solutions is evident.

The deflection curve at time $t^* = 10T_1^*$ is depicted in Figure 6.16a; that shows some differences between the exact and the finite element solutions. Using a finer finite-element mesh of 40 elements and smaller time step $\Delta t^* = T_1^*/1000 = 4.075172833 \times 10^{-04}$, we obtain a good comparison with the exact solution as shown in Figure 6.16b.

6.5 PROBLEMS

(1) A uniform bar of length L, cross-sectional area A, and mass density ρ is fixed at $x = 0$ and free at $x = L$. The bar is initially at rest and with zero displacement. At time $t = 0$, an axial force $F(t) = F_o sin(\Omega t)$ is applied at $x = L$. Model the

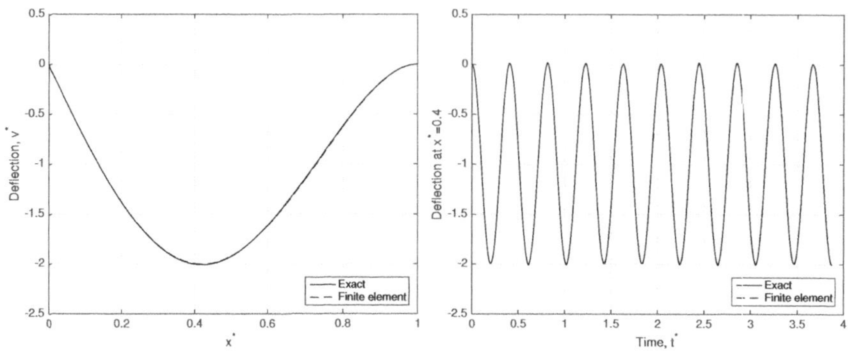

(a) (b)

Figure 6.15 Forced beam vibration. (a) Deflection at $t^* = 9.5T_1^*$, and (b) deflection at $x^* = 0.4$. A mesh of 10 elements is used and $\Delta t^* = 4.075172833 \times 10^{-03}$.

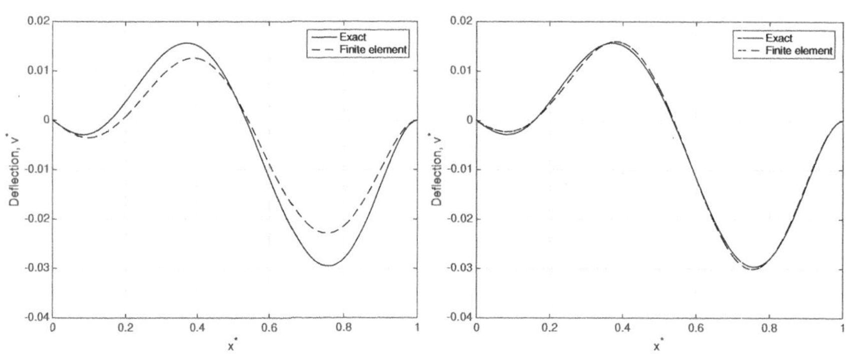

(a) (b)

Figure 6.16 Forced beam vibration, deflection at $t^* = 10T_1^*$. (a) A mesh of 10 elements is used and $\Delta t^* = 4.075172833 \times 10^{-03}$, (b) a mesh of 40 elements is used and $\Delta t^* = 4.075172833 \times 10^{-04}$.

bar by one cubic element, and use Newmark method with $\alpha = \gamma = 1/2$, to compute the time response for $0 \leq t \leq 16L/C$, where $C = \sqrt{E/\rho}$. Use F_o as a reference force and $u_o = F_oL/EA$ as a reference displacement. For $\Omega = \pi C/4L$, plot the nondimensional reaction at the fixed end R_1/F_o and the tip displacement $u(L,t)/u_o$ against time.

(2) Investigate the effects of mesh refinement on the results found in Problem (1) by comparing with results for two and three cubic elements.

(3) A uniform bar of length L, cross-sectional area A, and mass density ρ is fixed at $x = 0$ and is rigidly attached to concentrated mass $M_o = \rho AL$ at $x = L$ as shown

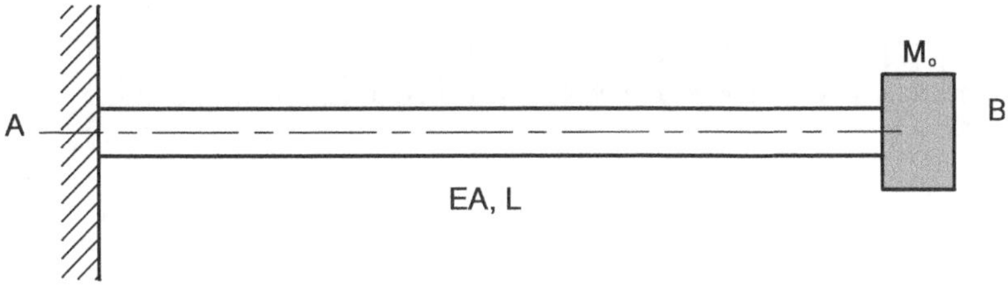

Figure 6.17 Longitudinal vibrations of a bar with concentrated mass at tip.

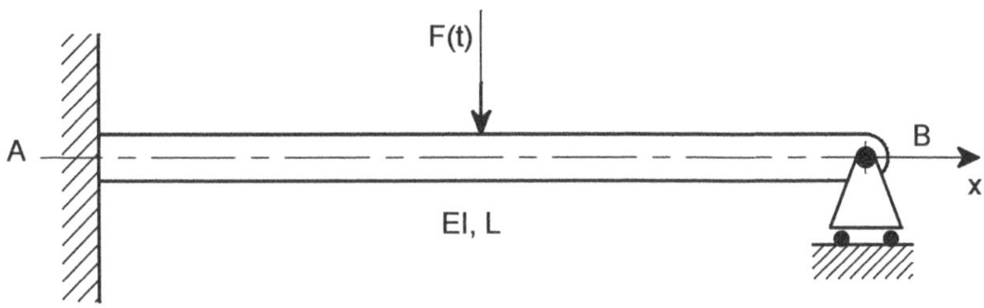

Figure 6.18 A beam forced by a concentrated force at midpoint.

in Figure 6.17. The bar is given an initial displacement $u(x,0) = u_o x/L$, and then released from rest. Model the bar by one cubic element, and use Newmark method with $\alpha = \gamma = 1/2$, to compute the time response for $0 \leq t \leq 16L/C$, where $C = \sqrt{E/\rho}$. Use $F_o = u_o EA/L$ as a reference force and u_o as a reference displacement, plot the nondimensional reaction at the fixed end R_1/F_o and tip displacement $u(L,t)/u_o$ against time.

(4) Investigate the effects of mesh refinement on the results found in Problem (3) by comparing with results for two and three cubic elements.

(5) A uniform beam AB (length L, mass density ρ, cross-sectional area A, and flexural rigidity EI) is fixed at A and simply supported at B as shown in Figure 6.18. It is forced by a concentrated lateral force $F(t) = (EI/2L^2)sin(\Omega t)$ applied at the midpoint ($x = L/2$), where $\Omega = 10\sqrt{EI/\rho AL^4}$. The motion starts from rest with zero deflection. Model the beam by two Hermite cubic elements, and use Newmark method with $\alpha = \gamma = 1/2$, integrate the global system in time, and plot the deflection at the midpoint $x = L/2$ against time. To simplify the analysis, introduce nondimensional parameters,

$$x^* = x/L$$

$$t^* = \sqrt{EI/\rho AL^4 t}$$

$$v^* = v/L$$

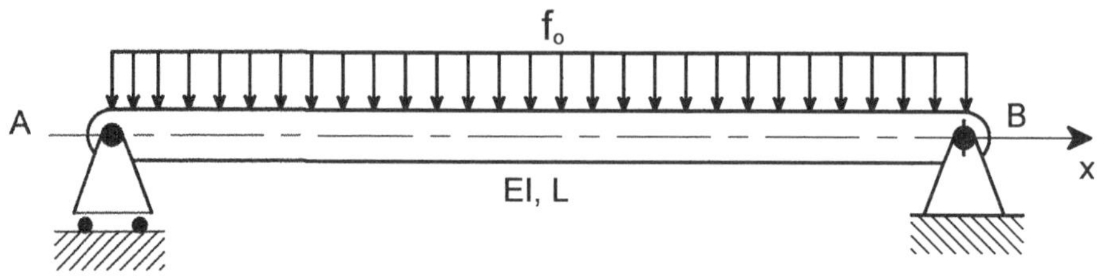

Figure 6.19 A simply supported beam forced by suddenly applied uniformly distributed load.

and show that the equation of motion takes the form

$$\frac{\partial^2 v^*}{\partial t^{*2}} + \frac{\partial^4 v^*}{\partial x^{*4}} = 0.$$

The concentrated force must also be made nondimensional. Plot v^* at $x^* = 1/2$ against t^*.

(6) We consider a uniform beam (cross-sectional area A, mass density ρ, length L, flexural rigidity EI) that is simply supported at $x = 0$ and $x = L$ as shown in Figure 6.19. The beam is initially at rest when a uniformly distributed downward load f_o is suddenly applied and maintained thereafter. We want to analyze the resulting beam vibration by direct time integration of the finite element equations. To that end, we introduce nondimensional variables $x^* = x/L$, $t^* = t\sqrt{EI/\rho AL^4}$, $v^* = EIv/f_oL^4$, into the beam equation of motion to obtain

$$\frac{\partial^2 v^*}{\partial t^{*2}} + \frac{\partial^4 v^*}{\partial x^{*4}} + 1 = 0. \tag{6.110}$$

The boundary conditions are: at $x^* = 0$ and $x^* = 1$ we have deflection $v^* = 0$ and moment $\partial^2 v^*/\partial x^{*2} = 0$. The initial conditions at $t^* = 0$ are zero deflection and velocity, $v^*(x^*, 0) = 0$ and $\partial v^*/\partial t^* = 0$.
Model the beam by one quintic three-node element, and plot the deflection v^* at $x^* = 0.5$ against time for $0 \leq t^* \leq 16/\pi$. Also, plot the reaction at one of the supports against time. Compare the maximum values of deflection and reaction to the quasi-static case in which the distributed load increases from zero to f_o very slowly. Quintic element matrices are given in Chapter 3 and 7.

(7) Consider the free transverse vibration of a cantilever, which is free at $x = 0$ and fixed at $x = L$. At time $t = 0$ the cantilever is given an initial deflection

$$v(x,0) = v_o[1 - sin(\pi x/2L)]$$

and then released from rest. According to the Euler-Bernoulli beam theory, the equation of motion is

$$\rho A \frac{\partial^2 v}{\partial t^2} + \frac{\partial^2}{\partial x^2}(EI\frac{\partial^2 v}{\partial x^2}) = 0 \quad \text{for} \quad 0 < x < L.$$

(a) For uniform ρA and EI, use the nondimensional variables $v^* = v/v_o, x^* = x/L$ and $t^* = t/\sqrt{\rho A L^4/EI}$ and write the differential equation and boundary and initial conditions in nondimensional forms.

(b) Using two Hermite elements and Newmark method for time integration ($\alpha = 0.5$ and $\gamma = 0.5$), plot the nondimensional reaction moment at the fixed end against time.

7 Differential Eigenvalue Problems

Differential eigenvalue problems are encountered in the analysis of free vibration and stability of structures and fluid motion, among other fields in engineering sciences. Examples include longitudinal and torsional vibration of bars and transverse vibration of beams. Knowing the natural frequencies of structures facilitates the analysis of free vibrations. For design purposes, knowing the natural frequencies is essential to safeguard against resonance conditions that can lead to large deflections and damage of the structures. The later problem is also related to the critical speed of a rotating shaft. Columns in equilibrium under axial compression force become unstable if the load exceeds a certain value. Fluid motions that are exact, or approximate, solutions to the Navier–Stokes equations may not be observable in laboratory experiments because small disturbances may grow in time or space and drive the flow into a different state. Analysis of the stability of fluid motions can shed light on such phenomena. The analysis of such problems gives rise to homogeneous differential equations with homogeneous boundary conditions. Thus, non-trivial solutions are obtained only if certain parameters in the equations take specific values that are called the eigenvalues. In this chapter, we use the finite element method in space to replace the differential eigenvalue problem by an algebraic one whose solution provides approximations to the eigenvalues and eigenfunctions. The eigenvalues may be the natural frequencies of structures, critical loads of columns, or complex frequencies of disturbance waves superimposed on an otherwise laminar fluid motion.

7.1 NATURAL FREQUENCIES OF LONGITUDINAL VIBRATION OF BARS

The equation of motion of longitudinal free oscillations of a bar (slender rod) is

$$\rho A\frac{\partial^2 u}{\partial t^2}-\frac{\partial}{\partial x}[EA\frac{\partial u}{\partial x}]=0, \qquad 0<x<L \tag{7.1}$$

where u is the axial (longitudinal) displacement, ρ is mass density (kg/m^3), A is cross-sectional area, L is the bar length, and E is Young's modulus. For a bar that is fixed at $x=0$ and free at $x=L$ the boundary conditions are

$$\begin{aligned} x=0 &: \quad u=0, && t>0 \\ x=L &: \quad EA\frac{\partial u}{\partial x}=0, && t>0 \end{aligned} \tag{7.2}$$

which are homogeneous; the second condition is zero stress at the free end. The coefficients in the differential equation are independent of time, hence we assume a

DOI: 10.1201/9781003323150-7

harmonic solution

$$u(x,t) = \hat{u}(x)e^{-\hat{i}\omega t}. \tag{7.3}$$

Substituting into the equation of motion and boundary conditions, we get

$$-\frac{d}{dx}[EA\frac{d\hat{u}}{dx}] - \omega^2 \rho A\hat{u} = 0, \qquad 0 < x < L \tag{7.4}$$

$$\begin{aligned} x = 0 &: \quad \hat{u} = 0 \\ x = L &: \quad EA\frac{d\hat{u}}{dx} = 0. \end{aligned} \tag{7.5}$$

The problem defined by Equation (7.4) with boundary conditions Equation (7.5) is a differential eigenvalue problem; a nontrivial solution is possible only if ω takes certain values, which are the natural frequencies of the bar. Using the Galerkin method, we find the element equations,

$$\left[[K^e] - \omega^2 [M^e]\right] \{\hat{U}^e\} = \{\hat{R}^e\}. \tag{7.6}$$

Example 7.1. Free Longitudinal Vibration of a Uniform Bar

Exact Solution

We consider a bar of uniform properties ρ, A, and E, and let $C = \sqrt{E/\rho}$ be the speed of sound in the bar. The exact solution to Equation (7.1) with boundary conditions Equation (7.2) is

$$u(x,t) = \sum_{n=0}^{\infty} \hat{B}_n sin\left(\frac{\omega_n x}{C}\right) e^{-\hat{i}\omega_n t} \tag{7.7}$$

where

$$\omega_n = (2n+1)\frac{\pi}{2}\frac{C}{L} \qquad n = 0, 1, 2, ... \tag{7.8}$$

are the natural frequencies, and $\hat{B}_n$ is complex amplitude of the n-th mode that can be found to satisfy certain initial conditions.

Finite element solution

We want to determine the lowest frequency of the bar. First, we solve the problem Equation (7.4) with boundary conditions Equation (7.5) using one linear element. The stiffness and mass matrices are given by Equation (2.218) and Equation (2.219), respectively. The finite element model for one linear element is

$$\left[\frac{EA}{L}\begin{bmatrix} 1 & -1 \\ -1 & 1 \end{bmatrix} - \omega^2 \frac{\rho AL}{6}\begin{bmatrix} 2 & 1 \\ 1 & 2 \end{bmatrix}\right]\left\{\begin{array}{c} \hat{U}_1 \\ \hat{U}_2 \end{array}\right\} = \left\{\begin{array}{c} \hat{R}_1 \\ \hat{R}_2 \end{array}\right\}. \tag{7.9}$$

The boundary conditions give $\hat{U}_1 = 0$ and $\hat{R}_2 = 0$. Applying the boundary conditions, we find the reduced system to be the second equation of the system Equation (7.9),

$$(1-2\lambda^2)\hat{U}_2 = 0 \tag{7.10}$$

where $\lambda^2 = \omega^2\rho L^2/6E$. For a nontrivial solution we must have $\lambda = \pm 1/\sqrt{2}$, which gives the frequency $\omega L/C = \sqrt{3} = 1.7321$, which is 10% higher than the exact value $\omega L/C = \pi/2 = 1.5708$.

Next, we solve the problem using one quadratic element for which the stiffness and mass matrices are given by Equation (2.268) and Equation (2.269), respectively. The finite element model is

$$\left[\frac{EA}{3L}\begin{bmatrix} 7 & -8 & 1 \\ -8 & 16 & -8 \\ 1 & -8 & 7 \end{bmatrix} - \omega^2\frac{\rho AL}{30}\begin{bmatrix} 4 & 2 & -1 \\ 2 & 16 & 2 \\ -1 & 2 & 4 \end{bmatrix}\right]\begin{Bmatrix} \hat{U}_1 \\ \hat{U}_2 \\ \hat{U}_3 \end{Bmatrix} = \begin{Bmatrix} \hat{R}_1 \\ 0 \\ \hat{R}_3 \end{Bmatrix}. \tag{7.11}$$

Applying the boundary conditions $\hat{U}_1 = 0$ and $\hat{R}_3 = 0$, we get the reduced system as the last two equations

$$\left[\frac{EA}{3L}\begin{bmatrix} 16 & -8 \\ -8 & 7 \end{bmatrix} - \omega^2\frac{\rho AL}{30}\begin{bmatrix} 16 & 2 \\ 2 & 4 \end{bmatrix}\right]\begin{Bmatrix} \hat{U}_2 \\ \hat{U}_3 \end{Bmatrix} = \begin{Bmatrix} 0 \\ 0 \end{Bmatrix}. \tag{7.12}$$

For nontrivial solution, we set the determinant of the coefficient matrix equal to zero,

$$\begin{Vmatrix} 16-16\lambda^2 & -8-2\lambda^2 \\ -8-2\lambda^2 & 7-4\lambda^2 \end{Vmatrix} = 0 \tag{7.13}$$

where $\lambda^2 = \omega^2\rho L^2/10E$. The roots are $\lambda^2 = (26 \pm \sqrt{496})/15$. The lowest frequency is $\omega L/C = 1.5767$, which is 0.4% higher than the exact value.

7.2 NATURAL FREQUENCIES OF BEAMS AND FRAMES

The equation of motion of Euler–Bernoulli beam is Equation (6.17), which is repeated here after neglecting the rotary inertia term,

$$\rho A\frac{\partial^2 v}{\partial t^2} + \frac{\partial^2}{\partial x^2}\left[D\frac{\partial^2 v}{\partial x^2}\right] + qv - f(x,t) = 0, \qquad 0 < x < L \tag{7.14}$$

where $D = EI$ is the flexural rigidity, which may vary with x, ρ is the mass density (kg/m^3), and A is the beam cross-sectional area. For free vibrations $f(x,t) = 0$ and the boundary conditions are homogeneous. The coefficients in the differential equation are independent of time, hence we assume a harmonic solution

$$v(x,t) = \hat{v}(x)e^{-\hat{\imath}\omega t} \tag{7.15}$$

Substituting into Equation (7.14), we obtain an equation for the complex amplitude $\hat{v}$,

$$-\omega^2\rho A\hat{v} + \frac{d^2}{dx^2}\left[D\frac{d^2\hat{v}}{dx^2}\right] + q\hat{v} = 0 \qquad 0 < x < L. \tag{7.16}$$

When subjected to homogeneous boundary conditions, Equation (7.16) is a differential eigenvalue problem. The finite element method can be used to approximate it with an algebraic eigenvalue problem.

Alternatively, we can also apply the finite element method to Equation (7.14), as presented in Chapter 6, and obtain the element equations Equation (6.53),

$$[M^e]\left\{\ddot{U}^e\right\}+[C^e]\left\{\dot{U}^e\right\}+[A^e]\left\{U^e\right\}=\left\{F^e\right\}+\left\{R^e\right\}. \tag{7.17}$$

Assemble all elements into a global system,

$$[M]\left\{\ddot{U}\right\}+[C]\left\{\dot{U}\right\}+[K]\left\{U\right\}=\left\{F\right\}+\left\{R\right\}. \tag{7.18}$$

For undamped free vibrations we have $[C]=0$, and no external forcing is applied $\{F\}=0$. To determine natural frequencies, we assume harmonic response for the nodal primary and secondary variables

$$\begin{aligned}\{U\} &= \left\{\hat{U}\right\}e^{-\hat{\imath}\omega t}\\ \{R\} &= \left\{\hat{R}\right\}e^{-\hat{\imath}\omega t}\end{aligned} \tag{7.19}$$

where ω is a natural frequency and $\hat{\imath}=\sqrt{-1}$, $\left\{\hat{U}\right\}$, and $\left\{\hat{R}\right\}$ are complex amplitudes of the nodal primary and secondary variables, respectively.

$$\left[-\omega^2[M]+[K]\right]\left\{\hat{U}\right\}=\left\{\hat{R}\right\}. \tag{7.20}$$

For a natural boundary condition, corresponding secondary variable is zero (free vibrations). For a homogeneous mixed boundary condition, the secondary variable is expressed in terms of the corresponding primary variable and substituted into the system of equations. After applying the boundary conditions, we obtain the reduced system of equations.

$$\left[[\bar{K}]-\omega^2[\bar{M}]\right]\left\{\hat{U}\right\}=0. \tag{7.21}$$

This is a homogeneous algebraic linear system of equations; a generalized eigenvalue problem. The eigenvalues are ω^2, and the eigenvectors are the mode shapes. The natural frequencies are the square root of the eigenvalues. We use the MATLAB function [Eigenvectors, Eigenvalues] = eig($[\bar{K}],[\bar{M}]$) to compute eigenvalues and eigenvectors.

Example 7.2. Natural Frequencies of a Cantilever Beam

We want to determine the natural frequencies of a cantilever beam of length L, uniform EI, and ρA. It is fixed at $x=0$ and free at $x=L$. The boundary conditions are

$$\begin{aligned} x &= 0: v=0 && \text{and} \quad \frac{dv}{dx}=0\\ x &= L: \frac{d}{dx}\left(D\frac{d^2v}{dx^2}\right)=0 && \text{and} \quad D\frac{d^2v}{dx^2}=0. \end{aligned} \tag{7.22}$$

We use one element ($h_e = h = L$) and determine two lowest natural frequencies. The mass and stiffness matrices are given by Equations (6.50) and (6.51), respectively,

$$\left[\frac{2EI}{h^3}\begin{bmatrix} 6 & 3h & -6 & 3h \\ 3h & 2h^2 & -3h & h^2 \\ -6 & -3h & 6 & -3h \\ 3h & h^2 & -3h & 2h^2 \end{bmatrix} - \omega^2\frac{\rho Ah}{420}\begin{bmatrix} 156 & 22h & 54 & -13h \\ 22h & 4h^2 & 13h & -3h^2 \\ 54 & 13h & 156 & -22h \\ -13h & -3h^2 & -22h & 4h^2 \end{bmatrix}\right]\begin{Bmatrix} \hat{U}_1 \\ \hat{U}_2 \\ \hat{U}_3 \\ \hat{U}_4 \end{Bmatrix} = \begin{Bmatrix} \hat{R}_1 \\ \hat{R}_2 \\ \hat{R}_3 \\ \hat{R}_4 \end{Bmatrix}. \quad (7.23)$$

The boundary conditions give $\hat{U}_1 = \hat{U}_2 = 0$, and $\hat{R}_3 = \hat{R}_4 = 0$. The reduced system is given by the last two equations

$$\left[\frac{2EI}{h^3}\begin{bmatrix} 6 & -3h \\ -3h & 2h^2 \end{bmatrix} - \omega^2\frac{\rho Ah}{420}\begin{bmatrix} 156 & -22h \\ -22h & 4h^2 \end{bmatrix}\right]\begin{Bmatrix} \hat{U}_3 \\ \hat{U}_4 \end{Bmatrix} = \begin{Bmatrix} 0 \\ 0 \end{Bmatrix}. \quad (7.24)$$

We define the nondimensional parameter $\lambda = \rho Ah^4\omega^2/EI$.

$$\left[420\begin{bmatrix} 6 & -3h \\ -3h & 2h^2 \end{bmatrix} - \lambda\begin{bmatrix} 78 & -11h \\ -11h & 2h^2 \end{bmatrix}\right]\begin{Bmatrix} \hat{U}_3 \\ \hat{U}_4 \end{Bmatrix} = \begin{Bmatrix} 0 \\ 0 \end{Bmatrix} \quad (7.25)$$

$$\begin{bmatrix} 2520 - 78\lambda & (-1260 + 11\lambda)h \\ (-1260 + 11\lambda)h & (840 - 2\lambda)h^2 \end{bmatrix}\begin{Bmatrix} \hat{U}_3 \\ \hat{U}_4 \end{Bmatrix} = \begin{Bmatrix} 0 \\ 0 \end{Bmatrix}. \quad (7.26)$$

For non-trivial solutions, we set the determinant of the coefficient matrix to zero, and obtain the characteristic equation.

$$\lambda^2 - 1224\lambda + 15120 = 0. \quad (7.27)$$

The roots are $\lambda_1 = 12.48$ and $\lambda_2 = 1211.52$. The lowest natural nondimensional frequencies are $\omega_1^* = \pm\sqrt{\lambda_1} = \pm 3.5327$ and $\omega_2^* = \pm\sqrt{\lambda_2} = \pm 34.8069$. The exact solution gives $\omega_1^* = 3.5160$ and $\omega_2^* = 22.0345$. The frequencies in radians per second are $\omega_1 = 3.5327\sqrt{EI/\rho AL^4}$ and $\omega_2 = 34.8069\sqrt{EI/\rho AL^4}$.

To determine the cigenvectors, we use the first equation of the system (7.26),

$$\hat{U}_4 = \frac{6}{L}\left(\frac{420 - 13\lambda}{1260 - 11\lambda}\right)\hat{U}_3. \quad (7.28)$$

Choose $\hat{U}_3 = 1$,

For $\lambda = \lambda_1$, we get $\hat{U}_4 = 1.3776/L$

For $\lambda = \lambda_2$, we get $\hat{U}_4 = 7.6220/L$

To plot the mode shape we use the element shape functions (Hermite cubic polynomials). $\hat{V}(x) = \psi_3(x)\hat{U}_3 + \psi_4(x)\hat{U}_4$.

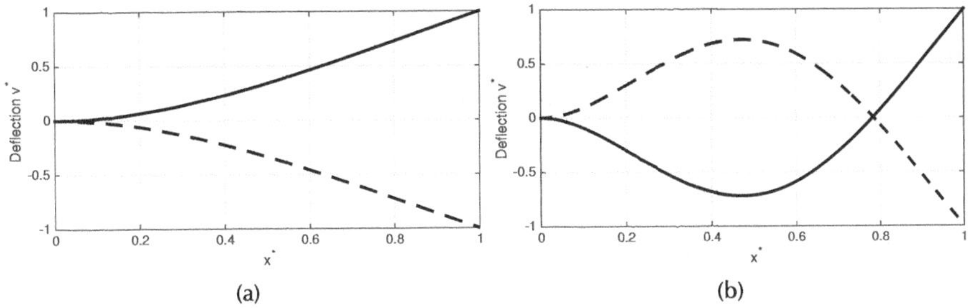

Figure 7.1 Free vibration of a cantilever beam. (a) First mode, and (b) second mode.

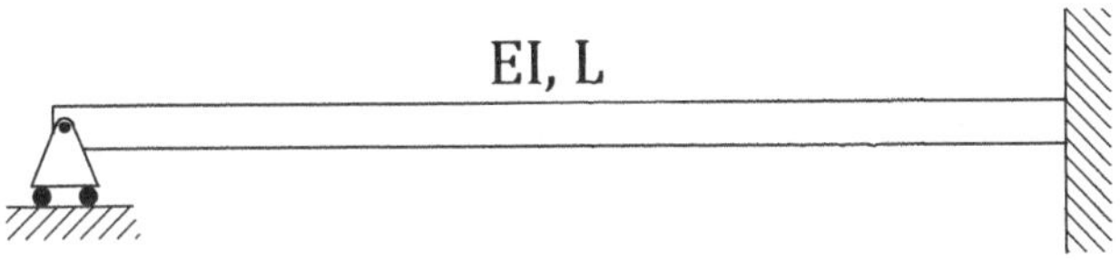

Figure 7.2 Free vibrations of a simply supported-fixed beam.

The exact natural frequencies of a cantilever beam are $\omega_n^* = \gamma_n^2$, where γ_n are the consecutive positive roots of the characteristic equation $cos\gamma cosh\gamma + 1 = 0$ (see, for example, [119]). A Newton–Raphson iteration is used to determine the lowest eight natural frequencies of a cantilever beam, $\omega_n^* =$

```
3.516015269e+00, 2.203449156e+01, 6.169721441e+01, 1.209019161e+02
1.998595301e+02, 2.985555310e+02, 4.169907861e+02, 5.551652476e+02
```

Using 15 elements, we find the the percentage errors to be:

$1.7e-05, 6.6e-04, 5.1e-03, 2.0e-02, 5.3e-02, 1.2e-01, 2.2e-01, 3.8e-01.$

The mode shapes of the first two modes are plotted in Figures 7.1a and 7.1b.

Example 7.3. Natural Frequencies of a Beam using Hermite Quintic Elements

We want to determine the natural frequencies of a uniform beam of length L and flexural rigidity EI, which is simply supported at $x = 0$ and fixed at $x = L$ as shown in Figure 7.2. We use Hermite quintic elements, each element has three local nodes and six degrees of freedom. The shape functions are given by Equation (3.72). The

Table 7.1
Natural Frequencies of Simply Supported-fixed Beam

Mode	4 elements	8 elements	Exact	% error
1	1.541820595e+01	1.541820572e+01	1.541820572e+01	6.5e-09
2	4.996493673e+01	4.996486238e+01	4.996486203e+01	7.0e-07
3	1.042499731e+02	1.042477097e+02	1.042476965e+02	1.3e-05
4	1.783338158e+02	1.782699153e+02	1.782697295e+02	1.0e-04
5	2.723506392e+02	2.720324243e+02	2.720309713e+02	5.3e-04
6	3.870741288e+02	3.855390815e+02	3.855314219e+02	2.0e-03
7	5.249298559e+02	5.187989126e+02	5.187710813e+02	5.4e-03
8	6.793652518e+02	6.719497032e+02	6.717499495e+02	3.0e-02

element mass and stiffness matrices are

$$[M^e] = \frac{(\rho A)_e h_e}{13860}\begin{bmatrix} 2092 & 114h_e & 880 & -160h_e & 262 & -29h_e \\ 114h_e & 8h_e^2 & 88h_e & -12h_e^2 & 29h_e & -3h_e^2 \\ 880 & 88h_e & 5632 & 0 & 880 & -88h_e \\ -160h_e & -12h_e^2 & 0 & 128h_e^2 & 160h_e & -12h_e^2 \\ 262 & 29h_e & 880 & 160h_e & 2092 & -114h_e \\ -29h_e & -3h_e^2 & -88h_e & -12h_e^2 & -114h_e & 8h_e^2 \end{bmatrix} \tag{7.29}$$

$$[K^e] = \frac{2(EI)_e}{35h_e^3}\begin{bmatrix} 2546 & 569h_e & -1792 & 960h_e & -754 & 121h_e \\ 569h_e & 166h_e^2 & -448h_e & 160h_e^2 & -121h_e & 19h_e^2 \\ -1792 & -448h_e & 3584 & 0 & -1792 & 448h_e \\ 960h_e & 160h_e^2 & 0 & 640h_e^2 & -960h_e & 160h_e^2 \\ -754 & -121h_e & -1792 & -960h_e & 2546 & -569h_e \\ 121h_e & 19h_e^2 & 448h_e & 160h_e^2 & -569h_e & 166h_e^2 \end{bmatrix}. \tag{7.30}$$

We note that the exact natural frequencies of the beam are $\omega_m = \lambda_m^2\sqrt{EI/\rho AL^4}$ where λ_m are consecutive positive roots of the characteristic equation $tan(\lambda_m) = tanh(\lambda_m)$ (see, for example, [119]), which we solve using a Newton–Raphson iteration. The finite-element results for the first eight frequencies are compared to the exact solution in Table 7.1 using 4 and 8 elements. The percentage errors shown in the table are for a mesh of 8 elements.

Example 7.4. Free Vibrations of a Cantilever Beam with Concentrated Mass at the Tip

Figure 7.3 shows a small body of mass m_o and moment of inertia I_o about a z-axis through its mass center attached to the tip of a cantilever beam. As the beam

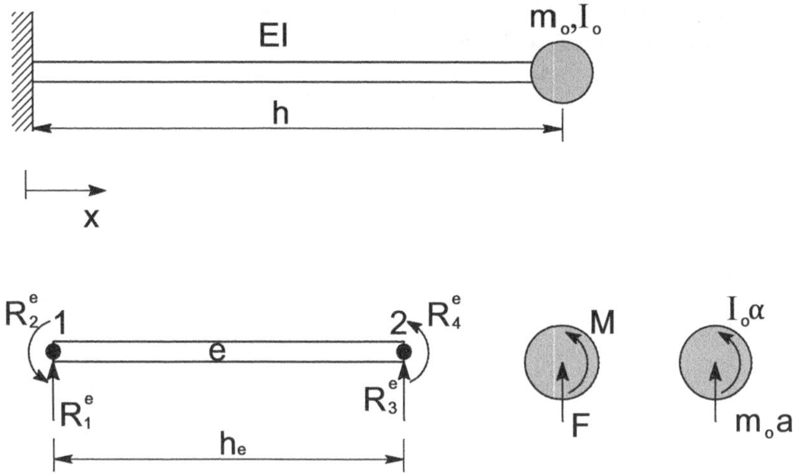

Figure 7.3 Free vibrations of a cantilever beam with concentrated mass at the tip.

vibrates the body also vibrates with the beam's tip. Thus, the body vertical acceleration is equal to the acceleration of the beam $\partial^2 v/\partial t^2$ evaluated at the tip. Because the beam tip also rotates about the z-axis, the body angular acceleration is equal to $\partial^2(\partial v/\partial x)/\partial t^2$ evaluated at the tip. The equations of motion of the body are

$$F = m_o a \tag{7.31}$$

$$M = I_o \alpha. \tag{7.32}$$

We discretize the beam into finite elements, and let element # e be the element attached to the body. The force F and moment M acting on the body are exerted by the beam tip, therefore

$$F = -R_3^e \tag{7.33}$$

and

$$M = -R_4^e. \tag{7.34}$$

The acceleration of the body is equal to the acceleration of node # 2 of that element,

$$a = \ddot{U}_3^e \tag{7.35}$$

and

$$\alpha = \ddot{U}_4^e. \tag{7.36}$$

The equations of motion of the body provide the boundary conditions at the beam tip,

$$R_3^e + m_o \ddot{U}_3^e = 0 \tag{7.37}$$

and

$$R_4^e + I_o \ddot{U}_4^e = 0. \tag{7.38}$$

These are mixed boundary conditions. To determine the natural frequencies, we assume harmonic oscillations, and obtain

$$\hat{R}_3^e - m_o\omega^2\hat{U}_3^e = 0 \tag{7.39}$$

and

$$\hat{R}_4^e - I_o\omega^2\hat{U}_4^e = 0. \tag{7.40}$$

Using one element, we have

$$\left[\frac{2EI}{h^3}\begin{bmatrix} 6 & 3h & -6 & 3h \\ 3h & 2h^2 & -3h & h^2 \\ -6 & -3h & 6 & -3h \\ 3h & h^2 & -3h & 2h^2 \end{bmatrix} - \omega^2\frac{\rho Ah}{420}\begin{bmatrix} 156 & 22h & 54 & -13h \\ 22h & 4h^2 & 13h & -3h^2 \\ 54 & 13h & 156 & -22h \\ -13h & -3h^2 & -22h & 4h^2 \end{bmatrix}\right]\begin{Bmatrix} \hat{U}_1 \\ \hat{U}_2 \\ \hat{U}_3 \\ \hat{U}_4 \end{Bmatrix} = \begin{Bmatrix} \hat{R}_1 \\ \hat{R}_2 \\ \hat{R}_3 \\ \hat{R}_4 \end{Bmatrix}. \tag{7.41}$$

Boundary conditions give $\hat{U}_1 = \hat{U}_2 = 0$, $\hat{R}_3 = \omega^2 m_o\hat{U}_3$, and $\hat{R}_4 = \omega^2 I_o\hat{U}_4$. The reduced system is given by the last two equations

$$\left[\frac{2EI}{h^3}\begin{bmatrix} 6 & -3h \\ -3h & 2h^2 \end{bmatrix} - \omega^2\frac{\rho Ah}{420}\begin{bmatrix} 156+\mu_1 & -22h \\ -22h & (4+\mu_2)h^2 \end{bmatrix}\right]\begin{Bmatrix} \hat{U}_3 \\ \hat{U}_4 \end{Bmatrix} = \begin{Bmatrix} 0 \\ 0 \end{Bmatrix}. \tag{7.42}$$

where $\mu_1 = 420m_o/(\rho Ah)$ and $\mu_2 = 420I_o/(\rho Ah^3)$. We define the nondimensional parameter $\lambda = \rho Ah^4\omega^2/EI$, and obtain the eigenvalue problem,

$$\begin{bmatrix} 5040-(156+\mu_1)\lambda & (-2520+22\lambda)h \\ (-2520+22\lambda)h & \{1680-(4+\mu_2)\lambda\}h^2 \end{bmatrix}\begin{Bmatrix} \hat{U}_3 \\ \hat{U}_4 \end{Bmatrix} = \begin{Bmatrix} 0 \\ 0 \end{Bmatrix}. \tag{7.43}$$

For given μ_1 and μ_2, we set the determinant equal to zero and solve for the eigenvalues λ.

Example 7.5. Critical Speed of Spinning Shaft: Whirling of Shafts

Large amplitude lateral vibrations can develop on a rotating shaft if its angular velocity approaches the natural frequency of transverse vibration of the spinning body [119]. We consider a steel shaft of diameter $d = 50$ mm and length $2h = 2$ m that carries a disc of mass $m_d = 40$ kg at the midpoint as shown in Figure 7.4. The bearings at A and B provide negligible moments, and the one at A may be treated as a thrust bearing. Determine the critical speed in rpm.

Solution using two Hermite cubic elements

We model the shaft by two cubic Hermite beam elements. The assembled equations before application of boundary conditions are

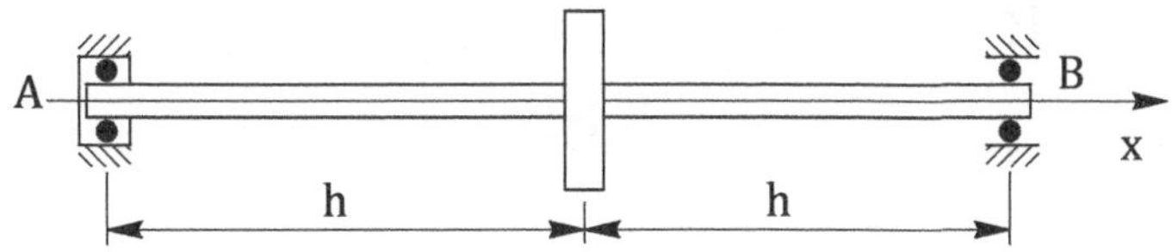

Figure 7.4 A rotating shaft with a disc at the midpoint.

$$\left[\frac{2EI}{h^3}\begin{bmatrix} 6 & 3h & -6 & 3h & 0 & 0 \\ 3h & 2h^2 & -3h & h^2 & 0 & 0 \\ -6 & -3h & 12 & 0 & -6 & 3h \\ 3h & h^2 & 0 & 4h^2 & -3h & h^2 \\ 0 & 0 & -6 & -3h & 6 & -3h \\ 0 & 0 & 3h & h^2 & -3h & 2h^2 \end{bmatrix} - \omega^2 \frac{\rho A h}{420}\begin{bmatrix} 156 & 22h & 54 & -13h & 0 & 0 \\ 22h & 4h^2 & 13h & -3h^2 & 0 & 0 \\ 54 & 13h & 312 & 0 & 54 & -13h \\ -13h & -3h^2 & 0 & 8h^2 & 13h & -3h^2 \\ 0 & 0 & 54 & 13h & 156 & -22h \\ 0 & 0 & -13h & -3h^2 & -22h & 4h^2 \end{bmatrix}\right]\begin{Bmatrix} \hat{U}_1 \\ \hat{U}_2 \\ \hat{U}_3 \\ \hat{U}_4 \\ \hat{U}_5 \\ \hat{U}_6 \end{Bmatrix} = \begin{Bmatrix} \hat{R}_1^1 \\ \hat{R}_2^1 \\ \hat{R}_3^1 + \hat{R}_1^2 \\ \hat{R}_4^1 + \hat{R}_2^2 \\ \hat{R}_3^2 \\ \hat{R}_4^2 \end{Bmatrix}. \tag{7.44}$$

The boundary conditions at the bearings A and B are $\hat{U}_1 = \hat{U}_5 = \hat{R}_2^1 = \hat{R}_4^2 = 0$. We treat the disc as a concentrated mass and neglect its rotary inertia. Thus, the conditions at the common node between the two elements are $\hat{R}_3^1 + \hat{R}_1^2 = \omega^2 m_d \hat{U}_3$ and $\hat{R}_4^1 + \hat{R}_2^2 = 0$. The reduced system is given by Equations 2, 3, 4, and 6 of the system Equation (7.44).

$$\left[840\begin{bmatrix} 2h^2 & -3h & h^2 & 0 \\ -3h & 12 & 0 & 3h \\ h^2 & 0 & 4h^2 & h^2 \\ 0 & 3h & h^2 & 2h^2 \end{bmatrix} - \lambda\begin{bmatrix} 4h^2 & 13h & -3h^2 & 0 \\ 13h & 312+\mu & 0 & -13h \\ -3h^2 & 0 & 8h^2 & -3h^2 \\ 0 & -13h & -3h^2 & 4h^2 \end{bmatrix}\right]\begin{Bmatrix} \hat{U}_2 \\ \hat{U}_3 \\ \hat{U}_4 \\ \hat{U}_6 \end{Bmatrix} = \begin{Bmatrix} 0 \\ 0 \\ 0 \\ 0 \end{Bmatrix} \tag{7.45}$$

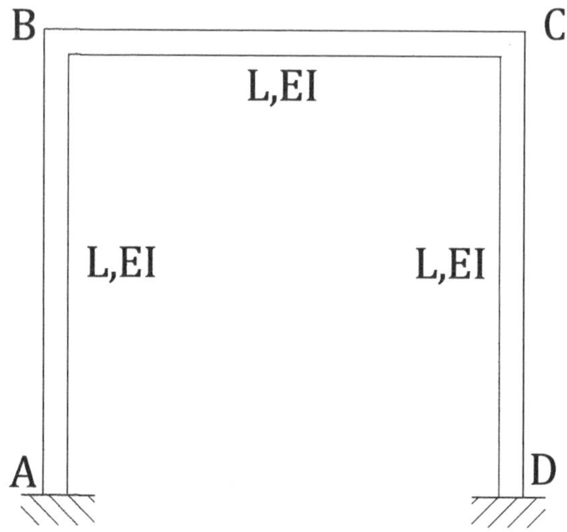

Figure 7.5 Π-frame fixed at A and D.

where $\mu = 420m_d/(\rho Ah)$, and $\lambda = \rho Ah^4\omega^2/EI$. To solve for the eigenvalue λ, we set the determinant of the coefficient matrix equal to zero. For the given data: $d =$ 0.05m, $h = 1$ m, $A = \pi d^2/4$, $I = \pi(d/2)^4/4$, $E = 200$ GPa, $\rho = 8000$ kg/m^3, and $m_d = 40$ kg, the lowest eigenvalue is $\lambda = 1.70471$. Thus, $\omega = \sqrt{\lambda EI/\rho Ah^4}$, and hence the critical speed is $60\omega/2\pi = 779.25$ rpm. If we approximate the shaft-disc assembly by a spring-mass system, where the spring constant is $k = 48EI/(2h)^3$, and the mass is m_d, we get an approximation of the critical speed of $\omega_a = \sqrt{k/m}$, which is 916.13 rpm.

Example 7.6. Natural Frequencies of a Frame

The mass matrix for the local displacements of a frame element as shown in Figure 3.20 is

$$[\bar{M}^e] = \frac{(\rho A)_e h_e}{420}\begin{bmatrix} 140 & 0 & 0 & 70 & 0 & 0 \\ 0 & 156 & 22h_e & 0 & 54 & -13h_e \\ 0 & 22h_e & 4h_e^2 & 0 & 13h_e & -3h_e^2 \\ 70 & 0 & 0 & 140 & 0 & 0 \\ 0 & 54 & 13h_e & 0 & 156 & -22h_e \\ 0 & -13h_e & -3h_e^2 & 0 & -22h_e & 4h_e^2 \end{bmatrix}. \tag{7.46}$$

Projecting the displacement vector onto global coordinates, we transform the mass matrix

$$[M^e] = [T^e]^T\,[\bar{M}^e]\,[T^e] \tag{7.47}$$

where $[T^e]$ is the transformation matrix given by Equation (3.82). The stiffness matrix for a frame element is given by Equation (3.79).

Now, we consider a Π-frame as shown in Figure 7.5 made of steel ($E = 200$ GPa, $\rho = 8000$ kg/m^3). The length $L = 4$ m, and the cross-sectional area is a square of

Table 7.2
Natural Frequencies of Π-frame in Hz

Mode	12 elements	24 elements	48 elements	96 elements
1	5.520090549e+00	5.520029626e+00	5.520024840e+00	5.520024305e+00
2	2.176533554e+01	2.175668459e+01	2.175613591e+01	2.175610149e+01
3	3.556814770e+01	3.553670775e+01	3.553461087e+01	3.553446286e+01
4	3.847185944e+01	3.842440455e+01	3.842138479e+01	3.842119525e+01
5	7.775421884e+01	7.739034776e+01	7.736615516e+01	7.736461380e+01
6	9.549677610e+01	9.483378593e+01	9.478911886e+01	9.478628359e+01
7	1.103243495e+02	1.093971114e+02	1.093319269e+02	1.093275989e+02
8	1.675436999e+02	1.644342745e+02	1.642080617e+02	1.641933997e+02

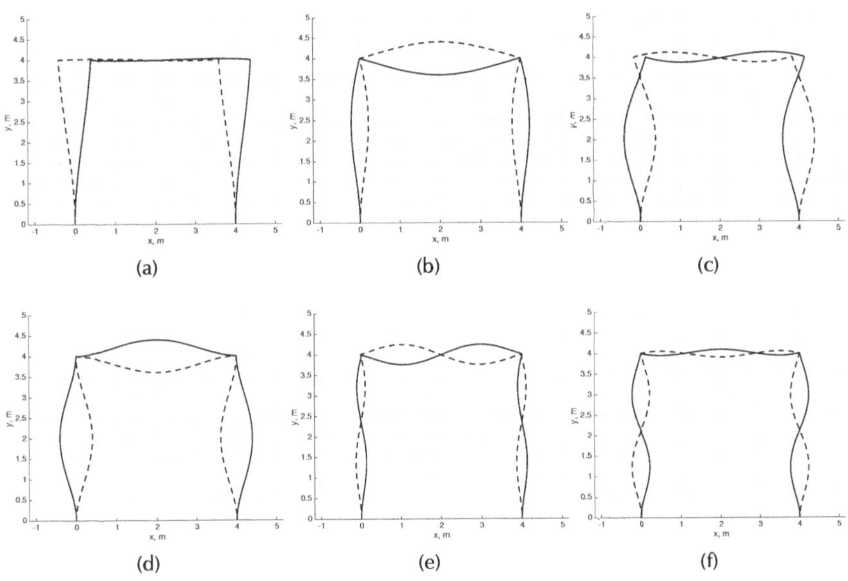

Figure 7.6 Mode shapes of the first six modes of free vibration of a Π−frame. (a) Mode 1, $f = 5.5200$ Hz, (b) mode 2, $f = 21.756$ Hz, (c) mode 3, $f = 35.534$ Hz, (d) mode 4, $f = 38.421$ Hz, (e) mode 5, $f = 77.365$ Hz, and (f) mode 6, $f = 94.786$ Hz.

side of 0.12 m. We want to determine the natural frequencies and mode shapes of the frame. We use a cubic Hermite element, and divide each side into equal elements. The first eight frequencies are shown in Table 7.2 as we refine the mesh. The mode shapes of the first six modes are shown in Figures 7.6a–7.6f. Modes 2, 4, and 6 are symmetric while modes 1, 3, and 5 are antisymmetric.

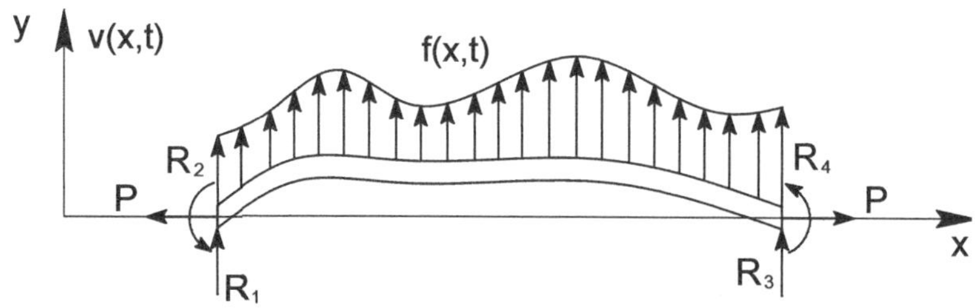

Figure 7.7 A beam subjected to lateral and axial forces.

7.3 EFFECTS OF AXIAL FORCE ON BEAM DEFLECTION

The equation of motion of Euler–Bernoulli beam subjected to lateral distributed force $f(x,t)$ and two equal but opposite axial forces applied at its ends as shown in Figure 7.7 is (see [119])

$$\rho A\frac{\partial^2 v}{\partial t^2}+\frac{\partial^2}{\partial x^2}\left[D\frac{\partial^2 v}{\partial x^2}\right]-P\frac{\partial^2 v}{\partial x^2}-f(x,t)=0, \qquad 0<x<L \qquad (7.48)$$

where $D=EI$ is the flexural rigidity, and P is the magnitude of the axial force assumed to be tensile. A compressive force means that P is negative. We can use Equation (7.48) to study the effects of axial forces on natural frequencies of beams by setting $f=0$. We can also use the same equation to determine critical loads (buckling) of columns by setting $f=0$ and removing the inertia term. The weak form for a two-node element is

$$\int_{x_1^e}^{x_2^e}\rho Aw\frac{\partial^2 v}{\partial t^2}dx+\int_{x_1^e}^{x_2^e}\frac{\partial^2 w}{\partial x^2}D\frac{\partial^2 v}{\partial x^2}dx+\int_{x_1^e}^{x_2^e}P\frac{\partial w}{\partial x}\frac{\partial v}{\partial x}dx=\int_{x_1^e}^{x_2^e}wfdx$$
$$+w(x_1^e)R_1^e+w(x_2^e)R_3^e+(\frac{dw}{dx}|_{x_1^e})R_2^e+(\frac{dw}{dx}|_{x_2^e})R_4^e \qquad (7.49)$$

where $w(x)$ is a test (weight) function, and

$$\begin{aligned}
R_1^e &= \left[\frac{\partial}{\partial x}(D\frac{\partial^2 v}{\partial x^2})-P\frac{\partial v}{\partial x}\right]_{x_1^e}\\
R_2^e &= -\left(D\frac{\partial^2 v}{\partial x^2}\right)_{x_1^e}\\
R_3^e &= -\left[\frac{\partial}{\partial x}\left(D\frac{\partial^2 v}{\partial x^2}\right)-P\frac{\partial v}{\partial x}\right]_{x_2^e}\\
R_4^e &= \left(D\frac{\partial^2 v}{\partial x^2}\right)_{x_2^e}
\end{aligned} \qquad (7.50)$$

are the secondary variables at the element's nodes.

Finite Element Model

Using Hermite cubic polynomials to approximate the deflection over the element, we obtain the Galerkin finite-element model,

$$[M^e]\{\ddot{U}^e\}+[[K^e]+P[S^e]]\{U^e\}=\{F^e\}+\{R^e\}. \tag{7.51}$$

The mass and stiffness matrices are given by Equations (6.50) and (6.51), respectively. The stability matrix $[S^e]$, also called the geometric stiffness matrix, is given by

$$[S^e]=\int_{x_1^e}^{x_2^e}[H]^T[H]dx \tag{7.52}$$

where

$$[H]=\begin{bmatrix}\frac{d\psi_1}{dx} & \frac{d\psi_2}{dx} & \frac{d\psi_3}{dx} & \frac{d\psi_4}{dx}\end{bmatrix} \tag{7.53}$$

or in index notation,

$$S_{ij}^e=\int_{x_1^e}^{x_2^e}\frac{\partial\psi_i}{\partial x}\frac{\partial\psi_j}{\partial x}dx. \tag{7.54}$$

For the cubic Hermite element, we have

$$[S^e]=\frac{1}{30h_e}\begin{bmatrix}36 & 3h_e & -36 & 3h_e\\ 3h_e & 4h_e^2 & -3h_e & -h_e^2\\ -36 & -3h_e & 36 & -3h_e\\ 3h_e & -h_e^2 & -3h_e & 4h_e^2\end{bmatrix}. \tag{7.55}$$

For the quintic three-node Hermite element, we have

$$[S^e]=\frac{1}{630h_e}\begin{bmatrix}1668 & 39h_e & -1536 & 240h_e & -132 & -9h_e\\ 39h_e & 28h_e^2 & -48h_e & -8h_e^2 & 9h_e & -5h_e^2\\ -1536 & -48h_e & 3072 & 0 & -1536 & 48h_e\\ 240h_e & -8h_e^2 & 0 & 256h_e^2 & -240h_e & -8h_e^2\\ -132 & 9h_e & -1536 & -240h_e & 1668 & -39h_e\\ -9h_e & -5h_e^2 & 48h_e & -8h_e^2 & -39h_e & 28h_e^2\end{bmatrix}. \tag{7.56}$$

Example 7.7. Critical Load of a Beam

Consider a beam of length L, uniform EI, and ρA. It is build-in at $x=0$ and simply supported at $x=L$ as shown in Figure 7.8. The boundary conditions are

$$\begin{aligned} x= & \;0: v=0 \quad \text{and} \quad \frac{dv}{dx}=0\\ x= & \;L: v=0 \quad \text{and} \quad D\frac{d^2v}{dx^2}=0. \end{aligned} \tag{7.57}$$

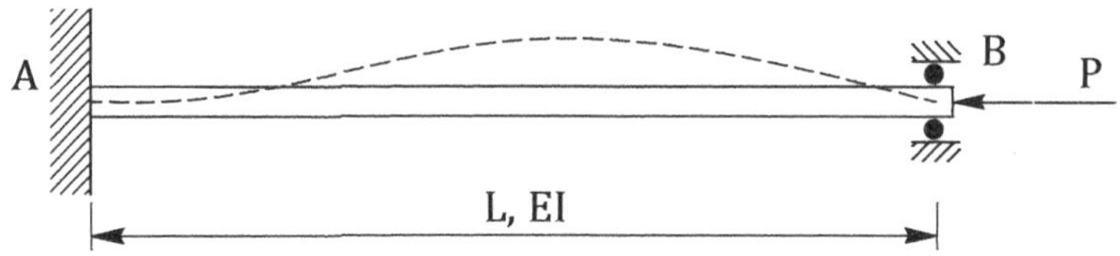

Figure 7.8 Buckling of a beam.

We want to determine the critical load using one three-node quintic element, and compare with the exact solution. Dropping the inertia and forcing terms from Equation (7.51), we have

$$[[K^e]-P[S^e]]\{U^e\}=\{R^e\}. \tag{7.58}$$

We changed the sign of P because it is compressive in this example. For one element, the element length is $h_e=L$, we get

$$\left[\frac{2EI}{35L^3}\begin{bmatrix} 2546 & 569L & -1792 & 960L & -754 & 121L \\ 569L & 166L^2 & -448L & 160L^2 & -121L & 19L^2 \\ -1792 & -448L & 3584 & 0 & -1792 & 448L \\ 960L & 160L^2 & 0 & 640L^2 & -960L & 160L^2 \\ -754 & -121L & -1792 & -960L & 2546 & -569L \\ 121L & 19L^2 & 448L & 160L^2 & -569L & 166L^2 \end{bmatrix}\right.$$

$$\left.-\frac{P}{630L}\begin{bmatrix} 1668 & 39L & -1536 & 240L & -132 & -9L \\ 39L & 28L^2 & -48L & -8L^2 & 9L & -5L^2 \\ -1536 & -48L & 3072 & 0 & -1536 & 48L \\ 240L & -8L^2 & 0 & 256L^2 & -240L & -8L^2 \\ -132 & 9L & -1536 & -240L & 1668 & -39L \\ -9L & -5L^2 & 48L & -8L^2 & -39L & 28L^2 \end{bmatrix}\right]\begin{Bmatrix} U_1 \\ U_2 \\ U_3 \\ U_4 \\ U_5 \\ U_6 \end{Bmatrix}$$

$$=\begin{Bmatrix} R_1 \\ R_2 \\ 0 \\ 0 \\ R_5 \\ R_6 \end{Bmatrix}. \tag{7.59}$$

The boundary conditions give $U_1 = U_2 = U_5 = 0$, and $R_6 = 0$. The reduced system is given by Equations 3, 4, and 6 of the system Equation (7.59).

$$\left[\begin{bmatrix} 3584 & 0 & 448L \\ 0 & 640L^2 & 160L^2 \\ 448L & 160L^2 & 166L^2 \end{bmatrix} - \lambda \begin{bmatrix} 3072 & 0 & 48L \\ 0 & 256L^2 & -8L^2 \\ 48L & -8L^2 & 28L^2 \end{bmatrix}\right] \begin{Bmatrix} U_3 \\ U_4 \\ U_6 \end{Bmatrix} = \begin{Bmatrix} 0 \\ 0 \\ 0 \end{Bmatrix} \quad (7.60)$$

where $\lambda = PL^2/36EI$. We set the determinant equal to zero, and obtain the characteristic equation

$$-162\lambda^3 + 1575\lambda^2 - 3010\lambda + 1225 = 0. \quad (7.61)$$

The roots are $PL^2/EI = 20.28578563415129$, 65.93098494071566, and 263.7832294251328. The exact critical load for the beam in this example is $P = \gamma^2 EI/L^2$, where γ is a solution to $\gamma = tan\gamma$. Thus, the exact critical load is $PL^2/EI = 20.19072855642663$, the error of one quintic element model is less that 0.5%.

Example 7.8. Effects of Axial Force on Natural Frequency of a Beam

Equation (7.51) shows that the beam stiffness matrix is enhanced by the presence of a tensile axial force P and is reduced if P is compressive. The higher (lower) stiffness implies higher (lower) natural frequencies. We consider uniform simply supported beam of length $2h$, subjected to compressive axial force P. We use two cubic Hermite beam elements. The assembled equations before application of boundary conditions are

$$\left[\frac{2EI}{h^3}\begin{bmatrix} 6 & 3h & -6 & 3h & 0 & 0 \\ 3h & 2h^2 & -3h & h^2 & 0 & 0 \\ -6 & -3h & 12 & 0 & -6 & 3h \\ 3h & h^2 & 0 & 4h^2 & -3h & h^2 \\ 0 & 0 & -6 & -3h & 6 & -3h \\ 0 & 0 & 3h & h^2 & -3h & 2h^2 \end{bmatrix}\right.$$

$$\left. - \frac{P}{30h}\begin{bmatrix} 36 & 3h & -36 & 3h & 0 & 0 \\ 3h & 4h^2 & -3h & -h^2 & 0 & 0 \\ -36 & -3h & 72 & 0 & -36 & 3h \\ 3h & -h^2 & 0 & 8h^2 & -3h & -h^2 \\ 0 & 0 & -36 & -3h & 36 & -3h \\ 0 & 0 & 3h & -h^2 & -3h & 4h^2 \end{bmatrix}\right.$$

$$
-\omega^2\frac{\rho Ah}{420}\begin{bmatrix} 156 & 22h & 54 & -13h & 0 & 0 \\ 22h & 4h^2 & 13h & -3h^2 & 0 & 0 \\ 54 & 13h & 312 & 0 & 54 & -13h \\ -13h & -3h^2 & 0 & 8h^2 & 13h & -3h^2 \\ 0 & 0 & 54 & 13h & 156 & -22h \\ 0 & 0 & -13h & -3h^2 & -22h & 4h^2 \end{bmatrix}\Bigg]\begin{Bmatrix} \hat{U}_1 \\ \hat{U}_2 \\ \hat{U}_3 \\ \hat{U}_4 \\ \hat{U}_5 \\ \hat{U}_6 \end{Bmatrix}
$$

$$
= \begin{Bmatrix} \hat{R}_1^1 \\ \hat{R}_2^1 \\ \hat{R}_3^1+\hat{R}_1^2 \\ \hat{R}_4^1+\hat{R}_2^2 \\ \hat{R}_3^2 \\ \hat{R}_4^2 \end{Bmatrix}. \tag{7.62}
$$

The boundary conditions at the supports are $\hat{U}_1 = \hat{U}_5 = \hat{R}_2^1 = \hat{R}_4^2 = 0$. The conditions at the common node between the two elements are $\hat{R}_3^1 + \hat{R}_1^2 = 0$ and $\hat{R}_4^1 + \hat{R}_2^2 = 0$. The reduced system is given by Equations 2, 3, 4, and 6 of the system Equation (7.62).

$$
\Bigg[840\begin{bmatrix} 2h^2 & -3h & h^2 & 0 \\ -3h & 12 & 0 & 3h \\ h^2 & 0 & 4h^2 & h^2 \\ 0 & 3h & h^2 & 2h^2 \end{bmatrix} - \frac{7\pi^2 r}{2}\begin{bmatrix} 4h^2 & -3h & -h^2 & 0 \\ -3h & 72 & 0 & 3h \\ -h^2 & 0 & 8h^2 & -h^2 \\ 0 & 3h & -h^2 & 4h^2 \end{bmatrix}
$$

$$
-\lambda\begin{bmatrix} 4h^2 & 13h & -3h^2 & 0 \\ 13h & 312 & 0 & -13h \\ -3h^2 & 0 & 8h^2 & -3h^2 \\ 0 & -13h & -3h^2 & 4h^2 \end{bmatrix}\Bigg]\begin{Bmatrix} \hat{U}_2 \\ \hat{U}_3 \\ \hat{U}_4 \\ \hat{U}_6 \end{Bmatrix} = \begin{Bmatrix} 0 \\ 0 \\ 0 \\ 0 \end{Bmatrix} \tag{7.63}
$$

where $\lambda = \rho Ah^4\omega^2/EI$, and $r = P/P_{cr}$ is the ratio of the force P to Euler's critical load $P_{cr} = \pi^2 EI/(2h)^2$. The exact frequencies are $\omega_m\sqrt{\rho AL^4/EI} = (m\pi)^2\sqrt{1-r/m^2}$ [119]. For $r = 0.2$, the two-element model gives the lowest frequency of 8.870759263, which is lower than that for $r = 0$ (no axial force) of 9.908558712.

7.4 HYDRODYNAMIC STABILITY: ORR–SOMMERFELD EQUATION

Analysis of the stability of fluid motions is an important branch of fluid mechanics. If a laminar motion is subjected to small disturbances, different physical mechanisms may transfer energy from the basic laminar motion to the disturbances. And if the disturbances grow in space or time, the laminar motion can take a different form or may even lose its organized structure and become turbulent.

To study the stability of a laminar fluid motion (the basic state), we superimpose small disturbances and determine the evolution of the disturbances in space and time. If any form of the disturbance grows, the basic state is declared unstable; otherwise, it is stable or neutral. The basic state may be an exact solution to the Navier–Stokes equations obtained under certain simplifying assumptions [108]. For example, assuming steady flow and the stream lines to be straight parallel lines, we obtain Poiseuille flow in a long pipe or between two parallel plates (channel flow), among many other flows. Exact solutions can also be obtained for steady motions for concentric circular streamlines. An example is the flow between two rotating long concentric cylinders. The basic state may be an approximate solution such as a laminar boundary-layer flow past a flat plate at high Reynolds number; that is, Blasius flow. All the mentioned flows exhibit instabilities depending on the Reynolds number of the flow or other operating parameters.

The basic state and the disturbed fluid motion, which is the sum of the basic state and disturbances, must satisfy the equations of fluid motion albeit with different assumptions. The linear momentum (Navier–Stokes) and continuity equations are

$$\rho \frac{D\underline{V}}{Dt} = -\nabla p + \mu \nabla^2 \underline{V} \tag{7.64}$$

$$\nabla \cdot \underline{V} = 0 \tag{7.65}$$

where $\underline{V}$, ρ, p, and μ are the velocity, density, pressure, and dynamic viscosity. A basic state is a solution to the Navier–Stokes equations, hence

$$\rho \frac{D\underline{V}_B}{Dt} = -\nabla p_B + \mu \nabla^2 \underline{V}_B \tag{7.66}$$

$$\nabla \cdot \underline{V}_B = 0 \tag{7.67}$$

where the subscript B indicates a basic state quantity. The disturbed flow (= basic state + disturbance) is written as

$$\tilde{\underline{V}} = \underline{V}_B + \underline{v}', \quad \tilde{p} = p_B + p' \tag{7.68}$$

where the $(\prime)$ indicates a disturbance quantity, must also satisfy the Navier–Stokes equations,

$$\rho \left[\frac{\partial \underline{V}_B}{\partial t} + \frac{\partial \underline{v}'}{\partial t} + (\underline{V}_B + \underline{v}') \cdot \nabla (\underline{V}_B + \underline{v}') \right] = -\nabla p_B - \nabla p' + \mu \nabla^2 \underline{V}_B + \mu \nabla^2 \underline{v}' \tag{7.69}$$

$$\nabla \cdot (\underline{V}_B + \underline{v}') = 0. \tag{7.70}$$

Subtracting equations satisfied by the basic state Equations (7.66) and (7.67), we obtain the governing equations for the disturbance field,

$$\rho \left(\frac{\partial \underline{v}'}{\partial t} + \underline{V}_B \cdot \nabla \underline{v}' + \underline{v}' \cdot \nabla \underline{V}_B + \underline{v}' \cdot \nabla \underline{v}' \right) = -\nabla p' + \mu \nabla^2 \underline{v}' \tag{7.71}$$

$$\nabla \cdot \underline{v}' = 0. \tag{7.72}$$

No limitations have been imposed on the disturbance field yet. Therefore, Equations (7.71) and (7.72) are exact equations. For a given basic state $(\underline{V}_B, p_B)$, the only nonlinear term in these equations is $\underline{v}' \cdot \nabla \underline{v}'$. In *Linear Stability Theory*, we neglect quadratic and higher-order terms in disturbances, and obtain the linearized disturbance equations

$$\rho\left(\frac{\partial \underline{v}'}{\partial t}+\underline{V}_B\cdot\nabla\underline{v}'+\underline{v}'\cdot\nabla\underline{V}_B\right)=-\nabla p'+\mu\nabla^2\underline{v}' \tag{7.73}$$

$$\nabla\cdot\underline{v}'=0. \tag{7.74}$$

For certain flows, such as velocity profiles with inflection points, the effects of viscosity on the development of disturbances can be neglected, and the analysis is called *inviscid stability analysis*. We neglect the viscous term $\mu\nabla^2\underline{v}'$, and obtain

$$\rho\left(\frac{\partial \underline{v}'}{\partial t}+\underline{V}_B\cdot\nabla\underline{v}'+\underline{v}'\cdot\nabla\underline{V}_B\right)=-\nabla p' \tag{7.75}$$

$$\nabla\cdot\underline{v}'=0. \tag{7.76}$$

7.4.1 ORR–SOMMERFELD EQUATION

We assume incompressible flow, and consider a basic state that is a parallel shear flow along the x-axis. Thus, the velocity field is $(U(y),0,0)$, and the pressure in the basic state is denoted by P. The disturbance field is two-dimensional with Cartesian velocity components $u(x,y,t)$ and $v(x,y,t)$, and pressure $p(x,y,t)$. The total flow (basic state + disturbance) is now

$$\begin{aligned}\tilde{u} &= U(y)+u(x,y,t)\\ \tilde{v} &= v(x,y,t)\\ \tilde{p} &= P+p(x,y,t)\end{aligned} \tag{7.77}$$

We substitute the total flow into the continuity and Navier–Stokes equations, subtract the equations satisfied by the basic state, drop the nonlinear terms in disturbance velocity, and obtain the disturbance equations,

Continuity Equation

$$\frac{\partial u}{\partial x}+\frac{\partial v}{\partial y}=0. \tag{7.78}$$

Momentum Equations

$$\rho\left(\frac{\partial u}{\partial t}+U\frac{\partial u}{\partial x}+v\frac{dU}{dy}\right)=-\frac{\partial p}{\partial x}+\mu\left[\frac{\partial^2 u}{\partial x^2}+\frac{\partial^2 u}{\partial y^2}\right] \tag{7.79}$$

$$\rho\left(\frac{\partial v}{\partial t}+U\frac{\partial v}{\partial x}\right)=-\frac{\partial p}{\partial y}+\mu\left[\frac{\partial^2 v}{\partial x^2}+\frac{\partial^2 v}{\partial y^2}\right]. \tag{7.80}$$

Using reference length L_r and velocity U_r, we introduce nondimensional variables $t^* = U_r t/L_r$, $(x^*, y^*) = (x/L_r, y/L_r)$, $(u^*, v^*) = (u/U_r, v/U_r)$, $U^* = U/U_r$, and $p^* = p/\rho U_r^2$, and write the disturbance equations in nondimensional form. For simplicity of notation, we drop the asterisks,

$$\frac{\partial u}{\partial x} + \frac{\partial v}{\partial y} = 0 \tag{7.81}$$

$$\frac{\partial u}{\partial t} + U\frac{\partial u}{\partial x} + v\frac{dU}{dy} = -\frac{\partial p}{\partial x} + \frac{1}{Re}\left[\frac{\partial^2 u}{\partial x^2} + \frac{\partial^2 u}{\partial y^2}\right] \tag{7.82}$$

$$\frac{\partial v}{\partial t} + U\frac{\partial v}{\partial x} = -\frac{\partial p}{\partial y} + \frac{1}{Re}\left[\frac{\partial^2 v}{\partial x^2} + \frac{\partial^2 v}{\partial y^2}\right] \tag{7.83}$$

where $Re = \rho U_r L_r/\mu$ is the Reynolds number based on the reference length and velocity. The basic state data are the velocity profile $U(y)$ and Reynolds number Re. Equations (7.81)–(7.83) are three equations to be solved for the disturbance u, v, and p. The flow domain is assumed to be unbounded in the main stream direction x-axis, but it can be bounded or unbounded in the transverse direction y-axis. We note that the coefficients of these equations are independent of t and x; they depend on y or constant. Therefore, exponential solutions in x and t are admissible; they are called normal mode solutions. We try a solution of the form,

$$\begin{aligned} u(x,y,t) &= \hat{u}(y)e^{i(\alpha x - \omega t)} \\ v(x,y,t) &= \hat{v}(y)e^{i(\alpha x - \omega t)} \\ p(x,y,t) &= \hat{p}(y)e^{i(\alpha x - \omega t)} \end{aligned} \tag{7.84}$$

where α and ω are constants, with the understanding that we take the real part of the complex-valued solution. Substituting into the disturbance equations (7.81–7.83), we obtain the equations for $\hat{u}$, $\hat{v}$, and $\hat{p}$, for simplicity we drop the hat,

$$i\alpha u + \frac{dv}{dy} = 0 \tag{7.85}$$

$$i(\alpha U - \omega)u + v\frac{dU}{dy} + i\alpha p - \frac{1}{Re}\left[-\alpha^2 u + \frac{d^2 u}{dy^2}\right] = 0 \tag{7.86}$$

$$i(\alpha U - \omega)v + \frac{dp}{dy} - \frac{1}{Re}\left[-\alpha^2 v + \frac{d^2 v}{dy^2}\right] = 0. \tag{7.87}$$

We eliminate u and p from Equations (7.85)–(7.87), and obtain an equation for v,

$$\frac{d^4 v}{dy^4} - 2\alpha^2\frac{d^2 v}{dy^2} + \alpha^4 v - iRe(\alpha U - \omega)\left[\frac{d^2 v}{dy^2} - \alpha^2 v\right] + i\alpha Re\frac{d^2 U}{dy^2}v = 0. \tag{7.88}$$

This is the Orr–Sommerfeld equation. Subjected to homogeneous boundary conditions, Equation (7.88) is a differential eigenvalue problem. Because $U(y)$ and Re are specified by the basic state, the eigenvalues are α and ω. For a nontrivial solution for v, a relationship $\Lambda(\alpha,\omega) = 0$ must be satisfied; such an equation is called a *dispersion relation*. That is to say, we cannot prescribe arbitrary values for α and ω. In general, these parameters are complex numbers; we can specify one of them and consider the other as the eigenvalue.

Temporal Stability Analysis

For temporal stability analysis, we specify a real value for the wavenumber α, and solve for the complex frequency $\omega = \omega_r + i\omega_i$. We write the normal mode solution Equation (7.84) for the v component as

$$v(x,y,t) = \parallel \hat{v}(y) \parallel e^{i\theta(y)} e^{i(\alpha x - \omega_r t)} e^{\omega_i t} \tag{7.89}$$

which is a traveling vorticity wave of length $\lambda = 2\pi/\alpha$, and phase speed of $C_{ph} = \omega_r/\alpha$, and $\theta(y)$ is the phase of the complex eigenfunction $\hat{v}(y)$. The factor $exp(\omega_i t)$ shows that if $\omega_i > 0$ the disturbance magnitude grows in time and hence the basic state is unstable. If $\omega_i < 0$ the disturbance will be damped for the selected wave number α. We call ω_i the growth rate. The basic state is stable if all possible disturbances (all values of α) are damped. A neutral point is obtained if $\omega_i = 0$. A complex phase speed is defined by $c = c_r + ic_i = \omega/\alpha$, hence $c_r = \omega_r/\alpha$ and $c_i = \omega_i/\alpha$.

Spatial Stability Analysis

For spatial stability analysis, we specify a real value for the frequency ω, and solve for the complex wavenumber $\alpha = \alpha_r + i\alpha_i$. We write the normal mode solution Equation (7.84) for the v component as

$$v(x,y,t) = \parallel \hat{v}(y) \parallel e^{i\theta(y)} e^{i(\alpha_r x - \omega t)} e^{-\alpha_i x} \tag{7.90}$$

which is a traveling vorticity wave of length $\lambda = 2\pi/\alpha_r$, and phase speed of $C_{ph} = \omega/\alpha_r$, and $\theta(y)$ is the phase of the complex eigenfunction $\hat{v}(y)$. The factor $exp(-\alpha_i x)$ shows that if $\alpha_i < 0$ the disturbance magnitude grows in space and hence the basic state is unstable. If $\alpha_i > 0$ the disturbance will be damped for the selected frequency ω. The basic state is stable if all possible disturbances (all values of ω) are damped. A neutral point is obtained if $\alpha_i = 0$. The neutral point of temporal analysis will be the same as that of spatial analysis since both α and ω are real.

7.4.2 WEAK FORM OF ORR–SOMMERFELD EQUATION

We rearrange terms and rewrite Equation (7.88) as

$$\beta \frac{d^2}{dy^2}\left(\frac{d^2 v}{dy^2} - 2\alpha^2 v\right) - (U-c)\frac{d^2 v}{dy^2} + \frac{d^2 U}{dy^2} v + \alpha^2 (U-c) v + \beta\alpha^4 v = 0 \tag{7.91}$$

where $\beta = 1/(i\alpha Re)$, and $c = \omega/\alpha$. We consider an element defined by two nodes with global coordinates y_1^e and y_2^e. To derive the weak form, we multiply Equation

(7.91) by a test (weight) function $w(y)$, integrate over the element domain. It suffices to integrate the fourth-order derivative term by parts twice as we have done for the Euler–Bernoulli beam equation; however, the resulting secondary variables do not have a clear physical meaning. We anticipate the primary variables to be the velocity components u and v, which are kinematic in nature. The continuity Equation (7.85) gives $u = (i/\alpha)dv/dy$. Thus, we may take the primary variables to be v and dv/dy. We also anticipate the secondary variables to be kinetic quantities; which are stresses. The relevant stress components are σ_{yy} and σ_{yx},

$$\sigma_{yy} = -p + \mu\frac{\partial v}{\partial y} \tag{7.92}$$

$$\sigma_{yx} = \mu\left(\frac{\partial u}{\partial y} + \frac{\partial v}{\partial x}\right). \tag{7.93}$$

For the normal mode solution Equation (7.84), and using Equations (7.85) and (7.86), we have

$$\begin{aligned} i\alpha\sigma_{yy} &= \beta\frac{d}{dy}\left(\frac{d^2v}{dy^2} - 2\alpha^2 v\right) - (U-c)\frac{dv}{dy} + \frac{dU}{dy}v \\ \sigma_{yx} &= -\beta\left(\frac{d^2v}{dy^2} + \alpha^2 v\right). \end{aligned} \tag{7.94}$$

We multiply Equation (7.91) by w and integrate the first three terms by parts, and obtain

$$\begin{aligned} &-\int_{y_1^e}^{y_2^e}\beta\frac{dw}{dy}\frac{d}{dy}\left(\frac{d^2v}{dy^2} - 2\alpha^2 v\right)dy + \int_{y_1^e}^{y_2^e}\frac{d}{dy}[w(U-c)]\,dy \\ &-\int_{y_1^e}^{y_2^e}\frac{dU}{dy}\frac{d}{dy}[wv]\,dy + \int_{y_1^e}^{y_2^e}\alpha^2 w(U-c)v dy + \int_{y_1^e}^{y_2^e}\beta\alpha^4 wv dy \\ &+\left[\beta\frac{d}{dy}\left(\frac{d^2v}{dy^2} - 2\alpha^2 v\right)w - (U-c)\frac{dv}{dy}w + \frac{dU}{dy}vw\right]_{y_1^e}^{y_2^e} = 0. \end{aligned} \tag{7.95}$$

Before we perform the second integration by part of the first term we add $\alpha^2 v$ and subtract the same term, to get

$$\begin{aligned} &\int_{y_1^e}^{y_2^e}\beta\frac{d^2w}{dy^2}\left(\frac{d^2v}{dy^2} + \alpha^2 v\right)dy + \int_{y_1^e}^{y_2^e}3\beta\alpha^2\frac{dw}{dy}\frac{dv}{dy}dy + \int_{y_1^e}^{y_2^e}\frac{d}{dy}[w(U-c)]\,dy \\ &\quad -\int_{y_1^e}^{y_2^e}\frac{dU}{dy}\frac{d}{dy}[wv]\,dy + \int_{y_1^e}^{y_2^e}\alpha^2 w(U-c)v dy + \int_{y_1^e}^{y_2^e}\beta\alpha^4 wv dy \\ &\quad +\left[\left(\beta\frac{d}{dy}\left(\frac{d^2v}{dy^2} - 2\alpha^2 v\right) - (U-c)\frac{dv}{dy} + \frac{dU}{dy}v\right)w\right]_{y_1^e}^{y_2^e} \\ &\quad +\left[-\beta\left(\frac{d^2v}{dy^2} + \alpha^2 v\right)\frac{dw}{dy}\right]_{y_1^e}^{y_2^e} = 0. \end{aligned} \tag{7.96}$$

Substituting Equation (7.94) into the boundary terms, we get

$$\int_{y_1^e}^{y_2^e} \beta \frac{d^2w}{dy^2}\left(\frac{d^2v}{dy^2}+\alpha^2 v\right)dy + \int_{y_1^e}^{y_2^e} 3\beta\alpha^2 \frac{dw}{dy}\frac{dv}{dy}dy + \int_{y_1^e}^{y_2^e} \frac{d}{dy}[w(U-c)]\,dy$$
$$-\int_{y_1^e}^{y_2^e} \frac{dU}{dy}\frac{d}{dy}[wv]\,dy + \int_{y_1^e}^{y_2^e} \alpha^2 w(U-c)v dy + \int_{y_1^e}^{y_2^e} \beta\alpha^4 wv dy$$
$$+\left[i\alpha\left(w\sigma_{yy} + \frac{1}{i\alpha}\frac{dw}{dy}\sigma_{yx}\right)\right]_{y_1^e}^{y_2^e} = 0. \tag{7.97}$$

Looking at the boundary term, we define the secondary variables to be σ_{yy} and σ_{yx}, and the corresponding primary variables to be v and $u = (1/i\alpha)dv/dy$. More simply, we define the primary variables to be v and dv/dy, and define the corresponding secondary variables at the element nodes by

$$\begin{aligned}
R_1^e &= \left[\beta\frac{d}{dy}\left(\frac{d^2v}{dy^2} - 2\alpha^2 v\right) - (U-c)\frac{dv}{dy} + \frac{dU}{dy}v\right]_{y_1^e} \\
R_2^e &= -\beta\left(\frac{d^2v}{dy^2} + \alpha^2 v\right)_{y_1^e} \\
R_3^e &= -\left[\beta\frac{d}{dy}\left(\frac{d^2v}{dy^2} - 2\alpha^2 v\right) - (U-c)\frac{dv}{dy} + \frac{dU}{dy}v\right]_{y_2^e} \\
R_4^e &= \beta\left(\frac{d^2v}{dy^2} + \alpha^2 v\right)_{y_2^e}.
\end{aligned} \tag{7.98}$$

Comparing with Equation (7.94), we see that the secondary variables R_1^e and R_2^e are $i\alpha\sigma_{yy}$ and σ_{yx} at the first node of the element. The secondary variables at the second node have the same physical meaning but with a negative sign. Now, we write the weak form of the Orr–Sommerfeld equation for an element as

$$\beta\int_{y_1^e}^{y_2^e}\left[\frac{d^2w}{dy^2}\left(\frac{d^2v}{dy^2}+\alpha^2 v\right) + 3\alpha^2\frac{dw}{dy}\frac{dv}{dy} + \alpha^4 wv\right]dy$$
$$+\int_{y_1^e}^{y_2^e}\left(U\frac{dw}{dy}\frac{dv}{dy} + \alpha^2 Uwv - \frac{dU}{dy}\frac{dw}{dy}v\right)dy - c\int_{y_1^e}^{y_2^e}\left(\frac{dw}{dy}\frac{dv}{dy} + wv\right)dy$$
$$= w(y_1^e)R_1^e + \left(\frac{dw}{dy}|_{y_1^e}\right)R_2^e + w(y_2^e)R_3^e + \left(\frac{dw}{dy}|_{y_2^e}\right)R_4^e. \tag{7.99}$$

Approximation and Hermite Quintic Shape Functions

In the present case of a fourth-order differential equation, the weak form contains the second derivative d^2v/dx^2, and hence it requires continuity of the trial solution $v(x)$ and its first derivative dv/dx. Lagrange interpolation (shape) functions are not admissible in this case because they do not allow us to enforce continuity of the slope at the inter-element nodes even if we use quadratic or cubic elements. The Hermite cubic shape functions used for the beam element can be used for the weak form of the Orr–Sommerfeld equation. Mamou and Khalid [70] developed a quintic Hermite element and used it for finite-element modeling of the Orr–Sommerfeld equation. They showed that a quintic element is computationally more efficient than the cubic one. The quintic element has two nodes with three degrees of freedom at each node, namely the value of the function and its first and second derivatives. The approximation over the element is expressed in terms of the six degrees of freedom. With the quintic element, we can enforce continuity of v, dv/dy, and d^2v/dy^2 throughout the domain. We note that the three-node quintic element presented in Chapter 3 for the beam equation is different from the present quintic element.

We write the approximation $V^e(\xi)$ in terms of the U^e's, which are the degrees of freedom of the element,

$$V^e(\xi) = \psi_1(\xi)U_1^e + \psi_2(\xi)U_2^e + \psi_3(\xi)U_3^e + \psi_4(\xi)U_4^e + \psi_5(\xi)U_5^e + \psi_6(\xi)U_6^e \tag{7.100}$$

where U_1^e, U_2^e, and U_3^e are approximations of v, dv/dy, and d^2v/dy^2 at node # 1, respectively. Similarly, U_4^e, U_5^e, and U_6^e are approximations of v, dv/dy, and d^2v/dy^2 at node # 2, respectively. We define $\xi = [2y - (y_1^e + y_2^e)]/h_e$, where $h_e = y_2^e - y_1^e$ is the element length. The element domain is mapped into $-1 \leq \xi \leq 1$. The shape functions are [70]

$$\begin{aligned}
\psi_1(\xi) &= (1-\xi)^3(8+9\xi+3\xi^2)/16 \\
\psi_2(\xi) &= (1-\xi)^3(1+\xi)(5+3\xi)(h_e/2)/16 \\
\psi_3(\xi) &= (1-\xi)^3(1+\xi)^2(h_e/2)^2/16 \\
\psi_4(\xi) &= (1+\xi)^3(8-9\xi+3\xi^2)/16 \\
\psi_5(\xi) &= (1+\xi)^3(-1+\xi)(5-3\xi)(h_e/2)/16 \\
\psi_6(\xi) &= (1+\xi)^3(1-\xi)^2(h_e/2)^2/16
\end{aligned} \tag{7.101}$$

and are plotted in Figures 7.9a–7.9f for $h_e = 1$. They satisfy the properties:

$$\begin{aligned}
&\psi_i(y_1^e) = \delta_{i1}, \quad \frac{d\psi_i}{dy}\Big|_{y_1^e} = \delta_{i2}, \quad \frac{d^2\psi_i}{dy^2}\Big|_{y_1^e} = \delta_{i3} \\
&\psi_i(y_2^e) = \delta_{i4}, \quad \frac{d\psi_i}{dy}\Big|_{y_2^e} = \delta_{i5}, \quad \frac{d^2\psi_i}{dy^2}\Big|_{y_2^e} = \delta_{i6}
\end{aligned} \tag{7.102}$$

where $\delta_{ij} = 1$ if $i = j$, and $\delta_{ij} = 0$ if $i \neq j$

Galerkin

To obtain the element equations, we substitute the approximation Equation (7.100) into the weak form Equation (7.99), and use $w_i = \psi_i$ for $i = 1, 2, ..., 6$.

$$[[A^e] - c\,[B^e]]\,\{U^e\} = \{R^e\}. \tag{7.103}$$

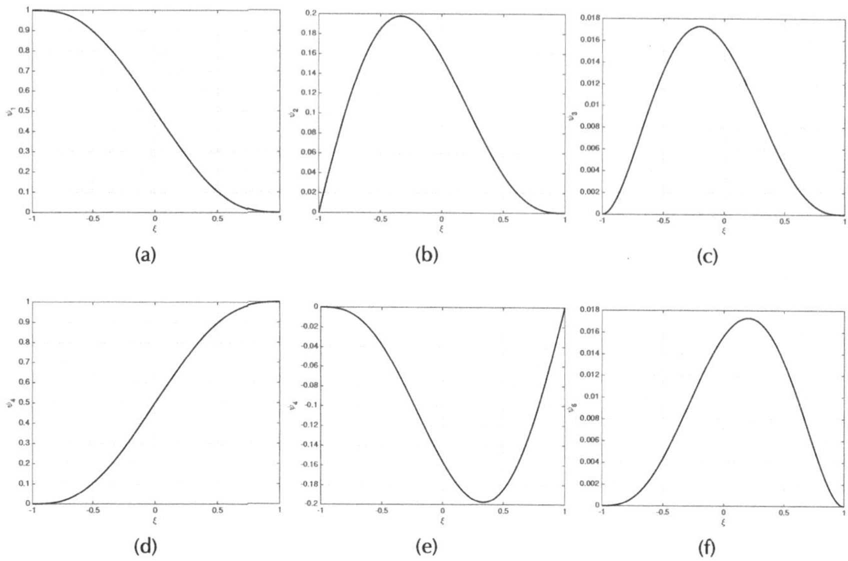

Figure 7.9 Shape functions of a two-node quintic element; three degrees of freedom at each node, namely the function and its first and second derivatives.

The entries of the element matrices in index notation for $(i = 1,2,...,6)$ and $(j = 1,2,...,6)$ are

$$a_{ij} = \int_{y_1^e}^{y_2^e} \beta \left[\frac{d^2\psi_i}{dy^2} \left(\frac{d^2\psi_j}{dy^2} + \alpha^2 \psi_j \right) + 3\alpha^2 \frac{d\psi_i}{dy}\frac{d\psi_j}{dy} + \alpha^4 \psi_i \psi_j \right] dy$$
$$+ \int_{y_1^e}^{y_2^e} \left(U \frac{d\psi_i}{dy}\frac{d\psi_j}{dy} + \alpha^2 U \psi_i \psi_j - \frac{dU}{dy}\frac{d\psi_i}{dy} \psi_j \right) dy \qquad (7.104)$$

$$b_{ij} = \int_{y_1^e}^{y_2^e} \left(\frac{d\psi_i}{dy}\frac{d\psi_j}{dy} + \psi_i \psi_j \right) dy \qquad (7.105)$$

$$R_i^e = \delta_{i1} R_1^e + \delta_{i2} R_2^e + \delta_{i4} R_3^e + \delta_{i5} R_4^e. \qquad (7.106)$$

Gauss–Legendre quadrature can be used to evaluate the integrals. However, for simple velocity profiles $U(y)$ such as plane Poiseuille flow, we evaluate the integrals analytically.

Example 7.9. Temporal Stability Analysis of Poiseuille Channel Flow

In this section, we study the stability of flow between two parallel plates, which is called a channel flow. The flow is driven by streamwise pressure gradient. Using half the distance between the walls (d) as a reference length and the maximum velocity (U_o) as a reference velocity, the velocity profile is

$$U(y) = 1 - y^2 \qquad -1 \leq y \leq 1. \tag{7.107}$$

The Reynolds number is $Re = \rho U_o d/\mu$. The boundary conditions on the disturbance velocity are no-slip $u = 0$ and no-flow through $v = 0$ on the channel walls at $y = \pm 1$, hence

$$\begin{aligned} y = -1 \quad &: \quad v = 0, \ \frac{dv}{dy} = 0 \\ y = 1 \quad &: \quad v = 0, \ \frac{dv}{dy} = 0. \end{aligned} \tag{7.108}$$

For temporal stability analysis, we prescribe a real wavenumber α, and determine the complex phase speed $c = \omega/\alpha$ as the eigenvalue. We divide the domain $-1 \leq y \leq 1$ into $N-1$ elements. The number of global nodes is N. We assemble all elements into a global system of $3N$ degrees of freedom,

$$[[A] - c[B]]\{U\} = \{R\}. \tag{7.109}$$

The entries of global vector $\{R\}$ are zero at all interior nodes, and are unknown at the wall nodes $i = 1$ and $i = N$. To apply the EBC Equation (7.108), we modify the matrices as follows:

(1) First we set $A(i,j) = B(i,j) = 0$ for $i = 1, 2, N-1, N$ and $j = 1, ..., N$.

(2) Then we set $A(i,i) = 1$ for $i = 1, 2, N-1, N$.

The modified system is the generalized eigenvalue problem

$$[[\bar{A}] - c[\bar{B}]]\{U\} = \{0\} \tag{7.110}$$

which we solve using MATLAB function $[H, c] = \text{eig}([\bar{A}], [\bar{B}])$ that computes eigenvalues c and eigenvectors H.

First we investigate the accuracy of the finite element method for the standard case $Re = 10^4$ and $\alpha = 1$. We focus our attention on the most amplified mode (largest c_i) for which highly accurate values have been reported by Orszag [81], Kirchner [60], and Mamou and Khalid [70]. We use Kirchner's values of $c_r = 0.23752648882047$ and $c_i = 0.00373967062979582$ for comparison and error estimates. First, we use uniform mesh and compute the error as we increase the number of elements from 10 to 100. The errors in c_r and c_i are listed in Table 7.3, and plotted in Figure 7.10a. The results suggest that the error drops as N^{-7}. Mamou and Khalid [70] found that, for the same number of elements, a nonuniform mesh gives higher accuracy than uniform mesh because the nonuniform mesh resolves the complex structure of the eigenfunctions near solid walls. We use a mildly stretched mesh where the grid points are distributed as $y_j = (5\xi_j - \xi_j^3)/4$, where $\xi_j = -1 + 2(j-1)/(N-1)$. The errors in c_r and c_i are listed in Table 7.4, and plotted in Figure 7.10b. The

Table 7.3
Errors in c for Poiseuille Channel Flow at $Re = 10^4$ and $\alpha = 1$

N-1	c_r	c_i	$\lvert\Delta c_r\rvert$	$\lvert\Delta c_i\rvert$
10	2.321009532e-01	7.547964182e-03	5.4255e-03	3.8083e-03
20	2.373776055e-01	3.655478379e-03	1.4888e-04	8.4192e-05
30	2.375202777e-01	3.732781878e-03	6.2111e-06	6.8887e-06
40	2.375261993e-01	3.739365164e-03	2.8949e-07	3.0546e-07
50	2.375264745e-01	3.739667878e-03	1.4348e-08	2.7454e-09
60	2.375264909e-01	3.739676379e-03	2.1218e-09	5.7560e-09
70	2.375264905e-01	3.739673196e-03	1.7035e-09	2.5729e-09
80	2.375264896e-01	3.739671631e-03	7.8140e-10	1.0077e-09
90	2.375264892e-01	3.739670984e-03	3.9901e-10	3.6124e-10
100	2.375264891e-01	3.739670832e-03	2.7799e-10	2.0914e-10

Uniform mesh of N nodes.

Table 7.4
Errors in c for Poiseuille Channel Flow at $Re = 10^4$ and $\alpha = 1$

N-1	c_r	c_i	$\lvert\Delta c_r\rvert$	$\lvert\Delta c_i\rvert$
10	2.370997492e-01	3.829672715e-03	4.2674e-04	9.0002e-05
20	2.375218405e-01	3.734255075e-03	4.6483e-06	5.4155e-06
30	2.375265121e-01	3.739617243e-03	2.3256e-08	5.3380e-08
40	2.375264864e-01	3.739666721e-03	2.3983e-09	3.9018e-09
50	2.375264884e-01	3.739670052e-03	4.1147e-10	5.7063e-10
60	2.375264887e-01	3.739670495e-03	1.0120e-10	1.2758e-10
70	2.375264888e-01	3.739670586e-03	2.2996e-11	3.7066e-11
80	2.375264888e-01	3.739670618e-03	1.5159e-11	4.7590e-12
90	2.375264888e-01	3.739670636e-03	2.8488e-11	1.2972e-11
100	2.375264888e-01	3.739670566e-03	1.5147e-12	5.6651e-11

Nonuniform mesh of N nodes.

error drops as N^{-7} but its magnitude is an order of magnitude smaller than with the uniform mesh. There is loss in accuracy if the number of elements exceeds 80, indicating dominance of round off errors. A stronger stretching is obtained with $y_j = sin(\pi\xi_j/2))$ where $\xi_j = -1+2(j-1)/(N-1)$. The best results are obtained with $N = 40$, which gives $c = 2.375264877e-01+3.739671221e-03i$. The errors are $|\Delta c_r| = 1.1100e-09$ and $|\Delta c_i| = 5.9791e-10$. The real and imaginary parts of the eigenfunction v are shown in Figure 7.11. It is of interest to investigate the existence of a critical Reynolds number Re_c below which all disturbances are damped.

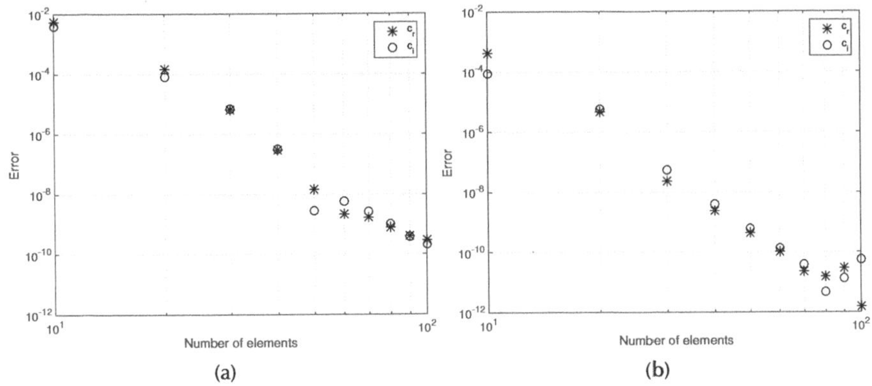

Figure 7.10 Error in the eigenvalue $c = c_r + ic_i$ of Poiseuille channel flow for $\alpha = 1$ and $Re = 10^4$. (a) Uniform mesh, and (b) nonuniform mesh.

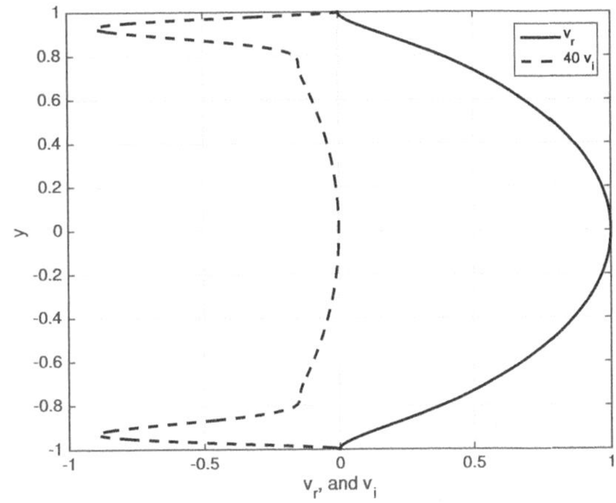

Figure 7.11 Eigenfunction $v = v_r(y) + iv_i(y)$ for $\alpha = 1$ and $Re = 10^4$.

Using linear stability analysis, Orszag [81] computed the critical Reynolds number for channel flow $Re_c = 5777.22$. The corresponding wavenumber is $\alpha_c = 1.02056$ and phase speed is $c_r = 0.26400174$. In Figure 7.12a we depict c_i against α for four Reynolds numbers: 4000, 5777.22, 10,000, and 20,000. For low $Re = 4000$ we see that $c_i < 0$ indicating damped waves. For $Re = Re_c$, the $c_i - \alpha$ curve is below $c_i = 0$, but it touches that line at $\alpha = \alpha_c$. For the higher Re of 10,000 and 20,000 there is a finite range of amplified wavenumbers, and two neutral points where the growth rate (or c_i) is zero. The wavenumber at a neutral point depends on the Reynolds number. We plot in Figure 7.12b the wavenumber α of neutral points against the Reynolds number. For $Re > Re_c$, there are two values of α; a low value α_l and a high value α_u. For $\alpha_l \leq \alpha \leq \alpha_u$, the disturbance grows in time, whereas for α outside that range,

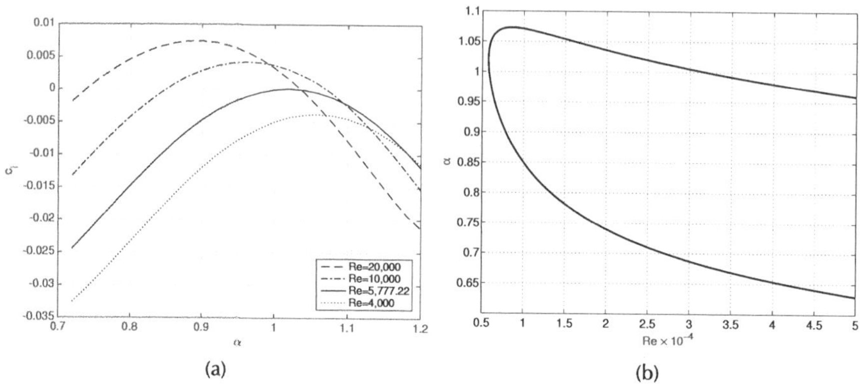

Figure 7.12 Stability analysis of Poiseuille plane channel flow. (a) Growth rate as a function of wavenumber for different Reynolds numbers, and (b) marginal stability curve.

disturbances decay. The curve shown in Figure (7.12b) is called the marginal stability curve. In-depth analysis of hydrodynamic stability can be found in [31].

7.5 PROBLEMS

(1) A uniform slender elastic bar is fixed at one end and is free at the other end. The displacement of longitudinal vibrations, $u(x,t)$, is governed by the wave equation

$$\rho A \frac{\partial^2 u}{\partial t^2} - \frac{\partial}{\partial x}(EA\frac{\partial u}{\partial x}) = 0$$

where ρ is the mass density, A is the cross-sectional area, and E is Young's modulus. Model the bar by two linear elements, and determine the first two natural frequencies, and compute the percentage errors.

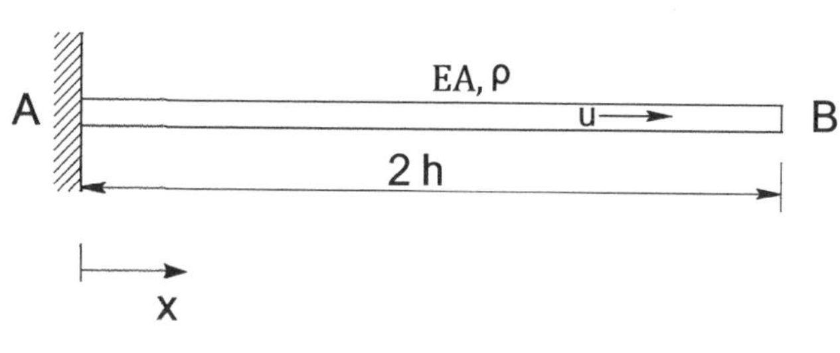

Figure 7.13 Longitudinal vibration of a bar.

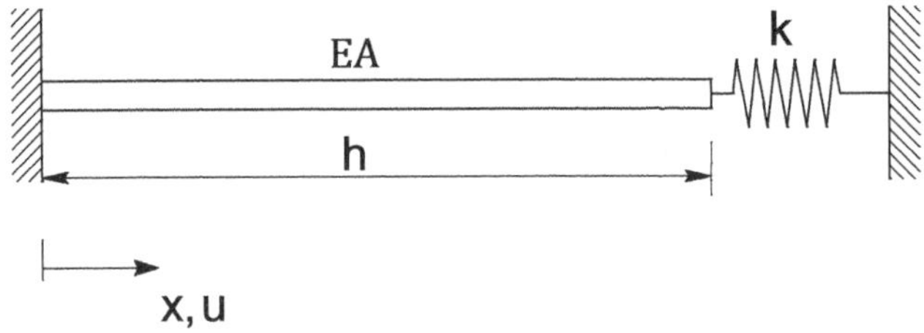

Figure 7.14 Longitudinal vibration of a bar restrained by a spring.

(2) Solve Problem (1) using one cubic element. Determine the lowest natural frequency and compare it with the exact frequency $\omega = \pi\sqrt{E/\rho}/2L$.

(3) A uniform slender elastic bar is fixed at $x = 0$ and is restrained by a linear spring of stiffness $k = 2EA/3h$ at $x = h$ as shown in Figure 7.14. Let ρ denote the mass density, A the cross-sectional area, and E Young's modulus. Model the bar by one quadratic element, and determine the first two natural frequencies.

(4) A circular shaft of torsional rigidity GJ is fixed at end $x = 0$ and rigidly attached to a thin disc at end $x = L$ as shown in Figure 7.15. The disc mass moment of inertia about the shaft axis is denoted by I_o. The disc is twisted about the x-axis and then released resulting in free torsional oscillations of the disc and shaft. The equation of motion of free torsional vibrations is given by

$$\rho J \frac{\partial^2 \theta}{\partial t^2} - \frac{\partial}{\partial x}\left(GJ\frac{\partial \theta}{\partial x}\right) = 0 \tag{7.111}$$

where θ is the twist angle, ρ is mass density, J is polar area moment of inertia, and G is shear modulus. Note that the primary variable is θ, and the secondary variable is $GJ\partial\theta/\partial x$, which is the torque. Assume $I_o = 0.1\rho LJ$, model the shaft by one quadratic element, and determine the lowest natural frequency of torsional vibrations.

(5) Using Euler–Bernoulli's beam theory, determine the lowest natural frequency of transverse vibrations of a beam that is built-in at both ends. The beam has uniform properties: mass density ρ, cross-sectional area A, length L, and flexural rigidity EI.

 (a) Use two Hermite cubic elements, and express the answer in nondimensional form

$$\omega^* = \omega\sqrt{\rho AL^4/EI}$$

 (b) Use one three-node quintic element.

(6) Using Euler–Bernoulli's beam theory, determine the lowest natural frequency of transverse vibrations of the continuous beam shown in Figure 3.36. You may assume $\rho A = 25$ kg/m. Use at least three Hermite cubic elements, and express

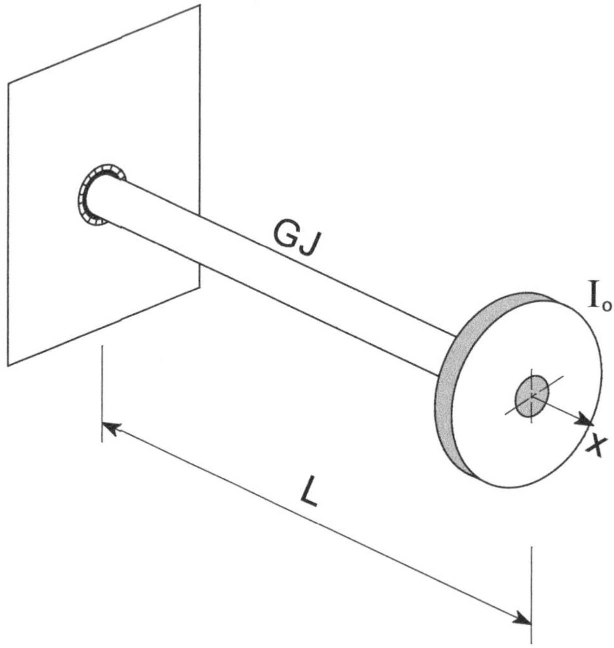

Figure 7.15 Free torsional vibrations of a circular shaft with a disc at tip.

the answer in nondimensional form

$$\omega^* = \omega\sqrt{\rho A L^4/EI}.$$

Plot the deflection curve (elastic line) of the first mode of free vibration (lowest frequency).

(7) We want to determine the lowest four natural frequencies of a cantilever beam of length L and uniform ρA and EI. To simplify the analysis, we use Euler–Bernoulli beam theory, and introduce nondimensional parameters,

$$x^* = x/L, \; t^* = t\sqrt{EI/\rho A L^4}, \; w^* = w/L$$

Show that the equation of motion for transverse vibrations takes the form

$$\frac{\partial^2 w^*}{\partial t^{*2}} + \frac{\partial^4 w^*}{\partial x^{*4}} = 0.$$

In nondimensional form, the frequency becomes

$$\omega^* = \omega\sqrt{\rho A L^4/EI}.$$

Use the finite-element method, and model the beam by 3, 4, 5, 6, 7, and 8 elements. Plot the percentage error in the frequencies against the number of elements. What is the minimum number of elements if the error is to be less than 1% in any of the lowest four frequencies?

(8) Consider a cantilever beam of length L and uniform ρA and EI. A body of mass M_s is rigidly attached to the free end of the beam. We want to determine the lowest four natural frequencies of the beam-body assembly. To simplify the analysis, we neglect rotary inertia of beam and body, and assume the mass M_s to be concentrated at the tip. We introduce nondimensional parameters,

$$x^* = x/L, \; t^* = t\sqrt{EI/\rho AL^4}, \; w^* = w/L$$

and obtain the equation of motion for transverse vibrations

$$\frac{\partial^2 w^*}{\partial t^{*2}} + \frac{\partial^4 w^*}{\partial x^{*4}} = 0$$

In nondimensional form, the frequency becomes

$$\omega^* = \omega\sqrt{\rho AL^4/EI}$$

The boundary conditions for this case are

$$x = 0: \; w^* = 0, \; \frac{\partial w^*}{\partial x^*} = 0$$

$$x^* = 1: \; -\frac{\partial^3 w^*}{\partial x^{*3}} = -\mu\frac{\partial^2 w^*}{\partial t^{*2}}, \; \frac{\partial^2 w^*}{\partial x^{*2}} = 0$$

where $\mu = M_s/\rho AL$ is the ratio of the concentrated mass at the tip to the mass of beam. Use the finite-element method, and model the beam by eight elements. For $\mu = 0.3$, determine the percentage changes in frequencies when compared to those of the case of $\mu = 0$.

(9) Using Euler–Bernoulli's beam theory, determine the lowest natural frequency of the transverse vibrations of the beam shown in Figure 3.37. Determine the percentage change in frequency when compared to that of the case of $k = 0$ (a cantilever without spring at the free end).

(10) A beam AB (length L, cross-sectional area A, mass density ρ, and flexural rigidity EI) is continuously restrained by an elastic foundation of modulus k_f (force per unit area) as shown in Figure 7.16. The beam ends may be treated as simple supports. The equation of motion for transverse vibrations $w(x,t)$ is given by

$$\rho A\frac{\partial^2 w}{\partial t^2} + \frac{\partial^2}{\partial x^2}(EI\frac{\partial^2 w}{\partial x^2}) + k_f w - f(x,t) = 0$$

where k_f is the foundation elastic modulus assumed constant in this problem. The boundary conditions are

$$x = 0: \; w = 0 \qquad \text{and} \qquad EI\frac{\partial^2 w}{\partial x^2} = 0$$

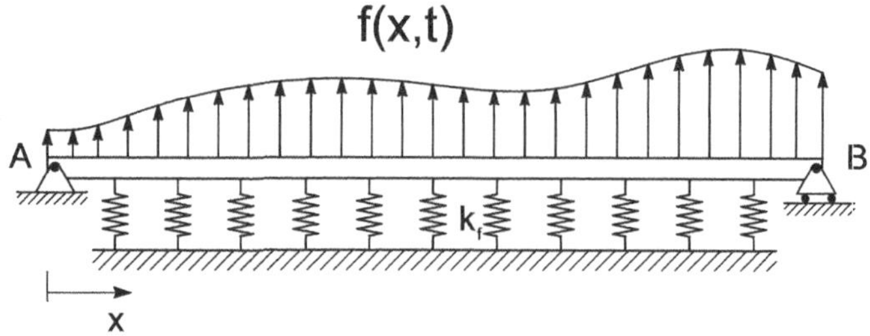

Figure 7.16 Simply supported beam on an elastic foundation.

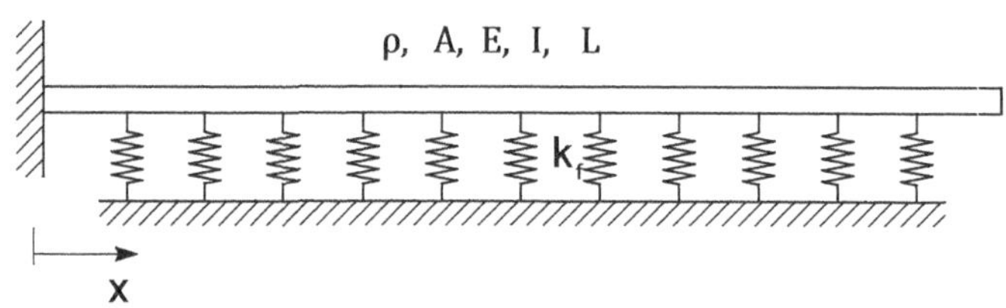

Figure 7.17 A cantilever beam on an elastic foundation.

$$x = L: \; w = 0 \qquad \text{and} \qquad EI\frac{\partial^2 w}{\partial x^2} = 0.$$

(a) Derive the weak form and classify the boundary conditions.

(b) Model the beam by one Euler–Bernoulli beam element, and determine the lowest natural frequency in terms of k_f, L, ρ, A, and EI.

(11) A cantilever beam (length L, cross-sectional area A, mass density ρ, and flexural rigidity EI) is attached to an elastic foundation of constant k_f as shown in Figure 7.17. The equation of motion for free transverse vibrations is

$$\rho A\frac{\partial^2 w}{\partial t^2} + \frac{\partial^2}{\partial x^2}\left(EI\frac{\partial^2 w}{\partial x^2}\right) + k_f w = 0.$$

(a) Derive the weak form, and classify the boundary conditions.

(b) The finite element method is used to study the effects of the elastic foundation on the natural frequencies of the beam. The Euler–Bernoulli element is used for, which we obtain the equation

$$[M^e]\{\ddot{U}^e\} + [K^e]\{U^e\} = \{R^e\}$$

Write down the expressions for $[M^e]$ and $[K^e]$.

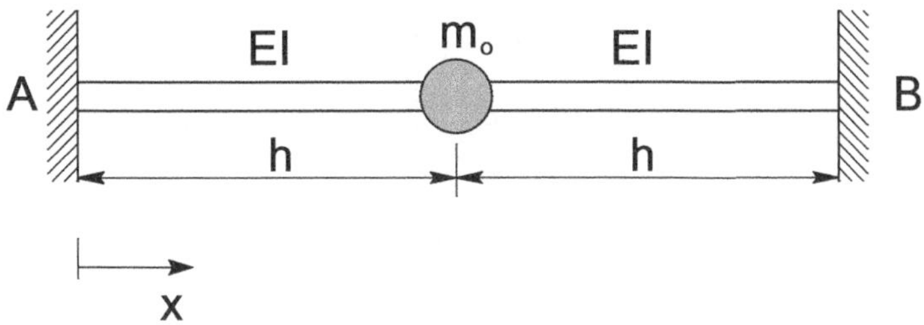

Figure 7.18 Problem 12.

(c) Model the beam by one Euler–Bernoulli beam element, and determine the lowest natural frequency in terms of k_f, L, ρ, A, and EI.

(12) The beam AB (uniform EI and ρA) is fixed at A and B. A concentrated mass ($m_o = \rho Ah$) is rigidly attached to the midsection of the beam as shown in Figure 7.18. Given EI, h, and ρA, use the finite element method to determine an approximation to the lowest transverse natural frequency of the beam-mass assembly.

(13) Using Euler–Bernoulli's beam theory, determine the lowest natural frequency of the transverse vibration of a beam AB whose end A is fixed and end B is subjected to four different conditions as shown in Figures 7.19a–7.19d. The beam has uniform properties: mass density ρ, cross-sectional area A, length L, and flexural rigidity EI. Also given are $m_o = 0.1\rho AL$, and $k_o = 1.2363EI/L^3$. Neglect the rotary inertia of the concentrated mass (i.e., neglect the effects of I_o). Use at least two Hermite cubic elements, and express the answer in nondimensional form

$$\omega^* = \omega\sqrt{\rho AL^4/EI}$$

(14) The cantilever beam (length $2h$ and flexural rigidity EI, and mass per unit length $= \rho A$) is attached to a linear spring of stiffness k at its tip. A body of mass m is rigidly attached at the midsection as shown in Figure 7.20. We want to determine the lowest natural frequency using the finite-element method. We model the beam by two Euler–Bernoulli elements, and assume harmonic response

$$\begin{aligned} \{U\} &= \{\hat{U}\} e^{-\hat{i}\omega t} \\ \{R\} &= \{\hat{R}\} e^{-\hat{i}\omega t} \end{aligned} \tag{7.112}$$

where ω is a natural frequency and $\hat{i} = \sqrt{-1}$, $\{\hat{U}\}$, and $\{\hat{R}\}$ are complex amplitudes of the nodal primary and secondary variables, respectively. The assembled global system before application of the boundary condition takes the form

$$\left[[K] - \omega^2 [M]\right] \{\hat{U}\} = \{\hat{R}\}. \tag{7.113}$$

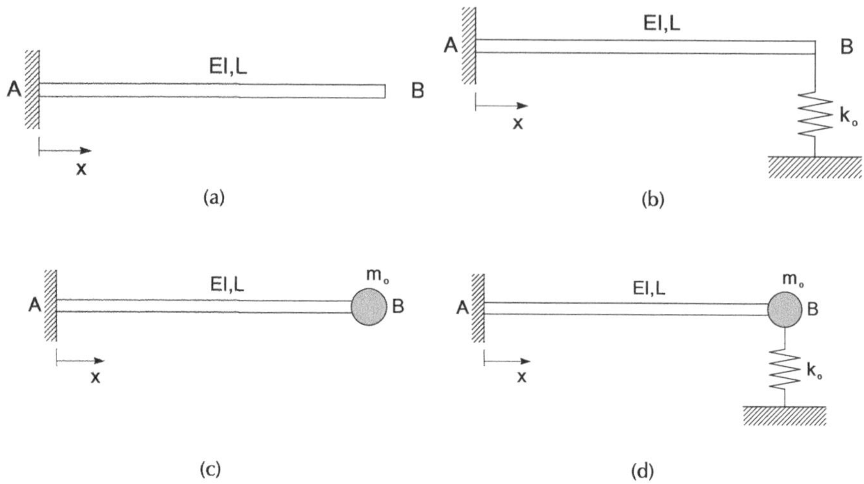

Figure 7.19 Free vibrations of a cantilever beam with various free-end loading, (a) cantilever beam, (b) end B is attached to a linear spring of stiffness k_o, (c) end B is attached to a small mass m_o, and (d) end B is attached to a small mass m_o and a linear spring of stiffness k_o.

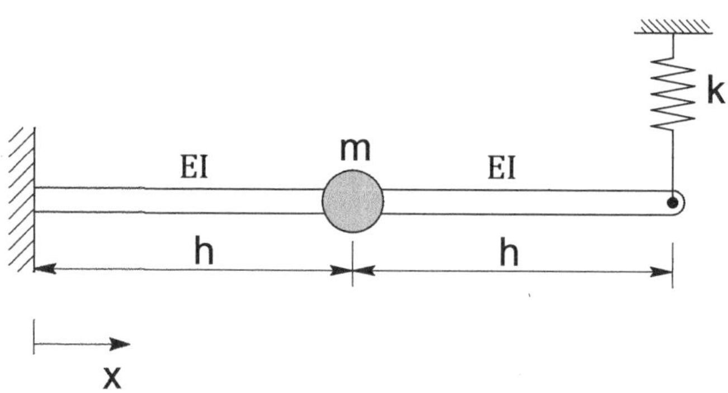

Figure 7.20 Problem 14.

Apply the boundary conditions and obtain a reduced system in the form

$$\left[[\bar{K}] - \omega^2 [\bar{M}]\right] \{\hat{U}\} = \{0\}. \tag{7.114}$$

Provide expressions for $[\bar{K}]$ and $[\bar{M}]$ in terms of EI, ρA, h, k, and m, only.

(15) Using Euler–Bernoulli's beam theory, determine the lowest natural frequency of transverse vibrations of a beam AB whose end A is fixed and end B is attached to a mass-spring oscillator as shown in Figure 7.21. The beam has uniform properties: mass density ρ, cross-sectional area A, length L, and flexural rigidity EI. Also given $m_o = 0.1\rho AL$, and $k_o = 1.2363EI/L^3$. Use at least two Hermite

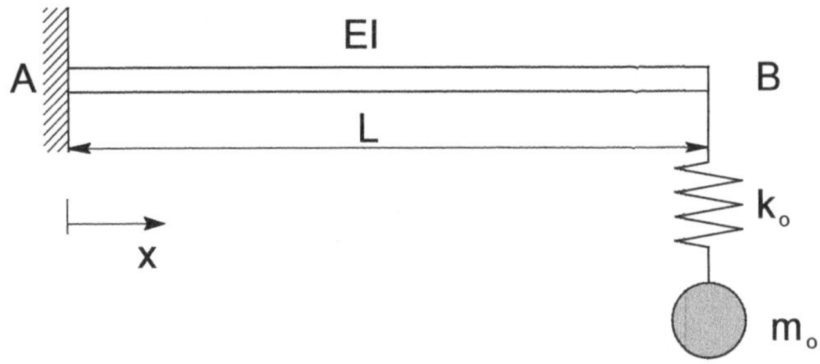

Figure 7.21 Free vibrations of a cantilever beam attached to a mass-spring oscillator.

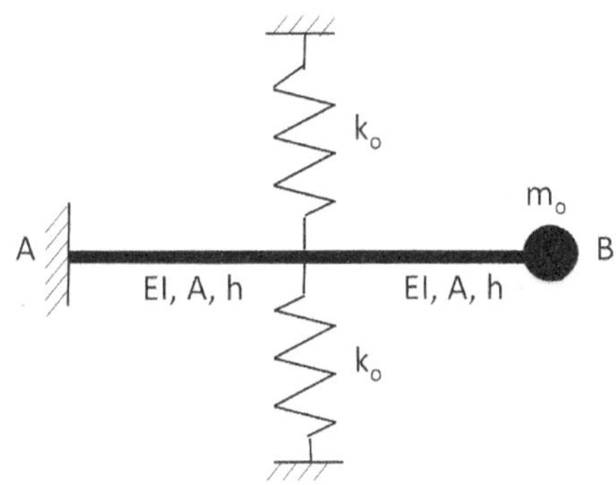

Figure 7.22 Beam *AB*, fixed at *A* and attached to two springs at midpoint.

cubic elements, and express the answer in nondimensional form

$$\omega^* = \omega\sqrt{\rho A L^4/EI}.$$

(16) A uniform beam *AB* of length $2h$, flexural rigidity EI, cross-sectional area A, and mass density ρ is fixed at A and carries a concentrated mass m_o at B as shown in Figure 7.22. It is also attached to two linear springs each of stiffness k_o at the midpoint. Take $k_o = 2EI/h^3$ and $m_o = \rho Ah/21$. Model the beam by two Euler–Bernoulli elements, and neglect the moment of inertia of m_o about mass center. Determine the two lowest natural frequencies.

(17) A uniform beam *AB* of length h and flexural rigidity EI is fixed at A and attached to a torsional spring of stiffness $k = 2EI/h$ at B as shown in Figure 7.23a. Model the beam by one Euler–Bernoulli element.

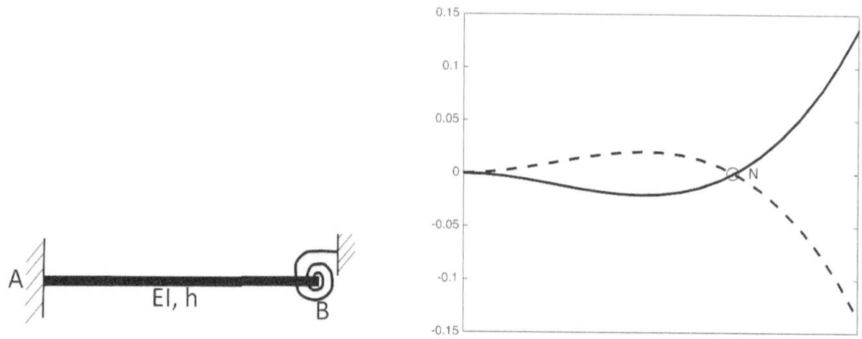

(a) (b)

Figure 7.23 (a) Beam AB, fixed at A and attached to a torsional spring at B. (b) Mode shape of the second mode.

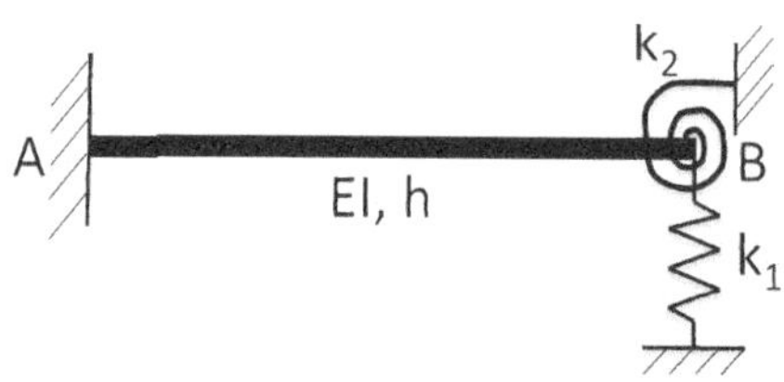

Figure 7.24 Problem 18.

(a) Determine the frequency of the second mode of free vibrations.

(b) The mode shape of the second mode is sketched in Figure 7.23b. Determine the x-coordinate of the nodal point "N".

(18) A uniform beam AB of length h and flexural rigidity EI is fixed at A as shown in Figure 7.24. At end B, a linear spring exerts a force opposite to displacement ($F = k_1\delta$) and a torsional spring exerts a moment opposite to rotation ($M = k_2\theta$). Model the beam by one Euler–Bernoulli element, and take $k_1 = 4EI/h^3$, and $k_2 = 2EI/h$.
Calculate approximations to the lowest two natural frequencies of transverse vibrations.

(19) The Γ-frame is fixed at A and simply supported at C as shown in Figure 7.25. It is made of steel ($E = 200$ GPa, $\rho = 8000$ kg/m^3). The length AB is 4m, and BC is 3 m. The cross-sectional area is a square of side of 0.12 m. (a) Use two frame elements, and determine the two lowest natural frequencies, and sketch the corresponding mode shapes. (b) Use seven elements each of length 1 m, and plot the first two mode shapes.

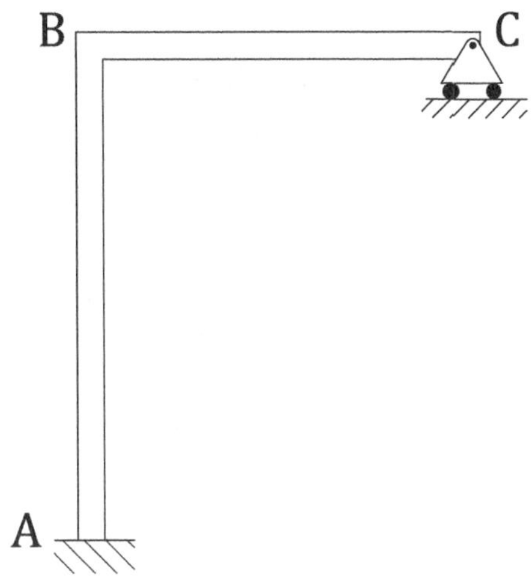

Figure 7.25 Γ-frame fixed at A and simply supported at C.

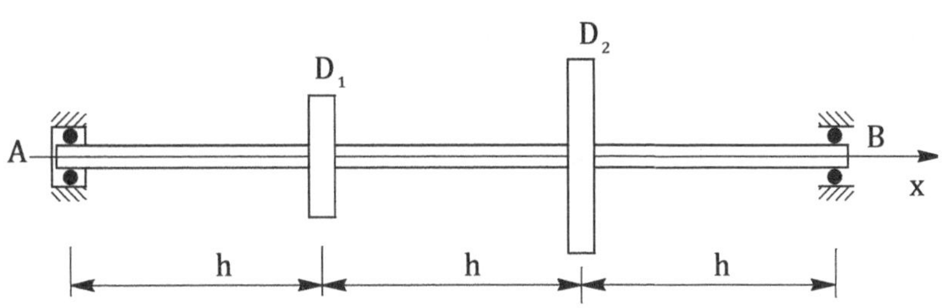

Figure 7.26 Spinning shaft with two discs.

(20) Determine the two lowest natural frequencies of the VT-frame shown in Figure 3.51, and sketch the corresponding mode shapes.

(21) Determine the two lowest natural frequencies of the frame shown in Figure 3.19, and sketch the corresponding mode shapes.

(22) Determine the two lowest natural frequencies of the traffic light frame shown in Figure 3.53, and sketch the corresponding mode shapes. Treat the light signals as concentrated masses and neglect rotary inertia.

(23) We consider a steel shaft of diameter $d = 50$ mm and length $3h = 2.1$ m that carries two discs as shown in Figure 7.26. The masses of D_1 and D_2 are 30 and and 40 kg, respectively. The bearings at A and B provide negligible moments, and the one at A may be treated as a thrust bearing. Determine the critical speed in rpm.

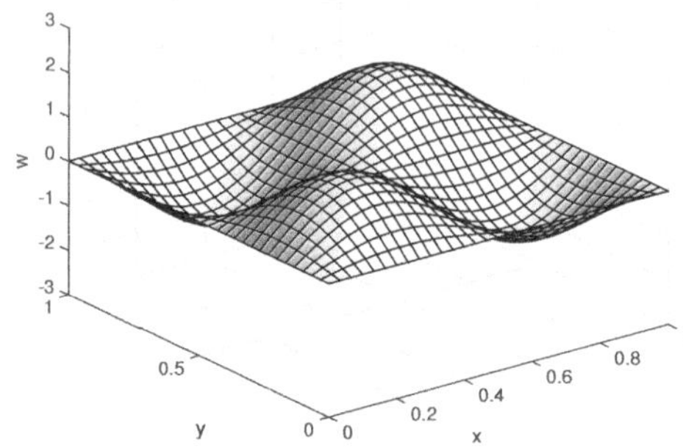

Figure 7.27 Transverse vibration of a rectangular membrane.

(24) A membrane is a very thin plate (negligible bending resistance) stretched and affixed at its boundary. The equation of motion for transverse vibrations is

$$\rho \frac{\partial^2 w}{\partial t^2} - T\left(\frac{\partial^2 w}{\partial x^2} + \frac{\partial^2 w}{\partial y^2}\right) - f(x,y,t) = 0$$

where $w(x,y,t)$ is the deflection, $f(x,y,t)$ is externally applied pressure, ρ is mass per unit area, and T is the tension per unit length (constant). The deflection is zero on the boundary.

(a) Consider a rectangular membrane ($0 \leq x \leq a$ and $0 \leq y \leq b$) and derive the weak form of the equation of motion.

(b) The membrane is discretized into small rectangular elements of sides h_x and h_y. Each element has four nodes, and the deflection is approximated using the shape functions of the bilinear rectangular element. Write down the mass and stiffness matrices ($[M^e]$ and $[K^e]$) for the element.

(25) The exact natural frequencies of a rectangular membrane of sides (a,b) are [119]

$$\omega_{mn} = \pi \sqrt{\frac{T}{\rho}\left(\frac{m^2}{a^2} + \frac{n^2}{b^2}\right)} \tag{7.115}$$

for $m = 1,2,...$ and $n = 1,2,....$ The mode shapes are

$$\hat{w}_{mn} = sin\left(\frac{m\pi x}{a}\right) sin\left(\frac{n\pi y}{b}\right). \tag{7.116}$$

Consider a square membrane of side a, and determine the lowest natural frequency using (a) four bilinear square elements, and (b) nine bilinear square elements. Compare the results to the exact frequency.

(26) **Fluid-Structure Interaction: One-Dimensional Panel Flutter**

Flutter of thin elastic panels has been briefly introduced in Problem (45) of Chapter 3. In the current exercise, we determine the critical dynamic pressure at the onset of flutter. We assume that when the elastic panel is undisturbed (i.e., flat), the fluid pressure of the supersonic flow above the panel is equal to the pressure on the lower side where the flow is stagnant, hence $p_\ell = p_\infty$. The plate is in equilibrium and should remain flat and coincident with the x-axis. If a small transverse force is applied and then removed, the equilibrium will be disturbed, the plate will oscillate, and the pressure on the upper surface will vary in space and time. However, the oscillations amplitude may damp or grow in time. The purpose of this problem is to determine the critical condition at which the oscillation amplitude grows in time; this condition is known as the flutter condition. The equation of motion of the plate is

$$\rho_p A\frac{\partial^2 v}{\partial t^2}+\frac{\partial^2}{\partial x^2}\left(D\frac{\partial^2 v}{\partial x^2}\right)-f(x,t)=0 \tag{7.117}$$

where ρ_p is the plate mass density, h is plate thickness, and $A = hb$ is the cross-sectional area. We assume unit width, hence $b = 1$. And f is the distributed load, which is the pressure difference $f(x,t) = p_\ell - p_u(x,t)$; it is an unknown function of position and time. Unsteady aerodynamic models establish a relationship of the pressure on plate and its motion. In this example, we use the simplest unsteady supersonic flow model in which the pressure at a point on the plate is proportional to the instantaneous local vertical velocity of the point. This model is known as the linear piston theory, which is a good approximation for Mach numbers in the range 1.5 to 5

$$p_u - p_\infty = \frac{\rho_\infty U_\infty}{\sqrt{M_\infty^2-1}}\left(\frac{M_\infty^2-2}{M_\infty^2-1}\frac{\partial v}{\partial t}+U_\infty\frac{\partial v}{\partial x}\right). \tag{7.118}$$

Noting that $p_\ell = p_\infty$, we write the equation of motion as

$$\rho_p h\frac{\partial^2 v}{\partial t^2}+\frac{\partial^2}{\partial x^2}\left(D\frac{\partial^2 v}{\partial x^2}\right)+\frac{\rho_\infty U_\infty}{\sqrt{M_\infty^2-1}}\left(\frac{M_\infty^2-2}{M_\infty^2-1}\frac{\partial v}{\partial t}+U_\infty\frac{\partial v}{\partial x}\right)=0. \tag{7.119}$$

(a) Introduce nondimensional variables

$$x^* = \frac{x}{L},\ t^* = \frac{t\sqrt{D}}{\sqrt{\rho_p h L^4}},\ v^* = \frac{v}{h} \tag{7.120}$$

and show that Equation (7.119) becomes

$$\frac{\partial^2 v^*}{\partial t^{*2}}+\alpha_o\sqrt{\mu_o\Lambda}\frac{\partial v^*}{\partial t^*}+\frac{\partial^4 v^*}{\partial x^{*4}}+\Lambda\frac{\partial v^*}{\partial x^*}=0 \tag{7.121}$$

where

$$\alpha_o = \frac{M_\infty^2-2}{M_\infty^2-1} \tag{7.122}$$

$$\mu_o = \frac{1}{\sqrt{M_\infty^2-1}}\frac{\rho_\infty L}{\rho_p h} \tag{7.123}$$

$$\Lambda = \frac{1}{\sqrt{M_\infty^2 - 1}} \frac{\rho_\infty U_\infty^2 L^3}{D_o}. \tag{7.124}$$

For simply supported panels, the boundary conditions are

$$\begin{aligned} x^* &= 0 \quad v^* = 0 \quad \text{and} \quad \frac{\partial^2 v^*}{\partial x^{*2}} = 0 \\ x^* &= 1 \quad v^* = 0 \quad \text{and} \quad \frac{\partial^2 v^*}{\partial x^{*2}} = 0. \end{aligned} \tag{7.125}$$

(b) Derive the weak form of Equation (7.121) for an element.

(c) For a two-node cubic Hermitian element (length h_e), use the Galerkin method to derive the element equations in the form:

$$[M^e]\{\ddot{U}^e\} + \alpha_o\sqrt{\mu_o\Lambda}[H_1^e]\{\dot{U}^e\} + \left[[K^e] + \Lambda[H_2^e]\right]\{U^e\} = \{R^e\} \tag{7.126}$$

and provide analytic expressions for the matrices $[K^e]$, $[H_1^e]$, and $[H_2^e]$. These matrices are functions of the element length only.
We note that the interaction of fluid with the vibrating structure modifies the equation of motion in two ways. The second term, which is proportional to velocity, may have a damping effect. More importantly, the structure stiffness is modified due to the fluid loading term $\Lambda[H_2^e]$. The effective stiffness of the structure is now the sum of structure stiffness matrix $[K^e]$ and another stiffness matrix $\Lambda[H_2^e]$ due to the induced fluid pressure on the structure. If Λ exceeds certain value the modified stiffness can help the structure absorb energy from the fluid causing it to flutter and reach sustained vibrations. Our objective is to determine the value of Λ at the onset of flutter. To that end, we assume a harmonic response for the deflection and secondary variables,

$$\{U^e\} = \{\hat{U}^e\}e^{-i\omega t}, \ \{R^e\} = \{\hat{R}^e\}e^{-i\omega t} \tag{7.127}$$

and substitute into Equation (7.126),

$$\left[-\omega^2[M^e] - i\omega\alpha_o\sqrt{\mu_o\Lambda}[H_1^e] + \left[[K^e] + \Lambda[H_2^e]\right]\right]\{\hat{U}^e\} = \{\hat{R}^e\}. \tag{7.128}$$

Use at least two elements, assemble the elements, apply the boundary conditions, and obtain the eigenvalue problem for ω

$$\left[-\omega^2[M] - i\omega\alpha_o\sqrt{\mu_o\Lambda}[H_1] + \left[[K] + \Lambda[H_2]\right]\right]\{\hat{U}\} = \{0\}. \tag{7.129}$$

(d) We consider flow of air with properties: $M_\infty = \sqrt{5}$, $\rho_\infty = 0.405$ kg m^{-3} over an aluminum panel with properties: $h = 3$ mm, $L = 1$ m, and $\rho_p = 2700$ kg m^{-3}, and compute $\mu_o = 0.025$ and $\alpha_o = 0.75$. Fix these values of α_o and

μ_o. To simplify the eigenvalue problem, we neglect the damping term, the eigenvalue problem becomes

$$\Big[\big[[K] + \Lambda [H_2]\big] - \omega^2 [M]\Big] \{\hat{U}\} = \{0\}. \tag{7.130}$$

Choose values of Λ in the range $0 \leq \Lambda \leq 600$ and determine the frequencies and growth rates of the first (lowest) and second modes of vibrations. Plot the frequencies and growth rates against Λ. It will be observed that the first-mode frequency increases as Λ increases from the zero, while the second-mode frequency will decrease slightly. When the frequencies of the two modes coalesce, flutter begins. Determine Λ at the onset of flutter. MATLAB: [V,D]=eig(A,B) solves the eigenvalue problem $[A]\{x\} = \lambda [B]\{x\}$. The eigenvalues are given by the diagonal of matrix $[D]$.

(e) For $E = 70$ GPa and $\nu = 0.35$, determine the critical air velocity (m/s) and flutter frequency (Hz).

(f) Determine the critical air velocity (m/s) and flutter frequency (Hz) if the damping term is not neglected.

8 Plane Elasticity

In this chapter, finite-element analysis is presented for elastic deformations of a continuum. The analysis is limited to plane elasticity problems for which certain components of stresses or strains are identically zero while the nonzero terms depend on two space Cartesian coordinates x and y only. Externally applied forces are parallel to the xy-plane. The analysis also includes strains due to temperature variation in space but the coupling terms in the equilibrium and energy equations are neglected. The temperature distribution in the continuum is either prescribed or determined by solving a heat conduction problem. The stress analysis for the prescribed temperature field is then performed. Throughout the chapter, the assumption of small displacement is adopted and the equations of equilibrium are linearized. Material properties are assumed to be constant and independent of temperature. The deformation is assumed to be in the elastic range and issues of material failure are not considered here. Since the problem domain is an area in the xy-plane, all the isoparametric elements and related transformations developed in Chapter 4 can be used for plane elasticity problems. However, in the present chapter, there are two degrees of freedom at each node given by the x and y components of the displacement field. Deformation and stresses in thin plates subjected to transverse loads are presented in Chapter 9. Geometrically nonlinear plates and buckling due to in-plane load are addressed in Chapter 10.

8.1 CONSTITUTIVE EQUATIONS FOR LINEAR ELASTICITY

Relative to Cartesian coordinates (x,y,z), the displacements components are denoted by (u,v,w). In index notation, the coordinates are denoted by x_i and the displacement vector by u_i, $i = 1,2,3$. For very small displacements, the linear strain tensor is

$$\varepsilon_{ij} = \frac{1}{2}\left(\frac{\partial u_i}{\partial x_j} + \frac{\partial u_j}{\partial x_i}\right) \tag{8.1}$$

it is symmetric $\varepsilon_{ij} = \varepsilon_{ji}$. The constitutive equation for isotropic linear elastic material gives the components of the stress tensor

$$\sigma_{ij} = \lambda \varepsilon_{kk}\delta_{ij} + 2\mu\varepsilon_{ij} \tag{8.2}$$

λ and μ are called the Lame elastic constants, which are related to the Young's modulus E, the shear modulus G, and Poisson's ratio ν by

$$\mu \equiv G = \frac{E}{2(1+\nu)}, \quad \lambda = \frac{\nu E}{(1+\nu)(1-2\nu)}. \tag{8.3}$$

The strain tensor can also be written in terms of stress tensor as

$$\varepsilon_{ij} = \frac{1+\nu}{E}\sigma_{ij} - \frac{\nu}{E}\sigma_{kk}\delta_{ij}. \tag{8.4}$$

DOI: 10.1201/9781003323150-8

Matrix Form for 3D Linear Elasticity

It will be convenient in the ensuing analysis if the constitutive equation is expressed in matrix form. To this end, we define stress and strain vectors

$$\{\sigma\} = \begin{Bmatrix} \sigma_{xx} \\ \sigma_{yy} \\ \sigma_{zz} \\ \sigma_{xy} \\ \sigma_{yz} \\ \sigma_{zx} \end{Bmatrix}, \quad \{\varepsilon\} = \begin{Bmatrix} \varepsilon_{xx} \\ \varepsilon_{yy} \\ \varepsilon_{zz} \\ \gamma_{xy} \\ \gamma_{yz} \\ \gamma_{zx} \end{Bmatrix} = \begin{Bmatrix} \frac{\partial u}{\partial x} \\ \frac{\partial v}{\partial y} \\ \frac{\partial w}{\partial z} \\ \frac{\partial u}{\partial y} + \frac{\partial v}{\partial x} \\ \frac{\partial v}{\partial z} + \frac{\partial w}{\partial y} \\ \frac{\partial w}{\partial x} + \frac{\partial u}{\partial z} \end{Bmatrix} \tag{8.5}$$

where $\gamma_{xy} = 2\varepsilon_{xy}$, $\gamma_{xz} = 2\varepsilon_{xz}$, $\gamma_{yz} = 2\varepsilon_{yz}$ are the engineering shear strains. The constitutive relation Equation (8.2) can be written in matrix form

$$\{\sigma\} = [D]\{\varepsilon\} \tag{8.6}$$

where

$$[D] = \frac{E}{(1+\nu)(1-2\nu)} \begin{bmatrix} 1-\nu & \nu & \nu & 0 & 0 & 0 \\ \nu & 1-\nu & \nu & 0 & 0 & 0 \\ \nu & \nu & 1-\nu & 0 & 0 & 0 \\ 0 & 0 & 0 & \frac{1-2\nu}{2} & 0 & 0 \\ 0 & 0 & 0 & 0 & \frac{1-2\nu}{2} & 0 \\ 0 & 0 & 0 & 0 & 0 & \frac{1-2\nu}{2} \end{bmatrix}. \tag{8.7}$$

The inverse relation gives strain components in terms of stresses,

$$\{\varepsilon\} = [D]^{-1}\{\sigma\} \tag{8.8}$$

where

$$[D]^{-1} = \frac{1}{E} \begin{bmatrix} 1 & -\nu & -\nu & 0 & 0 & 0 \\ -\nu & 1 & -\nu & 0 & 0 & 0 \\ -\nu & -\nu & 1 & 0 & 0 & 0 \\ 0 & 0 & 0 & 2(1+\nu) & 0 & 0 \\ 0 & 0 & 0 & 0 & 2(1+\nu) & 0 \\ 0 & 0 & 0 & 0 & 0 & 2(1+\nu) \end{bmatrix} \tag{8.9}$$

Constitutive Equations for Plane Linear Elasticity

We consider two well-known plane elasticity problems, the first is plane stress and the second is plane strain. We define

$$\{\sigma\} = \left\{ \begin{array}{c} \sigma_{xx} \\ \sigma_{yy} \\ \sigma_{xy} \end{array} \right\}, \quad \text{and} \quad \{\varepsilon\} = \left\{ \begin{array}{c} \varepsilon_{xx} \\ \varepsilon_{yy} \\ \gamma_{xy} \end{array} \right\} \tag{8.10}$$

Plane Stress

The case of plane stress is defined by

$$\sigma_{zz} = 0, \ \sigma_{xz} = 0, \ \sigma_{yz} = 0. \tag{8.11}$$

The non-zero components of stress are σ_{xx}, σ_{yy}, and σ_{xy}; they are functions of x and y (and possibly of time for dynamic problems), but independent of z. This state of stress can be realized for thin bodies (body thickness in the z-direction is much smaller than that in the x- or y-direction) and the body is loaded by forces parallel to the xy-plane (i.e., no loading in the z-direction). The inplane loading does not vary in the z-direction. Enforcing the conditions Equation (8.11) on the 3D constitutive Equation (8.8), we get for plane stress

$$\{\varepsilon\} = [D]^{-1}\{\sigma\} \tag{8.12}$$

where

$$[D]^{-1} = \frac{1}{E}\begin{bmatrix} 1 & -\nu & 0 \\ -\nu & 1 & 0 \\ 0 & 0 & 2(1+\nu) \end{bmatrix} \tag{8.13}$$

and

$$\varepsilon_{zz} = -\frac{\nu}{E}(\sigma_{xx} + \sigma_{yy}), \ \gamma_{xz} = 0, \ \gamma_{yz} = 0. \tag{8.14}$$

The inverse of Equation (8.12) gives

$$\{\sigma\} = [D]\{\varepsilon\} \tag{8.15}$$

where

$$[D] = \frac{E}{1-\nu^2}\begin{bmatrix} 1 & \nu & 0 \\ \nu & 1 & 0 \\ 0 & 0 & \frac{1-\nu}{2} \end{bmatrix} \tag{8.16}$$

Plane Strain

A state of plane strain is defined by

$$\varepsilon_{zz} = 0, \ \gamma_{xz} = 0, \ \gamma_{yz} = 0. \tag{8.17}$$

The non-zero components of strain are ε_{xx}, ε_{yy}, and γ_{xy}. Enforcing the conditions Equation (8.17) on the 3D constitutive Equation (8.6), we get for plane strain

$$\{\sigma\} = [D]\{\varepsilon\} \tag{8.18}$$

where

$$[D] = \frac{E}{(1+\nu)(1-2\nu)} \begin{bmatrix} 1-\nu & \nu & 0 \\ \nu & 1-\nu & 0 \\ 0 & 0 & \frac{1-2\nu}{2} \end{bmatrix} \tag{8.19}$$

and

$$\sigma_{zz} = \frac{\nu E}{(1+\nu)(1-2\nu)} (\varepsilon_{xx} + \varepsilon_{yy}), \ \sigma_{xz} = 0, \ \sigma_{yz} = 0. \tag{8.20}$$

The inverse of Equation (8.18) gives

$$\{\varepsilon\} = [D]^{-1}\{\sigma\} \tag{8.21}$$

where

$$[D]^{-1} = \frac{1+\nu}{E} \begin{bmatrix} 1-\nu & -\nu & 0 \\ -\nu & 1-\nu & 0 \\ 0 & 0 & 2 \end{bmatrix}. \tag{8.22}$$

The constitutive equations for plane stress and strain are summarized in Table 8.1.

8.2 PRINCIPLE OF VIRTUAL DISPLACEMENTS: PLANE ELASTICITY

The principle of virtual displacements for 3D problems has been derived from the equations of equilibrium in Chapter 3. To obtain the principle of virtual displacements for plane elasticity we start with Equation (3.130), which is repeated here,

$$\int_\Omega \sigma_{ji}\bar{\varepsilon}_{ij} d\Omega = \oint_S \bar{u}_i t_i dS + \int_\Omega \bar{u}_i f_i d\Omega \tag{8.23}$$

where σ_{ij} is the stress tensor at equilibrium, t_i is surface traction vector, and f_i is body force per unit volume. We apply the principle to a continuous body as shown in Figure 8.1a. Its boundary is defined by a cylindrical surface and two planes $z = 0$ and $z = \ell_z$; we refer to ℓ_z as the body thickness. The intersection of the body with

Table 8.1
Conditions and Constitutive Equations for Plane Stress and Strain

Plane stress	Plane strain
$\sigma_{zz}=0,\ \sigma_{xz}=0,\ \sigma_{yz}=0$	$\varepsilon_{zz}=0,\ \varepsilon_{xz}=0,\ \varepsilon_{yz}=0$
$\varepsilon_{zz}=-\frac{\nu}{E}(\sigma_{xx}+\sigma_{yy})$	$\sigma_{zz}=\frac{\nu E}{(1+\nu)(1-2\nu)}(\varepsilon_{xx}+\varepsilon_{yy})$
$\varepsilon_{xz}=0,\ \varepsilon_{yz}=0$	$\sigma_{xz}=0,\ \sigma_{yz}=0$
$\begin{Bmatrix}\sigma_{xx}\\ \sigma_{yy}\\ \sigma_{xy}\end{Bmatrix}=[D]\begin{Bmatrix}\varepsilon_{xx}\\ \varepsilon_{yy}\\ \gamma_{xy}\end{Bmatrix}$	$\begin{Bmatrix}\sigma_{xx}\\ \sigma_{yy}\\ \sigma_{xy}\end{Bmatrix}=[D]\begin{Bmatrix}\varepsilon_{xx}\\ \varepsilon_{yy}\\ \gamma_{xy}\end{Bmatrix}$
$[D]=\frac{E}{1-\nu^2}\begin{bmatrix}1 & \nu & 0\\ \nu & 1 & 0\\ 0 & 0 & \frac{1-\nu}{2}\end{bmatrix}$	$[D]=\frac{E}{(1+\nu)(1-2\nu)}\begin{bmatrix}1-\nu & \nu & 0\\ \nu & 1-\nu & 0\\ 0 & 0 & \frac{1-2\nu}{2}\end{bmatrix}$

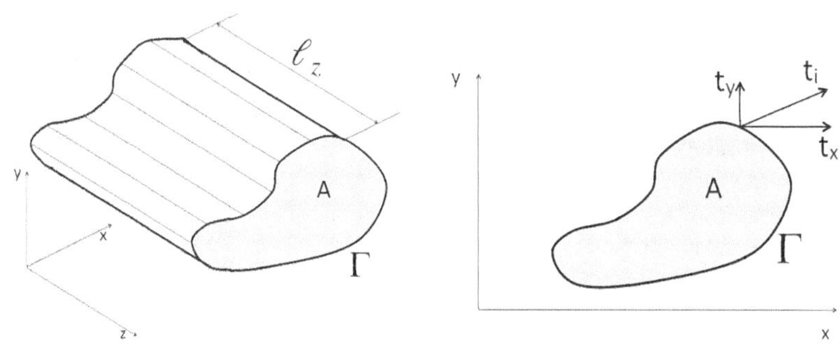

(a) (b)

Figure 8.1 (a) Plane elasticity, a prismatic body of cross-sectional area A and length ℓ_z. (b) Cross-section A, and surface traction $\vec{t}=t_x\hat{\imath}+t_y\hat{\jmath}$ on cylindrical surface defined by contour Γ and length ℓ_z.

any plane $z=z_o$, for $0\le z_o\le \ell_z$, is an area A whose contour is denoted by Γ, see Figure 8.1b. The integrand of the first term in Equation (8.23) is

$$\begin{aligned}\sigma_{ji}\bar{\varepsilon}_{ij} &= \sigma_{xx}\bar{\varepsilon}_{xx}+\sigma_{yy}\bar{\varepsilon}_{yy}+\sigma_{zz}\bar{\varepsilon}_{zz}+2\sigma_{xy}\bar{\varepsilon}_{xy}+2\sigma_{xz}\bar{\varepsilon}_{xz}+2\sigma_{yz}\bar{\varepsilon}_{yz}\\ &= \sigma_{xx}\bar{\varepsilon}_{xx}+\sigma_{yy}\bar{\varepsilon}_{yy}+\sigma_{zz}\bar{\varepsilon}_{zz}+\sigma_{xy}\bar{\gamma}_{xy}+\sigma_{xz}\bar{\gamma}_{xz}+\sigma_{yz}\bar{\gamma}_{yz}. \end{aligned} \tag{8.24}$$

For plane elasticity, the displacement components in the x and y directions (u,v) do not depend on z. Therefore, virtual displacements $\bar{u}$ and $\bar{v}$ must also be independent of z. Hence, the corresponding virtual strains

$$\bar{\varepsilon}_{xx}=\frac{\partial\bar{u}}{\partial x},\ \bar{\varepsilon}_{yy}=\frac{\partial\bar{v}}{\partial y},\ \bar{\gamma}_{xy}=\frac{\partial\bar{u}}{\partial y}+\frac{\partial\bar{v}}{\partial x} \tag{8.25}$$

are independent of z. For plane stress, as defined by Equation (8.11), $\sigma_{zz} = \sigma_{xz} = \sigma_{yz} = 0$. And for plane strain, as defined by Equation (8.17), and for consistency between the kinematics of the actual and virtual displacements, we must have $\bar{\varepsilon}_{zz} = \bar{\gamma}_{xz} = \bar{\gamma}_{yz} = 0$. Therefore, for plane stress or strain, we get

$$\begin{aligned}\sigma_{ji}\bar{\varepsilon}_{ij} &= \sigma_{xx}\bar{\varepsilon}_{xx} + \sigma_{yy}\bar{\varepsilon}_{yy} + \sigma_{xy}\bar{\gamma}_{xy} \\ &= \left\{ \begin{array}{ccc} \bar{\varepsilon}_{xx} & \bar{\varepsilon}_{yy} & \bar{\gamma}_{xy} \end{array} \right\} \left\{ \begin{array}{c} \sigma_{xx} \\ \sigma_{yy} \\ \sigma_{xy} \end{array} \right\} \\ &= \{ \ \bar{\varepsilon} \ \}^T \{ \ \sigma \ \}. \end{aligned} \tag{8.26}$$

For plane stress or strain, we have $t_x = t_y = 0$ on the planes $z = 0$ and $z = \ell_z$. For plane stress, the surface traction $t_z = 0$ on those planes. And for plane strain the virtual displacement $\bar{w} = 0$, as will be justified shortly. Thus, the boundary integrals on the planes $z = 0$ and $z = \ell_z$ are zero because $\sigma_{zz} = 0$ for plane stress, and $\bar{w} = 0$ for plane strain. The other components of traction are zero. We justify the condition $\bar{w} = 0$ for plane strain by writing the conditions Equation (8.17) in terms of displacements

$$\frac{\partial w}{\partial z} = 0, \quad \frac{\partial u}{\partial z} + \frac{\partial w}{\partial x} = 0, \quad \frac{\partial v}{\partial z} + \frac{\partial w}{\partial y} = 0. \tag{8.27}$$

Since u and v are independent of z, it follows that w must be a constant. We take $w = 0$ through out the domain for plane strain. The virtual displacement field must be kinematically consistent with the actual field. Thus, $\bar{w} = 0$ throughout the domain. On the cylindrical surface, traction t_i is parallel to the xy-plane with the two components t_x and t_y independent of z, and $t_z = 0$. The body force f_i is a two-dimensional vector field (f_x and f_y are independent of z, and $f_z = 0$.) We take an infinitesimal element of volume $d\Omega = \ell_z dA$ and an element of surface $dS = \ell_z ds$ where ds is an arc length on the contour Γ. Now, we can adapt Equation (8.23) for plane elasticity as

$$\begin{aligned}\int_A \int_0^{\ell_z} \left\{ \begin{array}{ccc} \bar{\varepsilon}_{xx} & \bar{\varepsilon}_{yy} & \bar{\gamma}_{xy} \end{array} \right\} \left\{ \begin{array}{c} \sigma_{xx} \\ \sigma_{yy} \\ \sigma_{xy} \end{array} \right\} dz dA &= \oint_\Gamma \int_0^{\ell_z} \left\{ \begin{array}{cc} \bar{u} & \bar{v} \end{array} \right\} \left\{ \begin{array}{c} t_x \\ t_y \end{array} \right\} dz ds \\ &+ \int_A \int_0^{\ell_z} \left\{ \begin{array}{cc} \bar{u} & \bar{v} \end{array} \right\} \left\{ \begin{array}{c} f_x \\ f_y \end{array} \right\} dz dA. \end{aligned} \tag{8.28}$$

The integrands do not depend on z, hence

$$\begin{aligned}\int_A \ell_z \left\{ \begin{array}{ccc} \bar{\varepsilon}_{xx} & \bar{\varepsilon}_{yy} & \bar{\gamma}_{xy} \end{array} \right\} \left\{ \begin{array}{c} \sigma_{xx} \\ \sigma_{yy} \\ \sigma_{xy} \end{array} \right\} dA &= \oint_\Gamma \ell_z \left\{ \begin{array}{cc} \bar{u} & \bar{v} \end{array} \right\} \left\{ \begin{array}{c} t_x \\ t_y \end{array} \right\} ds \\ &+ \int_A \ell_z \left\{ \begin{array}{cc} \bar{u} & \bar{v} \end{array} \right\} \left\{ \begin{array}{c} f_x \\ f_y \end{array} \right\} dA. \end{aligned} \tag{8.29}$$

The first integral on the right-hand-side is a boundary term, which indicates that the secondary variables are surface traction (t_x, t_y) while the displacements (u, v) are the corresponding primary variables. Further development of the principle requires specification and application of the boundary conditions, which affects the boundary integral. If the area A is divided into finite elements, Equation (8.29) can be applied to an element

$$\int_{A_e} \ell_z \begin{Bmatrix} \bar{\varepsilon}_{xx} \\ \bar{\varepsilon}_{yy} \\ \bar{\gamma}_{xy} \end{Bmatrix}^T \begin{Bmatrix} \sigma_{xx} \\ \sigma_{yy} \\ \sigma_{xy} \end{Bmatrix} dA = \oint_{\Gamma_e} \ell_z \begin{Bmatrix} \bar{u} \\ \bar{v} \end{Bmatrix}^T \begin{Bmatrix} t_x \\ t_y \end{Bmatrix} ds + \int_{A_e} \ell_z \begin{Bmatrix} \bar{u} \\ \bar{v} \end{Bmatrix}^T \begin{Bmatrix} f_x \\ f_y \end{Bmatrix} dA \quad (8.30)$$

where A_e is the element area and Γ_e is its boundary. For the typical element whose boundary is completely embedded within the domain, the traction components t_x and t_y are unknown and their integration around the element boundary will be represented by unknown weighted secondary variables at the element nodes.

8.3 ELEMENT EQUATIONS AND MATRICES

The element equations are relations among the primary and weighted secondary variables at the element nodes. We solve for the displacement field (u,v), which is the primary variable. Therefore, there are two primary variables at each node, namely the displacement components in the x- and y- directions. We consider them to be the two degrees of freedom of the node. For an element with n nodes, the number of degrees of freedom per element is $2n$. They are denoted by u_i^e and v_i^e, for $i = 1,...,n$, and shown in Figure 8.2a for a linear triangular element and in Figure 8.2b for a biquadratic quadrilateral element. The principle of virtual displacements will be used to derive a system of algebraic equations that relates the element's primary variables (nodal displacements) to the weighted secondary variables (nodal forces).

Approximation over an Element

The displacements at any point (x,y) within the element including points on the element boundary are expressed in terms of the element's degrees of freedom and shape functions.

$$\begin{aligned} u(x,y) \approx u^e(x,y) &= \sum_{j=1}^{n} u_j^e \psi_j(x,y) \\ v(x,y) \approx v^e(x,y) &= \sum_{j=1}^{n} v_j^e \psi_j(x,y) \end{aligned} \quad (8.31)$$

where $\psi_j(x,y)$, $j = 1,...,n$ are the element shape functions as derived in Chapter 4, including isoparametric elements. The functions $u^e(x,y)$ and $v^e(x,y)$ are the trial solution over an element, which can be written in matrix form

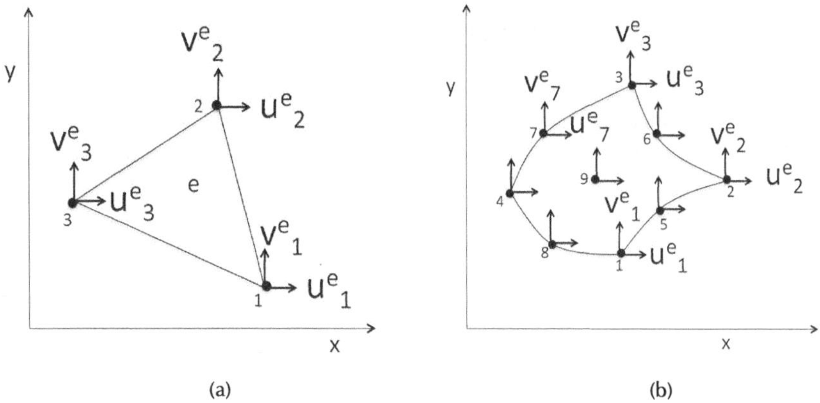

Figure 8.2 (a) Bilinear triangular element (constant strain), 3 nodes and 6 degrees of freedom. (b) Biquadratic quadrilateral element, 9 nodes and 18 degrees of freedom.

$$\begin{Bmatrix} u^e(x,y) \\ v^e(x,y) \end{Bmatrix} = \begin{bmatrix} \psi_1 & 0 & \psi_2 & 0 & . & \psi_j & 0 & . & \psi_n & 0 \\ 0 & \psi_1 & 0 & \psi_2 & . & 0 & \psi_j & . & 0 & \psi_n \end{bmatrix} \begin{Bmatrix} u^e_1 \\ v^e_1 \\ u^e_2 \\ v^e_2 \\ . \\ u^e_j \\ v^e_j \\ . \\ u^e_n \\ v^e_n \end{Bmatrix}. \tag{8.32}$$

We define the arrays

$$\{U^e\} = \begin{Bmatrix} u^e_1 \\ v^e_1 \\ u^e_2 \\ v^e_2 \\ . \\ u^e_j \\ v^e_j \\ . \\ u^e_n \\ v^e_n \end{Bmatrix} \tag{8.33}$$

and

$$[\psi] = \begin{bmatrix} \psi_1 & 0 & \psi_2 & 0 & . & \psi_j & 0 & . & \psi_n & 0 \\ 0 & \psi_1 & 0 & \psi_2 & . & 0 & \psi_j & . & 0 & \psi_n \end{bmatrix} \tag{8.34}$$

and write the trial solution as

$$\left\{ \begin{array}{c} u^e(x,y) \\ v^e(x,y) \end{array} \right\} = [\psi]\{U^e\}. \tag{8.35}$$

Note that $[\psi]$ is a function of x and y and of dimensions $(2,2n)$, while $\{U^e\}$ is the values of displacements at the element's nodes and of dimensions $(2n,1)$.

Strains and Stresses in an Element

We express the strains at any point in an element in terms of the displacements at the nodes and shape functions.

$$\begin{aligned} \left\{ \begin{array}{c} \varepsilon^e_{xx}(x,y) \\ \varepsilon^e_{yy}(x,y) \\ \gamma^e_{xy}(x,y) \end{array} \right\} &= \left\{ \begin{array}{c} \frac{\partial u^e}{\partial x} \\ \frac{\partial v^e}{\partial y} \\ \frac{\partial u^e}{\partial y} + \frac{\partial v^e}{\partial x} \end{array} \right\} \\ &= \left[\begin{array}{cccccccccc} \frac{\partial \psi_1}{\partial x} & 0 & \frac{\partial \psi_2}{\partial x} & 0 & . & \frac{\partial \psi_j}{\partial x} & 0 & . & \frac{\partial \psi_n}{\partial x} & 0 \\ 0 & \frac{\partial \psi_1}{\partial y} & 0 & \frac{\partial \psi_2}{\partial y} & . & 0 & \frac{\partial \psi_j}{\partial y} & . & 0 & \frac{\partial \psi_n}{\partial y} \\ \frac{\partial \psi_1}{\partial y} & \frac{\partial \psi_1}{\partial x} & \frac{\partial \psi_2}{\partial y} & \frac{\partial \psi_2}{\partial x} & . & \frac{\partial \psi_j}{\partial y} & \frac{\partial \psi_j}{\partial x} & . & \frac{\partial \psi_n}{\partial y} & \frac{\partial \psi_n}{\partial x} \end{array} \right] \left\{ \begin{array}{c} u^e_1 \\ v^e_1 \\ u^e_2 \\ v^e_2 \\ . \\ u^e_j \\ v^e_j \\ . \\ u^e_n \\ v^e_n \end{array} \right\}. \end{aligned} \tag{8.36}$$

We define the gradient matrix

$$[B^e] = \left[\begin{array}{cccccccccc} \frac{\partial \psi_1}{\partial x} & 0 & \frac{\partial \psi_2}{\partial x} & 0 & . & \frac{\partial \psi_j}{\partial x} & 0 & . & \frac{\partial \psi_n}{\partial x} & 0 \\ 0 & \frac{\partial \psi_1}{\partial y} & 0 & \frac{\partial \psi_2}{\partial y} & . & 0 & \frac{\partial \psi_j}{\partial y} & . & 0 & \frac{\partial \psi_n}{\partial y} \\ \frac{\partial \psi_1}{\partial y} & \frac{\partial \psi_1}{\partial x} & \frac{\partial \psi_2}{\partial y} & \frac{\partial \psi_2}{\partial x} & . & \frac{\partial \psi_j}{\partial y} & \frac{\partial \psi_j}{\partial x} & . & \frac{\partial \psi_n}{\partial y} & \frac{\partial \psi_n}{\partial x} \end{array} \right] \tag{8.37}$$

and write the strains as

$$\left\{ \begin{array}{c} \varepsilon^e_{xx}(x,y) \\ \varepsilon^e_{yy}(x,y) \\ \gamma^e_{xy}(x,y) \end{array} \right\} = [B^e]\{U^e\}. \tag{8.38}$$

Note that $[B^e]$ is a function of x and y and of dimensions $(3,2n)$. Multiplying Equation (8.38) by the appropriate elastic coefficient matrix $[D^e]$, we obtain the stresses at any point in the element domain,

$$\left\{\begin{array}{c}\sigma_{xx}^{e}(x,y)\\ \sigma_{yy}^{e}(x,y)\\ \sigma_{xy}^{e}(x,y)\end{array}\right\} = [D^e]\,[B^e]\,\{U^e\}. \tag{8.39}$$

The matrix $[D^e]$ is given by Equation (8.16) for plane stress, and by (8.19) for plane strain.

Virtual Displacements and Strains for an Element: Galerkin

The virtual displacement field is an arbitrary field that we select; it is a completely known field. In the Galerkin method, we construct the virtual displacement over the element by using the same shape functions of the trial solution, which is given by Equation (8.31) or (8.35). Thus, we choose the virtual displacements of element # e as

$$\left\{\begin{array}{c}\bar{u}^e(x,y)\\ \bar{v}^e(x,y)\end{array}\right\} = \left[\begin{array}{cccccccccccc}\psi_1 & 0 & \psi_2 & 0 & . & \psi_j & 0 & . & \psi_n & 0\\ 0 & \psi_1 & 0 & \psi_2 & . & 0 & \psi_j & . & 0 & \psi_n\end{array}\right]\left\{\begin{array}{c}d_{x1}\\ d_{y1}\\ d_{x2}\\ d_{y2}\\ .\\ d_{xj}\\ d_{yj}\\ .\\ d_{xn}\\ d_{yn}\end{array}\right\}. \tag{8.40}$$

We define the array

$$\{d\} = \left\{\begin{array}{c}d_{x1}\\ d_{y1}\\ d_{x2}\\ d_{y2}\\ .\\ d_{xj}\\ d_{yj}\\ .\\ d_{xn}\\ d_{yn}\end{array}\right\} \tag{8.41}$$

and write the virtual displacements in matrix form

$$\left\{\begin{array}{c}\bar{u}^e(x,y)\\ \bar{v}^e(x,y)\end{array}\right\} = [\psi]\,\{d\}. \tag{8.42}$$

The array $\{d\}$ is of dimensions $(2n,1)$, it gives the two components of virtual displacements at the element's n nodes. We can construct $2n$ linearly independent virtual displacement fields for the element by choosing one at-a-time of the components of $\{d\}$ to be non-zero, say 1, while the remaining $(2n-1)$ components to be zero. Any other virtual displacement field can be obtained by linear superposition of those $2n$ fields. The virtual strains follow directly from Equation (8.38),

$$\left\{ \begin{array}{c} \bar{\varepsilon}^e_{xx}(x,y) \\ \bar{\varepsilon}^e_{yy}(x,y) \\ \bar{\gamma}^e_{xy}(x,y) \end{array} \right\} = [B^e]\{d\}. \tag{8.43}$$

Element Equations

Substituting Equations (8.39), (8.42), and (8.43) into the principle of virtual displacement for an element Equation (8.30), we get

$$\begin{aligned} \int_{A_e} \ell_z \{d\}^T [B^e]^T [D^e][B^e]\{U^e\}\, dA &= \oint_{\Gamma_e} \ell_z \{d\}^T [\psi]^T \left\{ \begin{array}{c} t_x \\ t_y \end{array} \right\} ds \\ &+ \int_{A_e} \ell_z \{d\}^T [\psi]^T \left\{ \begin{array}{c} f_x \\ f_y \end{array} \right\} dA. \end{aligned} \tag{8.44}$$

Since $\{d\}$ and $\{U^e\}$ are constant arrays, we rearrange terms and obtain

$$\begin{aligned} \{d\}^T \Bigg[\left(\int_{A_e} \ell_z [B^e]^T [D^e][B^e]\, dA \right) \{U^e\} &- \oint_{\Gamma_e} \ell_z [\psi]^T \left\{ \begin{array}{c} t_x \\ t_y \end{array} \right\} ds \\ &- \int_{A_e} \ell_z [\psi]^T \left\{ \begin{array}{c} f_x \\ f_y \end{array} \right\} dA \Bigg] = 0. \end{aligned} \tag{8.45}$$

We define,

Stiffness Matrix

$$[K^e] = \int_{A_e} \ell_z [B^e]^T [D^e][B^e]\, dA. \tag{8.46}$$

Forcing Vector

$$\{F^e\} = \int_{A_e} \ell_z [\psi]^T \left\{ \begin{array}{c} f_x \\ f_y \end{array} \right\} dA. \tag{8.47}$$

Weighted Secondary Variables Vector

$$\{R^e\} = \oint_{\Gamma_e} \ell_z [\psi]^T \left\{ \begin{array}{c} t_x \\ t_y \end{array} \right\} ds \tag{8.48}$$

and rewrite Equation (8.45),

$$\{d\}^T \Big[[K^e]\{U^e\} - \{F^e\} - \{R^e\} \Big] = 0. \tag{8.49}$$

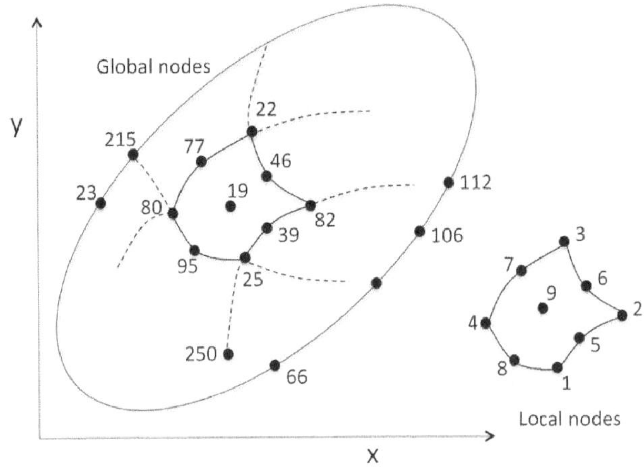

Figure 8.3 Global nodes for a biquadratic quadrilateral element and corresponding local nodes. Element's local nodes (1, 2, 3, 4, 5, 6, 7, 8, 9) correspond to global nodes (25, 82, 22, 80, 39, 46, 77, 95, 19).

Since $\{d\}^T$ is an arbitrary array of dimensions (1, 2n), the multiplying bracket must be zero, and we obtain the element equations,

$$[K^e]\{U^e\} = \{F^e\} + \{R^e\}. \tag{8.50}$$

This is a system of $2n$ algebraic equations for each element, where n is the number of nodes per element. For an element completely embedded within the area A, both of the arrays of primary variables $\{U^e\}$ and weighted secondary variables $\{R^e\}$ are unknown, hence there are $4n$ unknowns. Every term in Equation (8.50) has the dimension of force $[F]$ or units of Newtons (N). Stiffness matrix $[K^e]$ and forcing vector $\{F^e\}$ are domain integrals, given by integrals over the area A_e of the element, whereas weighted secondary variables vector $\{R^e\}$ is a boundary integral, given by a line integral around the element boundary Γ_e.

8.4 ELEMENTS ASSEMBLY AND GLOBAL SYSTEM

Let's assume that the problem domain is divided into elements using N global nodes. Figure 8.3 shows a sketch of such a domain and one of the elements that is a 9-node biquadratic element, its global nodes are (25, 82, 22, 80, 39, 46, 77, 95, 19), and the corresponding local nodes are $(j = 1,2,3,4,5,6,7,8,9)$. For the sake of discussion, let's assume that this is element number $ie = 37$. Then row number 37 of the connectivity matrix $[CM]$ should read

$$[CM](37,1:9) = \{25\ 82\ 22\ 80\ 39\ 46\ 77\ 95\ 19\}. \tag{8.51}$$

Since there are two degrees of freedom at each node, we arrange the $2N$ degrees of freedom of the global system in an array $\{U\}$ of dimensions $(2N,1)$.

$$\{U\} = \left\{ \begin{array}{c} u_1 \\ v_1 \\ u_2 \\ v_2 \\ \cdot \\ u_j \\ v_j \\ \cdot \\ u_N \\ v_N \end{array} \right\}. \tag{8.52}$$

At global node j, the x-displacement is $u_j = U_{2j-1}$ and the y-displacement is $v_j = U_{2j}$.

Displacements at the local nodes of a specified element can be extracted from the global array as follows. For local node number k of element # ie, we extract the corresponding global node from connectivity matrix $i = [MC](ie,k)$. We obtain $u_k^e = U_{2i-1}$ and $v_k^e = U_{2i}$. As an example, for element $ie = 37$ shown in Figure 8.3, the correspondence between the element local degrees of freedom and those of the global system is given by

$$\left\{ \begin{array}{c} u_1^e \\ v_1^e \\ u_2^e \\ v_2^e \\ \cdot \\ \cdot \\ \cdot \\ \cdot \\ u_9^e \\ v_9^e \end{array} \right\} = \left\{ \begin{array}{c} U_{49} \\ U_{50} \\ U_{163} \\ U_{164} \\ \cdot \\ \cdot \\ \cdot \\ \cdot \\ U_{37} \\ U_{38} \end{array} \right\}. \tag{8.53}$$

Each element has a system of equations of the form Equation (8.50), where the stiffness matrix $[K^e]$ and forcing vector $\{ F^e \}$ can be readily computed for the known mesh, material properties, and applied body loads. The arrays $\{ U^e \}$ of primary variables and of weighted secondary variables $\{ R^e \}$ are unknown since no boundary conditions are applied when deriving the element equations. Boundary conditions are specified on the outer boundary of the entire domain. Therefore, the element equations are assembled into a global system of equations that contains all primary and weighted secondary variables at all global nodes. Thus, the global system before application of the boundary conditions is a system of $2N$ equations of the general form

$$[K]\{U\} \quad = \quad \{F\} + \{R\} \tag{8.54}$$

where the coefficient matrix $[K]$ is of dimensions $(2N, 2N)$, and the forcing vector $\{F\}$ and the vector of weighted secondary variables $\{R\}$ are of dimension $(2N, 1)$.

The entries of the stiffness matrix $k_{i_g j_g}$ are obtained by accumulating contributions from the stiffness matrices of all the elements whose local degrees of freedom coincide with global DOF i_g and j_g. The connectivity (or identity) matrix is very helpful in the assembly of elements. Once the stiffness matrix and forcing vector are computed for an element, they are assembled into the global system and the system is updated. Assembly of the elements can be accomplished by the following MATLAB algorithm. We define the parameters:

$nnpe =$ Number of nodes per element

$nvpn =$ Number of degrees of freedom per node

$nvpe = nnpe \times nvpn =$ Number of degrees of freedom per element

$nvg = nng \times nvpn =$ Number of degrees of freedom in the global system

Dimensions of element stiffness matrix is $(nvpe, nvpe)$

Dimensions of element forcing vector is $(nvpe, 1)$

Dimensions of global stiffness matrix is (nvg, nvg)

Dimensions of global forcing vector is $(nvg, 1)$

For element # *ie*, compute an array of integers ijg of dimensions $(nvpe, 1)$ that gives the row (or column) in the global matrix that corresponds to row i (or column j) of the element matrix.

```
%% The connectivity matrix gives the global nodes of element ie:
for jnd=1:nnpe                 % for each local node (jnd) of
                               % the element
nodes(jnd)=mconn(ie,jnd);      % nodes(jnd) is the corresponding
                               % global node
end
%%%%% Compute ijg
m=0;                           % a counter
for jnd=1:nnpe                 % for each local node of the element
mprev=(nodes(jnd)-1)*nvpn;     % mprev is the number of rows
                               % (or columns) preceeding the global
                               % row (or column) corresponding to
                               % the first degree of freedom of the
                               % node jnd
for i=1:nvpn                   % for each degree of freedom of the
                               % node
m=m+1;                         % increase the counter by 1
ijg(m)=mprev+i;                % this is the row (or column) in the
                               % global matrix that corresponds to
                               % degree of freedom number i of node
                               % jnd of element ie
```

```
end
end
```

Then assemble element matrices into global system.

```
for i=1:nvpe
ig=ijg(i);                      % ig = row in global matrix that
                                % corresponds  to row i in element
                                % matrix
fg(ig)=fg(ig)+fe(i);            % assemble element forcing vector
                                % fe(i) into global forcing vector
                                % fg(ig)
for j=1:nvpe
jg=ijg(j);                        % jg = column in global matrix
                                  % that corresponds
                                  % to column j in element matrix
kg(ig,jg)=kg(ig,jg)+ke(i,j);      % assemble element stiffness
                                  % matrix ke(i,j) into global
                                  % stiffness matrix kg(ig,jg)
end
end
```

We remark that some of the components of the global array of weighted secondary variables $\{R\}$ may be known while the others are unknown. As a result of the balance of weighted secondary variables, the entries of the global $\{R\}$ for all interior nodes are either zero or equal to externally applied concentrated forces not accounted for by the distributed body force (f_x, f_y). For boundary nodes where a natural boundary condition is prescribed, the entries of the weighted secondary variables are computed and added to the global forcing vector. For boundary nodes where an essential boundary condition is prescribed, the entries of $\{R\}$ are unknown whereas the corresponding entries of $\{U\}$ are known. We can either remove the corresponding equations that contain the unknown weighted secondary variables thereby obtaining a reduced system for the unknown primary variables, or replace those equations with equations that assign the known values of primary variables and hence keep the original dimension of the matrices and arrays. We use the second method in the MATLAB codes developed for this book. For boundary nodes where mixed boundary conditions are prescribed, we replace the unknown secondary variable in terms of the unknown primary variable. Thus, all the unknown entries of $\{R\}$ are either removed by applying EBC or replaced in terms of primary variables at MBC. After applying the boundary conditions, the modified (or reduced) system takes the form

$$[\bar{K}]\{U\} = \{\bar{F}\} \tag{8.55}$$

which is solved for the array of primary variables at all global nodes $\{U\}$. Afterward, we compute the unknown weighted secondary variables and other quantities of interest. If we save the coefficient matrix and forcing vector before applications of boundary conditions, the array of weighted secondary variables (known and unknown) can be found from Equation (8.54) rewritten here as

$$\{R\} = [K]\{U\} - \{F\}. \tag{8.56}$$

An alternative treatment of a natural boundary condition is to compute $\{R^e\}$ given by Equation (8.48) for elements whose edges fall on an NBC. In this case, the integral is performed only on those edges where traction components are prescribed. The contributions of edges embedded in the domain or subjected to an EBC are assigned zero values. The computed $\{R^e\}$ is added to $\{F^e\}$ before assembly into the global forcing vector $\{F\}$.

Integration Rules and Computation of Stresses: Barlow Points

Displacements and stresses at any point can be computed in terms of the element shape functions and displacements at the element's nodes. Strains and stresses can be computed by Equations (8.36), and (8.39), which requires the space derivatives of the shape functions. Lagrange interpolation functions do not enforce continuity of space derivatives at a common edge between two adjacent elements, as such, the stresses may exhibit a discontinuity as the edge is crossed. The discontinuity is expected to decrease with mesh refinement. It is recommended that stresses to be computed at the Barlow points [8] and [9], which are the base points of reduced integration rule of the element. For the bilinear quadrilateral element, the full integration rule is (2×2) Gauss–Legendre quadrature, which is used for the stiffness matrix and load vector. The Barlow point for the element is the base point for the (1×1) Gauss–Legendre quadrature; which is the reduced integration rule. Similarly, for the biquadratic element, the full integration rule is (3×3) Gauss–Legendre quadrature, and Barlow points are base points of the (2×2) Gauss–Legendre quadrature. Similar definitions are used for higher-order elements.

Example 8.1. A Contrived Exact Solution for Plane Elasticity

The objective of this example is to construct a simple yet exact analytic solution to the equations of equilibrium for plane elasticity. In the next example, we use such a solution to assess errors and verify finite-element solutions. We propose analytic expressions for displacement field $u(x,y)$, and $v(x,y)$, and determine the corresponding strains ε_{xx}, ε_{yy}, and γ_{xy}. Stresses follow from the appropriate constitutive relations, and the divergence of stress tensor is then evaluated. The equations of equilibrium are then used to determine the body forces that balance the divergence of stress tensor.

$$f_x = -\left(\frac{\partial \sigma_{xx}}{\partial x} + \frac{\partial \sigma_{yx}}{\partial y}\right), \; f_y = -\left(\frac{\partial \sigma_{xy}}{\partial x} + \frac{\partial \sigma_{yy}}{\partial y}\right). \tag{8.57}$$

To demonstrate this procedure, we propose a displacement field

$$\begin{aligned} u(x,y) &= \gamma a \left(\frac{x}{a}\right)\left(\frac{y}{b}\right)\left[\left(\frac{x}{a}\right)\left(\frac{y}{b}\right)+1\right] \\ v(x,y) &= \gamma b \left(\frac{x}{a}\right)\left(\frac{y}{b}\right)\left[\left(\frac{x}{a}\right)\left(\frac{y}{b}\right)-1\right] \end{aligned} \tag{8.58}$$

which gives strains

$$\{\varepsilon\} = \left\{\begin{array}{c}\varepsilon_{xx}\\ \varepsilon_{yy}\\ \gamma_{xy}\end{array}\right\} = \gamma\left\{\begin{array}{c}\left(\frac{y}{b}\right)\left[2\left(\frac{x}{a}\right)\left(\frac{y}{b}\right)+1\right]\\ \left(\frac{x}{a}\right)\left[2\left(\frac{x}{a}\right)\left(\frac{y}{b}\right)-1\right]\\ \left(\frac{x}{b}\right)\left[2\left(\frac{x}{a}\right)\left(\frac{y}{b}\right)+1\right]+\left(\frac{y}{a}\right)\left[2\left(\frac{x}{a}\right)\left(\frac{y}{b}\right)-1\right]\end{array}\right\}. \tag{8.59}$$

Assuming a state of plane stress, we use Equation (8.15) and obtain

$$\{\sigma\} \equiv \left\{\begin{array}{c}\sigma_{xx}\\ \sigma_{yy}\\ \sigma_{xy}\end{array}\right\} = \frac{\gamma E}{1-\nu^2}\left\{\begin{array}{c}\left(\frac{y}{b}\right)\left[2\left(\frac{x}{a}\right)\left(\frac{y}{b}\right)+1\right]+\nu\left(\frac{x}{a}\right)\left[2\left(\frac{x}{a}\right)\left(\frac{y}{b}\right)-1\right]\\ \left(\frac{x}{a}\right)\left[2\left(\frac{x}{a}\right)\left(\frac{y}{b}\right)-1\right]+\nu\left(\frac{y}{b}\right)\left[2\left(\frac{x}{a}\right)\left(\frac{y}{b}\right)+1\right]\\ \frac{1-\nu}{2}\left[\left(\frac{x}{b}\right)\left(2\left(\frac{x}{a}\right)\left(\frac{y}{b}\right)+1\right)+\left(\frac{y}{a}\right)\left(2\left(\frac{x}{a}\right)\left(\frac{y}{b}\right)-1\right)\right]\end{array}\right\}. \tag{8.60}$$

The body forces that satisfy the equations of equilibrium are

$$\begin{aligned} f_x &= \frac{\gamma E}{a(1-\nu^2)}\left\{(1+\nu)\left[\frac{1}{2}-2\left(\frac{x}{a}\right)\left(\frac{y}{b}\right)\right]\right. \\ &\quad \left.-(1-\nu)\left(\frac{a}{b}\right)^2\left(\frac{x}{a}\right)^2-2\left(\frac{y}{b}\right)^2\right\} \\ f_y &= -\frac{\gamma E}{b(1-\nu^2)}\left\{(1+\nu)\left[\frac{1}{2}+2\left(\frac{x}{a}\right)\left(\frac{y}{b}\right)\right]\right. \\ &\quad \left.+(1-\nu)\left(\frac{b}{a}\right)^2\left(\frac{y}{b}\right)^2+2\left(\frac{x}{a}\right)^2\right\}. \end{aligned} \tag{8.61}$$

One can select a domain of any shape in the xy-plane, and use Equations (8.58) and (8.60) to specify displacements and/or traction on the boundary of that domain. Then, use the finite-element method to solve the equations of equilibrium with body forces given by Equation (8.61). The accuracy of finite-element solution can be evaluated by comparison with the exact solution for unknown displacements and weighted secondary variables.

Example 8.2. Error Assessment of Finite Element Solution

We consider a rectangular plate $ABCD$ ($0 \le x \le a, 0 \le y \le b$) as shown in Figure 8.4 in a state of plane stress under the action of body forces given by Equation (8.61). The exact solution of Example 8.1 is used to prescribe the following boundary conditions. On AB, $y = 0$ and $0 \le x \le a$:

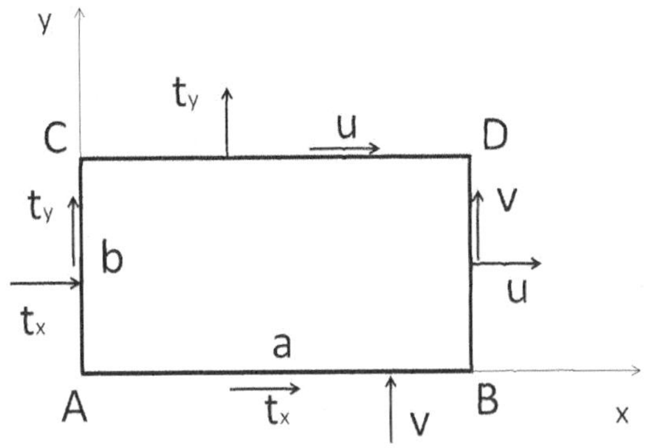

Figure 8.4 Example 8.2. Boundary conditions for a plane elasticity problem.

$$\begin{aligned} v &= 0 \\ t_x &= -\sigma_{yx} \\ &= -\frac{\gamma E}{1-\nu^2}\frac{1-\nu}{2}\left(\frac{x}{b}\right). \end{aligned} \tag{8.62}$$

On CD, $y = b$ and $0 \le x \le a$:

$$\begin{aligned} u &= \gamma a\left(\frac{x}{a}\right)\left[\left(\frac{x}{a}\right)+1\right] \\ t_y &= \sigma_{yy} \\ &= \frac{\gamma E}{1-\nu^2}\left\{\left(\frac{x}{a}\right)\left[2\left(\frac{x}{a}\right)-1\right]+\nu\left[2\left(\frac{x}{a}\right)+1\right]\right\}. \end{aligned} \tag{8.63}$$

On AC, $x = 0$ and $0 \le y \le b$:

$$\begin{aligned} t_x &= -\sigma_{xx} \\ &= -\frac{\gamma E}{1-\nu^2}\left(\frac{y}{b}\right) \\ t_y &= -\sigma_{xy} \\ &= \frac{\gamma E}{1-\nu^2}\frac{1-\nu}{2}\left(\frac{y}{a}\right). \end{aligned} \tag{8.64}$$

On BD, $x = a$ and $0 \le y \le b$:

$$u = \gamma a\left(\frac{y}{b}\right)\left[\left(\frac{y}{b}\right)+1\right],\; v = \gamma b\left(\frac{y}{b}\right)\left[\left(\frac{y}{b}\right)-1\right]. \tag{8.65}$$

Since the exact solution for displacements is biquadratic in x and y, we anticipate that the finite-element solution to be identical to the exact solution if we use one biquadratic element. The global and local node numbers are shown in Figures 8.5a and 8.5b, respectively. The shape functions are given by Equation (4.207). Our objective is to asses the error in the finite element method, so we assign convenient numerical

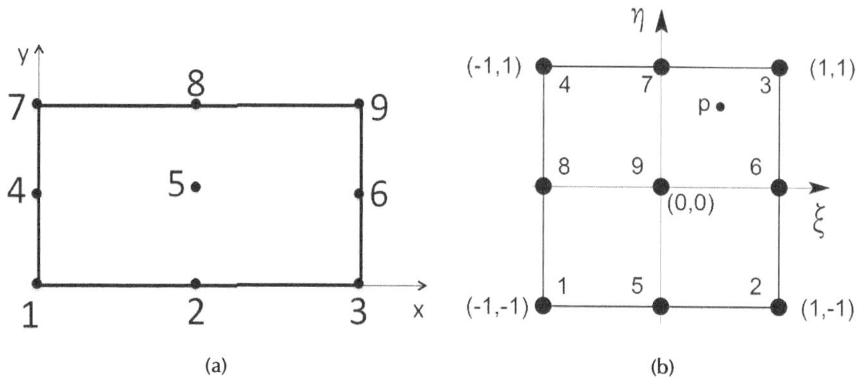

Figure 8.5 (a) Global node numbers for one element. (b) Canonical biquadratic quadrilateral element.

values for the parameters $\gamma = 9$, $a = 2$, $b = 1$, $E = 1$, and $\nu = 0.25$. The solution for displacements, as given in Equation (8.58), is independent of ν and E. We also choose unit thickness $\ell_z = 1$.

First, we compute the contributions of prescribed surface traction to the array of weighted secondary variables. The contributions of unspecified traction remain unknown and will be found from the finite-element solution.

$$\begin{aligned} \{R^e\} &= \oint [\psi]^T \left\{ \begin{array}{c} t_x \\ t_y \end{array} \right\} ds \\ &= \oint \begin{bmatrix} \psi_1 & 0 \\ 0 & \psi_1 \\ . & . \\ . & . \\ . & . \\ \psi_9 & . \\ . & \psi_9 \end{bmatrix} \left\{ \begin{array}{c} t_x \\ t_y \end{array} \right\} ds \end{aligned} \tag{8.66}$$

$$\{R^e\} = \oint \left\{ \begin{array}{c} \psi_1 t_x \\ \psi_1 t_y \\ . \\ . \\ . \\ \psi_9 t_x \\ \psi_9 t_y \end{array} \right\} ds. \tag{8.67}$$

For a 9-node biquadratic element, the vector $\{R^e\}$ is of dimensions $(18,1)$. The 9 odd-numbered components R_1^e, R_3^e, ..., R_{17}^e are nodal forces in the x-direction equivalent to the distributed traction t_x all around the element boundary. Similarly, the 9 even-numbered components R^e_2, R^e_4, ..., R^e_{18} are nodal forces in the y-direction equivalent to the distributed traction t_y all around the element boundary. The prescribed traction components are shown in Figure 8.4. For the example at hand, the boundary integral is divided into four integrals,

$$
\begin{aligned}
\{R^e\} &= \int_{AB} \begin{Bmatrix} \psi_1 t_x \\ \psi_1 t_y \\ \cdot \\ \cdot \\ \cdot \\ \psi_9 t_x \\ \psi_9 t_y \end{Bmatrix} ds + \int_{BD} \begin{Bmatrix} \psi_1 t_x \\ \psi_1 t_y \\ \cdot \\ \cdot \\ \cdot \\ \psi_9 t_x \\ \psi_9 t_y \end{Bmatrix} ds \\
&+ \int_{DC} \begin{Bmatrix} \psi_1 t_x \\ \psi_1 t_y \\ \cdot \\ \cdot \\ \cdot \\ \psi_9 t_x \\ \psi_9 t_y \end{Bmatrix} ds + \int_{CA} \begin{Bmatrix} \psi_1 t_x \\ \psi_1 t_y \\ \cdot \\ \cdot \\ \cdot \\ \psi_9 t_x \\ \psi_9 t_y \end{Bmatrix} ds. \qquad (8.68)
\end{aligned}
$$

For the interior node of the biquadratic element, shape function ψ_9 is identically zero on the element boundary. Hence, $R^e_{17} = 0$ and $R^e_{18} = 0$; that is to say, the system of nodal forces equivalent to distributed traction on the element boundary does not have components at node number 9. Let's consider the contribution of each edge to the other components of the array $\{R^e\}$. Edge AB coincides with edge 1-5-2 of the canonical element in the $\xi - \eta$ plane. We recall that on edge 1-5-2 of the biquadratic element all shape functions are zero except ψ_1, ψ_2, and ψ_5. Thus, the contribution of the traction on AB to $\{R^e\}$ becomes

$$
\begin{Bmatrix} R^e_1 \\ R^e_2 \\ R^e_3 \\ R^e_4 \\ R^e_9 \\ R^e_{10} \end{Bmatrix}_{AB} = \int_{AB} \begin{Bmatrix} \psi_1 t_x \\ \psi_1 t_y \\ \psi_2 t_x \\ \psi_2 t_y \\ \psi_5 t_x \\ \psi_5 t_y \end{Bmatrix} ds. \qquad (8.69)
$$

In this example, t_x is prescribed on AB, but t_y is unknown; we can compute R^e_1, R^e_3, and R^e_9, whereas R^e_2, R^e_4, and R^e_{10} remain unknown. Using isoparametric-element variables, we find $\eta = -1, x = a(\xi+1)/2 = \xi + 1$, and $ds = dx = d\xi$. We also know $t_x = -3.6x = -3.6(\xi+1)$.

$$
\begin{aligned}
(R^e_1)_{AB} &= \int_{-1}^{+1} \psi_1(\xi,-1)[-3.6(\xi+1)]d\xi \\
&= \int_{-1}^{+1} \frac{1}{2}(\xi^2 - \xi)[-3.6(\xi+1)]d\xi \\
&= 0
\end{aligned}
$$

$$\begin{aligned}(R_3^e)_{AB} &= \int_{-1}^{+1} \psi_2(\xi,-1)[-3.6(\xi+1)]d\xi \\ &= \int_{-1}^{+1} \frac{1}{2}(\xi^2+\xi)[-3.6(\xi+1)]d\xi \\ &= -2.4 \\ (R_9^e)_{AB} &= \int_{-1}^{+1} \psi_5(\xi,-1)[-3.6(\xi+1)]d\xi \\ &= \int_{-1}^{+1} (1-\xi^2)[-3.6(\xi+1)]d\xi \\ &= -4.8. \end{aligned} \tag{8.70}$$

Thus, the prescribed distributed traction t_x on edge AB is replaced by equivalent weighted secondary variables $R_1^e = 0$ at node #1, $R_3^e = -2.4$ at node #2, and $R_9^e = -4.8$ at node #5 of the biquadratic element. These are forces in the x-direction. The contribution of t_x at all other nodes is zero ($R_{2j-1}^e = 0$, $j = 3,4,6,7,8,9$). The traction t_y on AB is unknown, so it is represented by unknown weighted secondary variables R_2^e, R_4^e, and R_{10}^e, while its contributions to the other nodes are zero ($R_{2j}^e = 0$, $j = 3,4,6,7,8,9$).

Edge DC coincides with edge 3-7-4 of the canonical element in the $\xi - \eta$ plane. We recall that on edge 3-7-4 of the biquadratic element all shape functions are zero except ψ_3, ψ_4, and ψ_7. Thus, the contribution of the traction on DC to $\{R^e\}$ becomes

$$\begin{Bmatrix} R_5^e \\ R_6^e \\ R_7^e \\ R_8^e \\ R_{13}^e \\ R_{14}^e \end{Bmatrix}_{DC} = \int_{DC} \begin{Bmatrix} \psi_3 t_x \\ \psi_3 t_y \\ \psi_4 t_x \\ \psi_4 t_y \\ \psi_7 t_x \\ \psi_7 t_y \end{Bmatrix} ds. \tag{8.71}$$

In this example, t_y is prescribed on DC, but t_x is unknown; we can compute R_6^e, R_8^e, and R_{14}^e, whereas R_5^e, R_7^e, and R_{13}^e remain unknown. On edge DC, we find $\eta = 1$, $x = a(\xi+1)/2 = \xi+1$, and $-ds = dx = d\xi$. We also know $t_y = 2.4(2x^2 - x + 1)$, hence

$$\begin{aligned}(R_6^e)_{DC} &= \int_{-1}^{+1} \psi_3(\xi,1)[4.8(\xi+1)^2 - 2.4\xi]d\xi \\ &= \int_{-1}^{+1} \frac{1}{2}(\xi^2+\xi)[4.8(\xi+1)^2 - 2.4\xi]d\xi \\ &= 4.96 \\ (R_8^e)_{DC} &= \int_{-1}^{+1} \psi_4(\xi,1)[4.8(\xi+1)^2 - 2.4\xi]d\xi \\ &= \int_{-1}^{+1} \frac{1}{2}(\xi^2-\xi)[4.8(\xi+1)^2 - 2.4\xi]d\xi \\ &= 0.16 \end{aligned}$$

$$\begin{aligned}(R^e_{14})_{DC} &= \int_{-1}^{+1} \psi_7(\xi,1)[4.8(\xi+1)^2 - 2.4\xi]d\xi \\ &= \int_{-1}^{+1} (1-\xi^2)[4.8(\xi+1)^2 - 2.4\xi]d\xi \\ &= 7.68. \end{aligned} \tag{8.72}$$

Thus, the prescribed distributed traction t_y on edge DC is replaced by equivalent weighted secondary variables $R^e_6 = 4.96$ at node #3, $R^e_8 = 0.16$ at node #4, and $R^e_{14} = 7.68$ at node #7 of the biquadratic element. These are forces in the y-direction. The contribution of t_y at all other nodes is zero ($R^e_{2j} = 0$, $j = 1,2,5,6,8,9$). The traction t_x on DC is unknown, so it is represented by unknown weighted secondary variables R^e_5, R^e_7, and R^e_{13}, while its contributions to the other nodes are zero ($R^e_{2j-1} = 0$, $j = 1,2,5,6,8,9$).

Edge CA coincides with edge 4-8-1 of the canonical element in the $\xi - \eta$ plane. We recall that on edge 4-8-1 of the biquadratic element all shape functions are zero except ψ_4, ψ_8, and ψ_1. Thus, the contribution of the traction on CA to $\{R^e\}$ becomes

$$\begin{Bmatrix} R^e_1 \\ R^e_2 \\ R^e_7 \\ R^e_8 \\ R^e_{15} \\ R^e_{16} \end{Bmatrix}_{CA} = \int_{CA} \begin{Bmatrix} \psi_1 t_x \\ \psi_1 t_y \\ \psi_4 t_x \\ \psi_4 t_y \\ \psi_8 t_x \\ \psi_8 t_y \end{Bmatrix} ds. \tag{8.73}$$

In this example, both t_x and t_y are prescribed on CA, where $t_x = -9.6y$, and $t_y = 1.8y$. On that edge, we have $\xi = -1$, $y = b(\eta+1)/2 = (\eta+1)/2$, and $-ds = dy = d\eta/2$, hence

$$\begin{aligned}(R^e_1)_{CA} &= \int_{-1}^{+1} \psi_1(-1,\eta)[-2.4(\eta+1)]d\eta \\ &= \int_{-1}^{+1} \frac{1}{2}(\eta^2-\eta)[-2.4(\eta+1)]d\eta \\ &= 0 \\ (R^e_7)_{CA} &= \int_{-1}^{+1} \psi_4(-1,\eta)[-2.4(\eta+1)]d\eta \\ &= \int_{-1}^{+1} \frac{1}{2}(\eta^2+\eta)[-2.4(\eta+1)]d\eta \\ &= -1.6 \\ (R^e_{15})_{CA} &= \int_{-1}^{+1} \psi_8(-1,\eta)[-2.4(\eta+1)]d\eta \\ &= \int_{-1}^{+1} (1-\eta^2)[-2.4(\eta+1)]d\eta \\ &= -3.2. \end{aligned} \tag{8.74}$$

Thus, the prescribed distributed traction t_x on edge CA is replaced by equivalent weighted secondary variables $R^e_1 = 0$ at node #1, $R^e_7 = -1.6$ at node #4, and

$R^e_{15} = -3.2$ at node #8 of the biquadratic element. These are forces in the x-direction. The contribution of t_x at all other nodes is zero ($R^e_{2j-1} = 0,\ j = 2,3,5,6,7,9$). The integrals for t_y are

$$\begin{aligned}
(R^e_2)_{CA} &= \int_{-1}^{+1} \psi_1(-1,\eta)[0.45(\eta+1)]d\eta \\
&= \int_{-1}^{+1} \frac{1}{2}(\eta^2-\eta)[0.45(\eta+1)]d\eta \\
&= 0 \\
(R^e_8)_{CA} &= \int_{-1}^{+1} \psi_4(-1,\eta)[0.45(\eta+1)]d\eta \\
&= \int_{-1}^{+1} \frac{1}{2}(\eta^2+\eta)[0.45(\eta+1)]d\eta \\
&= 0.3 \\
(R^e_{16})_{CA} &= \int_{-1}^{+1} \psi_8(-1,\eta)[0.45(\eta+1)]d\eta \\
&= \int_{-1}^{+1} (1-\eta^2)[0.45(\eta+1)]d\eta \\
&= 0.6. \qquad (8.75)
\end{aligned}$$

Thus, the prescribed distributed traction t_y on edge CA is replaced by equivalent weighted secondary variables $R^e_2 = 0$ at node #1, $R^e_8 = 0.3$ at node #4, and $R^e_{16} = 0.6$ at node #8 of the biquadratic element. These are forces in the y-direction. The contribution of t_y at all other nodes is zero ($R^e_{2j} = 0,\ j = 2,3,5,6,7,9$).
Finally, we consider edge BD which coincides with edge 2-6-3 of the canonical element in the $\xi - \eta$ plane. We recall that on edge 2-6-3 of the biquadratic element all shape functions are zero except ψ_2, ψ_6, and ψ_3. Thus, the contribution of the traction on BD to $\{R^e\}$ becomes

$$\left\{\begin{array}{c} R^e_3 \\ R^e_4 \\ R^e_5 \\ R^e_6 \\ R^e_{11} \\ R^e_{12} \end{array}\right\}_{BD} = \int_{BD} \left\{\begin{array}{c} \psi_2 t_x \\ \psi_2 t_y \\ \psi_3 t_x \\ \psi_3 t_y \\ \psi_6 t_x \\ \psi_6 t_y \end{array}\right\} ds. \qquad (8.76)$$

In this example, both t_x and t_y are unknown on BD. Thus, there are unknown forces in the x-direction R^e_3, R^e_5, and R^e_{11}, and unknown forces in the y-direction R^e_4, R^e_6, and R^e_{12}, at nodes number 2, 3, and 6, respectively. The contribution of the unknown traction t_x and t_y at all other nodes is zero ($R^e_{2j-1} = R^e_{2j} = 0,\ j = 1,4,5,7,8,9$).
Summing the results of Equations (8.70), (8.72), (8.74), and (8.75), we replace the prescribed surface traction on AB, DC, and CA by equivalent weighted secondary

Table 8.2

Example 8.2, Prescribed and Unknown Primary Variables (u, v), and Weighted Secondary Variables (R_x, R_y)

Global node	local node	u	v	R_x	R_y	F_x	F_y
1	1	?	0	0	?	22.5	−16.25
2	5	?	0	−4.8	?	−6	−193
3	2	0	0	−2.4+?	?	−37.5	−96.25
4	8	?	?	−3.2	0.6	26	−77
5	9	?	?	0	0	−480	−1220
6	6	13.5	−2.25	?	?	−314	−597
7	4	0	?	−1.6+?	0.46	−17.5	−23.75
8	7	13.5	?	?	7.68	−266	−423
9	3	36	0	?	4.96+?	−127.5	−203.75

F_x and F_y are (forcing vectors equivalent to body forces) ×75, and "?" indicates unknown quantity.

variables at the element's nodes

$$\begin{aligned}
&R_1^e = 0,\ R_2^e = 0,\ R_3^e = -2.4,\ R_4^e = 0,\ R_5^e = 0,\ R_6^e = 4.96,\\
&R_7^e = -1.6,\ R_8^e = 0.46,\ R_9^e = -4.8,\ R_{10}^e = 0,\ R_{11}^e = 0,\ R_{12}^e = 0,\\
&R_{13}^e = 0,\ R_{14}^e = 7.68,\ R_{15}^e = -3.2,\ R_{16}^e = 0.6,\ R_{17}^e = 0,\ R_{18}^e = 0. \quad (8.77)
\end{aligned}$$

If the element is subjected to any virtual displacement field $(\bar{u}, \bar{v})$ constructed by linear superposition of the element shape functions, then the virtual work of the concentrated forces given by Equation (8.77) will be equal to the virtual work of the known distributed traction on the element boundary. We note that the values given in Equation (8.77) are not the total values of weighted secondary variables because, for some of the nodes, unknown surface traction may still contribute additional forces to those values. We recall that at a point where the primary variable is prescribed, the corresponding secondary variable is unknown. Thus, if one of the pair (u, R_x) is given the other is unknown. Similarly, if one of the pair (v, R_y) is given, the other is unknown.

The distributed body forces given by Equation (8.61) are substituted into Equation (8.47), which gives the forcing vector at the nodes. We evaluated the integrals by (3×3) Gauss–Legendre quadrature formula, and the results are shown in Table 8.2 for one biquadratic element. It also shows the local and global nodes, and the known and unknown primary (u, v) and weighted secondary variables (R_x, R_y). For a mesh of one element, there are 18 primary variables and 18 corresponding weighted secondary variables. As shown in the table, 18 of the 36 variables are known while the other 18 are unknown. Since we are using one element, we solve the system (8.50) for the 18 unknowns. Stiffness matrix $[K^e]$ is also evaluated by (3×3) Gauss–Legendre quadrature formula. The finite-element solution is summarized in

Table 8.3
Example 8.2, Finite Element Solution by One Biquadratic Element

Global node	local node	u	v	R_x	R_y	F_x	F_y
1	1	0	0	0	0	22.5	−16.25
2	5	0	0	−4.8	6.4	−6	−193
3	2	0	0	−3.12	4.34	−37.5	−96.25
4	8	0	0	−3.2	0.6	26	−77
5	9	5.625	−1.6875	0	0	−480	−1220
6	6	13.5	−2.25	7.04	9.72	−314	−597
7	4	0	0	−2.68	0.46	−17.5	−23.75
8	7	13.5	−2.25	10.56	7.68	−266	−423
9	3	36	0	12.2	8.8	−127.5	−203.75

F_x and F_y are (forcing vectors equivalent to body forces) $\times 75$

Table 8.3. The computed displacements at the element's nodes are exact. And since the trial solution is biquadratic in ξ and η, it is also biquadratic in x and y for the present orientation of the rectangular domain. Thus, the finite-element solution is exact at any point within the element. One can also verify that all weighted secondary variables are exact. We also note that $\sum R_x + \sum F_x = 0$, and $\sum R_y + \sum F_y = 0$, as it should be for a body in static equilibrium. Also, sum of moments about any point is zero.

Example 8.3. Stresses in a Strip with a Central Circular Hole

We want to determine stress concentration around a circular hole drilled in a rectangular plate loaded by uniform uniaxial tensile stress as shown in Figure 8.6. Howland [52] developed an analytic solution using successive approximations for the case of a long strip in the x-direction but finite width in the y-direction. The origin of the coordinate system is placed at the center of the circular hole. The rectangle height is $2b$, hence the strip occupies the domain from $y = -b$ to $y = b$. We use b as a reference length and solve the problem in nondimensional variables. The nondimensional hole radius is $\lambda = r_o/b$. We assume a strip of finite length in the x-direction, and impose symmetry conditions on the x- and y- axes. We solve the problem on the first quadrant of the coordinate system. The boundary conditions are

$$\begin{aligned} &\text{ABG- symmetry}: && v = 0,\ t_x = 0 \\ &\text{CD- symmetry}: && u = 0,\ t_y = 0 \\ &\text{AC- traction-free surface}: && t_x = 0,\ t_y = 0 \\ &\text{DEF- traction-free surface}: && t_x = 0,\ t_y = 0 \\ &\text{FG- prescribed traction}: && t_x = \sigma_o,\ t_y = 0. \end{aligned} \tag{8.78}$$

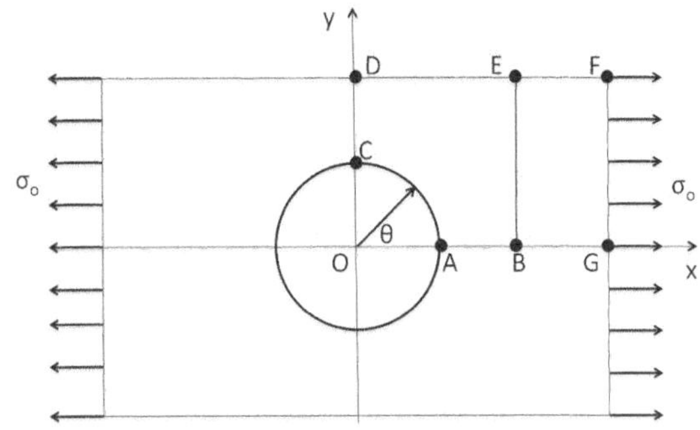

Figure 8.6 Howland problem.

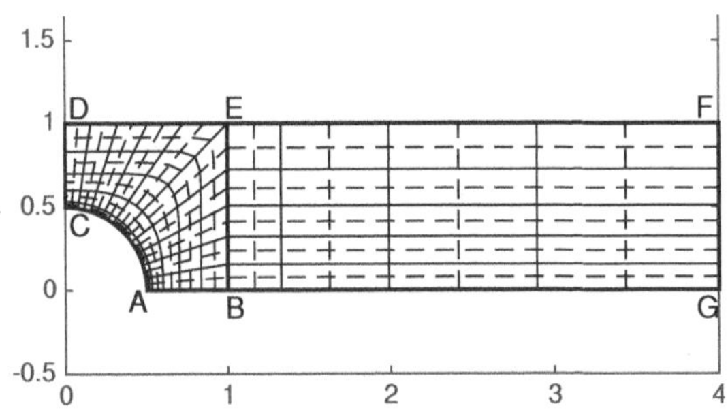

Figure 8.7 Howland problem, a coarse mesh of biquadratic elements.

A coarse mesh of quadrilateral biquadratic elements is shown in Figure 8.7 for $\lambda = 0.5$. The finite-element results presented here are obtained on a finer mesh of 1140 elements. The element matrices are evaluated by the (3×3) Gauss–Legendre quadrature formula. A thin plate of thickness $\ell_z = 0.01$ and a state of plane stress are assumed with properties $\nu = 0.3$ and $E = 210$ GPa. Uniform tensile stress of $\sigma_o = 1$ MPa is used.

For $b = 1$ m, the normal stress σ_{xx} on the line $x = 0$ and $0.5 \leq y \leq 1$ is shown in Figure 8.8a, and the hoop stress $\sigma_{\theta\theta}$ on the circular hole is shown in Figure 8.8b. The figures show excellent agreement between the finite-element solution and Howland analytic solution [52]. For this case of $r_o/b = 0.5$, the stress concentration factor is approximately 4.3. The displacements u on the x-axis and v on the y-axis are shown in Figures 8.9a and 8.9b, respectively. We recall that λ is the ratio of the hole diameter to the strip width. The normal stress σ_{xx} on the line $x = 0$ for $\lambda \leq y \leq 1$ is shown in Figure 8.10 for different values of λ. The finite-element solution is in agreement with Howland analytic solution [52]. A stress concentration factor of 3 is obtained if the hole radius is small in comparison with lateral dimensions of the plate.

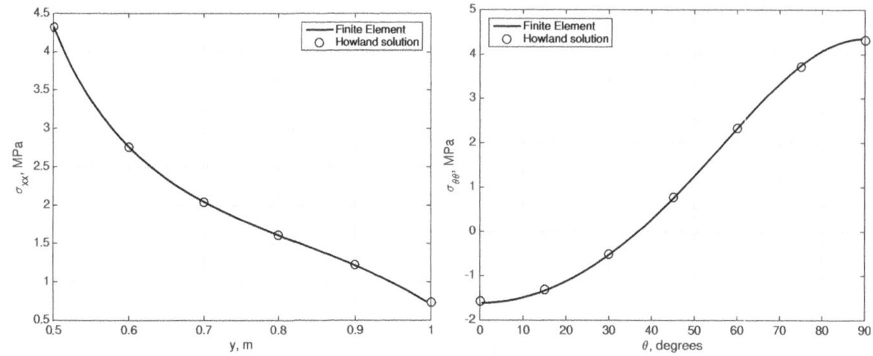

(a) (b)

Figure 8.8 Comparison of finite-element results to analytic solution [52]. (a) Stress σ_{xx} on line CD. (b) Hoop stress $\sigma_{\theta\theta}$ on circular arc AC. $b = 1$ m, $\lambda = 0.5$.

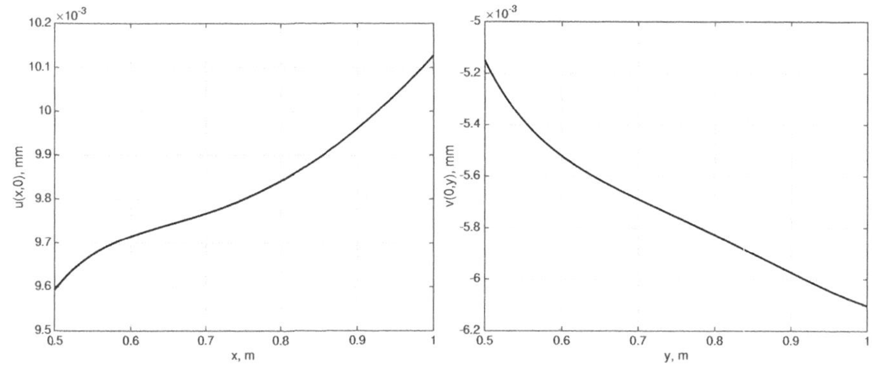

(a) (b)

Figure 8.9 (a) Displacement $u(x,0)$ on line AB. (b) Displacement $v(0,y)$ on line CD. $b = 1$ m, $\lambda = 0.5$.

Example 8.4. A Circular Cylinder Pressurized by an Eccentric Circular Conduit

A circular conduit is eccentrically drilled in a long circular cylinder as depicted in Figure 8.11. The inner cylindrical surface is subjected to uniform pressure $\sigma_{rr} = -\sigma_o$, while the outer cylindrical surface is free of traction. A state of plane strain is assumed, and symmetry conditions are imposed on a line passing through the circle centers. A mesh of quadrilateral biquadratic elements on half the domain is generated and depicted in Figure 8.11. The mesh is 180 elements and 777 global nodes, each

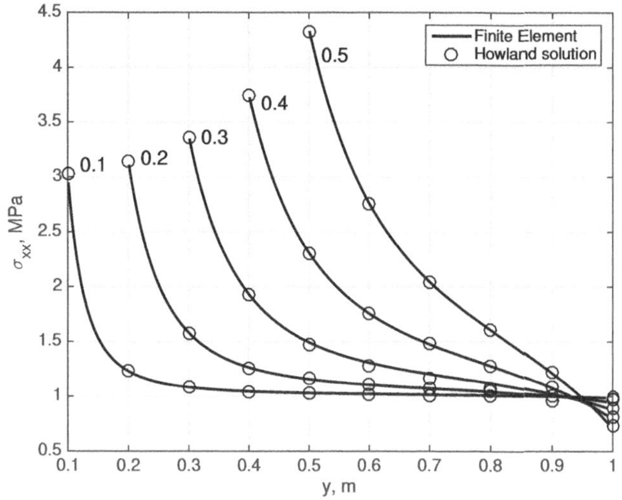

Figure 8.10 Comparison of finite-element results to analytic solution [52]. Stress σ_{xx} on line CD for hole-diameter to strip-width ratio $\lambda = (0.1, 0.2, 0.3, 0.4, 0.5)$.

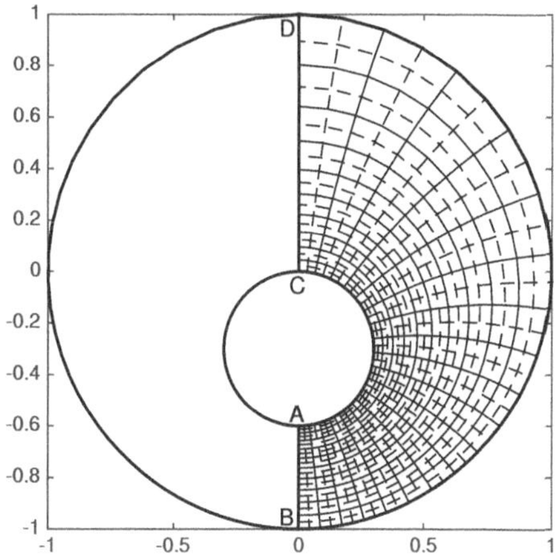

Figure 8.11 A circular cylinder pressurized by an eccentric circular conduit. A mesh of biquadratic elements on half domain.

element has 9 local nodes. The boundary conditions are

$$\begin{aligned}
\text{AB- symmetry}: &\quad u = 0,\ t_y = 0 \\
\text{CD- symmetry}: &\quad u = 0,\ t_y = 0 \\
\text{AC- prescribed traction}: &\quad \sigma_{rr} = -\sigma_o,\ \sigma_{r\theta} = 0 \\
\text{Cylinder (r=1)- traction-free surface}: &\quad \sigma_{rr} = 0,\ \sigma_{r\theta} = 0 \\
\text{Point B}: &\quad v = 0.
\end{aligned} \tag{8.79}$$

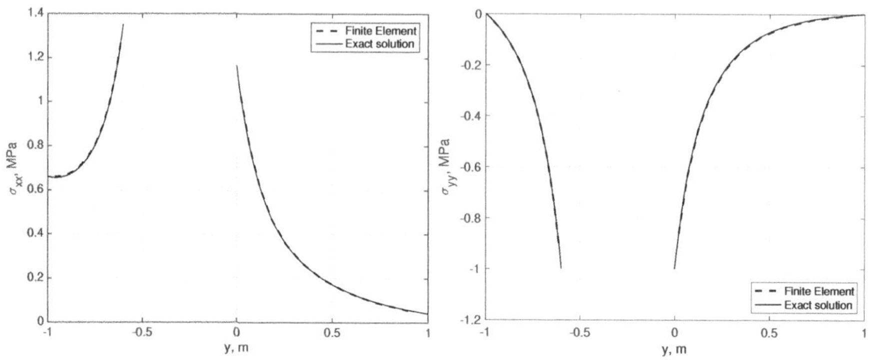

(a) (b)

Figure 8.12 Comparison of finite-element and exact analytic solution [55]. (a) Stress σ_{xx} on lines *AB* and *CD*. (b) Stress σ_{yy} on lines *AB* and *CD*.

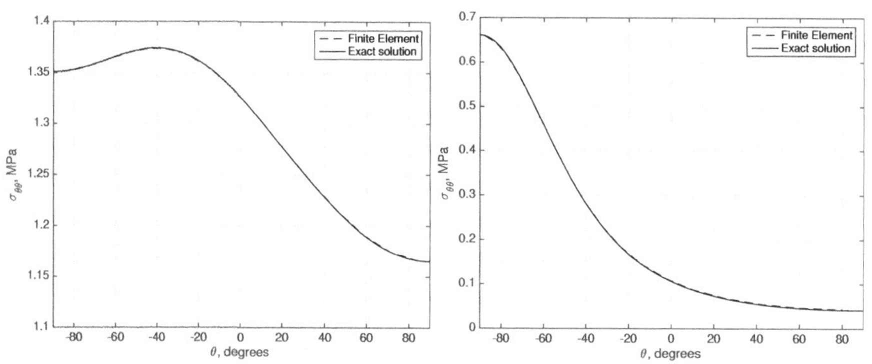

(a) (b)

Figure 8.13 Comparison of finite-element and exact analytic solution [55]. (a) Hoop stress $\sigma_{\theta\theta}$ on inner circular boundary *AC*. (b) Hoop stress $\sigma_{\theta\theta}$ on outer circular boundary *BD*.

The condition at point B is needed to prevent rigid body translation in the y-direction; any other point could have been selected. We compare the finite element solution to Jeffery's [55] analytic exact solution. Results are obtained for $\nu = 0.3$, $E = 210$ GPa, and $\sigma_o = 1$ MPa. The cylinder outer radius is 1 m, and the circular conduit radius is 0.3 m and center at $(x, y) = (0, -0.3)$. Stresses σ_{xx} and σ_{yy} on the y-axis are shown in Figures 8.12a and 8.12b, respectively. We note that σ_{xx} is tensile whereas σ_{yy} is compressive. The hoop stress $\sigma_{\theta\theta}$ is plotted in Figures 8.13a and 8.13b on the inner and outer circles, respectively. Figure 8.14 shows the displacement v on the y-axis.

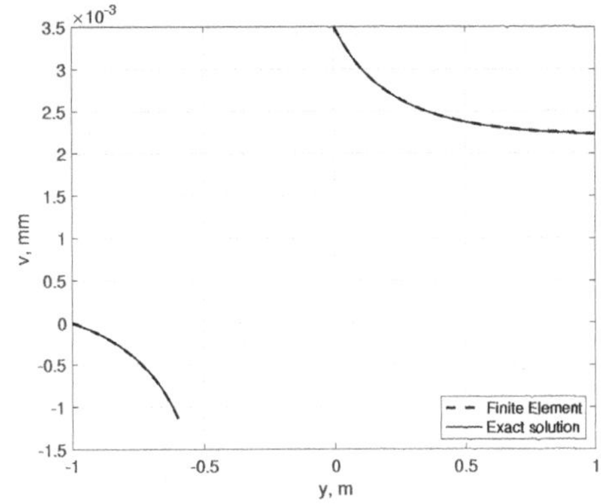

Figure 8.14 Comparison of finite-element and exact analytic solution [55]. Displacement v on lines BA and CD.

8.5 UNCOUPLED LINEAR THERMOELASTICITY

Large magnitude stresses develop in solids as a result of temperature variations with time or gradients in space. Stresses can also develop due to boundary constraints even with uniform temperature changes. Theory of thermal stresses is well presented in the book by Boley and Weiner [13] among other books. In this chapter, we formulate the finite-element method to compute thermal stresses in the simple case of plane uncoupled thermoelasticity. Inertia terms in the equation of motion and internal energy generation by time varying strains in the thermal energy equation are neglected. We limit the analysis here to isotropic linear elastic materials.

8.5.1 CONSTITUTIVE EQUATIONS

Love [66] remarked that constitutive equations for linear elastic materials were developed by Duhamel and Neumann (see also [106]),

$$\sigma_{ij} = C_{ijkl}\varepsilon_{kl} - \beta_{ij}(T - T_o) \tag{8.80}$$

where T is the current temperature and T_o is reference temperature at which the material is stress-free [13], and C_{ijkl} are material elastic coefficients; it is a fourth-order tensor. For isotropic materials, $\beta_{ij} = \beta\delta_{ij}$, and hence

$$\sigma_{ij} = \lambda\varepsilon_{kk}\delta_{ij} + 2\mu\varepsilon_{ij} - \beta\Delta T\delta_{ij} \tag{8.81}$$

where $\Delta T = T - T_o$ is the temperature change from the reference state. This equation can be inverted to obtain the strain tensor,

$$\varepsilon_{ij} = \frac{1+\nu}{E}\sigma_{ij} - \frac{\nu}{E}\sigma_{kk}\delta_{ij} + \alpha\Delta T\delta_{ij} \tag{8.82}$$

where

$$\alpha = \frac{\beta(1-2\nu)}{E}, \ \beta = \frac{\alpha E}{1-2\nu} \tag{8.83}$$

α is the coefficient of thermal expansion. With $\{\sigma\}$ and $\{\varepsilon\}$ as defined in Equations (8.5), we write Equation (8.81) in matrix form,

$$\{\sigma\} = [D]\{\varepsilon\} - \beta\Delta T \begin{Bmatrix} 1 \\ 1 \\ 1 \\ 0 \\ 0 \\ 0 \end{Bmatrix}. \tag{8.84}$$

The six components of strain $\{\varepsilon\}$ are defined in terms of actual displacements in Equation (8.5). Multiplying Equation (8.84) by $[D]^{-1}$, we get

$$\{\varepsilon\} = [D]^{-1}\{\sigma\} + \alpha\Delta T \begin{Bmatrix} 1 \\ 1 \\ 1 \\ 0 \\ 0 \\ 0 \end{Bmatrix}. \tag{8.85}$$

Thermal Strains

The second term in Equation (8.85) is called thermal strain,

$$\{\varepsilon_T\} = \alpha\Delta T \begin{Bmatrix} 1 \\ 1 \\ 1 \\ 0 \\ 0 \\ 0 \end{Bmatrix}. \tag{8.86}$$

For isotropic materials, thermal strains are normal strains. We rewrite Equation (8.85) as

$$\{\varepsilon\} = [D]^{-1}\{\sigma\} + \{\varepsilon_T\} \tag{8.87}$$

and hence we obtain the stress as

$$\{\sigma\} = [D][\{\varepsilon\} - \{\varepsilon_T\}]. \tag{8.88}$$

Plane Stress

Imposing the conditions of plane stress Equation (8.11) on Equation (8.87), we find

$$\{\sigma\} = [D][\{\varepsilon\} - \{\varepsilon_T\}] \tag{8.89}$$

where $[D]$ is now given by Equation (8.16), and

$$\{\varepsilon_T\} = \alpha\Delta T \begin{Bmatrix} 1 \\ 1 \\ 0 \end{Bmatrix}. \tag{8.90}$$

Plane Strain

Imposing the conditions of plane strain Equation (8.17) on Equation (8.88), we find

$$\{\sigma\} = [D][\{\varepsilon\} - \{\varepsilon_T\}] \tag{8.91}$$

where $[D]$ is now given by Equation (8.19), and

$$\{\varepsilon_T\} = (1+\nu)\alpha\Delta T \left\{ \begin{array}{c} 1 \\ 1 \\ 0 \end{array} \right\}. \tag{8.92}$$

8.5.2 ELEMENT EQUATIONS AND MATRICES, THERMAL LOADING

The principle of virtual displacements Equation (8.30) is independent of the constitutive equations, hence we can apply it to the case of thermal stresses as well. To derive the element equations, we repeat the analysis of Section 8.3, but instead of the stress-strain relation given by Equation (8.39), we use

$$\begin{aligned} \left\{ \begin{array}{c} \sigma^e_{xx}(x,y) \\ \sigma^e_{yy}(x,y) \\ \sigma^e_{xy}(x,y) \end{array} \right\} &= [D^e][\{\varepsilon^e\} - \{\varepsilon^e_T\}] \\ &= [D^e]\left\{[B^e]\{U^e\} - \{\varepsilon^e_T\}\right\}. \end{aligned} \tag{8.93}$$

Substituting Equations (8.93), (8.42), and (8.43) into the principle of virtual displacements for an element Equation (8.30), we get,

$$\begin{aligned} &\int_{A_e} \ell_z \{d\}^T [B^e]^T [D^e] \left\{[B^e]\{U^e\} - \{\varepsilon^e_T\}\right\} dA = \\ &\oint_{\Gamma_e} \ell_z \{d\}^T [\psi]^T \left\{ \begin{array}{c} t_x \\ t_y \end{array} \right\} ds + \int_{A_e} \ell_z \{d\}^T [\psi]^T \left\{ \begin{array}{c} f_x \\ f_y \end{array} \right\} dA. \end{aligned} \tag{8.94}$$

Factoring out $\{d\}^T$, we obtain

$$\begin{aligned} \{d\}^T \Big[&\int_{A_e} \ell_z [B^e]^T [D^e][B^e]\{U^e\} dA - \int_{A_e} \ell_z [B^e]^T [D^e]\{\varepsilon^e_T\} dA \\ &- \oint_{\Gamma_e} \ell_z [\psi]^T \left\{ \begin{array}{c} t_x \\ t_y \end{array} \right\} ds - \int_{A_e} \ell_z [\psi]^T \left\{ \begin{array}{c} f_x \\ f_y \end{array} \right\} dA \Big] = 0. \end{aligned} \tag{8.95}$$

We define thermal loading vector

$$[F^e_T] = \int_{A_e} \ell_z [B^e]^T [D^e]\{\varepsilon^e_T\} dA. \tag{8.96}$$

With the definitions of element stiffness matrix, forcing vector, and weighted secondary variables vector given by Equations (8.46), (8.47), and (8.48), respectively, we rewrite Equation (8.95),

$$\{d\}^T \{[K^e]\{U^e\} - \{F^e\} - \{F^e_T\} - \{R^e\}\} = 0 \tag{8.97}$$

Since $\{d\}^T$ is an arbitrary array of dimensions (1, 2n), the multiplying bracket must be zero, and we obtain the element equations,

$$[K^e]\{U^e\} = \{F^e\} + \{F_T^e\} + \{R^e\} \tag{8.98}$$

We can compute thermal loading if the temperature fields $T(x,y)$ and $T_o(x,y)$ are known. Let's assume that $T(x,y)$ is determined by solving a heat conduction problem, and that the same isoparametric element is used for both the temperature and stress analysis problems. The temperature field can be written as

$$\begin{aligned} T^e(x,y) &= \{\ \psi_1 \quad \psi_2 \quad . \quad \psi_j \quad . \quad \psi_n\ \} \left\{ \begin{array}{c} T_1^e \\ T_2^e \\ . \\ T_j^e \\ . \\ T_n^e \end{array} \right\} \\ &= \{\psi\}\{T^e\} \end{aligned} \tag{8.99}$$

where T_j^e are temperature at the element's nodes, and ψ_j are the shape functions. Similarly, we write the reference temperature,

$$\begin{aligned} T_o^e(x,y) &= \{\ \psi_1 \quad \psi_2 \quad . \quad \psi_j \quad . \quad \psi_n\ \} \left\{ \begin{array}{c} T_{o1}^e \\ T_{o2}^e \\ . \\ T_{oj}^e \\ . \\ T_{on}^e \end{array} \right\} \\ &= \{\psi\}\{T_o^e\} \end{aligned} \tag{8.100}$$

where T_{oj}^e are the reference temperature at the element's nodes; it is usually uniform. The change in temperature needed for computation of thermal loading becomes,

$$\Delta T^e(x,y) = \{\psi\}\Big\{\{T^e\} - \{T_o^e\}\Big\} \tag{8.101}$$

$$[F_T^e] = \int_{A_e} \ell_z \alpha_1 [B^e]^T [D^e] \left\{ \begin{array}{c} 1 \\ 1 \\ 0 \end{array} \right\} \{\psi\}\Big\{\{T^e\} - \{T_o^e\}\Big\} dA \tag{8.102}$$

where $\alpha_1 = \alpha$ for plane stress, and $\alpha_1 = (1+\nu)\alpha$ for plane strain.

Example 8.5. Thermal Stresses Around an Insulated Circular Hole in an Infinite Plate subjected to steady heat flow

We consider a large plate thermally loaded by uniform heat flux in the negative y-direction. The temperature distribution in the plate is given by $T = (\tilde{q}/k)y$, where $\tilde{q}$

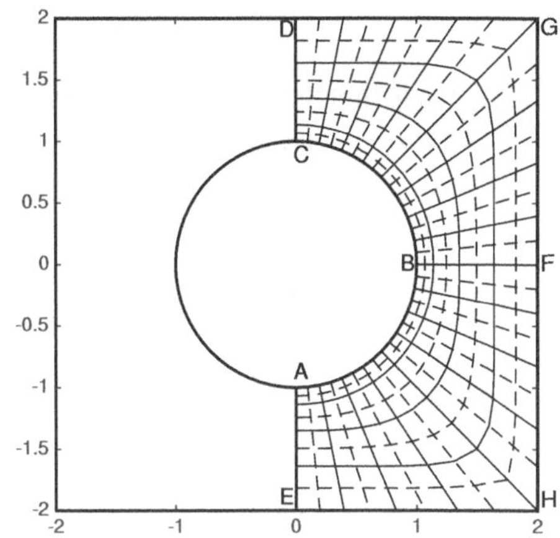

Figure 8.15 An insulated circular hole in a square plate subjected to steady heat flow. A mesh of biquadratic elements on half domain.

is the heat flux, and k is thermal conductivity. Now, a circular hole of radius a and center $(0,0)$ is drilled in the plate. The inner surface of the hole is insulated, hence the boundary condition is $\partial T/\partial r = 0$ at $r = a$. The temperature in the plate with the hole becomes,

$$T = \frac{\tilde{q}a}{k}\left(\frac{r}{a} + \frac{a}{r}\right) sin\theta. \tag{8.103}$$

We want to find the stresses around the circular hole due to temperature gradient. Florence and Goodier [37] (see also Sadd [106]) provided an analytic solution to this problem,

$$\begin{aligned} \sigma_{rr} &= -\frac{\tilde{q}ac}{2k}\left(\frac{a}{r} - \frac{a^3}{r^3}\right) sin\theta \\ \sigma_{\theta\theta} &= -\frac{\tilde{q}ac}{2k}\left(\frac{a}{r} + \frac{a^3}{r^3}\right) sin\theta \\ \sigma_{r\theta} &= \frac{\tilde{q}ac}{2k}\left(\frac{a}{r} - \frac{a^3}{r^3}\right) cos\theta \end{aligned} \tag{8.104}$$

$c = E\alpha$ for plane stress. We solve the problem by the finite-element method and compare with the exact solution. The finite-element solution is obtained on a square domain with a hole at its center as shown in Figure 8.15. A mesh of biquadratic elements (64 elements and 297 nodes) is generated on half the domain, and symmetry conditions are imposed on AE and CD. The inner surface is free of traction. The boundary conditions are

$$\begin{aligned}
\text{AE- symmetry}: &\quad u=0,\ t_y=0\\
\text{CD- symmetry}: &\quad u=0,\ t_y=0\\
\text{ABC- traction-free surface}: &\quad t_x=0,\ t_y=0\\
\text{EHFGD- prescribed traction}: &\quad t_x=\hat{t}_x,\ t_y=\hat{t}_y\\
\text{Point B}: &\quad v=0.
\end{aligned} \tag{8.105}$$

The exact analytic stresses Equation (8.104) are used to prescribe traction components $\hat{t}_x$ and $\hat{t}_y$ on the outer boundary. Stress components in cylindrical and Cartesian coordinates are related by (e.g., Sadd [106])

$$\begin{aligned}
\sigma_{rr} &= \sigma_{xx}cos^2\theta+\sigma_{yy}sin^2\theta+2\sigma_{xy}sin\theta cos\theta\\
\sigma_{\theta\theta} &= \sigma_{xx}sin^2\theta+\sigma_{yy}cos^2\theta-2\sigma_{xy}sin\theta cos\theta\\
\sigma_{r\theta} &= (\sigma_{yy}-\sigma_{xx})\,sin\theta cos\theta+\sigma_{xy}\left(cos^2\theta-sin^2\theta\right)
\end{aligned} \tag{8.106}$$

$$\begin{aligned}
\sigma_{xx} &= \sigma_{rr}cos^2\theta+\sigma_{\theta\theta}sin^2\theta-2\sigma_{r\theta}sin\theta cos\theta\\
\sigma_{yy} &= \sigma_{rr}sin^2\theta+\sigma_{\theta\theta}cos^2\theta+2\sigma_{r\theta}sin\theta cos\theta\\
\sigma_{xy} &= (\sigma_{rr}-\sigma_{\theta\theta})\,sin\theta cos\theta+\sigma_{r\theta}\left(cos^2\theta-sin^2\theta\right).
\end{aligned} \tag{8.107}$$

A state of plane stress is assumed with properties: $E=200$ GPa, $\alpha=12\times10^{-6}$ °C^{-1}, $k=44$ Wm^{-1}°C^{-1}, $\tilde{q}=2200$ Wm^{-2}, $a=1$ m, $\nu=0.25$, and $\ell_z=10$ mm. Far from the insulated hole, the temperature gradient is $\partial T/\partial y=\tilde{q}/k=50$ °C/m. The temperature at the points A, B, and C are -100, 0, 100°C, respectively. Because the hole is insulated, the temperature difference between the cold point A and the hot point C is 200 °C. We define nondimensional hoop stress $\sigma^*_{\theta\theta}$ and displacements u^* and v^* by

$$\sigma^*_{\theta\theta}=\frac{\sigma_{\theta\theta}}{E\alpha\tilde{q}a/k},\quad u^*=\frac{u}{\alpha\tilde{q}a^2/k},\quad v^*=\frac{v}{\alpha\tilde{q}a^2/k}. \tag{8.108}$$

The nondimensional variables are independent of the specific values used for E, α, $\tilde{q}$, and k. The finite-element solution is in excellent agreement with the exact analytic solution as shown in Figures 8.16a and 8.16b.

8.6 PROBLEMS

(1) A rectangular plate ($a=0.4$ m, $b=0.2$ m, thickness $\ell_z=10$ mm, mechanical properties $E=70$ GPa, and $\nu=0.30$) $ABCD$ is fixed along sides AD and CD as shown in Figure 8.17. The plate is subjected to surface traction $t_x=\sigma_o(1-x/a)$ on side AB and uniform traction $t_x=\sigma_o$ on side BC, where $\sigma_o=0.5$ MPa. For simplicity, the plate is modeled by one bilinear 4-node element. A state of plane stress is assumed, and the stiffness matrix is given in Table 8.5. Calculate:

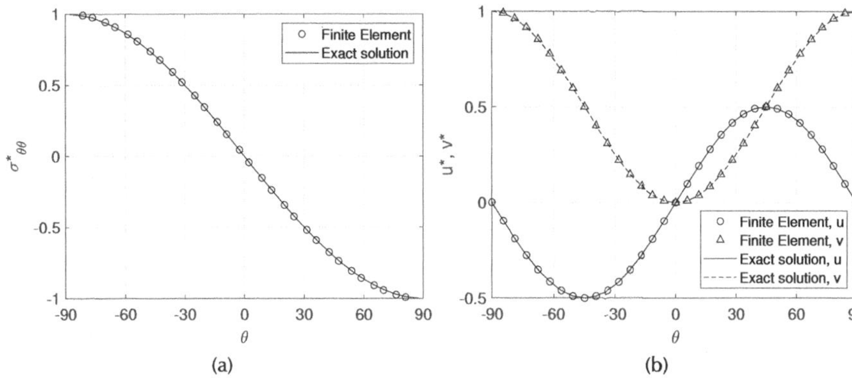

Figure 8.16 Comparison of finite-element and exact analytic solutions. (a) Hoop stress on the circular hole. (b) Displacements u and v on the hole.

Table 8.4
Material Properties

Property	Units	Steel	Aluminum	Copper	Gold	Silver	Bronze
ρ	kg/m^3	7860	2700	8900	19300	10524	8800
C_p	J/kg°C	470	963	375	129	234	420
k	W/m°C	44	204	400	317	419	52
E	GPa	200	73	120	79	0	26
ν		0.25	0.35	0.36	0	0	0
$\alpha \times 10^6$	$°C^{-1}$	12	23	17	9.5	17	0

ρ Density, Cp Specific heat, k Thermal conductivity, E modulus of elasticity, ν Poisson's ratio, α Coefficient of thermal expansion.

(a) the displacement (u, v) at point B.

(b) the strain component ε_{xx} at the center of the plate.

(2) Assuming a state of plane stress with properties E and ν and uniform body force (f_x, f_y), derive the stiffness matrix $[K^e]$ and forcing vector $\{F^e\}$ for the bilinear triangular element shown in Figure 8.18 for $a = b$. This is known as a constant-strain triangular (CST) element. Evaluate integrals analytically as numerical integration is unnecessary for this problem.

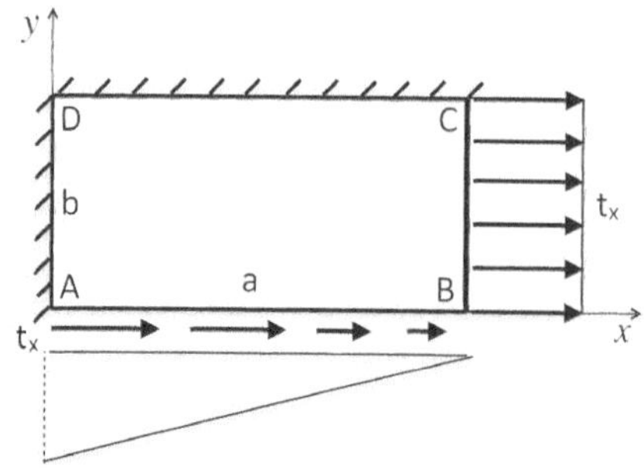

Figure 8.17 Plane elasticity problem on a rectangular plate.

Table 8.5
Stiffness Matrix: $10^{-8}[K]$

3.0769	1.2500	−0.3846	−0.0962	−1.1538	0.0962	−1.5385	−1.2500
1.2500	5.5769	0.0962	2.1154	−0.0962	−4.9038	−1.2500	−2.7885
−0.3846	0.0962	3.0769	−1.2500	−1.5385	1.2500	−1.1538	−0.0962
−0.0962	2.1154	−1.2500	5.5769	1.2500	−2.7885	0.0962	−4.9038
−1.1538	−0.0962	−1.5385	1.2500	3.0769	−1.2500	−0.3846	0.0962
0.0962	−4.9038	1.2500	−2.7885	−1.2500	5.5769	−0.0962	2.1154
−1.5385	−1.2500	−1.1538	0.0962	−0.3846	−0.0962	3.0769	1.2500
−1.2500	−2.7885	−0.0962	−4.9038	0.0962	2.1154	1.2500	5.5769

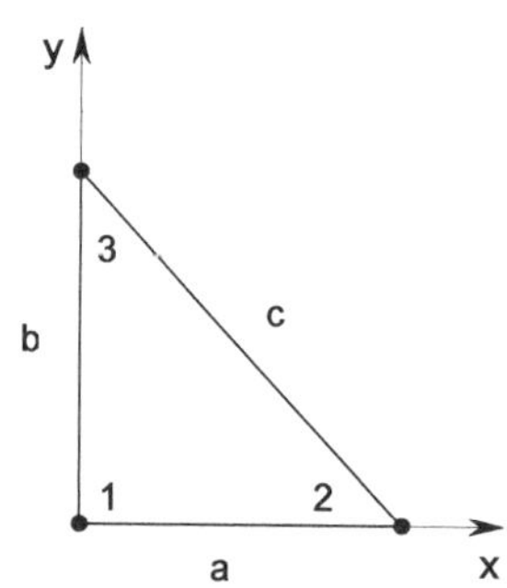

Figure 8.18 Bilinear triangular element.

(3) Consider the equilibrium of a thin elastic plate in a state of plane stress. The equations of equilibrium are

$$\begin{aligned}\frac{\partial \sigma_{xx}}{\partial x}+\frac{\partial \sigma_{yx}}{\partial y}+f_x &= 0\\ \frac{\partial \sigma_{xy}}{\partial x}+\frac{\partial \sigma_{yy}}{\partial y}+f_y &= 0.\end{aligned} \qquad (8.109)$$

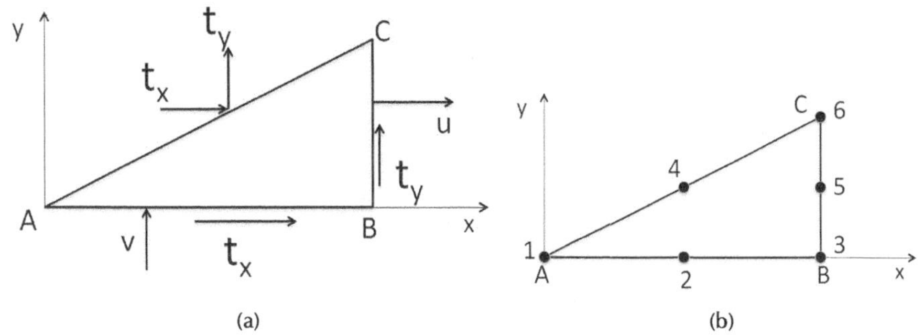

Figure 8.19 (a) Prescribed boundary conditions for triangular plate. (b) A biquadratic triangular element, and 6 global nodes.

Assume linearly elastic isotropic material with constants E and ν. If the body force is given by

$$\begin{aligned} f_x &= \frac{\gamma E}{a(1-\nu^2)}\left\{(1+\nu)\left[\frac{1}{2}-2\left(\frac{x}{a}\right)\left(\frac{y}{b}\right)\right]\right. \\ &\quad \left. -(1-\nu)\left(\frac{a}{b}\right)^2\left(\frac{x}{a}\right)^2-2\left(\frac{y}{b}\right)^2\right\} \\ f_y &= -\frac{\gamma E}{b(1-\nu^2)}\left\{(1+\nu)\left[\frac{1}{2}+2\left(\frac{x}{a}\right)\left(\frac{y}{b}\right)\right]\right. \\ &\quad \left. +(1-\nu)\left(\frac{b}{a}\right)^2\left(\frac{y}{b}\right)^2+2\left(\frac{x}{a}\right)^2\right\} \end{aligned} \tag{8.110}$$

verify that the displacement field

$$\begin{aligned} u(x,y) &= \gamma a\left(\frac{x}{a}\right)\left(\frac{y}{b}\right)\left[\left(\frac{x}{a}\right)\left(\frac{y}{b}\right)+1\right] \\ v(x,y) &= \gamma b\left(\frac{x}{a}\right)\left(\frac{y}{b}\right)\left[\left(\frac{x}{a}\right)\left(\frac{y}{b}\right)-1\right] \end{aligned} \tag{8.111}$$

is an exact solution to the equations of equilibrium. γ, a, and b are arbitrary constants.

(4) A state of plane stress is to be analyzed in a triangular plate shown in Figure 8.19a. The exact solution given in Problem (3) is used to prescribe boundary conditions on the plate sides. On side AB, the vertical displacement v and shear stress t_x are prescribed. On side BC, the horizontal displacement u and shear stress t_y are prescribed. On side AC, the traction (t_x, t_y) are prescribed.
The triangular plate is modeled by one biquadratic isoparametric element as shown in Figure 8.19b, where the coordinates of B and C are (2,0) and (2,1), respectively. Nodes 2, 4, and 5 are in the middle of the respected side. For the nominal data: $a=2$, $b=1$, $\gamma=9$, $E=1$, $\nu=0.25$, determine:

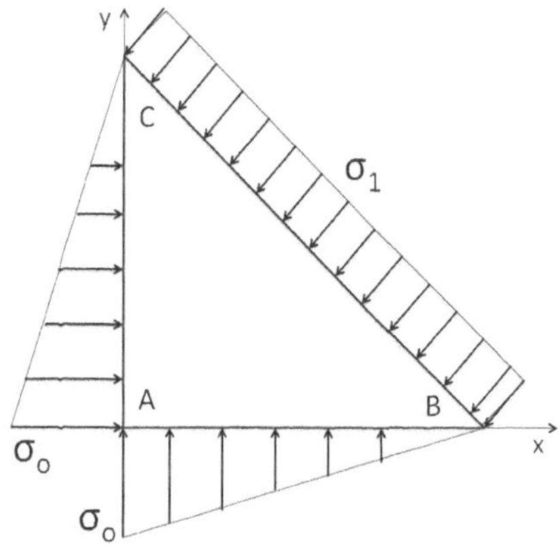

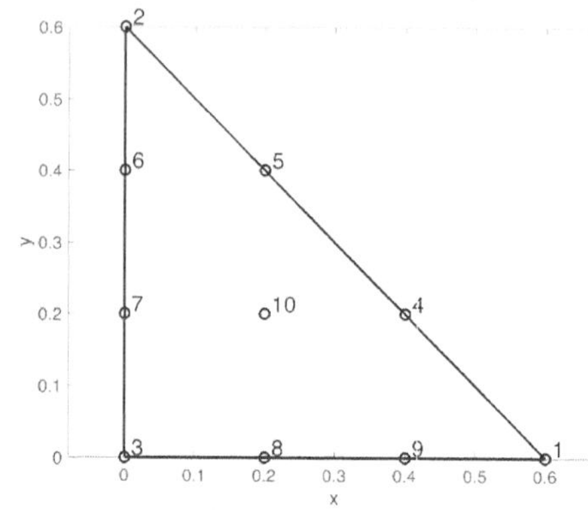

Figure 8.20 (a) Triangular plate loaded by normal traction. (b) A bicubic triangular element, and 10 global nodes.

(a) The element stiffness matrix $[K^e]$.

(b) The element forcing vector $\{F^e\}$.

(c) The displacement (u,v) and stresses $(\sigma_{xx},\sigma_{yy},\sigma_{xy})$ at the point $(x,y) = (1.5,0.25)$, and compare those values with the exact solution. Explain any discrepancy between exact and finite-element solutions.

(5) A steel triangular plate (thickness $\ell_z = 0.02$ m, $E = 200$ GPa, $\nu = 0.25$) is in static equilibrium under the action of the applied normal traction as shown in Figure 8.20a. Neglecting body forces and assuming a state of plane stress, model the plate by one bicubic triangular element as shown in Figure 8.20b. Take $AB = AC = 0.6$ m, $\sigma_o = 4$ MPa, and $\sigma_1 = 2$ MPa.

(a) Replace the distributed traction on side AB by equivalent forces at the element's nodes. Provide the array of equivalent forces.

(b) Replace the distributed traction on side BC by equivalent forces at the element's nodes. Provide the array of equivalent forces.

(c) Compute displacements (u,v) at all nodes in units of μm.

(d) Compute stresses σ_{xx}, σ_{yy}, and σ_{xy} at the centroid of the plate in units of MPa.

(6) Consider a cantilever rectangular plate ($a = 200$ mm, $b = 40$ mm, $\ell_z = 10$ mm) with mechanical properties $E = 200$ GPa, and $\nu = 0.25$. The lower edge is subjected to uniform vertical traction $t_y = 1$ MPa as shown in Figure 8.21. Assume plane stress conditions.

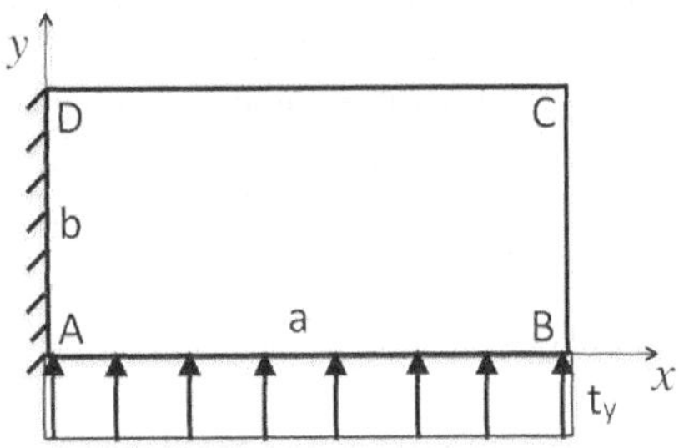

Figure 8.21 Cantilever plate.

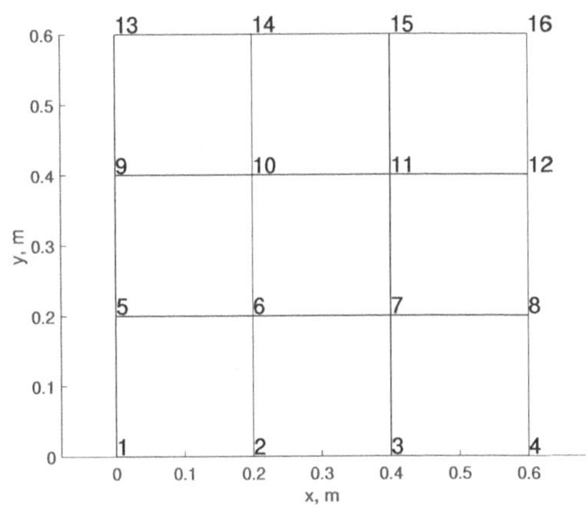

Figure 8.22 One bicubic square element.

(a) Plot the vertical displacement component v on the plate central plane and compare with the deflection predicted by the Euler–Bernoulli beam theory. Report results for bilinear and biquadratic rectangular elements. Make sure that the finite-element results are grid independent, or at least refine the mesh as much as your computer resources allow.

(b) Choose a vertical line near $x = 100$ mm and plot the stress components σ_{xx} and σ_{xy} on that line.

(c) Compare the results of part (b) with "mechanics of materials" solution or the Euler–Bernoulli beam theory.

(7) Repeat Problem (6) for a plate with dimensions ($a = 200$ mm, $b = 80$ mm, $\ell_z = 10$ mm).

(8) (a) We want to determine the steady temperature distribution in an Aluminum square plate of side $a = 0.6$ m and uniform thickness $\ell_z = 0.01$m as shown in Figure 8.22. The left and bottom sides are insulated. The right side

$x = 0.6$ m is at 0 °C, while the temperature on the top side $y = 0.6$ m varies linearly with x, $T(x, 0.6) = 100(1 - x/a)$ °C. No energy is generated within the plate. The temperature field is assumed to be two-dimensional. The thermal conductivity of plate is $k = 204$ W/m°C. Use one bicubic element (16 nodes), and evaluate the stiffness matrix by the (4×4) Gauss–Legendre integration rule.

(a.i) Determine the unknown nodal temperatures.

(a.ii) Determine the weighted secondary variables R_j at the boundary nodes and verify conservation of energy.

(b) We want to analyze the thermal stresses for the temperature distribution found in Part (a). The plate is free of stresses at the uniform temperature $T_o = 20$ °C. The left and bottom edges are fixed (displacements $u = 0$, and $v = 0$) while the other two edges are free of traction ($t_x = 0$, and $t_y = 0$). We assume a state of plane stress, and model the plate by one bicubic element. Evaluate the stiffness matrix and thermal loading vector by the (4×4) Gauss–Legendre integration rule. Given material properties $E = 73$ GPa, $\alpha = 23 \times 10^{-6}$ °C^{-1}, and $\nu = 0.35$, determine:

(b.i) Displacements (u, v) at the center $(x, y) = (0.3, 0.3)$m in units of mm.

(b.ii) Stresses $(\sigma_{xx}, \sigma_{yy}, \sigma_{xy})$ at the center $(x, y) = (0.3, 0.3)$m in units of Pa.

(c) The nodal forces (weighted secondary variables R_j) in units of N at the nodes on the fixed sides, and verify that the plate is in equilibrium.

(9) We wish to use the finite-element method to analyze linear elastic deformation of the plate shown in Figure 8.23. The plate boundary is denoted by $ABCDEFA$. Given the dimensions $AB = BC = CD = DE = 100$ mm, plate thickness $\ell_z = 10$ mm, and mechanical properties $E = 200$ GPa, and $\nu = 0.25$. Side AB is fixed, and side EF is subjected to a piecewise linear traction t_x as shown with maximum value of 2 MPa at the midpoint of EF, and no traction in the y-direction ($t_y = 0$). The remainder of the boundary is free of traction. Assume plane stress conditions, and divide the plate into three biquadratic elements (9-node element). Determine:

(a) The stiffness matrix $[K^e]$ of element # 3. Use (3×3) Gauss–Legendre integration rule.

(b) The array of secondary variables $\{R^e\}$ of element # 3.

(c) The stress components σ_{xx}, σ_{yy}, and σ_{xy} in element # 3 at the Barlow point nearest to corner C. For the biquadratic element, the Barlow points are the base points of the (2×2) Gauss–Legendre quadrature formula.

(10) We wish to use the finite-element method to analyze deformation and stresses in a culvert modeled as an isotropic linear elastic material with Poisson's ratio $\nu = 0.15$ and modulus of elasticity $E = 30$ GPa. The cross-section of the

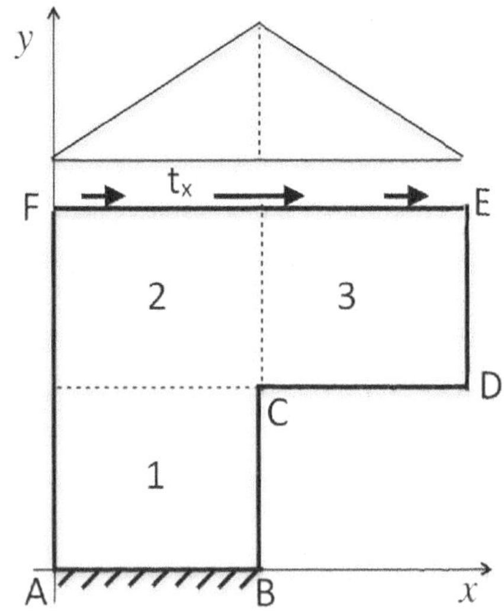

Figure 8.23 Angle-shape domain divided into three biquadratic elements. Piece-wise linear shear applied on side FE.

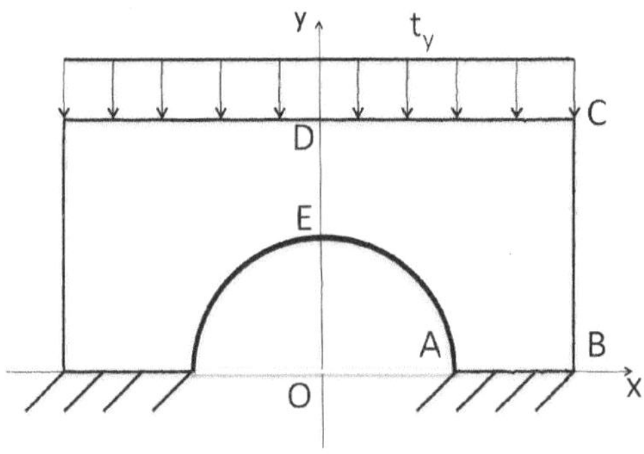

Figure 8.24 Culvert.

culvert is shown in Figure 8.24 (AE is semicircle of radius $OA = 2.5$ m, $OB = 4$m, and $BC = 4$m). The top boundary is loaded by uniform surface traction $t_y = -1$ MPa. We assume a state of plane strain. We also assume symmetry about the $yz-$plane, and solve the problem on the domain $ABCDEA$. The boundary conditions are: (1) on AB essential boundary conditions are applied with displacements $u = 0$, and $v = 0$, (2) on BC and EA natural boundary conditions are applied with surface traction $t_x = 0$ and $t_y = 0$, (3) on CD natural boundary conditions are applied with surface traction $t_x = 0$ and $t_y = -1$ MPa, and (4) on DE symmetry conditions are used where $u = 0$, and $t_y = 0$.
A MATLAB code is provided that can be used to generate a 2D mesh of quadrilateral biquadratic (or bilinear) elements. It also computes the shape

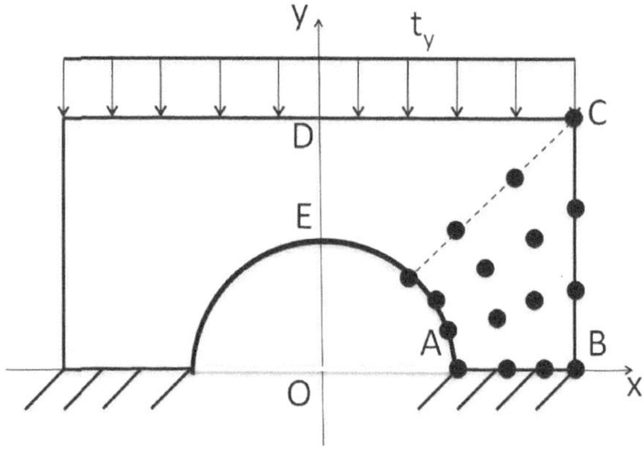

Figure 8.25 *ABCDEA* is modeled by two bicubic elements. The nodes of one of the elements are shown.

functions (ψ_i) and their first derivatives with respect to the natural coordinates ($\partial\psi_i/\partial\xi, \partial\psi_i/\partial\eta$). Use this code to generate a grid of 4 isoparametric biquadratic elements ($Q9$). The relevant mesh parameters in MATLAB code are $nxe = 2; nye = 2;\ sr = 1.2; sn1 = 2; sn2 = 12; snr = 4$. Evaluate the element matrices by using Gauss–Legendre (3×3) quadrature rule, the base points and corresponding weight factors are also provided by the MATLAB code.
Provide:

(a) The displacements (in millimeters) at C, D, and E.

(b) The stress tensor at the Barlow point nearest to point E. For the biquadratic element, the Barlow points are the base points of the (2×2) Gauss–Legendre quadrature formula.

(11) We reconsider the culvert of Problem (10), and use a mesh of two bicubic elements. The 16 nodes of one of the elements are shown in Figure 8.25. The boundary nodes are equally spaced on the circular arc AE, and on lines AB, BC, CD, and DE. The interior nodes are equally spaced on lines connecting nodes on AE to the outer boundary. Neglect body forces, and evaluate the element matrices by using Gauss–Legendre (4×4) quadrature rule.

(a) Specify and classify boundary conditions on $ABCDEA$.

(b) Find the displacements (in millimeters) at C, D, and E.

(c) Plot σ_{xx} in MPa as a function of y on line ED.

(12) A circular hole of radius b is drilled at the center of a thin rectangular plate of height $4b$ and length $12b$. A tensile stress t_x is applied to edges IK and JH as shown in Figure 8.26 while the other edges are unloaded and free to deform. A state of plane stress is assumed. Assuming symmetry about the x- and y-axes, we solve for the displacements $u(x,y)$, and $v(x,y)$ in one quadrant; say the second

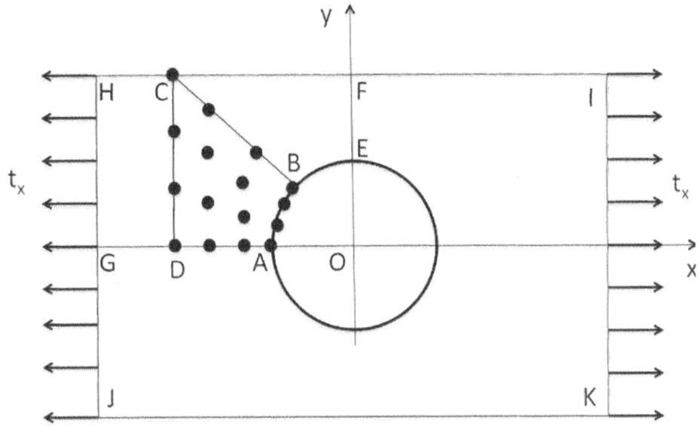

Figure 8.26 A strip with a circular hole under uniaxial tension.

quadrant as shown in Figure 8.26. The domain is divided into three bicubic elements labeled *ABCD, BEFC*, and *DCHG*. Each element has 16 local nodes, and the total number of global nodes is 40. The coordinates of points *B*, *C*, *D*, *G*, and *H* are $(-b/\sqrt{2}, b/\sqrt{2})$, $(-2b, 2b)$, $(-2b, 0)$, $(-6b, 0)$, and $(-6b, 2b)$, respectively. The nodes of element *ABCD* are shown in the figure; the nodes are equally spaced on *AB* and on element boundary. Given $b = 25$ mm, $E = 73$ GPa, $\nu = 0.33$, and $t_x = 1$ MPa. The plate thickness is uniform.

(a) Specify and classify (essential, natural, mixed) boundary conditions on *GDA*, *ABE*, *HCF*, *GH*, and *EF*.

(b) Evaluate the element matrices by the (4×4) Gauss–Legendre quadrature formula.

(c) Report the values of global weighted secondary variables R_j at the 4 nodes on *GH* and *EF*.

(d) Report the values of displacements at the nodes of element *BEFC*.

(e) Report the values of the stresses σ_{xx}, σ_{yy}, and σ_{xy} at the Barlow points of element *BEFC*. For the bicubic element, the Barlow points are the base points of the (3×3) Gauss–Legendre quadrature formula.

(13) Hot liquid flows steadily in a circular conduit drilled in a long column of rectangular cross-section *JKIH* as shown in Figure 8.27. The conduit radius is b and the rectangular height is $4b$ and length $6b$. We assume that the temperature of the circular conduit surface to be uniform at 100°C and that the outer surface of the column also to be at uniform temperature of 0°C. We want to use the finite-element method to analyze the temperature field and thermal stresses in the column. The state of stress in the column is assumed to be plane strain. Assuming symmetry about the *x*- and *y*- axes, we solve for the temperature $T(x, y)$ and displacements $u(x, y)$, and $v(x, y)$ in the second quadrant as shown in

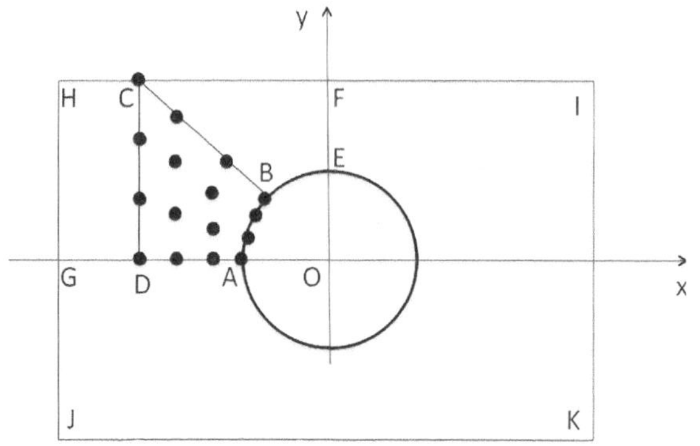

Figure 8.27 Circular conduit drilled in a rectangular column.

Figure 8.27. The domain is divided into three bicubic elements labeled *ABCD*, *BEFC*, and *DCHG*. Each element has 16 local nodes, and the total number of global nodes is 40. The coordinates of points *B*, *C*, *D*, *G*, and *H* are $(-b/\sqrt{2}, b/\sqrt{2})$, $(-2b, 2b)$, $(-2b, 0)$, $(-3b, 0)$, and $(-3b, 2b)$, respectively. The nodes of element *ABCD* are shown in the figure; the nodes are equally spaced on *AB* and on the element boundary. Take $b = 0.2$ m, $E = 73$ GPa, $\nu = 0.33$, thermal conductivity $k = 204$ W m^{-1} °C^{-1}, and coefficient of thermal expansion $\alpha = 23.2 \times 10^{-6}$ °C^{-1}. The reference temperature at which there is no stresses in the column is 20°C.

(a) Specify and classify (essential, natural, mixed) boundary conditions on *GDA*, *ABE*, *HCF*, *GH*, and *EF*.

(b) Prove that the rate of heat flow to the environment per unit length of the column is equal to the sum of weighted secondary variables on the outer boundary of the rectangle. Compute that rate. Ignore the change in the liquid temperature along conduit length.

(c) Report the values of the thermal loading vector for element *BEFC*.

(d) Report the values of displacements at the nodes of element *BEFC* (in units of mm).

(e) Report the values of the stresses σ_{xx}, σ_{yy}, and σ_{xy} (in units of MPa) at the Barlow points of element *BEFC*. For the bicubic element, the Barlow points are the base points of the (3×3) Gauss–Legendre quadrature formula.

(14) We want to determine thermal stresses around an insulated circular hole drilled in an infinite plate subjected to uniform heat flux $\vec{q} = -\tilde{q}\hat{j}$ W/m^2, and compare the finite-element results with the exact analytic solution as a way to verify the FEM computations. For a state of plane stress and polar coordinates (r, θ)

with origin at the center of circular hole (*It is important to remember where the origin is*), the exact solutions for temperature and stresses [106] are given by Equations (8.103) and (8.104), respectively.
We solve this thermoelastic problem by the finite-element method on the domain bounded by the circular hole and an outer eccentric circle as shown in Figure 8.11, and specify the plate temperature by Equation (8.103). The exact solution Equation (8.104) is used to prescribe traction on the circular boundaries.
A MATLAB code (*eccentric_cylinders_Mesh.m*) for mesh generation is provided. It generates a mesh of biquadratic quadrilateral elements such as that shown in Figure 8.11, and provides the connectivity matrix. The number of elements is $nxe = 12$ in the radial direction and $nye = 32$ in the circumferential direction.

Develop a finite-element code for solving this problem and obtain results for the following data (steel plate): hole radius $a = r_1 = 0.5$ m, outer domain radius $r_2 = 1$ m, distance between centers 0.3 m, thermal conductivity $k = 44$ W m^{-1} °C^{-1}, Poisson's ratio $\nu = 0.25$, elastic modules $E = 200$ GPa, coefficient of linear thermal expansion $\alpha = 12 \times 10^{-6}$ °C^{-1}, and heat flux $\tilde{q} = 2200$ W/m^2. Recognize the symmetry of temperature and displacement filed, and solve the problem on half the domain.

(a) Specify proper boundary conditions for the stress problem.

(b) Take the origin at the center of the inner cylinder. Compute and plot the hoop stress $\sigma_{\theta\theta}$ on AC for $-\pi/2 \leq \theta \leq \pi/2$. On the same graph, compare the finite element solution to the exact stress given by Equation (8.104). Report results in units of MPa.

(c) Plot the displacement components u and v on AC for $-\pi/2 \leq \theta \leq \pi/2$. Report results in units of μm.

(15) Reconsider Problem (14), and make one change: Specify zero traction on the outer circle. Provide the plot of part (b). Summarize lessons learned from problems (14) and (15).

(16) An eccentric cylinder of radius $r_1 = 0.5$ m is drilled through a larger cylinder of radius $r_2 = 1$ m, as shown in Figure 8.11. The distance between centers is 0.3 m. The initial temperature of the cylinder is 10 °C, corresponding to zero stresses. During operation, the inner surface is at uniform temperature of 30 °C, while the temperature of the outer surface is given by $T = 80 cos(\theta/2)$ °C, where θ is the angle a radial line from the center of the larger circle makes with line CD, so that $\theta = 0$ at point D and π at point B. The material properties (aluminum) are: thermal conductivity $k = 204$ W m^{-1} °C^{-1}, Poisson's ratio $\nu = 0.35$, elastic modules $E = 73$ GPa, and coefficient of linear thermal expansion $\alpha = 23 \times 10^{-6}$ °C^{-1}. A MATLAB code (*eccentric_cylinders_Mesh.m*) for mesh generation is provided. The number of elements is $nxe = 12$ in the radial direction and $nye = 32$ in the circumferential direction.

Assume a state of plane strain, and solve for thermal stresses in the cylinder due to the nonuniform temperature distribution, which is to be determined first by solving the heat conduction problem.

(a) Take the origin at the center of the inner cylinder, and plot the hoop stress $\sigma_{\theta\theta}$ on AC for $-\pi/2 \leq \theta \leq \pi/2$. Report results in units of MPa.

(b) Take the origin at the center of the outer cylinder, and plot the hoop stress $\sigma_{\theta\theta}$ on BD for $-\pi/2 \leq \theta \leq \pi/2$. Report results in units of MPa.

(c) Plot the displacement components u and v on AC for $-\pi/2 \leq \theta \leq \pi/2$. Report results in units of μm.

9 Kirchhoff–Love and Reissner–Mindlin Plates

A plate is a thin flat solid structure carrying transverse load. In this chapter, we develop the finite-element method for structural components that can be modeled as thin or moderately thick plates. The starting point of the formulation is the principle of virtual displacements for three-dimensional continuum. Certain geometric constraints on the plate thickness and transverse deflection permit postulating simple forms of the displacement field, and enable reduction of a three-dimensional problem to a two-dimensional one.

Ground breaking works on theory of plates date back to Lagrange (1811) and Kirchhoff (1850) while further advancements have been due to Hencky [49], Reissner, [98–100], and Mindlin [76]. Two well-known plate theories are presented in this book. The first is Kirchhoff–Love theory, also known as the classical plate theory (*CPT*), in which transverse shear strains in planes normal to the plate middle plane are neglected. It is applicable to thin plates where the plate thickness is less than one-hundredth of a characteristic length in the plate plane. For example, a square plate is considered thin if its thickness is less than 1% of the length of its side. The second plate theory is Reissner–Mindlin theory, also known as the first-order shear deformation theory. Generally speaking, it can be used for thin and moderately thick plates where the plate thickness could be as large as one-tenth of a typical length in the plate plane. The thickness of a circular plate can be as large as 10% of its diameter. Applications in this chapter include static deflection and free vibrations of plates in vacuum. The chapter closes with fluid-structure interactions such as vibrations of plates submerged in heavy fluids and hydroelastic panel divergence and flutter. MATLAB codes for all applications are provided. In Chapter 10, the finite-element method is presented for geometrically nonlinear Reissner–Mindlin theory using von Karman approximations. For comprehensive coverage of various theories of plates, the reader is referred to the books by Timoshenko and Krieger [118] and Reddy [95, 96].

9.1 CLASSICAL PLATE THEORY, *CPT*

In a Cartesian coordinate system, xyz, the plate geometry is defined by two surfaces $z = \pm h/2$, and its intersection with the xy-plane defines the middle plane with area A and contour Γ as shown in Figures 9.1a–9.1b. The plate thickness h is assumed to be small in comparison to the lateral dimensions, but it may be a function of x and y. However, its slopes are very small, that is $\partial h/\partial x << 1$ and $\partial h/\partial y << 1$. First, we exclude inplane loading ($F_x = 0$, $F_y = 0$, $M_z = 0$), and limit the loading to transverse loads (F_z, M_x, M_y) that may be distributed or concentrated, where (F_x, F_y, F_z) and

DOI: 10.1201/9781003323150-9

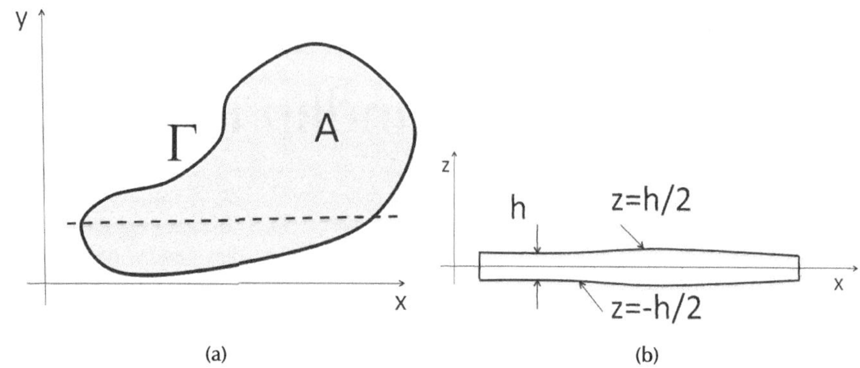

Figure 9.1 (a) Plate middle plane of area A and contour Γ. (b) Intersection of plate with a plane parallel to xz-plane, h is the plate thickness.

(M_x, M_y, M_z) are force, and moment vectors, respectively. A plate is usually loaded by external transverse forces (F_z) on the upper and/or lower surfaces $z = \pm h/2$, other forces and moments are exerted on the plate by supports. In this chapter, we also assume that the deflection is small in comparison with the plate thickness and neglect strains of the middle plane.

Kinematics of ***CPT***

The displacements of a point p in the middle plane with coordinates (x, y) is denoted by $u_o(x,y)$, $v_o(x,y)$, and $w_o(x,y)$. In classical plate theory, certain kinematic constraints enable expressing the displacements of any point (x,y,z) in the plate in terms of u_o, v_o, w_o, and their gradients. The first fundamental assumption of *CPT* is that a material straight line normal to the middle plane before deformation (deflection) remains straight and normal to the deformed (deflected) middle plane. The second assumption is that changes in length of that line are neglected. Now, we consider a point q with coordinates (x,y,z) on the normal line at point p as shown in Figure 9.2. After deformation, point p is displaced to p', and point q is displaced to q'. Let $\hat{n}$ denote the unit normal to the deformed middle plane at p'. The two adopted assumptions give

$$\begin{aligned} \overrightarrow{p'q'} &= z\hat{n} \\ &= z\left(n_x\hat{i} + n_y\hat{j} + n_z\hat{k}\right). \end{aligned} \tag{9.1}$$

The displacement of point q is

$$\overrightarrow{qq'} = \overrightarrow{pp'} + \overrightarrow{p'q'} - \overrightarrow{pq} \tag{9.2}$$

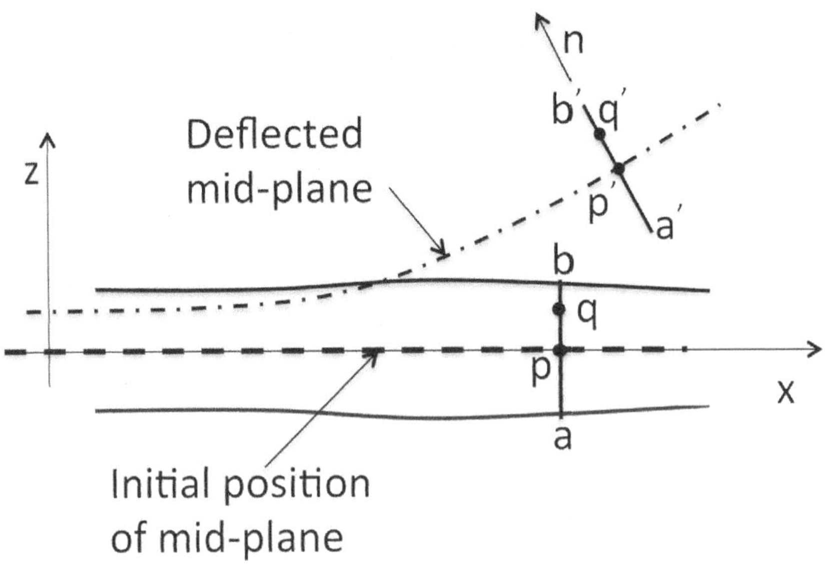

Figure 9.2 *CPT* kinematics. A filament *ab* initially normal to the middle plane remains normal to the plane after deflection.

whose components are

$$\begin{aligned} u(x,y,z) &= u_o(x,y) + zn_x \\ v(x,y,z) &= v_o(x,y) + zn_y \\ w(x,y,z) &= w_o(x,y) + zn_z - z. \end{aligned} \tag{9.3}$$

The equation of the deformed middle plane is

$$z - w_o(x,y) = 0. \tag{9.4}$$

Thus, the unit normal is

$$\hat{n} = \frac{-\frac{\partial w_o}{\partial x}\hat{\imath} - \frac{\partial w_o}{\partial y}\hat{\jmath} + \hat{k}}{\sqrt{\left(\frac{\partial w_o}{\partial x}\right)^2 + \left(\frac{\partial w_o}{\partial y}\right)^2 + 1}}. \tag{9.5}$$

Assuming small slopes, $|\partial w_o/\partial x| << 1$, $|\partial w_o/\partial y| << 1$, we approximate $\hat{n}$ by

$$\hat{n} = -\frac{\partial w_o}{\partial x}\hat{\imath} - \frac{\partial w_o}{\partial y}\hat{\jmath} + \hat{k} \tag{9.6}$$

and write the displacements of point q as

$$\begin{aligned} u(x,y,z) &= u_o(x,y) - z\frac{\partial w_o}{\partial x} \\ v(x,y,z) &= v_o(x,y) - z\frac{\partial w_o}{\partial y} \\ w(x,y,z) &= w_o(x,y). \end{aligned} \tag{9.7}$$

For these displacements, the corresponding linear strains are

$$\begin{aligned} &\varepsilon_{xx} = \frac{\partial u_o}{\partial x} - z\frac{\partial^2 w_o}{\partial x^2}, \quad \varepsilon_{yy} = \frac{\partial v_o}{\partial y} - z\frac{\partial^2 w_o}{\partial y^2}, \quad \gamma_{xy} = \frac{\partial u_o}{\partial y} + \frac{\partial v_o}{\partial x} - 2z\frac{\partial^2 w_o}{\partial x \partial y} \\ &\varepsilon_{zz} = 0, \quad \gamma_{xz} = 0, \quad \gamma_{yz} = 0. \end{aligned} \tag{9.8}$$

Plate Bending Problem

For deflection smaller than the plate thickness and in the absence of inplane loading, the strains due to displacements (u_o, v_o) of the middle plane are negligible (Timoshenko and Krieger [118]). Thus the displacements, strains, and stresses can be found by solving one problem for $w_o(x,y)$, which we call the bending problem.

Change Notation: We want to determine the middle plane deflection $w_o(x,y)$, in the following we drop the subscript "o", and denote it by $w(x,y)$.

Strains

$$\begin{aligned} \varepsilon_{xx} &= -z\frac{\partial^2 w}{\partial x^2}, \; \varepsilon_{yy} = -z\frac{\partial^2 w}{\partial y^2}, \; \gamma_{xy} = -2z\frac{\partial^2 w}{\partial x \partial y}, \\ \varepsilon_{zz} &= 0, \; \gamma_{xz} = 0, \; \gamma_{yz} = 0. \end{aligned} \tag{9.9}$$

In matrix form,

$$\left\{ \begin{array}{c} \varepsilon_{xx} \\ \varepsilon_{yy} \\ \gamma_{xy} \end{array} \right\} = -z \left\{ \begin{array}{c} \frac{\partial^2 w}{\partial x^2} \\ \frac{\partial^2 w}{\partial y^2} \\ 2\frac{\partial^2 w}{\partial x \partial y} \end{array} \right\}, \quad \{\varepsilon\} = -z\{\kappa\} \tag{9.10}$$

where the curvature $\{\kappa\}$ is independent of z,

$$\{\kappa\} = \left\{ \begin{array}{c} \frac{\partial^2 w}{\partial x^2} \\ \frac{\partial^2 w}{\partial y^2} \\ 2\frac{\partial^2 w}{\partial x \partial y} \end{array} \right\}. \tag{9.11}$$

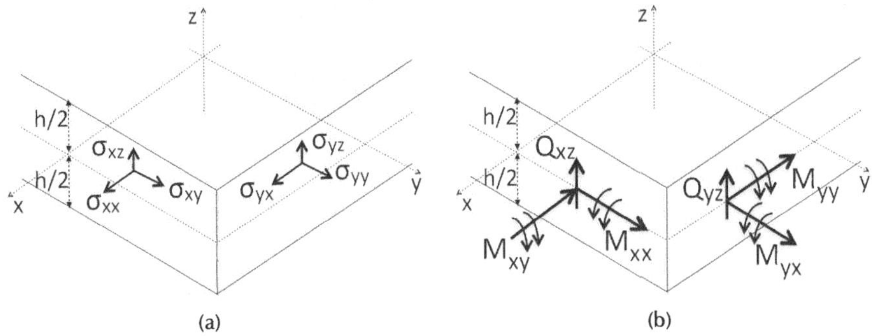

Figure 9.3 (a) Stresses at a point in a plate. (b) Resultant transverse shear and bending and twisting moments.

Stresses

Stresses on vertical planes through a point in the plate are shown in Figure 9.3a. The resultant bending and twisting moments, and transverse shear forces are defined by

$$M_{xx} = \int_{-h/2}^{h/2} z\sigma_{xx}dz, \quad M_{yy} = \int_{-h/2}^{h/2} z\sigma_{yy}dz, \quad M_{yx} = M_{xy} = \int_{-h/2}^{h/2} z\sigma_{xy}dz$$
$$Q_{xz} = \int_{-h/2}^{h/2} \sigma_{xz}dz, \quad Q_{yz} = \int_{-h/2}^{h/2} \sigma_{yz}dz \tag{9.12}$$

and shown in Figure 9.3b. Because the plate thickness is small, the applied transverse load on the upper and/or lower surfaces is resisted by bending moment M_{nn}, which is the resultant of normal stress σ_{nn}, twisting moment M_{ns}, which is the resultant of shear stress σ_{ns}, and transverse shear force Q_{nz}, which is the resultant of shear stress σ_{nz}, all per unit length of the plate edge, as shown in Figure 9.4. The strains given by Equation (9.9) imply a state of plane strain; however, in the theory of plates we use a constitutive equation for plane stress. We tolerate the inconsistency of the assumption of plane stress $\sigma_{zz} = 0$ with the applied pressure on the plate top or bottom surfaces. The stresses $(\sigma_{xx}, \sigma_{yy}, \sigma_{xy})$ are much higher than σ_{zz}. Thus, the constitutive equation for plates is

$$\begin{Bmatrix} \sigma_{xx} \\ \sigma_{yy} \\ \sigma_{xy} \end{Bmatrix} = \frac{E}{1-\nu^2} \begin{bmatrix} 1 & \nu & 0 \\ \nu & 1 & 0 \\ 0 & 0 & \frac{1-\nu}{2} \end{bmatrix} \begin{Bmatrix} \varepsilon_{xx} \\ \varepsilon_{yy} \\ \gamma_{xy} \end{Bmatrix}, \quad \{\sigma\} = [D]\{\varepsilon\} \tag{9.13}$$

where

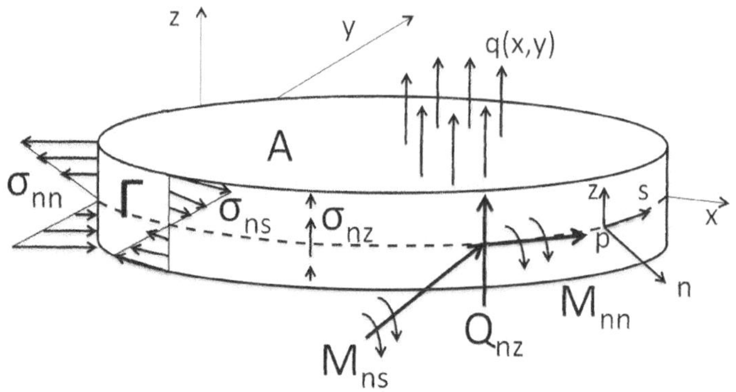

Figure 9.4 Stresses σ_{nn}, σ_{ns}, σ_{nz} on cylindrical surface, and resultant bending and twisting moments M_{nn}, M_{ns} and shear force Q_{nz}. $q(x,y)$ distributed load on the plate upper surface.

$$[D] = \frac{E}{1-\nu^2}\begin{bmatrix} 1 & \nu & 0 \\ \nu & 1 & 0 \\ 0 & 0 & \frac{1-\nu}{2} \end{bmatrix}. \tag{9.14}$$

Substituting $\{\varepsilon\}$ from Equation (9.10), we get

$$\{\sigma\} = -z[D]\{\kappa\}. \tag{9.15}$$

9.1.1 PRINCIPLE OF VIRTUAL DISPLACEMENTS

The principle of virtual displacements for a 3D continuum has been derived from the equations of equilibrium in Chapter 3. To obtain the principle of virtual displacements for plates, we start with Equation (3.130), which is repeated here,

$$\int_\Omega \bar{\varepsilon}_{ij}\sigma_{ji}d\Omega = \oint_S \bar{u}_i t_i dS + \int_\Omega \bar{u}_i f_i d\Omega. \tag{9.16}$$

A virtual displacement field is an arbitrary field but it must be consistent with the kinematic constraints imposed on the actual displacement field. Thus, in assuming a virtual displacement field for the bending problem, we choose a virtual deflection $\bar{w}(x,y)$ of the middle plane, and impose the condition that a normal line on that plane before deflection remains normal on the deflected plane and its length remains constant. Hence, the other components of virtual displacements are

$$\bar{u} = -z\frac{\partial\bar{w}}{\partial x}, \quad \bar{v} = -z\frac{\partial\bar{w}}{\partial y}. \tag{9.17}$$

The virtual strains are

$$\left\{\begin{array}{c}\bar{\varepsilon}_{xx}\\ \bar{\varepsilon}_{yy}\\ \bar{\gamma}_{xy}\end{array}\right\} = -z\left\{\begin{array}{c}\frac{\partial^2\bar{w}}{\partial x^2}\\ \frac{\partial^2\bar{w}}{\partial y^2}\\ 2\frac{\partial^2\bar{w}}{\partial x\partial y}\end{array}\right\}, \quad \{\bar{\varepsilon}\} = -z\{\bar{\kappa}\} \tag{9.18}$$

where $\{\bar{\kappa}\}$ is independent of z,

$$\{\bar{\kappa}\} = \left\{\begin{array}{c}\frac{\partial^2\bar{w}}{\partial x^2}\\ \frac{\partial^2\bar{w}}{\partial y^2}\\ 2\frac{\partial^2\bar{w}}{\partial x\partial y}\end{array}\right\}. \tag{9.19}$$

The bounding surface of the plate consists of three parts: The upper and lower surfaces $z = \pm h$, and a cylindrical (not necessarily circular) surface with axis parallel to the z-axis; the intersection of the cylinder with the middle plane is the contour Γ of area A. The principle of virtual displacements Equation (9.16) takes the form

$$\int_A\int_{-h/2}^{h/2}\left\{\begin{array}{ccc}\bar{\varepsilon}_{xx} & \bar{\varepsilon}_{yy} & \bar{\gamma}_{xy}\end{array}\right\}\left\{\begin{array}{c}\sigma_{xx}\\ \sigma_{yy}\\ \sigma_{xy}\end{array}\right\}dzdA =$$

$$\int_A\left\{\begin{array}{ccc}\bar{u} & \bar{v} & \bar{w}\end{array}\right\}\left\{\begin{array}{c}t_x\\ t_y\\ t_z\end{array}\right\}_{z=+h/2} dA + \int_A\left\{\begin{array}{ccc}\bar{u} & \bar{v} & \bar{w}\end{array}\right\}\left\{\begin{array}{c}t_x\\ t_y\\ t_z\end{array}\right\}_{z=-h/2} dA$$

$$+\oint_\Gamma\int_{-h/2}^{h/2}\left\{\begin{array}{ccc}\bar{u} & \bar{v} & \bar{w}\end{array}\right\}\left\{\begin{array}{c}t_x\\ t_y\\ t_z\end{array}\right\}dzds + \int_A\int_{-h/2}^{h/2}\left\{\begin{array}{ccc}\bar{u} & \bar{v} & \bar{w}\end{array}\right\}\left\{\begin{array}{c}f_x\\ f_y\\ f_z\end{array}\right\}dzdA. \tag{9.20}$$

The lower and upper surfaces $z = \pm h/2$ are assumed to be free of traction parallel to the middle plane ($t_x = t_y = 0$), and only distributed load $t_z = q(x,y)$ or concentrated forces normal to the middle plane are applied on those surfaces. A concentrated force of magnitude P at (x_p, y_p) can be represented as a distribution (force per unit area) using the Dirac delta function δ; $q(x,y) = P\delta(x - x_p)\delta(y - y_p)$. We also assume that the body force is parallel to the z-axis, hence $f_x = f_y = 0$. Equation (9.20) reduces to

$$\int_A\int_{-h/2}^{h/2}\left\{\bar{\varepsilon}_{xx}\ \bar{\varepsilon}_{yy}\ \bar{\gamma}_{xy}\right\}\left\{\begin{array}{c}\sigma_{xx}\\ \sigma_{yy}\\ \sigma_{xy}\end{array}\right\}dzdA = \int_A \bar{w}qdA + \int_A\int_{-h/2}^{h/2}\bar{w}f_z dzdA$$

$$+\oint_\Gamma\int_{-h/2}^{h/2}\left\{\bar{u}\ \bar{v}\ \bar{w}\right\}\left\{\begin{array}{c}t_x\\ t_y\\ t_z\end{array}\right\}dzds. \tag{9.21}$$

In Equation (9.21), q is the net of distributed load on the upper and lower surfaces. Substituting stresses and virtual strains from Equations (9.15) and (9.18), respectively, we get

$$\int_A \int_{-h/2}^{h/2} z^2 \{\bar{\kappa}\}^T [D] \{\kappa\} dz dA = \int_A \bar{w} q dA + \int_A \int_{-h/2}^{h/2} \bar{w} f_z dz dA + \oint_\Gamma \int_{-h/2}^{h/2} \left\{ \bar{u} \;\; \bar{v} \;\; \bar{w} \right\} \begin{Bmatrix} t_x \\ t_y \\ t_z \end{Bmatrix} dz ds. \tag{9.22}$$

We evaluate the integral

$$\int_{-h/2}^{h/2} z^2 dz = \frac{h^3}{12} \tag{9.23}$$

define

$$\begin{aligned} [\tilde{D}] &= \frac{h^3}{12} [D] \\ &= \frac{Eh^3}{12(1-\nu^2)} \begin{bmatrix} 1 & \nu & 0 \\ \nu & 1 & 0 \\ 0 & 0 & \frac{1-\nu}{2} \end{bmatrix} \end{aligned} \tag{9.24}$$

$$\tilde{f}_z = \int_{-h/2}^{h/2} f_z dz \tag{9.25}$$

and rewrite Equation (9.22)

$$\int_A \{\bar{\kappa}\}^T [\tilde{D}] \{\kappa\} dA = \int_A \bar{w} (q + \tilde{f}_z) dA + \oint_\Gamma \int_{-h/2}^{h/2} \left\{ \bar{u} \;\; \bar{v} \;\; \bar{w} \right\} \begin{Bmatrix} t_x \\ t_y \\ t_z \end{Bmatrix} dz ds. \tag{9.26}$$

The quantity

$$D = \frac{Eh^3}{12(1-\nu^2)} \tag{9.27}$$

is called the plate flexural rigidity. Note that $\tilde{f}_z$ is a body force per unit area of the plate.

At a point p on the contour Γ, we define local coordinate system (n, s, z), where $\hat{n}$ is the unit normal to Γ in the middle plane, $\hat{s}$ is unit tangent such that $\hat{n} = \hat{s} \times \hat{k}$ as shown in Figure 9.4. Let us project the surface traction on the cylindrical surface S into normal $t_n = \sigma_{nn}$ and tangential components $t_s = \sigma_{ns}$ and $t_z = \sigma_{nz}$ as shown in Figure 9.4. Similarly, we project the virtual displacements

$$\bar{u}_n = -z\frac{\partial \bar{w}}{\partial n}, \quad \bar{u}_s = -z\frac{\partial \bar{w}}{\partial s}, \quad \bar{u}_z = \bar{w}. \tag{9.28}$$

We note that

$$\begin{aligned}\bar{u}t_x + \bar{v}t_y + \bar{w}t_z &= \vec{\bar{u}} \cdot \vec{t} \\ &= \bar{u}_n \sigma_{nn} + \bar{u}_s \sigma_{ns} + \bar{w}\sigma_{nz} \end{aligned} \tag{9.29}$$

and consider the integral

$$\begin{aligned}\int_{-h/2}^{h/2} \left\{ \bar{u} \; \bar{v} \; \bar{w} \right\} \left\{ \begin{matrix} t_x \\ t_y \\ t_z \end{matrix} \right\} dz &= \int_{-h/2}^{h/2} \vec{\bar{u}} \cdot \vec{t}\, dz \\ &= \int_{-h/2}^{h/2} (\bar{u}_n \sigma_{nn} + \bar{u}_s \sigma_{ns} + \bar{w}\sigma_{nz})\, dz \\ &= \int_{-h/2}^{h/2} \left(-z\sigma_{nn}\frac{\partial \bar{w}}{\partial n} - z\sigma_{ns}\frac{\partial \bar{w}}{\partial s} + \bar{w}\sigma_{nz} \right) dz \\ &= -\frac{\partial \bar{w}}{\partial n} M_{nn} - \frac{\partial \bar{w}}{\partial s} M_{ns} + \bar{w} Q_{nz} \end{aligned} \tag{9.30}$$

where

$$M_{nn} = \int_{-h/2}^{h/2} z\sigma_{nn} dz, \quad M_{ns} = \int_{-h/2}^{h/2} z\sigma_{ns} dz, \quad Q_{nz} = \int_{-h/2}^{h/2} \sigma_{nz} dz. \tag{9.31}$$

Substituting Equation (9.30) into Equation (9.26), we write the principle of virtual displacements for the *CPT*,

$$\int_A \{\bar{\kappa}\}^T [\tilde{D}] \{\kappa\} dA = \int_A \bar{w}\left(q + \tilde{f}_z\right) dA + \oint_\Gamma \left(-\frac{\partial \bar{w}}{\partial n} M_{nn} - \frac{\partial \bar{w}}{\partial s} M_{ns} + \bar{w} Q_{nz} \right) ds. \tag{9.32}$$

Primary and Secondary Variables

Looking at the boundary term given by the contour integral on the right-hand-side, we define secondary variables as the coefficients of virtual displacement and its derivatives (Reddy [97]). Thus, the secondary variables are the shear force Q_{nz}, twisting moment M_{ns}, and bending moment M_{nn}, all per unit length of contour Γ. The corresponding primary variables are deflection w, slope $\partial w/\partial s$, and slope $\partial w/\partial n$, respectively. So, it appears that we can prescribe three boundary conditions at each point of the plate edge. On the plate edge Γ (or part of it), we can prescribe plate deflection w and independently prescribe its slope $\partial w/\partial n$. However, on the plate edge, we would not be able to prescribe w and $\partial w/\partial s$ independently. If w is prescribed then, by differentiation, we also prescribed $\partial w/\partial s$. Similarly, if $\partial w/\partial s$ is prescribed then, by integration, we find w aside from a constant. The terms $-M_{ns}\partial \bar{w}/\partial s$ and $Q_{nz}\bar{w}$ are the virtual work of shear stresses σ_{ns} and σ_{nz}, respectively. Aside from rigid body translation $\bar{w} = c$, both of shear stress components will simultaneously do virtual work through any virtual displacement $\bar{w} \neq 0$. This means that we only have

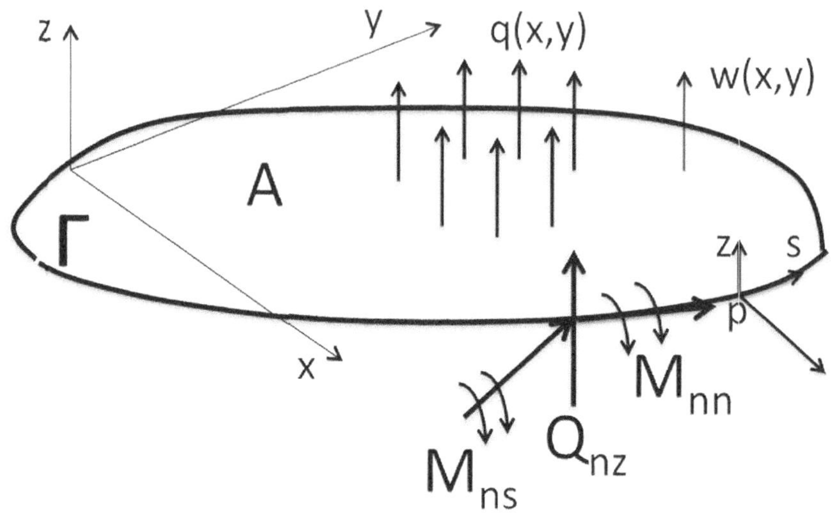

Figure 9.5 Resultant bending and twisting moments M_{nn}, M_{ns}, and shear force Q_{nz}. $q(x,y)$ net distributed load on the plate upper and lower surfaces.

two primary variables w and $\partial w/\partial n$. Kirchhoff provided the secondary variable corresponding to w as the effective shear force

$$Q^*_{nz} = Q_{nz} + \frac{\partial M_{ns}}{\partial s} \tag{9.33}$$

and, hence, the principle of virtual displacements for *CPT* becomes

$$\int_A \{\bar{\kappa}\}^T [\tilde{D}] \{\kappa\} dA = \int_A \bar{w}(q + \tilde{f}_z)\, dA + \oint_\Gamma \left(-\frac{\partial \bar{w}}{\partial n} M_{nn} + \bar{w} Q^*_{nz}\right) ds. \tag{9.34}$$

Boundary Conditions

At each point of the boundary Γ, we need to specify two boundary conditions. For the first condition, we must specify one of the pair (w, Q^*_{nz}), or a relation between the two elements of this pair. For the second condition, we must specify one of the pair $(\partial w/\partial n, M_{nn})$, or a relation between the two elements of this pair. The notions of essential, natural, and mixed boundary conditions applies.

Simply Supported (Hinged) Edge

$$w = 0, \quad M_{nn} = 0. \tag{9.35}$$

The first condition is an *EBC* while the second is an *NBC*. The secondary variable Q^*_{nz} and primary variable $\partial w/\partial n$ are unknown.

Clamped (Fixed) Edge

$$w = 0, \quad \frac{\partial w}{\partial n} = 0. \tag{9.36}$$

Both conditions are *EBCs*. The secondary variables Q^*_{nz} and M_{nn} are unknown.

Free Edge

$$Q^*_{nz} = 0, \quad M_{nn} = 0. \tag{9.37}$$

Both conditions are *NBCs*. The primary variables w and $\partial w/\partial n$ are unknown.

Line of Symmetry
A line in the xy-plane is a line of symmetry if the plate geometry, material properties, supports, and loading are symmetric about that line. Let $\hat{n}$ be the unit normal to that line in the plate middle plane.

$$\frac{\partial w}{\partial n} = 0, \quad Q^*_{nz} = 0. \tag{9.38}$$

The first condition is an *EBC* while the second is an *NBC*. The secondary variable M_{nn} and primary variable w are unknown.

Remarks on the Geometric Boundary Conditions for *CPT*:
In addition to applying the essential boundary conditions, we should also impose any geometric constraints that are consistent with those conditions as needed by the parametrization of the displacement field. Consider a clamped edge $y = c$; a constant. On that edge the boundary conditions are $w = 0$, and $\partial w/\partial y = 0$. These conditions are true for all values of x on the edge. It follows that $\partial w/\partial x = 0$ and $\partial^2 w/\partial x\partial y = 0$ on that edge. Similarly, for a simply supported edge $x = c$; a constant, the geometric boundary condition is $w = 0$, that is true for all values of y on the edge. Hence, $\partial w/\partial y = 0$ is also valid on that edge. Finally, we consider a line of symmetry $y = c$; a constant. On that line, the geometric condition is $\partial w/\partial y = 0$, that is valid for all values of x on the line of symmetry. Hence $\partial^2 w/\partial x\partial y = 0$ is also true on that line. In cases where the deflection over the element is expressed in terms of the geometric parameters w, $\partial w/\partial x$, $\partial w/\partial y$, and $\partial^2 w/\partial x\partial y$ at the element's nodes, the known values of those parameters should be imposed at the boundary nodes.

9.1.2 APPROXIMATIONS AND ELEMENT SHAPE FUNCTIONS

A Non-Conforming Rectangular Element
A rectangular element with 4 nodes and 3 degrees of freedom at each node namely deflection and slopes $(w, \partial w/\partial x, \partial w/\partial y)$ is non-conforming. Such an element will not be used in this book in favor of conforming elements.

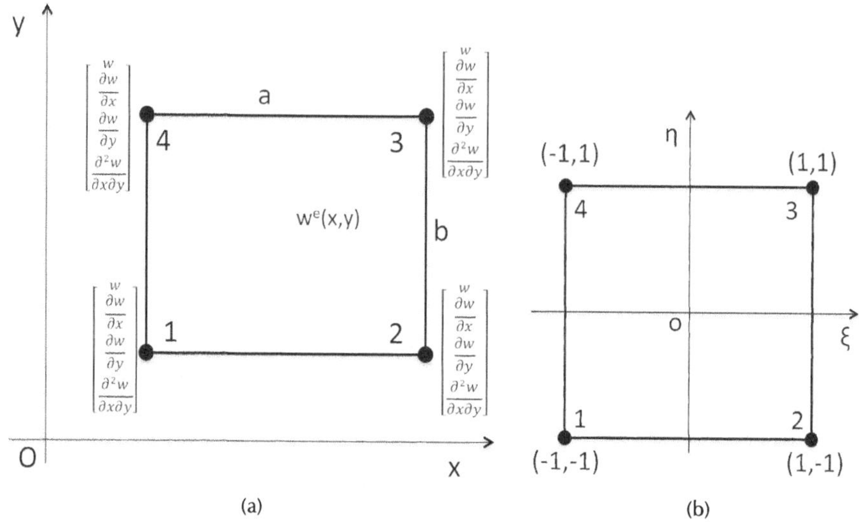

Figure 9.6 (a) Bogner-Fox-Schmit rectangular element, four degrees of freedom at each node. Deflection $w^e(x,y)$ at any point of the element is expressed in terms of the 16 geometric parameters at the element's nodes. (b) Canonical element in the $\xi\eta$-plane.

A Conforming Rectangular Element: Bogner–Fox–Schmit Element

This is a 4-node rectangular element with sides parallel to Cartesian coordinates x and y axes as shown in Figure 9.6. The coordinates of the 4 nodes are denoted by (x_i^e, y_i^e), for $i = 1,2,3,4$. The element has 16 degrees of freedom; four at each node, namely the deflection w, slopes $\partial w/\partial x$, $\partial w/\partial y$, and mixed derivative $\partial^2 w/\partial x \partial y$. We define,

$$\text{for node 1: } U_1^e = w_1^e, \quad U_2^e = \left(\frac{\partial w}{\partial x}\right)_1, \quad U_3^e = \left(\frac{\partial w}{\partial y}\right)_1, \quad U_4^e = \left(\frac{\partial^2 w}{\partial x \partial y}\right)_1 \tag{9.39}$$

$$\text{for node 2: } U_5^e = w_2^e, \quad U_6^e = \left(\frac{\partial w}{\partial x}\right)_2, \quad U_7^e = \left(\frac{\partial w}{\partial y}\right)_2, \quad U_8^e = \left(\frac{\partial^2 w}{\partial x \partial y}\right)_2 \tag{9.40}$$

$$\text{for node 3: } U_9^e = w_3^e, \quad U_{10}^e = \left(\frac{\partial w}{\partial x}\right)_3, \quad U_{11}^e = \left(\frac{\partial w}{\partial y}\right)_3, \quad U_{12}^e = \left(\frac{\partial^2 w}{\partial x \partial y}\right)_3 \tag{9.41}$$

y^4	xy^4	x^2y^4	x^3y^4	x^4y^4
y^3	xy^3	x^2y^3	x^3y^3	x^4y^3
y^2	xy^2	x^2y^2	x^3y^2	x^4y^2
y	xy	x^2y	x^3y	x^4y
1	x	x^2	x^3	x^4

Figure 9.7 Polynomial selection. For rectangular element of 16 (or 12) DOF, choose the 16 (or 12) terms of lowest-power that keep symmetry in x and y.

$$\text{for node 4: } U_{13}^e = w_4^e, \quad U_{14}^e = \left(\frac{\partial w}{\partial x}\right)_4, \quad U_{15}^e = \left(\frac{\partial w}{\partial y}\right)_4, \quad U_{16}^e = \left(\frac{\partial^2 w}{\partial x \partial y}\right)_4. \tag{9.42}$$

We arrange the 16 degrees of freedom in an array $\{U^e\}$ of dimensions $(16,1)$.

$$\{U^e\} = \{U_1^e\ U_2^e\ U_3^e\ U_4^e\ U_5^e\ U_6^e\ U_7^e\ U_8^e\ U_9^e\ U_{10}^e\ U_{11}^e\ U_{12}^e\ U_{13}^e\ U_{14}^e\ U_{15}^e\ U_{16}^e\}^T. \tag{9.43}$$

The deflection at any point (x,y) in the element is a polynomial in x and y. Figure 9.7 shows the different terms that can form polynomials in x and y. We select the 16 terms with the lowest power in x and lowest power in y but keep the symmetry in x and y.

$$\begin{aligned} W^e(x,y) \quad &= c_1 + c_2x + c_3y + c_4xy \\ &+ c_5x^2 + c_6y^2 + c_7x^2y + c_8xy^2 + c_9x^2y^2 \\ &+ c_{10}x^3 + c_{11}y^3 + c_{12}x^3y + c_{13}xy^3 + c_{14}x^3y^2 + c_{15}x^2y^3 + c_{16}x^3y^3. \end{aligned} \tag{9.44}$$

The 16 coefficients $c_1, \cdots, c_{16}$ can be determined in terms of the 16 degrees of freedom of the element U_j^e, $j = 1, \cdots, 16$, and of course the coordinates of the element's nodes. Thus, the approximation over the element can now be written in the form

$$W^e(x,y) = \sum_{j=1}^{16} \psi_j(x,y) U_j^e \tag{9.45}$$

where ψ_j, $j = 1, \cdots, 16$, are the element shape functions. We consider a rectangular element of sides $a = x_2^e - x_1^e$, $b = y_4^e - y_1^e$ as shown in Figure 9.6. Introducing the linear transformation

$$\begin{aligned} x &= x_1^e + a(\xi+1)/2 \\ y &= y_1^e + b(\eta+1)/2 \\ \xi &= -1+2(x-x_1^e)/a \\ \eta &= -1+2(y-y_1^e)/b \end{aligned} \tag{9.46}$$

where $-1 \leq \xi \leq 1$, and $-1 \leq \eta \leq 1$, we obtain the shape functions in terms of ξ and η. We developed a MATLAB symbolic code for generating the functions.

$$\begin{aligned}
\psi_1 &= (1/16)(\eta+2)(\eta-1)(\eta-1)(\xi+2)(\xi-1)(\xi-1) \\
\psi_2 &= (1/16)(\eta+2)(\eta-1)(\eta-1)(\xi+1)(\xi-1)(\xi-1) \\
\psi_3 &= (1/16)(\eta+1)(\eta-1)(\eta-1)(\xi+2)(\xi-1)(\xi-1) \\
\psi_4 &= (1/16)(\eta+1)(\eta-1)(\eta-1)(\xi+1)(\xi-1)(\xi-1) \\
\psi_5 &= (-1/16)(\eta+2)(\eta-1)(\eta-1)(\xi-2)(\xi+1)(\xi+1) \\
\psi_6 &= (1/16)(\eta+2)(\eta-1)(\eta-1)(\xi-1)(\xi+1)(\xi+1) \\
\psi_7 &= (-1/16)(\eta+1)(\eta-1)(\eta-1)(\xi-2)(\xi+1)(\xi+1) \\
\psi_8 &= (1/16)(\eta+1)(\eta-1)(\eta-1)(\xi-1)(\xi+1)(\xi+1) \\
\psi_9 &= (1/16)(\eta-2)(\eta+1)(\eta+1)(\xi-2)(\xi+1)(\xi+1) \\
\psi_{10} &= (-1/16)(\eta-2)(\eta+1)(\eta+1)(\xi-1)(\xi+1)(\xi+1) \\
\psi_{11} &= (-1/16)(\eta-1)(\eta+1)(\eta+1)(\xi-2)(\xi+1)(\xi+1) \\
\psi_{12} &= (1/16)(\eta-1)(\eta+1)(\eta+1)(\xi-1)(\xi+1)(\xi+1) \\
\psi_{13} &= (-1/16)(\eta-2)(\eta+1)(\eta+1)(\xi+2)(\xi-1)(\xi-1) \\
\psi_{14} &= (-1/16)(\eta-2)(\eta+1)(\eta+1)(\xi+1)(\xi-1)(\xi-1) \\
\psi_{15} &= (1/16)(\eta-1)(\eta+1)(\eta+1)(\xi+2)(\xi-1)(\xi-1) \\
\psi_{16} &= (1/16)(\eta-1)(\eta+1)(\eta+1)(\xi+1)(\xi-1)(\xi-1).
\end{aligned} \tag{9.47}$$

At the element's nodes, each of the 16 functions satisfies the conditions
$\psi_j = 0$, $\partial\psi_j/\partial\xi = 0$, $\partial\psi_j/\partial\eta = 0$, and $\partial^2\psi_j/\partial\xi\partial\eta = 0$, except:
At node # 1, $\psi_1 = 1$, $\partial\psi_2/\partial\xi = 1$, $\partial\psi_3/\partial\eta = 1$, and $\partial^2\psi_4/\partial\xi\partial\eta = 1$.
At node # 2, $\psi_5 = 1$, $\partial\psi_6/\partial\xi = 1$, $\partial\psi_7/\partial\eta = 1$, and $\partial^2\psi_8/\partial\xi\partial\eta = 1$.
At node # 3, $\psi_9 - 1$, $\partial\psi_{10}/\partial\xi = 1$, $\partial\psi_{11}/\partial\eta = 1$, and $\partial^2\psi_{12}/\partial\xi\partial\eta = 1$.
At node # 4, $\psi_{13} = 1$, $\partial\psi_{14}/\partial\xi = 1$, $\partial\psi_{15}/\partial\eta = 1$, and $\partial^2\psi_{16}/\partial\xi\partial\eta = 1$.

For the rectangular element, the x and y derivatives are

$$\begin{aligned}
\frac{\partial\psi_j}{\partial x} &= \frac{2}{a}\frac{\partial\psi_j}{\partial\xi}, \quad \frac{\partial^2\psi_j}{\partial x^2} = \frac{4}{a^2}\frac{\partial^2\psi_j}{\partial\xi^2} \\
\frac{\partial\psi_j}{\partial y} &= \frac{2}{b}\frac{\partial\psi_j}{\partial\eta}, \quad \frac{\partial^2\psi_j}{\partial y^2} = \frac{4}{b^2}\frac{\partial^2\psi_j}{\partial\eta^2} \\
\frac{\partial^2\psi_j}{\partial x\partial y} &= \frac{4}{ab}\frac{\partial\psi_j}{\partial\xi\partial\eta}.
\end{aligned} \tag{9.48}$$

9.1.3 ELEMENT EQUATIONS AND MATRICES

The principle of virtual displacements Equation (9.34) for an element becomes,

$$\int_{A_e} \{\bar{\kappa}\}^T \left[\tilde{D}\right] \{\kappa\} dA = \int_{A_e} \bar{w}\left(q+\tilde{f}_z\right) dA + \oint_{\Gamma_e} \left(-\frac{\partial \bar{w}}{\partial n} M_{nn} + \bar{w} Q^*_{nz}\right) ds. \quad (9.49)$$

The approximation over the element can be written as

$$W^e(x,y) = \left\{ \begin{array}{ccccc} \psi_1 & \psi_2 & \cdot & \psi_j & \cdot & \psi_{16} \end{array} \right\} \left\{ \begin{array}{c} U_1^e \\ U_2^e \\ \cdot \\ U_j^e \\ \cdot \\ U_{16}^e \end{array} \right\}. \quad (9.50)$$

We define the array

$$\{\psi\} = \left\{ \begin{array}{cccccc} \psi_1 & \psi_2 & \cdot & \psi_j & \cdot & \psi_{16} \end{array} \right\} \quad (9.51)$$

and write the trial solution in matrix form

$$W^e(x,y) = \{\psi\}\{U^e\}. \quad (9.52)$$

The curvature of the plate becomes,

$$\left\{ \begin{array}{c} \frac{\partial^2 W^e(x,y)}{\partial x^2} \\ \frac{\partial^2 W^e(x,y)}{\partial y^2} \\ 2\frac{\partial^2 W^e(x,y)}{\partial x \partial y} \end{array} \right\} = \left[\begin{array}{cccc} \frac{\partial^2 \psi_1(x,y)}{\partial x^2} & \frac{\partial^2 \psi_2(x,y)}{\partial x^2} & \cdots & \frac{\partial^2 \psi_{16}(x,y)}{\partial x^2} \\ \frac{\partial^2 \psi_1(x,y)}{\partial y^2} & \frac{\partial^2 \psi_2(x,y)}{\partial y^2} & \cdots & \frac{\partial^2 \psi_{16}(x,y)}{\partial y^2} \\ 2\frac{\partial^2 \psi_1(x,y)}{\partial x \partial y} & 2\frac{\partial^2 \psi_2(x,y)}{\partial x \partial y} & \cdots & 2\frac{\partial^2 \psi_{16}(x,y)}{\partial x \partial y} \end{array} \right] \left\{ \begin{array}{c} U_1^e \\ U_2^e \\ \cdot \\ \cdot \\ \cdot \\ U_{16}^e \end{array} \right\}. \quad (9.53)$$

We define the curvature matrix $[B^e]$ by

$$[B^e] = \left[\begin{array}{cccc} \frac{\partial^2 \psi_1(x,y)}{\partial x^2} & \frac{\partial^2 \psi_2(x,y)}{\partial x^2} & \cdots & \frac{\partial^2 \psi_{16}(x,y)}{\partial x^2} \\ \frac{\partial^2 \psi_1(x,y)}{\partial y^2} & \frac{\partial^2 \psi_2(x,y)}{\partial y^2} & \cdots & \frac{\partial^2 \psi_{16}(x,y)}{\partial y^2} \\ 2\frac{\partial^2 \psi_1(x,y)}{\partial x \partial y} & 2\frac{\partial^2 \psi_2(x,y)}{\partial x \partial y} & \cdots & 2\frac{\partial^2 \psi_{16}(x,y)}{\partial x \partial y} \end{array} \right] \quad (9.54)$$

$$\{\kappa\} = [B^e]\{U^e\}. \quad (9.55)$$

Galerkin

The virtual deflection (which is the weight or test function in the weak form) is a linear superposition of the element shape functions,

$$\begin{aligned} \bar{W}^e(x,y) &= \{\psi\}\{d\} \\ &= \{d\}^T\{\psi\}^T \end{aligned} \quad (9.56)$$

$$\{\bar{\kappa}\} = [B^e]\{d\}, \; \{\bar{\kappa}\}^T = \{d\}^T [B^e]^T \tag{9.57}$$

Since $u_n = un_x + vn_y$, we have

$$\begin{aligned}
\frac{\partial \bar{W}^e}{\partial n} &= \frac{\partial \bar{W}^e}{\partial x} n_x + \frac{\partial \bar{W}^e}{\partial y} n_y \\
&= \{d\}^T \left[\frac{\partial \{\psi\}^T}{\partial x} n_x + \frac{\partial \{\psi\}^T}{\partial y} n_y \right] \\
&= \{d\}^T \begin{bmatrix} \frac{\partial \psi_1(x,y)}{\partial x} & \frac{\partial \psi_2(x,y)}{\partial x} & \cdots & \frac{\partial \psi_{16}(x,y)}{\partial x} \\ \frac{\partial \psi_1(x,y)}{\partial y} & \frac{\partial \psi_2(x,y)}{\partial y} & \cdots & \frac{\partial \psi_{16}(x,y)}{\partial y} \end{bmatrix}^T \left\{ \begin{matrix} n_x \\ n_y \end{matrix} \right\} \\
&= \{d\}^T [G^e]^T \{n\}
\end{aligned} \tag{9.58}$$

where

$$[G^e] = \begin{bmatrix} \frac{\partial \psi_1(x,y)}{\partial x} & \frac{\partial \psi_2(x,y)}{\partial x} & \cdots & \frac{\partial \psi_{16}(x,y)}{\partial x} \\ \frac{\partial \psi_1(x,y)}{\partial y} & \frac{\partial \psi_2(x,y)}{\partial y} & \cdots & \frac{\partial \psi_{16}(x,y)}{\partial y} \end{bmatrix} \tag{9.59}$$

and

$$\{n\} = \left\{ \begin{matrix} n_x \\ n_y \end{matrix} \right\}. \tag{9.60}$$

The principle of virtual displacements gives

$$\begin{aligned}
\int_{A_e} \{d\}^T [B^e]^T [\tilde{D}^e] [B^e] \{U^e\} dA &= \int_{A_e} \{d\}^T \{\psi\}^T (q + \tilde{f}_z) dA \\
&+ \oint_{\Gamma_e} \{d\}^T [G^e]^T \{n\} (-M_{nn}) ds + \oint_{\Gamma_e} \{d\}^T \{\psi\}^T Q^*_{nz} ds.
\end{aligned} \tag{9.61}$$

Since $\{d\}$ and $\{U^e\}$ are constant arrays, we rearrange terms and obtain

$$\begin{aligned}
\{d\}^T \Big[\Big(\int_{A_e} [B^e]^T [\tilde{D}^e] [B^e] dA \Big) \{U^e\} - \int_{A_e} \{\psi\}^T (q + \tilde{f}_z) dA \\
- \oint_{\Gamma_e} [G^e]^T \{n\} (-M_{nn}) ds - \oint_{\Gamma_e} \{\psi\}^T Q^*_{nz} ds \Big] = 0.
\end{aligned} \tag{9.62}$$

We define,

Stiffness Matrix

$$[K^e] = \int_{A_e} [B^e]^T [\tilde{D}^e] [B^e] dA. \tag{9.63}$$

Forcing Vector

$$\{F^e\} = \int_{A_e} \{\psi\}^T (q + \tilde{f}_z) dA. \tag{9.64}$$

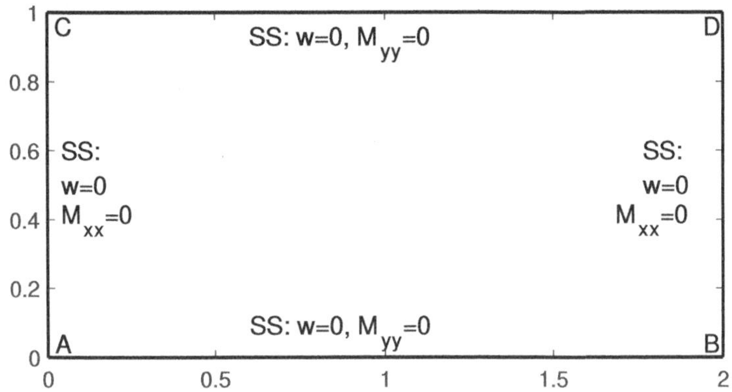

Figure 9.8 Boundary conditions for a simply supported rectangular plate.

Weighted Secondary Variables Vector

$$\{R^e\} = \oint_{\Gamma_e} [G^e]^T \{n\} (-M_{nn}) \, ds + \oint_{\Gamma_e} \{\psi\}^T Q^*_{nz} ds \tag{9.65}$$

and rewrite Equation (9.62),

$$\{d\}^T \{[K^e]\{U^e\} - \{F^e\} - \{R^e\}\} = 0. \tag{9.66}$$

Since $\{d\}^T$ is an arbitrary array of dimensions (1,16), the multiplying bracket must be zero, and we obtain the element equations,

$$[K^e]\{U^e\} = \{F^e\} + \{R^e\}. \tag{9.67}$$

The element equations are assembled into a global system, and boundary conditions are imposed. The system is solved for the unknown primary and weighted secondary variables at the global nodes.

Example 9.1. Uniformly Loaded Rectangular Plate

Consider a rectangular plate ($a = 2, b = 1, h = 0.02$) with all sides simply supported as shown in Figure 9.8. Uniform distributed pressure ($q = -1$) is assumed. The entire plate is modeled by uniform mesh of square elements ($nxe \times nye = 64 \times 32$). The distributed reactions (components of the weighted secondary variable corresponding to resultant of transverse shear) on the plate edges are compared with the exact analytic solution in Figure 9.9a. The maximum deflection and weighted reactions at the plate corners are compared with the exact solution in Table 9.1, which indicates convergence of the finite-element solution as the mesh is refined. In the table, the corner force is downward. It is verified that the sum of the weighted effective shear forces R_j at all nodes on the plate boundary is exactly equal to the net pressure

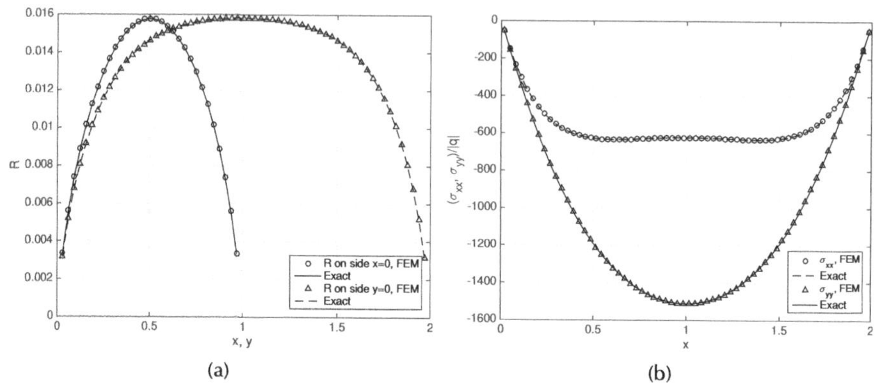

Figure 9.9 Comparison of finite element solution to exact analytic solution for a simply supported rectangular plate, $a = 2$, $b = 1$, $h = 0.02$, $\nu = 0.25$. Mesh of $nxe = 64$ and $nye = 32$. (a) Weighted reaction at the nodes on the plate edges; the corner nodes are not shown, see Table 9.1. (b) Maximum stresses σ_{xx} and σ_{yy} at $z = h/2$ on the plate center-line $y = b/2$. $|q|$ is applied pressure on the upper surface.

Table 9.1
Example 9.1, Maximum Deflection w^* and Corner Reaction R_c, Exact Solution $w^* = 6.33041441 \times 10^{-4}$

nxe, nye	w^* FEM	% error	R_c^* FEM	R_c^* Exact	% error
4,2	6.3475237e-04	2.70e-01	3.498515e-02	2.622623e-02	33.4
8,4	6.3314446e-04	1.63e-02	-5.426225e-02	-5.837536e-02	7.046
16,8	6.3304782e-04	1.01e-03	-8.508153e-02	-8.659564e-02	1.748
32,16	6.3304184e-04	6.29e-05	-9.491726e-02	-9.541960e-02	5.26e-01
64,32	6.3304147e-04	3.92e-06	-9.790934e-02	-9.806637e-02	1.60e-01

Nondimensional maximum deflection, $w^* = w(a/2, b/2)D/|q|a^4$. Nondimensional corner reaction R_c, $R_c^* = R_c/|q|a^2$.

force $-qA$. However, if we exclude the weighted effective shear forces at the four corner nodes of the plate, then the sum of R_j is greater than $-qA$. The maximum normal stresses σ_{xx} and σ_{yy} on the plate centerline $y = b/2$, $z = h/2$ are compared with the exact solution in Figure 9.9b. A mesh convergence study for the stresses at a point on the plate centerline is given in Table 9.2.

Dimensional Example:
Consider a steel ($\rho = 7860$ kg/m^3, $E = 200$ GPa, $\nu = 0.25$) rectangular plate ($a =$ 1m, $b = 0.5$m) of thickness $h = 10$ mm. If we take the distributed pressure to be

Table 9.2
Example 9.1, Normal Stresses σ_{xx} and σ_{yy} at a Point on the Plate Centerline $y = b/2, z = h/2$

nxe, nye	x	σ^*_{xx} FEM	% error	σ^*_{yy} FEM	% error
4,2	0.75	-6.939767e+02	10.1	-1.674762e+03	16.5
8,4	0.875	-6.414011e+02	2.62	-1.551131e+03	3.84
16,8	0.9375	-6.275665e+02	6.59e-01	-1.521830e+03	9.43e-01
32,16	0.96875	-6.240760e+02	1.65e-01	-1.514592e+03	2.35e-01
64,32	0.984375	-6.232016e+02	4.13e-02	-1.512788e+03	5.87e-02

Nondimensional stresses, $\sigma^* = \sigma/|q|$.

the weight of unit area of the plate, $q = \rho g h = 771.066$ Pa, we find the maximum deflection at the plate center $w_{max} = 0.027456566$ mm. If the pressure is $q = 1$ MPa, the maximum deflection will be $w_{max} = 35.608581$ mm; much greater than the plate thickness. For such a loading nonlinear effects must be considered, see Chapter 10.

9.1.4 ISOPARAMETRIC HERMITE ELEMENTS

Petera and Pittman [84] developed an isoparametric Hermite plate element and demonstrated its use for plates with curved boundaries. It is a 4-node quadrilateral element with 4 degrees of freedom (DOF) at each node for a total of 16 DOF per element.

Mapping

$$x^e(\xi,\eta) = \sum_{j=1}^{16} \psi_j(\xi,\eta) X_j^e, \quad y^e(\xi,\eta) = \sum_{j=1}^{16} \psi_j(\xi,\eta) Y_j^e. \tag{9.68}$$

where $\psi_j, j = 1, \quad , 16$ are given by Equation (9.47). The geometric degrees of freedom (X_j^e, Y_j^e) are defined as follows. For node # 1,

$$X_1^e = x_1, \quad X_2^e = \left(\frac{\partial x}{\partial \xi}\right)_1, \quad X_3^e = \left(\frac{\partial x}{\partial \eta}\right)_1, \quad X_4^e = \left(\frac{\partial^2 x}{\partial \xi \partial \eta}\right)_1$$
$$Y_1^e = y_1, \quad Y_2^e = \left(\frac{\partial y}{\partial \xi}\right)_1, \quad Y_3^e = \left(\frac{\partial y}{\partial \eta}\right)_1, \quad Y_4^e = \left(\frac{\partial^2 y}{\partial \xi \partial \eta}\right)_1 \tag{9.69}$$

with similar definitions at the other 3 nodes resulting in 16 geometric parameters for mapping of each of the coordinates x and y. For mapping of x, we arrange them in an array $\{X^e\}$ of dimensions $(16,1)$.

$$\{X^e\} = \{X_1^e\, X_2^e\, X_3^e\, X_4^e\, X_5^e\, X_6^e\, X_7^e\, X_8^e\, X_9^e\, X_{10}^e\, X_{11}^e\, X_{12}^e\, X_{13}^e\, X_{14}^e\, X_{15}^e\, X_{16}^e\}^T \tag{9.70}$$

and for mapping of y, we have

$$\{Y^e\} = \{Y_1^e\, Y_2^e\, Y_3^e\, Y_4^e\, Y_5^e\, Y_6^e\, Y_7^e\, Y_8^e\, Y_9^e\, Y_{10}^e\, Y_{11}^e\, Y_{12}^e\, Y_{13}^e\, Y_{14}^e\, Y_{15}^e\, Y_{16}^e\}^T. \tag{9.71}$$

Isoparametric Hermite mapping is now written as

$$\{x^e(\xi,\eta)\; y^e(\xi,\eta)\} = \{\psi\}\left[\{X^e\}\; \{Y^e\}\right]. \tag{9.72}$$

Computations of element matrices usually require values of the first and second partial derivatives of the trial solution $W^e(x,y)$ with respect to x and y. However, the trial solution is expressed explicitly in terms of ξ and η rather than x and y. We first compute first and second derivatives of the trial solution with respect to ξ and η. Those derivatives are then transformed to derivatives with respect to x and y.

$$W_x^e = W_\xi^e \xi_x + W_\eta^e \eta_x, \quad W_y^e = W_\xi^e \xi_y + W_\eta^e \eta_y \tag{9.73}$$

$$\begin{aligned}
W_{xx}^e &= W_{\xi\xi}^e \xi_x \xi_x + W_{\eta\eta}^e \eta_x \eta_x + 2W_{\xi\eta}^e \xi_x \eta_x + W_\xi^e \xi_{xx} + W_\eta^e \eta_{xx} \\
W_{yy}^e &= W_{\xi\xi}^e \xi_y \xi_y + W_{\eta\eta}^e \eta_y \eta_y + 2W_{\xi\eta}^e \xi_y \eta_y + W_\xi^e \xi_{yy} + W_\eta^e \eta_{yy} \\
W_{xy}^e &= W_{\xi\xi}^e \xi_x \xi_y + W_{\eta\eta}^e \eta_x \eta_y + W_{\xi\eta}^e (\xi_x \eta_y + \xi_y \eta_x) + W_\xi^e \xi_{xy} + W_\eta^e \eta_{xy}
\end{aligned} \tag{9.74}$$

with subscripts indicate partial differentiation. At any selected point (ξ,η) in the canonical element, we compute the coordinates of the corresponding point (x,y) in the physical space using the mapping Equation (9.68). By direct differentiation of the shape functions, we also compute the following 10 parameters

$$a_1 = x_\xi,\; a_2 = x_\eta,\; a_3 = x_{\xi\xi},\; a_4 = x_{\xi\eta},\; a_5 = x_{\eta\eta} \tag{9.75}$$

$$b_1 = y_\xi,\; b_2 = y_\eta,\; b_3 = y_{\xi\xi},\; b_4 = y_{\xi\eta},\; b_5 = y_{\eta\eta}. \tag{9.76}$$

The required derivatives of ξ and η with respect to x and y, can be computed as follows.

$$J = a_1 b_2 - a_2 b_1 \tag{9.77}$$

$$J\xi_x = b_2,\; J\xi_y = -a_2,\; J\eta_x = -b_1,\; J\eta_y = a_1 \tag{9.78}$$

$$\begin{aligned}
\xi_{xx} &= \frac{\partial}{\partial x}(\xi_x) \\
&= \frac{\partial}{\partial \xi}(\xi_x)\,\xi_x + \frac{\partial}{\partial \eta}(\xi_x)\,\eta_x \\
&= \frac{\partial}{\partial \xi}\left(\frac{y_\eta}{J}\right)\xi_x + \frac{\partial}{\partial \eta}\left(\frac{y_\eta}{J}\right)\eta_x \\
&= \left(\frac{y_{\xi\eta}}{J} - \frac{y_\eta J_\xi}{J^2}\right)\frac{y_\eta}{J} + \left(\frac{y_{\eta\eta}}{J} - \frac{y_\eta J_\eta}{J^2}\right)\frac{-y_\xi}{J} \\
J^3 \xi_{xx} &= \left(y_{\xi\eta} J - y_\eta J_\xi\right) y_\eta - \left(y_{\eta\eta} J - y_\eta J_\eta\right) y_\xi
\end{aligned} \tag{9.79}$$

$$
\begin{aligned}
J_\xi &= y_{\xi\eta}x_\xi + y_\eta x_{\xi\xi} - y_{\xi\xi}x_\eta - y_\xi x_{\xi\eta} \\
&= b_4a_1 + b_2a_3 - b_3a_2 - b_1a_4 \\
J_\eta &= y_{\eta\eta}x_\xi + y_\eta x_{\xi\eta} - y_{\xi\eta}x_\eta - y_\xi x_{\eta\eta} \\
&= b_5a_1 + b_2a_4 - b_4a_2 - b_1a_5
\end{aligned} \tag{9.80}
$$

$$
\begin{aligned}
J^3\xi_{xx} &= -a_5b_1^2b_2 + a_2b_5b_1^2 + 2a_4b_1b_2^2 - 2a_2b_4b_1b_2 - a_3b_2^3 + a_2b_3b_2^2 \\
J^3\xi_{yy} &= b_5a_1^2a_2 - a_5b_2a_1^2 - 2b_4a_1a_2^2 + 2a_4b_2a_1a_2 + b_3a_2^3 - a_3b_2a_2^2 \\
J^3\xi_{xy} &= a_2a_3b_2^2 - a_1a_4b_2^2 + a_2^2b_1b_4 - a_2^2b_2b_3 - a_1a_2b_1b_5 + a_1a_2b_2b_4 \\
&\quad + a_1a_5b_1b_2 - a_2a_4b_1b_2 \\
J^3\eta_{xx} &= a_5b_1^3 - 2a_4b_1^2b_2 - a_1b_5b_1^2 + a_3b_1b_2^2 + 2a_1b_4b_1b_2 - a_1b_3b_2^2 \\
J^3\eta_{yy} &= -b_5a_1^3 + 2b_4a_1^2a_2 + a_5b_1a_1^2 - b_3a_1a_2^2 - 2a_4b_1a_1a_2 + a_3b_1a_2^2 \\
J^3\eta_{xy} &= a_2a_4b_1^2 - a_1a_5b_1^2 + a_1^2b_1b_5 - a_1^2b_2b_4 - a_1a_2b_1b_4 + a_1a_2b_2b_3 \\
&\quad + a_1a_4b_1b_2 - a_2a_3b_1b_2.
\end{aligned} \tag{9.81}
$$

Stiffness Matrix

For isoparametric elements, stiffness matrix is given by

$$
[K^e] = \int_{-1}^{+1}\int_{-1}^{+1} [B^e]^T \left[\tilde{D}^e\right][B^e] J d\xi d\eta \tag{9.82}
$$

where

$$
[B^e] = [H_1^e]\left[\tilde{B}_1^e\right] + [H_2^e]\left[\tilde{B}_2^e\right] \tag{9.83}
$$

$$
[H_1^e] = \begin{bmatrix} \xi_x\xi_x & \eta_x\eta_x & \xi_x\eta_x \\ \xi_y\xi_y & \eta_y\eta_y & \xi_y\eta_y \\ 2\xi_x\xi_y & 2\eta_x\eta_y & \xi_x\eta_y + \xi_y\eta_x \end{bmatrix} \tag{9.84}
$$

$$
\left[\tilde{B}_1^e\right] = \begin{bmatrix} \frac{\partial^2\psi_1}{\partial\xi^2} & \frac{\partial^2\psi_2}{\partial\xi^2} & \cdots & \frac{\partial^2\psi_{16}}{\partial\xi^2} \\ \frac{\partial^2\psi_1}{\partial\eta^2} & \frac{\partial^2\psi_2}{\partial\eta^2} & \cdots & \frac{\partial^2\psi_{16}}{\partial\eta^2} \\ 2\frac{\partial^2\psi_1}{\partial\xi\partial\eta} & 2\frac{\partial^2\psi_2}{\partial\xi\partial\eta} & \cdots & 2\frac{\partial^2\psi_{16}}{\partial\xi\partial\eta} \end{bmatrix} \tag{9.85}
$$

$$
[H_2^e] = \begin{bmatrix} \xi_{xx} & \eta_{xx} \\ \xi_{yy} & \eta_{yy} \\ 2\xi_{xy} & 2\eta_{xy} \end{bmatrix} \tag{9.86}
$$

$$
\left[\tilde{B}_2^e\right] = \begin{bmatrix} \frac{\partial\psi_1}{\partial\xi} & \frac{\partial\psi_2}{\partial\xi} & \cdots & \frac{\partial\psi_{16}}{\partial\xi} \\ \frac{\partial\psi_1}{\partial\eta} & \frac{\partial\psi_2}{\partial\eta} & \cdots & \frac{\partial\psi_{16}}{\partial\eta} \end{bmatrix}. \tag{9.87}
$$

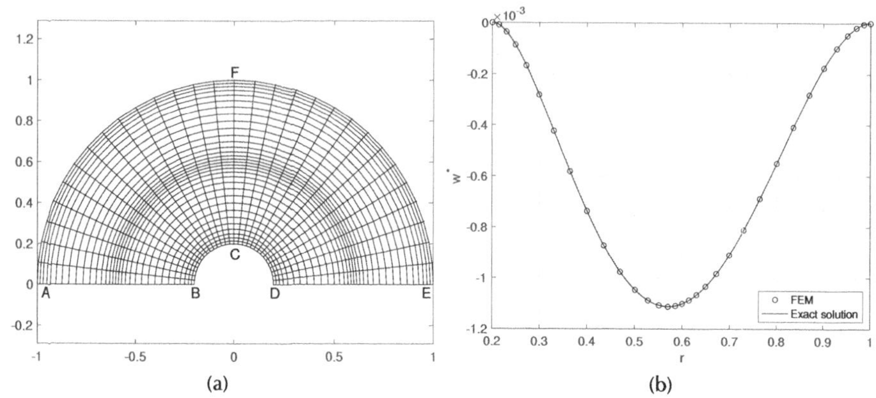

Figure 9.10 Comparison of finite element solution to exact analytic solution for uniformly loaded annular plate. (a) A mesh of 32×30 Hermite elements. AB and DE are simply supported, BCD and AFE are clamped. (b) Deflection on line $\theta = \pi/2$, $w^* = wD/|q|b^4$, $\nu = 0.25$.

Forcing Vector

$$\{F^e\} = \int_{-1}^{+1} \int_{-1}^{+1} \{\psi\}^T (q + \tilde{f}_z) J d\xi d\eta \tag{9.88}$$

$$\left\{ \begin{array}{c} W_x^e \\ W_y^e \end{array} \right\} = \begin{bmatrix} \xi_x & \eta_x \\ \xi_y & \eta_y \end{bmatrix} \left[\tilde{B}_2^e\right] \{U^e\}. \tag{9.89}$$

Example 9.2. Uniformly Loaded Semi-circular Annular Plate

We consider an annular plate of inner and outer radii a and b shown in Figure 9.10a. The straight edges AB, and DE (radial lines) are simply supported, and the semi-circular arcs EFA and DCB are clamped. Using Cartesian coordinates, we apply the boundary conditions

$$\text{on } AB \text{ and } DE: \; w = 0, \; M_{yy} = 0 \tag{9.90}$$

$$\text{on } EFA \text{ and } DCB: \; w = 0, \; \frac{\partial w}{\partial x} = 0, \; \frac{\partial w}{\partial y} = 0. \tag{9.91}$$

We generate a mesh in polar coordinates made of radial and semi-circular lines. The mesh is uniform in the circumferential direction with $\Delta\theta = \pi/n_\theta$, where n_θ is the number of intervals in the θ-direction. The grid is nonuniform in the radial direction and the number of intervals is n_r. The radial coordinate is given by

$$r = \frac{b+a}{2} + \frac{b-a}{2} g(\zeta) \tag{9.92}$$

where $-1 \leq \zeta \leq 1$ is uniformly distributed, $\Delta\zeta = 2/n_r$, and $g(-1) = -1$ and $g(1) = 1$. A 4-node quadrilateral element in polar coordinates is bounded by two radial lines $\theta = \theta_1^e (= \theta_2^e)$ and $\theta = \theta_4^e (= \theta_3^e)$, and two circular arcs $r = r_1^e (= r_4^e)$ and $r = r_2^e (= r_3^e)$ or equivalently ζ_1 and ζ_2. A point (ξ, η) in the canonical square element is mapped to a point with polar coordinates (ζ, θ) by

$$\zeta = \zeta_1^e + \frac{1}{2}\Delta\zeta\,(\xi + 1)\,,\ \theta = \theta_1^e + \frac{1}{2}\Delta\theta\,(\eta + 1) \tag{9.93}$$

where

$$\Delta\zeta = \zeta_2^e - \zeta_1^e,\quad \Delta\theta = \theta_4^e - \theta_1^e. \tag{9.94}$$

The Cartesian coordinates (x, y) of the point are given by

$$x = rcos\theta,\quad y = rsin\theta \tag{9.95}$$

and can be expressed in terms of (ξ, η), thus we obtain

$$\begin{aligned} x_\xi &= r_\xi cos\theta,\quad x_\eta = -\frac{1}{2}\Delta\theta rsin\theta \\ y_\xi &= r_\xi sin\theta,\quad y_\eta = \frac{1}{2}\Delta\theta rcos\theta \end{aligned} \tag{9.96}$$

$$\begin{aligned} x_{\xi\xi} &= r_{\xi\xi} cos\theta,\quad x_{\eta\eta} = -\frac{1}{4}(\Delta\theta)^2 rcos\theta,\quad x_{\xi\eta} = -\frac{1}{2} r_\xi \Delta\theta sin\theta \\ y_{\xi\xi} &= r_{\xi\xi} sin\theta,\quad y_{\eta\eta} = -\frac{1}{4}(\Delta\theta)^2 rsin\theta,\quad y_{\xi\eta} = \frac{1}{2} r_\xi \Delta\theta cos\theta \end{aligned} \tag{9.97}$$

where

$$r_\xi = \frac{b-a}{2}\frac{\Delta\zeta}{2}\frac{dg}{d\zeta},\ r_{\xi\xi} = \frac{b-a}{2}\left(\frac{\Delta\zeta}{2}\right)^2 \frac{d^2g}{d\zeta^2}. \tag{9.98}$$

Uniform mesh in the radial direction, if desired, can be obtained by putting $g(\zeta) = \zeta$, or using $\Delta r = (b-a)/n_r$, and hence

$$r = r_1^e + \frac{1}{2}\Delta r\,(\xi + 1)\,,\ \theta = \theta_1^e + \frac{1}{2}\Delta\theta\,(\eta + 1) \tag{9.99}$$

where

$$\Delta r = r_2^e - r_1^e,\quad \Delta\theta = \theta_4^e - \theta_1^e \tag{9.100}$$

$$\begin{aligned} x_\xi &= \frac{1}{2}\Delta rcos\theta,\quad x_\eta = -\frac{1}{2}\Delta\theta rsin\theta \\ y_\xi &= \frac{1}{2}\Delta rsin\theta,\quad y_\eta = \frac{1}{2}\Delta\theta rcos\theta \end{aligned} \tag{9.101}$$

Table 9.3
Deflection of a Semi-circular Annular Plate, $a = 0.2$, $b = 1$. Percentage Error in $w^*(0.6, \pi/2)$. $w^*_{exact} = -1.1019329e-03$

$n_r \parallel n_\theta =$	30	60	120
8	5.57e-02	6.00e-02	6.01e-02
16	4.75e-03	3.78e-03	4.34e-03
32	1.15e-02	8.47e-04	2.09e-04
64	1.37e-02	1.43e-03	1.24e-04
Non-uniform radial mesh			
32	6.10e-04	1.61e-04	2.09e-04
64	2.84e-03	2.03e-05	2.23e-05

(n_r, n_θ), Number of elements in r and θ directions.

$$x_{\xi\xi} = 0, \quad x_{\eta\eta} = -\frac{1}{4}(\Delta\theta)^2 rcos\theta, \quad x_{\xi\eta} = -\frac{1}{4}\Delta r \Delta\theta sin\theta$$
$$y_{\xi\xi} = 0, \quad y_{\eta\eta} = -\frac{1}{4}(\Delta\theta)^2 rsin\theta, \quad y_{\xi\eta} = \frac{1}{4}\Delta r \Delta\theta cos\theta. \tag{9.102}$$

In this example, the stretching function $g(\zeta)$ concentrates points near the clamped edges $r = a$ and $r = b$, and in the middle of the radial domain. It is given by

$$g(\zeta) = \zeta - \alpha sin(2\pi\zeta) \tag{9.103}$$

and $\alpha = 0.07$. A mesh of 960 elements in polar coordinates is shown in Figure 9.10a. The dimensionless deflection is defined by $w^* = wD/|q|b^4$. Comparison with the exact solution (e.g., [64]) for deflection on the line $\theta = \pi/2$ is shown in Figure 9.10b. A mesh convergence study for the deflection at the point $(r, \theta) = (0.6, \pi/2)$ is shown in Table 9.3. Uniform mesh ($\alpha = 0$) is used for the first four lines in the table, and non-uniform mesh with optimized α is used for the last two lines.

9.2 REISSNER–MINDLIN FIRST-ORDER SHEAR DEFORMATION PLATE THEORY

Kirchhoff–Love plate theory assumes that a material filament initially normal to the middle plane remains normal to that plane after deflection. The assumption is valid for very thin plates. Thus, strain energy due to transverse shear is neglected, and the model exhibits higher rigidity than the physical plate. Reissner ([98–100]), and Mindlin [76] plate theory allows arbitrary rotations of the filament relative to the normal direction to the midplane after deflection. As shown in Figure 9.11, a filament *ab* initially normal to the middle plane before deflection takes the position *a′b′* after deflection. It rotates through an angle θ_x about the negative y-axis. A positive value of θ_x produces negative x-displacement for a point q on the filament. The filament

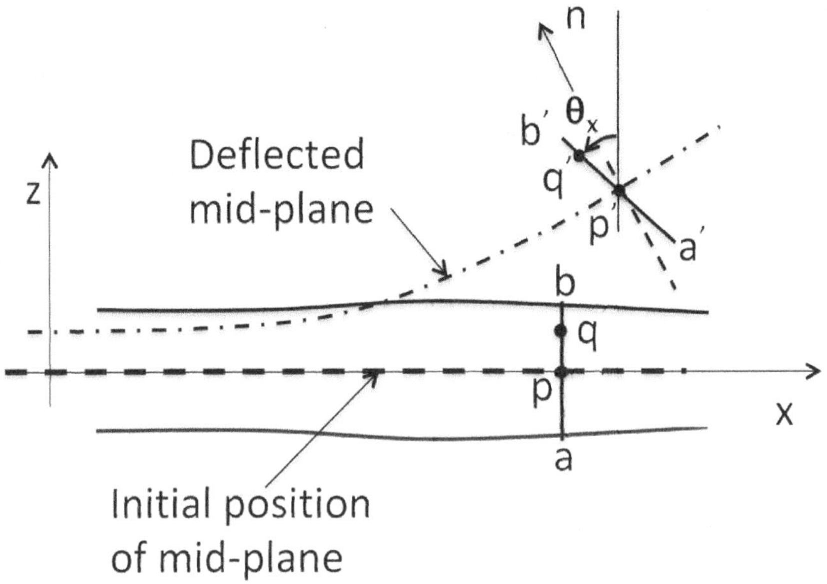

Figure 9.11 Kinematics of shear deformation plate theory. The y-axis is into the page. A filament ab initially normal to the middle plane rotates through an angle θ_x about the (negative) y-axis. Its rotation about the x-axis is not shown.

also rotates about the x-axis through an angle θ_y, but such a rotation is not shown in the view of this figure. In addition to the displacements of a point p in the middle plane (u_o, v_o, w_o), the theory introduces two more degrees of freedom representing the filament rotations about the x- and y-axes, and the displacement field is given by

$$\begin{aligned} u(x,y,z) &= u_o(x,y) - z\theta_x \\ v(x,y,z) &= v_o(x,y) - z\theta_y \\ w(x,y,z) &= w_o(x,y) \end{aligned} \tag{9.104}$$

where $\theta_x(x,y)$ is rotation of a filament initially normal to the middle plane about the (negative) y-axis, and $\theta_y(x,y)$ is its rotation about the (positive) x-axis. For these displacements, the linear strains are

$$\begin{aligned} \varepsilon_{xx} &= \frac{\partial u_o}{\partial x} - z\frac{\partial \theta_x}{\partial x}, \ \varepsilon_{yy} = \frac{\partial v_o}{\partial y} - z\frac{\partial \theta_y}{\partial y}, \ \varepsilon_{zz} = 0, \\ \gamma_{xy} &= \frac{\partial u_o}{\partial y} + \frac{\partial v_o}{\partial x} - z\left(\frac{\partial \theta_x}{\partial y} + \frac{\partial \theta_y}{\partial x}\right), \\ \gamma_{xz} &= -\theta_x + \frac{\partial w_o}{\partial x}, \ \gamma_{yz} = -\theta_y + \frac{\partial w_o}{\partial y}. \end{aligned} \tag{9.105}$$

Plate Bending Problem

We assume that there are no externally applied inplane loading and that the supports do not produce such loading, and that the deflection is smaller than the plate thickness. With these assumptions, the inplane displacements (u_o, v_o) due to transverse loading are negligible (Timoshenko and Krieger [118]). Thus, the displacements, strains, and stresses can be found by solving for three functions, namely, the deflection of middle plane $w_o(x,y)$ and rotations $\theta_x(x,y)$ and $\theta_y(x,y)$.

Change Notation: We want to determine the middle plane deflection $w_o(x,y)$, in the following we drop the subscript "o", and denote it by $w(x,y)$. The displacement field is

$$u(x,y,z) = -z\theta_x,\ v(x,y,z) = -z\theta_y,\ w(x,y,z) = w(x,y). \tag{9.106}$$

Strains

$$\begin{gathered}\varepsilon_{xx} = -z\frac{\partial \theta_x}{\partial x},\ \varepsilon_{yy} = -z\frac{\partial \theta_y}{\partial y},\ \varepsilon_{zz} = 0 \\ \gamma_{xy} = -z\left(\frac{\partial \theta_x}{\partial y} + \frac{\partial \theta_y}{\partial x}\right),\ \gamma_{xz} = -\theta_x + \frac{\partial w}{\partial x},\ \gamma_{yz} = -\theta_y + \frac{\partial w}{\partial y}.\end{gathered} \tag{9.107}$$

We note that for the special case of *CPT*: $\theta_x = \partial w/\partial x$, and $\theta_y = \partial w/\partial y$, hence $\gamma_{xz} = 0$ and $\gamma_{yz} = 0$. The five components of strains are divided into two groups. The first group is the flexural group

$$\left\{\varepsilon^b\right\} \equiv \left\{\begin{array}{c}\varepsilon_{xx}\\ \varepsilon_{yy}\\ \gamma_{xy}\end{array}\right\} = -z\left\{\begin{array}{c}\frac{\partial \theta_x}{\partial x}\\ \frac{\partial \theta_y}{\partial y}\\ \frac{\partial \theta_x}{\partial y} + \frac{\partial \theta_y}{\partial x}\end{array}\right\},\quad \left\{\varepsilon^b\right\} = -z\{\kappa\} \tag{9.108}$$

where

$$\{\kappa\} = \left\{\begin{array}{c}\frac{\partial \theta_x}{\partial x}\\ \frac{\partial \theta_y}{\partial y}\\ \frac{\partial \theta_x}{\partial y} + \frac{\partial \theta_y}{\partial x}\end{array}\right\} \tag{9.109}$$

and the second group is the transverse shear strains,

$$\{\varepsilon^s\} \equiv \left\{\begin{array}{c}\gamma_{xz}\\ \gamma_{yz}\end{array}\right\} = \left\{\begin{array}{c}-\theta_x + \frac{\partial w}{\partial x}\\ -\theta_y + \frac{\partial w}{\partial y}\end{array}\right\}. \tag{9.110}$$

We note that $\{\kappa\}$ and $\{\varepsilon^s\}$ are independent of z.

Stresses

The corresponding two groups of stresses are

$$\begin{aligned}\left\{\sigma^b\right\} \equiv \left\{\begin{array}{c}\sigma_{xx}\\ \sigma_{yy}\\ \sigma_{xy}\end{array}\right\} &= \frac{E}{1-\nu^2}\left[\begin{array}{ccc}1 & \nu & 0\\ \nu & 1 & 0\\ 0 & 0 & \frac{1-\nu}{2}\end{array}\right]\left\{\begin{array}{c}\varepsilon_{xx}\\ \varepsilon_{yy}\\ \gamma_{xy}\end{array}\right\}\\ \left\{\sigma^b\right\} &= [D]\left\{\varepsilon^b\right\}\\ &= -z[D]\{\kappa\}\end{aligned} \tag{9.111}$$

$$\begin{aligned}\{\sigma^s\} \equiv \left\{\begin{array}{c}\sigma_{xz}\\ \sigma_{yz}\end{array}\right\} &= \frac{E}{2(1+\nu)}\left[\begin{array}{cc}1 & 0\\ 0 & 1\end{array}\right]\left\{\begin{array}{c}-\theta_x+\frac{\partial w}{\partial x}\\ -\theta_y+\frac{\partial w}{\partial y}\end{array}\right\}\\ \{\sigma^s\} &= [D^s]\{\varepsilon^s\}\end{aligned} \tag{9.112}$$

we note that $E/2(1+\nu) = G$ is the shear modulus.

9.2.1 PRINCIPLE OF VIRTUAL DISPLACEMENTS

A virtual displacement field is an arbitrary field but it has to be consistent with the kinematic constraints imposed on the actual displacement field. Thus, the virtual displacement field must include three generalized displacements given by the deflection of the middle plane $\bar{w}(x,y)$, and two independent rotations $\bar{\theta}_x$ and $\bar{\theta}_y$. Hence, the Cartesian components of virtual displacements are

$$\bar{u}(x,y,z) = -z\bar{\theta}_x, \quad \bar{v}(x,y,z) = -z\bar{\theta}_y, \quad \bar{w}(x,y,z) = \bar{w}(x,y) \tag{9.113}$$

and virtual strains are

$$\left\{\bar{\varepsilon}^b\right\} \equiv \left\{\begin{array}{c}\bar{\varepsilon}_{xx}\\ \bar{\varepsilon}_{yy}\\ \bar{\gamma}_{xy}\end{array}\right\} = -z\left\{\begin{array}{c}\frac{\partial\bar{\theta}_x}{\partial x}\\ \frac{\partial\bar{\theta}_y}{\partial y}\\ \frac{\partial\bar{\theta}_x}{\partial y}+\frac{\partial\bar{\theta}_y}{\partial x}\end{array}\right\}, \quad \left\{\bar{\varepsilon}^b\right\} = -z\{\bar{\kappa}\} \tag{9.114}$$

where

$$\{\bar{\kappa}\} = \left\{\begin{array}{c}\frac{\partial\bar{\theta}_x}{\partial x}\\ \frac{\partial\bar{\theta}_y}{\partial y}\\ \frac{\partial\bar{\theta}_x}{\partial y}+\frac{\partial\bar{\theta}_y}{\partial x}\end{array}\right\} \tag{9.115}$$

$$\{\bar{\varepsilon}^s\} \equiv \left\{\begin{array}{c}\bar{\gamma}_{xz}\\ \bar{\gamma}_{yz}\end{array}\right\} = \left\{\begin{array}{c}-\bar{\theta}_x+\frac{\partial\bar{w}}{\partial x}\\ -\bar{\theta}_y+\frac{\partial\bar{w}}{\partial y}\end{array}\right\} \tag{9.116}$$

We note that $\{\bar{\kappa}\}$ and $\{\bar{\varepsilon}^s\}$ are independent of z.

The bounding surface of the plate consists of three parts: The upper and lower surfaces $z = \pm h$, and a cylindrical (not necessarily circular) surface with axis parallel to the z-axis; the intersection of the cylinder with the middle plane is the contour Γ of area A. The principle of virtual displacements Equation (9.16) takes the form

$$\int_A \int_{-h/2}^{h/2} \left\{ \bar{\varepsilon}_{xx}\ \bar{\varepsilon}_{yy}\ \bar{\gamma}_{xy} \right\} \left\{ \begin{array}{c} \sigma_{xx} \\ \sigma_{yy} \\ \sigma_{xy} \end{array} \right\} dz dA + \int_A \int_{-h/2}^{h/2} \left\{ \bar{\gamma}_{xz}\ \bar{\gamma}_{yz} \right\} \left\{ \begin{array}{c} \sigma_{xz} \\ \sigma_{yz} \end{array} \right\} dz dA =$$

$$\int_A \left\{ \bar{u}\ \bar{v}\ \bar{w} \right\} \left\{ \begin{array}{c} t_x \\ t_y \\ t_z \end{array} \right\}_{z=+h/2} dA + \int_A \left\{ \bar{u}\ \bar{v}\ \bar{w} \right\} \left\{ \begin{array}{c} t_x \\ t_y \\ t_z \end{array} \right\}_{z=-h/2} dA$$

$$+ \oint_\Gamma \int_{-h/2}^{h/2} \left\{ \bar{u}\ \bar{v}\ \bar{w} \right\} \left\{ \begin{array}{c} t_x \\ t_y \\ t_z \end{array} \right\} dz ds + \int_A \int_{-h/2}^{h/2} \left\{ \bar{u}\ \bar{v}\ \bar{w} \right\} \left\{ \begin{array}{c} f_x \\ f_y \\ f_z \end{array} \right\} dz dA. \tag{9.117}$$

The lower and upper surfaces $z = \pm h/2$ are assumed to be free of traction parallel to the middle plane ($t_x = t_y = 0$), and only distributed load $t_z = q(x,y)$ or concentrated forces normal to the middle plane are applied on those surfaces. A concentrated force of magnitude P at (x_p, y_p) can be represented as a distribution (force per unit area) using the Dirac delta function δ; $q(x,y) = P\delta(x - x_p)\,\delta(y - y_p)$. We also assume that the body force is parallel to the z-axis, hence $f_x = f_y = 0$.

$$\int_A \int_{-h/2}^{h/2} \left\{ \bar{\varepsilon}^b \right\}^T \left\{ \sigma^b \right\} dz dA + \int_A \int_{-h/2}^{h/2} \left\{ \bar{\varepsilon}^s \right\}^T \left\{ \sigma^s \right\} dz dA =$$

$$\int_A \bar{w} q dA + \int_A \int_{-h/2}^{h/2} \bar{w} f_z dz dA + \oint_\Gamma \int_{-h/2}^{h/2} \left\{ \bar{u}\ \bar{v}\ \bar{w} \right\} \left\{ \begin{array}{c} t_x \\ t_y \\ t_z \end{array} \right\} dz ds. \tag{9.118}$$

Substituting stresses from Equations (9.111)–(9.112), and virtual strains from Equations (9.114)–(9.116), we get

$$\int_A \int_{-h/2}^{h/2} z^2 \left\{ \bar{\kappa} \right\}^T [D] \left\{ \kappa \right\} dz dA + \int_A \int_{-h/2}^{h/2} \left\{ \bar{\varepsilon}^s \right\}^T [D^s] \left\{ \varepsilon^s \right\} dz dA =$$

$$\int_A \bar{w} q dA + \int_A \int_{-h/2}^{h/2} \bar{w} f_z dz dA + \oint_\Gamma \int_{-h/2}^{h/2} \left\{ \bar{u}\ \bar{v}\ \bar{w} \right\} \left\{ \begin{array}{c} t_x \\ t_y \\ t_z \end{array} \right\} dz ds. \tag{9.119}$$

We evaluate the integral

$$\int_{-h/2}^{h/2} z^2 dz = \frac{h^3}{12} \tag{9.120}$$

define

$$
\begin{aligned}
[\tilde{D}] &= \frac{h^3}{12}[D] \\
&= \frac{Eh^3}{12(1-\nu^2)}\begin{bmatrix} 1 & \nu & 0 \\ \nu & 1 & 0 \\ 0 & 0 & \frac{1-\nu}{2} \end{bmatrix}
\end{aligned} \tag{9.121}
$$

$$
\begin{aligned}
[\tilde{D}^s] &= \kappa h [D^s] \\
&= \frac{\kappa E h}{2(1+\nu)}\begin{bmatrix} 1 & 0 \\ 0 & 1 \end{bmatrix}
\end{aligned} \tag{9.122}
$$

where $\kappa = 5/6$, is a shear correction factor,

$$
\tilde{f}_z = \int_{-h/2}^{h/2} f_z dz \tag{9.123}
$$

and rewrite Equation (9.119)

$$
\int_A \{\bar{\kappa}\}^T [\tilde{D}]\{\kappa\} dA + \int_A \{\bar{\varepsilon}^s\}^T [\tilde{D}^s]\{\varepsilon^s\} dA =
$$

$$
\int_A \bar{w}\,(q+\tilde{f}_z)\,dA + \oint_\Gamma \int_{-h/2}^{h/2} \left\{ \bar{u} \;\; \bar{v} \;\; \bar{w} \right\} \begin{Bmatrix} t_x \\ t_y \\ t_z \end{Bmatrix} dz ds. \tag{9.124}
$$

We note that $\tilde{f}_z$ is a body force per unit area of the middle plane. At a point p on the contour Γ, we define local coordinate system (n,s,z), where $\hat{n}$ is the unit normal to Γ in the middle plane, $\hat{s}$ is unit tangent such that $\hat{n} = \hat{s} \times \hat{k}$. Let us project the surface traction on the cylindrical surface S into normal $t_n = \sigma_{nn}$ and tangential components $t_s = \sigma_{ns}$ and $t_z = \sigma_{nz}$ as shown in Figure 9.4. Similarly, we project the virtual displacements

$$
\bar{u}_n = -z\bar{\theta}_n, \; \bar{u}_s = -z\bar{\theta}_s, \; \bar{u}_z = \bar{w} \tag{9.125}
$$

where $\bar{\theta}_n$ is virtual rotation of a filament initially normal to the middle plane about the (negative) s-axis, and $\bar{\theta}_s$ is its virtual rotation about the (positive) n-axis. The actual rotations of the filament are denoted by θ_n and θ_s. We note that

$$
\begin{aligned}
\bar{u}t_x + \bar{v}t_y + \bar{w}t_z &= \vec{\bar{u}} \cdot \vec{t} \\
&= \bar{u}_n\sigma_{nn} + \bar{u}_s\sigma_{ns} + \bar{w}\sigma_{nz}
\end{aligned} \tag{9.126}
$$

and consider the integral

$$\begin{aligned}
\int_{-h/2}^{h/2} \left\{ \bar{u}\ \bar{v}\ \bar{w} \right\} \left\{ \begin{array}{c} t_x \\ t_y \\ t_z \end{array} \right\} dz &= \int_{-h/2}^{h/2} \vec{\bar{u}} \cdot \vec{t}\, dz \\
&= \int_{-h/2}^{h/2} \left(\bar{u}_n \sigma_{nn} + \bar{u}_s \sigma_{ns} + \bar{w} \sigma_{nz} \right) dz \\
&= \int_{-h/2}^{h/2} \left[\left(-z\bar{\theta}_n\right) \sigma_{nn} + \left(-z\bar{\theta}_s\right) \sigma_{ns} + \bar{w}\sigma_{nz} \right] dz \\
&= -\bar{\theta}_n M_{nn} - \bar{\theta}_s M_{ns} + \bar{w} Q_{nz} \qquad (9.127)
\end{aligned}$$

where

$$M_{nn} = \int_{-h/2}^{h/2} z\sigma_{nn} dz, \ M_{ns} = \int_{-h/2}^{h/2} z\sigma_{ns} dz, \ Q_{nz} = \int_{-h/2}^{h/2} \sigma_{nz} dz. \qquad (9.128)$$

The principle of virtual displacements for the Reissner–Mindlin plate is

$$\begin{aligned}
&\int_A \{\bar{\kappa}\}^T \left[\tilde{D}\right] \{\kappa\} dA + \int_A \{\bar{\varepsilon}^s\}^T \left[\tilde{D}^s\right] \{\varepsilon^s\} dA = \\
&\int_A \bar{w} \left(q + \tilde{f}_z\right) dA + \oint_\Gamma \left(-\bar{\theta}_n M_{nn} - \bar{\theta}_s M_{ns} + \bar{w} Q_{nz}\right) ds. \qquad (9.129)
\end{aligned}$$

Primary and Secondary Variables

Looking at the boundary term given by the contour integral on the right of Equation (9.129), we define secondary variables as the coefficients of virtual generalized displacements (Reddy [97]). Thus, the secondary variables are the shear force Q_{nz}, twisting moment M_{ns}, and bending moment M_{nn}, all per unit length of boundary Γ. The corresponding primary variables are deflection w, rotation θ_s, and rotation θ_n, respectively. Thus, according to Reissner plate theory ([98], [99]), it is necessary that we prescribe three boundary conditions at each point on the plate edge.

Boundary Conditions

At each point of the boundary Γ, we need to specify three boundary conditions. We may specify one member of the three pairs of primary and secondary variables: (w, Q_{nz}), (θ_s, M_{ns}), and (θ_n, M_{nn}), or a relation between the two members of any of the pairs. The notions of essential, natural, and mixed boundary conditions applies. Prescribing a primary variable (a kinematic quantity) such as w, θ_s, and θ_n constitutes an essential boundary condition. Whereas prescribing a secondary variable (kinetic quantity) constitutes a natural boundary condition. A mixed boundary condition prescribes a relation between a primary and a secondary variable of the same pair; but neither variable is known.

Simply Supported (Hinged) Edge

$$w = 0, \quad \theta_s = 0, \quad M_{nn} = 0. \tag{9.130}$$

The first and second conditions are *EBCs* while the third is an *NBC*. The secondary variables Q_{nz}, M_{ns} and primary variable θ_n are unknown. This set of boundary conditions is denoted by *SS*2. Alternatively, the second condition may be replaced by $M_{ns} = 0$, in which case θ_s is unknown. The alternative set is denoted by *SS*1. Unless otherwise indicated, simply supported edge means *SS*2.

Clamped (Fixed) Edge

$$w = 0, \quad \theta_s = 0, \quad \theta_n = 0. \tag{9.131}$$

All three conditions are *EBCs*. The secondary variables Q_{nz}, M_{ns}, and M_{nn} are unknown.

Free Edge

$$Q_{nz} = 0, \quad M_{ns} = 0, \quad M_{nn} = 0. \tag{9.132}$$

All three conditions are *NBCs*. The primary variables w, θ_s, and θ_n are unknown.

A Line of Symmetry

If a straight line in the xy-plane is a line of (geometry, material, and load) symmetry, we may choose to solve the problem on half the problem domain (on either side of the line). In this case, on that line we apply the conditions:

$$\theta_n = 0, \quad Q_{nz} = 0, \quad M_{ns} = 0. \tag{9.133}$$

For example, if the line of symmetry is parallel to the x-axis, we have

$$\theta_y = 0, \quad Q_{yz} = 0, \quad M_{yx} = 0 \tag{9.134}$$

and if it is parallel to the $y-$axis, we have

$$\theta_x = 0, \quad Q_{xz} = 0, \quad M_{xy} = 0. \tag{9.135}$$

In all of the above boundary conditions Equations (9.130–9.135), weighted secondary variables are either zero because of the *homogeneous NBCs* or unknown because of the *EBCs* (e.g., Hughes et al. [54].)

9.2.2 APPROXIMATIONS AND ELEMENT EQUATIONS

In Reissner–Mindlin plate theory only first derivatives of the unknown displacements and rotations appear in the curvature terms, and thus continuity of those derivatives across elements is not essential. Lagrange interpolation polynomials are admissible for approximating the displacements and rotations for the element. This is a major simplification over the *CPT*. For an element with n local nodes, the approximations over the element are

$$\begin{aligned}\theta_x^e(x,y) &= \sum_{j=1}^{n}\theta_{xj}^e\psi_j(x,y)\\ \theta_y^e(x,y) &= \sum_{j=1}^{n}\theta_{yj}^e\psi_j(x,y)\\ w^e(x,y) &= \sum_{j=1}^{n}w_j^e\psi_j(x,y)\end{aligned} \tag{9.136}$$

where $\psi_j(x,y)$, $j=1,\ldots,n$ are the element shape functions as derived in Chapter 4, including isoparametric elements. In matrix form, we have

$$\left\{\begin{array}{c}\theta_x^e(x,y)\\ \theta_y^e(x,y)\\ w^e(x,y)\end{array}\right\} = \begin{bmatrix}\psi_1 & 0 & 0 & \psi_2 & . & . & . & \psi_j & . & . & . & \psi_n & 0 & 0\\ 0 & \psi_1 & 0 & 0 & \psi_2 & . & . & . & \psi_j & . & . & 0 & \psi_n & 0\\ 0 & 0 & \psi_1 & 0 & 0 & \psi_2 & . & . & . & \psi_j & . & 0 & 0 & \psi_n\end{bmatrix}\left\{\begin{array}{c}\theta_{x1}^e\\ \theta_{y1}^e\\ w_1^e\\ .\\ \theta_{xj}^e\\ \theta_{yj}^e\\ w_j^e\\ .\\ \theta_{xn}^e\\ \theta_{yn}^e\\ w_n^e\end{array}\right\}. \tag{9.137}$$

We define the arrays

$$\{U^e\} = \left\{\theta_{x1}^e\ \theta_{y1}^e\ w_1^e\ \cdots\ \theta_{xj}^e\ \theta_{yj}^e\ w_j^e\ \cdots\ \theta_{xn}^e\ \theta_{yn}^e\ w_n^e\right\}^T \tag{9.138}$$

and

$$[\psi] = \begin{bmatrix}\psi_1 & 0 & 0 & \psi_2 & . & . & . & \psi_j & . & . & . & \psi_n & 0 & 0\\ 0 & \psi_1 & 0 & 0 & \psi_2 & . & . & . & \psi_j & . & . & 0 & \psi_n & 0\\ 0 & 0 & \psi_1 & 0 & 0 & \psi_2 & . & . & . & \psi_j & . & 0 & 0 & \psi_n\end{bmatrix} \tag{9.139}$$

and write the trial solution as

$$\left\{\begin{array}{c}\theta_x^e(x,y)\\ \theta_y^e(x,y)\\ w^e(x,y)\end{array}\right\} = [\psi]\{U^e\}. \tag{9.140}$$

Note that $[\psi]$ is a function of x and y and of dimensions $(3,3n)$, while $\{U^e\}$ are the values of generalized displacements at the element's nodes and of dimensions

$(3n,1)$. For later use, we also define the arrays

$$\{\psi\}_1 = \left\{ \psi_1\ 0\ 0\ \psi_2\ \ldots\ \psi_j\ \ldots\ \psi_n\ 0\ 0 \right\} \tag{9.141}$$

$$\{\psi\}_2 = \left\{ 0\ \psi_1\ 0\ 0\ \psi_2\ \ldots\ \psi_j\ ..\ 0\ \psi_n\ 0 \right\} \tag{9.142}$$

$$\{\psi\}_3 = \left\{ 0\ 0\ \psi_1\ 0\ 0\ \psi_2\ \ldots\ \psi_j\ .\ 0\ 0\ \psi_n \right\} \tag{9.143}$$

each is of dimensions $(1,3n)$. All of the terms in the statement of virtual displacements can now be written in terms of the element's shape functions and their derivatives.

Flexural Strains and Stresses

$$\left\{ \begin{array}{c} \frac{\partial \theta_x^e}{\partial x} \\ \frac{\partial \theta_y^e}{\partial y} \\ \frac{\partial \theta_x^e}{\partial y} + \frac{\partial \theta_y^e}{\partial x} \end{array} \right\} = \begin{bmatrix} \frac{\partial \psi_1}{\partial x} & 0 & 0 & \frac{\partial \psi_2}{\partial x} & . & .\,. & \frac{\partial \psi_j}{\partial x} & . & .\,. & \frac{\partial \psi_n}{\partial x} & 0 & 0 \\ 0 & \frac{\partial \psi_1}{\partial y} & 0 & 0 & \frac{\partial \psi_2}{\partial y} & .\,. & . & \frac{\partial \psi_j}{\partial y} & .\,. & 0 & \frac{\partial \psi_n}{\partial y} & 0 \\ \frac{\partial \psi_1}{\partial y} & \frac{\partial \psi_1}{\partial x} & 0 & \frac{\partial \psi_2}{\partial y} & \frac{\partial \psi_2}{\partial x} & 0\,. & \frac{\partial \psi_j}{\partial y} & \frac{\partial \psi_j}{\partial x} & 0\,. & \frac{\partial \psi_n}{\partial y} & \frac{\partial \psi_n}{\partial x} & 0 \end{bmatrix} \left\{ \begin{array}{c} \theta_{x1}^e \\ \theta_{y1}^e \\ w_1^e \\ . \\ \theta_{xj}^e \\ \theta_{yj}^e \\ w_j^e \\ . \\ \theta_{xn}^e \\ \theta_{yn}^e \\ w_n^e \end{array} \right\}. \tag{9.144}$$

We define the gradient matrix

$$\left[B^b\right] = \begin{bmatrix} \frac{\partial \psi_1}{\partial x} & 0 & 0 & \frac{\partial \psi_2}{\partial x} & . & .\,. & \frac{\partial \psi_j}{\partial x} & . & .\,. & \frac{\partial \psi_n}{\partial x} & 0 & 0 \\ 0 & \frac{\partial \psi_1}{\partial y} & 0 & 0 & \frac{\partial \psi_2}{\partial y} & .\,. & . & \frac{\partial \psi_j}{\partial y} & .\,. & 0 & \frac{\partial \psi_n}{\partial y} & 0 \\ \frac{\partial \psi_1}{\partial y} & \frac{\partial \psi_1}{\partial x} & 0 & \frac{\partial \psi_2}{\partial y} & \frac{\partial \psi_2}{\partial x} & 0\,. & \frac{\partial \psi_j}{\partial y} & \frac{\partial \psi_j}{\partial x} & 0\,. & \frac{\partial \psi_n}{\partial y} & \frac{\partial \psi_n}{\partial x} & 0 \end{bmatrix} \tag{9.145}$$

and write

$$\{\kappa\} = \left[B^b\right]\{U^e\} \tag{9.146}$$

$$\begin{aligned} \left\{\varepsilon^b\right\} &= -z\{\kappa\} \\ &= -z\left[B^b\right]\{U^e\} \end{aligned} \tag{9.147}$$

$$\begin{aligned}\left\{\sigma^b\right\} &= [D]\left\{\varepsilon^b\right\}\\ &= -z[D]\left[B^b\right]\{U^e\}\end{aligned} \tag{9.148}$$

Transverse Shear Strains and Stresses

$$\left\{\begin{matrix}-\theta_x^e+\frac{\partial w^e}{\partial x}\\ -\theta_y^e+\frac{\partial w^e}{\partial y}\end{matrix}\right\}=\begin{bmatrix}-\psi_1 & 0 & \frac{\partial \psi_1}{\partial x} & -\psi_2 & 0 & \frac{\partial \psi_2}{\partial x} & \ldots & -\psi_n & 0 & \frac{\partial \psi_n}{\partial x}\\ 0 & -\psi_1 & \frac{\partial \psi_1}{\partial y} & 0 & -\psi_2 & \frac{\partial \psi_2}{\partial y} & \ldots & 0 & -\psi_n & \frac{\partial \psi_n}{\partial y}\end{bmatrix}\left\{\begin{matrix}\theta_{x1}^e\\ \theta_{y1}^e\\ w_1^e\\ \cdot\\ \theta_{xj}^e\\ \theta_{yj}^e\\ w_j^e\\ \cdot\\ \theta_{xn}^e\\ \theta_{yn}^e\\ w_n^e\end{matrix}\right\} \tag{9.149}$$

$$[B^s]=\begin{bmatrix}-\psi_1 & 0 & \frac{\partial \psi_1}{\partial x} & -\psi_2 & 0 & \frac{\partial \psi_2}{\partial x} & \ldots & -\psi_n & 0 & \frac{\partial \psi_n}{\partial x}\\ 0 & -\psi_1 & \frac{\partial \psi_1}{\partial y} & 0 & -\psi_2 & \frac{\partial \psi_2}{\partial y} & \ldots & 0 & -\psi_n & \frac{\partial \psi_n}{\partial y}\end{bmatrix} \tag{9.150}$$

$$\{\varepsilon^s\}=[B^s]\{U^e\} \tag{9.151}$$

$$\begin{aligned}\{\sigma^s\} &= [D^s]\{\varepsilon^s\}\\ &= [D^s][B^s]\{U^e\}.\end{aligned} \tag{9.152}$$

Virtual Displacements and Strains: Galerkin

$$\left\{\begin{matrix}\bar{\theta}_x^e(x,y)\\ \bar{\theta}_y^e(x,y)\\ \bar{w}^e(x,y)\end{matrix}\right\}=[\psi]\{d\} \tag{9.153}$$

where $\{d\}$ is an array of arbitrary constants giving the virtual generalized displacements at the element's nodes.

$$\bar{\theta}_x^e=\{d\}^T\{\psi\}_1^T,\ \bar{\theta}_y^e=\{d\}^T\{\psi\}_2^T,\ \bar{w}=\{d\}^T\{\psi\}_3^T \tag{9.154}$$

$$\{\bar{\kappa}\}=\left[B^b\right]\{d\},\ \{\bar{\kappa}\}^T=\{d\}^T\left[B^b\right]^T \tag{9.155}$$

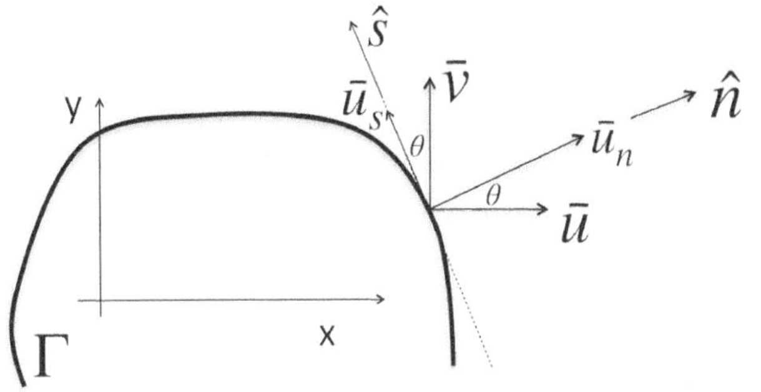

Figure 9.12 Projection of virtual displacements $(\bar{u}, \bar{v})$ into normal and tangential components $\bar{u}_n$ and $\bar{u}_s$. Unit normal $\hat{n} = cos\theta\hat{i} + sin\theta\hat{j}$.

$$\left\{\bar{\varepsilon}^b\right\}^T = -z\{d\}^T\left[B^b\right]^T, \{\bar{\varepsilon}^s\}^T = \{d\}^T[B^s]^T. \tag{9.156}$$

We transform $\bar{\theta}_n$ and $\bar{\theta}_s$ to $\bar{\theta}_x$ and $\bar{\theta}_y$. With reference to Figure 9.12, we write

$$\begin{aligned} -z\bar{\theta}_n &= \bar{u}_n = \bar{u}cos\theta + \bar{v}sin\theta = \bar{u}n_x + \bar{v}n_y \\ &= \left(-z\bar{\theta}_x\right)n_x + \left(-z\bar{\theta}_y\right)n_y \end{aligned} \tag{9.157}$$

$$\bar{\theta}_n = \bar{\theta}_x n_x + \bar{\theta}_y n_y \tag{9.158}$$

$$\begin{aligned} -z\bar{\theta}_s &= \bar{u}_s = \bar{v}cos\theta - \bar{u}sin\theta = \bar{v}n_x - \bar{u}n_y \\ &= \left(-z\bar{\theta}_y\right)n_x - \left(-z\bar{\theta}_x\right)n_y \end{aligned} \tag{9.159}$$

$$\bar{\theta}_s = \bar{\theta}_y n_x - \bar{\theta}_x n_y \tag{9.160}$$

$$\begin{aligned} \bar{\theta}_n M_{nn} + \bar{\theta}_s M_{ns} &= \left(\bar{\theta}_x n_x + \bar{\theta}_y n_y\right)M_{nn} + \left(\bar{\theta}_y n_x - \bar{\theta}_x n_y\right)M_{ns} \\ &= \bar{\theta}_x M_1 + \bar{\theta}_y M_2 \end{aligned} \tag{9.161}$$

$$M_1 = M_{nn}n_x - M_{ns}n_y, \; M_2 = M_{nn}n_y + M_{ns}n_x \tag{9.162}$$

$$M_{nn} = M_1 n_x + M_2 n_y, \; M_{ns} = M_2 n_x - M_1 n_y. \tag{9.163}$$

The statement of virtual displacements now becomes

$$\begin{aligned} &\int_{A_e}\{d\}^T\left[B^b\right]^T[\tilde{D}]\left[B^b\right]\{U^e\}dA + \int_{A_e}\{d\}^T[B^s]^T\left[\tilde{D}^s\right][B^s]\{U^e\}dA \\ &\qquad = \int_{A_e}\{d\}^T\{\psi\}_3^T(q + \tilde{f}_z)dA \\ &+ \oint_{\Gamma_e}\{d\}^T\left[\{\psi\}_1^T(-M_1) + \{\psi\}_2^T(-M_2) + \{\psi\}_3^T Q_{nz}\right]ds. \end{aligned} \tag{9.164}$$

We define,

Stiffness Matrix

$$[K^e] = \int_{A_e} \left[B^b\right]^T [\tilde{D}] \left[B^b\right] dA + \int_{A_e} [B^s]^T \left[\tilde{D}^s\right] [B^s] dA \tag{9.165}$$

$$[K^e] = \left[K^b\right] + [K^s] \tag{9.166}$$

$$\left[K^b\right] = \int_{A_e} \left[B^b\right]^T [\tilde{D}] \left[B^b\right] dA \tag{9.167}$$

$$[K^s] = \int_{A_e} [B^s]^T \left[\tilde{D}^s\right] [B^s] dA. \tag{9.168}$$

Forcing Vector

$$\{F^e\} = \int_{A_e} \{\psi\}_3^T (q + \tilde{f}_z) dA. \tag{9.169}$$

Weighted Secondary Variables Vector

$$\{R^e\} = \oint_{\Gamma_e} \left[\{\psi\}_1^T (-M_1) + \{\psi\}_2^T (-M_2) + \{\psi\}_3^T Q_{nz}\right] ds \tag{9.170}$$

and rewrite Equation (9.164),

$$\{d\}^T \{[K^e]\{U^e\} - \{F^e\} - \{R^e\}\} = 0. \tag{9.171}$$

Since $\{d\}^T$ is an arbitrary array of constants, the multiplying bracket must be zero, and we obtain the element equations,

$$[K^e]\{U^e\} = \{F^e\} + \{R^e\}. \tag{9.172}$$

Element equations are assembled into a global system, and boundary conditions are applied. The modified (or reduced) system is solved for the unknown primary and weighted secondary variables.

Implementation of Boundary Conditions on a Simply Supported Edge

There are three geometric variables arranged in the order (θ_x, θ_y, w) at each global node. Implementation of the boundary conditions on a simply supported edge that is parallel to the coordinate axes (x or y) is straightforward since the prescribed geometric or bending moment are aligned with the axes. However, implementation of the geometric $\theta_s = 0$ and moment $M_{nn} = 0$ conditions [see Equation (9.130)] at a global node on a curved boundary or inclined edge requires combining the first and second equations at that node. Equation (9.160) is used for θ_s and Equation (9.163) is used for M_{nn}. First, the local unit normal (n_x, n_y) is computed at all global nodes on the curved boundary. Second, the first equation in the assembled global system at a node on the boundary is multiplied by n_x and the second equation is multiplied by n_y. A combined equation is created by the sum of the two new equations and setting the sum of weighted secondary variables to zero. Third, the combined equation replaces the second equation of the global system at the node. Fourth, the first equation in global system at the node is replaced by $-n_y\theta_x + n_x\theta_y = 0$, which corresponds to $\theta_s = 0$.

Selective Integration Rules

The Reissner–Mindlin plate theory may experience shear lock for low order elements such as the 4-node bilinear element. Selective integration rules can be used to circumvent shear lock (Hughes et al. [54], Hughes [53]). Reduced-order integration is used for computing the shear stiffness matrix Equation (9.168), whereas full integration rules are used for all other matrices and vectors. Using Gauss–Legendre quadrature, the shear stiffness matrix for the 4-node bilinear element is computed by (1×1) rule while the 2×2 rule is used for the other matrices. For the biquadratic 9-node or 8-node element, we use 3×3 rule for full integration and 2×2 rule for reduced integration. For the bicubic 16-node element, we use 4×4 rule for full integration and 3×3 rule for reduced integration. The problem of shear lock is detrimental for the 4-node bilinear element and is much less so for other higher-order elements.

Example 9.3. Uniformly Loaded Rectangular Plate

We consider a simply supported ($SS2$) rectangular plate $-a/2 \le x \le a/2$, $-b/2 \le y \le b/2$. Assume $a = 2$ and $b = 1$, take $h = 0.02$, and $\nu = 0.25$. The plate is loaded by uniformly distributed pressure q. With the origin at the plate center, using double symmetry of load and deformation about the x and y axes, we solve the problem on the first quadrant $0 \le x \le 1$ and $0 \le y \le 0.5$.

Boundary Conditions:

Symmetry about the x-axis, $y = 0$

$$0 \le x \le 1: \quad \theta_y = 0, \quad Q_{yz} = 0, \quad M_{yx} = 0. \tag{9.173}$$

Symmetry about the y-axis, $x = 0$

$$0 \le y \le 0.5: \quad \theta_x = 0, \quad Q_{xz} = 0, \quad M_{xy} = 0. \tag{9.174}$$

Simply supported edge $SS2$, $x = 1$

$$0 \le y \le 0.5: \quad w = 0, \quad \theta_y = 0, \quad M_{xx} = 0. \tag{9.175}$$

Simply supported edge $SS2$, $y = 0.5$

$$0 \le x \le 1: \quad w = 0, \quad \theta_x = 0, \quad M_{yy} = 0. \tag{9.176}$$

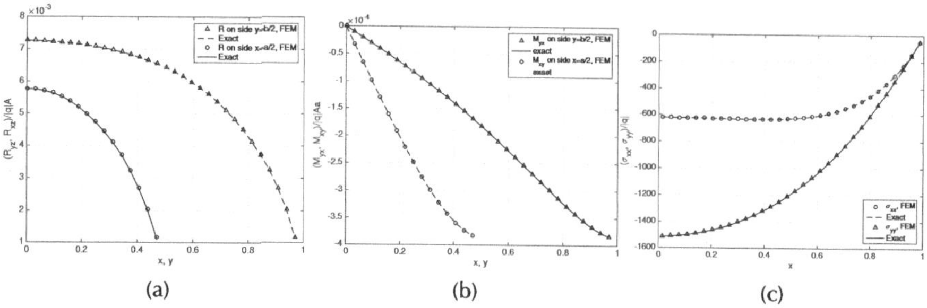

Figure 9.13 Selective integration rules. Comparison of finite-element and exact analytic solution for Reissner–Mindlin plate. Rectangular plate with *SS*2 boundary conditions. Mesh $nxe \times nye = 32 \times 16$ bilinear quadrilateral elements. Weighted reactions, (a) Forces R_{xz} on edge $x = a/2$, and R_{yz} on edge $y = b/2$. (b) Twist moments M_{yx} on edge $y = b/2$, and M_{xy} on edge $x = a/2$. (c) Normal stresses on plate centerline $y = 0$ and $z = h/2$.

The 4-node bilinear rectangular element is used. There are four nodes and three degrees of freedom at each node, namely deflection w, and rotations θ_x, and θ_y.

Using Selective Integration Rules:
Flexural group: Full integration, Gauss–Legendre 2×2 rule.
Transverse shear stiffness matrix: Reduced integration, Gauss–Legendre 1×1 rule.
We define nondimensional deflection $w^* = wD/qa^4$, scaled rotation $\theta_x^* = \theta_x D/qa^3$, and normal stresses $\sigma_{xx}^* = \sigma_{xx}/|q|$, and $\sigma_{yy}^* = \sigma_{yy}/|q|$. Comparisons to the exact analytic solutions are depicted in Figures 9.13a, 9.13b, and 9.13c. Results of the bilinear element with selective integration rules are in excellent agreement with the exact solutions. Table 9.4 shows grid convergence of the maximum deflection and rotation. Table 9.5 also shows grid convergence for the normal stresses at a point on the plate centerline $y = 0$.

Using Full Integration Rules:
Flexural group: Full integration, Gauss–Legendre 2×2 rule.
Transverse shear group: Full integration, Gauss–Legendre 2×2 rule.
As shown in Figures 9.14a-9.14b, full integration for the transverse shear stiffness matrix produces incorrect deflection and rotation for the 4-node bilinear element if the plate is thin. The corresponding stresses are shown in Figure 9.14c. For a thick plate, with $h = 0.1$, results for the bilinear element are shown in Figures 9.15a, 9.15b, and 9.15c. Comparisons with the exact solution still indicate some errors albeit less than that for thin plates. Results obtained with full integration of the shear stiffness matrix for high-order elements such as the biquadratic 9-node and bicubic 16-node elements show good agreement with the exact solution even for a thin plate. An example is shown in Figures 9.16a, 9.15b, and 9.15c for $h = 0.002$. Based on these

Table 9.4
Example 9.3, Maximum Deflection w^* and Rotation θ_x^*, Exact Solution $w^* = 6.33800586 \times 10^{-4}$, $\theta_x^* = 2.46855054 \times 10^{-3}$

nxe, nye	w^* FEM	% error	θ_x^* FEM	% error
4,2	5.9630416e-04	5.92	2.3325508e-03	5.51
8,4	6.2479172e-04	1.42	2.4352467e-03	1.35
16,8	6.3157010e-04	3.52e-01	2.4602910e-03	3.35e-01
32,16	6.3324431e-04	8.78e-02	2.4664902e-03	8.35e-02
64,32	6.3366160e-04	2.19e-02	2.4680357e-03	2.09e-02

Nondimensional maximum deflection, $w^* = w(0,0)D/|q|a^4$.
Scaled rotation $\theta_x^* = \theta_x(a/2,0)D/|q|a^3$.

Table 9.5
Example 9.3, Normal Stresses σ_{xx}^* and σ_{yy}^* at a Point on the Plate Centerline $y = 0$

nxe, nye	x	σ_{xx}^* FEM	% error	σ_{yy}^* FEM	% error
4,2	0.125	-5.7039547e+02	8.74	-1.3581992e+03	9.08
8,4	6.25e-02	-6.1015180e+02	2.13	-1.4747992e+03	2.18
16,8	3.125e-02	-6.1974253e+02	5.31e-01	-1.5029055e+03	5.39e-01
32,16	1.5625e-02	-6.2211946e+02	1.32e-01	-1.5098706e+03	1.34e-01
64,32	7.8125e-03	-6.2271241e+02	3.31e-02	-1.5116081e+03	3.36e-02

Nondimensional stresses, $\sigma^* = \sigma/|q|$.
Exact solution for $nxe = 64$, $nye = 32$: At $x = 7.8125 \times 10^{-3}$,
$\sigma_{xx}^* = -6.22918598 \times 10^2$, $\sigma_{yy}^* = -1.51211544 \times 10^3$.

examples, biquadratic or higher-order elements are recommended for the Reissner–Mindlin plate with selective integration.

9.3 DYNAMIC RESPONSE OF PLATES

In this section, we develop finite-element models for dynamic response of plates due to externally applied transverse loads. We start with Cauchy's equation of motion

$$\frac{\partial \sigma_{ji}}{\partial x_j} + f_i = \rho \frac{\partial^2 u_i}{\partial t^2} \tag{9.177}$$

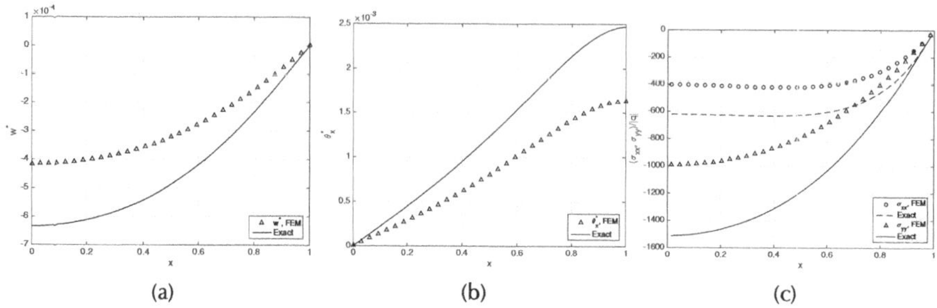

Figure 9.14 Full integration for flexural and shear stiffness matrices. Comparison of finite-element and exact analytic solution for Reissner–Mindlin plate. Rectangular plate with $SS2$ boundary conditions. Thin plate $h = 0.02$. Mesh $nxe = 32$, $nye = 16$, 4-node bilinear element. (a) Deflection, w^* on plate centerline $y = 0$. (b) Rotation θ_x^* on plate centerline $y = 0$. (c) Normal stresses on plate centerline $y = 0$ and $z = h/2$.

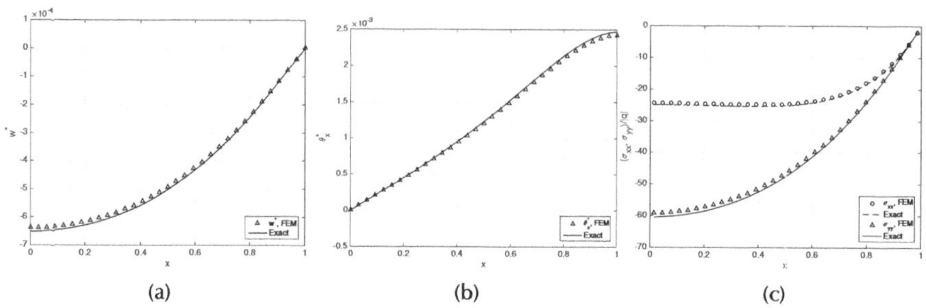

Figure 9.15 Full integration for flexural and shear stiffness matrices. Comparison of finite-element and exact analytic solution for Reissner–Mindlin plate. Rectangular plate with $SS2$ boundary conditions. Thick plate $h = 0.1$. Mesh $nxe = 32$, $nye = 16$ 4-node bilinear elements. (a) Deflection, w^* on plate centerline $y = 0$. (b) Rotation θ_x^* on plate centerline $y = 0$. (c) Normal stresses on plate centerline $y = 0$ and $z = h/2$.

where σ_{ij} is the stress tensor, f_i is body force (force per unit volume), and ρ is mass density. In component form, we have three equations of motion,

$$\frac{\partial \sigma_{xx}}{\partial x} + \frac{\partial \sigma_{yx}}{\partial y} + \frac{\partial \sigma_{zx}}{\partial z} + f_x = \rho \frac{\partial^2 u}{\partial t^2} \tag{9.178}$$

$$\frac{\partial \sigma_{xy}}{\partial x} + \frac{\partial \sigma_{yy}}{\partial y} + \frac{\partial \sigma_{zy}}{\partial z} + f_y = \rho \frac{\partial^2 v}{\partial t^2} \tag{9.179}$$

$$\frac{\partial \sigma_{xz}}{\partial x} + \frac{\partial \sigma_{yz}}{\partial y} + \frac{\partial \sigma_{zz}}{\partial z} + f_z = \rho \frac{\partial^2 w}{\partial t^2}. \tag{9.180}$$

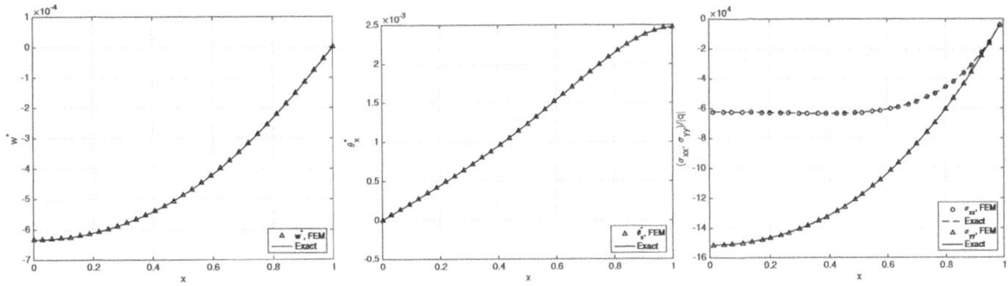

(a) (b) (c)

Figure 9.16 Full integration for flexural and shear stiffness matrices. Comparison of finite-element and exact analytic solution for Reissner–Mindlin plate. Rectangular plate with $SS2$ boundary conditions. Thin plate $h = 0.002$. Mesh $nxe = 16$, $nye = 8$ biquadratic 9-node quadrilateral elements. (a) Deflection, w^* on plate centerline $y = 0$. (b) Rotation θ_x^* on plate centerline $y = 0$. (c) Normal stresses on plate centerline $y = 0$ and $z = h/2$.

We recall the kinematics of CPT and RM plates, the deflection w is independent of z. We integrate Equation (9.180) with respect to z over the plate thickness, $-h/2 \leq z \leq h/2$,

$$\frac{\partial Q_{xz}}{\partial x} + \frac{\partial Q_{yz}}{\partial y} + q + \tilde{f}_z = \rho h \frac{\partial^2 w}{\partial t^2} \tag{9.181}$$

where

$$Q_{xz} = \int_{-h/2}^{h/2} \sigma_{xz} dz, \quad Q_{yz} = \int_{-h/2}^{h/2} \sigma_{yz} dz, \quad \tilde{f}_z = \int_{-h/2}^{h/2} f_z dz$$
$$q = \sigma_{zz}(x,y,h/2) - \sigma_{zz}(x,y,-h/2). \tag{9.182}$$

We multiply Equations (9.178) and (9.179) by z and integrate with respect to z over the plate thickness.

$$\frac{\partial M_{xx}}{\partial x} + \frac{\partial M_{yx}}{\partial y} + z\sigma_{zx} \Big|_{z=-h/2}^{z=h/2} - Q_{zx} + \int_{-h/2}^{h/2} f_x z dz = \int_{-h/2}^{h/2} \rho \frac{\partial^2 u}{\partial t^2} z dz \tag{9.183}$$

$$\frac{\partial M_{xy}}{\partial x} + \frac{\partial M_{yy}}{\partial y} + z\sigma_{zy} \Big|_{z=-h/2}^{z=h/2} - Q_{zy} + \int_{-h/2}^{h/2} f_y z dz = \int_{-h/2}^{h/2} \rho \frac{\partial^2 v}{\partial t^2} z dz \tag{9.184}$$

where

$$M_{xx} = \int_{-h/2}^{h/2} z\sigma_{xx} dz, \; M_{yy} = \int_{-h/2}^{h/2} z\sigma_{yy} dz, \; M_{yx} = \int_{-h/2}^{h/2} z\sigma_{yx} dz.$$
$$M_{xy} = M_{yx} \tag{9.185}$$

The resultant shear forces Equation (9.182) and moments Equation (9.185) are depicted in Figure 9.3b.
For the bending problem, $f_x = 0$, $f_y = 0$, and the shear stresses σ_{zx} and σ_{zy} are zero on $z = \pm h/2$.

$$\frac{\partial M_{xx}}{\partial x} + \frac{\partial M_{yx}}{\partial y} - Q_{zx} = \int_{-h/2}^{h/2} \rho \frac{\partial^2 u}{\partial t^2} z dz \tag{9.186}$$

$$\frac{\partial M_{xy}}{\partial x} + \frac{\partial M_{yy}}{\partial y} - Q_{zy} = \int_{-h/2}^{h/2} \rho \frac{\partial^2 v}{\partial t^2} z dz. \tag{9.187}$$

9.3.1 EQUATION OF MOTION FOR *CPT*

Equations for the *CPT* are obtained by substituting $u = -z\partial w/\partial x$, and $v = -z\partial w/\partial y$ into Equations (9.186) and (9.187), respectively. We get

$$\frac{\partial M_{xx}}{\partial x} + \frac{\partial M_{yx}}{\partial y} - Q_{zx} = -\frac{\rho h^3}{12} \frac{\partial^3 w}{\partial t^2 \partial x} \tag{9.188}$$

$$\frac{\partial M_{xy}}{\partial x} + \frac{\partial M_{yy}}{\partial y} - Q_{zy} = -\frac{\rho h^3}{12} \frac{\partial^3 w}{\partial t^2 \partial y}. \tag{9.189}$$

To obtain an equation for w, we differentiate Equation (9.188) with respect to x and Equation (9.189) with respect to y, and add the results to Equation (9.181),

$$\begin{aligned} \frac{\partial^2 M_{xx}}{\partial x^2} + 2\frac{\partial^2 M_{xy}}{\partial x \partial y} + \frac{\partial^2 M_{yy}}{\partial y^2} + (q + \tilde{f}_z) = \rho h \frac{\partial^2 w}{\partial t^2} \\ - \frac{\partial}{\partial x}\left(\frac{\rho h^3}{12} \frac{\partial^3 w}{\partial t^2 \partial x}\right) - \frac{\partial}{\partial y}\left(\frac{\rho h^3}{12} \frac{\partial^3 w}{\partial t^2 \partial y}\right). \end{aligned} \tag{9.190}$$

Next, we use the constitutive equations to express the moments in terms of curvature.

$$\begin{aligned} \left\{ \begin{array}{c} \sigma_{xx} \\ \sigma_{yy} \\ \sigma_{xy} \end{array} \right\} &= \frac{E}{1-\nu^2} \begin{bmatrix} 1 & \nu & 0 \\ \nu & 1 & 0 \\ 0 & 0 & \frac{1-\nu}{2} \end{bmatrix} \left\{ \begin{array}{c} \varepsilon_{xx} \\ \varepsilon_{yy} \\ \gamma_{xy} \end{array} \right\} \\ \{\sigma\} &= [D]\{\varepsilon\} \\ &= -z[D] \left\{ \begin{array}{c} \frac{\partial^2 w}{\partial x^2} \\ \frac{\partial^2 w}{\partial y^2} \\ 2\frac{\partial^2 w}{\partial x \partial y} \end{array} \right\}. \end{aligned} \tag{9.191}$$

Substituting into Equation (9.185), we get

$$\begin{Bmatrix} M_{xx} \\ M_{yy} \\ M_{xy} \end{Bmatrix} = -D \begin{bmatrix} 1 & \nu & 0 \\ \nu & 1 & 0 \\ 0 & 0 & \frac{1-\nu}{2} \end{bmatrix} \begin{Bmatrix} \frac{\partial^2 w}{\partial x^2} \\ \frac{\partial^2 w}{\partial y^2} \\ 2\frac{\partial^2 w}{\partial x \partial y} \end{Bmatrix} \tag{9.192}$$

where $D = Eh^3/12(1-\nu^2)$. For a uniform plate (h is constant), Equation (9.190) gives

$$\begin{aligned} -D\left(\frac{\partial^4 w}{\partial x^4} + 2\frac{\partial^4 w}{\partial x^2 \partial y^2} + \frac{\partial^4 w}{\partial y^4}\right) + (q + \tilde{f}_z) = \rho h \frac{\partial^2 w}{\partial t^2} \\ - \frac{\partial}{\partial x}\left(\frac{\rho h^3}{12}\frac{\partial^3 w}{\partial t^2 \partial x}\right) - \frac{\partial}{\partial y}\left(\frac{\rho^3}{12}\frac{\partial^3 w}{\partial t^2 \partial y}\right). \end{aligned} \tag{9.193}$$

The last two terms on the right are rotary inertia, neglecting those terms we obtain

$$\rho h \frac{\partial^2 w}{\partial t^2} + D\left(\frac{\partial^4 w}{\partial x^4} + 2\frac{\partial^4 w}{\partial x^2 \partial y^2} + \frac{\partial^4 w}{\partial y^4}\right) - (q + \tilde{f}_z) = 0 \tag{9.194}$$

$$\rho h \frac{\partial^2 w}{\partial t^2} + D\nabla^4 w - (q + \tilde{f}_z) = 0. \tag{9.195}$$

For a plate in static equilibrium, we get the well known biharmonic equation,

$$\nabla^4 w = \frac{q + \tilde{f}_z}{D}. \tag{9.196}$$

9.3.2 EQUATIONS OF MOTION FOR REISSNER–MINDLIN PLATE

Kinematics of the Reissner–Mindlin plate theory gives $u = -z\theta_x(x,y,t)$, and $v = -z\theta_y(x,y,t)$, hence Equations (9.186) and (9.187) become,

$$\frac{\partial M_{xx}}{\partial x} + \frac{\partial M_{yx}}{\partial y} - Q_{zx} = -\frac{\rho h^3}{12}\frac{\partial^2 \theta_x}{\partial t^2} \tag{9.197}$$

$$\frac{\partial M_{xy}}{\partial x} + \frac{\partial M_{yy}}{\partial y} - Q_{zy} = -\frac{\rho h^3}{12}\frac{\partial^2 \theta_y}{\partial t^2}. \tag{9.198}$$

Again, the constitutive equations give

$$\begin{aligned}
\begin{Bmatrix} \sigma_{xx} \\ \sigma_{yy} \\ \sigma_{xy} \end{Bmatrix} &= \frac{E}{1-\nu^2}\begin{bmatrix} 1 & \nu & 0 \\ \nu & 1 & 0 \\ 0 & 0 & \frac{1-\nu}{2} \end{bmatrix}\begin{Bmatrix} \varepsilon_{xx} \\ \varepsilon_{yy} \\ \gamma_{xy} \end{Bmatrix} \\
\{\sigma\} &= [D]\{\varepsilon\} \\
&= -z[D]\begin{Bmatrix} \frac{\partial \theta_x}{\partial x} \\ \frac{\partial \theta_y}{\partial y} \\ \frac{\partial \theta_x}{\partial y} + \frac{\partial \theta_y}{\partial x} \end{Bmatrix}
\end{aligned} \tag{9.199}$$

and hence the moments follow from Equation (9.185),

$$\begin{Bmatrix} M_{xx} \\ M_{yy} \\ M_{xy} \end{Bmatrix} = -D\begin{bmatrix} 1 & \nu & 0 \\ \nu & 1 & 0 \\ 0 & 0 & \frac{1-\nu}{2} \end{bmatrix}\begin{Bmatrix} \frac{\partial \theta_x}{\partial x} \\ \frac{\partial \theta_y}{\partial y} \\ \frac{\partial \theta_x}{\partial y} + \frac{\partial \theta_y}{\partial x} \end{Bmatrix} \tag{9.200}$$

where $D = Eh^3/12(1-\nu^2)$. The transverse shear stresses are

$$\begin{aligned}
\{\sigma^s\} \equiv \begin{Bmatrix} \sigma_{xz} \\ \sigma_{yz} \end{Bmatrix} &= \frac{E}{2(1+\nu)}\begin{bmatrix} 1 & 0 \\ 0 & 1 \end{bmatrix}\begin{Bmatrix} -\theta_x + \frac{\partial w}{\partial x} \\ -\theta_y + \frac{\partial w}{\partial y} \end{Bmatrix} \\
\{\sigma^s\} &= [D^s]\{\varepsilon^s\}
\end{aligned} \tag{9.201}$$

where $E/2(1+\nu) = G$ is the shear modulus. The resultant transverse shear forces are obtained from Equation (9.182),

$$\begin{Bmatrix} Q_{xz} \\ Q_{yz} \end{Bmatrix} = \kappa G\begin{bmatrix} 1 & 0 \\ 0 & 1 \end{bmatrix}\begin{Bmatrix} -\theta_x + \frac{\partial w}{\partial x} \\ -\theta_y + \frac{\partial w}{\partial y} \end{Bmatrix} \tag{9.202}$$

where $\kappa = 5/6$ is the shear correction factor. Substituting Equations (9.200), and (9.202) into Equations (9.197), (9.198), and (9.181), we get

$$\begin{aligned}
\frac{\rho h^3}{12}\frac{\partial^2 \theta_x}{\partial t^2} - D\frac{\partial}{\partial x}\left(\frac{\partial \theta_x}{\partial x} + \nu\frac{\partial \theta_y}{\partial y}\right) - D\frac{1-\nu}{2}\frac{\partial}{\partial y}\left(\frac{\partial \theta_x}{\partial y} + \frac{\partial \theta_y}{\partial x}\right)& \\
-\kappa G\left(-\theta_x + \frac{\partial w}{\partial x}\right) = 0&
\end{aligned} \tag{9.203}$$

$$\begin{aligned}
\frac{\rho h^3}{12}\frac{\partial^2 \theta_y}{\partial t^2} - D\frac{\partial}{\partial y}\left(\frac{\partial \theta_y}{\partial y} + \nu\frac{\partial \theta_x}{\partial x}\right) - D\frac{1-\nu}{2}\frac{\partial}{\partial x}\left(\frac{\partial \theta_x}{\partial y} + \frac{\partial \theta_y}{\partial x}\right)& \\
-\kappa G\left(-\theta_y + \frac{\partial w}{\partial y}\right) = 0&
\end{aligned} \tag{9.204}$$

$$\rho h\frac{\partial^2 w}{\partial t^2}-\kappa G\frac{\partial}{\partial x}\left(-\theta_x+\frac{\partial w}{\partial x}\right)-\kappa G\frac{\partial}{\partial y}\left(-\theta_y+\frac{\partial w}{\partial y}\right)-(q+\tilde{f}_z)=0. \tag{9.205}$$

Equations (9.203), (9.204), and (9.205) are the equations of motion for Reissner–Mindlin plate as obtained by Mindlin [76].

9.3.3 FINITE ELEMENT MODEL

Starting with the Cauchy's equation of motion Equation (9.177), we derive the principle of virtual displacements for 3D dynamical problems. The derivation follows the same steps as for deriving the weak form of the equations of equilibrium given in Chapter 3. The principle of virtual displacements for dynamical problems is

$$\int_\Omega \rho\bar{u}_i\ddot{u}_i d\Omega+\int_\Omega \bar{\varepsilon}_{ij}\sigma_{ji}d\Omega=\int_\Omega \bar{u}_i f_i d\Omega+\oint_S \bar{u}_i t_i dS \tag{9.206}$$

where $\ddot{u}_i$ is the acceleration. The first term is the virtual work done by the inertia ($\rho\ddot{u}_i$) through a virtual displacement field $\bar{u}_i$. The procedure used for deriving the finite-element models for the static deflection problem can be used here for the dynamic case. The inertia term poses no difficulties, and the element equation of motion takes the form

$$[M^e]\{\ddot{U}^e\}+[K^e]\{U^e\}=\{F^e\}+\{R^e\} \tag{9.207}$$

where $[K^e]$ is the stiffness matrix, $\{F^e\}$ is the forcing vector, and $\{R^e\}$ is the array of weighted secondary variables. The matrix $[M^e]$ is the consistent mass matrix.

Consistent Mass Matrix: Classical Plate Theory *CPT*

For the classical plate theory, the term $\bar{u}_i\ddot{u}_i$ can be written in terms of the plate deflection w,

$$\begin{aligned}\bar{u}_i\ddot{u}_i &= \bar{u}\ddot{u}+\bar{v}\ddot{v}+\bar{w}\ddot{w}\\ &= \left(-z\frac{\partial\bar{w}}{\partial x}\right)\left(-z\frac{\partial\ddot{w}}{\partial x}\right)+\left(-z\frac{\partial\bar{w}}{\partial y}\right)\left(-z\frac{\partial\ddot{w}}{\partial y}\right)+\bar{w}\ddot{w}\\ &= z^2\left(\frac{\partial\bar{w}}{\partial x}\frac{\partial\ddot{w}}{\partial x}+\frac{\partial\bar{w}}{\partial y}\frac{\partial\ddot{w}}{\partial y}\right)+\bar{w}\ddot{w}.\end{aligned} \tag{9.208}$$

The inertia term for an element becomes

$$\begin{aligned}\int_{\Omega^e}\rho\bar{u}_i\ddot{u}_i d\Omega &= \int_{A_e}\int_{-h/2}^{h/2}\rho\left[z^2\left(\frac{\partial\bar{w}}{\partial x}\frac{\partial\ddot{w}}{\partial x}+\frac{\partial\bar{w}}{\partial y}\frac{\partial\ddot{w}}{\partial y}\right)+\bar{w}\ddot{w}\right]dzdA\\ &= \int_{A_e}\rho\left[\frac{h^3}{12}\left(\frac{\partial\bar{w}}{\partial x}\frac{\partial\ddot{w}}{\partial x}+\frac{\partial\bar{w}}{\partial y}\frac{\partial\ddot{w}}{\partial y}\right)+h\bar{w}\ddot{w}\right]dA.\end{aligned} \tag{9.209}$$

Substituting the trial solution for w, and invoking the Galerkin method for selecting the virtual displacement, we get

$$\begin{aligned}
\int_{\Omega^e} \rho \bar{u}_i \ddot{u}_i d\Omega &= \{d\}^T \left[\int_{A_e} \rho \frac{h^3}{12} \left(\frac{\partial \{\psi\}^T}{\partial x} \frac{\partial \{\psi\}}{\partial x} + \frac{\partial \{\psi\}^T}{\partial y} \frac{\partial \{\psi\}}{\partial y} \right) dA \right] \{\ddot{U}^e\} \\
&\quad + \{d\}^T \left[\int_{A_e} \rho h \{\psi\}^T \{\psi\} dA \right] \{\ddot{U}^e\} \\
&= \{d\}^T [M^e] \{\ddot{U}^e\}
\end{aligned} \tag{9.210}$$

where $\{\psi\}$ is given by Equation (9.51), and where

$$[M^e] = \int_{A_e} \rho \frac{h^3}{12} \left(\frac{\partial \{\psi\}^T}{\partial x} \frac{\partial \{\psi\}}{\partial x} + \frac{\partial \{\psi\}^T}{\partial y} \frac{\partial \{\psi\}}{\partial y} \right) dA + \int_{A_e} \rho h \{\psi\}^T \{\psi\} dA. \tag{9.211}$$

The first integral is the rotary inertia matrix, and is neglected for very thin plates. Thus, for the *CPT*, the consistent mass matrix is

$$[M^e] = \int_{A_e} \rho h \{\psi\}^T \{\psi\} dA. \tag{9.212}$$

Consistent Mass Matrix: Reissner–Mindlin Plate

For Reissner–Mindlin plate theory, the term $\bar{u}_i \ddot{u}_i$ can be written in terms of the plate deflection w, and rotations θ_x and θ_y,

$$\begin{aligned}
\bar{u}_i \ddot{u}_i &= \bar{u}\ddot{u} + \bar{v}\ddot{v} + \bar{w}\ddot{w} \\
&= \left(-z\bar{\theta}_x\right)\left(-z\ddot{\theta}_x\right) + \left(-z\bar{\theta}_y\right)\left(-z\ddot{\theta}_y\right) + \bar{w}\ddot{w} \\
&= z^2 \left(\bar{\theta}_x \ddot{\theta}_x + \bar{\theta}_y \ddot{\theta}_y\right) + \bar{w}\ddot{w}.
\end{aligned} \tag{9.213}$$

The inertia term for an element becomes

$$\begin{aligned}
\int_{\Omega^e} \rho \bar{u}_i \ddot{u}_i d\Omega &= \int_{A_e} \int_{-h/2}^{h/2} \rho \left[z^2 \left(\bar{\theta}_x \ddot{\theta}_x + \bar{\theta}_y \ddot{\theta}_y\right) + \bar{w}\ddot{w} \right] dz dA \\
&= \int_{A_e} \rho \left[\frac{h^3}{12} \left(\bar{\theta}_x \ddot{\theta}_x + \bar{\theta}_y \ddot{\theta}_y\right) + h\bar{w}\ddot{w} \right] dA \\
&= \int_{A_e} \frac{\rho h}{12} \left\{ \begin{array}{c} \bar{\theta}_x \\ \bar{\theta}_y \\ \bar{w} \end{array} \right\}^T \left[\begin{array}{ccc} h^2 & 0 & 0 \\ 0 & h^2 & 0 \\ 0 & 0 & 12 \end{array} \right] \left\{ \begin{array}{c} \ddot{\theta}_x \\ \ddot{\theta}_y \\ \ddot{w} \end{array} \right\}.
\end{aligned} \tag{9.214}$$

Substituting the trial solution for θ_x, θ_y, and w, and invoking the Galerkin method for selecting the virtual displacement, we get

$$\begin{aligned}
\int_{\Omega^e} \rho \bar{u}_i \ddot{u}_i d\Omega &= \{d\}^T \left[\int_{A_e} \rho [\psi]^T \left[\begin{array}{ccc} h^3/12 & 0 & 0 \\ 0 & h^3/12 & 0 \\ 0 & 0 & h \end{array} \right] [\psi] dA \right] \{\ddot{U}^e\} \\
&= \{d\}^T [M^e] \{\ddot{U}^e\}.
\end{aligned} \tag{9.215}$$

For Reissner–Mindlin plate, the consistent mass matrix is

$$[M^e] = \int_{A_e} \rho\, [\psi]^T \begin{bmatrix} h^3/12 & 0 & 0 \\ 0 & h^3/12 & 0 \\ 0 & 0 & h \end{bmatrix} [\psi]\, dA \tag{9.216}$$

where $[\psi]$ is given by Equation (9.139).

9.3.4 FREE VIBRATIONS: NATURAL FREQUENCIES

To study the plate dynamics, we first assemble the equations of motion of all elements into a global system. Before application of the boundary conditions, the system is of the form

$$[M]\{\ddot{U}\} + [K]\{U\} = \{F\} + \{R\}. \tag{9.217}$$

In this section, we study free vibrations of plates, which is the time response of the plate to initial disturbances in displacement and/or velocity. The plate motion is unforced, and hence $\{F\} = 0$. At interior nodes, all components of the weighted secondary variable $\{R\}$ are zero. Natural boundary conditions are homogeneous, and hence the corresponding components of $\{R\}$ are zero. Also, for mixed boundary conditions, the corresponding components of $\{R\}$ can be written in terms of the unknown $\{U\}$. After applying the boundary conditions, we obtain a homogeneous system for the unknown generalized displacements,

$$[\bar{M}]\{\ddot{U}\} + [\bar{K}]\{U\} = \{0\}. \tag{9.218}$$

In this section we are interested only in determining the natural frequencies of plates rather than the plate response to initial disturbances. The mass and stiffness matrices are independent of time. Thus, we assume harmonic response,

$$\{U\} = \{\hat{U}\} e^{i\omega t}. \tag{9.219}$$

and substitute into Equation (9.218). The natural frequencies ω and mode shapes of free vibrations are determined by solving the algebraic eigenvalue problem

$$\left[[\bar{K}] - \omega^2 [\bar{M}]\right] \{\hat{U}\} = \left\{0\right\}. \tag{9.220}$$

We use the eigenvalue problem solver in MATLAB,

$$eig(\mathscr{V}, \mathscr{D}) = (\bar{K}, \bar{M})$$

that returns the eigenvalues on the diagonal matrix $\mathscr{D}$, and the eigenvectors in the matrix $\mathscr{V}$.

Table 9.6
Example 9.4, *CPT*. Nondimensional Natural Frequencies of a Square Plate, all Sides are Simply Supported

Mode m, n	Mesh 4×4	Mesh 8×8	Mesh 16×16
1,1	1.9741843e+01	1.9739372e+01	1.9739219e+01
1,2	4.9476259e+01	4.9356314e+01	4.9348544e+01
2,1	4.9476259e+01	4.9356314e+01	4.9348544e+01
2,2	7.9127053e+01	7.8967372e+01	7.8957489e+01
1,3	1.0018640e+02	9.8799597e+01	9.8702675e+01
3,1	1.0018640e+02	9.8799597e+01	9.8702675e+01
2,3	1.2956987e+02	1.2838940e+02	1.2831019e+02
3,2	1.2956987e+02	1.2838940e+02	1.2831019e+02
1,4	1.7952488e+02	1.6837349e+02	1.6782194e+02

Example 9.4. *CPT*: Square Plate with All Sides Simply Supported

For the *CPT*, the exact solution for free vibrations of simply supported rectangular plate of sides (a,b) can be written as

$$w(x,y,t) = \sum_{n=1}^{\infty}\sum_{m=1}^{\infty} a_{mn}(t) sin\frac{m\pi x}{a} sin\frac{n\pi y}{b} \tag{9.221}$$

$$a_{mn}(t) = A_{mn} cos\omega t + B_{mn} sin\omega t \tag{9.222}$$

where A_{mn} and B_{mn} are arbitrary constants. We define the nondimensional frequency

$$\omega^* = \omega a^2 \sqrt{\rho h/D}. \tag{9.223}$$

The natural frequency of mode (m,n) is

$$\omega^* = \pi^2 \left[m^2 + n^2 (b/a)^2\right]. \tag{9.224}$$

We consider a square plate $a = 1$, $b = 1$, $h = 0.1$, and $\nu = 0.3$. The full plate is discretized using the Bogner–Fox–Schmit 4-node conforming element. The natural frequencies of low-order modes are given in Table 9.6, and the percentage errors are given in Table 9.7 for different mesh resolutions. The fundamental mode $(m,n) = (1,1)$ is well predicted by a mesh of $4 \times 4 = 16$ elements. Finer meshes are needed to predict accurately the frequencies of higher-order modes. Because of the plate and boundary conditions symmetries, modes $(m,n) = (1,2)$ and $(m,n) = (2,1)$ have equal frequencies. The mode shapes are shown in Figures 9.17a and 9.17b, respectively, and they are obtained using a slightly asymmetric mesh of $(nxe \times nye) =$

Table 9.7
Example 9.4, Percentage Error in Natural Frequencies

Mode m, n	Mesh 4×4	Mesh 8×8	Mesh 16×16
1,1	1.33e-02	8.28e-04	5.16e-05
1,2	2.60e-01	1.68e-02	1.06e-03
2,1	2.60e-01	1.68e-02	1.06e-03
2,2	2.16e-01	1.33e-02	8.28e-04
1,3	1.51e+00	1.05e-01	6.72e-03
3,1	1.51e+00	1.05e-01	6.72e-03
2,3	9.86e-01	6.59e-02	4.15e-03
3,2	9.86e-01	6.59e-02	4.15e-03
1,4	7.00e+00	3.52e-01	2.30e-02

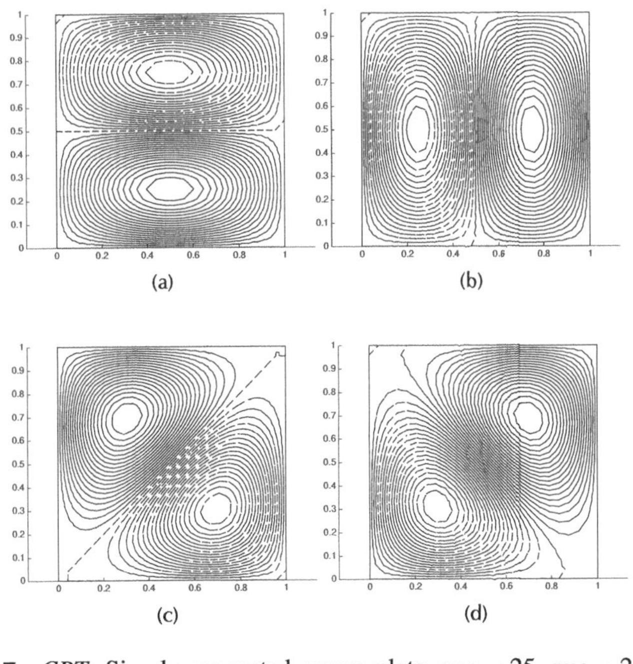

Figure 9.17 *CPT*, Simply supported square plate. $nxe = 25$, $nye = 24$: (a) mode (1,2), and (b) mode (2,1). $nxe = 24$, $nye = 24$: (c) mode (2,1) + mode (1,2), and (d) mode (2,1) − mode (1,2).

25×24 elements. If a symmetric mesh of $(nxe \times nye) = 24 \times 24$ elements is used, the numerical eigenvalue solver gives the modes shapes shown in Figures 9.17c and 9.17d, respectively. These mode shapes are linear superposition of those shown in Figures 9.17a and 9.17b.

Table 9.8

Example 9.5, Reissner–Mindlin plate. Nondimensional Natural Frequencies of a Square Plate, *SFSC*

Mode m,n	Mesh 4×4	Mesh 8×8	Mesh 16×16
1,1	1.2268751e+01	1.2261279e+01	1.2260658e+01
1,2	3.0557011e+01	3.0480104e+01	3.0474674e+01
2,1	3.9095229e+01	3.8739587e+01	3.8714654e+01
2,2	5.6391137e+01	5.6004860e+01	5.5975854e+01
1,3	6.4179839e+01	6.3035368e+01	6.2958088e+01
3,1	8.2241940e+01	7.9061286e+01	7.8828769e+01
2,3	8.7611330e+01	8.6368844e+01	8.6278095e+01
3,2	9.7247128e+01	9.4325420e+01	9.4106762e+01
1,4	1.1409537e+02	1.0666817e+02	1.0619865e+02

Table 9.9

Example 9.5, Reissner–Mindlin Plate. Percentage Error in Natural Frequencies of Square Plate, *SFSC*

Mode m,n	Mesh 4×4	Mesh 8×8	Mesh 16×16
1,1	6.65e-02	5.54e-03	4.76e-04
1,2	2.71e-01	1.90e-02	1.23e-03
2,1	9.88e-01	6.92e-02	4.79e-03
2,2	7.46e-01	5.58e-02	4.03e-03
1,3	1.96e+00	1.31e-01	8.56e-03
3,1	4.35e+00	3.16e-01	2.13e-02
2,3	1.55e+00	1.13e-01	7.88e-03
3,2	3.35e+00	2.50e-01	1.72e-02
1,4	7.47e+00	4.73e-01	3.11e-02

Example 9.5. Reissner–Mindlin Plate: Square Plate with Two Opposite Sides Simply Supported and the Other Sides Clamped and Free, *SFSC*

We consider a square plate $a = 1$, $b = 1$, $h = 0.1$, and $\nu = 0.3$. The two sides $x = 0$ and $x = 1$ are simply supported, and the side $y = 1$ is clamped while $y = 0$ is free. We use the biquadratic quadrilateral Lagrange element, and compare the finite-element results with the exact solutions obtained by Hashemi and Arsanjani [47]. The shear correction factor is $\kappa = 0.86667$. We define the nondimensional frequency

$$\omega^* = \omega a^2 \sqrt{\rho h/D}. \tag{9.225}$$

The natural frequencies of low-order modes are given in Table 9.8 and the percentage errors are given in Table 9.9 for different mesh resolutions. The lowest mode is well

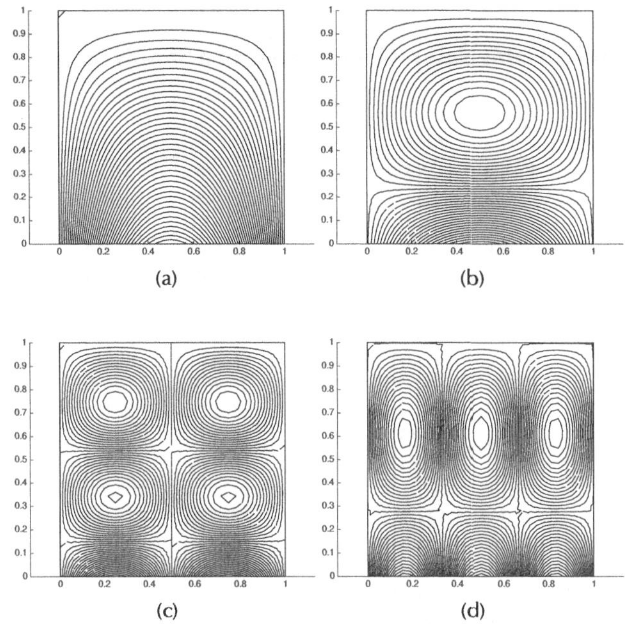

Figure 9.18 Reissner–Mindlin plate, $S-F-S-C$ square plate, $nxe = 16$, $nye = 16$. (a) mode (1,1), (b) mode (1,2), (c) mode (2,3), and (d) mode (3,2).

predicted by a mesh of $4 \times 4 = 16$ elements. Finer meshes are needed to predict accurately the frequencies of higher-order modes.

The mode shapes for the lowest-order modes (1,1) and (1,2) are depicted in Figures 9.18a–9.18b. Higher-order mode shapes (2,3) and (3,2) are also shown in Figures 9.18c–9.18d.

Example 9.6. Reissner–Mindlin Plate: Natural Frequencies of a Simply Supported Circular Plate

We consider a simply supported circular plate of radius $R = 1$, thickness $h = 0.001$, Poisson's ratio $\nu = 0.3$, and shear correction factor $\kappa = 5/6$. Finite element results are obtained for four meshes. Three meshes use the bilinear 4-node element, and the number of global nodes and elements are: Mesh 1 (nodes = 2185, elements = 2128), Mesh 2 (nodes = 4561, elements = 4480), Mesh 3 (nodes = 7801, elements = 7696). The fourth mesh (Mesh 4) uses bicubic 16-node quadrilateral element, and the number of nodes and elements are (nodes = 1009, elements = 108); the mesh is depicted in Figure 9.19a. We define the dimensionless frequency

$$\omega^* = \omega R^2 \sqrt{\rho h / D}. \tag{9.226}$$

Table 9.10
Reissner–Mindlin Plate: Natural Frequencies of a Simply Supported Circular Plate

(m_r, m_θ)	Mesh 1	Mesh 2	Mesh 3	Mesh 4	Ref. [109]	$\Delta\omega^*$
(1,0)	4.9408	4.9379	4.9367	4.935	4.935	0.000
(1,1)	13.933	13.914	13.908	13.898	13.898	0.000
(1,2)	25.700	25.652	25.636	25.613	25.613	0.000
(2,0)	29.869	29.789	29.760	29.720	29.720	0.000
(1,3)	40.161	40.054	40.014	39.957	39.957	0.000
(2,1)	48.836	48.645	48.575	48.480	48.478	0.002
(1,4)	57.238	57.032	56.953	56.842	56.841	0.001
(2,2)	70.685	70.379	70.267	70.118	70.116	0.002
(3,0)	75.015	74.558	74.388	74.161	74.155	0.006
(1,5)	76.949	76.561	76.412	76.209	76.202	0.007
(2,3)	95.663	95.072	94.851	94.562	94.547	0.015
(1,6)	99.278	98.611	98.355	98.016		
(3,1)	104.405	103.529	103.208	102.808	102.772	0.036
(2,4)	123.364	122.485	122.156	121.729	121.699	0.030
(1,7)	124.280	123.188	122.770	122.251		
(3,2)	136.498	135.312	134.880	134.362	134.294	0.068
(4,0)	141.349	139.722	139.126	138.410	138.315	0.095
(1,8)	151.991	150.294	149.645	148.917		
(2,5)	154.065	152.714	152.210	151.581	151.514	0.067
(3,3)	172.550	170.474	169.712	168.856	168.669	0.187
(4,1)	181.674	179.043	178.088	177.177	176.795	0.382
(1,9)	182.519	179.964	178.989	178.075		
(2,6)	187.575	185.652	184.935	184.058		
(3,4)	211.145	208.323	207.278	206.074	205.843	0.231
(1,10)	215.908	212.213	210.805	209.939		
(4,2)	224.087	220.900	219.747	218.538	218.194	0.344
(2,7)	224.052	221.349	220.340	219.180		

The dimensionless natural frequencies of different modes are given in Table 9.10, where m_r is the number of points of zero deflection on the radius, and m_θ is the number of full waves in the circumferential direction. Modes of $m_\theta = 0$ are symmetric modes where deflection is independent of θ. The results of Mesh 4 are in excellent agreement with the exact values obtained by Senjanovic et al. [109]. Even with very fine meshes, the results of the bilinear 4-node elements are less accurate than those of the bicubic element. The fundamental mode shape is shown in Figure 9.19b, and two non-axisymmetric mode shapes are shown in Figures 9.19c and 9.19d.

Example 9.7. Reissner–Mindlin Plate: Natural Frequencies of a Clamped Circular Plate

We reconsider the plate of Example 9.6 and change the boundary conditions to a clamped edge. The nondimensional natural frequencies of different modes are given

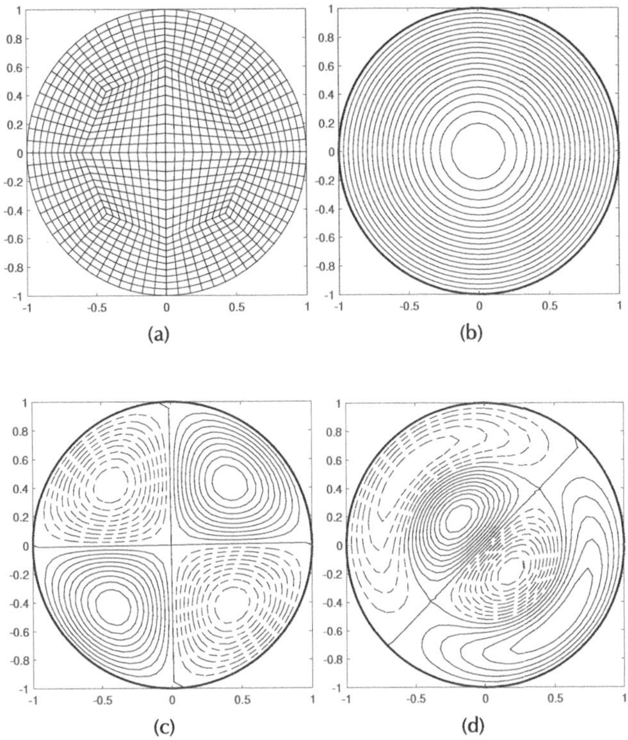

Figure 9.19 Free vibrations of a simply supported circular plate. (a) Mesh 4: Bicubic quadrilateral elements; nodes = 1009 and elements = 108. Mode shapes (m_r, m_θ): (b) Fundamental mode shape (1,0); contour levels step 0.05, (c) (1,2), and (d) (2,1); contour levels step 0.1.

in Table 9.11, where m_r is the number of points of zero deflection on the radius, and m_θ is the number of full waves in the circumferential direction. Modes of $m_\theta = 0$ are symmetric modes where deflection is independent of θ. Results of mesh 4 are in excellent agreement with the exact values obtained by Senjanovic et al. [109]. As in the case of simply supported plate, the results demonstrate the high accuracy achieved by the bicubic 16-node element in comparison to the low-order bilinear 4-node element.

Example 9.8. Reissner–Mindlin Plate: Natural Frequencies of Right Triangular Plates

Next, we consider a right-angle triangular plate of sides a, b, and $c = \sqrt{a^2+b^2}$, and use a as a reference length. For the purpose of comparing with thin plate results,

Table 9.11
Reissner–Mindlin Plate: Natural Frequencies of a Clamped Circular Plate

(m_r, m_θ)	Mesh 1	Mesh 2	Mesh 3	Mesh 4	Ref. [109]	$\Delta\omega^*$
(1,0)	10.232	10.223	10.220	10.216	10.216	0.000
(1,1)	21.328	21.292	21.279	21.260	21.260	0.000
(1,2)	35.021	34.942	34.914	34.877	34.877	0.000
(2,0)	40.033	39.893	39.841	39.771	39.771	0.000
(1,3)	51.342	51.177	51.115	51.031	51.029	0.002
(2,1)	61.409	61.096	60.982	60.832	60.828	0.004
(1,4)	70.213	69.925	69.816	69.668	69.665	0.003
(2,2)	85.441	84.977	84.809	84.586	84.581	0.005
(3,0)	90.355	89.687	89.440	89.115	89.102	0.013
(1,5)	91.704	91.196	91.005	90.749	90.737	0.012
(2,3)	112.635	111.772	111.454	111.042	111.019	0.023
(1,6)	115.783	114.960	114.648	114.240		
(3,1)	122.315	121.116	120.676	120.127	120.076	0.051
(2,4)	142.423	141.192	140.734	140.137	140.104	0.033
(1,7)	142.538	141.238	140.745	140.155		
(3,2)	156.760	155.167	154.590	153.926	153.810	0.116
(4,0)	162.194	160.038	159.250	158.341	158.179	0.162
(1,8)	171.993	170.032	169.288	168.437		
(2,5)	175.235	173.405	172.728	171.908	171.797	0.111
(3,3)	195.359	192.652	191.657	190.657	190.296	0.361
(1,9)	204.264	201.382	200.290	199.237		
(4,1)	205.393	201.955	200.714	199.925	199.045	0.880
(2,6)	210.804	208.281	207.347	206.236		
(3,4)	236.318	232.689	231.348	229.846	229.507	0.339
(1,10)	239.436	235.319	233.759	232.705		
(2,7)	249.313	245.885	244.616	243.198		
(4,2)	250.229	246.149	244.680	243.236	242.708	0.528

we take the plate thickness $h/a = 0.001$, and Poisson's ratio $\nu = 0.3$. The plate is divided into 75 quadrilateral bicubic elements with 721 global nodes; similar to the mesh shown in Figure 9.22a. Gorman used classical plate theory, and computed highly accurate natural frequencies for a simply supported plate (3 sides) [42], and a clamped plate (3 sides) [43]. Natural frequencies are shown in Tables 9.12 and 9.13 for the two types of boundary conditions, respectively. The conditions $\theta_x = 0$, $\theta_y = 0$, $w = 0$ are also applied at the three vertices. The tables show excellent agreement between the present finite-element computations and Gorman's analytical results.

9.4 FLUID-STRUCTURE INTERACTION: LINEAR ANALYSIS

In this section, we formulate the finite-element method for vibrations of plates submerged in a fluid. In the absence of plate motion, the fluid may be stagnant or

Table 9.12

Natural Frequencies in Vacuum of a Right Triangular Plate with Simply Supported Sides SSS; $h/a = 0.001$, $\nu = 0.3$

Mode	$b/a = 1, FEM$	$b/a = 1$, Ref. [42]	$b/a = 2, FEM$	$b/a = 2$, Ref. [42]
1	49.3476	49.35	27.7590	27.76
2	98.6947	98.70	49.8847	49.89
3	128.302	128.3	74.6736	74.67
4	167.781	167.8	81.3205	81.32

Table 9.13

Natural Frequencies in Vacuum of a Right Triangular Plate with Clamped Sides CCC; $h/a = 0.001$, $\nu = 0.3$

Mode	$b/a = 1, FEM$	$b/a = 1$, Ref. [43]	$b/a = 2, FEM$	$b/a = 2$, Ref. [43]
1	93.7864	93.86	53.4446	53.45
2	157.781	157.7	82.4277	82.42
3	194.757	194.8	113.493	113.5
4	242.790	242.8	121.840	121.9

streaming uniformly parallel to the plate. Flexible thin elastic panels may lose stability when exposed to high speed parallel flow; usually on one side of the panel. Such panels may be used on hulls of light fast ships, glass windows/curtains on high rise buildings, casings of submersible vehicles, and wings/bodies of unmanned and micro aerial vehicles (UAVs and MAVs, [15]). Other engineering applications can be in energy harvesting [104, 121, 122], and flow control [25]. The emphasis of the analysis presented here is on the onset of instability, and hence a linear theory is sufficient. We assume linear vibrations and incompressible flow. Divergence and flutter of geometrically nonlinear panels are presented in Chapter 10 for incompressible and supersonic flows.

The literature is very rich on theoretical, computational, and experimental investigations of two-dimensional (infinite span) and rectangular panels as can be found in ([28], [29], [32], [35], [44], [61], [68], [74], [86], [114], [113]) among many other investigations, and no attempt is made here for a comprehensive review. Dowel has provided comprehensive reviews of panel flutter in [27] and [29]. Dugundji et al. [32] conducted experiments and theoretical analysis of two-dimensional panels on elastic foundation in subsonic flow. They found traveling-wave instability to be the dominant mode of flutter, which is preceded by a mild divergence-type instability as the fluid velocity increases from zero. Kornecki et al. [61] investigated the effects of boundary conditions on instability of 2D panels, and found that divergence prevails for a panel fixed at both ends whereas flutter occurs for a cantilevered panel with free trailing edge. Ellen [35] studied stability of rectangular panels at subsonic Mach numbers. He developed an analytic expression for the pressure, and simplified it for low frequency to determine divergence velocity in the limits of high and low aspect ratio of the panel. Lucey and Carpenter [68] investigated hydroelastic stability of rectangular compliant walls (panels supported by elastic foundation). They analyzed the effects of panel aspect ratio and elastic foundation constant on critical velocity and mode shape at divergence and coalescence flutter. Merrett and Hilton [74] developed theoretical and computational models for elastic and viscoelastic panel flutter. They used approximate aerodynamic models and predicted panel flutter for incompressible flow as well as for subsonic and supersonic Mach numbers, and presented numerical results for a square panel. Tan et al. [114] considered square panels and promoted the use of localized stiffening as a mean to delay divergence and suppress flutter. Tan et al. [113] conducted computational study of stability of two-dimensional panels and identified the effects of mass ratio (or density ratio) on the flutter mode; particularly the switch from modal coalescence divergence at low mass ratio to modal coalescence flutter at high mass ratio.

In previous numerical analyses of hydroelastic stability, structure model is the classical plate theory, and either modal expansion in conjunction with Galerkin method or a finite-difference method is used for the space operators, which is adequate for rectangular domains. It is the thin plate that is prone to instability and flutter, and Kirchhoff–Love theory should be sufficient. However, in the present section, the plate dynamics is modeled by Reissner–Mindlin plate theory, which is valid for thin and moderately thick plates. Small disturbance potential flow theory is used to compute hydrodynamic pressure in terms of the plate motion. Although modal expansions can be adapted to regular shapes such as circles and triangles, the finite-element method is more versatile for irregular shapes. The finite-element method is used to discretize the continuous fluid-structure problem into a finite dynamical homogeneous linear system for the degrees of freedom of the structure. The Lagrange interpolation functions are used for the plate structural dynamics as well for the hydrodynamic model, which is given by a panel method. Pitman and Lucey [86] put forward a method for direct determination of the eigenmodes of finite flow-structure systems. Their method is used here to analyze the hydroelastic divergence and flutter of elastic panels of arbitrary shapes.

9.4.1 COUPLED FLUID-STRUCTURAL MODEL

Let L denote a characteristic length of the panel (radius of a circle or side of a triangle), and h its thickness assumed uniform. The panel response to fluid pressure fluctuations is determined by two nondimensional parameters namely: mass loading μ (also called density ratio), and hydrodynamic loading Λ (also called velocity ratio) defined by

$$\mu = \frac{\rho_f}{\rho_p}\frac{L}{h}, \quad \Lambda = \frac{\rho_f V_o^2 L^3}{D}, \quad D = \frac{Eh^3}{12(1-\nu^2)} \tag{9.227}$$

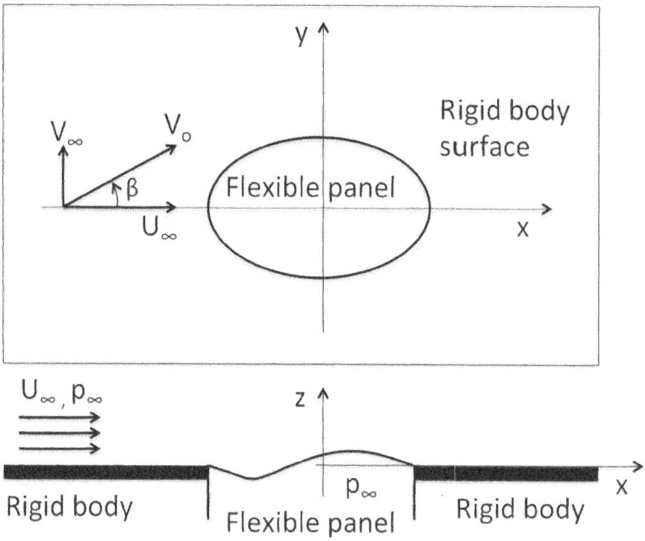

Figure 9.20 A flexible panel surrounded by a flat rigid body. Far field flow velocity (U_∞, V_∞) and pressure p_∞ on upper surface, and uniform pressure p_∞ on the lower surface of panel with negligible velocity.

where ρ_f and ρ_p are mass densities of fluid and plate, respectively, V_o is the fluid velocity magnitude, and D is the flexural rigidity, where E is elastic modulus and ν is Poisson's ratio. Figure 9.20 define panel geometry and flow parameters. In Reissner–Mindlin plate theory, Poisson's ratio ν and plate thickness ratio $h^* = h/L$ are two more parameters. In addition to these four parameters, the flow direction relative to the plate given by an angle β may also influence the response. Three parameters μ, Λ, and β combine flow and plate properties, whereas the parameters h^* and ν are plate properties. We note that in the classical plate theory, only three parameters are needed μ, Λ, and β; whereas h^* and ν do not appear explicitly.

Element Equation of Motion

The equations of motion for the typical element # e is given by Equation (9.207)

$$[M^e]\{\ddot{U}^e\} + [K^e]\{U^e\} = \{F^e\} + \{R^e\}. \tag{9.228}$$

The stiffness matrix $[K^e]$ and consistent mass matrix $[M^e]$ are given by Equations (9.165) and (9.216), respectively. The forcing vector due to distributed transverse load $f(x,y,t)$ on the panel is

$$\{F^e\} = \int_{A_e} \{\psi\}_3^T f dA. \tag{9.229}$$

For the present case of hydroelastic analysis, the distributed load is the fluid pressure jump across the panel.

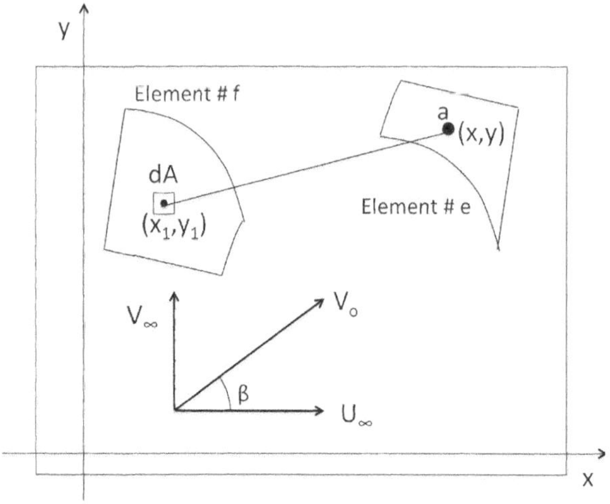

Figure 9.21 Two arbitrary elements in the xy-plane that coincides with the undeformed plate. Velocity potential and velocity components at an arbitrary point a due to source distribution on element # f are found by integration over the element area A_f.

Hydrodynamic Model

The forcing vector in the equation of motion of the plate element is nodal forces equivalent to the distributed fluid pressure. A hydrodynamic model is used to express that vector in terms of deflection, velocity, and acceleration at the element nodes. In the absence of plate motion, the upper side of the panel is exposed to uniform flow as shown in Figure 9.21 with velocity components U_∞ and V_∞ in the x and y directions, respectively, and pressure p_∞. The lower side of panel is at stagnation pressure p_∞. Inviscid flow theory is used to model velocity and pressure disturbances on the upper side due to plate vibration, while disturbances on the lower side are neglected and hence the pressure remains at p_∞. Incompressible, irrotational small disturbance assumptions are also used here. Thus, fluid motion satisfies the linearized boundary condition

$$\frac{\partial \phi}{\partial z} = \frac{\partial w}{\partial t} + U_\infty \frac{\partial w}{\partial x} + V_\infty \frac{\partial w}{\partial y} \quad on \;\; z = 0 \tag{9.230}$$

where ϕ is the disturbance velocity potential, $w(x,y,t)$ is the deflection of the mid-plane of plate. This condition is satisfied by a source distribution in the $z = 0$ plane over the vibrating panel area. The source strength (q) is zero over the rigid baffle surrounding the panel, and for points in the panel area it is given by

$$q(x,y,t) = 2\frac{\partial \phi}{\partial z} \quad on \;\; z = 0. \tag{9.231}$$

Using Equation (9.230), we obtain

$$\frac{1}{2}q(x,y,t) = \frac{\partial w}{\partial t} + U_\infty \frac{\partial w}{\partial x} + V_\infty \frac{\partial w}{\partial y}. \tag{9.232}$$

With the source strength given in terms of plate deflection $w(x,y,t)$, we want to write the pressure disturbance at any point on the plate in terms of the yet unknown deflection and its space and time derivatives. To that end, we consider two elements labeled e and f in the finite-element mesh as shown in Figure 9.21. Approximation of the deflection over element # f is

$$w^f(x,y,t) = \{\psi\}\{W^f(t)\} \tag{9.233}$$

where

$$\{\psi\} = \{\ \psi_1\ \psi_2\ \cdots\ \psi_j\ \cdots\ \psi_n\ \} \tag{9.234}$$

and

$$\{W^f(t)\} = \left\{\ w_1^f, \ldots, w_j^f, \ldots, w_n^f\ \right\}^T. \tag{9.235}$$

We obtain space and time derivatives of $w^f(x,y,t)$ from Equation (9.233), and write the source strength over element f as

$$\frac{1}{2}q^f(x,y,t) = \{\psi\}\{\dot{W}^f\} + \{\tilde{U}_\infty\}[S]\{W^f\} \tag{9.236}$$

$$\{\tilde{U}_\infty\} = \{U_\infty\ V_\infty\} \tag{9.237}$$

$$[S] = \begin{bmatrix} \frac{\partial \psi_1}{\partial x} & \frac{\partial \psi_2}{\partial x} & \cdots & \frac{\partial \psi_j}{\partial x} & \cdots & \frac{\partial \psi_n}{\partial x} \\ \frac{\partial \psi_1}{\partial y} & \frac{\partial \psi_2}{\partial y} & \cdots & \frac{\partial \psi_j}{\partial y} & \cdots & \frac{\partial \psi_n}{\partial y} \end{bmatrix}. \tag{9.238}$$

Here, we use the notation $\dot{Q} \equiv dQ/dt$. The velocity potential ϕ at any point (x,y) on the plane $z = 0$ due to the distributed source q^f on element # f is

$$\phi(x,y,t) = -\frac{1}{4\pi}\int_{A_f} \frac{q^f(x_1,y_1,t)}{[(x-x_1)^2+(y-y_1)^2]^{1/2}}dA. \tag{9.239}$$

Using the source strength given in Equation (9.236), we obtain

$$\phi(x,y,t) = -\frac{1}{2\pi}\int_{A_f} \frac{\{\psi\}\{\dot{W}^f\} + \{\tilde{U}_\infty\}[S]\{W^f\}}{[(x-x_1)^2+(y-y_1)^2]^{1/2}}dA. \tag{9.240}$$

and hence the disturbance velocity components and the local time derivative of ϕ are

$$\frac{\partial \phi}{\partial x} = \frac{1}{2\pi}\int_{A_f} \frac{(x-x_1)\left[\{\psi\}\{\dot{W}^f\} + \{\tilde{U}_\infty\}[S]\{W^f\}\right]}{[(x-x_1)^2+(y-y_1)^2]^{3/2}}dA \tag{9.241}$$

$$\frac{\partial \phi}{\partial y} = \frac{1}{2\pi} \int_{A_f} \frac{(y-y_1)\left[\{\psi\}\{\dot{W}^f\} + \{\tilde{U}_\infty\}[S]\{W^f\}\right]}{[(x-x_1)^2+(y-y_1)^2]^{3/2}} dA \tag{9.242}$$

$$\frac{\partial \phi}{\partial t} = -\frac{1}{2\pi} \int_{A_f} \frac{\{\psi\}\{\ddot{W}^f\} + \{\tilde{U}_\infty\}[S]\{\dot{W}^f\}}{[(x-x_1)^2+(y-y_1)^2]^{1/2}} dA. \tag{9.243}$$

Defining the matrices

$$\{E_o\} = -\frac{1}{2\pi} \int_{A_f} \frac{\{\psi\}}{[(x-x_1)^2+(y-y_1)^2]^{1/2}} dA \tag{9.244}$$

$$\{E_1\} = \frac{1}{2\pi} \int_{A_f} \frac{(x-x_1)\{\psi\}}{[(x-x_1)^2+(y-y_1)^2]^{3/2}} dA \tag{9.245}$$

$$\{E_2\} = \frac{1}{2\pi} \int_{A_f} \frac{(y-y_1)\{\psi\}}{[(x-x_1)^2+(y-y_1)^2]^{3/2}} dA \tag{9.246}$$

$$[F_o] = -\frac{1}{2\pi} \int_{A_f} \frac{[S]}{[(x-x_1)^2+(y-y_1)^2]^{1/2}} dA \tag{9.247}$$

$$[F_1] = \frac{1}{2\pi} \int_{A_f} \frac{(x-x_1)[S]}{[(x-x_1)^2+(y-y_1)^2]^{3/2}} dA \tag{9.248}$$

$$[F_2] = \frac{1}{2\pi} \int_{A_f} \frac{(y-y_1)[S]}{[(x-x_1)^2+(y-y_1)^2]^{3/2}} dA \tag{9.249}$$

we write

$$\frac{\partial \phi}{\partial t} = \{E_o\}\{\ddot{W}^f\} + \{\tilde{U}_\infty\}[F_o]\{\dot{W}^f\} \tag{9.250}$$

$$\frac{\partial \phi}{\partial x} = \{E_1\}\{\dot{W}^f\} + \{\tilde{U}_\infty\}[F_1]\{W^f\} \tag{9.251}$$

$$\frac{\partial \phi}{\partial y} = \{E_2\}\{\dot{W}^f\} + \{\tilde{U}_\infty\}[F_2]\{W^f\}. \tag{9.252}$$

Pressure disturbance is obtained from Bernoulli's equation linearized for small disturbances,

$$-\frac{p-p_\infty}{\rho_f} = \frac{\partial \phi}{\partial t} + U_\infty \frac{\partial \phi}{\partial x} + V_\infty \frac{\partial \phi}{\partial y}. \tag{9.253}$$

When the induced velocity of one element on another element is computed, the integrals in Equations (9.240–9.249) are regular because the point of evaluation (x,y) is not in the range of integration. However, for self-induced velocity by an element, proper limit is taken to avoid the singularity. Pressure disturbance at any point on the plane $z=0$ due to the source distribution on element # f is

$$-\frac{p-p_\infty}{\rho_f} = [G_o]\{\ddot{W}^f\} + V_o\{\Gamma_1\}[G_1]\{\dot{W}^f\} + V_o^2\{\Gamma_2\}[G_2]\{W^f\} \tag{9.254}$$

where

$$[G_o] = \{E_o\},\ [G_1] = \left[[F_o] + \begin{bmatrix} \{E_1\} \\ \{E_2\} \end{bmatrix}\right],\ [G_2] = \begin{bmatrix} [F_1] \\ [F_2] \end{bmatrix} \tag{9.255}$$

$$\begin{aligned} \{\Gamma_1\} &= \{cos\beta \quad sin\beta\} \\ \{\Gamma_2\} &= \{cos^2\beta \quad cos\beta sin\beta \quad cos\beta sin\beta \quad sin^2\beta\} \end{aligned} \tag{9.256}$$

$$V_o = \sqrt{U_\infty^2 + V_\infty^2}, \qquad cos\beta = \frac{U_\infty}{V_o}, \qquad sin\beta = \frac{V_\infty}{V_o}. \tag{9.257}$$

We note that the hydrodynamic matrices involve integrals of the element shape functions and first derivatives only. Thus, Lagrange interpolation functions, which are used for the structure model, are also admissible for the hydrodynamic model.

Fluid-Structural Model

Forcing vector for element number # e is given by Equation (9.229) for distributed load f, which is the pressure jump $f = -(p-p_\infty)$, hence

$$\{F^e\} = \int_{A_e} \{\psi\}^T [-(p-p_\infty)] dA. \tag{9.258}$$

Using Equation (9.254), we find the contribution of the source distribution on element # f to the forcing vector of element # e to be

$$\begin{aligned} \{F^e\}^f &= \rho_f \int_{A_e} \{\psi\}^T \left[[G_o]\{\ddot{W}^f\} + V_o\{\Gamma_1\}[G_1]\{\dot{W}^f\} + V_o^2\{\Gamma_2\}[G_2]\{W^f\}\right] dA \\ &= \rho_f [H_o]\{\ddot{W}^f\} + \rho_f V_o [H_1]\{\dot{W}^f\} + \rho_f V_o^2 [H_2]\{W^f\} \end{aligned} \tag{9.259}$$

where

$$\begin{aligned} [H_o] &= \int_{A_e} \{\psi\}^T [G_o] dA \\ [H_1] &= \int_{A_e} \{\psi\}^T \{\Gamma_1\}[G_1] dA \\ [H_2] &= \int_{A_e} \{\psi\}^T \{\Gamma_2\}[G_2] dA. \end{aligned} \tag{9.260}$$

The forcing vector of element # e is the sum of the contributions of all elements including the self-induced pressure. It is obtained by assembling the matrices $[H_o]$, $[H_1]$, and $[H_2]$ into global matrices $[H_o^g]$, $[H_1^g]$, and $[H_2^g]$ and expressing the vector for element # e in terms of the global array $\{U\}$ and its time derivatives.

$$\{F^e\} = \rho_f [H_o^g] \{\ddot{U}\} + \rho_f V_o [H_1^g] \{\dot{U}\} + \rho_f V_o^2 [H_2^g] \{U\}. \tag{9.261}$$

The equation of motion for an element is

$$[M^e] \{\ddot{U}^e\} + [C^e] \{\dot{U}^e\} + [K^e] \{U^e\} = \{F^e\} + \{R^e\} \tag{9.262}$$

we added a structural damping matrix $[C^e]$ if needed, however, the results presented in this chapter assume zero structural damping. We assemble all elements into a global system

$$[M] \{\ddot{U}\} + [C] \{\dot{U}\} + [K] \{U\} = \{F\} + \{R\}. \tag{9.263}$$

Finite-element model of the fluid-structural problem becomes

$$[[M] - \rho_f [H_o^g]] \{\ddot{U}\} + [[C] - \rho_f V_o [H_1^g]] \{\dot{U}\} + [[K] - \rho_f V_o^2 [H_2^g]] \{U\} = \{R\}. \tag{9.264}$$

Nondimensional Variables

Note that the physical dimensions of $[H_o^g]$, $[H_1^g]$, and $[H_2^g]$ are volume (L^3), area (L^2), and length (L), respectively. These matrices contribute only to the equation of motion of the transverse deflection w, and have no contributions to the moment equations for θ_x or θ_y. Introducing nondimensional space and time variables

$$(x^*, y^*, z^*) = \frac{(x, y, z)}{L}, \quad t^* = \frac{t}{L^2} \sqrt{\frac{D}{\rho_p h}} \tag{9.265}$$

into Equation (9.264), we get

$$[[M] - \mu [H_o^g]] \{\ddot{U}\} + \left[[C] - \sqrt{\mu \Lambda} [H_1^g]\right] \{\dot{U}\} + [[K] - \Lambda [H_2^g]] \{U\} = \{R\}. \tag{9.266}$$

All matrices are nondimensional. To compute mass $[M]$ and stiffness $[K]$ matrices we need to specify values for ν and h^* only. No parameters are need for computing $[H_o^g]$, whereas only flow direction β is needed for computing $[H_1^g]$ and $[H_2^g]$. The hydrodynamic loading matrices depend only on the panel geometry, and are independent of the physical properties of the fluid and plate material. Also, they can be computed and stored before applications of boundary conditions.

Two special cases are of interest. The first case is computation of the effects of added mass on natural frequencies of one-side wetted panels. In this case case $\Lambda = 0$, and assuming zero structural damping, the fluid-structural system is

$$[[M] - \mu [H_o^g]] \{\ddot{U}\} + [K] \{U\} = \{R\}. \tag{9.267}$$

The added mass matrix is $-\mu\left[H_o^g\right]$. The second special case is computation of velocity ratio Λ at divergence-onset and recovery-onset if it happens at zero frequency, which are given by static condition, hence we drop the time derivatives; the relevant fluid-structural system is

$$\left[[K]-\Lambda\left[H_2^g\right]\right]\{U\}=\{R\}. \tag{9.268}$$

In this case Λ is an eigenvalue. The array of assembled secondary variables $\{R\}$ is decided by the boundary conditions. Applying boundary conditions to the general system Equation (9.264), we obtain a reduced/modified dynamical system for the coupled fluid-structure interaction problem

$$[M_*]\{\ddot{U}\}+[C_*]\{\dot{U}\}+[K_*]\{U\}=0 \tag{9.269}$$

where $\{U\}$ is the array of deflection and rotations (slopes) at the global nodes of the finite-element mesh. The mass and stiffness matrices include the added mass and stiffness due to fluid. In this section, we neglect structural damping, and hence the damping matrix is due to the fluid only. We use Pitman and Lucey's [86] method for direct determination of the eigenmodes of finite flow-structure systems. Define state variables $\{Q_1\}=\{U\}$, and $\{Q_2\}=\{\dot{U}\}$, and rewrite the system as

$$\begin{bmatrix} [I] & [0] \\ [0] & [M_*] \end{bmatrix}\begin{bmatrix} \{\dot{Q}_1\} \\ \{\dot{Q}_2\} \end{bmatrix}=\begin{bmatrix} [0] & [I] \\ -[K_*] & -[C_*] \end{bmatrix}\begin{bmatrix} \{Q_1\} \\ \{Q_2\} \end{bmatrix} \tag{9.270}$$

or

$$[B]\ \{\dot{Q}\}\ =[A]\{Q\}. \tag{9.271}$$

We assume a normal mode solution,

$$\{Q\}=e^{s^*t^*}\{\hat{Q}\} \tag{9.272}$$

and obtain the eigenvalue problem

$$[[A]-s^*[B]]\{\hat{Q}\}=0. \tag{9.273}$$

In a mesh of N global nodes, dimension of matrices $[A]$ and $[B]$ is $(6N,6N)$, and dimension of $\{\hat{Q}\}$ is $(6N,1)$. The generalized eigenvalue problem is solved for eigenvalues s_j^* and corresponding eigenvectors $\{\hat{Q}\}_j$ for $j=1,\ldots,6N$. The complex eigenvalue of mode j is written as $s_j^*=s_r^*+is_i^*$. The real part s_r^* is growth rate of mode j and the imaginary part s_i^* is its frequency. The mode shape is given by the array $\{\hat{Q}\}_j$.

Numerical Procedures

Isoparametric elements are used. All integrals of hydrodynamic matrices are transformed and evaluated numerically by Gauss–Legendre quadrature on the canonical element as given in Section 4.8 (see also [54]). Gauss–Legendre rules of integration are 4×4, 8×8, and 12×12 for the bilinear, biquadratic, and bicubic elements, respectively. In computing integrals for self-induced velocity potential and velocity components of an element, the element is divided into (16, 64, or 144) sub-elements and the integrals are evaluated using analytic expressions for a quadrilateral panel with uniform source strength ([59]). The same Gauss–Legendre rules are used for computing bending stiffness matrix, but reduced integration is used for the shear stiffness matrix ([54], [97]). Generalized eigenvalue problems are solved using QZ algorithm, and all computations are programmed in MATLAB.

9.4.2 FREE VIBRATIONS OF SUBMERGED PLATES: NAVMI FACTORS

Surface pressure on plates vibrating in fluids is proportional to acceleration, and as such, it is equivalent to an increase in mass of the plate known as the virtual mass. Natural frequencies of a plate submerged in fluid are reduced due to the added (virtual) mass. The natural frequencies for a flexible plate surrounded by a rigid plane and in contact with unbounded fluid on one side are denoted by ω_f, and the corresponding frequencies in vacuum are denoted by ω. For the same mode shape, we expect $\omega > \omega_f$. Amabili and Kwak [6] expressed frequency reduction in terms of a nondimensional added virtual mass incremental ($NAVMI$) factor,

$$\Gamma = \frac{(\omega/\omega_f)^2 - 1}{\mu} \tag{9.274}$$

where μ is the mass ratio defined by Equation (9.227).

Example 9.9. Reissner–Mindlin: *NAVMI* Factors for Circular Plates

We consider a circular plate of radius R, thickness $h^* = h/R = 0.001$, Poisson's ratio $\nu = 0.3$, shear correction factor $\kappa = 5/6$, and mass ratio $\mu = 1$. Nondimensional frequency is defined by

$$\omega^* = \omega R^2 \sqrt{\frac{\rho_p h}{D}}. \tag{9.275}$$

A mesh of 192 bicubic elements with 1777 global nodes finer than that depicted in Figure 9.19a is used. As shown in Table 9.14, natural frequencies of different radial (m_r) and circumferential (m_θ) modes in vacuum are in excellent agreement with the data reported by Senjanovic et al. [109] who used a modified Reissner–Mindlin plate theory. Amabili and Kwak [6], used classical plate theory, and computed $NAVMI$ factors. In one approach, they expanded the deflection of a wet plate in terms of the mode shapes of the plate in vacuum. Their results for this approach are compared with the present finite-element computations in Table 9.14 for mass loading factor $\mu = 1$; for this value changes in mode shapes due to fluid loading are negligible. The maximum difference of $NAVMI$ factors is less than 5% for all presented modes.

Finite-element computations are repeated for a clamped plate with the same geometric and material parameters as the simply supported plate, and using the same

Table 9.14
Natural Frequencies in Vacuum and *NAVMI* Factors for Simply Supported Circular Plate; $h^* = 0.001$, $\nu = 0.3$, $\mu = 1$

(m_r, m_θ)	$\omega^* - FEM$	ω^* Ref. [109]	$\Gamma - FEM$	$\Gamma-$ Ref. [6]
(1,0)	4.93514e+00	4.93500e+00	7.55879e-01	7.5539e-01
(1,1)	1.38981e+01	1.38980e+01	3.32468e-01	3.3225e-01
(1,2)	2.56131e+01	2.56130e+01	2.26915e-01	2.2675e-01
(2,0)	2.97199e+01	2.97200e+01	2.44084e-01	2.5680e-01
(1,3)	3.99570e+01	3.99570e+01	1.75346e-01	1.7521e-01
(2,1)	4.84787e+01	4.84780e+01	1.68380e-01	1.6951e-01
(1,4)	5.68410e+01	5.68410e+01	1.44016e-01	1.4389e-01
(2,2)	7.01166e+01	7.01160e+01	1.33666e-01	1.3390e-01
(3,0)	7.41562e+01	7.41550e+01	1.43264e-01	1.5041e-01
(1,5)	7.62027e+01	7.62020e+01	1.22714e-01	
(2,3)	9.45494e+01	9.45470e+01	1.12294e-01	1.1232e-01
(1,6)	9.79960e+01		1.07187e-01	
(3,1)	1.02777e+02	1.02772e+02	1.12336e-01	1.1321e-01
(2,4)	1.21705e+02	1.21699e+02	9.74529e-02	9.7410e-02
(1,7)	1.22188e+02		9.53172e-02	
(3,2)	1.34302e+02	1.34294e+02	9.50698e-02	9.5280e-02
(4,0)	1.38331e+02	1.38315e+02	1.00736e-01	1.0513e-01
(1,8)	1.48753e+02		8.59228e-02	
(2,5)	1.51525e+02	1.51514e+02	8.64109e-02	
(3,3)	1.68698e+02	1.68669e+02	8.32391e-02	8.3270e-02
(4,1)	1.76841e+02	1.76795e+02	8.40721e-02	8.4690e-02
(1,9)	1.77680e+02		7.82835e-02	
(2,6)	1.83963e+02		7.78180e-02	
(3,4)	2.05890e+02	2.05843e+02	7.44163e-02	7.4380e-02
(1,10)	2.08965e+02		7.19425e-02	
(4,2)	2.18239e+02	2.18194e+02	7.37613e-02	7.3920e-02
(2,7)	2.18985e+02		7.09047e-02	

mesh. As shown in Table 9.15, natural frequencies of different radial (m_r) and circumferential (m_θ) modes in vacuum are in excellent agreement with the data reported by Senjanovic et al. [109]. The *NAVMI* factors are also compared with Amabili and Kwak [6] results in Table 9.15; the maximum difference is less than 8% for all considered modes.

Example 9.10. Reissner–Mindlin: *NAVMI* Factors for Simply Supported Triangular Panels

An equilateral triangular panel is considered with all sides being simply supported. The plate side L is used as reference length. The values $\nu = 0.35$, $h/L \equiv h^* = 0.01$, and $\mu = 37.037$ are used. A mesh of quadrilateral bicubic elements as shown in Figure 9.22a is used, the number of elements and global nodes are 108 and 1027,

Table 9.15
Natural Frequencies in Vacuum and *NAVMI* Factors for Clamped Circular Plate; $h^* = 0.001$, $\nu = 0.3$, $\mu = 1$

(m_r, m_θ)	$\omega^* - FEM$	ω^* Ref. [109]	$\Gamma - FEM$	$\Gamma-$ Ref. [6]
(1,0)	1.02158e+01	1.02160e+00	6.55553e-01	6.5381e-01
(1,1)	2.12603e+01	2.12600e+01	2.99278e-01	2.9883e-01
(1,2)	3.48768e+01	3.48770e+01	2.08598e-01	2.0834e-01
(2,0)	3.97709e+01	3.97710e+01	2.54662e-01	2.7613e-01
(1,3)	5.10296e+01	5.10290e+01	1.63403e-01	1.6321e-01
(2,1)	6.08285e+01	6.08280e+01	1.67008e-01	1.6914e-01
(1,4)	6.96653e+01	6.96650e+01	1.35506e-01	1.3535e-01
(2,2)	8.45819e+01	8.45810e+01	1.30915e-01	1.3123e-01
(3,0)	8.91047e+01	8.91020e+01	1.51871e-01	1.6513e-01
(1,5)	9.07390e+01	9.07370e+01	1.16297e-01	
(2,3)	1.11023e+02	1.11019e+02	1.09531e-01	1.0968e-01
(1,6)	1.14215e+02		1.02153e-01	
(3,1)	1.20084e+02	1.20076e+02	1.14092e-01	1.1591e-01
(2,4)	1.40067e+02	1.40104e+02	9.42272e-02	9.4960e-02
(1,7)	1.40113e+02		9.19664e-02	
(3,2)	1.53822e+02	1.53810e+02	9.51122e-02	9.5650e-02
(4,0)	1.58206e+02	1.58179e+02	1.06909e-01	1.1541e-01
(1,8)	1.68274e+02		8.25619e-02	
(2,5)	1.71815e+02	1.71797e+02	8.41924e-02	
(3,3)	1.90343e+02	1.90296e+02	8.26841e-02	8.2860e-02
(4,1)	1.98828e+02	1.99045e+02	8.29134e-02	8.7480e-02
(1,9)	1.99122e+02		7.86394e-02	
(2,6)	2.06094e+02		7.58652e-02	
(3,4)	2.29583e+02	2.29507e+02	7.36401e-02	7.3680e-02
(1,10)	2.29583e+02		6.95256e-02	
(4,2)	2.42780e+02	2.42708e+02	7.45012e-02	7.4940e-02
(2,7)	2.42928e+02		6.91808e-02	

respectively. Nondimensional frequency is defined by

$$\omega^* = \omega L^2 \sqrt{\frac{\rho_p h}{D}}. \tag{9.276}$$

Natural frequencies in vacuum and *NAVMI* factors for the first ten free vibration modes are given in Table 9.16. The first mode frequency in vacuum is 52.5838, and the wet mode shape is shown in Figure 9.22b. The (wet) mode shapes for modes 2 and 3 are shown in Figures 9.22c and 9.22d, respectively. Modes 2 and 3 have different shapes but equal frequencies.

9.4.3 LINEAR HYDROELASTIC STABILITY

The linear fluid-structure model given by Equation (9.273) is used to determine the hydrodynamic loading at onset of divergence and modal coalescence flutter for a

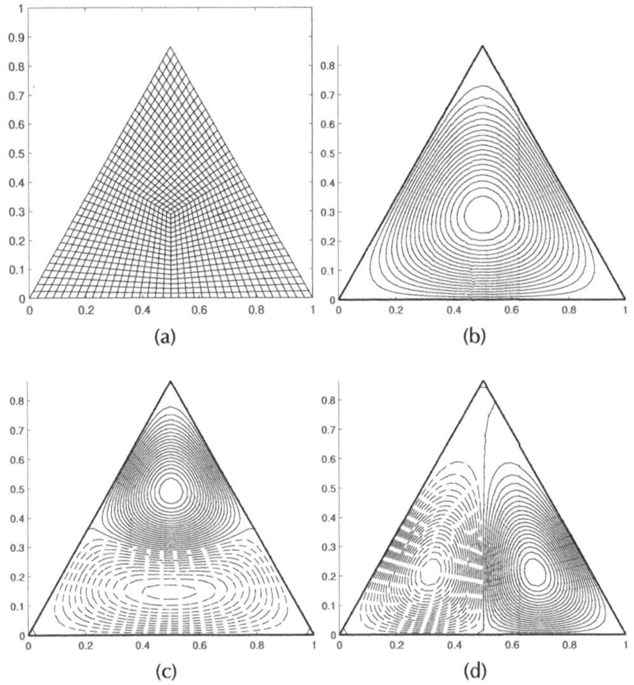

Figure 9.22 Free vibration of an equilateral simply supported triangular panel. (a) Mesh of quadrilateral bicubic elements; 108 elements and 1027 global nodes. Deflection contours (wet) mode shapes: (b) Fundamental (first) mode, (c) second mode, and (d) third mode. Step in contour levels is 0.05.

Table 9.16

Natural Frequencies in Vacuum and *NAVMI* Factors of Equilateral Triangular Plate with Simply Supported Sides *SSS*; $h^* = 0.01$, $\nu = 0.35$

Mode	ω^*	Γ	Mode	ω^*	Γ
1	5.25838e+01	2.58076e-01	6	2.27089e+02	7.86771e-02
2	1.22528e+02	1.18111e-01	7	3.31233e+02	6.54163e-02
3	1.22528e+02	1.18111e-01	8	3.31233e+02	6.54163e-02
4	2.09692e+02	9.20141e-02	9	3.65850e+02	5.94661e-02
5	2.27089e+02	7.86771e-02	10	3.65852e+02	5.89495e-02

thin elastic panel surrounded by a rigid plane. However, an analysis that neglects geometric nonlinear deformation of plates is not valid for dynamic pressure greater than that at static divergence. Once divergence occurs, the plate develops large deflection and the inplane strains become effective in increasing the structure stiffness.

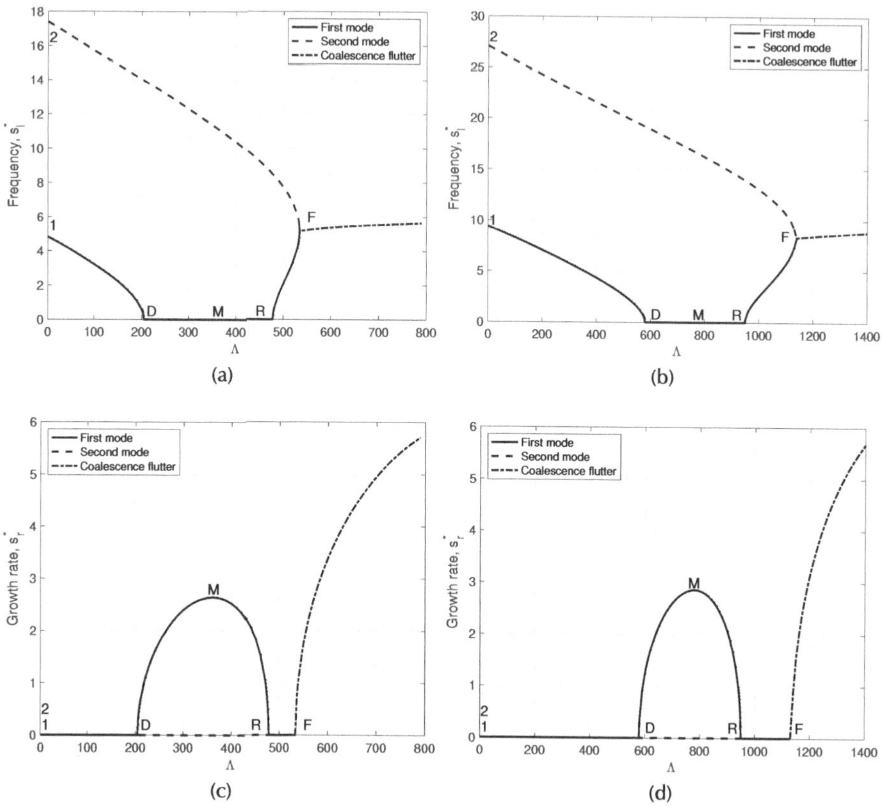

Figure 9.23 Divergence and flutter of a square plate. (a) Simply supported plate frequency. (b) Clamped plate frequency. (a) Simply supported plate growth rate. (b) Clamped plate growth rate.

The effects of geometric non-linearity on the hydroelastic response of elastic panels will be presented in Chapter 10.

Example 9.11. Divergence and Flutter of Square Panels

We consider a square panel of side a, thickness $h/a = 0.01$, Poisson ratio $\nu = 0.35$, and shear correction factor $\kappa = 5/6$. The plate and fluid mass densities $\rho_p = 2600$ and $\rho_f = 1000$ kg/m^3 give $\mu = \rho_f a/\rho_p h = 38.46$. A mesh of $16 \times 16 = 256$ bicubic elements and 2401 global nodes is used. The eigenvalue problem Equation (9.273) is solved for prescribed values of Λ. The growth rate and frequency are given by the real and imaginary parts of the eigenvalue $s^* = s_r^* + is_i^*$, respectively. If the growth rate of any mode is positive, the panel is unstable and a small disturbance from the flat position grows exponentially in time. Nonlinear effects, when considered, can limit such a growth.

For a simply supported plate, frequencies of the first two modes are plotted in Figure 9.23a, and growth rates are shown in Figure 9.23c. Static divergence begins at $\Lambda_D =$

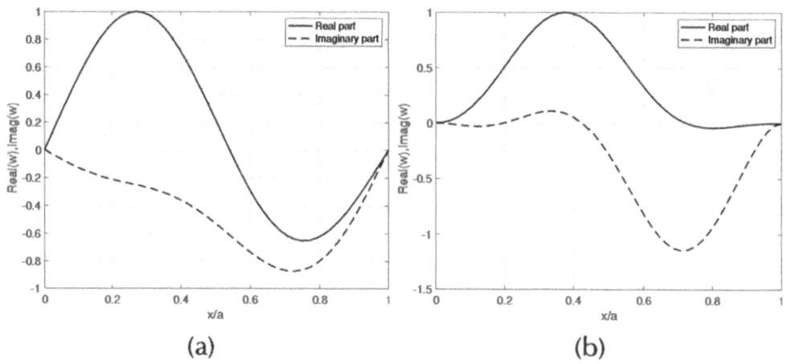

Figure 9.24 Flutter mode shape for a square plate, $h/a = 0.01$, $\nu = 0.35$, $\mu = 38.46$. (a) Simply supported at $\Lambda = 540$. (b) Clamped plate at $\Lambda = 1320$.

204, and recovery at $\Lambda_R = 478$. At $\Lambda = 540$, the complex frequency of the amplified mode is $s^* = 1.0203 + 5.170i$ and its complex conjugate. The flutter mode shape (eigenfunction) is shown in Figure 9.24a. For the complex conjugate frequency, the eigenfunction is the complex conjugate of that shown in the figure.

Similarly, for the clamped square plate static divergence begins at $\Lambda_D = 579$, and recovery at $\Lambda_R = 950$. At $\Lambda = 1320$, the complex frequency of amplified mode is $s^* = 4.992 + 8.701i$ and its complex conjugate. The flutter mode shape (eigenfunction) is shown in Figure 9.24b. For the complex conjugate frequency, the eigenfunction is the complex conjugate of that shown in the figure.

9.5 PROBLEMS

(1) We want to determine the deflection and stresses in a square aluminum plate ($E = 73$ GPa, $\nu = 0.35$) of side $a = 1.2$ m and thickness $h = 10$ mm that is loaded by uniform pressure $q = 5$ kPa on the upper surface $z = 5$ mm. Consider the two different cases of boundary conditions: Case(1) all four sides are clamped, and case (2) the sides $x = \pm a/2$ are simply supported and the sides $y = \pm a/2$ are clamped, see Figure 9.25a. We assume double symmetry about the x- and y-axes, and solve the problem on the first quadrant. The classical plate theory is used, and the first quadrant is modeled by one Bogner–Fox–Schmit conforming element.

   (a) For the two cases: determine the deflection $w(0,0)$ at the center in mm. Also report the nondimensional deflection $w(0,0)/h$ and $w(0,0)D/qa^4$, where $D = Eh^3/12(1-\nu^2)$.

   (b) For case (1): In a plane of constant $x = 0.6$ m, determine the stresses ($\sigma_{xx}, \sigma_{xy}, \sigma_{xz}$ in MPa) at a point p with coordinates $(y,z) = (0,-5)$ mm, see Figure 9.25b.

(2) The bending problem of a rectangular plate ($2a = 300$ mm, $2b = 200$ mm, $h = 10$ mm) with mechanical properties ($E = 70$ GPa, $\nu = 0.30$) is to be analyzed

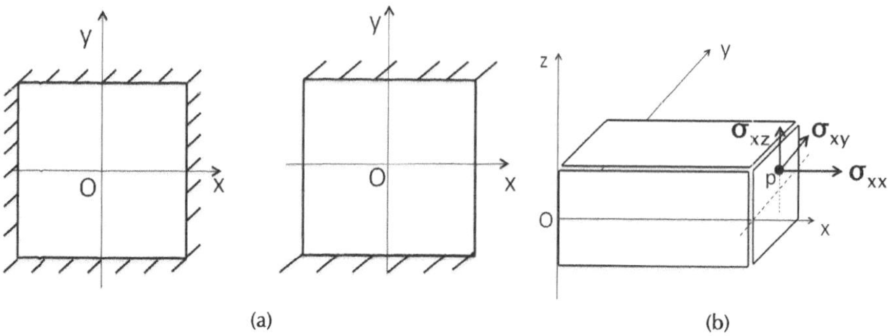

Figure 9.25 Square plate with different boundary conditions

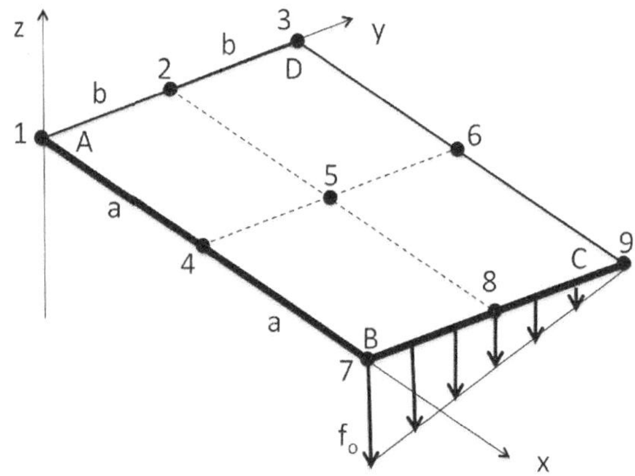

Figure 9.26 Bending of a rectangular plate by distributed load on edge *BC*.

according to the Reissner–Mindlin shear deformation plate theory. As shown in Figure 9.26, side *AD* is treated as clamped and side *CD* is simply supported (*SS*2). Sides *AB* and *BC* are free. A linearly varying transverse load is applied to edge *BC*, $f_o = 400$ N/m. The plate is divided into four elements, and the global nodes are shown in the figure. The assembled system is written in the general form

$$[K]\{U\} = \{F\} + \{R\}$$

(a) From the given boundary conditions and loading, provide numerical values for all the known elements of the primary variables $\{U\}$ and weighted secondary variables $\{R\}$.

(b) Determine the deflection at corner *B*.

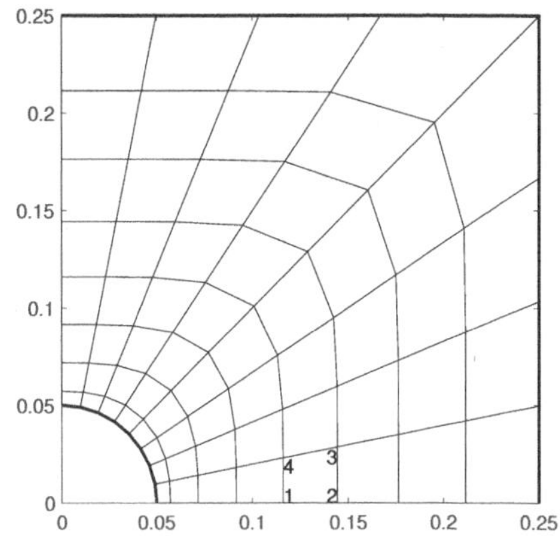

Figure 9.27 A mesh of bilinear 4-node elements on one quarter of a square plate with circular hole at the center.

(3) Reissner–Mindlin shear deformation plate theory is used to analyze the bending of a thick square plate of dimensions of ($a = 0.5$ m, $b = 0.5$ m, $h = 20$ mm) and mechanical properties of ($E = 70$ GPa, $\nu = 0.30$). A hole of radius $r_o = 0.05$ m is drilled at the center of the plate. The plate is simply supported ($SS2$) at its four edges, and uniformly distributed transverse load ($q = -0.5$ MPa) is applied to the upper surface. The first quadrant of the plate is modeled by imposing symmetry conditions on the x and y axes. The isoparametric bilinear 4-node element is used and a grid of ($nxe = 8$, $nye = 8$) is generated and shown in Figure 9.27.
The following data pertain to the element with local nodes $1-2-3-4$ as shown in the figure. Cartesian coordinates of element's nodes:

$$\{x^e\} = \{0.115975, 0.144284, 0.144283, 0.115969\} \text{ m}$$
$$\{y^e\} = \{0, 0, 0.0286996, 0.0230676\} \text{ m}$$

(a) Determine the stiffness matrix $[K^e]$ of this element. Use a shear correction factor $\kappa = 5/6$, and state the integration rules you used for the numerical evaluation of the two integrals in the stiffness matrix Equation (9.165).

(b) Given the primary variables [rotations and deflection]:

$$\{\theta_x^e\} = 0.01\{1.1970, 1.3407, 1.3136, 1.1783\} \text{ radians}$$
$$\{\theta_y^e\} = 0.001\{0, 0, 1.8869, 1.9920\} \text{ radians}$$
$$\{w^e\} = \{-2.0164, -1.6460, -1.6305, -1.9811\} \text{ mm.}$$

Calculate the stress components σ_{xx}, σ_{yy}, and σ_{xy} at the Barlow points of the element and $z = h/2$. For the bilinear element, the Barlow points are the base points of the (1×1) Gauss–Legendre quadrature formula.

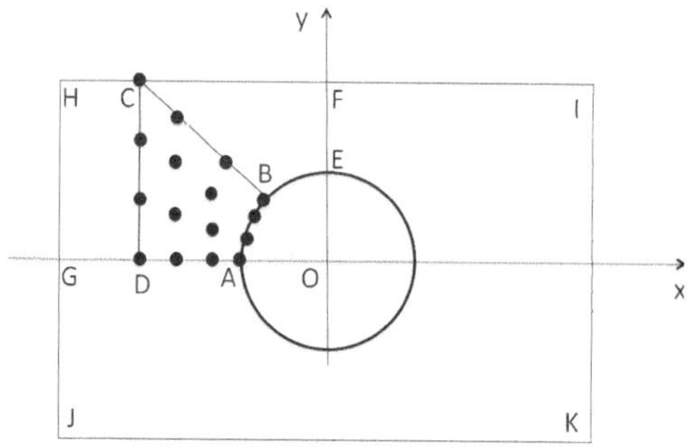

Figure 9.28 Bending of a rectangular plate with circular hole at the center.

(4) A circular hole of radius b is drilled at the center of a thin rectangular plate of height $4b$ and length $6b$ as shown in Figure 9.28. Edges IK and JH are clamped while the other edges are free to deform. Uniform pressure distribution q is applied to the upper surface of the plate. The plate thickness is h. The bending problem is to be analyzed according to the shear deformation plate theory. Assuming symmetry about the x- and y- axes, we solve for the deflection $w(x,y)$ and rotations $\theta_x(x,y)$ and $\theta_y(x,y)$ in one quadrant; say the second quadrant as shown in Figure 9.28. The domain is divided into three bicubic elements labeled $ABCD$, $BEFC$, and $DCHG$. Each element has 16 local nodes, and the total number of global nodes is 40. The coordinates of points B, C, D, G, and H are $(-b/\sqrt{2}, b/\sqrt{2})$, $(-2b, 2b)$, $(-2b, 0)$, $(-3b, 0)$, and $(-3b, 2b)$, respectively. The nodes of element $ABCD$ are shown in the figure; the nodes are equally spaced on AB, BC, AD, and CD.
Take $b = 0.2$ m, $h = 12$ mm, $E = 73$ GPa, $\nu = 0.33$, and $q = 5$ kPa.

(a) Specify and classify (essential, natural, mixed) boundary conditions on GDA, ABE, HCF, GH, and EF.

(b) Use full integration (4x4) Gauss–Legendre quadrature rule for the bending stiffness matrix and reduced integration rule (3x3) for the shear stiffness matrix.

(i) Provide a table that shows the x-coordinates of the global nodes on edge HCF and the values of w in mm, θ_x and θ_y in degrees at those nodes. Plot the three functions against x.

(ii) Provide a table that shows the (x,y) coordinates of the Barlow points of element $BEFC$, and the values of σ_{xx}, σ_{yy} and σ_{xy} in units of MPa at those points and $z = h/2$.

(c) Use full integration (4x4) Gauss–Legendre quadrature rule for both bending and shear stiffness matrices. Compare the results with those found in parts (i) and (ii).

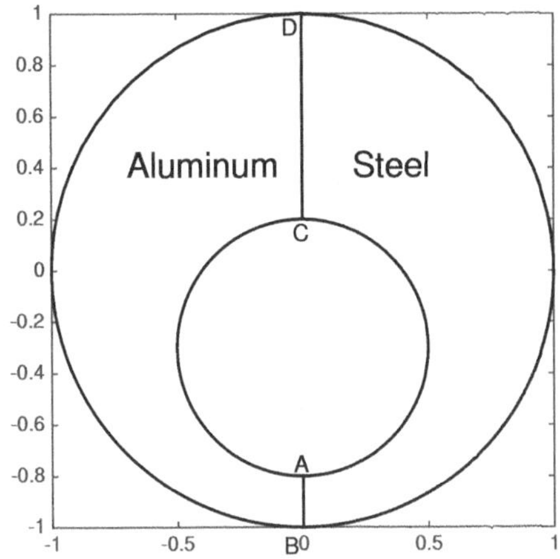

Figure 9.29 A steel-aluminum disc with eccentric hole.

(5) Two semicircular plates one made of steel and the other of aluminum are welded to form a disc of diameter 2 m, then a circular hole of diameter 1 m is drilled with eccentricity 0.3 m, as shown in Figure 9.29. The disc thickness is 30 mm. Its outer circular edge is simply supported ($SS1 : w = 0,\ M_{nn} = 0,\ M_{ns} = 0$) while the inner circle is a free edge. The disc is loaded by uniform pressure of 50 kPa. Neglect the plate weight, and use Reissner–Mindlin shear deformation plate theory to determine deflection and reactions. A MATLAB code (*eccentric_cylinders_Mesh_full.m*) for mesh generation is provided. It generates a mesh of biquadratic quadrilateral elements, and provides the connectivity matrix. The number of elements is $nxe = 20$ in the radial direction and $nye = 60$ in the circumferential direction.

For steel: Poisson's ratio $\nu = 0.25$, and elastic modulus $E = 200$ GPa.

For aluminum: Poisson's ratio $\nu = 0.35$, and elastic modulus $E = 73$ GPa.

(a) Take the origin at the center of the small circle, and plot deflection w as a function of θ on AC for steel and on CA for aluminum. Report results in units of mm. For plotting, take $\theta = 0$ at point A on the steel side, $\theta = 180°$ at point C, and $\theta = 360°$ at point A on the aluminum side.

(b) Take the origin at the center of the small circle, plot rotations θ_n and θ_s as a function of θ on AC for steel and on CA for aluminum. Report results in units of degrees. Note that $\hat{n}$ is the outward unit normal on the boundary, and $\hat{s} = \hat{k} \times \hat{n}$.

(c) Take the origin at the center of the large circle, plot (weighted) reaction R_{nz} as a function of θ on BD for steel and on DB for aluminum. Report results in units of kN. Verify that the obtained reactions satisfy the equations of static equilibrium, and explain how you did the verification.

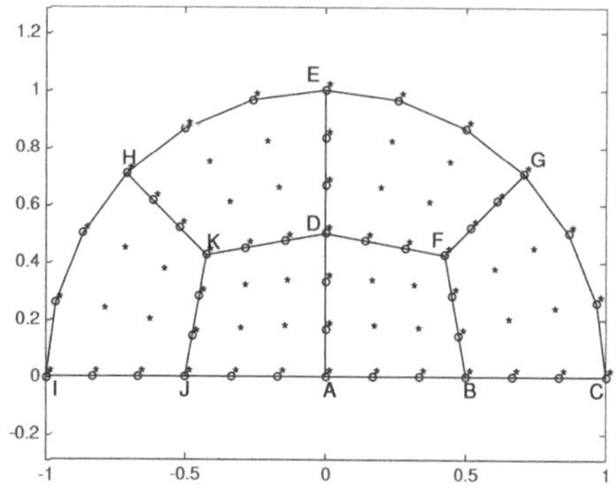

Figure 9.30 A mesh of 6 elements on semicircular plate. Point F at $(x,y) = (4/3,4/3)/\pi$.

(6) We wish to use the finite-element method to determine the natural frequencies of a semicircular plate of radius $R = 1$ and thickness $h/R = 0.1$, see Figure 9.30. The diameter $IJABC$ is clamped and the circumference $CGEHI$ is a free edge. The plate is divided into six bicubic elements. Use Reissner–Mindlin shear deformation plate theory, and evaluate mass and bending stiffness matrices by Gauss-Legendre 4×4 rule and the transverse shear stiffness matrix by the 3×3 rule. Take Poisson's ratio $\nu = 0.25$, and shear correction factor $\kappa = 5/6$. Define a nondimensional frequency by

$$\omega^* = \omega R^2 \sqrt{\rho h/D}. \tag{9.277}$$

where ρ is mass density and D is the plate flexural rigidity.

(a) Report the first (lowest) four natural frequencies (nondimensional).

(b) Take the origin at the semicircle center, and plot deflection w of the first four modes as a function of θ on the free edge. For each mode, normalize the deflection so that the maximum deflection on the edge is 1.

(c) Plot the rotation θ_x as a function of y on line ADE for the four modes.

(7) We wish to use the finite-element method to determine the natural frequencies of an equilateral triangular plate of side $a = 1$ and thickness $h/a = 0.01$. The plate is divided into three bicubic Lagrange elements as shown in Figure 9.31. The nodes are equally spaced on the edges of the plate and on the sides of each element. All sides of the plate are simply supported $SS2$. Use Reissner–Mindlin shear deformation plate theory, and evaluate mass and bending stiffness matrices by Gauss–Legendre 4×4 rule and the transverse shear stiffness matrix by the 3×3 rule. Take Poisson's ratio $\nu = 0.3$, and shear correction factor $\kappa = 5/6$. Define a nondimensional frequency by

$$\omega^* = \omega a^2 \sqrt{\rho h/D} \tag{9.278}$$

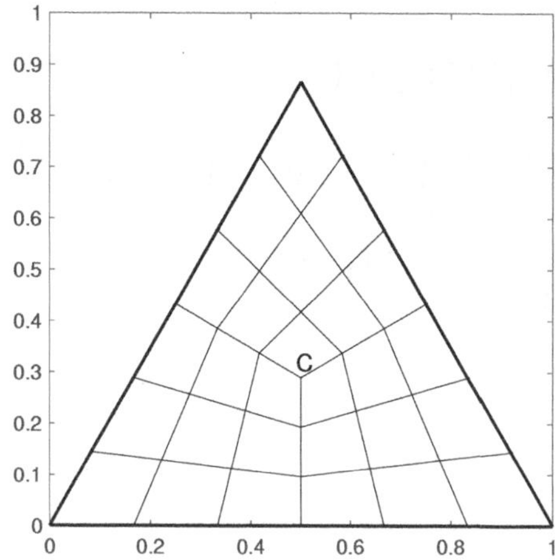

Figure 9.31 A mesh of three bicubic elements on an equilateral triangular plate. Point C is the centroid.

where ρ is mass density and D is the plate flexural rigidity.

(a) Report the first (lowest) four natural frequencies (nondimensional).

(b) Plot the mode shapes for the first and second modes.

10 Nonlinear Reissner–Mindlin Plate and Applications

In this chapter, we formulate the finite-element method for geometrically nonlinear Reissner–Mindlin plate. The von Karman assumptions for large-deflection are adopted for the nonlinear strain-displacement relations. The chapter starts with the equations of equilibrium for a continuum with finite displacements. The principle of virtual displacements is then derived and used to formulate finite-element models for geometrically nonlinear moderately thick plates. Applications include static deflection under transverse loads, and stability analysis (buckling and post-buckling) of plates due to inplane loading or temperature rise. The finite-element method is also formulated for nonlinear vibrations. The methods of harmonic balance and direct numerical integration are used to determine natural frequencies of periodic motions. Linear and nonlinear panel flutter at supersonic speeds are presented, and limit cycle oscillations amplitudes are determined. The chapter closes with a brief study of hydroelastic stability of geometrically nonlinear panels.

10.1 EQUATIONS OF EQUILIBRIUM OF FINITE DISPLACEMENTS

Green's Strain Tensor

The position of a point in the undeformed body is given by the rectangular Cartesian coordinates (x,y,z), and in index notation by x_i, $i = 1,2,3$. After deformation its coordinates are $(x+u, y+v, z+w)$, where (u,v,w) are Cartesian components of the displacement vector, also denoted by u_i. In index notation, Green's strain tensor (e.g., Fung [39], and Washizu [124]) is defined by

$$\varepsilon_{ij} = \frac{1}{2}\left(\frac{\partial u_i}{\partial x_j} + \frac{\partial u_j}{\partial x_i}\right) + \frac{1}{2}\frac{\partial u_k}{\partial x_i}\frac{\partial u_k}{\partial x_j}. \tag{10.1}$$

The expanded form is

$$\begin{aligned}
\varepsilon_{xx} &= \frac{\partial u}{\partial x} + \frac{1}{2}\left[\left(\frac{\partial u}{\partial x}\right)^2 + \left(\frac{\partial v}{\partial x}\right)^2 + \left(\frac{\partial w}{\partial x}\right)^2\right],\\
\varepsilon_{yy} &= \frac{\partial v}{\partial y} + \frac{1}{2}\left[\left(\frac{\partial u}{\partial y}\right)^2 + \left(\frac{\partial v}{\partial y}\right)^2 + \left(\frac{\partial w}{\partial y}\right)^2\right],
\end{aligned}$$

DOI: 10.1201/9781003323150-10

$$
\begin{aligned}
\varepsilon_{zz} &= \frac{\partial w}{\partial z} + \frac{1}{2}\left[\left(\frac{\partial u}{\partial z}\right)^2 + \left(\frac{\partial v}{\partial z}\right)^2 + \left(\frac{\partial w}{\partial z}\right)^2\right], \\
\varepsilon_{xy} &= \frac{1}{2}\left(\frac{\partial u}{\partial y} + \frac{\partial v}{\partial x}\right) + \frac{1}{2}\left[\frac{\partial u}{\partial x}\frac{\partial u}{\partial y} + \frac{\partial v}{\partial x}\frac{\partial v}{\partial y} + \frac{\partial w}{\partial x}\frac{\partial w}{\partial y}\right], \\
\varepsilon_{xz} &= \frac{1}{2}\left(\frac{\partial u}{\partial z} + \frac{\partial w}{\partial x}\right) + \frac{1}{2}\left[\frac{\partial u}{\partial x}\frac{\partial u}{\partial z} + \frac{\partial v}{\partial x}\frac{\partial v}{\partial z} + \frac{\partial w}{\partial x}\frac{\partial w}{\partial z}\right], \\
\varepsilon_{yz} &= \frac{1}{2}\left(\frac{\partial w}{\partial y} + \frac{\partial v}{\partial z}\right) + \frac{1}{2}\left[\frac{\partial u}{\partial y}\frac{\partial u}{\partial z} + \frac{\partial v}{\partial y}\frac{\partial v}{\partial z} + \frac{\partial w}{\partial y}\frac{\partial w}{\partial z}\right],
\end{aligned}
\tag{10.2}
$$

It is a symmetric tensor, $\varepsilon_{xy} = \varepsilon_{yx}$, $\varepsilon_{xz} = \varepsilon_{zx}$, and $\varepsilon_{yz} = \varepsilon_{zy}$. We also define $\gamma_{xy} = 2\varepsilon_{xy}$, $\gamma_{xz} = 2\varepsilon_{xz}$, and $\gamma_{yz} = 2\varepsilon_{yz}$.

Equations of Equilibrium

In terms of the second Piola–Kirchhoff stress tensor σ_{ij}, the equations of equilibrium of an infinitesimally small element are (Novozhilov [80], Fung [39], and Washizu [124])

$$
\begin{aligned}
&\frac{\partial}{\partial x}\left[\left(1+\frac{\partial u}{\partial x}\right)\sigma_{xx} + \frac{\partial u}{\partial y}\sigma_{xy} + \frac{\partial u}{\partial z}\sigma_{xz}\right] \\
&\frac{\partial}{\partial y}\left[\left(1+\frac{\partial u}{\partial x}\right)\sigma_{yx} + \frac{\partial u}{\partial y}\sigma_{yy} + \frac{\partial u}{\partial z}\sigma_{yz}\right] \\
&\frac{\partial}{\partial z}\left[\left(1+\frac{\partial u}{\partial x}\right)\sigma_{zx} + \frac{\partial u}{\partial y}\sigma_{zy} + \frac{\partial u}{\partial z}\sigma_{zz}\right] \qquad + f_x = 0
\end{aligned}
\tag{10.3}
$$

$$
\begin{aligned}
&\frac{\partial}{\partial x}\left[\frac{\partial v}{\partial x}\sigma_{xx} + \left(1+\frac{\partial v}{\partial y}\right)\sigma_{xy} + \frac{\partial v}{\partial z}\sigma_{xz}\right] \\
&\frac{\partial}{\partial y}\left[\frac{\partial v}{\partial x}\sigma_{yx} + \left(1+\frac{\partial v}{\partial y}\right)\sigma_{yy} + \frac{\partial v}{\partial z}\sigma_{yz}\right] \\
&\frac{\partial}{\partial z}\left[\frac{\partial v}{\partial x}\sigma_{zx} + \left(1+\frac{\partial v}{\partial y}\right)\sigma_{zy} + \frac{\partial v}{\partial z}\sigma_{zz}\right] \qquad + f_y = 0
\end{aligned}
\tag{10.4}
$$

$$
\begin{aligned}
&\frac{\partial}{\partial x}\left[\frac{\partial w}{\partial x}\sigma_{xx} + \frac{\partial w}{\partial y}\sigma_{xy} + \left(1+\frac{\partial w}{\partial z}\right)\sigma_{xz}\right] \\
&\frac{\partial}{\partial y}\left[\frac{\partial w}{\partial x}\sigma_{yx} + \frac{\partial w}{\partial y}\sigma_{yy} + \left(1+\frac{\partial w}{\partial z}\right)\sigma_{yz}\right] \\
&\frac{\partial}{\partial z}\left[\frac{\partial w}{\partial x}\sigma_{zx} + \frac{\partial w}{\partial y}\sigma_{zy} + \left(1+\frac{\partial w}{\partial z}\right)\sigma_{zz}\right] \qquad + f_z = 0
\end{aligned}
\tag{10.5}
$$

where (f_x, f_y, f_z) is the body force per unit volume of the undeformed body. In index notation, the equations of equilibrium are

$$
\frac{\partial}{\partial x_k}\left[\left(\delta_{ij} + \frac{\partial u_i}{\partial x_j}\right)\sigma_{kj}\right] + f_i = 0 \qquad (i = 1,2,3). \tag{10.6}
$$

10.2 PRINCIPLE OF VIRTUAL DISPLACEMENTS

A constitutive equation provides a relation between the stress tensor σ_{ij} and the strain tensor ε_{ij}. Such an equation allows us to write the equations of equilibrium in terms of the displacement u_i. Our objective is to determine the unknown displacement field u_i that satisfies the equations of equilibrium subjected to certain boundary conditions. Following Washizu [124], we use the equations of equilibrium to derive the principle of virtual displacements. The body in equilibrium is given an infinitesimally small virtual displacement filed $\bar{u}_i$, that is an arbitrary field but we select it to be zero at points on the body surface where the actual displacement u_i is prescribed. It is a continuous field and differentiable as needed by the final form of the principle. We multiply the equations of equilibrium Equation (10.6) by $\bar{u}_i$, and integrate the resulting equation over the initial (undeformed) configuration.

$$\int_\Omega \bar{u}_i \frac{\partial}{\partial x_k}\left[\left(\delta_{ij}+\frac{\partial u_i}{\partial x_j}\right)\sigma_{kj}\right]d\Omega+\int_\Omega \bar{u}_i f_i d\Omega=0. \tag{10.7}$$

Substituting the identity

$$\frac{\partial}{\partial x_k}\left[\bar{u}_i\left(\delta_{ij}+\frac{\partial u_i}{\partial x_j}\right)\sigma_{kj}\right]=\bar{u}_i\frac{\partial}{\partial x_k}\left[\left(\delta_{ij}+\frac{\partial u_i}{\partial x_j}\right)\sigma_{kj}\right]+\frac{\partial \bar{u}_i}{\partial x_k}\left(\delta_{ij}+\frac{\partial u_i}{\partial x_j}\right)\sigma_{kj} \tag{10.8}$$

into Equation (10.7), we get

$$\int_\Omega \frac{\partial \bar{u}_i}{\partial x_k}\left(\delta_{ij}+\frac{\partial u_i}{\partial x_j}\right)\sigma_{kj}d\Omega=\int_\Omega \frac{\partial}{\partial x_k}\left[\bar{u}_i\left(\delta_{ij}+\frac{\partial u_i}{\partial x_j}\right)\sigma_{kj}\right]d\Omega+\int_\Omega \bar{u}_i f_i d\Omega. \tag{10.9}$$

Using Gauss theorem, we replace the domain integral on the right side by a boundary integral,

$$\int_\Omega \frac{\partial \bar{u}_i}{\partial x_k}\left(\delta_{ij}+\frac{\partial u_i}{\partial x_j}\right)\sigma_{kj}d\Omega=\oint_{\partial\Omega} \bar{u}_i\left(\delta_{ij}+\frac{\partial u_i}{\partial x_j}\right)\sigma_{kj}\hat{n}_k dS+\int_\Omega \bar{u}_i f_i d\Omega \tag{10.10}$$

where $\hat{n}_k$ is the outward unit normal on the surface element dS of the undeformed body. As shown by Washizu ([124]), the Cartesian components of the surface traction t_i (defined per unit surface area of the undeformed body) are related to the second Piola–Kirchhoff stress tensor by

$$t_i=\left(\delta_{ij}+\frac{\partial u_i}{\partial x_j}\right)\sigma_{kj}\hat{n}_k \tag{10.11}$$

and hence Equation (10.10) becomes

$$\int_\Omega \frac{\partial \bar{u}_i}{\partial x_k}\left(\delta_{ij}+\frac{\partial u_i}{\partial x_j}\right)\sigma_{kj}d\Omega=\oint_{\partial\Omega} \bar{u}_i t_i dS+\int_\Omega \bar{u}_i f_i d\Omega. \tag{10.12}$$

Let's denote the integrand on the left side by Λ,

$$\begin{aligned}\Lambda &= \frac{\partial \bar{u}_i}{\partial x_k}\left(\delta_{ij} + \frac{\partial u_i}{\partial x_j}\right)\sigma_{kj} \\ &= \left(\frac{\partial \bar{u}_i}{\partial x_k}\delta_{ij} + \frac{\partial \bar{u}_i}{\partial x_k}\frac{\partial u_i}{\partial x_j}\right)\sigma_{kj} \\ &= \left(\frac{\partial \bar{u}_j}{\partial x_k} + \frac{\partial \bar{u}_i}{\partial x_k}\frac{\partial u_i}{\partial x_j}\right)\sigma_{kj}. \end{aligned} \tag{10.13}$$

We note that Λ is a scalar and that the stress tensor is symmetric $\sigma_{kj} = \sigma_{jk}$, and hence we can write Λ also as

$$\Lambda = \left(\frac{\partial \bar{u}_k}{\partial x_j} + \frac{\partial \bar{u}_i}{\partial x_j}\frac{\partial u_i}{\partial x_k}\right)\sigma_{kj}. \tag{10.14}$$

Taking the average of the expressions given by Equations (10.13) and (10.14), we write Λ as

$$\Lambda = \left[\frac{1}{2}\left(\frac{\partial \bar{u}_j}{\partial x_k} + \frac{\partial \bar{u}_k}{\partial x_j}\right) + \frac{1}{2}\left(\frac{\partial \bar{u}_i}{\partial x_j}\frac{\partial u_i}{\partial x_k} + \frac{\partial \bar{u}_i}{\partial x_k}\frac{\partial u_i}{\partial x_j}\right)\right]\sigma_{kj} \tag{10.15}$$

We define

$$\bar{\varepsilon}_{jk} = \frac{1}{2}\left(\frac{\partial \bar{u}_j}{\partial x_k} + \frac{\partial \bar{u}_k}{\partial x_j}\right) + \frac{1}{2}\left(\frac{\partial \bar{u}_i}{\partial x_j}\frac{\partial u_i}{\partial x_k} + \frac{\partial \bar{u}_i}{\partial x_k}\frac{\partial u_i}{\partial x_j}\right). \tag{10.16}$$

Note that $\bar{\varepsilon}_{jk}$ depends on both the virtual displacement $\bar{u}_i$ and the actual displacement u_i that exits in the deformed body at equilibrium. We rewrite Equation (10.12),

$$\int_\Omega \bar{\varepsilon}_{jk}\sigma_{kj} d\Omega = \oint_{\partial\Omega} \bar{u}_i t_i dS + \int_\Omega \bar{u}_i f_i d\Omega. \tag{10.17}$$

Further development of the surface integral in Equation (10.17) depends on the boundary conditions imposed on the body in equilibrium. On parts of the surface denoted by Γ_E where the actual displacement $u_i = \hat{u}_i$ is prescribed the surface traction t_i is unknown, and we select the virtual displacement $\bar{u}_i$ to be zero on Γ_E. This is called an essential or geometric boundary condition. On parts of the surface denoted by Γ_N where the surface traction is prescribed $t_i = \hat{t}_i$, the surface integral can be evaluated for the selected virtual displacement. The principle of virtual displacement becomes,

$$\int_\Omega \bar{\varepsilon}_{jk}\sigma_{kj} d\Omega = \int_{\Gamma_N} \bar{u}_i \hat{t}_i dS + \int_\Omega \bar{u}_i f_i d\Omega. \tag{10.18}$$

Identifying the virtual displacement with a variation of the actual displacement, $\bar{u}_i \equiv \delta u_i$, we rewrite Equation (10.16) in terms of δu_i,

$$
\begin{aligned}
\bar{\varepsilon}_{jk} &= \frac{1}{2}\left(\frac{\partial \delta u_j}{\partial x_k}+\frac{\partial \delta u_k}{\partial x_j}\right)+\frac{1}{2}\left(\frac{\partial \delta u_i}{\partial x_j}\frac{\partial u_i}{\partial x_k}+\frac{\partial \delta u_i}{\partial x_k}\frac{\partial u_i}{\partial x_j}\right) \\
&= \frac{1}{2}\left[\delta\left(\frac{\partial u_j}{\partial x_k}\right)+\delta\left(\frac{\partial u_k}{\partial x_j}\right)\right]+\frac{1}{2}\left[\delta\left(\frac{\partial u_i}{\partial x_j}\right)\frac{\partial u_i}{\partial x_k}+\delta\left(\frac{\partial u_i}{\partial x_k}\right)\frac{\partial u_i}{\partial x_j}\right] \\
&= \delta\left[\frac{1}{2}\left(\frac{\partial u_j}{\partial x_k}+\frac{\partial u_k}{\partial x_j}\right)+\frac{1}{2}\frac{\partial u_i}{\partial x_j}\frac{\partial u_i}{\partial x_k}\right] \\
&= \delta\varepsilon_{jk}. \qquad (10.19)
\end{aligned}
$$

Thus, $\bar{\varepsilon}_{jk}$ is the variation of Green's strain tensor due to virtual displacement $\bar{u}_i$. In terms of the variations δu_i and $\delta\varepsilon_{jk}$, the principle of virtual displacements takes the form

$$\int_{\Omega} \delta\varepsilon_{jk}\sigma_{kj}d\Omega = \int_{\Gamma_N} \delta u_i \hat{t}_i dS + \int_{\Omega} \delta u_i f_i d\Omega. \qquad (10.20)$$

We note that the principle of virtual displacements is linear in $\bar{u}_i \equiv \delta u_i$. In this chapter, we limit the presentation to linear isotropic elastic materials; thus, the stresses are linear functions of strains. Geometrical non-linearity stems from the quadratic terms in the strain-displacement relation Equation (10.1). Since $\bar{u}_i$ is prescribed, Equation (10.18) or (10.20) will have linear, quadratic, and cubic terms in the unknown displacement u_i.

10.3 GEOMETRICALLY NONLINEAR REISSNER–MINDLIN PLATE

In this section, we use the principle of virtual displacements to formulate the large deflection plate problem taking into consideration the effects of transverse shear. We start with the static deflection problem and consider dynamic effects in later sections.

Displacements and Strains

In the formulation of Reissner–Mindlin plate theory (Reissner [100], Mindlin [76], Reddy [97], and Hughes [53]), the displacement components u, v, and w of an arbitrary point (x,y,z) is expressed in terms of three geometric quantities, namely the displacement of a point in the middle plane and rotations about the x and y axes of a filament initially normal to the middle plane at the point,

$$
\begin{aligned}
u &= u_o(x,y) - z\theta_x(x,y), \\
v &= v_o(x,y) - z\theta_y(x,y), \\
w &= w_o(x,y) \qquad (10.21)
\end{aligned}
$$

where (u_o, v_o, w_o) is the displacement of a point in the middle plane, $\theta_x(x,y)$ is rotation of the filament about the (negative) y-axis, and $\theta_y(x,y)$ is its rotation about the (positive) x-axis. Substituting the assumed displacement field Equation (10.21) into the Green's strain tensor, we get

$$\begin{aligned}
\varepsilon_{xx} &= \frac{\partial u_o}{\partial x} - z\frac{\partial \theta_x}{\partial x} + \frac{1}{2}\left(\frac{\partial w_o}{\partial x}\right)^2 + \frac{1}{2}z^2\left[\left(\frac{\partial \theta_x}{\partial x}\right)^2 + \left(\frac{\partial \theta_y}{\partial x}\right)^2\right], \\
\varepsilon_{yy} &= \frac{\partial v_o}{\partial y} - z\frac{\partial \theta_y}{\partial y} + \frac{1}{2}\left(\frac{\partial w_o}{\partial y}\right)^2 + \frac{1}{2}z^2\left[\left(\frac{\partial \theta_x}{\partial y}\right)^2 + \left(\frac{\partial \theta_y}{\partial y}\right)^2\right], \\
\gamma_{xy} &= \frac{\partial u_o}{\partial y} + \frac{\partial v_o}{\partial x} - z\left(\frac{\partial \theta_x}{\partial y} + \frac{\partial \theta_y}{\partial x}\right) + \frac{\partial w_o}{\partial x}\frac{\partial w_o}{\partial y} \\
&\quad + z^2\left[\frac{\partial \theta_x}{\partial x}\frac{\partial \theta_x}{\partial y} + \frac{\partial \theta_y}{\partial x}\frac{\partial \theta_y}{\partial y}\right], \\
\gamma_{xz} &= \frac{\partial w_o}{\partial x} - \theta_x + z\left[\theta_x\frac{\partial \theta_x}{\partial x} + \theta_y\frac{\partial \theta_y}{\partial x}\right], \\
\gamma_{yz} &= \frac{\partial w_o}{\partial y} - \theta_y + z\left[\theta_x\frac{\partial \theta_x}{\partial y} + \theta_y\frac{\partial \theta_y}{\partial y}\right].
\end{aligned} \tag{10.22}$$

We neglected the nonlinear parts of strains involving the membrane stretching u_o and v_o. In the theory of plates, we also neglect ε_{zz} and ignore σ_{zz}. The linear and nonlinear strain terms are separated into arrays that show the dependence on z explicitly,

$$\begin{Bmatrix} \varepsilon_{xx} \\ \varepsilon_{yy} \\ \varepsilon_{xy} \end{Bmatrix} = \{\ \varepsilon_o\ \} - z\{\ \varepsilon_b\ \} + \{\ \varepsilon_o^{NL}\ \} + z^2\{\ \varepsilon_b^{NL}\ \} \tag{10.23}$$

$$\begin{Bmatrix} \gamma_{xz} \\ \gamma_{yz} \end{Bmatrix} = \{\ \gamma_b\ \} + z\{\ \gamma_b^{NL}\ \} \tag{10.24}$$

where

$$\begin{aligned}
\left\{\varepsilon_o\right\} &= \begin{Bmatrix} \frac{\partial u_o}{\partial x} \\ \frac{\partial v_o}{\partial y} \\ \frac{\partial u_o}{\partial y} + \frac{\partial v_o}{\partial x} \end{Bmatrix}, \quad \left\{\varepsilon_b\right\} = \begin{Bmatrix} \frac{\partial \theta_x}{\partial x} \\ \frac{\partial \theta_y}{\partial y} \\ \frac{\partial \theta_x}{\partial y} + \frac{\partial \theta_y}{\partial x} \end{Bmatrix} \\
\left\{\varepsilon_o^{NL}\right\} &= \frac{1}{2}\begin{Bmatrix} \left(\frac{\partial w_o}{\partial x}\right)^2 \\ \left(\frac{\partial w_o}{\partial y}\right)^2 \\ 2\frac{\partial w_o}{\partial x}\frac{\partial w_o}{\partial y} \end{Bmatrix}, \quad \left\{\varepsilon_b^{NL}\right\} = \frac{1}{2}\begin{Bmatrix} \left(\frac{\partial \theta_x}{\partial x}\right)^2 + \left(\frac{\partial \theta_y}{\partial x}\right)^2 \\ \left(\frac{\partial \theta_x}{\partial y}\right)^2 + \left(\frac{\partial \theta_y}{\partial y}\right)^2 \\ 2\frac{\partial \theta_x}{\partial x}\frac{\partial \theta_x}{\partial y} + 2\frac{\partial \theta_y}{\partial x}\frac{\partial \theta_y}{\partial y} \end{Bmatrix}
\end{aligned} \tag{10.25}$$

$$\{\ \gamma_b\ \} = \begin{Bmatrix} \frac{\partial w_o}{\partial x} - \theta_x \\ \frac{\partial w_o}{\partial y} - \theta_y \end{Bmatrix}, \quad \{\ \gamma_b^{NL}\ \} = \begin{Bmatrix} \theta_x\frac{\partial \theta_x}{\partial x} + \theta_y\frac{\partial \theta_y}{\partial x} \\ \theta_x\frac{\partial \theta_x}{\partial y} + \theta_y\frac{\partial \theta_y}{\partial y} \end{Bmatrix}. \tag{10.26}$$

The superscript "NL" indicates nonlinear strains. The von Karman approximations are frequently adopted in the analysis of large defection of plates, in which case all nonlinear strains are neglected except $\{\varepsilon_o^{NL}\}$ is retained.

Constitutive Equation: Isotropic Linear Elastic Material

We consider isotropic linear elastic material with constant Poisson's ratio ν and Young's modulus E. A state of plane stress is assumed, hence

$$\left\{\begin{array}{c}\sigma_{xx}\\ \sigma_{yy}\\ \sigma_{xy}\end{array}\right\} = [D]\left\{\begin{array}{c}\varepsilon_{xx}\\ \varepsilon_{yy}\\ \gamma_{xy}\end{array}\right\} \tag{10.27}$$

where

$$[D] = \frac{E}{1-\nu^2}\begin{bmatrix} 1 & \nu & 0 \\ \nu & 1 & 0 \\ 0 & 0 & \frac{1-\nu}{2} \end{bmatrix} \tag{10.28}$$

and

$$\left\{\begin{array}{c}\sigma_{xz}\\ \sigma_{yz}\end{array}\right\} = [D^s]\left\{\begin{array}{c}\gamma_{xz}\\ \gamma_{yz}\end{array}\right\} \tag{10.29}$$

where

$$[D^s] = \frac{E}{2(1+\nu)}\begin{bmatrix} 1 & 0 \\ 0 & 1 \end{bmatrix} \tag{10.30}$$

$$\left\{\begin{array}{c}\sigma_{xx}\\ \sigma_{yy}\\ \sigma_{xy}\end{array}\right\} = [D]\{\ \varepsilon_o\ \} - z[D]\{\ \varepsilon_b\ \} + [D]\{\ \varepsilon_o^{NL}\ \} + z^2[D]\{\ \varepsilon_b^{NL}\ \} \tag{10.31}$$

$$\left\{\begin{array}{c}\sigma_{xz}\\ \sigma_{yz}\end{array}\right\} = [D^s]\{\ \gamma_b\ \} + z[D^s]\{\ \gamma_b^{NL}\ \}. \tag{10.32}$$

Virtual Displacements and Strains

Virtual displacements $(\bar{u}, \bar{v}, \bar{w})$ are infinitesimally small and arbitrary but must be consistent with the kinematic constraints imposed on the actual displacements. For the Reissner–Mindlin plate, we take,

$$\begin{aligned}\bar{u} &= \bar{u}_o(x,y) - z\bar{\theta}_x(x,y),\\ \bar{v} &= \bar{v}_o(x,y) - z\bar{\theta}_y(x,y),\\ \bar{w} &= \bar{w}_o(x,y).\end{aligned} \tag{10.33}$$

We obtain the virtual strains by substituting Equations (10.21) and (10.33) into Equation (10.16),

$$\left\{\begin{array}{c}\bar{\varepsilon}_{xx}\\ \bar{\varepsilon}_{yy}\\ \bar{\varepsilon}_{xy}\end{array}\right\} = \{\ \bar{\varepsilon}_o\ \}-z\{\ \bar{\varepsilon}_b\ \}+\{\ \bar{\varepsilon}_o^{NL}\ \}+z^2\{\ \bar{\varepsilon}_b^{NL}\ \} \tag{10.34}$$

$$\left\{\begin{array}{c}\bar{\gamma}_{xz}\\ \bar{\gamma}_{yz}\end{array}\right\} = \{\ \bar{\gamma}_b\ \}+z\{\ \bar{\gamma}_b^{NL}\ \} \tag{10.35}$$

where

$$\begin{aligned}
\{\ \bar{\varepsilon}_o\ \} &= \left\{\begin{array}{c}\dfrac{\partial \bar{u}_o}{\partial x}\\ \dfrac{\partial \bar{v}_o}{\partial y}\\ \dfrac{\partial \bar{u}_o}{\partial y}+\dfrac{\partial \bar{v}_o}{\partial x}\end{array}\right\},\quad \{\ \bar{\varepsilon}_b\ \}=\left\{\begin{array}{c}\dfrac{\partial \bar{\theta}_x}{\partial x}\\ \dfrac{\partial \bar{\theta}_y}{\partial y}\\ \dfrac{\partial \bar{\theta}_x}{\partial y}+\dfrac{\partial \bar{\theta}_y}{\partial x}\end{array}\right\},\\
\{\ \bar{\varepsilon}_o^{NL}\ \} &= \left\{\begin{array}{c}\dfrac{\partial \bar{w}_o}{\partial x}\dfrac{\partial w_o}{\partial x}\\ \dfrac{\partial \bar{w}_o}{\partial y}\dfrac{\partial w_o}{\partial y}\\ \dfrac{\partial \bar{w}_o}{\partial x}\dfrac{\partial w_o}{\partial y}+\dfrac{\partial \bar{w}_o}{\partial y}\dfrac{\partial w_o}{\partial x}\end{array}\right\},\\
\left\{\ \bar{\varepsilon}_b^{NL}\ \right\} &= \left\{\begin{array}{c}\dfrac{\partial \bar{\theta}_x}{\partial x}\dfrac{\partial \theta_x}{\partial x}+\dfrac{\partial \bar{\theta}_y}{\partial x}\dfrac{\partial \theta_y}{\partial x}\\ \dfrac{\partial \bar{\theta}_x}{\partial y}\dfrac{\partial \theta_x}{\partial y}+\dfrac{\partial \bar{\theta}_y}{\partial y}\dfrac{\partial \theta_y}{\partial y}\\ \dfrac{\partial \bar{\theta}_x}{\partial x}\dfrac{\partial \theta_x}{\partial y}+\dfrac{\partial \bar{\theta}_x}{\partial y}\dfrac{\partial \theta_x}{\partial x}+\dfrac{\partial \bar{\theta}_y}{\partial x}\dfrac{\partial \theta_y}{\partial y}+\dfrac{\partial \bar{\theta}_y}{\partial y}\dfrac{\partial \theta_y}{\partial x}\end{array}\right\}
\end{aligned} \tag{10.36}$$

$$\begin{aligned}
\left\{\ \bar{\gamma}_b\ \right\} &= \left\{\begin{array}{c}\dfrac{\partial \bar{w}_o}{\partial x}-\bar{\theta}_x\\ \dfrac{\partial \bar{w}_o}{\partial y}-\bar{\theta}_y\end{array}\right\},\\
\{\ \bar{\gamma}_b^{NL}\ \} &= \left\{\begin{array}{c}\bar{\theta}_x\dfrac{\partial \theta_x}{\partial x}+\theta_x\dfrac{\partial \bar{\theta}_x}{\partial x}+\bar{\theta}_y\dfrac{\partial \theta_y}{\partial x}+\theta_y\dfrac{\partial \bar{\theta}_y}{\partial x}\\ \bar{\theta}_x\dfrac{\partial \theta_x}{\partial y}+\theta_x\dfrac{\partial \bar{\theta}_x}{\partial y}+\bar{\theta}_y\dfrac{\partial \theta_y}{\partial y}+\theta_y\dfrac{\partial \bar{\theta}_y}{\partial y}\end{array}\right\}.
\end{aligned} \tag{10.37}$$

Internal Virtual Work

The internal virtual work per unit volume for the plate is

$$\begin{aligned}
\bar{\varepsilon}_{ij}\sigma_{ji} &= \bar{\varepsilon}_{xx}\sigma_{xx}+\bar{\varepsilon}_{yy}\sigma_{yy}+\bar{\gamma}_{xy}\sigma_{xy}+\bar{\gamma}_{xz}\sigma_{xz}+\bar{\gamma}_{yz}\sigma_{yz}\\
&= \left\{\begin{array}{c}\bar{\varepsilon}_{xx}\\ \bar{\varepsilon}_{yy}\\ \bar{\gamma}_{xy}\end{array}\right\}^T\left\{\begin{array}{c}\sigma_{xx}\\ \sigma_{yy}\\ \sigma_{xy}\end{array}\right\}+\left\{\begin{array}{c}\bar{\gamma}_{xz}\\ \bar{\gamma}_{yz}\end{array}\right\}^T\left\{\begin{array}{c}\sigma_{xz}\\ \sigma_{yz}\end{array}\right\}.
\end{aligned} \tag{10.38}$$

Substituting Equations (10.31), (10.32), (10.34), and (10.35), we obtain

$$
\begin{aligned}
\bar{\varepsilon}_{ij}\sigma_{ji} = & \left[\left\{\bar{\varepsilon}_o\right\}^T - z\left\{\bar{\varepsilon}_b\right\}^T + \left\{\bar{\varepsilon}_o^{NL}\right\}^T + z^2\left\{\bar{\varepsilon}_b^{NL}\right\}^T\right] \times \\
& \left[[D]\left\{\varepsilon_o\right\} - z[D]\left\{\varepsilon_b\right\} + [D]\left\{\varepsilon_o^{NL}\right\} + z^2[D]\left\{\varepsilon_b^{NL}\right\}\right] \\
& + \left[\left\{\bar{\gamma}_b\right\}^T + z\left\{\bar{\gamma}_b^{NL}\right\}^T\right]\left[[D^s]\left\{\gamma_b\right\} + z[D^s]\left\{\gamma_b^{NL}\right\}\right].
\end{aligned}
\tag{10.39}
$$

The internal virtual work per unit area of the midplane is obtained by integrating Equation (10.39) with respect to z over the plate thickness h,

$$
\begin{aligned}
& \int_{-h/2}^{h/2} \begin{Bmatrix} \bar{\varepsilon}_{xx} \\ \bar{\varepsilon}_{yy} \\ \bar{\gamma}_{xy} \end{Bmatrix}^T \begin{Bmatrix} \sigma_{xx} \\ \sigma_{yy} \\ \sigma_{xy} \end{Bmatrix} dz + \int_{-h/2}^{h/2} \begin{Bmatrix} \bar{\gamma}_{xz} \\ \bar{\gamma}_{yz} \end{Bmatrix}^T \begin{Bmatrix} \sigma_{xz} \\ \sigma_{yz} \end{Bmatrix} dz = \\
& \{\ \bar{\varepsilon}_o\ \}^T\left[[D_o]\{\ \varepsilon_o\ \} + [D_o]\{\ \varepsilon_o^{NL}\ \} + [D_b]\{\ \varepsilon_b^{NL}\ \}\right] \\
& \qquad + \{\ \bar{\varepsilon}_b\ \}^T[D_b]\{\ \varepsilon_b\ \} \\
& + \{\ \bar{\varepsilon}_o^{NL}\ \}^T\left[[D_o]\{\ \varepsilon_o\ \} + [D_o]\{\ \varepsilon_o^{NL}\ \} + [D_b]\{\ \varepsilon_b^{NL}\ \}\right] \\
& + \{\ \bar{\varepsilon}_b^{NL}\ \}^T\left[[D_b]\{\ \varepsilon_o\ \} + [D_b]\{\ \varepsilon_o^{NL}\ \} + [D_{bN}]\{\ \varepsilon_b^{NL}\ \}\right] \\
& \qquad + \{\ \bar{\gamma}_b\ \}^T[D_b^s]\{\ \gamma_b\ \} + \{\ \bar{\gamma}_b^{NL}\ \}^T[D_N^s]\{\ \gamma_b^{NL}\ \}
\end{aligned}
\tag{10.40}
$$

where

$$
\begin{aligned}
& [D_o] = h[D],\ [D_b] = \frac{h^3}{12}[D],\ [D_{bN}] = \frac{h^5}{80}[D], \\
& [D_b^s] = \kappa h[D^s],\ [D_N^s] = \kappa\frac{h^3}{12}[D^s]
\end{aligned}
\tag{10.41}
$$

where $\kappa = 5/6$ is a shear correction factor.

Principle of Virtual Displacements for the Plate

Following the same steps for treating the virtual work of externally applied surface traction and body forces leading to Equation (9.129) for the linear case, we rewrite the principle of virtual displacements Equation (10.17) for the geometrically nonlinear Reissner–Mindlin plate as

$$\int_A \{\ \bar{\varepsilon}_o\ \}^T [[D_o]\{\ \varepsilon_o\ \} + [D_o]\{\ \varepsilon_o^{NL}\ \} + [D_b]\{\ \varepsilon_b^{NL}\ \}]\, dA$$
$$+ \int_A \{\ \bar{\varepsilon}_b\ \}^T [D_b]\{\ \varepsilon_b\ \}\, dA$$
$$+ \int_A \{\ \bar{\varepsilon}_o^{NL}\ \}^T [[D_o]\{\ \varepsilon_o\ \} + [D_o]\{\ \varepsilon_o^{NL}\ \} + [D_b]\{\ \varepsilon_b^{NL}\ \}]\, dA$$
$$+ \int_A \{\ \bar{\varepsilon}_b^{NL}\ \}^T [[D_b]\{\ \varepsilon_o\ \} + [D_b]\{\ \varepsilon_o^{NL}\ \} + [D_{bN}]\{\ \varepsilon_b^{NL}\ \}]\, dA$$
$$+ \int_A \{\ \bar{\gamma}_b\ \}^T [D_b^s]\{\ \gamma_b\ \}\, dA + \int_A \{\ \bar{\gamma}_b^{NL}\ \}^T [D_N^s]\{\ \gamma_b^{NL}\ \}\, dA$$
$$= \int_A \bar{w}\left(q + \tilde{f}_z\right) dA + \oint_\Gamma \left(-\bar{\theta}_n M_{nn} - \bar{\theta}_s M_{ns} + \bar{w} Q_{nz}\right) ds. \quad (10.42)$$

The nonlinear terms $\{\ \varepsilon_b^{NL}\ \}$, and $\{\ \gamma_b^{NL}\ \}$ are usually neglected, and the principle simplifies to

$$\int_A \{\ \bar{\varepsilon}_o\ \}^T [D_o] [\{\ \varepsilon_o\ \} + \{\ \varepsilon_o^{NL}\ \}]\, dA + \int_A \{\ \bar{\varepsilon}_b\ \}^T [D_b]\{\ \varepsilon_b\ \}\, dA$$
$$+ \int_A \{\ \bar{\varepsilon}_o^{NL}\ \}^T [D_o] [\{\ \varepsilon_o\ \} + \{\ \varepsilon_o^{NL}\ \}]\, dA + \int_A \{\ \bar{\gamma}_b\ \}^T [D_b^s]\{\ \gamma_b\ \}\, dA$$
$$= \int_A \bar{w}\left(q + \tilde{f}_z\right) dA + \oint_\Gamma \left(-\bar{\theta}_n M_{nn} - \bar{\theta}_s M_{ns} + \bar{w} Q_{nz}\right) ds. \quad (10.43)$$

The finite-element results reported in this chapter are based on Equation (10.43). Some results are based on a reduced form of Equation (10.42) and will be explicitly identified.

10.4 FINITE ELEMENT MODEL

Interpolation of Displacements and Strains Over an Element

To proceed with finite-element analysis of the theory, we divide the plate into small quadrilateral elements; each is defined by n local nodes. Each node has five degrees of freedom; three displacements and two rotations. Isoparametric elements are used to map the element into a canonical element and express displacement and rotations at an arbitrary point within the element in terms of their values at the element's nodes. Over the typical element # e, the interpolated solution in matrix form is written as

$$\begin{Bmatrix} u_o^e(x,y) \\ v_o^e(x,y) \\ w_o^e(x,y) \\ \theta_x^e(x,y) \\ \theta_y^e(x,y) \end{Bmatrix} = [\psi]\{U^e\} \quad (10.44)$$

where

$$[\psi]=\Big[[\psi]_1\ \ldots\ [\psi]_j\ \ldots\ [\psi]_n\Big],\ [\psi]_j=\begin{bmatrix}\psi_j & 0 & 0 & 0 & 0\\ 0 & \psi_j & 0 & 0 & 0\\ 0 & 0 & \psi_j & 0 & 0\\ 0 & 0 & 0 & \psi_j & 0\\ 0 & 0 & 0 & 0 & \psi_j\end{bmatrix} \tag{10.45}$$

and

$$\{U^e\}=\{u^e_{o1},v^e_{o1},w^e_{o1},\theta^e_{x1},\theta^e_{y1},...,u^e_{oj},v^e_{oj},w^e_{oj},\theta^e_{xj},\theta^e_{yj},...,u^e_{on},v^e_{on},w^e_{on},\\ \theta^e_{xn},\theta^e_{yn}\}^T \tag{10.46}$$

where ψ_j $(j=1,2,...,n)$ are the two-dimensional Lagrange interpolation functions presented in Chapter 4. Note that $[\psi]$ is of dimensions $(5,5n)$, while $\{U^e\}$ are the values of generalized displacements at the element's nodes and of dimensions $(5n,1)$. Next, we obtain the strains at any point in the element in terms of the generalized displacements at the element's nodes.

$$\{\ \varepsilon_o\ \}=\begin{Bmatrix}\frac{\partial u_o}{\partial x}\\ \frac{\partial v_o}{\partial y}\\ \frac{\partial u_o}{\partial y}+\frac{\partial v_o}{\partial x}\end{Bmatrix}=[B_o]\{U^e\} \tag{10.47}$$

$$[B_o]=\Big[[B_o]_1\ \ldots\ [B_o]_j\ \ldots\ [B_o]_n\Big],\ [B_o]_j=\begin{bmatrix}\frac{\partial\psi_j}{\partial x} & 0 & 0\ 0\ 0\\ 0 & \frac{\partial\psi_j}{\partial y} & 0\ 0\ 0\\ \frac{\partial\psi_j}{\partial y} & \frac{\partial\psi_j}{\partial x} & 0\ 0\ 0\end{bmatrix} \tag{10.48}$$

$$\{\ \varepsilon_b\ \}=\begin{Bmatrix}\frac{\partial\theta_x}{\partial x}\\ \frac{\partial\theta_y}{\partial y}\\ \frac{\partial\theta_x}{\partial y}+\frac{\partial\theta_y}{\partial x}\end{Bmatrix}=[B_b]\{U^e\} \tag{10.49}$$

$$[B_b]=\Big[[B_b]_1\ \ldots\ [B_b]_j\ \ldots\ [B_b]_n\Big],\ [B_b]_j=\begin{bmatrix}0\ 0\ 0 & \frac{\partial\psi_j}{\partial x} & 0\\ 0\ 0\ 0 & 0 & \frac{\partial\psi_j}{\partial y}\\ 0\ 0\ 0 & \frac{\partial\psi_j}{\partial y} & \frac{\partial\psi_j}{\partial x}\end{bmatrix}. \tag{10.50}$$

Note that $[B_o]$ and $[B_b]$ are of dimensions $(3,5n)$. The deflection w_o at any point within the element can be written in terms of the array $\{U^e\}$

$$w^e_o(x,y)=\{\psi\}_3\{U^e\} \tag{10.51}$$

where

$$\{\psi\}_3=\Big[\{\psi\}_1\ \ldots\ \{\psi\}_j\ \ldots\ \{\psi\}_n\Big],\ \{\psi\}_j=\{0\ 0\ \psi_j\ 0\ 0\} \tag{10.52}$$

hence the derivatives become

$$\frac{\partial w_o^e}{\partial x} = \{B_x^{NL}\}\{U^e\}, \ \frac{\partial w_o^e}{\partial y} = \{B_y^{NL}\}\{U^e\} \tag{10.53}$$

where

$$\begin{aligned} \{B_x^{NL}\} &= \left[\{B_x^{NL}\}_1 \{B_x^{NL}\}_2 \ldots \{B_x^{NL}\}_j \ldots \{B_x^{NL}\}_n\right] \\ \{B_y^{NL}\} &= \left[\{B_y^{NL}\}_1 \{B_y^{NL}\}_2 \ldots \{B_y^{NL}\}_j \ldots \{B_y^{NL}\}_n\right] \end{aligned} \tag{10.54}$$

$$\{B_x^{NL}\}_j = \left\{0 \ 0 \ \tfrac{\partial \psi_j}{\partial x} \ 0 \ 0\right\}, \ \{B_y^{NL}\}_j = \left\{0 \ 0 \ \tfrac{\partial \psi_j}{\partial y} \ 0 \ 0\right\}. \tag{10.55}$$

We define

$$\left[B^{NL}\right] = \left[\begin{array}{c} \{B_x^{NL}\} \\ \{B_y^{NL}\} \end{array}\right] \tag{10.56}$$

and write the derivatives in matrix form

$$\left\{\begin{array}{c} \frac{\partial w_o^e}{\partial x} \\ \frac{\partial w_o^e}{\partial y} \end{array}\right\} = \left[B^{NL}\right]\{U^e\}. \tag{10.57}$$

The inplane nonlinear strains are

$$\begin{aligned} \{\ \varepsilon_o^{NL}\ \} &= \frac{1}{2}\left\{\begin{array}{c} \left(\frac{\partial w_o}{\partial x}\right)^2 \\ \left(\frac{\partial w_o}{\partial y}\right)^2 \\ 2\frac{\partial w_o}{\partial x}\frac{\partial w_o}{\partial y} \end{array}\right\} = \frac{1}{2}\left[\begin{array}{cc} \frac{\partial w_o}{\partial x} & 0 \\ 0 & \frac{\partial w_o}{\partial y} \\ \frac{\partial w_o}{\partial y} & \frac{\partial w_o}{\partial x} \end{array}\right]\left\{\begin{array}{c} \frac{\partial w_o}{\partial x} \\ \frac{\partial w_o}{\partial y} \end{array}\right\} \\ &= \frac{1}{2}\left[\ A_o\ \right]\left\{\begin{array}{c} \frac{\partial w_o}{\partial x} \\ \frac{\partial w_o}{\partial y} \end{array}\right\} = \frac{1}{2}\left[\ A_o\ \right]\left[B^{NL}\right]\{U^e\} \\ &= \frac{1}{2}\left[\ G^{NL}\ \right]\{U^e\} \end{aligned} \tag{10.58}$$

where

$$\begin{aligned} [A_o] &= \left[\begin{array}{cc} \frac{\partial w_o}{\partial x} & 0 \\ 0 & \frac{\partial w_o}{\partial y} \\ \frac{\partial w_o}{\partial y} & \frac{\partial w_o}{\partial x} \end{array}\right] \\ &= \left[\begin{array}{cc} \{B_x^{NL}\}\{U^e\} & 0 \\ 0 & \{B_y^{NL}\}\{U^e\} \\ \{B_y^{NL}\}\{U^e\} & \{B_x^{NL}\}\{U^e\} \end{array}\right] \end{aligned} \tag{10.59}$$

$$\left[G^{NL}\right] = [A_o]\left[B^{NL}\right]. \tag{10.60}$$

Note that $\left[B^{NL}\right]$ and $\left[G^{NL}\right]$ are of dimensions $(2,5n)$, and $(3,5n)$, respectively.

$$\begin{aligned}\left\{\ \varepsilon_b^{NL}\ \right\} &= \frac{1}{2}\left\{\begin{array}{c}\left(\frac{\partial\theta_x}{\partial x}\right)^2 \\ \left(\frac{\partial\theta_x}{\partial y}\right)^2 \\ 2\frac{\partial\theta_x}{\partial x}\frac{\partial\theta_x}{\partial y}\end{array}\right\} + \frac{1}{2}\left\{\begin{array}{c}\left(\frac{\partial\theta_y}{\partial x}\right)^2 \\ \left(\frac{\partial\theta_y}{\partial y}\right)^2 \\ 2\frac{\partial\theta_y}{\partial x}\frac{\partial\theta_y}{\partial y}\end{array}\right\} \\ &= \frac{1}{2}\begin{bmatrix}\frac{\partial\theta_x}{\partial x} & 0 \\ 0 & \frac{\partial\theta_x}{\partial y} \\ \frac{\partial\theta_x}{\partial y} & \frac{\partial\theta_x}{\partial x}\end{bmatrix}\left\{\begin{array}{c}\frac{\partial\theta_x}{\partial x} \\ \frac{\partial\theta_x}{\partial y}\end{array}\right\} + \frac{1}{2}\begin{bmatrix}\frac{\partial\theta_y}{\partial x} & 0 \\ 0 & \frac{\partial\theta_y}{\partial y} \\ \frac{\partial\theta_y}{\partial y} & \frac{\partial\theta_y}{\partial x}\end{bmatrix}\left\{\begin{array}{c}\frac{\partial\theta_y}{\partial x} \\ \frac{\partial\theta_y}{\partial y}\end{array}\right\}.\end{aligned} \tag{10.61}$$

Defining the matrices

$$\begin{gathered}\left[\ A_1\ \right] = \begin{bmatrix}\frac{\partial\theta_x}{\partial x} & 0 \\ 0 & \frac{\partial\theta_x}{\partial y} \\ \frac{\partial\theta_x}{\partial y} & \frac{\partial\theta_x}{\partial x}\end{bmatrix},\ \left[\ A_2\ \right] = \begin{bmatrix}\frac{\partial\theta_y}{\partial x} & 0 \\ 0 & \frac{\partial\theta_y}{\partial y} \\ \frac{\partial\theta_y}{\partial y} & \frac{\partial\theta_y}{\partial x}\end{bmatrix} \\ \left\{\begin{array}{c}\frac{\partial\theta_x}{\partial x} \\ \frac{\partial\theta_x}{\partial y}\end{array}\right\} = \left[B_1^{NL}\right]\{U^e\},\ \left\{\begin{array}{c}\frac{\partial\theta_y}{\partial x} \\ \frac{\partial\theta_y}{\partial y}\end{array}\right\} = \left[B_2^{NL}\right]\{U^e\}, \\ \left[\ G_1^{NL}\ \right] = [A_1]\left[B_1^{NL}\right],\ \left[\ G_2^{NL}\ \right] = [A_2]\left[B_2^{NL}\right] \\ \left[\ G_3^{NL}\ \right] = \left[\ G_1^{NL}\ \right] + \left[\ G_2^{NL}\ \right]\end{gathered} \tag{10.62}$$

where

$$\left[B_1^{NL}\right] = \begin{bmatrix}\left\{B_{1x}^{NL}\right\} \\ \left\{B_{1y}^{NL}\right\}\end{bmatrix},\ \left[B_2^{NL}\right] = \begin{bmatrix}\left\{B_{2x}^{NL}\right\} \\ \left\{B_{2y}^{NL}\right\}\end{bmatrix} \tag{10.63}$$

$$\begin{gathered}\{B_{1x}^{NL}\}_j = \left\{0\ 0\ 0\ \frac{\partial\psi_j}{\partial x}\ 0\right\},\ \{B_{1y}^{NL}\}_j = \left\{0\ 0\ 0\ \frac{\partial\psi_j}{\partial y}\ 0\right\} \\ \{B_{2x}^{NL}\}_j = \left\{0\ 0\ 0\ 0\ \frac{\partial\psi_j}{\partial x}\right\},\ \{B_{2y}^{NL}\}_j = \left\{0\ 0\ 0\ 0\ \frac{\partial\psi_j}{\partial y}\right\}\end{gathered} \tag{10.64}$$

we obtain

$$\left\{\ \varepsilon_b^{NL}\ \right\} = \frac{1}{2}\left[\ G_3^{NL}\ \right]\{U^e\} \tag{10.65}$$

We define

$$\begin{aligned}\left\{\varepsilon_s\right\} &= \begin{Bmatrix} \frac{\partial w_o}{\partial x} - \theta_x \\ \frac{\partial w_o}{\partial y} - \theta_y \end{Bmatrix} \\ &= [B_s]\{U^e\} \end{aligned} \tag{10.66}$$

$$[B_s] = \begin{bmatrix} [B_s]_1 & \ldots & [B_s]_j & \ldots & [B_s]_n \end{bmatrix},\ [B_s]_j = \begin{bmatrix} 0 & 0 & \frac{\partial \psi_j}{\partial x} & -\psi_j & 0 \\ 0 & 0 & \frac{\partial \psi_j}{\partial y} & 0 & -\psi_j \end{bmatrix}. \tag{10.67}$$

Interpolation of Virtual Displacements and Strains Over an Element

Next, consider the virtual displacements and strain for an element. The virtual displacements at the element's node are given by an array $\{d\}$. The element shape functions [see Equations (10.44) and (10.45)] are used to find the virtual displacements at any point (x,y) in the element. This is equivalent to using the Galerkin method for selecting the weight functions in the weak formulation.

$$\begin{Bmatrix} \bar{u}_o^e(x,y) \\ \bar{v}_o^e(x,y) \\ \bar{w}_o^e(x,y) \\ \bar{\theta}_x^e(x,y) \\ \bar{\theta}_y^e(x,y) \end{Bmatrix} = [\psi]\{d\}. \tag{10.68}$$

The virtual strains can be easily expressed in therm of the array $\{d\}$ and the matrices that have already been defined for the actual strains.

$$\{\ \bar{\varepsilon}_o\ \} = [B_o]\{d\} \tag{10.69}$$

$$\{\ \bar{\varepsilon}_b\ \} = [B_b]\{d\} \tag{10.70}$$

$$\begin{aligned}\left\{\bar{\varepsilon}_o^{NL}\right\} &= \begin{Bmatrix} \frac{\partial w_o}{\partial x}\frac{\partial \bar{w}_o}{\partial x} \\ \frac{\partial w_o}{\partial y}\frac{\partial \bar{w}_o}{\partial y} \\ \frac{\partial w_o}{\partial y}\frac{\partial \bar{w}_o}{\partial x} + \frac{\partial w_o}{\partial x}\frac{\partial \bar{w}_o}{\partial y} \end{Bmatrix} = \begin{bmatrix} \frac{\partial w_o}{\partial x} & 0 \\ 0 & \frac{\partial w_o}{\partial y} \\ \frac{\partial w_o}{\partial y} & \frac{\partial w_o}{\partial x} \end{bmatrix} \begin{Bmatrix} \frac{\partial \bar{w}_o}{\partial x} \\ \frac{\partial \bar{w}_o}{\partial y} \end{Bmatrix} \\ &= \begin{bmatrix} A_o \end{bmatrix} \begin{Bmatrix} \frac{\partial \bar{w}_o}{\partial x} \\ \frac{\partial \bar{w}_o}{\partial y} \end{Bmatrix} = \begin{bmatrix} A_o \end{bmatrix} [B^{NL}]\{d\} \\ &= [G^{NL}]\{d\} \end{aligned} \tag{10.71}$$

$$\{\ \bar{\varepsilon}_b^{NL}\ \} = \begin{bmatrix} G_3^{NL} \end{bmatrix}\{d\} \tag{10.72}$$

$$\left\{ \bar{\gamma}_b \right\} = [B_s]\{d\}. \tag{10.73}$$

Element Equations and Matrices

The element equations are obtained from the principle of virtual displacements Equation (10.42) following the usual procedure given in Chapter 9.

$$[K^e]\{U^e\} = \{F^e\} + \{R^e\}. \tag{10.74}$$

The stiffness matrix is

$$\begin{aligned}[K^e] = &\int_{A_e} \{ [B_o] + [G^{NL}] \}^T [D_o] \{ [B_o] + \tfrac{1}{2}[G^{NL}] \} dA \\ &+ \int_{A_e} [B_b]^T [D_b] [B_b] dA + \int_{A_e} [B_s]^T [D_b^s] [B_s] dA \\ &+ \int_{A_e} \{ [B_o] + [G^{NL}] \}^T [D_b] \{ \tfrac{1}{2}[G_3^{NL}] \} dA \\ &+ \int_{A_e} \{ [G_3^{NL}] \}^T [D_b] \{ [B_o] + \tfrac{1}{2}[G^{NL}] \} dA \\ &+ \int_{A_e} \{ [G_3^{NL}] \}^T [D_{bN}] \{ \tfrac{1}{2}[G_3^{NL}] \} dA. \end{aligned} \tag{10.75}$$

We neglected nonlinear transverse shear strain $\{ \gamma_b^{NL} \}$. For the simplified principle of virtual displacements Equation (10.43), the stiffness matrix is

$$\begin{aligned}[K^e] = &\int_{A_e} \{ [B_o] + [G^{NL}] \}^T [D_o] \{ [B_o] + \tfrac{1}{2}[G^{NL}] \} dA \\ &+ \int_{A_e} [B_b]^T [D_b] [B_b] dA + \int_{A_e} [B_s]^T [D_b^s] [B_s] dA. \end{aligned} \tag{10.76}$$

For the purpose of evaluating the virtual work of surface traction, we define the arrays

$$\{\psi\}_1 = \{ \psi_1 \ 0\ 0\ 0\ 0 \ \ldots \ \psi_j \ 0\ 0\ 0\ 0 \ \ldots \ \psi_n \ 0\ 0\ 0\ 0 \} \tag{10.77}$$

$$\{\psi\}_2 = \{ 0 \ \psi_1 \ 0\ 0\ 0 \ \ldots \ 0 \ \psi_j \ 0\ 0\ 0 \ \ldots \ 0 \ \psi_n \ 0\ 0\ 0 \} \tag{10.78}$$

$$\{\psi\}_3 = \{ 0\ 0 \ \psi_1 \ 0\ 0 \ \ldots \ 0\ 0 \ \psi_j \ 0\ 0 \ \ldots \ 0\ 0 \ \psi_n \ 0\ 0 \} \tag{10.79}$$

$$\{\psi\}_4 = \{ 0\ 0\ 0 \ \psi_1 \ 0 \ \ldots \ 0\ 0\ 0 \ \psi_j \ 0 \ \ldots \ 0\ 0\ 0 \ \psi_n \ 0 \} \tag{10.80}$$

$$\{\psi\}_5 = \{ 0\ 0\ 0\ 0 \ \psi_1 \ \ldots \ 0\ 0\ 0\ 0 \ \psi_j \ \ldots \ 0\ 0\ 0\ 0 \ \psi_n \}. \tag{10.81}$$

Each array is of dimensions $(1, 5n)$. The forcing vector due to the jump in distributed load $f(x,y)$ normal to the upper and lower surfaces of the panel $z = \pm h/2$, is

$$\{F^e\} = \int_{A_e} \{\psi\}_3^T f dA. \tag{10.82}$$

The weighted secondary variables vector is given by

$$\{R^e\} = \oint_{\Gamma_e} \left[\{\psi\}_1^T Q_{nx} + \{\psi\}_2^T Q_{ny} + \{\psi\}_3^T Q_{nz} \right] ds + \oint_{\Gamma_e} \left[\{\psi\}_4^T (-M_1) + \{\psi\}_5^T (-M_2) \right] ds \tag{10.83}$$

$$\begin{aligned} Q_{nx} &= Q_{nn}\hat{n}_x - Q_{ns}\hat{n}_y, \\ Q_{ny} &= Q_{nn}\hat{n}_y + Q_{ns}\hat{n}_x, \\ M_1 &= M_{nn}\hat{n}_x - M_{ns}\hat{n}_y, \\ M_2 &= M_{nn}\hat{n}_y + M_{ns}\hat{n}_x \end{aligned} \tag{10.84}$$

where Γ_e is the element boundary in the middle plane, and $(\hat{n}_x, \hat{n}_y, 0)$ is unit normal to the element boundary. M_{nn} and M_{ns} are bending and twisting moments, Q_{nn}, Q_{ns}, and Q_{nz} are edge forces in the normal, tangential, and transverse directions, respectively, all per unit length of contour Γ_e. The components of the weighted secondary variables array $\{R^e\}$ are determined by the boundary conditions on the plate edges. Unknown components at essential (geometric) boundary conditions will be removed from the assembled system by application of the boundary conditions. Natural boundary conditions for free vibration and flutter are homogeneous, and hence the corresponding components of $\{R^e\}$ are zero.

Boundary Conditions

Two types of boundary conditions are of interest:

Immovable simply supported edge:

$$u_o = 0, \quad v_o = 0, \quad w_o = 0, \quad \theta_s = 0, \quad M_{nn} = 0. \tag{10.85}$$

Clamped (fixed) edge:

$$u_o = 0, \quad v_o = 0, \quad w_o = 0, \quad \theta_s = 0, \quad \theta_n = 0 \tag{10.86}$$

where

$$\theta_n = \theta_x \hat{n}_x + \theta_y \hat{n}_y, \quad \theta_s = \theta_y \hat{n}_x - \theta_x \hat{n}_y. \tag{10.87}$$

Integration Rules

The Reissner–Mindlin plate theory may experience shear lock for low-order elements such as the 4-node bilinear element. Selective integration rules can be used to circumvent shear lock. Reduced-order integration is used for computing the shear stiffness matrix given by the third integral in Equation (10.76) [also given by Equation (9.168)], whereas full integration rules are used for all other matrices and vectors. Using Gauss-Legendre quadrature, the shear stiffness matrix for the 4-node bilinear element is computed by (1×1) rule while the 2×2 rule is used for the other matrices. For the biquadratic 9-node or 8-node element, we use 3×3 rule for full integration and 2×2 rule for reduced integration. For the bicubic 16-node element, we use 4×4 rule for full integration and 3×3 rule for reduced integration. In certain applications, higher-order integration rules may be necessary. The problem of shear lock is prevalent for the 4-node bilinear element and is much less severe for biquadratic and higher-order elements.

10.5 STATIC DEFLECTION OF GEOMETRICALLY NONLINEAR PLATES

In this section, we analyze static deflection of plates due to distributed or concentrated loads. We note that the matrix $[G^{NL}]$, given by Equation (10.60), is not a constant matrix; it depends on the deflection w_o. Hence, the stiffness matrix $[K^e]$, given by Equation (10.76), depends on the solution $\{U^e\}$. The equations of equilibrium for an element

$$[K^e]\{U^e\} = \{F^e\} + \{R^e\} \tag{10.88}$$

are nonlinear. Newton–Raphson iteration will be used to solve the algebraic nonlinear system.

Newton–Raphson Method

Let $\{U^e\}_k$ denote the nodal solution at the kth iteration. Thus,

$$\{U^e\}_{(k+1)} = \{U^e\}_k + \Delta\{U^e\}_k \tag{10.89}$$

where $\Delta\{U^e\}_k$ is an increment to be determined. The stiffness matrix at the $(k+1)$th iteration can be written as

$$\begin{aligned}[K^e]_{(k+1)} = \int_{A_e} \{\ [B_o] + [G^{NL}]_{(k+1)}\ \}^T [D_o] \{\ [B_o] + \tfrac{1}{2}[G^{NL}]_{(k+1)}\ \} dA \\ + \int_{A_e} [B_b]^T [D_b][B_b] dA + \int_{A_e} [B_s]^T [D_b^s][B_s] dA\end{aligned} \tag{10.90}$$

and by Equations (10.59) and (10.60), we have

$$\begin{aligned}
[G^{NL}]_{(k+1)} &= \begin{bmatrix} \{B_x^{NL}\}\{U^e\}_{(k+1)} & 0 \\ 0 & \{B_y^{NL}\}\{U^e\}_{(k+1)} \\ \{B_y^{NL}\}\{U^e\}_{(k+1)} & \{B_x^{NL}\}\{U^e\}_{(k+1)} \end{bmatrix} \begin{bmatrix} \{B_x^{NL}\} \\ \{B_y^{NL}\} \end{bmatrix} \\
&= \begin{bmatrix} \{B_x^{NL}\}\{U^e\}_k & 0 \\ 0 & \{B_y^{NL}\}\{U^e\}_k \\ \{B_y^{NL}\}\{U^e\}_k & \{B_x^{NL}\}\{U^e\}_k \end{bmatrix} \begin{bmatrix} \{B_x^{NL}\} \\ \{B_y^{NL}\} \end{bmatrix} \\
&+ \begin{bmatrix} \{B_x^{NL}\}\Delta\{U^e\}_k & 0 \\ 0 & \{B_y^{NL}\}\Delta\{U^e\}_k \\ \{B_y^{NL}\}\Delta\{U^e\}_k & \{B_x^{NL}\}\Delta\{U^e\}_k \end{bmatrix} \begin{bmatrix} \{B_x^{NL}\} \\ \{B_y^{NL}\} \end{bmatrix} \\
&= [G^{NL}]_k + \Delta[G^{NL}]_k
\end{aligned} \tag{10.91}$$

where

$$\Delta\left[G^{NL}\right]_k = \begin{bmatrix} \{B_x^{NL}\}\Delta\{U^e\}_k & 0 \\ 0 & \{B_y^{NL}\}\Delta\{U^e\}_k \\ \{B_y^{NL}\}\Delta\{U^e\}_k & \{B_x^{NL}\}\Delta\{U^e\}_k \end{bmatrix} \begin{bmatrix} \{B_x^{NL}\} \\ \{B_y^{NL}\} \end{bmatrix} \tag{10.92}$$

$$\begin{aligned} [K^e]_{(k+1)} &= \int_{A_e} \{\, [B_o] + [G^{NL}]_k \,\}^T [D_o] \{\, [B_o] + \tfrac{1}{2}[G^{NL}]_k \,\} dA \\ &+ \int_{A_e} [B_b]^T [D_b] [B_b] dA + \int_{A_e} [B_s]^T [D_b^s] [B_s] dA \\ &+ \int_{A_e} \{\ [B_o] + [G^{NL}]_k\ \}^T [D_o] \frac{1}{2} \Delta\left[G^{NL}\right]_k dA \\ &+ \int_{A_e} \Delta\left[G^{NL}\right]_k^T [D_o] \{\ [B_o] + \tfrac{1}{2}[G^{NL}]_k\ \} dA \\ &= [K^e]_k + \Delta[K^e]_k \end{aligned} \tag{10.93}$$

where

$$\begin{aligned} \Delta[K^e]_k &= \int_{A_e} \{\ [B_o] + [G^{NL}]_k\ \}^T [D_o] \frac{1}{2} \Delta\left[G^{NL}\right]_k dA \\ &+ \int_{A_e} \Delta\left[G^{NL}\right]_k^T [D_o] \{\ [B_o] + \tfrac{1}{2}[G^{NL}]_k\ \} dA. \end{aligned} \tag{10.94}$$

We neglected quadratic terms in the increment $\Delta\{U^e\}_k$. The element equations can be written in terms of the increment

$$([K^e]_k + [\Delta K^e]_k)(\{U_k^e\} + \Delta\{U_k^e\}) = \{F^e\} + \{R^e\}. \tag{10.95}$$

Neglecting the product of increment in solution and increment in stiffness matrix, we obtain

$$[K^e]_k \Delta\{U^e\}_k + \Delta[K^e]_k \{U^e\}_k = \{F^e\} + \{R^e\} - [K^e]_k \{U^e\}_k. \tag{10.96}$$

We will show that

$$[\Delta K^e]_k \{U^e\}_k = [S^e]_k \Delta\{U^e\}_k \tag{10.97}$$

where $[S^e]_k$ depends on $\{U^e\}_k$ only [see Equation (10.105)], hence the linearized system for the increment becomes

$$([K^e]_k + [S^e]_k)\Delta\{U^e\}_k = \{F^e\} + \{R^e\} - [K^e]_k \{U^e\}_k. \tag{10.98}$$

We define the tangent stiffness matrix by

$$[K_T^e]_k = [K^e]_k + [S^e]_k \tag{10.99}$$

and obtain the linearized element equations at iteration k,

$$[K_T^e]_k \Delta\{U^e\}_k = \{F^e\} + \{R^e\} - [K^e]_k \{U^e\}_k. \tag{10.100}$$

Now, we prove Equation (10.97).

$$\Delta[K^e]_k\{U^e\}_k = \int_{A_e} \{\ [B_o]+[G^{NL}]_k\ \}^T [D_o]\frac{1}{2}\Delta\left[G^{NL}\right]_k\{U^e\}_k dA$$
$$+\int_{A_e}\Delta\left[G^{NL}\right]_k^T [D_o]\{\ [B_o]+\tfrac{1}{2}[G^{NL}]_k\ \}\{U^e\}_k dA. \quad (10.101)$$

To simplify the notation, we define $X = \{B_x^{NL}\}$, $Y = \{B_y^{NL}\}$, $\Delta = \Delta\{U^e\}_k$, and $U = \{U^e\}_k$.

$$\begin{aligned}
\Delta\left[G^{NL}\right]_k\{U^e\}_k &= \begin{bmatrix} X\Delta & 0 \\ 0 & Y\Delta \\ Y\Delta & X\Delta \end{bmatrix}\begin{bmatrix} X \\ Y \end{bmatrix}U = \begin{bmatrix} X\Delta & 0 \\ 0 & Y\Delta \\ Y\Delta & X\Delta \end{bmatrix}\begin{bmatrix} XU \\ YU \end{bmatrix} \\
&= \begin{bmatrix} (XU)(X\Delta) \\ (YU)(Y\Delta) \\ (XU)(Y\Delta)+(YU)(X\Delta) \end{bmatrix} \\
&= \begin{bmatrix} XU & 0 \\ 0 & YU \\ YU & XU \end{bmatrix}\begin{bmatrix} X\Delta \\ Y\Delta \end{bmatrix} = \begin{bmatrix} XU & 0 \\ 0 & YU \\ YU & XU \end{bmatrix}\begin{bmatrix} X \\ Y \end{bmatrix}\Delta \\
&= \left[G^{NL}\right]_k\Delta\{U^e\}_k. \qquad (10.102)
\end{aligned}$$

We define $\Lambda = [D_o]\{\ [B_o]+\frac{1}{2}[G^{NL}]_k\ \}\{U^e\}_k = [\Lambda_1\ \Lambda_2\ \Lambda_3]^T$.

$$\begin{aligned}
\Delta\left[G^{NL}\right]_k^T\Lambda &= [\ X^T\ \ Y^T\ ]\begin{bmatrix} X\Delta & 0 & Y\Delta \\ 0 & Y\Delta & X\Delta \end{bmatrix}\begin{bmatrix} \Lambda_1 \\ \Lambda_2 \\ \Lambda_3 \end{bmatrix} \\
&= [\ X^T\ \ Y^T\ ]\begin{bmatrix} \Lambda_1(X\Delta)+\Lambda_3(Y\Delta) \\ \Lambda_2(Y\Delta)+\Lambda_3(X\Delta) \end{bmatrix} \\
&= \begin{bmatrix} X \\ Y \end{bmatrix}^T\begin{bmatrix} \Lambda_1X+\Lambda_3Y \\ \Lambda_2Y+\Lambda_3X \end{bmatrix}\Delta. \qquad (10.103)
\end{aligned}$$

Substituting Equations (10.102) and (10.103) into Equation (10.101), we get

$$[\Delta K^e]_k\{U^e\}_k = [S^e]_k\Delta\{U^e\}_k \qquad (10.104)$$

where

$$\begin{aligned}
[S^e]_k &= \int_{A_e}\{\ [B_o]+[G^{NL}]_k\ \}^T[D_o]\frac{1}{2}\left[G^{NL}\right]_k dA \\
&\quad +\int_{A_e}\begin{bmatrix} X \\ Y \end{bmatrix}^T\begin{bmatrix} \Lambda_1X+\Lambda_3Y \\ \Lambda_2Y+\Lambda_3X \end{bmatrix}dA. \qquad (10.105)
\end{aligned}$$

The matrices in Equation (10.100) for each element are assembled into global matrices, and boundary conditions are applied. The global system is solved for the increments at all nodes, and the solution is updated.

$$\{U\}_{(k+1)} = \{U\}_k+\Delta\{U\}_k. \qquad (10.106)$$

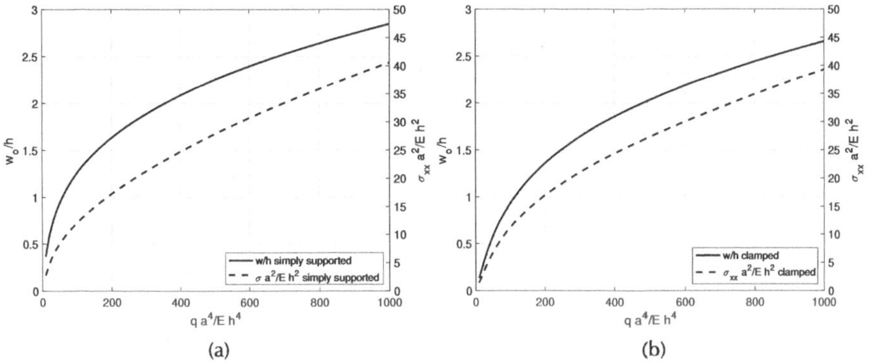

Figure 10.1 Square plate: Central deflection $w_o(0,0)/h$ and normal stress $\sigma_{xx}(0,0,h/2)a^2/Eh^2$. (a) Simply supported plate. (b) Clamped plate.

A problem-dependent convergence criterion is used for the termination of the iteration process. For static deflection problems, the convergence criterion is

$$\frac{\Delta\{U\}_k^T \Delta\{U\}_k}{F^T F} < \varepsilon \tag{10.107}$$

where F is the forcing vector, and ε is a prescribed tolerance on the order of 10^{-10}.

Example 10.1. Square Plate Carrying Uniform Distributed Load

We consider a square plate subjected to uniformly distributed load q. The plate side is $a = 300$ cm, its thickness $h = 3$ cm, and material properties $\nu = 0.316$, and $E = 0.3 \times 10^8$ Pa. This is one of the many examples solved by Pica et al. [85] and Panda and Barik [82]. The origin $(x,y) = (0,0)$ is at the center of the square. The first quadrant is modeled by $6 \times 6 = 36$ bicubic elements (16 nodes per element and 361 global nodes), and symmetry conditions are imposed on the x- and y-axes. Results are presented for the central deflection $w_o(0,0)/h$ and the maximum normal stress $\sigma_{xx}(0,0,h/2)a^2/Eh^2$ for a range of dimensionless load qa^4/Eh^4. Central deflection and stress are shown in Figures 10.1a and 10.1b for the simply supported, and clamped plates, respectively. Results of the linear plate analysis, not shown in the figures, would be given by the tangent at the origin. Results of the linear analysis would deviate significantly from the nonlinear results for deflections greater than $0.1h$. Table 10.1 shows grid convergence of the deflection and maximum normal stress at the plate center for the load $qa^4/Eh^4 = 402$ for the simply supported plate. Three types of elements are tested: Bilinear $Q4$, biquadratic $Q9$, and bicubic $Q16$. The convergence for the bilinear element is much slower than the biquadratic or bicubic element; it requires a much finer mesh. Tables 10.2, and 10.3 give the deflection, and maximum normal stress at the plate center, respectively, for a range of the distributed load. The number of global nodes is 49 for $Q4$, $Q9$, and $Q16$. It

Table 10.1
Simply Supported Square Plate: Grid Convergence at Load $qa^4/Eh^4 = 402$, Central Deflection $w_o(0,0)/h$ and Maximum Normal Stress $\sigma_{xx}(0,0,h/2)a^2/Eh^2$

Nodes	$w-Q4$	$w-Q9$	$w-Q16$	$\sigma-Q4$	$\sigma-Q9$	$\sigma-Q16$
49	2.10389	2.08674	2.08682	25.28349	25.06706	24.42638
169	2.09101	2.08679	2.08679	24.90791	24.82262	24.76119
361	2.08866	2.08679	2.08679	24.84000	24.79951	24.78100
625	2.08784	2.08679	2.08679	24.81630	24.79303	24.78438
961	2.08746	2.08679	2.08679	24.80536	24.79031	24.78530
1369	2.08726	2.08679	2.08679	24.79943	24.78892	24.78564

Table 10.2
Simply Supported Square Plate: Central Deflection $w_o(0,0)/h$

qa^4/Eh^4	$Q4$	$Q9$	$Q16$	$Q16$-fine	Reference [82]	Δ
17.79	0.54678	0.54477	0.54463	0.54465	0.5450	0.000
38.3	0.83272	0.82767	0.82750	0.82753	0.8281	0.001
63.4	1.04848	1.04101	1.04084	1.04088	1.0416	0.001
95	1.23969	1.23016	1.23000	1.23005	1.2309	0.001
134.9	1.42078	1.40939	1.40927	1.40931	1.4102	0.001
184	1.59484	1.58181	1.58173	1.58176	1.5826	0.001
245	1.76854	1.75400	1.75397	1.75399	1.7547	0.001
318	1.93920	1.92328	1.92331	1.92330	1.9239	0.001
402	2.10389	2.08674	2.08682	2.08679	2.0871	0.000

is 961 nodes for the fine mesh labeled $Q16$-fine. The tables also show the difference (Δ) between the present results for $Q16$-fine and those by Panda and Barik [82]. Grid convergence for the clamped plate for the load $qa^4/Eh^4 = 402$ is given in Table 10.4 for the bilinear $Q4$, biquadratic $Q9$, and bicubic $Q16$ elements. The results for the clamped plate are more sensitive to grid resolution than those of the simply supported plate. Tables 10.5 and 10.6 give the deflection and maximum normal stress at the plate center, respectively, for a range of the distributed load. The number of global nodes is 49 for $Q4$, $Q9$, and $Q16$. It is 961 nodes for the fine mesh labeled $Q16$-fine. The tables also show the difference (Δ) between the present results for $Q16$-fine and those by Pica et al. [85].

Example 10.2. Circular Plate Carrying Uniform Distributed Load

We consider a circular plate subjected to uniformly distributed load q. The plate radius is $R = 100$ cm, its thickness $h = 2$ cm, and material properties $\nu = 0.3$, and

Table 10.3
Simply Supported Square Plate: Central Normal Stress $\sigma_{xx}(0,0,h/2)a^2/Eh^2$

qa^4/Eh^4	$Q4$	$Q9$	$Q16$	$Q16$-fine	Reference [82]	Δ
17.79	4.43271	4.46305	4.36481	4.38448	4.3881	0.004
38.3	7.29043	7.29346	7.12105	7.17229	7.1796	0.007
63.4	9.68655	9.65138	9.41738	9.50536	9.5160	0.011
95	12.00392	11.92871	11.63606	11.76448	11.7789	0.014
134.9	14.38681	14.27333	13.92063	14.09241	14.1106	0.018
184	16.86633	16.71900	16.30325	16.51995	16.5418	0.022
245	19.53808	19.36176	18.87666	19.14040	19.1660	0.026
318	22.36337	22.16400	21.60354	21.91507	21.9443	0.029
402	25.28349	25.06706	24.42638	24.78530	24.8184	0.033

Table 10.4
Clamped Square Plate: Grid Convergence at Load $qa^4/Eh^4 = 402$, Central Deflection $w_o(0,0)/h$, and Normal Stress $\sigma_{xx}(0,0,h/2)a^2/Eh^2$

Nodes	$w-Q4$	$w-Q9$	$w-Q16$	$\sigma-Q4$	$\sigma-Q9$	$\sigma-Q16$
49	1.87942	1.85439	1.85409	25.19780	25.21251	23.64581
169	1.86271	1.85661	1.85692	24.60364	24.55243	24.35712
361	1.85972	1.85713	1.85727	24.49657	24.47020	24.40080
625	1.85868	1.85727	1.85733	24.45905	24.44342	24.40785
961	1.85820	1.85731	1.85734	24.44174	24.43148	24.40974
1369	1.85794	1.85733	1.85735	24.43235	24.42513	24.41042

Table 10.5
Clamped Square Plate: Central Deflection $w_o(0,0)/h$

qa^4/Eh^4	$Q4$	$Q9$	$Q16$	$Q16$-fine	Reference [85]	Δ
17.79	0.23585	0.23693	0.23655	0.23662	0.2368	0.000
38.3	0.46977	0.46981	0.46913	0.46943	0.4699	0.000
63.4	0.69434	0.69132	0.69047	0.69116	0.6915	0.000
95	0.90960	0.90255	0.90163	0.90280	0.9029	0.000
134.9	1.11716	1.10586	1.10494	1.10662	1.1063	0.000
184	1.31565	1.30031	1.29947	1.30163	1.3009	0.001
245	1.51103	1.49198	1.49127	1.49387	1.4928	0.001
318	1.69992	1.67761	1.67710	1.68007	1.6786	0.001
402	1.87942	1.85439	1.85409	1.85734	1.8555	0.002

Table 10.6
Clamped Square Plate: Central Normal Stress $\sigma_{xx}(0,0,h/2)a^2/Eh^2$

qa^4/Eh^4	$Q4$	$Q9$	$Q16$	$Q16$-fine	Reference [85]	Δ
17.79	2.59460	2.68667	2.54197	2.57007	2.6319	0.062
38.3	5.42718	5.60131	5.28367	5.34052	5.4816	0.141
63.4	8.30023	8.51051	8.00719	8.11079	8.3258	0.215
95	11.14406	11.34764	10.65445	10.83038	11.103	0.273
134.9	13.95381	14.12259	13.24069	13.51363	13.827	0.313
184	16.71272	16.83478	15.77142	16.15847	16.497	0.339
245	19.52295	19.59798	18.35769	18.87099	19.225	0.354
318	22.36062	22.39821	20.98945	21.63116	21.994	0.363
402	25.19780	25.21251	23.64581	24.40974	24.780	0.370

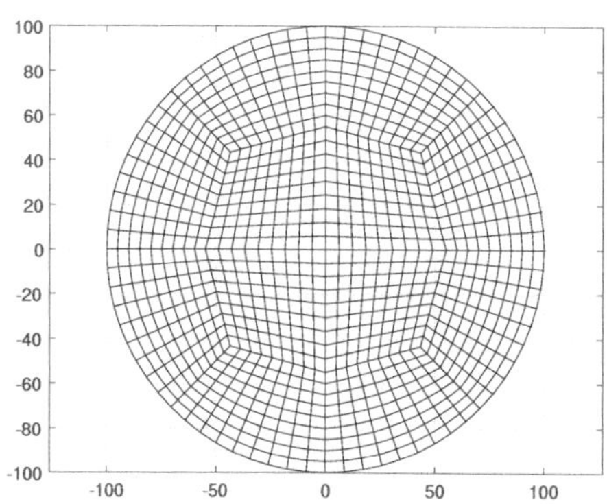

Figure 10.2 A mesh of 108 bicubic elements for a circular plate, 1009 nodes.

$E = 0.1 \times 10^8$ Pa. This problem has been solved also by Pica et al. [85], and Panda and Barik [82]. The origin $(x,y) = (0,0)$ is at the circle center. The full circle is modeled by 108 bicubic elements (16 nodes per element and 1009 global nodes) as shown in Figure 10.2. Results are presented for the central deflection $w_o(0,0)/h$ and normal stress $\sigma_{xx}(0,0,h/2)R^2/Eh^2$ for a range of the dimensionless load qR^4/Eh^4. Central deflection and stress are shown in Figures 10.3a and 10.3b for the simply supported and clamped plates, respectively.

Results are also obtained for quadrilateral elements of lower order than bicubic. For the simply supported plate, the deflection $w_o(0,0)$ and normal stress $\sigma_{xx}(0,0,h/2)$ are computed using 457 global nodes for quadrilateral elements $Q4$, $Q9$, and $Q16$; the corresponding number of elements are 432, 108, and 48, respectively. The results are shown in Tables 10.7 and 10.8. Results are also obtained for a fine mesh of bicubic $Q16$ element using 192 elements and 1777 nodes. The differences Δ between the fine mesh results and Panda and Barik's [82] computations are also shown in the

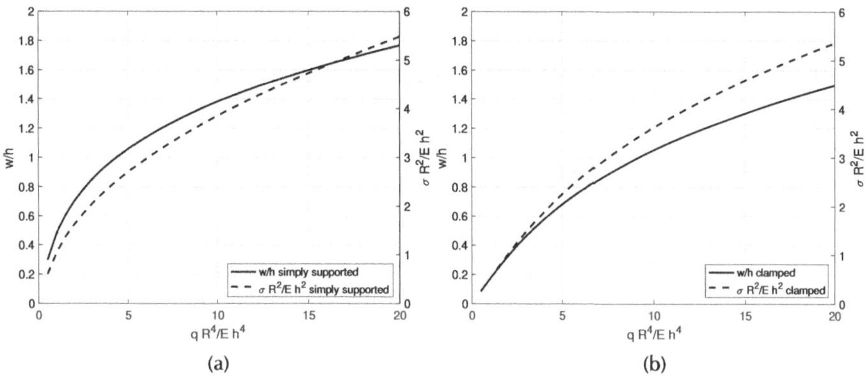

Figure 10.3 Circular plate: Central deflection $w_o(0,0)/h$ and normal stress $\sigma_{xx}(0,0,h/2)R^2/Eh^2$. (a) Simply supported plate. (b) Clamped plate.

Table 10.7
Simply Supported Circular Plate: Central Deflection $w_o(0,0)/h$

qR^4/Eh^4	$Q4$	$Q9$	$Q16$	$Q16$-fine	Reference [82]	Δ
1	0.4794	0.4808	0.4808	0.4808	0.4795	0.001
2	0.7027	0.7037	0.7038	0.7038	0.7035	0.000
3	0.8504	0.8512	0.8513	0.8513	0.8517	0.000
6	1.1363	1.1367	1.1368	1.1368	1.1385	0.002
10	1.3797	1.3799	1.3800	1.3800	1.3826	0.003
15	1.5974	1.5975	1.5976	1.5976	1.6008	0.003

Table 10.8
Simply Supported Circular Plate: Central Normal Stress $\sigma_{xx}(0,0,h/2)R^2/Eh^2$

qR^4/Eh^4	$Q4$	$Q9$	$Q16$	$Q16$-fine	Reference [82]	Δ
1	1.0385	1.0472	1.0389	1.0392	1.0348	0.004
2	1.6306	1.6422	1.6281	1.6288	1.6261	0.003
3	2.0582	2.0717	2.0533	2.0544	2.053	0.001
6	2.9736	2.9920	2.9647	2.9665	2.9678	0.001
10	3.8568	3.8811	3.8457	3.8480	3.8515	0.004
15	4.7381	4.7697	4.7264	4.7288	4.7342	0.005

Table 10.9
Clamped Circular Plate: Central Deflection $w_o(0,0)/h$

qR^4/Eh^4	$Q4$	$Q9$	$Q16$	$Q16$-fine	Reference [85]	Δ
1	0.1673	0.1683	0.1683	0.1683	0.1682	0.000
2	0.3217	0.3233	0.3233	0.3234	0.3231	0.000
3	0.4577	0.4595	0.4595	0.4595	0.4591	0.000
6	0.7697	0.7709	0.7710	0.7710	0.7702	0.001
10	1.0525	1.0526	1.0528	1.0528	1.0514	0.001
15	1.3035	1.3027	1.3029	1.3030	1.3007	0.002

Table 10.10
Clamped Circular Plate: Central Normal Stress $\sigma_{xx}(0,0,h/2)R^2/Eh^2$

qR^4/Eh^4	$Q4$	$Q9$	$Q16$	$Q16$-fine	Reference [85]	Δ
1	0.5018	0.5122	0.5051	0.5054	0.5304	0.025
2	1.0024	1.0233	1.0076	1.0081	1.0611	0.053
3	1.4653	1.4949	1.4702	1.4710	1.5524	0.081
6	2.5792	2.6247	2.5763	2.5789	2.7384	0.160
10	3.6311	3.6871	3.6157	3.6216	3.8657	0.244
15	4.6018	4.6668	4.5748	4.5844	4.9163	0.332

tables. Results are also obtained for the clamped plate using the same grids as for the simply supported plate. The central deflection and stress are shown in Tables 10.9, and 10.10, respectively. The differences between the fine mesh results and Pica et al.'s [85] computations are also shown in the tables.

10.6 BUCKLING AND POST-BUCKLING OF PLATES

As an introduction to the subject of plate buckling, we consider a square plate of side a and uniform thickness h. The middle plane is assumed to be perfectly flat. The plate is in static equilibrium under the action of uniformly distributed compressive stress $\sigma_o > 0$ applied to two parallel edges as shown in Figure 10.4a. The stress resultant per unit length is $N_{xx} = \sigma_o h$. Except for restricting rigid-body translation and rotation, the plate is free to deform; the edges are unrestrained for motion parallel to the midplane. The state of stress and displacements at any point can be determined by solving the equations of equilibrium assuming a state of plane stress as presented in Chapter 8. The solution is

$$\sigma_{xx} = -\sigma_o,\ \sigma_{yy} = 0,\ \sigma_{xy} = 0,$$
$$u_o = -(\sigma_o/E)x,\ v_o = \nu(\sigma_o/E)y,\ w_o = 0. \tag{10.108}$$

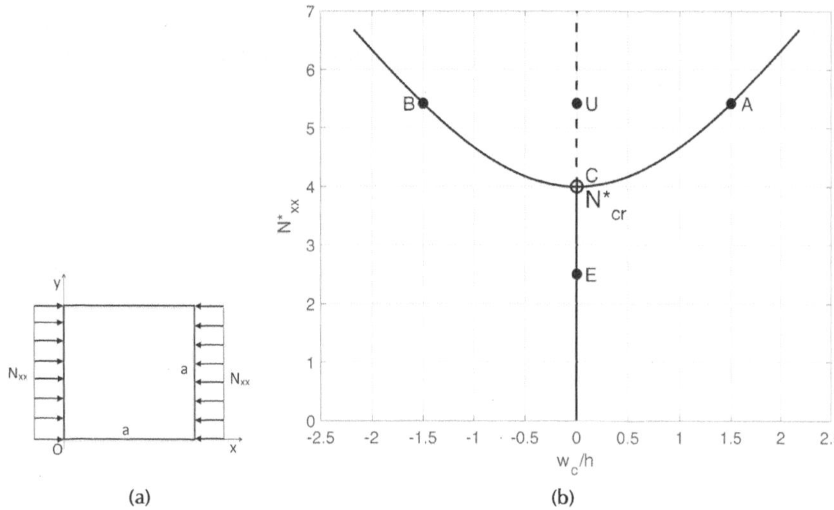

Figure 10.4 Post-buckling of a simply supported square plate, $h/a = 0.01, \nu = 0.3$. (a) Uniaxial load N_{xx} with unrestrained membrane displacements. (b) Post-buckling central deflection-load response.

The equilibrium state given by Equation (10.108) is called the *basic state*. If σ_o is increased from zero and the plate is perfectly flat, the deflection w_o remains zero for any value of σ_o. However, such an equilibrium state will be unstable if σ_o exceeds a critical value denoted by σ_{cr}. If a tiny lateral load (such as pressure distribution normal to the midplane) is applied to the plate and then removed, the plate returns to the basic state if $\sigma_o < \sigma_{cr}$, and the basic state is stable. But, if $\sigma_o > \sigma_{cr}$, the plate will transition to a different equilibrium state in which the deflection $w_o \neq 0$, and the basic state is unstable. The displacements and stresses in the new equilibrium state depend heavily on the boundary conditions imposed on the plate bending geometry (w, θ_x, θ_y) in addition to the conditions imposed on the inplane displacements (u_o, v_o).

We define nondimensional uniaxial load by

$$N_{xx}^* = \frac{\sigma_o h a^2}{\pi^2 D}, \quad N_{cr}^* = \frac{\sigma_{cr} h a^2}{\pi^2 D}, \quad D = \frac{E h^3}{12(1-\nu^2)}. \tag{10.109}$$

When the four edges are simply supported, the classical plate theory predicts the critical load $N_{cr}^* = 4$. Finite-element computations are obtained for Reissner–Mindlin plate ($h/a = 0.01$, $\nu = 0.3$, and a mesh of 64 bicubic elements). The plate central deflection w_c/h is shown in Figure 10.4b as a function of the loading N_{xx}^*. At a point such as E for which $N_{xx}^* < N_{cr}^*$, the plate is in stable equilibrium in the initial flat configuration $w_o = 0$. Whereas at a point such as U for which $N_{xx}^* > N_{cr}^*$, the equilibrium of the initial flat configuration $w_o = 0$ is unstable, and the plate snaps to the deflected position A or B depending on the type of disturbance. In Figure 10.4b, the dashed line represents unstable equilibrium whereas solid lines represent stable equilibrium. The critical point C is called a bifurcation point and the equilibrium path is called pitchfork supercritical bifurcation.

In this section, finite-element procedures will be presented for determining the critical loads and post-buckling paths of initially flat plates for various inplane loading and boundary conditions. To this end, we first extend the principle of virtual displacements to the case of initially stressed elastic body. We follow Washizu [124] and take the initially stressed state as the reference configuration. Cartesian components of the initial stress vectors on three mutually perpendicular planes at a point define the initial stress tensor σ^o_{ij}. The initial stresses are in equilibrium with externally applied loads. Incremental displacements u_i due to buckling or additional body and surface forces are measured from the reference configuration.

$$\int_\Omega \bar{\varepsilon}_{jk}\sigma_{kj}d\Omega + \int_\Omega \frac{\partial \bar{u}_m}{\partial x_j}\frac{\partial u_m}{\partial x_k}\sigma^o_{kj}d\Omega = \oint_{\partial\Omega} \bar{u}_i t_i dS + \int_\Omega \bar{u}_i f_i d\Omega. \tag{10.110}$$

Equation (10.110) is the principle of virtual displacements for an initially stressed body [124]. The second integral on the left hand side is the only difference between Equation (10.110) and Equation (10.17) for a body free of initial stresses. Because the virtual displacements $\bar{u}_m$ and the initial stresses σ^o_{kj} are prescribed, that term is linear in the unknown displacements u_m. In the study of stability (buckling) problems, the initial stresses are given by the state of stress in the basic state. For buckling analysis, the basic state stresses are determined first at prescribed *reference* values of the applied external loads, and those stresses are then multiplied by an unknown parameter, which is determined as an eigenvalue. To extend the finite-element model Equation (10.74) for the study of Reissner–Mindlin plate buckling, we substitute the assumed displacements Equations (10.21) and (10.33) into the term

$$\begin{aligned}\frac{\partial \bar{u}_m}{\partial x_j}\frac{\partial u_m}{\partial x_k}\sigma^o_{kj} &= \left(\frac{\partial \bar{u}}{\partial x}\frac{\partial u}{\partial x} + \frac{\partial \bar{v}}{\partial x}\frac{\partial v}{\partial x} + \frac{\partial \bar{w}}{\partial x}\frac{\partial w}{\partial x}\right)\sigma^o_{xx} \\ &+ \left(\frac{\partial \bar{u}}{\partial y}\frac{\partial u}{\partial y} + \frac{\partial \bar{v}}{\partial y}\frac{\partial v}{\partial y} + \frac{\partial \bar{w}}{\partial y}\frac{\partial w}{\partial y}\right)\sigma^o_{yy} \\ + \Bigg[\left(\frac{\partial \bar{u}}{\partial x}\frac{\partial u}{\partial y} + \frac{\partial \bar{v}}{\partial x}\frac{\partial v}{\partial y} + \frac{\partial \bar{w}}{\partial x}\frac{\partial w}{\partial y}\right) &+ \left(\frac{\partial \bar{u}}{\partial y}\frac{\partial u}{\partial x} + \frac{\partial \bar{v}}{\partial y}\frac{\partial v}{\partial x} + \frac{\partial \bar{w}}{\partial y}\frac{\partial w}{\partial x}\right)\Bigg]\sigma^o_{xy}\end{aligned} \tag{10.111}$$

and upon neglecting terms involving products of membrane stretching (u_o, v_o) and bending displacements (w_o, θ_x, θ_y), we get

$$\begin{aligned}\frac{\partial \bar{u}_m}{\partial x_j}\frac{\partial u_m}{\partial x_k}\sigma^o_{kj} \approx\ & \frac{\partial \bar{w}_o}{\partial x}\frac{\partial w_o}{\partial x}\sigma^o_{xx} + \frac{\partial \bar{w}_o}{\partial y}\frac{\partial w_o}{\partial y}\sigma^o_{yy} + \left(\frac{\partial \bar{w}_o}{\partial x}\frac{\partial w_o}{\partial y} + \frac{\partial \bar{w}_o}{\partial y}\frac{\partial w_o}{\partial x}\right)\sigma^o_{xy} \\ & + z^2\left(\frac{\partial \bar{\theta}_x}{\partial x}\frac{\partial \theta_x}{\partial x} + \frac{\partial \bar{\theta}_y}{\partial x}\frac{\partial \theta_y}{\partial x}\right)\sigma^o_{xx} + z^2\left(\frac{\partial \bar{\theta}_x}{\partial y}\frac{\partial \theta_x}{\partial y} + \frac{\partial \bar{\theta}_y}{\partial y}\frac{\partial \theta_y}{\partial y}\right)\sigma^o_{yy} \\ & + z^2\left[\left(\frac{\partial \bar{\theta}_x}{\partial x}\frac{\partial \theta_x}{\partial y} + \frac{\partial \bar{\theta}_y}{\partial x}\frac{\partial \theta_y}{\partial y}\right) + \left(\frac{\partial \bar{\theta}_x}{\partial y}\frac{\partial \theta_x}{\partial x} + \frac{\partial \bar{\theta}_y}{\partial y}\frac{\partial \theta_y}{\partial x}\right)\right]\sigma^o_{xy}.\end{aligned} \tag{10.112}$$

For the case in which the initial stresses do not change with z, that is to say they are uniform across the plate thickness, we obtain the integral

$$\begin{aligned}\int_{-h/2}^{h/2} \frac{\partial \bar{u}_m}{\partial x_j}\frac{\partial u_m}{\partial x_k}\sigma^o_{kj}dz &= \frac{\partial \bar{w}_o}{\partial x}\frac{\partial w_o}{\partial x}N^o_{xx} + \frac{\partial \bar{w}_o}{\partial y}\frac{\partial w_o}{\partial y}N^o_{yy} \\ &+ \left(\frac{\partial \bar{w}_o}{\partial x}\frac{\partial w_o}{\partial y} + \frac{\partial \bar{w}_o}{\partial y}\frac{\partial w_o}{\partial x}\right)N^o_{xy} \\ +\frac{h^2}{12}\left(\frac{\partial \bar{\theta}_x}{\partial x}\frac{\partial \theta_x}{\partial x} + \frac{\partial \bar{\theta}_y}{\partial x}\frac{\partial \theta_y}{\partial x}\right)N^o_{xx} &+ \frac{h^2}{12}\left(\frac{\partial \bar{\theta}_x}{\partial y}\frac{\partial \theta_x}{\partial y} + \frac{\partial \bar{\theta}_y}{\partial y}\frac{\partial \theta_y}{\partial y}\right)N^o_{yy} \\ +\frac{h^2}{12}\left[\left(\frac{\partial \bar{\theta}_x}{\partial x}\frac{\partial \theta_x}{\partial y} + \frac{\partial \bar{\theta}_y}{\partial x}\frac{\partial \theta_y}{\partial y}\right) \right. &+ \left.\left(\frac{\partial \bar{\theta}_x}{\partial y}\frac{\partial \theta_x}{\partial x} + \frac{\partial \bar{\theta}_y}{\partial y}\frac{\partial \theta_y}{\partial x}\right)\right]N^o_{xy} \qquad (10.113)\end{aligned}$$

where $N^o_{xx} = h\sigma^o_{xx}$, $N^o_{yy} = h\sigma^o_{yy}$, and $N^o_{xy} = h\sigma^o_{xy}$. In terms of the virtual strains defined by Equation (10.36), we have

$$\int_{-h/2}^{h/2} \frac{\partial \bar{u}_m}{\partial x_j}\frac{\partial u_m}{\partial x_k}\sigma^o_{kj}dz = \{\bar{\varepsilon}^{NL}_o\}^T \begin{Bmatrix} N^o_{xx} \\ N^o_{yy} \\ N^o_{xy}\end{Bmatrix} + \frac{h^2}{12}\{\bar{\varepsilon}^{NL}_b\}^T \begin{Bmatrix} N^o_{xx} \\ N^o_{yy} \\ N^o_{xy}\end{Bmatrix}. \qquad (10.114)$$

Introducing finite-element representations Equations (10.44) and (10.68) for displacements, we obtain

$$\begin{aligned}\int_{-h/2}^{h/2} \frac{\partial \bar{u}_m}{\partial x_j}\frac{\partial u_m}{\partial x_k}\sigma^o_{kj}dz &= \{d\}^T\left[B^{NL}\right]^T[A_o]^T\begin{Bmatrix} N^o_{xx} \\ N^o_{yy} \\ N^o_{xy}\end{Bmatrix} \\ +\frac{h^2}{12}\{d\}^T\left[B^{NL}_1\right]^T[A_1]^T\begin{Bmatrix} N^o_{xx} \\ N^o_{yy} \\ N^o_{xy}\end{Bmatrix} &+ \frac{h^2}{12}\{d\}^T\left[B^{NL}_2\right]^T[A_2]^T\begin{Bmatrix} N^o_{xx} \\ N^o_{yy} \\ N^o_{xy}\end{Bmatrix} \\ &= \{d\}^T\left[B^{NL}\right]^T[S]\left[B^{NL}\right]\{U^e\} \\ +\{d\}^T\left\{\frac{h^2}{12}\left[B^{NL}_1\right]^T[S]\left[B^{NL}_1\right] \right. &+ \left.\frac{h^2}{12}\left[B^{NL}_2\right]^T[S]\left[B^{NL}_2\right]\right\}\{U^e\} \qquad (10.115)\end{aligned}$$

where

$$[S] = \begin{bmatrix} N^o_{xx} & N^o_{xy} \\ N^o_{xy} & N^o_{yy}\end{bmatrix} \qquad (10.116)$$

The matrices $\left[B^{NL}_1\right]$ and $\left[B^{NL}_2\right]$ are given by Equation (10.63). Element equations for the initially stressed plate become,

$$[[K^e] + [K^e_o]]\{U^e\} = \{F^e\} + \{R^e\} \qquad (10.117)$$

where

$$\begin{aligned}[K^e_o] &= \int_{A_e}\left[B^{NL}\right]^T[S]\left[B^{NL}\right]dA \\ +\int_{A_e}\left\{\frac{h^2}{12}\left[B^{NL}_1\right]^T[S]\left[B^{NL}_1\right] \right. &+ \left.\frac{h^2}{12}\left[B^{NL}_2\right]^T[S]\left[B^{NL}_2\right]\right\}dA. \qquad (10.118)\end{aligned}$$

As mentioned previously, for the purpose of critical loads determination and post-buckling analysis, the initial stresses are determined first at reference values of the applied loads. Let $[S^*]$ denote the stress matrix at the reference values, hence

$$[S] = \lambda\, [S^*] \tag{10.119}$$

where λ is a parameter to be determined as an eigenvalue. The element equations become,

$$\left[[K^e] + \lambda \left[K_g^e\right]\right]\{U^e\} = \{F^e\} + \{R^e\} \tag{10.120}$$

where

$$\begin{aligned}\left[K_g^e\right] &= \int_{A_e} \left[B^{NL}\right]^T [S^*] \left[B^{NL}\right] dA \\ &+ \int_{A_e} \left\{\frac{h^2}{12}\left[B_1^{NL}\right]^T [S^*]\left[B_1^{NL}\right] + \frac{h^2}{12}\left[B_2^{NL}\right]^T [S^*]\left[B_2^{NL}\right]\right\} dA\end{aligned} \tag{10.121}$$

$\left[K_g^e\right]$ is called the geometric-stiffness matrix. Unless the plate is thick, the second integral being proportional to h^2 is usually neglected and the geometric-stiffness matrix simplifies to

$$\left[K_g^e\right] = \int_{A_e} \left[B^{NL}\right]^T [S^*] \left[B^{NL}\right] dA. \tag{10.122}$$

10.6.1 CRITICAL LOADS OF PLATES

A state of stress is created in a plate due to inplane loading applied to the plate edges. As an example, a rectangular plate may be subjected to uniaxial (N_{xx}), bi-axial (N_{xx}, N_{yy}), or pure shear (N_{xy}) loads distributed on the plate edges as depicted in Figure 10.5. A combination of shear and uniaxial or bi-axial loading is also important to consider. The critical loading is the least magnitude of applied loads at which the equilibrium becomes unstable (the bifurcation point). For the uniaxial case, we may take the reference load $N_{xx} = -Eh$ or any convenient value, and $N_{yy} = N_{xy} = 0$. Hence,

$$[S^*] = Eh\begin{bmatrix} -1 & 0 \\ 0 & 0 \end{bmatrix} \tag{10.123}$$

and the critical load will be $N_{xx,cr} = \lambda Eh$. Similarly, for bi-axial load we take $N_{xx} = -Eh$, $N_{yy} = -\beta Eh$, and $N_{xy} = 0$, where $\beta = N_{yy}/N_{xx}$ is a given parameter. Hence,

$$[S^*] = Eh\begin{bmatrix} -1 & 0 \\ 0 & -\beta \end{bmatrix}. \tag{10.124}$$

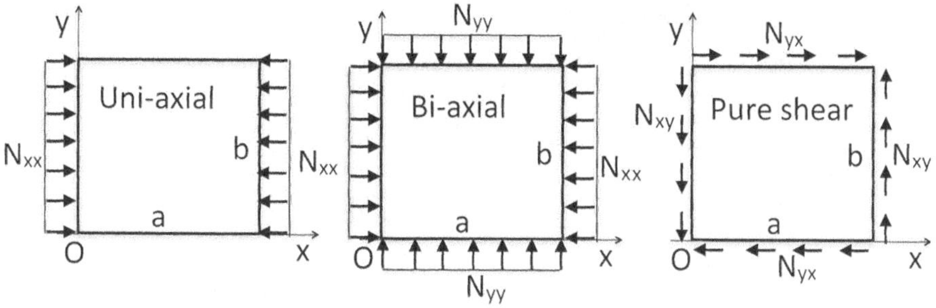

Figure 10.5 Uniaxial, biaxial, and pure shear inplane loading of a rectangular plate.

The critical bi-axial loading will be $N_{xx,cr} = \lambda Eh$. For pure shear, we have $N_{xx} = 0$, $N_{yy} = 0$, and $N_{xy} = Eh$, hence

$$[S^*] = Eh \begin{bmatrix} 0 & 1 \\ 1 & 0 \end{bmatrix}. \tag{10.125}$$

The critical pure shear loading will be $N_{xy,cr} = \lambda Eh$. At the critical load (bifurcation point), the bending geometric parameters $(w_o, \theta_x, \theta_y)$ are infinitesimally small. Hence the stiffness matrix Equation (10.75) simplifies to a linear matrix

$$[K^e] = \int_{A_e} [B_o]^T [D_o] [B_o] dA + \int_{A_e} [B_b]^T [D_b] [B_b] dA + \int_{A_e} [B_s]^T [D_b^s] [B_s] dA. \tag{10.126}$$

Thus, we have a linear eigenvalue problem, that can be further simplified by decoupling the additional flexural and membrane displacements. As such, there are only three degrees of freedom at each node $(w_o, \theta_x, \theta_y)$. However, for post-buckling analysis, the full coupled nonlinear system must be solved. For simple geometries such as a rectangular plate, the initial stress matrix $[S^*]$ (basic state stresses) is readily available for uniform uniaxial, biaxial, or pure shear. However, for more complex geometries or loading the initial stresses in the basic state can vary in space and may have to be determined numerically using for example the finite-element method.

Example 10.3. Critical Uniaxial Loading for a Square Plate

Consider a square plate of side a thickness h in uniaxial loading as shown in Figure 10.5. Equation (10.123) gives the initial stress matrix $[S^*]$. All sides are simply supported and boundary conditions are given by Equation (10.85). The square is modeled by uniform $8 \times 8 = 64$ bicubic elements. The variation of the nondimensional critical load $N_{cr}^* = N_{cr} a^2 / \pi^2 D$ with the plate thickness ratio h/a is shown in

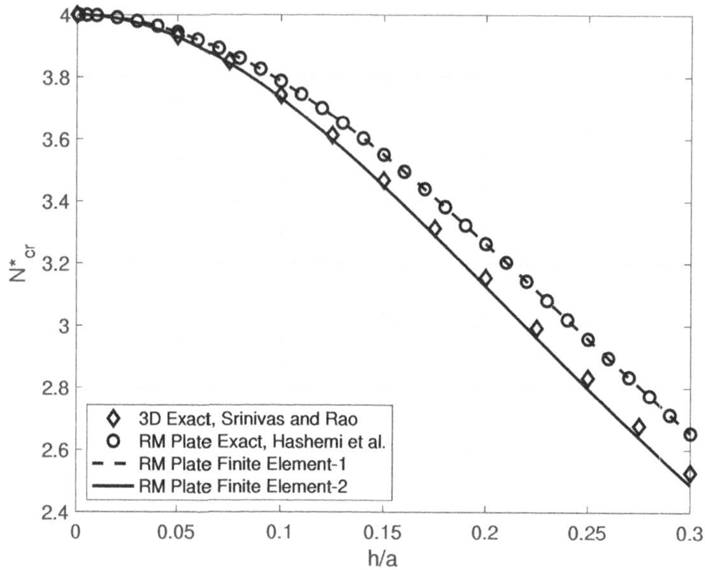

Figure 10.6 Effect of plate thickness on uniaxial critical load for simply supported square plate. $\nu = 0.3$.

Figure 10.6 for Poisson's ratio $\nu = 0.3$. For the geometric-stiffness matrix Equation (10.121), the finite-element results (finite-element−2) are in good agreement with Srinivas and Rao's [111] exact analytic solution to the three-dimensional elasticity problem without plate theory approximations. For the geometric-stiffness matrix Equation (10.122), the finite-element results (finite-element−1) are in perfect agreement with Hashemi et al.'s [48] exact analytic solution to the Reissner–Mindlin (*RM*) plate equations. Solution to the linear eigenvalue problem provides a spectrum of critical loads corresponding to different modes of buckling. The mode shapes in terms of deflection $w_o(x,y)$ for the lowest four eigenvalues are depicted in Figure 10.7. Modes 1, 2, and 3 have one, two, and three, respectively, half-waves in the direction parallel to the load, and one half-wave in the normal direction. Mode 4 has one wave in both directions. Modes 2 and 4 have zero deflection at the plate center. The first mode is the fundamental mode of buckling since it corresponds to the smallest load.

Example 10.4. Critical Biaxial Loading for a Square Plate

For biaxial loading, Equation (10.124) gives the initial stress matrix $[S^*]$; we consider a square plate and equal loads $N_{xx} = N_{yy}$, hence $\beta = 1$. The square is modeled by uniform $20 \times 20 = 400$ bicubic elements. The nondimensional critical load $N^*_{cr} = N_{cr}a^2/\pi^2 D$ for different values of the plate thickness ratio h/a is shown in Table 10.11 for Poisson's ratio $\nu = 0.3$. The geometric-stiffness matrix is given by

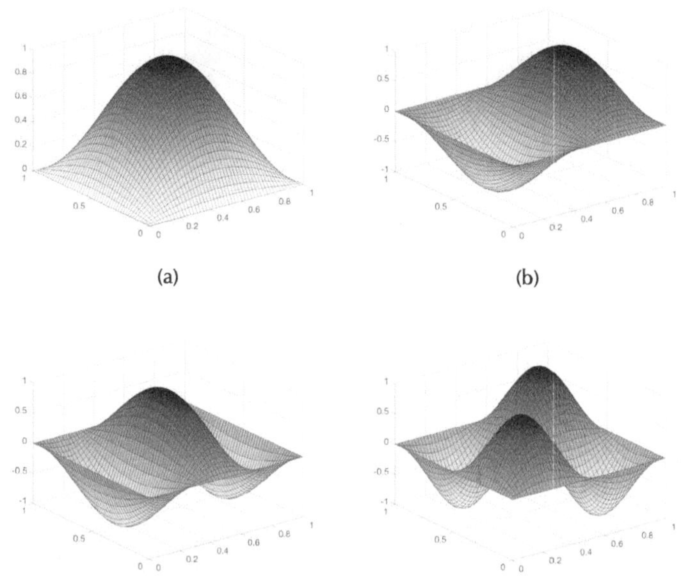

(a) (b) (c) (d)

Figure 10.7 Four buckling mode shapes of a simply supported square plate in uniaxial compression, $h/a = 0.01, \nu = 0.3$. (a) First mode, $N^*_{cr} = 3.9977$, (b) second mode, $N^*_{cr} = 6.2412$, (c) third mode, $N^*_{cr} = 11.0799$, and (d) fourth mode, $N^*_{cr} = 15.9640$.

Table 10.11
Critical Load for Square Plate under Equal Bi-axial Load
$N_{xx} = N_{yy}$. $\nu = 0.3$, $\kappa = 5/6$

Boundary conditions	h/a 0.001	h/a 0.01	h/a 0.05	h/a 0.1	h/a 0.15	h/a 0.2
S-S-S-S	1.99999	1.99887	1.97219	1.89323	1.77479	1.63187
S-C-S-F	1.14312	1.13874	1.11200	1.06415	1.00493	0.939019
S-C-S-C	3.82985	3.82463	3.70360	3.37609	2.95131	2.51863
C-C-C-C	5.30353	5.29441	5.08656	4.54785	3.88517	3.23994

Edge condition: simply supported S, clamped C, free F.

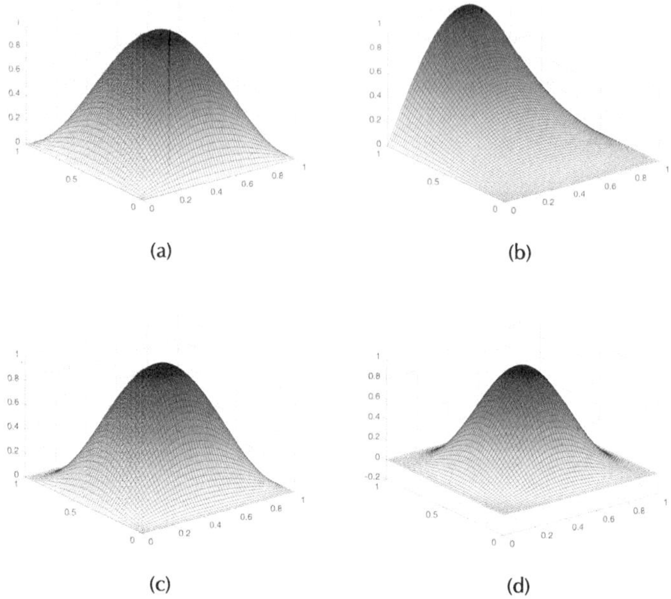

Figure 10.8 Fundamental buckling mode shape of a square plate in equal bi-axial compression, $h/a = 0.01, \nu = 0.3$. Boundary conditions: (a) SSSS, (b) SCSF, (c) SCSC, and (d) CCCC.

Equation (10.122). For the first three cases, two parallel edges of the plate are simply supported (S) while the other edges may be free (F) or clamped (C). The finite-element results are in perfect agreement with Hashemi et al.'s [48] exact analytic solution to the Reissner–Mindlin plate equations. The fourth case is for a clamped plate. The buckling mode shape for each of the four boundary conditions is shown in Figure 10.8.

Example 10.5. Rectangular Plate in Pure Shear

We consider a simply supported rectangular plate of sides (a, b) and thickness h in pure shear loading as shown in Figure 10.5. Equation (10.125) gives the basic-state stress matrix $[S^*]$. A uniform mesh of $16 \times 16 = 256$ bicubic elements $Q16$ is used. The critical load $N^*_{xy,cr} = N_{xy,cr} b^2/\pi^2 D$ for a thin plate of $b/h = 2000$ and $\nu = 0.3$ at different aspect ratios a/b is shown in Table 10.12. Timoshenko and Gere [117] provide exact analytic solution [Equation (e), Section 9.7, page 382] for thin plates, CPT. The finite-element results are in excellent agreement with the analytic solution. Effects of the thickness ratio on the critical load are shown in Table 10.13 for a square

Table 10.12
Critical Inplane Shear Load $N^*_{xy,cr}$ for Simply Supported Rectangular Thin Plate, $\nu = 0.3$, $b/h = 2000$

a/b	1.0	1.2	1.4	1.6	1.8	2.0
FEM	9.3245	7.9834	7.2874	6.9073	6.6882	6.5460
a/b	2.2	2.4	2.6	2.8	3.0	3.2
FEM	6.2933	6.1031	5.9760	5.8931	5.8401	5.8055

Table 10.13
Critical Inplane Shear Load $N^*_{xy,cr}$ for Simply Supported Square Plate, $\nu = 0.3$

b/h	1000	100	50	20	10	5
FEM	9.3243	9.3068	9.2541	8.9017	7.8368	5.2692

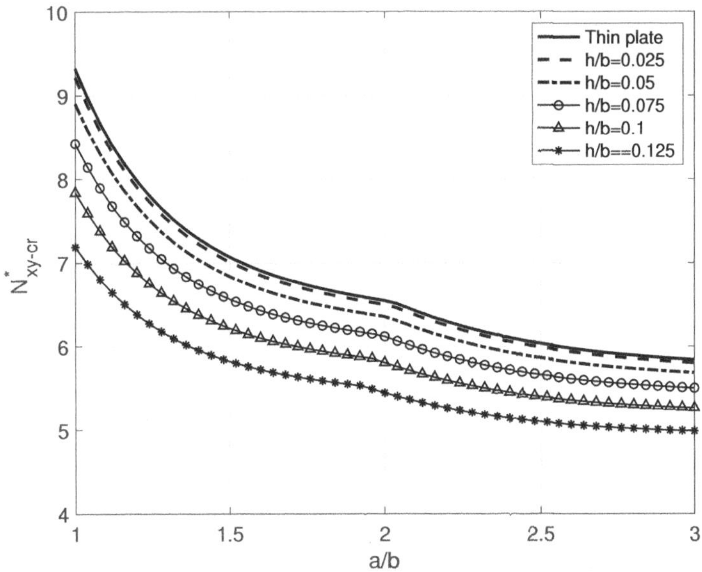

Figure 10.9 Effects of the aspect ratio a/b and plate thickness h/b on the critical load for a simply supported rectangular plate in pure shear. $\nu = 0.3$.

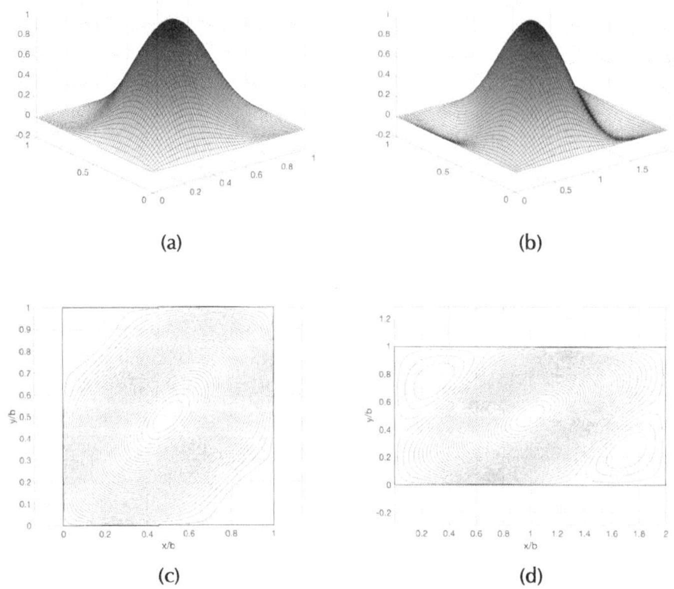

Figure 10.10 Buckling mode shapes of a simply supported plate in pure shear, $h/b = 0.01, \nu = 0.3$. (a, c) Square plate, $N^*_{xy,cr} = 9.306818$. (b, d) Rectangular plate $a/b = 2$, $N^*_{xy,cr} = 6.538126$.

plate $a/b = 1$, and for rectangular plates in Figure 10.9. Buckling mode shape for simply supported square and rectangular $a/b = 2$ plates are shown in Figures 10.10a and 10.10b, respectively. Because of the symmetry of plate geometry and boundary conditions, the eigenvalue solver provides two critical loads of equal magnitudes but opposite signs. The mode shapes in Figure 10.10 are for positive shear, and those for negative shear can be obtained by flipping the figures about a vertical edge.

10.6.2 POST-BUCKLING OF PLATES

If the applied load exceeds the critical value, the equilibrium basic state in which the plate is flat becomes unstable, and the plate takes a different equilibrium position in which the lateral deflection $w_o \neq 0$. The displacements and stresses in the new equilibrium position can be determined by solving the eigenvalue system given by Equation (10.120) or by direct solution of the nonlinear equilibrium system Equation (10.74).

Eigenvalue Solution for Post-Buckling

As in the case of determining the critical load, the applied load is assumed to be characterized by one parameter, which is treated as an eigenvalue. In this approach, the deflection w_o is prescribed at a certain point of the plate in the post-buckling equilibrium position, and the minimum load needed to achieve that deflection is determined as an eigenvalue of the system Equation (10.120). For example, in the case of a square plate, we want the deflection at the plate center in the new equilibrium position to be a prescribed fraction of the plate thickness. The eigenvalue system is nonlinear because the stiffness matrix depends on displacements of the bending problem (w_o, θ_x, θ_y) throughout the plate; which are unknown. Thus, an iterative process is followed. In the course of the iterative process, the eigenvectors corresponding to the eigenvalue having the smallest magnitude are scaled and used to compute the stiffness matrix. The scaling is such that the deflection w_o at a point in the plate will have the prescribed value. Upon solving the eigenvalue problem, the change in the eigenvalue of smallest magnitude is monitored, and if it exceeds certain tolerance the new eigenvectors are re-scaled. Then, the stiffness matrix is updated and the eigenvalue problem is solved again. This iteration is repeated until the change in the eigenvalue of smallest magnitude falls below a prescribed tolerance. The converged eigenvalue of smallest magnitude provides the required load. To construct the post-buckling deflection-load path, an incremental process is used. It starts with a very small target deflection for which the eigenvectors at the critical load provide a good initial guess. The target deflection is then incremented by a small step, and the initial displacements are taken as the converged eigenvectors of the previous step. Boundary conditions on membrane displacements (u_o, v_o) must be enforced in addition to conditions on the bending displacements $(w_o, \theta_x, \theta_y)$.

We note that the displacement field found by the eigenvalue approach gives the additional displacements measured relative to the basic state configuration. To find the total displacements measured relative to the unloaded state, we must add the basic state displacements to those predicted by the eigenvalue solver.

Post-Buckling Analysis by Direct Integration of the Equations of Equilibrium

The problem of finding equilibrium positions for post-buckling can be treated as a static equilibrium problem under the action of prescribed inplane loads such as uniaxial, biaxial, pure shear or combination loads. Two procedures can be followed. In the first approach, a tiny transverse load that may be concentrated or distributed over the plate area is applied in addition to the inplane loads responsible for the plate buckling. The inplane load is increased from zero in small steps. At each step, the equations of static equilibrium are solved using Newton–Raphson method. The initial guess for displacements is taken as the converged solution in the previous load step. If the load is smaller than the critical load, the plate deflection will be very small and remains proportional to the transverse tiny load. Once the prescribed inplane load approaches or exceeds the critical load, the plate develops large deflection and enters the post-buckling phase. Thereafter, the external tiny transverse load may be removed, and the process of increasing the applied inplane load continues to the desired level. In this approach, the load-deflection curve cannot pass through the bifurcation point because the stiffness matrix is singular at that point. In the second procedure, the applied inplane load starts with a value slightly greater than the critical load and the plate is initially deflected; no need to apply any lateral load. The eigenvalue procedure provides good initial displacements to start Newton–Raphson iteration. While the applied inplane load is fixed, the static equilibrium equations are solved and stiffness and tangent matrices are updated after each iteration. If the change in the plate deflection at a control point is smaller than certain tolerance the iteration is terminated. The converged solution for one load is used as an initial guess for the next load step. In the direct approach, the inplane load is given and the total displacements measured from the unloaded state are determined, whereas in the eigenvalue approach the deflection at a point is given and the inplane load is found, and the obtained displacements are additional displacements measured from the basic state.

Example 10.6. Post-Buckling of a Simply Supported Square Plate

We consider a simply supported square plate (side a, thickness $h/a = 0.01$, Poisson's ratio $\nu = 0.3$). The membrane displacements are unrestrained on the plate edges. The boundary conditions for the eigenvalue solver are

$$\begin{aligned}
x=0: &\quad w_o=0,\ M_{xx}=0,\ \theta_y=0,\ N_{xx}=0,\ N_{xy}=0\\
x=a: &\quad w_o=0,\ M_{xx}=0,\ \theta_y=0,\ N_{xx}=0,\ N_{xy}=0\\
y=0: &\quad w_o=0,\ M_{yy}=0,\ \theta_x=0,\ N_{yy}=0,\ N_{yx}=0\\
y=a: &\quad w_o=0,\ M_{yy}=0,\ \theta_x=0,\ N_{yy}=0,\ N_{yx}=0.
\end{aligned} \tag{10.127}$$

To prevent rigid body translation and rotation, we also apply the conditions: $(x,y) = (0,0) : u_o = 0, v_o = 0$, and $(x,y) = (a,0) : v_o = 0$. Starting from zero, the central deflection $w_o(0.5a, 0.5a)/h \equiv w_c/h$ is stepped by increments of 0.02 and the required load is found at each step by the eigenvalue solver. The post-buckling load-deflection is shown in Figure 10.4b.

For the direct approach of solving the nonlinear equations of equilibrium we apply the boundary conditions,

$$\begin{aligned}
x=0: &\quad w_o=0,\ M_{xx}=0,\ \theta_y=0,\ N_{xx}=\hat{N}_{xx},\ N_{xy}=0\\
x=a: &\quad w_o=0,\ M_{xx}=0,\ \theta_y=0,\ N_{xx}=-\hat{N}_{xx},\ N_{xy}=0\\
y=0: &\quad w_o=0,\ M_{yy}=0,\ \theta_x=0,\ N_{yy}=0,\ N_{yx}=0\\
y=a: &\quad w_o=0,\ M_{yy}=0,\ \theta_x=0,\ N_{yy}=0,\ N_{yx}=0
\end{aligned} \tag{10.128}$$

where $\hat{N}_{xx}$ is prescribed compressive force per unit length. To prevent rigid body translation and rotation, we also apply the conditions: $(x,y) = (a/2, a/2) : u_o = 0, v_o = 0$, and $(x,y) = (a, a/2) : v_o = 0$. The eigenvectors of the first mode at $w_c/h = 0.04$ are used to initialize the Newton–Raphson method, which predicts the same post-buckling curve shown in Figure 10.4b for $w_c/h \geq 0.04$. For the load

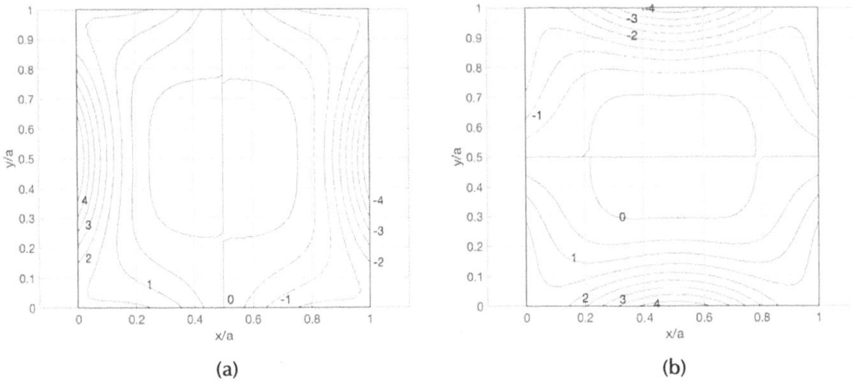

Figure 10.11 Middle plane displacements for post-buckling of simply supported square plate, $h/a = 0.01, \nu = 0.3, N^*_{xx} = 6.3337$. (a) $u_o a/h^2$, and (b) $v_o a/h^2$.

$N^*_{xx} = 6.3337$, contours of additional displacements $u_o a/h^2$ and $v_o a/h^2$ (measured relative to the basic state) are shown in Figure 10.11. Even though the applied load is uniform on two parallel edges while the other edges are unloaded, displacements are nonuniform on the edges with maximum values at the middle of the edges; $u_{max}a/h^2 = \pm 4.6122$ and $v_{max}a/h^2 = \pm 4.3541$. Contours of w_o/h are shown in Figure 10.12a; maximum deflection is at the plate center $w_{max}/h = 2.0041$. Displacements u_o and v_o on the plate edges and deflection w_o on lines through the plate center are shown in Figure 10.12b. The displacement u_o is zero at the center and two other points on the central line parallel to the x-axis. Similar distribution is predicted for v_o on a central line parallel to the y-axis.

Example 10.7. Post-Buckling by Edge-Shortening

Plate post-buckling can be studied by prescribing inplane displacements on the edges. Such an approach is commonly used in theoretical analysis using for example the Rayleigh–Ritz method. We consider a simply supported square plate of side a and thickness h. In the basic state, lines parallel to the x-axis are shortened by εa, where $\varepsilon = -\varepsilon_{xx}$ is a prescribed normal strain. The two edges $y = 0$ and $y = a$ are restrained, hence $v = 0$. All edges are free of shear stress $\sigma_{yx} = 0$. The basic state is given by

$$u = \varepsilon a\left(\frac{1}{2} - \frac{x}{a}\right), \quad v = 0,$$
$$\sigma_{xx} = -\frac{E\varepsilon}{1-\nu^2}, \quad \sigma_{yy} = \nu\sigma_{xx}, \quad \sigma_{xy} = 0. \tag{10.129}$$

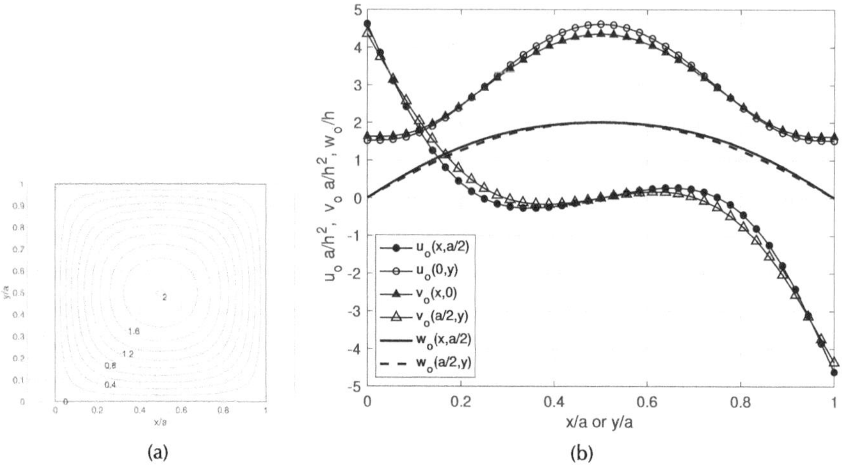

Figure 10.12 Post-buckling of a simply supported square plate, $h/a = 0.01, \nu = 0.3, N_{xx}^* = 6.3337$. (a) w_o/h, and (b) displacements on edges and central lines.

We define a scaled nondimensional edge shortening u^* as

$$u^* = \frac{12\left(1-\nu^2\right)}{\pi^2}\frac{a^2}{h^2}\varepsilon. \tag{10.130}$$

For the initial state of stress given by the basic state Equation (10.129), we determine the critical edge shortening $\varepsilon_{cr}a$ using the eigenvalue solver with boundary conditions

$$\begin{aligned}
x=0: &\quad w_o=0,\quad M_{xx}=0,\quad \theta_y=0,\quad u_o=0,\quad N_{xy}=0\\
x=a: &\quad w_o=0,\quad M_{xx}=0,\quad \theta_y=0,\quad u_o=0,\quad N_{xy}=0\\
y=0: &\quad w_o=0,\quad M_{yy}=0,\quad \theta_x=0,\quad v_o=0,\quad N_{yx}=0\\
y=a: &\quad w_o=0,\quad M_{yy}=0,\quad \theta_x=0,\quad v_o=0,\quad N_{yx}=0.
\end{aligned} \tag{10.131}$$

Using Rayleigh–Ritz method as presented by Dym [34] for thin plates (*CPT*) (see also Timoshenko and Gear [117]), we obtained $u_{cr}^* = 4(1-\nu)$. For $h/a = 0.005$ and $\nu = 0.3$, solution of eigenvalue problem for Reissner–Mindlin plate predicts $u_{cr}^* = 2.799605$ in agreement with the 2.8 for *CPT*. Next, Newton–Raphson method is used to predict the post-buckling deflection for prescribed edge shortening by solving the nonlinear equations of equilibrium and applying the boundary conditions,

$$\begin{aligned}
x=0: &\quad w_o=0,\quad M_{xx}=0,\quad \theta_y=0,\quad u_o=\varepsilon a/2,\quad N_{xy}=0\\
x=a: &\quad w_o=0,\quad M_{xx}=0,\quad \theta_y=0,\quad u_o=-\varepsilon a/2,\quad N_{xy}=0\\
y=0: &\quad w_o=0,\quad M_{yy}=0,\quad \theta_x=0,\quad v_o=0,\quad N_{yx}=0\\
y=a: &\quad w_o=0,\quad M_{yy}=0,\quad \theta_x=0,\quad v_o=0,\quad N_{yx}=0
\end{aligned} \tag{10.132}$$

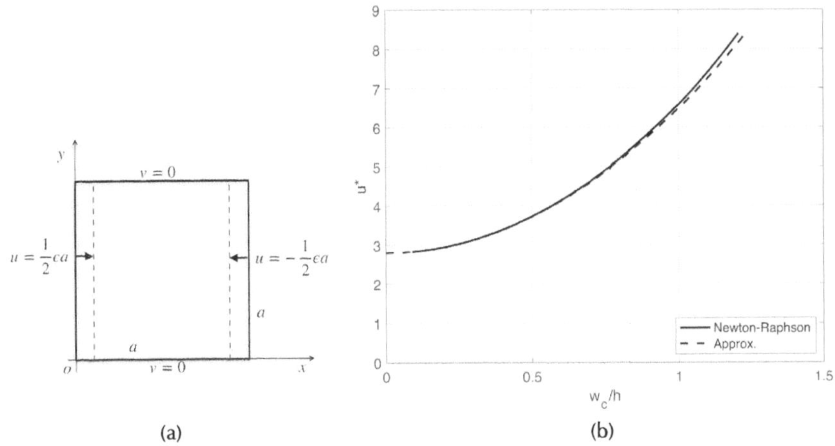

Figure 10.13 Post-buckling of a simply supported square plate, $h/a = 0.005$, $\nu = 0.3$. (a) Edge shortening in the x-direction $u_o = \pm\varepsilon a/2$ and restrained membrane displacements $v_o = 0$ on the edges $y = 0$ and $y = a$. (b) Central deflection.

where εa is greater than the critical value. Again, the eigenvalue solver provides initial displacements slightly away from the bifurcation point. Figure 10.13b shows the edge shortening u^* against central deflection w_c/h and comparison with approximate solution using Rayleigh–Ritz method for thin plates. For edge shortening of $u^* = 8.4$, contours of additional displacements (u_o, v_o, w_o) (relative to the basic state) are depicted in Figures 10.14a, 10.14b, and 10.15a, respectively. Maximum displacements are $|u_o|_{max}a/h^2 = 0.5501$, $|v_o|_{max}a/h^2 = 0.5101$, and $|w_o|_{max}/h = 1.2072$. Finite element solution for u_o and w_o on the line $y = a/2$ are compared with approximate solution by Rayleigh–Ritz method in Figure 10.15b. Distributions of normal stresses at $(z = 0)$ $\sigma^*_{xx} = \sigma_{xx}/(E\varepsilon_{cr})$ on the vertical edges ($x = 0$ and $x = a$) and on the restrained horizontal edges ($y = 0$, and $y = a$), $\sigma^*_{yy} = \sigma_{yy}/(E\varepsilon_{cr})$, are shown in Figure 10.16, and are in good agreement with the approximate Rayleigh–Ritz solution.

10.7 THERMAL BUCKLING OF REISSNER–MINDLIN PLATE

To study the effects of temperature variation on plate deflection and stability we have to account for thermal strains in the constitutive equations. To simplify the analysis we neglect the effects of temperature changes on the mechanical properties E and ν and assume them to be constants. We also assume a constant coefficient of thermal expansion α.

Constitutive Equation: Isotropic Linear Elastic Material

We consider isotropic linear elastic materials. A state of plane stress is also assumed, hence

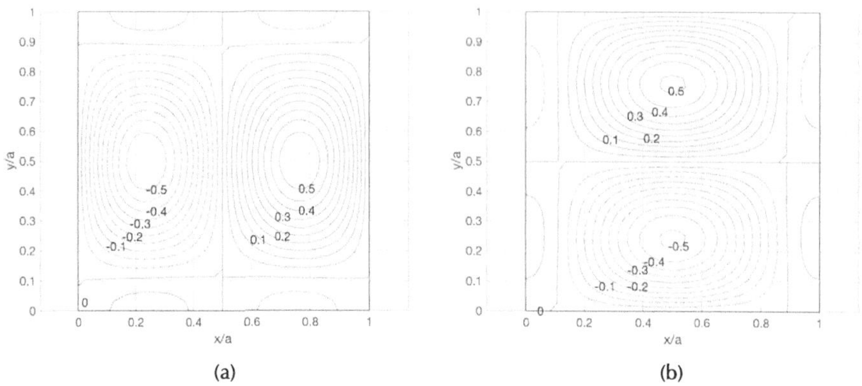

Figure 10.14 Post-buckling of a simply supported square plate, $h/a = 0.005, \nu = 0.3$, $u^* = 8.4$. (a) Contours of $u_o a/h^2$, and (b) contours of $v_o a/h^2$.

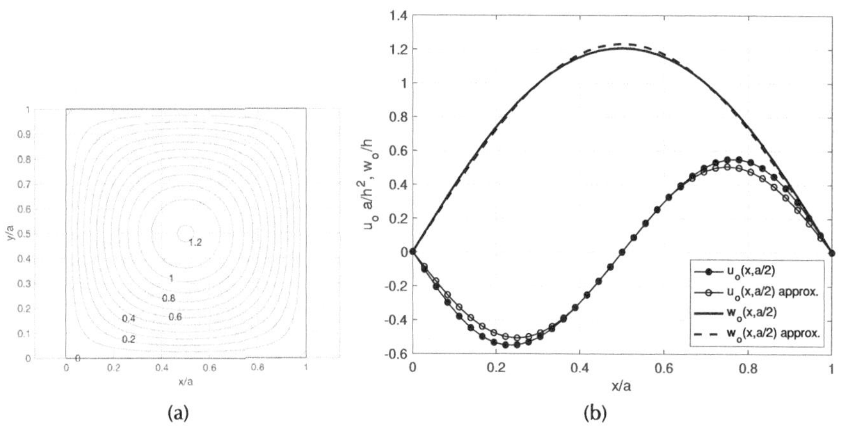

Figure 10.15 Post-buckling of simply supported square plate, $h/a = 0.005, \nu = 0.3$, $u^* = 8.4$. (a) Contours of w_o/h. (b) Displacements on $y = a/2$ and comparison with approximate Rayleigh–Ritz solution.

$$
\begin{aligned}
\sigma_{xx} &= \frac{E}{1-\nu^2}\left[\varepsilon_{xx} + \nu\varepsilon_{yy} - (1+\nu)\alpha T\right] \\
\sigma_{yy} &= \frac{E}{1-\nu^2}\left[\varepsilon_{yy} + \nu\varepsilon_{xx} - (1+\nu)\alpha T\right] \\
\sigma_{xy} &= \frac{E}{2(1+\nu)}\gamma_{xy}
\end{aligned}
\tag{10.133}
$$

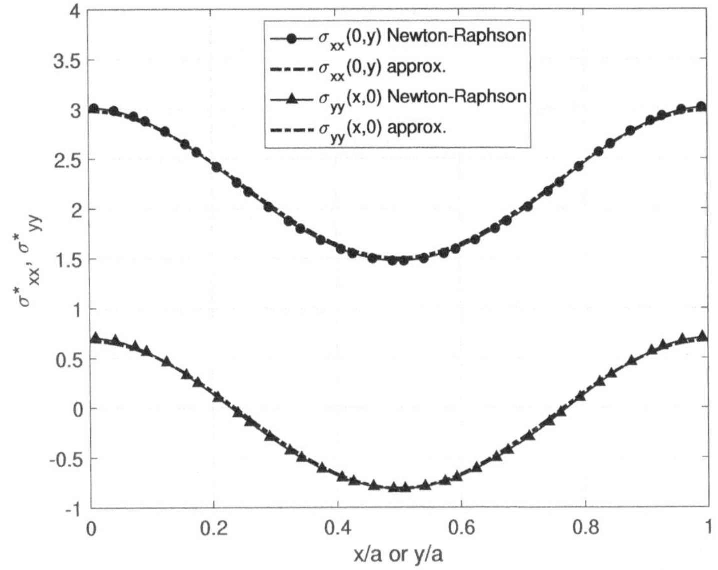

Figure 10.16 Post-buckling of simply supported square plate, $h/a = 0.005, \nu = 0.3$, $u^* = 8.4$. Stresses on plate edges at $z = 0$, and comparison with approximate Rayleigh–Ritz solution.

or in matrix form,

$$\left\{\begin{array}{c}\sigma_{xx}\\ \sigma_{yy}\\ \sigma_{xy}\end{array}\right\} = [D]\left[\left\{\begin{array}{c}\varepsilon_{xx}\\ \varepsilon_{yy}\\ \gamma_{xy}\end{array}\right\} - \left\{\ \varepsilon_T\ \right\}\right] \tag{10.134}$$

where

$$\left\{\ \varepsilon_T\ \right\} = \alpha T \left\{\begin{array}{c}1\\ 1\\ 0\end{array}\right\} \tag{10.135}$$

is thermal strain, and $[D]$ is given by Equation (10.28).

Thermoelastic Element Equations for Reissner–Mindlin Plate

The principle of virtual displacements Equation (10.42) has been used to derive the element equations of equilibrium Equations (10.74) for the constitutive Equation (10.27). We update those equations for the constitutive Equation (10.134) by adding the contribution of the stresses $-[D]\{\varepsilon_T\}$ to the internal virtual work Equation (10.39).

$$(\bar{\varepsilon}_{ij}\sigma_{ji})_T = -\left[\{\bar{\varepsilon}_o\}^T - z\{\ \bar{\varepsilon}_b\ \}^T + \{\bar{\varepsilon}_o^{NL}\}^T + z^2\{\bar{\varepsilon}_b^{NL}\}^T\right][D]\{\varepsilon_T\}. \tag{10.136}$$

The contribution to the internal virtual work per unit area of the middle plane is obtained by integrating Equation (10.136) with respect to z over the plate thickness h,

$$\int_{-h/2}^{h/2} (\bar{\varepsilon}_{ij}\sigma_{ji})_T\, dz = -\int_{-h/2}^{h/2} \{\bar{\varepsilon}_o\}^T [D] \{\varepsilon_T\} dz + \int_{-h/2}^{h/2} \{\bar{\varepsilon}_b\}^T [D] \{\varepsilon_T\} z dz$$
$$-\int_{-h/2}^{h/2} \{\bar{\varepsilon}_o^{NL}\}^T [D] \{\varepsilon_T\} dz - \int_{-h/2}^{h/2} \{\bar{\varepsilon}_b^{NL}\}^T [D] \{\varepsilon_T\} z^2 dz. \quad (10.137)$$

Define the resultant thermal stress forces and moments by

$$\{N_T\} \equiv \begin{Bmatrix} N_{xxT} \\ N_{yyT} \\ N_{xyT} \end{Bmatrix} = \int_{-h/2}^{h/2} [D] \begin{Bmatrix} \alpha T \\ \alpha T \\ 0 \end{Bmatrix} dz$$
$$\{M_T\} \equiv \begin{Bmatrix} M_{xxT} \\ M_{yyT} \\ M_{xyT} \end{Bmatrix} = \int_{-h/2}^{h/2} [D] \begin{Bmatrix} \alpha T \\ \alpha T \\ 0 \end{Bmatrix} z dz$$
$$\{J_T\} \equiv \begin{Bmatrix} J_{xxT} \\ J_{yyT} \\ J_{xyT} \end{Bmatrix} = \int_{-h/2}^{h/2} [D] \begin{Bmatrix} \alpha T \\ \alpha T \\ 0 \end{Bmatrix} z^2 dz. \quad (10.138)$$

If the temperature is uniform over the plate thickness (T is independent of z), then $\{M_T\} = 0$. The four integrals on the right-hand-side of Equation (10.137) are written explicitly as,

$$\int_{-h/2}^{h/2} \{\ \bar{\varepsilon}_o\ \}^T [D] \{\ \varepsilon_T\ \} dz = \{\ \bar{\varepsilon}_o\ \}^T \{\ N_T\ \}$$
$$= \begin{Bmatrix} \frac{\partial \bar{u}_o}{\partial x} \\ \frac{\partial \bar{v}_o}{\partial y} \\ \frac{\partial \bar{u}_o}{\partial y} + \frac{\partial \bar{v}_o}{\partial x} \end{Bmatrix}^T \begin{Bmatrix} N_{xxT} \\ N_{yyT} \\ N_{xyT} \end{Bmatrix}$$
$$\int_{-h/2}^{h/2} \{\ \bar{\varepsilon}_b\ \}^T [D] \{\ \varepsilon_T\ \} z dz = \{\ \bar{\varepsilon}_b\ \}^T \{\ M_T\ \}$$
$$= \begin{Bmatrix} \frac{\partial \bar{\theta}_x}{\partial x} \\ \frac{\partial \bar{\theta}_y}{\partial y} \\ \frac{\partial \bar{\theta}_x}{\partial y} + \frac{\partial \bar{\theta}_y}{\partial x} \end{Bmatrix}^T \begin{Bmatrix} M_{xxT} \\ M_{yyT} \\ M_{xyT} \end{Bmatrix} \quad (10.139)$$

$$\int_{-h/2}^{h/2}\left\{\bar{\varepsilon}_o^{NL}\right\}^T[D]\left\{\varepsilon_T\right\}dz = \left\{\bar{\varepsilon}_o^{NL}\right\}^T\left\{N_T\right\}$$
$$= \begin{Bmatrix}\frac{\partial \bar{w}_o}{\partial x}\\ \frac{\partial \bar{w}_o}{\partial y}\end{Bmatrix}^T \begin{bmatrix}\frac{\partial w_o}{\partial x} & 0\\ 0 & \frac{\partial w_o}{\partial y}\\ \frac{\partial w_o}{\partial y} & \frac{\partial w_o}{\partial x}\end{bmatrix}^T \begin{Bmatrix}N_{xxT}\\ N_{yyT}\\ N_{xyT}\end{Bmatrix}$$
$$= \begin{Bmatrix}\frac{\partial \bar{w}_o}{\partial x}\\ \frac{\partial \bar{w}_o}{\partial y}\end{Bmatrix}^T \begin{bmatrix}N_{xxT}\frac{\partial w_o}{\partial x}+N_{xyT}\frac{\partial w_o}{\partial y}\\ N_{xyT}\frac{\partial w_o}{\partial x}+N_{yyT}\frac{\partial w_o}{\partial y}\end{bmatrix}$$
$$= \begin{Bmatrix}\frac{\partial \bar{w}_o}{\partial x}\\ \frac{\partial \bar{w}_o}{\partial y}\end{Bmatrix}^T \begin{bmatrix}N_{xxT} & N_{xyT}\\ N_{xyT} & N_{yyT}\end{bmatrix} \begin{Bmatrix}\frac{\partial w_o}{\partial x}\\ \frac{\partial w_o}{\partial y}\end{Bmatrix} \tag{10.140}$$

$$\int_{-h/2}^{h/2}\{\ \bar{\varepsilon}_b^{NL}\ \}^T[D]\{\ \varepsilon_T\ \}z^2dz = \{\ \bar{\varepsilon}_b^{NL}\ \}^T\{\ J_T\ \}$$
$$= \begin{Bmatrix}\frac{\partial \bar{\theta}_x}{\partial x}\\ \frac{\partial \bar{\theta}_x}{\partial y}\end{Bmatrix}^T \begin{bmatrix}\frac{\partial \theta_x}{\partial x} & 0\\ 0 & \frac{\partial \theta_x}{\partial y}\\ \frac{\partial \theta_x}{\partial y} & \frac{\partial \theta_x}{\partial x}\end{bmatrix}^T \begin{Bmatrix}J_{xxT}\\ J_{yyT}\\ J_{xyT}\end{Bmatrix} + \begin{Bmatrix}\frac{\partial \bar{\theta}_y}{\partial x}\\ \frac{\partial \bar{\theta}_y}{\partial y}\end{Bmatrix}^T \begin{bmatrix}\frac{\partial \theta_y}{\partial x} & 0\\ 0 & \frac{\partial \theta_y}{\partial y}\\ \frac{\partial \theta_y}{\partial y} & \frac{\partial \theta_y}{\partial x}\end{bmatrix}^T \begin{Bmatrix}J_{xxT}\\ J_{yyT}\\ J_{xyT}\end{Bmatrix}$$
$$= \begin{Bmatrix}\frac{\partial \bar{\theta}_x}{\partial x}\\ \frac{\partial \bar{\theta}_x}{\partial y}\end{Bmatrix}^T \begin{bmatrix}J_{xxT} & J_{xyT}\\ J_{xyT} & J_{yyT}\end{bmatrix} \begin{Bmatrix}\frac{\partial \theta_x}{\partial x}\\ \frac{\partial \theta_x}{\partial y}\end{Bmatrix} + \begin{Bmatrix}\frac{\partial \bar{\theta}_y}{\partial x}\\ \frac{\partial \bar{\theta}_y}{\partial y}\end{Bmatrix}^T \begin{bmatrix}J_{xxT} & J_{xyT}\\ J_{xyT} & J_{yyT}\end{bmatrix} \begin{Bmatrix}\frac{\partial \theta_y}{\partial x}\\ \frac{\partial \theta_y}{\partial y}\end{Bmatrix}. \tag{10.141}$$

Using Equations (10.57), (10.62), (10.69), and (10.70), we write the four integrals for an element,

$$\int_{A_e}\int_{-h/2}^{h/2}\{\ \bar{\varepsilon}_o\ \}^T[D]\{\ \varepsilon_T\ \}dzdA = \{d\}^T\{\ F_{NT}^e\ \}$$
$$\int_{A_e}\int_{-h/2}^{h/2}\{\ \bar{\varepsilon}_b\ \}^T[D]\{\ \varepsilon_T\ \}zdzdA = \{d\}^T\{\ F_{MT}^e\ \}$$
$$\int_{A_e}\int_{-h/2}^{h/2}\{\ \bar{\varepsilon}_o^{NL}\ \}^T[D]\{\ \varepsilon_T\ \}dzdA = \{d\}^T[K_G^e]\{U^e\}$$
$$\int_{A_e}\int_{-h/2}^{h/2}\{\ \bar{\varepsilon}_b^{NL}\ \}^T[D]\{\ \varepsilon_T\ \}z^2dzdA = \{d\}^T[K_J^e]\{U^e\} \tag{10.142}$$

where

$$\{\ F_{NT}^e\ \} = \int_{A_e} [B_o]^T \{\ N_T\ \} dA$$

$$\{\ F_{MT}^e\ \} = \int_{A_e} [B_b]^T \{\ M_T\ \} dA$$

$$[K_G^e] = \int_{A_e} [B^{NL}]^T \begin{bmatrix} N_{xxT} & N_{xyT} \\ N_{xyT} & N_{yyT} \end{bmatrix} [B^{NL}]\, dA$$

$$[K_J^e] = \int_{A_e} [B_1^{NL}]^T \begin{bmatrix} J_{xxT} & J_{xyT} \\ J_{xyT} & J_{yyT} \end{bmatrix} [B_1^{NL}]\, dA + \int_{A_e} [B_2^{NL}]^T \begin{bmatrix} J_{xxT} & J_{xyT} \\ J_{xyT} & J_{yyT} \end{bmatrix} [B_2^{NL}]\, dA. \tag{10.143}$$

The equations for thermoelastic equilibrium of Reissner–Mindlin plate element are

$$([K^e] + [K_G^e] + [K_J^e])\{U^e\} = \{F_{NT}^e\} + \{F_{MT}^e\} + \{F^e\} + \{R^e\}. \tag{10.144}$$

The stiffness matrix $[K^e]$ is given by Equation (10.75) or (10.76). Newton–Raphson method and element Equation (10.144) can be used to study the effects of nonuniform temperature variation on the plate deflection and stability of equilibrium. In such a study, the displacements are measured from the undeformed state. If the plate temperature is uniform across the pate thickness, thermal buckling can be investigated by solving an eigenvalue problem, and the displacements are measured relative to the basic state that can be determined by solving a plane-stress thermoelastic problem as described in Chapter 8.

Example 10.8. Thermal Post-Buckling of a Rectangular Clamped Plate

Stresses developed in constrained plates due to temperature variations can lead to loss of stability. This example gives the basic features of plate buckling due to thermal loading. We consider a rectangular aluminum plate where all edges are restrained with respect to membrane displacement and clamped for lateral deflection. The boundary conditions are

$$\begin{aligned} x=0: &\quad w_o=0,\ \ \theta_x=0,\ \ \theta_y=0,\ \ u_o=0,\ \ v_o=0 \\ x=a: &\quad w_o=0,\ \ \theta_x=0,\ \ \theta_y=0,\ \ u_o=0,\ \ v_o=0 \\ y=0: &\quad w_o=0,\ \ \theta_x=0,\ \ \theta_y=0,\ \ u_o=0,\ \ v_o=0 \\ y=b: &\quad w_o=0,\ \ \theta_x=0,\ \ \theta_y=0,\ \ u_o=0,\ \ v_o=0. \end{aligned} \tag{10.145}$$

Plate dimensions and material properties are ($a = 1$ m, $b = 0.5$ m, $h = 6$ mm, $\nu = 0.35$, $E = 73$ GPa, $\alpha = 23 \times 10^{-6}$ °C^{-1}). A uniform mesh of $8 \times 8 = 64$ bicubic

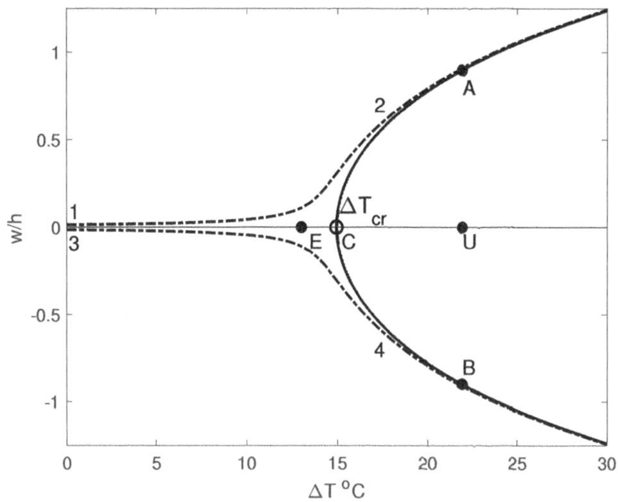

Figure 10.17 Thermal buckling of a clamped rectangular plate. $\Delta T_{cr} = 14.93665\,°C$.

elements (625 global nodes) is used. The plate is unstressed at the initial temperature. The plate is loaded by uniform distributed load $q = \pm 800$ Pa, which is kept constant as the plate temperature is increased. The temperature rise ΔT is uniform throughout the plate. The nonlinear equations of equilibrium are solved by the Newton–Raphson method. The central deflection $w_o(a/2, b/2)/h$ is plotted against the temperature increase in Figure 10.17. Line $1-2-A$ gives deflection for positive $q = 800$ Pa, while line $3-4-B$ is for negative $q = -800$ Pa. For temperature rise less than that of point E the deflection remains small, but we observe faster rate of deflection increase with temperature rise beyond point E. If no transverse load is applied to the plate $q = 0$, the plate will be in equilibrium with zero deflection as given by line $0-E-C-U$. However, there is a temperature rise ΔT_{cr} if exceeded such an equilibrium will be unstable. For the current example $\Delta T_{cr} = 14.93665\,°C$ at the bifurcation point C. If $\Delta T < \Delta T_{cr}$ such as point E, the equilibrium is stable. If we apply a small lateral force then remove it, the plate returns to the equilibrium position of zero deflection. However, for $\Delta T > \Delta T_{cr}$ such as point U, the equilibrium is unstable. A small lateral force will take the plate to a new equilibrium position at either point A or B for which the plate is buckled. Figure 10.18 shows deflection contours $w_o(x,y)/h$ for temperature rise of $\Delta T = 30°C$, and with maximum deflection $1.23799h$ at the center.

Example 10.9. Critical Uniform Temperature of a Square Plate

We consider a square plate (side a, thickness h, Poisson's ratio ν) subjected to uniform temperature rise ΔT. The plate edges are restrained. First, the stresses in the plate due to a unit temperature rise $\Delta T = 1$ are obtained by solving a plane stress

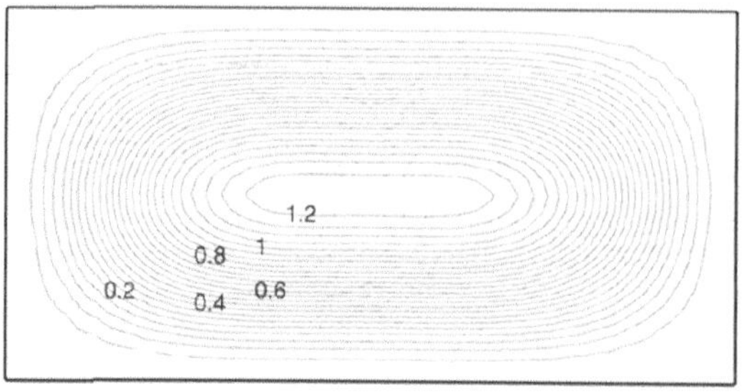

Figure 10.18 Contours of w_o/h, thermal post-buckling of a clamped rectangular plate at $\Delta T = 30\,^\circ\text{C}$. Central deflection $(w_o/h)_{max} = 1.23799$, $q = 0$.

problem with boundary conditions

$$\begin{aligned} x = 0: &\quad u = 0, \;\; N_{xy} = 0 \\ x = a: &\quad u = 0, \;\; N_{xy} = 0 \\ y = 0: &\quad v = 0, \;\; N_{yx} = 0 \\ y = a: &\quad v = 0, \;\; N_{yx} = 0. \end{aligned} \tag{10.146}$$

Second, the obtained stresses are used to compute the geometric stiffness matrix $\left[K_G^e\right]$ as given by Equation (10.143). Equation (10.144) for thermoelastic equilibrium of Reissner–Mindlin plate element is now written as

$$([K^e] + \lambda\, [K_G^e])\, \{U^e\} = \{R^e\} \tag{10.147}$$

where $\lambda \equiv \Delta T$ is unknown. At the critical condition, the plate deflection w_o is infinitesimally small, and hence, the linear stiffness matrix is used. The element equations are assembled into a global system and homogeneous boundary conditions are applied. The boundary conditions are given by Equation (10.131) for a simply supported plate, and by Equation (10.145) for a clamped plate. The linear eigenvalue problem is solved, and the lowest λ gives the critical temperature rise ΔT_{cr}. Because there is no coupling between the membrane displacements (u_o, v_o) and the transverse deflection w_o, the eigenvalue problem can be formulated in terms of flexural displacements $(w_o, \theta_x, \theta_y)$ only; the boundary conditions on u_o and v_o do not play a role when solver the linear eigenvalue problem. However, for post-buckling analysis, the coupled nonlinear system must be solved, and boundary conditions on u_o and v_o are important.

We define a nondimensional critical temperature rise by

$$\begin{aligned} \Delta T_{cr}^* &= \frac{E\alpha\Delta T_{cr} h a^2}{\pi^2 D} \\ \alpha\Delta T_{cr} &= \frac{\pi^2}{12(1-\nu^2)} \left(\frac{h}{a}\right)^2 \Delta T_{cr}^*. \end{aligned} \tag{10.148}$$

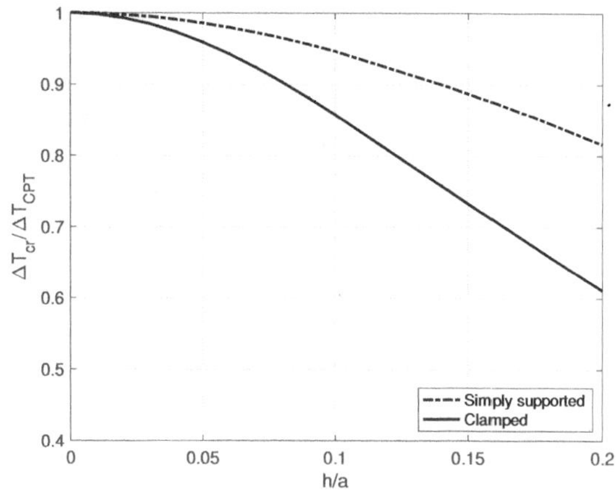

Figure 10.19 Thermal buckling of a square plate. Effects of thickness ratio (h/a) on the ratio of critical temperature to that of CPT. $\nu = 0.3$.

Taking $\nu = 0.3$, thin plate theory (CPT) predicts $\Delta T^*_{cr} = 1.4$ for simply supported square plate, and 3.7126 for a clamped plate. Being proportional to h^2, the critical temperature increases with plate thickness. The present Reissner–Mindlin finite-element model is used to determine the effects of plate thickness on the critical temperature rise. As shown in Figure 10.19, the ratio of critical temperature rise to that predicted by CPT decreases as the plate thickness increases. The thin plate theory predicts higher temperature rise for thermal buckling of thick plates than that of the Reissner–Mindlin theory; especially for the clamped plate.

Example 10.10. Thermal Buckling of a Simply Supported Rectangular Plate

We consider a rectangular plate (sides a, b, thickness h, Poisson's ratio ν) subjected to non-uniform temperature rise ΔT.

$$\Delta T(x,y) \quad = \quad -T_1 cos\left(\frac{2\pi y}{b}\right), \quad 0 \leq y \leq b. \tag{10.149}$$

Boley and Weiner [13] determined exact analytic solution for T_1 at onset of thermal buckling using thin plate theory CPT. They assumed that the plate edges remain straight.

$$\begin{aligned} u(0,y) = 0, \quad u(a,y) = u_1 \\ v(x,0) = 0, \quad v(x,b) = v_1 \end{aligned} \tag{10.150}$$

where u_1 and v_1 are constants. For the edges ($x = 0$, and $x = a$) to remain straight, axial distributed loads must be applied to the edges, however, the resultant axial force

Table 10.14
Critical Temperature Factor k_1 for Simply Supported Rectangular Thin Plate, $\nu = 0.3$, $h/a = 0.0001$

a/b	0.25	0.50	0.75	1.00	1.25	1.50
Reference [13]	13.3168	5.5250	4.0848	3.8516	4.0874	4.0848
Finite element	13.3160	5.5248	4.0845	3.8514	4.0872	4.0846
a/b	1.75	2.00	2.25	2.50	2.75	3.00
Reference [13]	3.8860	3.8516	3.9285	3.9303	3.8596	3.8516
Finite element	3.8858	3.8514	3.9283	3.9301	3.8594	3.8514

is zero

$$\int_0^b N_{xx}(0,y)dy = \int_0^b N_{xx}(a,y)dy = 0. \tag{10.151}$$

The analytic solution gives $u_1 = v_1 = 0$. Boley and Weiner [13] expressed the critical temperature T_{1cr} as

$$\frac{E\alpha T_{1cr}}{2} = k_1 \frac{\pi^2 E}{12(1-\nu^2)} \left(\frac{h}{b}\right)^2 \tag{10.152}$$

and provided a procedure for computing the factor k_1.

We use a uniform mesh of $8 \times 8 = 64$ bicubic elements. The stresses in the plate due to temperature rise Equation (10.149) with $T_1 = 1$ are obtained by solving a plane stress problem with boundary conditions

$$\begin{aligned} x = 0: &\quad u = 0, \quad N_{xy} = 0 \\ x = a: &\quad u = 0, \quad N_{xy} = 0 \\ y = 0: &\quad N_{yy} = 0, \quad N_{xy} = 0 \\ y = b: &\quad N_{yy} = 0, \quad N_{xy} = 0 \end{aligned} \tag{10.153}$$

and $\nu = 0$ at the corner $(x,y) = (0,0)$ to prevent rigid body translation. The finite-element solution also predicted that $v = 0$ on the edges ($y = 0$ and $y = b$). The geometric stiffness matrix is computed using the obtained stresses, and the critical temperature is determined by solving the eigenvalue problem Equation (10.147) for a simply supported plate. The factor k_1 obtained by the finite-element method is listed in Table 10.14 for selected values of the aspect ratio a/b, and it agrees very well with the analytic solution [13].

Example 10.11. Thermal Post-Buckling of a Clamped Skew Plate

We consider a 45°-skew plate in the shape of a rhombus of side a, thickness ratio $h/a = 0.01$, and Poisson's ratio $\nu = 0.3$. All edges are clamped, ($u_o = v_o = w_o =$

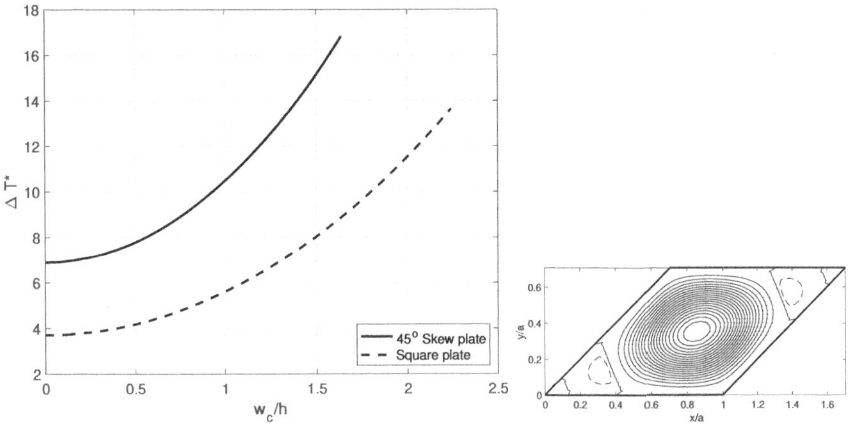

(a) (b)

Figure 10.20 Thermal buckling of a clamped plate, $h/a = 0.01$, $\nu = 0.3$. (a) Central deflection of buckled plate. (b) Deflection of 45°-skew plate, contours interval $= 0.05\ w_{max}/h$, dashed contour $= -0.006\ w_{max}/h$.

$\theta_x = \theta_y = 0$) on the boundary. We wish to determine the post-buckling equilibrium deflected shape due to uniform temperature rise ΔT. Using a perturbation method, Prabhu and Durvasula [87] solved this problem for von Karman plate. Prakash et al. [88] also solved this problem for Reissner–Mindlin nonlinear plate using a finite-element method. We use a mesh of $10 \times 10 = 100$ quadrilateral bicubic skew elements. The nondimensional temperature rise is defined by $\Delta T^* = E\alpha\Delta T h a^2/\pi^2 D$. An eigenvalue solver is used to determine the critical temperature rise ΔT^*_{cr} as well as $\Delta T^*(> \Delta T^*_{cr})$ for prescribed small central deflection. The first-mode eigenfunction is used to initialize Newton–Raphson method for larger prescribed ΔT^*. The critical temperature rise is $\Delta T^*_{cr} = 6.89597$ for the rhombus as compared to $\Delta T^*_{cr} = 3.70609$ for a square of equal side but of larger area. The equilibrium post-buckling central deflection is shown in Figure 10.20a as a function of ΔT^* for both the rhombus and square plate of side a. Contours of w_o/h are shown in Figure 10.20b at $\Delta T^* = 16.8184$, where the maximum deflection is $w_{max}/h = 1.63686$.

10.8 NONLINEAR VIBRATIONS: ELEMENT MATRICES

The equations of equilibrium for finite displacements are given by Equation (10.6), which states that sum of the forces per unit volume of the undeformed body is zero. The equations of motion can be obtained by equating the sum of forces to mass times acceleration. In index notation, the equations of motion are

$$\frac{\partial}{\partial x_k}\left[\left(\delta_{ij} + \frac{\partial u_i}{\partial x_j}\right)\sigma_{kj}\right] + f_i = \rho\ddot{u}_i \qquad (i = 1,2,3) \qquad (10.154)$$

where ρ is mass per unit volume of the undeformed body. We derive the principle of virtual displacements for dynamical problems (or the weak form of Equations (10.154)) by following the analysis leading to Equation (10.17),

$$\int_{\Omega} \rho \bar{u}_i \ddot{u}_i d\Omega + \int_{\Omega} \bar{\varepsilon}_{jk} \sigma_{kj} d\Omega = \oint_{\partial\Omega} \bar{u}_i t_i dS + \int_{\Omega} \bar{u}_i f_i d\Omega. \tag{10.155}$$

Element Equations and Matrices

The procedure for deriving the element equations and matrices for the static case can be followed to derive relevant equations for the dynamical case. For Reissner–Mindlin plate theory, the term $\bar{u}_i\ddot{u}_i$ can be written in terms of the plate deflection and rotations

$$\begin{aligned} \bar{u}_i\ddot{u}_i &= \bar{u}\ddot{u} + \bar{v}\ddot{v} + \bar{w}\ddot{w} \\ &= \left(\bar{u}_o - z\bar{\theta}_x\right)\left(\ddot{u}_o - z\ddot{\theta}_x\right) + \left(\bar{v}_o - z\bar{\theta}_y\right)\left(\ddot{v}_o - z\ddot{\theta}_y\right) + \bar{w}_o\ddot{w}_o \\ &= \bar{u}_o\ddot{u}_o + \bar{v}_o\ddot{v}_o + \bar{w}_o\ddot{w}_o + z^2\left(\bar{\theta}_x\ddot{\theta}_x + \bar{\theta}_y\ddot{\theta}_y\right) \\ &\quad -z\left(\bar{\theta}_x\ddot{u}_o + \bar{\theta}_y\ddot{v}_o + \bar{u}_o\ddot{\theta}_x + \bar{v}_o\ddot{\theta}_y\right). \end{aligned} \tag{10.156}$$

For an element, the inertia term becomes

$$\begin{aligned} \int_{\Omega^e} \rho \bar{u}_i\ddot{u}_i d\Omega &= \int_{A^e}\int_{-h/2}^{h/2} \rho\left[\bar{u}_o\ddot{u}_o + \bar{v}_o\ddot{v}_o + \bar{w}_o\ddot{w}_o + z^2\left(\bar{\theta}_x\ddot{\theta}_x + \bar{\theta}_y\ddot{\theta}_y\right)\right] dzdA \\ &\quad - \int_{A^e}\int_{-h/2}^{h/2} \rho z\left(\bar{\theta}_x\ddot{u}_o + \bar{\theta}_y\ddot{v}_o + \bar{u}_o\ddot{\theta}_x + \bar{v}_o\ddot{\theta}_y\right) dA \\ &= \int_{A^e} \rho\left[h\left(\bar{u}_o\ddot{u}_o + \bar{v}_o\ddot{v}_o + \bar{w}_o\ddot{w}_o\right) + \frac{h^3}{12}\left(\bar{\theta}_x\ddot{\theta}_x + \bar{\theta}_y\ddot{\theta}_y\right)\right] dA \\ &= \int_{A^e} \rho \begin{Bmatrix} \bar{u}_o \\ \bar{v}_o \\ \bar{w}_o \\ \bar{\theta}_x \\ \bar{\theta}_y \end{Bmatrix}^T \begin{bmatrix} h & 0 & 0 & 0 & 0 \\ 0 & h & 0 & 0 & 0 \\ 0 & 0 & h & 0 & 0 \\ 0 & 0 & 0 & \frac{h^3}{12} & 0 \\ 0 & 0 & 0 & 0 & \frac{h^3}{12} \end{bmatrix} \begin{Bmatrix} \ddot{u}_o \\ \ddot{v}_o \\ \ddot{w}_o \\ \ddot{\theta}_x \\ \ddot{\theta}_y \end{Bmatrix} dA. \end{aligned} \tag{10.157}$$

Substituting the trial solutions for u_o, v_o, w_o, θ_x, and θ_y, and invoking the Galerkin method for selecting the virtual displacements, we get

$$\begin{aligned} \int_{\Omega^e} \rho \bar{u}_i\ddot{u}_i d\Omega &= \{d\}^T \left[\int_{A^e} \rho [\psi]^T \begin{bmatrix} h & 0 & 0 & 0 & 0 \\ 0 & h & 0 & 0 & 0 \\ 0 & 0 & h & 0 & 0 \\ 0 & 0 & 0 & h^3/12 & 0 \\ 0 & 0 & 0 & 0 & h^3/12 \end{bmatrix} [\psi] dA\right] \{\ddot{U}^e\} \\ &= \{d\}^T [M^e] \{\ddot{U}^e\}. \end{aligned} \tag{10.158}$$

For Reissner–Mindlin plate, the consistent mass matrix is

$$[M^e] = \int_{A_e} \rho \, [\psi]^T \begin{bmatrix} h & 0 & 0 & 0 & 0 \\ 0 & h & 0 & 0 & 0 \\ 0 & 0 & h & 0 & 0 \\ 0 & 0 & 0 & h^3/12 & 0 \\ 0 & 0 & 0 & 0 & h^3/12 \end{bmatrix} [\psi] \, dA. \tag{10.159}$$

The element equation for nonlinear dynamical problems takes the usual form

$$[M^e]\{\ddot{U}^e\} + [K^e]\{U^e\} = \{F^e\} + \{R^e\}. \tag{10.160}$$

The stiffness matrix is given by Equation (10.75) or (10.76). The forcing vector $\{F^e\}$ and array of weighted secondary variables $\{R^e\}$ are defined by Equations (10.82) and (10.83), respectively, and they are functions of time. The element equations are assembled into a global system

$$[M]\{\ddot{U}\} + [K]\{U\} = \{F\} + \{R\}. \tag{10.161}$$

Given boundary and initial conditions for displacements and velocity, this system can be integrated in time using Newmark method as described in Section (6.3). The stiffness matrix $[K]$ depends on the unknown deflection w_o. At the beginning of a new time step, the stiffness matrix is evaluated using the converged solution of the previous step. The stiffness matrix is updated in a sub-iteration until convergence before the solution can be advanced to the next time step.

In-depth coverage of nonlinear response of plates is beyond the scope of this book, the reader is referred to Chia [23], Amabili [5]– [4], and Sathyamoorthy [107] for a comprehensive literature review. Here, we present simple examples to demonstrate the effects of oscillations amplitude on the natural frequency of geometrically nonlinear plates. In general, free or forced vibrations of nonlinear structures can exhibits non-harmonic responses. In the next subsection, we consider a simple one-degree-of-freedom oscillator to highlight this point and introduce certain definitions.

10.8.1 THE DUFFING EQUATION

The Duffing equation is a single-degree-of-freedom oscillator with a cubic nonlinear spring. The unforced equation without damping in scaled form is

$$\ddot{w} + w + \alpha_3 w^3 = 0. \tag{10.162}$$

We want to solve this equation for $w(t)$ with the initial conditions

$$t = 0: \; w = A, \; \text{ and } \; \dot{w} = 0. \tag{10.163}$$

The exact solution to Equation (10.162) as given by Rand [94] is

$$\begin{aligned} w &= a_1 \, cn(u,k), & u &= a_2 t + b, \\ a_2^2 &= 1 + \alpha_3 a_1^2, & k^2 &= \frac{a_1^2 \alpha_3}{2\left(1 + \alpha_3 a_1^2\right)} \end{aligned} \tag{10.164}$$

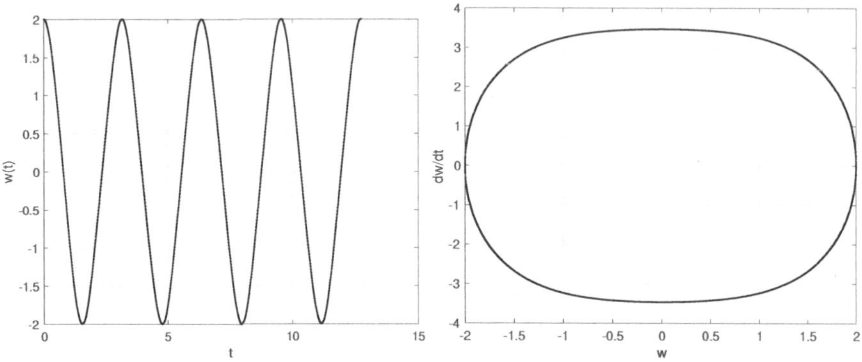

(a) (b)

Figure 10.21 Duffing equation, $\alpha_3 = 1$; exact solution for amplitude $A = 2$. (a) Time response. (b) Phase portrait.

where cn is the Jacobi elliptic cosine function, and a_1 and b are arbitrary constants. For the desired initial conditions Equation (10.163), we have $a_1 = A$, and $b = 0$. The solution is a periodic function of time t with period T_e and frequency $\omega_e = 2\pi/T_e$,

$$T_e = \frac{4K(k)}{a_2}, \qquad \omega_e = \frac{2\pi}{T_e} \tag{10.165}$$

where $K(k)$ is the complete elliptic integral of the first kind. As an example, we take $\alpha_3 = 1$, and $A = 2$. The time response $w - t$ is shown in Figure 10.21a, and the corresponding phase portrait $\dot{w} - w$ is shown in Figure 10.21b. The periodic solution is not a simple harmonic motion, it can be represented by a Fourier cosine series of odd harmonics,

$$w(t) = A_1 cos\omega_e t + A_3 cos3\omega_e t + A_5 cos5\omega_e t + \cdots \tag{10.166}$$

where the coefficients A_1, A_3, and A_5 etc. are real. Using FFT, the magnitude of the complex coefficients (half of the real coefficients) are depicted in Figure 10.22a. The frequency ω_e is a function of the vibration amplitude A. For very small amplitude, we neglect the cubic term $\alpha_3 w^3$ in Equation (10.162), and obtain the linear frequency $\omega_L = 1$. For finite amplitude A, the frequency is given by Equation (10.165). The plot of A versus the ratio ω_e/ω_L is known as the backbone curve and is shown in Figure 10.22b. The increase of frequency with vibration amplitude is due to the hardening effects of the spring or stiffness of the system. It is instructive to compare the exact solution to solutions by direct time integration of the equation of motion or approximate methods such as the method of harmonic balance, which will be used in this chapter with the finite-element method. Using Newmark scheme (see Section 6.3) with parameters $\alpha = 0.5$, and $\gamma = 0.5$, we integrate Equation (10.162) in time using a constant time step. Four sub-iterations are used to update the nonlinear term

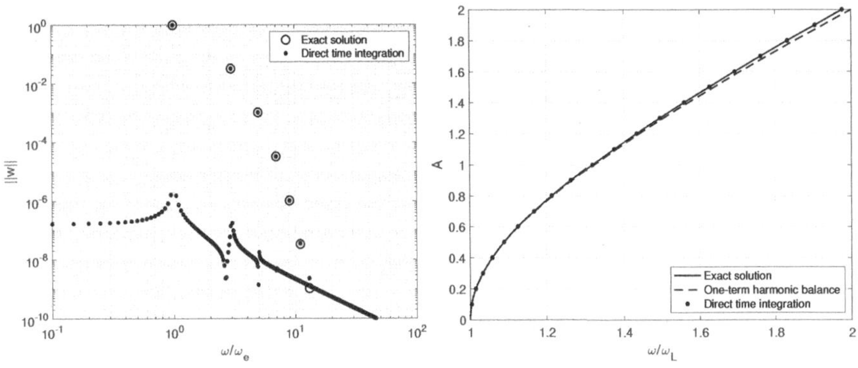

(a) (b)

Figure 10.22 Duffing equation, $\alpha_3 = 1$. (a) Fourier coefficients spectrum for amplitude $A = 2$; $\omega_e = 1.976016$. (b) Backbone curves; $\omega_L = 1$.

$\alpha_3 w^3$ at each time step. Computations are conducted for twenty periods $T_f = 20T_e$, and $\Delta t = T_e/2048$. For the example considered ($\alpha_3 = 1$, $A = 2$), the time response $w-t$ and phase portrait $\dot{w}-w$ are indistinguishable from the exact solution shown in Figures 10.21a and 10.21b, respectively. The Fourier coefficients spectrum is compared with the exact solution in Figure 10.22a. Excellent agreement is evident at the discrete frequencies $\omega/\omega_e = 1, 3, 5, 7$, and 9. The spectrum of direct numerical integration also includes a continuous spectrum but with very small amplitudes, which can be further reduced by using a smaller time step. To obtain the backbone curve, we repeat the time integration for different amplitudes A. The time instants t_j at which w is zero are used to determine the period of the direct time integration method, and hence the frequency. The backbone curve by direct time integration is in excellent agreement with the exact solution as shown in Figure 10.22b.

The Method of Harmonic Balance

Harmonic balance is an approximate method for solving nonlinear dynamical systems, (e.g., Nayfeh and Mook [78]). The periodic solution of Equation (10.162) is written as a sum of harmonic functions

$$w(t) = \sum_{m=0}^{M} A_m cos(m\omega t + m\beta_o). \tag{10.167}$$

We need $M+3$ equations for the unknown $M+1$ coefficients A_m, β_o, and frequency ω. Two equations are given by the initial conditions, and $M+1$ algebraic equations are obtained by substituting Equation (10.167) into Equation (10.162), and equating the coefficients of $cos(m\omega t)$ for $(m = 0, 1, 2, \cdots, M)$ to zero. Here, we propose a one-term solution

$$w(t) = A cos(\omega t + \beta_o) \equiv A cos\phi. \tag{10.168}$$

Substituting into Equation (10.162), and using the identity $cos^3\phi = (3cos\phi + cos3\phi)/4$, we get

$$(-\omega^2 + 1)Acos\phi + \frac{1}{4}\alpha_3 A^3(3cos\phi + cos3\phi) = 0. \tag{10.169}$$

Equating the coefficient of $cos\phi$ to zero, we get

$$\omega^2 = 1 + \frac{3}{4}\alpha_3 A^2; \qquad \omega \approx 1 + \frac{3}{8}\alpha_3 A^2 \tag{10.170}$$

ω is the nonlinear frequency according to the one-term harmonic balance solution. The backbone curve is compared to the exact and direct time integration solutions in Figure 10.22b.

10.8.2 FREE VIBRATION OF PLATES: HARMONIC BALANCE

We use the method of harmonic balance to study the effects of vibration amplitude on the natural frequency of geometrically nonlinear plates. To that end, we rewrite the element stiffness matrix Equation (10.76) as

$$\begin{aligned}[K^e] = \int_{A_e} [B_o]^T [D_o] [B_o] dA + \int_{A_e} [B_o]^T [D_o] \frac{1}{2}[G^{NL}] dA \\ + \int_{A_e} [B_b]^T [D_b] [B_b] dA + \int_{A_e} [B_s]^T [D_b^s] [B_s] dA \\ + \int_{A_e} [G^{NL}]^T [D_o] [B_o] dA + \int_{A_e} [G^{NL}]^T [D_o] \frac{1}{2}[G^{NL}] dA.\end{aligned} \tag{10.171}$$

The equations of motion of the membrane displacements (u_o, v_o) depend only on the first and second integrals. The first integral is a constant matrix and gives non-zero coefficients for u_o, and v_o only. The second integral produces a matrix with non-zero coefficients for the deflection w_o only. Because $[G^{NL}]$ is linear in the deflection w_o, there are quadratic terms in w_o in the equations of motions of u_o and v_o. The third and fourth integrals are constant matrices with non-zero coefficients only for the flexural degrees of freedom w_o, θ_x and θ_y. The fifth integral produces non-zero coefficients only for u_o, and v_o while the sixth integral produces non-zero coefficients only for w_o, both matrices contribute only to the equations of motion of w_o. Because $[G^{NL}]$ is linear in the deflection w_o, the sixth matrix is quadratic in w_o, hence the equation of motion for w_o contains cubic terms in w_o. Neglecting the inertia terms in the equations of motion for u_o, and v_o, we can write these displacements in terms of w_o. By eliminating u_o, and v_o from the equation of motion of w_o, we realize that the fifth integral also produces cubic terms in w_o.

We write two sets of equations of motion, the first set is for the middle plane displacements u_o and v_o, and the second set is for the flexural displacements w_o,

θ_x, and θ_y. For an element with n local nodes, we define the array of inplane nodal displacements,

$$\{U_p^e\} = \left\{ u_{o1}^e, v_{o1}^e, ..., u_{oj}^e, v_{oj}^e, ..., u_{on}^e, v_{on}^e \right\}^T \tag{10.172}$$

and the array of flexural nodal displacements,

$$\{U_b^e\} = \left\{ w_{o1}^e, \theta_{x1}^e, \theta_{y1}^e, ..., w_{oj}^e, \theta_{xj}^e, \theta_{yj}^e, ..., w_{on}^e, \theta_{xn}^e, \theta_{yn}^e \right\}^T. \tag{10.173}$$

The equations of motion for the inplane displacements are

$$\left[M_p^e\right]\{\ddot{U}_p^e\} + \left[K_p^e\right]\{U_p^e\} + \left[K_{pb}^e\right]\{U_b^e\} = \{F_p^e\} + \{R_p^e\}. \tag{10.174}$$

The equations of motion for the flexural displacements are

$$[M_b^e]\{\ddot{U}_b^e\} + [K_b^e]\{U_b^e\} + [K_{bb}^e]\{U_b^e\} + \left[K_{bp}^e\right]\{U_p^e\} = \{F_b^e\} + \{R_b^e\}. \tag{10.175}$$

To write expressions for the different coefficient matrices, we need to define new matrices with proper dimensions.

$$[\psi_p] = \left[[\psi]_1 \, [\psi]_2 \; \cdots \; [\psi]_j \; \cdots \; [\psi]_n \right], [\psi]_j = \begin{bmatrix} \psi_j & 0 \\ 0 & \psi_j \end{bmatrix} \tag{10.176}$$

$$[\psi_b] = \left[[\psi]_1 \, [\psi]_2 \; \cdots \; [\psi]_j \; \cdots \; [\psi]_n \right], [\psi]_j = \begin{bmatrix} \psi_j & 0 & 0 \\ 0 & \psi_j & 0 \\ 0 & 0 & \psi_j \end{bmatrix} \tag{10.177}$$

$$[B_o] = \left[[B_o]_1 \; \cdots \; [B_o]_j \; \cdots \; [B_o]_n \right], [B_o]_j = \begin{bmatrix} \frac{\partial \psi_j}{\partial x} & 0 \\ 0 & \frac{\partial \psi_j}{\partial y} \\ \frac{\partial \psi_j}{\partial y} & \frac{\partial \psi_j}{\partial x} \end{bmatrix} \tag{10.178}$$

$$[B_b] = \left[[B_b]_1 \; \cdots \; [B_b]_j \; \cdots \; [B_b]_n \right], [B_b]_j = \begin{bmatrix} 0 & \frac{\partial \psi_j}{\partial x} & 0 \\ 0 & 0 & \frac{\partial \psi_j}{\partial y} \\ 0 & \frac{\partial \psi_j}{\partial y} & \frac{\partial \psi_j}{\partial x} \end{bmatrix} \tag{10.179}$$

$$\left[B^{NL}\right] = \left[\begin{matrix} \{B_x^{NL}\} \\ \{B_y^{NL}\} \end{matrix} \right], \{B_x^{NL}\}_j = \left\{ \tfrac{\partial \psi_j}{\partial x} \, 0 \, 0 \right\}, \{B_y^{NL}\}_j = \left\{ \tfrac{\partial \psi_j}{\partial y} \, 0 \, 0 \right\} \tag{10.180}$$

$$[A_o] = \begin{bmatrix} \{B_x^{NL}\}\{U_b^e\} & 0 \\ 0 & \{B_y^{NL}\}\{U_b^e\} \\ \{B_y^{NL}\}\{U_b^e\} & \{B_x^{NL}\}\{U_b^e\} \end{bmatrix}, \left[G^{NL}\right] = [A_o]\left[B^{NL}\right] \tag{10.181}$$

$$[B_s] = \left[[B_s]_1 \, [B_s]_2 \ldots [B_s]_j \ldots [B_s]_n \right], [B_s]_j = \begin{bmatrix} \frac{\partial \psi_j}{\partial x} & -\psi_j & 0 \\ \frac{\partial \psi_j}{\partial y} & 0 & -\psi_j \end{bmatrix}. \tag{10.182}$$

Note that $[B_o]$, $[B_b]$, $[B^{NL}]$, $[G^{NL}]$, and $[B_s]$ are of dimensions $(3,2n)$, $(3,3n)$, $(2,3n)$, $(3,3n)$, and $(3,3n)$, respectively.
The consistent mass matrices are

$$\begin{aligned}
[M_p^e] &= \int_{A_e} \rho [\psi_p]^T \begin{bmatrix} h & 0 \\ 0 & h \end{bmatrix} [\psi_p]\, dA, \\
[M_b^e] &= \int_{A_e} \rho [\psi_b]^T \begin{bmatrix} h & 0 & 0 \\ 0 & h^3/12 & 0 \\ 0 & 0 & h^3/12 \end{bmatrix} [\psi_b]\, dA.
\end{aligned} \tag{10.183}$$

The linear stiffness matrices are

$$\begin{aligned}
[K_p^e] &= \int_{A_e} [B_o]^T [D_o] [B_o] dA, \\
[K_b^e] &= \int_{A_e} [B_b]^T [D_b] [B_b] dA + \int_{A_e} [B_s]^T [D_b^s] [B_s] dA
\end{aligned} \tag{10.184}$$

and geometric nonlinear matrices are

$$\begin{aligned}
[K_{bp}^e] &= \int_{A_e} [B_o]^T [D_o] \frac{1}{2} [G^{NL}] dA \\
[K_{bp}^e] &= \int_{A_e} [G^{NL}]^T [D_o] [B_o] dA \\
[K_{bb}^e] &= \int_{A_e} [G^{NL}]^T [D_o] \frac{1}{2} [G^{NL}] dA.
\end{aligned} \tag{10.185}$$

Because $[D_o]$ is symmetric, we have $\left[K_{bp}^e\right] = 2\left[K_{pb}^e\right]^T$.

The element equations are assembled into global systems and boundary conditions are applied for both the membrane and flexural displacements. For free vibrations, we obtain homogeneous systems.

$$[M_p]\{\ddot{U}_p\} + [K_p]\{U_p\} + [K_{pb}]\{U_b\} = 0 \tag{10.186}$$

$$[M_b]\{\ddot{U}_b\} + [K_b]\{U_b\} + [K_{bb}]\{U_b\} + [K_{bp}]\{U_p\} = 0. \tag{10.187}$$

In developing a one-term harmonic balance for free vibration of plates, we adopt the method presented by Ribeiro [102]. Neglecting the inertia of membrane motion, we obtain from Equation (10.186),

$$\{U_p\} = -[K_p]^{-1}[K_{pb}]\{U_b\} \tag{10.188}$$

and substituting into Equation (10.187), we obtain

$$[M_b]\{\ddot{U}_b\} + \left([K_b] + [K^{NL}]\right)\{U_b\} = 0 \tag{10.189}$$

where

$$\left[K^{NL}\right] = [K_{bb}] - \left[K_{bp}\right][K_p]^{-1}\left[K_{pb}\right]. \tag{10.190}$$

This matrix depends on w_o only, and the term $\left[K^{NL}\right]\{U_b\}$ produces cubic terms in w_o in the equation of w_o only. A one-term harmonic solution is assumed,

$$\{U_b\} = \left\{\hat{U}_b\right\} cos\phi,\ \phi = \omega t + \beta_o \tag{10.191}$$

where $\left\{\hat{U}_b\right\}$ is the vector of amplitudes of the nodal displacements. Substituting the solution Equation (10.191) into Equation (10.189), and equating the coefficient of $cos\phi$ to zero, we get

$$\left[-\omega^2[M_b] + [K_b] + \frac{3}{4}\left[K^{NL}(\hat{w}_o)\right]\right]\left\{\hat{U}_b\right\} = 0. \tag{10.192}$$

Nonlinear frequency of the first mode are obtained by solving this eigenvalue problem for prescribed deflection amplitude $\hat{w}_o$ at a reference point. An iteration is used in which the mode shape of the first mode is scaled such that $\hat{w}_o$ attains the prescribed value. The matrix $\left[K^{NL}(\hat{w}_o)\right]$ is updated after each iteration. The iteration is terminated if the condition $|\delta\omega|/\omega < 10^{-7}$ is satisfied.
Raju and Hinton [93] have presented periods of nonlinear vibrations of Reissner–Mindlin plates of different shapes and boundary conditions. They neglected the membrane displacements (u_o, v_o), and linearized the inplane strains as

$$\begin{aligned}
\varepsilon_{xx}^{NL} &= \frac{1}{2}\left(\frac{\partial w_o}{\partial x}\right)^2 = f_x\frac{\partial w_o}{\partial x} \\
\varepsilon_{yy}^{NL} &= \frac{1}{2}\left(\frac{\partial w_o}{\partial y}\right)^2 = f_y\frac{\partial w_o}{\partial y} \\
\gamma_{xy}^{NL} &= \frac{\partial w_o}{\partial x}\frac{\partial w_o}{\partial y} = f_x\frac{\partial w_o}{\partial y} + f_y\frac{\partial w_o}{\partial x}
\end{aligned} \tag{10.193}$$

where

$$f_x = \frac{1}{2}\frac{\partial w_o}{\partial x}, \quad \text{and} \quad f_y = \frac{1}{2}\frac{\partial w_o}{\partial y} \tag{10.194}$$

that are updated during the iterative scheme used for solving the eigenvalue problem. The results reported in [93] can be obtained by solving the eigenvalue problem

$$\left[-\omega^2[M_b] + [K_b] + \frac{1}{2}\left[K_{bb}(\hat{w}_o)\right]\right]\left\{\hat{U}_b\right\} = 0. \tag{10.195}$$

The linearization given by Equation (10.193) is not used in this book, and all numerical results presented in this chapter are based on the nonlinear system Equation (10.192) and the matrix given by Equation (10.190).

Table 10.15
First Mode Nondimensional Linear Frequency $\omega_o^* = \omega_o a^2 \sqrt{\rho h/D}$ for a Square Plate. $\nu = 0.3$, $\kappa = 5/6$

h/a	Simply supported	Clamped
0.001	19.7391	35.9848
0.1	19.0650	32.5243
0.2	17.4486	26.5246

Example 10.12. Backbone Curve for a Square Plate

We consider a square plate of side a and thickness h, and use $\nu = 0.3$ and shear correction factor $\kappa = 5/6$. A uniform mesh of $6 \times 6 = 36$ bicubic elements with 361 global nodes is used. Two sets of boundary conditions are applied.
Case 1: Simply supported plate with immovable edges, the boundary conditions are:

$$\begin{aligned} x = 0 \text{ and } x = a: \ & u_o = 0,\ N_{xy} = 0,\ w_o = 0,\ \theta_y = 0,\ M_{xx} = 0 \\ y = 0 \text{ and } y = a: \ & v_o = 0,\ N_{yx} = 0,\ w_o = 0,\ \theta_x = 0,\ M_{yy} = 0. \end{aligned} \tag{10.196}$$

Case 2: Fully clamped plate, the boundary conditions are:

$$\begin{aligned} x = 0 \text{ and } x = a: \ & u_o = 0,\ v_o = 0,\ w_o = 0,\ \theta_x = 0,\ \theta_y = 0 \\ y = 0 \text{ and } y = a: \ & u_o = 0,\ v_o = 0,\ w_o = 0,\ \theta_x = 0,\ \theta_y = 0. \end{aligned} \tag{10.197}$$

For very small vibration amplitude, the linear frequency of the first mode (ω_o) is given in Table 10.15 for three values of h/a. For large amplitude vibration, the effects of amplitude on the natural frequency is represented by the backbone curve, which shows the variation of the natural frequency (ω) with the maximum deflection w_c. The maximum deflection at the plate center $w_c(0.5a, 0.5a)/h$ is prescribed, and the frequencies are determined by solving the nonlinear eigenvalue problem. The ratio ω/ω_o of the first mode is plotted against w_c/h in Figure 10.23a for the simply supported plate. The results for $h/a = 0.001$ are in excellent agreement with Chu and Herrmann perturbation solution for thin plates [24]. Figure 10.23b shows the backbone curve for the clamped plate. For $h/a = 0.1$, the results are in very good agreement with Ribeiro [102] hierarchical finite-element method for Reissner–Mindlin plate. For both sets of boundary conditions, spring hardening is indicated by the increase of natural frequency with amplitude.

Example 10.13. Backbone Curve for a Circular Plate

We consider a circular plate of radius R and thickness ratio $h/R = 0.1$, and use $\nu = 0.3$ and shear correction factor $\kappa = 5/6$. The full circle is modeled by 48 bicubic

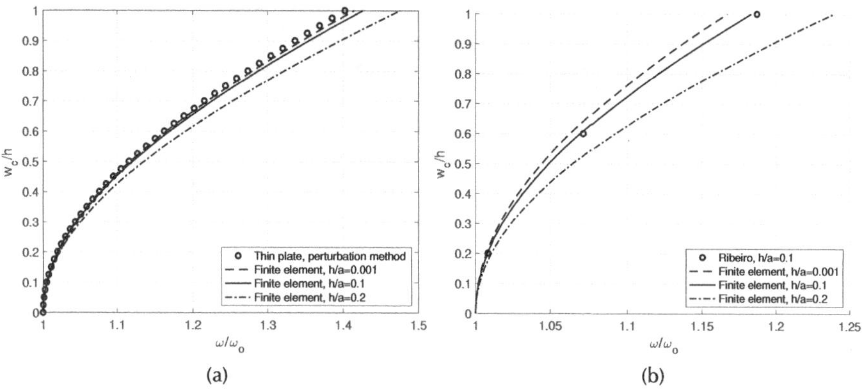

Figure 10.23 Backbone curves for a square plate, $\nu = 0.3$, $\kappa = 5/6$. (a) Simply supported plate with immovable edges. Comparison with perturbation method for thin plates [24]. (b) Clamped plate. Comparison with Ribeiro's finite-element solution at $h/a = 0.1$ [102].

elements (16 nodes per element and 457 global nodes) similar to but coarser than that shown in Figure 10.2. Two sets of boundary conditions are applied.
Case 1: Simply supported plate with immovable edges, the boundary conditions are:

$$r = R: \; u_o = 0, \; v_o = 0, \; w_o = 0, \; \theta_s = 0, \; M_{rr} = 0 \tag{10.198}$$

where θ_s is filament rotation around a radial line.
Case 2: Fully clamped plate, the boundary conditions are:

$$r = R: \; u_o = 0, \; v_o = 0, \; w_o = 0, \; \theta_x = 0, \; \theta_y = 0. \tag{10.199}$$

The linear frequency of the first mode $\omega_o^* = \omega_o R^2 \sqrt{\rho h/D}$ is 4.89417 for the simply supported plate, and 9.9438 for the clamped plate. The backbone curves are shown in Figure 10.24. Hardening spring effects are stronger for the simply supported plate than the clamped one.

Direct Time Integration

Direct time integration of the equations of motion requires initial values for the in-plane and flexural displacements (five geometric components). We choose the eigenvectors of the first mode (lowest frequency) of the eigenvalue problem:

$$\left(-\omega_a^2 [M] + [K]\right) \{U\}_o = 0 \tag{10.200}$$

to prescribe initial displacements, where the mass matrix and nonlinear stiffness matrix are the same matrices used with direct time integration. The stiffness matrix depends on the unknown deflection w_o, hence an iteration is used in which the eigenvectors are scaled after each iteration such that w_o at the circle center takes

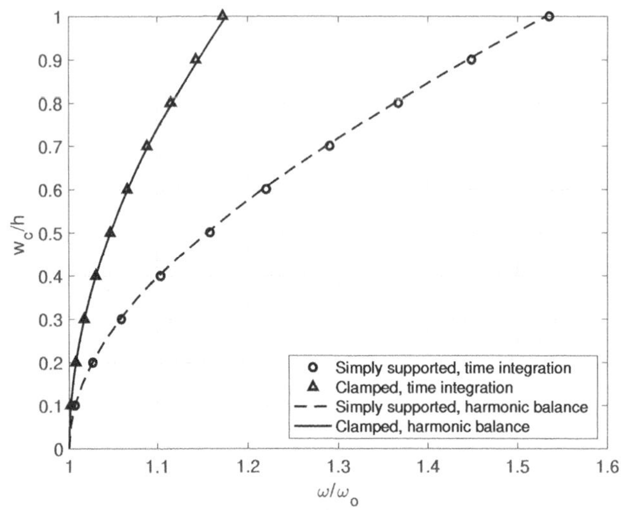

Figure 10.24 Backbone curves for circular plates, $h/R = 0.1$, $\nu = 0.3$, $\kappa = 5/6$. Comparison of harmonic balance and direct time integration.

a prescribed value. The iteration is terminated if the change in frequency satisfies the condition $\delta\omega_a/\omega_a < 10^{-7}$. The only purpose of solving the eigenvalue problem Equation (10.200) is to provide an initial displacement field for direct time integration. The obtained frequency ω_a is also used to estimate the period and to determine the time step Δt for a prescribed number of time steps pèr period, which is 300 or 400 steps. Using Newmark scheme (see Section 6.3) with parameters $\alpha = 0.5$, and $\gamma = 0.5$, we integrate the system Equation (10.161) in time using a constant time step. The initial velocity is zero. Two or four sub-iterations are used to account for the nonlinear stiffness matrix. Computations are conducted for roughly 20 estimated periods. The time instants t_j at which the deflection w_o at the circle center is zero are used to determine the period of the direct time integration method. Because the periods $T_j = t_{j+2} - t_j$ may change slightly in time, we averaged their values over the final ten periods. The backbone curves by direct time integration are in agreement with harmonic balance data as shown in Figure 10.24. A sample output of direct time integration is shown in Figure 10.25 for a simply supported plate and $w_c/h = 0.8$. Simulations are conducted for 32 periods using 400 time steps per period and two sub-iterations at each time step to account for the nonlinear stiffness matrix. The time history for the deflection at the plate center over the last 16 periods is shown in Figure 10.25a, and the corresponding phase portrait and Fourier expansion coefficients spectrum (FFT) are shown in Figures 10.25b and 10.25c, respectively. The two dominant frequencies in the response are the fundamental frequency ($\omega = 6.6912$) and the third harmonic (3ω).

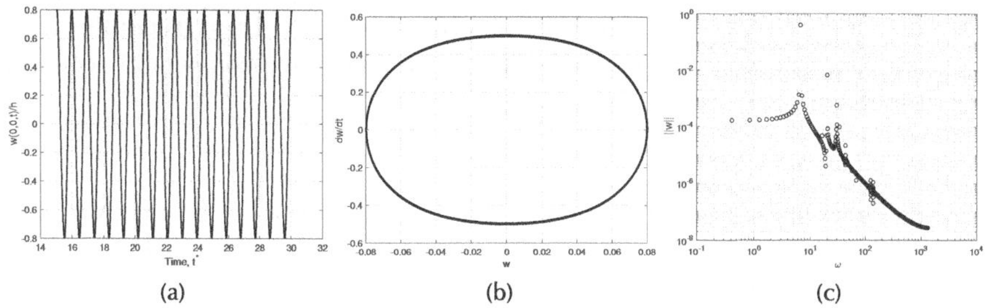

Figure 10.25 Direct time integration. Simply supported circular plate, $h/R = 0.1$, $\nu = 0.3$, $\kappa = 5/6$. (a) A sample of time variation of the deflection at the circle center $w_c(t) \equiv w_o(0,0,t)$; maximum defection $w_c/h = 0.8$, and $t^* = t/(R^2\sqrt{\rho h/D})$. (b) Phase portrait. (c) Fourier coefficients spectrum.

10.9 SUPERSONIC PANEL FLUTTER: LIMIT CYCLE OSCILLATIONS

Thin elastic panels are usually used on surfaces of supersonic airplanes and rockets. Because of the high dynamic pressure $(\rho_\infty U_\infty^2/2)$ on one side of the panel, a small transverse disturbance in the panel position may trigger sustained vibrations of the panel known as panel flutter, which can lead to fatigue and eventual failure of panel. Other engineering applications can be found in energy harvesting. Makihara and Shimose [69] proposed a theoretical model for energy harvesting from supersonic flutter using piezoelectric patches on rocket fins. Zhou et al. [127] studied supersonic flutter suppression of isotropic rectangular plates using piezoelectric actuators. Chai et al. [21] proposed to suppress supersonic flutter and thermal buckling of composite lattice sandwich panels by using elastic foundation. Dowell has provided comprehensive reviews of panel flutter ([27] and [29]). Panels of non-rectangular shapes are less investigated. Ragab and Fayed [91] determined divergence and flutter of circular and triangular panels at very low Mach numbers. Johns [56] was the first to study supersonic flutter of circular and elliptic panels. He used classical plate theory to model plate deflection and linear piston theory to compute the aerodynamic load. He used the Galerkin method and approximated the flutter mode by two assumed mode shapes that satisfy the clamped boundary conditions on the ellipse. Ragab and Fayed [92] studied linear and nonlinear supersonic flutter of plates of various shapes and determined limit cycle amplitudes (LCO). They found that when all conditions are fixed (dynamic pressure, panel area and thickness, material properties, etc.), panels of different shapes exhibit different flutter amplitudes and frequencies. The shape with the least LCO amplitude is equilateral triangle, while the circle shows the largest LCO amplitude. Such differences could be exploited by a designer either for enhancing flutter as in energy harvesting applications or flutter suppression for flight vehicles applications. In this section, we formulate the finite-element method for supersonic flutter of arbitrary shape panels. The plate dynamics is modeled by nonlinear Reissner–Mindlin plate theory, and the aerodynamic model is the linear piston theory (e.g., Dowell [28], Dowell and Bliss [30]). The emphasis of the analysis presented here is on the onset of coalescence flutter, and computation of limit cycle amplitudes.

Problem Formulation

We consider an infinite rigid plane with a flexible panel as depicted in Figure 9.20. In the absence of plate motion, the upper side of the panel is exposed to uniform flow with velocity components U_∞ and V_∞ in the x and y directions, respectively, and static pressure p_∞. Stagnation conditions with pressure p_∞ exist on the lower side of the panel. Gravitational effects are neglected, and hence the undeformed panel is in static equilibrium. If the equilibrium state is disturbed by an external lateral force and then removed, the panel vibrates and unsteady fluid velocity and pressure disturbances are generated on the upper side. Velocity and pressure disturbances are neglected on the lower side and the pressure on the panel lower surface remains p_∞. The equations of motion for the typical element are given by Equation (10.160). For the present case of aeroelastic analysis, the distributed load is the fluid pressure jump across the panel, which is given by the piston theory as will be discussed next.

Aerodynamic Model

The forcing vector in the equation of motion of the plate element is nodal forces equivalent to the distributed fluid pressure. An aerodynamic model is used to express that vector in terms of deflection, velocity, and acceleration at the element nodes. Inviscid flow theory is used to model velocity and pressure disturbances on the upper side due to plate vibrations. Thus, fluid motion satisfies the linearized boundary condition

$$v_z = \frac{\partial w_o}{\partial t} + U_\infty \frac{\partial w_o}{\partial x} + V_\infty \frac{\partial w_o}{\partial y}, \qquad on \quad z = 0 \tag{10.201}$$

where v_z is the fluid velocity normal to the plate, and $w_o(x,y,t)$ is the deflection of the middle plane of plate. The linearized piston theory gives the surface pressure by the simple relation

$$\begin{aligned} p - p_\infty &= \rho_\infty a_\infty v_z \\ &= \rho_\infty a_\infty \left(\frac{\partial w_o}{\partial t} + U_\infty \frac{\partial w_o}{\partial x} + V_\infty \frac{\partial w_o}{\partial y} \right) \end{aligned} \tag{10.202}$$

where ρ_∞ and a_∞ are the free-stream density and speed of sound, respectively. An analysis of the unsteady flow problem (Dowell [28]) gives a more accurate expression for small reduced frequencies, which is used in the present analysis.

$$p - p_\infty = \frac{\rho_\infty V_o}{\sqrt{M^2 - 1}} \left(\frac{M^2 - 2}{M^2 - 1} \frac{\partial w_o}{\partial t} + U_\infty \frac{\partial w_o}{\partial x} + V_\infty \frac{\partial w_o}{\partial y} \right) \tag{10.203}$$

where $V_o^2 = U_\infty^2 + V_\infty^2$, and $M = V_o/a_\infty$ is the Mach number. Using the approximation of deflection w_o over an element as given by Equations (10.51), we get

$$\frac{p - p_\infty}{\rho_\infty} = \alpha_1 V_o \{\psi\}_3 \{\dot{U}^e\} + \alpha_2 V_o^2 \{\Gamma\} [S] \{U^e\} \tag{10.204}$$

where

$$\alpha_1 = \frac{1}{\sqrt{M^2 - 1}} \frac{M^2 - 2}{M^2 - 1}, \qquad \alpha_2 = \frac{1}{\sqrt{M^2 - 1}} \tag{10.205}$$

$$\{\Gamma\} = \{\ cos\beta \quad sin\beta\ \}, \qquad cos\beta = \frac{U_\infty}{V_o}, \qquad sin\beta = \frac{V_\infty}{V_o} \tag{10.206}$$

$$[S] = \begin{bmatrix} 0\ 0\ \frac{\partial \psi_1}{\partial x}\ 0\ 0 \ .... \ \frac{\partial \psi_j}{\partial x}\ ...\ 0\ 0\ \frac{\partial \psi_n}{\partial x}\ 0\ 0 \\ 0\ 0\ \frac{\partial \psi_1}{\partial y}\ 0\ 0 \ .... \ \frac{\partial \psi_j}{\partial y}\ ...\ 0\ 0\ \frac{\partial \psi_n}{\partial y}\ 0\ 0 \end{bmatrix}. \tag{10.207}$$

Fluid-Structural Model

Forcing vector for element number # e is given by Equation (10.82) for distributed load f, which is the pressure jump $f = -(p - p_\infty)$, hence

$$\{F^e\} = \int_{A_e} \{\psi\}_3^T [-(p - p_\infty)] dA. \tag{10.208}$$

Using Equation (10.204), we find the forcing vector of element # e to be

$$\begin{aligned} \{F^e\} \quad &= \quad -\alpha_1 \rho_\infty V_o \int_{A_e} \{\psi\}_3^T \{\psi\}_3 \{\dot{U}^e\} dA \\ &\quad -\alpha_2 \rho_\infty V_o^2 \int_{A_e} \{\psi\}_3^T \{\Gamma\} [S] \{U^e\} dA \end{aligned} \tag{10.209}$$

which can be written as

$$\{F^e\} = \alpha_1 \rho_\infty V_o [H_1^e] \{\dot{U}^e\} + \alpha_2 \rho_\infty V_o^2 [H_2^e] \{U^e\} \tag{10.210}$$

where

$$[H_1^e] = -\int_{A_e} \{\psi\}_3^T \{\psi\}_3 dA, \qquad [H_2^e] = -\int_{A_e} \{\psi\}_3^T \{\Gamma\} [S] dA. \tag{10.211}$$

The equations of motion for element # e is

$$[M^e] \{\ddot{U}^e\} + [C^e] \{\dot{U}^e\} + [K^e] \{U^e\} = \{F^e\} + \{R^e\} \tag{10.212}$$

we added a structural damping matrix $[C^e]$ if needed, however, the results presented in this section assume zero structural damping. We assemble all elements into a global system

$$[M]\{\ddot{U}\}+[C]\{\dot{U}\}+[K]\{U\}=\{F\}+\{R\}. \qquad (10.213)$$

The global forcing vector $\{F\}$ is the assembly of element forcing vector given by Equation (10.210), hence

$$[M]\{\ddot{U}\}+[[C]-\alpha_1\rho_\infty V_o[H_1]]\{\dot{U}\}+\left[[K]-\alpha_2\rho_\infty V_o^2[H_2]\right]\{U\}=\{R\}. \qquad (10.214)$$

According to linearized piston theory [see Equation (10.203)], fluid pressure is independent of the local structural acceleration, and as such the mass matrix is unmodified. In other words, the added mass effects are neglected. Fluid-structure interaction produces linear damping proportional to the matrix $[H_1]$, and linear stiffness proportional to the matrix $[H_2]$. The effects of geometrically nonlinear strains are manifested only in the structural stiffness matrix $[K]$, which depends on the instantaneous transverse deflection w_o as given by matrix $\left[G^{NL}\right]$. In this section, we consider both linear ($\left[G^{NL}\right]=0$) and nonlinear flutter analysis.

Nondimensional Variables

Let L denote a characteristic length of the panel (radius of a circle or side of a square), and h its thickness assumed uniform. The panel response to fluid pressure fluctuations is determined by two nondimensional parameters namely: mass loading μ (also called density ratio), and aerodynamic loading Λ (also called velocity ratio) defined by

$$\mu=\frac{\rho_\infty}{\rho_p}\frac{L}{h},\quad \Lambda=\frac{\rho_\infty V_o^2 L^3}{D\sqrt{M^2-1}},\quad D=\frac{Eh^3}{12(1-\nu^2)} \qquad (10.215)$$

where D is the flexural rigidity. We note that Λ is usually written in terms of the dynamic pressure $q=\rho_\infty V_o^2/2$. Introducing nondimensional space coordinates and time variables

$$(x^*,y^*,z^*)=\frac{(x,y,z)}{L},\quad t^*=\frac{t}{L^2}\sqrt{\frac{D}{\rho_p h}} \qquad (10.216)$$

into Equation (10.214), we get

$$[M]\{\ddot{U}\}+\left[[C]-\alpha_o\sqrt{\mu_o\Lambda}[H_1]\right]\{\dot{U}\}+[[K]-\Lambda[H_2]]\{U\}=\{R\} \qquad (10.217)$$

where $\alpha_o=\alpha_1/\alpha_2$, and $\mu_o=\mu\alpha_2$. We note that for high Mach number $\alpha_1=\alpha_2=M^{-1}$, and hence $\mu_o=\mu/M$ and $\alpha_o=1$.

Linear Flutter Analysis

For low values of the aerodynamic loading parameter Λ, if a small lateral disturbance is applied to a panel and then removed, the panel vibrates but it returns to equilibrium.

However, if Λ exceeds a certain value, the vibration amplitude grows and reaches a limit cycle oscillations (LCO). Linear analysis is sufficient to determine the dynamic pressure Λ at the onset of flutter since at that condition the vibration amplitude may be assumed to be very small. With $\left[G^{NL}\right]=0$, the system of Equations (10.217) is linear, and a frequency domain analysis is sufficient for determining the dynamic pressure at onset of flutter and its frequency. We use Pitman and Lucey's [86] method for direct determination of the eigenmodes of finite flow-structure systems. Define state variables $\{Q_1\}=\{U\}$, and $\{Q_2\}=\{\dot{U}\}$, and rewrite the system as

$$[B]\ \{\dot{Q}\}\ =[A]\{Q\}. \tag{10.218}$$

Assume a normal mode solution,

$$\{Q\}=e^{s^*t^*}\{\hat{Q}\} \tag{10.219}$$

and solve the eigenvalue problem

$$[[A]-s^*[B]]\{\hat{Q}\}=0. \tag{10.220}$$

In general, the eigenvalues are complex, $s^*=s_r^*+is_i^*$, where s_r^* and s_i^* are nondimensional growth rate and frequency, respectively.

Nonlinear Flutter Analysis

To determine the LCO amplitude, we need to solve the nonlinear system of Equations (10.217). Given initial conditions, we integrate the system in the time domain using Newmark method (see Section 6.3). At each time step, the nonlinear element stiffness matrix is linearized, and sub-iteration is used for updating that matrix before advancing the solution to the next time step. Usually four sub-iterations are found to be sufficient for convergence. Let $\{\bar{U}^e\}$ denote a known solution at a sub-iteration, the stiffness matrix is computed as

$$\begin{aligned}[\bar{K}^e]=&\int_{A_e}\{\,[B_o]+[\bar{G}^{NL}]\,\}^T[D_o]\{\,[B_o]+\tfrac{1}{2}[\bar{G}^{NL}]\,\}dA\\&+\int_{A_e}[B_b]^T[D_b][B_b]dA+\int_{A_e}[B_s]^T[D_b^s][B_s]dA\end{aligned} \tag{10.221}$$

and is updated after each sub-iteration. Except when stated, a 16-node isoparametric bicubic quadrilateral element is used in all results presented here. Element matrices are numerically evaluated by the (4×4) Gauss–Legendre quadrature rule except the shear stiffness matrix which is evaluated by the reduced integration (3×3) rule.

Example 10.14. Flutter of a Simply Supported Square Plate

Panels of different shapes exhibit similar qualitative variations of vibration frequency and growth rate as the aerodynamic loading parameter Λ is increased from zero until onset of flutter. We consider a simply supported square panel as a typical example. The square side a, and free stream velocity U_∞ are used as reference parameters L,

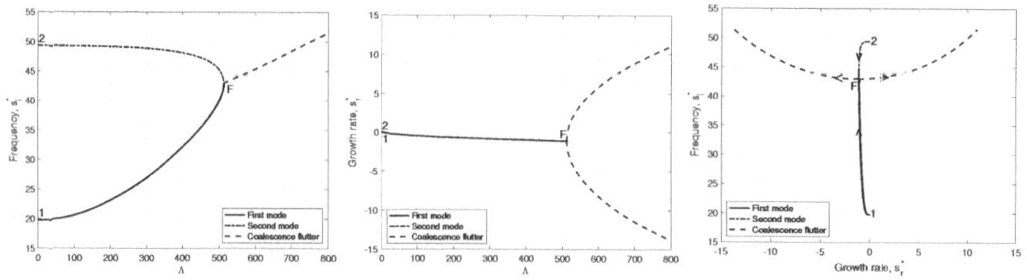

(a) (b) (c)

Figure 10.26 Supersonic flutter of a simply supported square plate, $h/a = 0.01$, $\nu = 0.35$, $\kappa = 5/6$, $\mu/M = 0.01$. First and second modes coalesce at point F. (a) Frequency. (b) Growth rate. (c) Root loci; arrows show direction of increasing Λ.

and V_o, respectively. A mesh of 36 bicubic elements with 361 nodes is used. The aeroelastic instability of the panel is investigated by fixing the parameters $\nu = 0.35$, $\kappa = 5/6$, $h/a = 0.01$, and the flow angle $\beta = 0$. The high Mach number approximation is used, hence $\alpha_o = 1$, and $\mu/M = 0.01$. We survey the eigenvalue s^* of system Equation (10.220) as a function of the parameter Λ. Frequencies of the first and second modes are depicted in Figure 10.26a, and growth rates are depicted in Figure 10.26b against the loading parameter Λ. At $\Lambda = 0$, frequencies of the first and second modes are 19.7316, and 49.3012, respectively, and the growth rate is zero. As Λ increases from zero, the frequency of the first mode increases while the frequency of the second mode decreases. For $\Lambda < \Lambda_F (= 512)$, the growth rates of the two modes are nearly equal and slightly negative indicating stability. The two modes coalesce at point F where $\Lambda_F = 512$ with frequency $s_i^* = 42.9357$, but there is a small damping as the growth rate is negative $s_r^* = -1.2693$. The value $\Lambda_F = 512$ is in excellent agreement with Guo and Mei [45] computations using 6 aeroelastic modes. The root loci of the two modes are shown in Figure 10.26c, which shows the coalescence of the two modes and the migration of one of the roots into the $s_r^* > 0$ half-plane signifying flutter. The value of Λ at which the growth rate changes from negative to positive is considered to be the onset of flutter. The root locus crosses to positive growth rate at $\Lambda_F = 514.22$ and frequency $s_i^* = 43.012$ and growth rate is nearly zero. At $\Lambda = 0$, the first mode shape is depicted in Figure 10.27a with maximum vibration amplitude at the plate center. The second mode shape is depicted in Figure 10.27b, where the line $x = a/2$ is a nodal line, $w_o = 0$. As Λ increases, the two mode shapes evolve and coalesce to the shape shown in Figure 10.27c at $\Lambda = 512$. The vibration amplitude near the upstream edge $x = 0$ is much smaller than that near the downstream edge $x/a = 1$.

Next we consider the nonlinear flutter problem and determine the limit cycle oscillations (*LCO*) amplitude of a simply supported square plate. We integrate the nonlinear system of Equations (10.217) using Newmark method (see Section 6.3) with parameters $\alpha = 0.5$, and $\gamma = 0.5$. Initial displacements are prescribed by eigenvectors

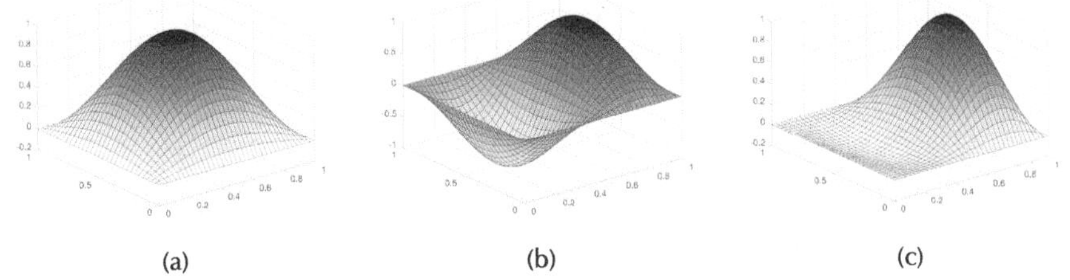

Figure 10.27 Supersonic flutter of a simply supported square plate, $h/a = 0.01$, $\nu = 0.35$, $\kappa = 5/6$, $\mu/M = 0.01$. (a) First mode, $\Lambda = 0$. (b) Second mode, $\Lambda = 0$. (c) Coalescence flutter mode, $\Lambda = 512$.

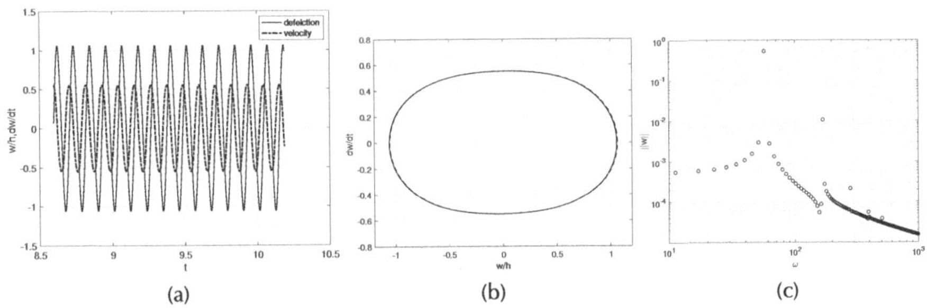

Figure 10.28 Supersonic flutter of a simply supported square plate, $h/a = 0.01$, $\nu = 0.35$, $\kappa = 5/6$, $\mu/M = 0.01$, $\Lambda = 820$. (a) Deflection and velocity at $(x,y) = (0.7778, 0.5)$. (b) Phase portrait. (c) Spectrum of FFT amplitudes.

of the system Equation (10.200) solved for prescribed small deflection at the plate center $w_o/h = 0.01$. The initial velocity is zero. For an example, we fix $\Lambda = 820$. A constant time step is prescribed $dt = 0.001592$, and three sub-iterations are used per time step. The deflection w_o and velocity dw_o/dt at the point $(x,y) = (0.7778, 0.5)$ are recorded at each time step until $t = 10.19$. The time signal and phase portrait for the final 14 periods are depicted in Figure 10.28a, and 10.28b, respectively. The nondimensional period is 0.111444 (frequency, 56.3798), and the *LCO* amplitude at the monitored point is $w_o/h = 1.059$. The spectrum of FFT amplitudes of the deflection is shown in Figure 10.28c, which shows the response is dominated by the fundamental frequency $\omega_o = 56.3798$; however, the response also includes the third super harmonic $3\omega_o$. The amplitudes of other frequencies are at least two orders of magnitude smaller than that of the fundamental frequency. The time integration is repeated for selected values of Λ, and the *LCO* amplitude at the $(x,y) = (0.75a, 0.5a)$ are plotted in Figure 10.29 for the simply supported plate. The onset of flutter for a

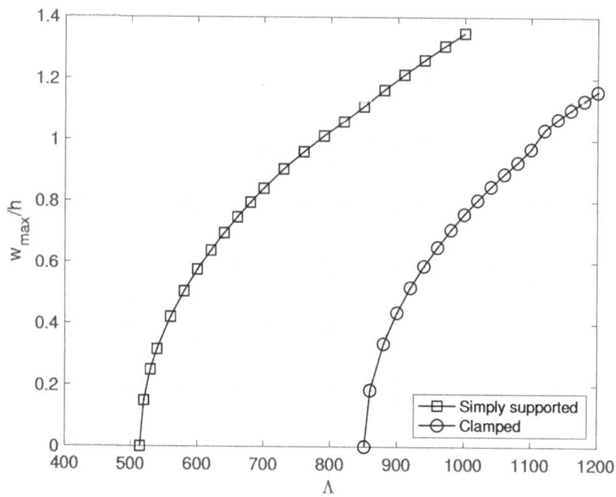

Figure 10.29 Supersonic flutter of simply supported and clamped square plates, $h/a = 0.01$, $\nu = 0.35$, $\kappa = 5/6$, $\mu/M = 0.01$. Limit cycle oscillations amplitude.

clamped square plate happens at $\Lambda = 851.6$, and the *LCO* amplitudes are also shown in the figure. Ragab and Fayed [92] have presented results for panels of various shapes including square, circular, semi-circular, elliptical, triangular, and hexagonal panels.

10.10 NONLINEAR HYDROELASTIC STABILITY OF PANELS

The linear hydroelastic stability of a square panel has been presented in Example 9.11. For hydrodynamic loading Λ greater than that at the onset of divergence, the panel deflection becomes large and geometric nonlinear effects become significant. For large deflection both the hydrodynamic pressure and structural stiffness are nonlinear functions of deflection. For incompressible potential flow, we still solve the Laplace's equation for the velocity potential, but the pressure follows from the Bernoulli's equation, which contains the square of the velocity magnitude. Moreover, the boundary conditions should be applied on the deformed surface rather than being transferred to the initial flat surface. Lucey et al. [67] presented a computational study of the nonlinear hydroelastic behavior of flexible walls on elastic foundation. They considered a panel of infinite span, and coupled a nonlinear potential flow model with a nonlinear structural model in which the plate deflection is a function of one-space dimension and time. Kapor [58] also studied the nonlinear infinite-span compliant-wall fluid-structure interaction problem using a Discrete-Vortex Method to account for the boundary-layer flow. Implementation of a nonlinear hydrodynamic model in structural vibration problems is more challenging than the linearized model. Fortunately, in most applications, the basic characteristics of the hydroelastic response of panels beyond divergence can be captured by a geometrically nonlinear structural model while keeping a linearized hydrodynamic model. The coupling of a linear hydrodynamic model with nonlinear structural model is common practice in aeroelasticity (e.g., Reynolds and Dowell [105], Boutet de Monvel and Chueshov [16]). Tubaldi et al. [123] considered a channel flow with one rigid wall and the other flexible, which is modeled as a periodically simply supported rectangular panel with immovable edges. They used von Karman nonlinear plate and a linear potential flow model to investigate vibrations and stability of the flexible wall. In the present analysis, we account for the structural geometric nonlinear effects, but use the linear hydrodynamic model presented in Chapter 9. The coupled fluid-structure equation of motion is obtained by substituting the fluid pressure loading given by Equation (9.261) into Equation (10.161) for nonlinear plate vibration,

$$\left[[M]-\rho_f\left[H_o^g\right]\right]\{\ddot{U}\}+\left[[C]-\rho_f V_o\left[H_1^g\right]\right]\{\dot{U}\}+\left[[K]-\rho_f V_o^2\left[H_2^g\right]\right]\{U\}=\{R\}. \tag{10.222}$$

Let L denote a characteristic length of the panel (radius of a circle or side of a triangle), and h its thickness assumed uniform. The panel response to fluid pressure fluctuations is determined by two nondimensional parameters namely: mass loading μ (also called density ratio), and hydrodynamic loading Λ (also called velocity ratio) defined by

$$\mu=\frac{\rho_f}{\rho_p}\frac{L}{h}, \quad \Lambda=\frac{\rho_f V_o^2 L^3}{D}, \quad D=\frac{Eh^3}{12(1-\nu^2)} \tag{10.223}$$

where ρ_f and ρ_p are mass densities of fluid and plate, respectively, V_o is the fluid velocity magnitude, and D is the flexural rigidity, where E is elastic modulus and ν is Poisson's ratio.

Example 10.15. Hydroelastic Stability of a Simply Supported Square Plate

The data of Example 9.11 are used here again (square side a, thickness $h/a=0.01$, Poisson's ratio $\nu=0.35$, shear correction factor $\kappa=5/6$, plate density $\rho_p=2600$ kg/m^3, fluid density $\rho_f=1000$ kg/m^3, mass ratio $\mu=38.46$, and Young's modulus $E=59$ GPa.) The panel is simply supported on the four edges. We want to investigate the nonlinear post-divergence response for different values of the hydrodynamic pressure parameter Λ. According to the linear theory, the panel becomes unstable and diverges at $\Lambda\approx 205$, and it recovers at $\Lambda\approx 479$ until $\Lambda\approx 537$ where it becomes unstable again and coalescence flutter begins. It is instructive to investigate the nonlinear response in the three stages: Divergence, recovery, and post flutter phases. To this end, we determine the nonlinear response for $\Lambda=360$, 500, and 700; values that fall in the stages of interest. We use uniform mesh of 64 bicubic elements with 625 global nodes, and integrate Equation (10.222) in time using Newmark method ($\alpha=\gamma=0.5$). The initial displacements and velocity are provided by the eigenfunctions of the linear problem at the specified value of Λ. The eigenvectors are scaled such that the deflection w_o/h at a point on the center line is given an initial value

Table 10.16
Hydroelastic Stability of a Square Plate. $h/a = 0.01$, $\nu = 0.3$, $\kappa = 5/6$, $\mu = 38.46$

Λ	Initial w_o/h	at x/a	Time step Δt
360	0.01	0.25	8.084×10^{-4}
500	0.01	0.25	1.354×10^{-3}
500	0.20	0.25	1.219×10^{-3}
700	0.01	0.25	3.855×10^{-4}

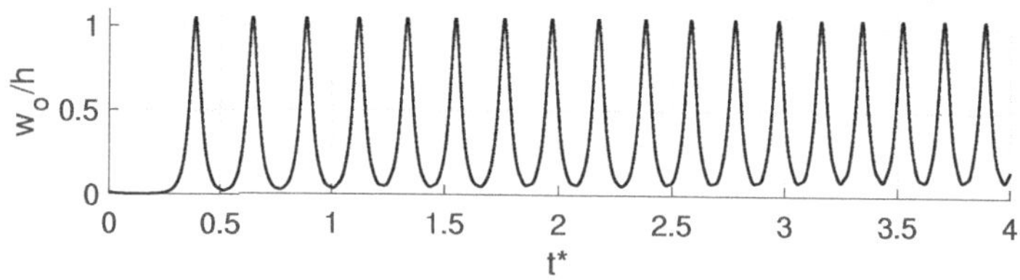

Figure 10.30 Central deflection w_c/h for hydroelastic divergence of simply supported square plate at $\Lambda = 360$.

as shown in Table 10.16. The table also shows the time step. The nonlinear stiffness matrix is updated in two sub-iterations at each time step.

At $\Lambda = 360$, the panel is linearly unstable, and a small deflection from the flat surface grows exponentially in time. Time integration of the nonlinear system without structural damping predicts an initial exponential growth but as the deflection at the center reaches a maximum of $w_o/h \approx 1$ it decreases to a minimum, and then the panel oscillates for a long time as shown in Figure 10.30. The minimum deflection increases with time but the maximum deflection appears to decrease very slowly. Figure 10.31a shows the deflection for a few cycles near the end of the simulation time, and Figure 10.31b depicts the phase portrait at the panel center $(x/a, y/a) = (0.5, 0.5)$. The response is not perfectly periodic. Sixteen consecutive deflection profiles on the plate center line $y/a = 0.5$ are shown in Figure 10.31c; the time interval between two consecutive profiles is $10\Delta t$. The preferred mode of divergence is the typical first mode in which the deflection is maximum at the center.

Pressure loading of $\Lambda = 500$ falls in the recovery phase, and the panel is linearly stable. According to the linear theory, the response should be small oscillations around the flat surface. Time integration of the nonlinear system without structural damping indicates that the response depends on the initial conditions. For a prescribed small initial deflection of $w_o/h = 0.01$ at $(x/a, y/a) = (0.25, 0.50)$, the response is oscillatory with an amplitude very near the prescribed initial deflection as shown in Figure 10.32a, and the frequency is equal to that of linear analysis. Figure 10.32b depicts

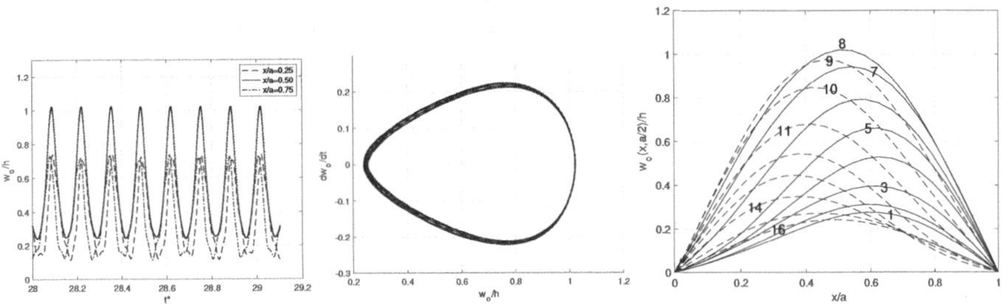

(a) (b) (c)

Figure 10.31 Hydroelastic divergence of a simply supported square plate, $\Lambda = 360$, and initial $w_o/h = 0.01$. (a) Deflection history at $y/a = 0.5$ and $x/a = 0.25, 0.50, 0.75$. (b) Phase portrait of time signal at $x/a = 0.5$. (c) Deflection profiles $w_o(x, a/2)$ at different times.

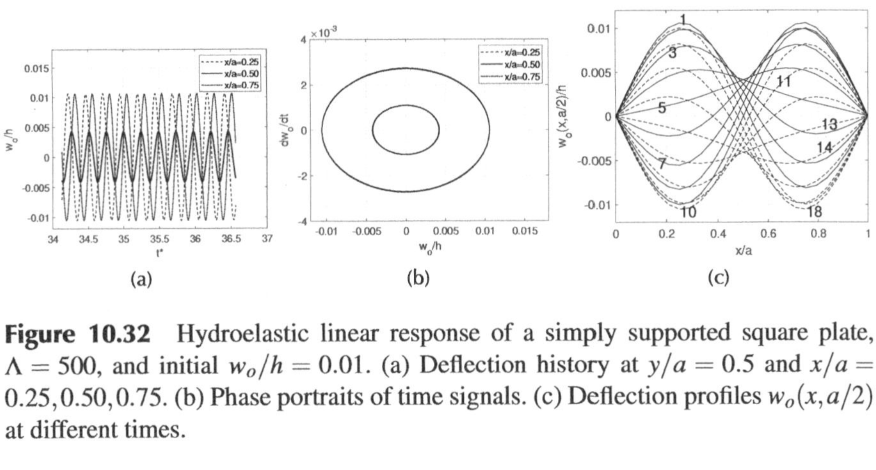

(a) (b) (c)

Figure 10.32 Hydroelastic linear response of a simply supported square plate, $\Lambda = 500$, and initial $w_o/h = 0.01$. (a) Deflection history at $y/a = 0.5$ and $x/a = 0.25, 0.50, 0.75$. (b) Phase portraits of time signals. (c) Deflection profiles $w_o(x, a/2)$ at different times.

the phase portraits at the three points $(x/a, y/a) = (0.25, 0.5), (0.5, 0.5), (0.75, 0.5)$. The response is periodic. Eighteen consecutive deflection profiles on the plate center line $y/a = 0.5$ over one period are shown in Figure 10.32c; the time interval between two consecutive profiles is $10\Delta t$. These results confirm the linear theory predictions in the recovery phase provided that the initial deflection is very small.

The response, however, is significantly different if the panel is given an initial large deflection. We prescribe an initial deflection of $w_o/h = 0.2$ at $(x/a, y/a) = (0.25, 0.50)$. Figure 10.33 shows the deflection at the panel center as a function of time. The amplitude grows initially and attains a maximum value, but then the panel appears to oscillate around a mean position. The oscillations amplitude

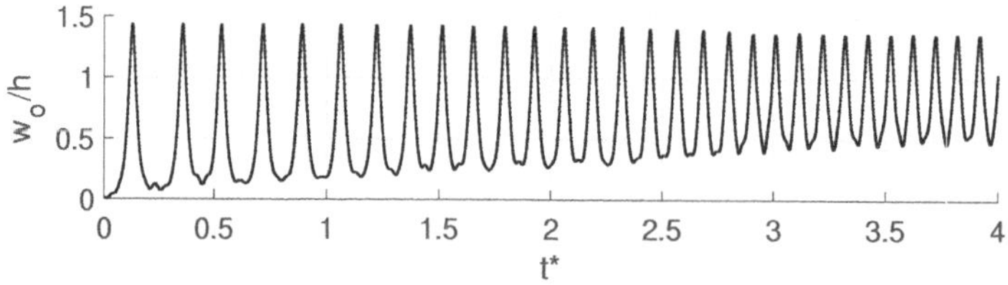

Figure 10.33 Central deflection w_c/h for hydroelastic divergence of simply supported square plate at $\Lambda = 500$, and initial $w_o/h = 0.2$ at $(x/a, y/a) = (0.25, 0.50)$.

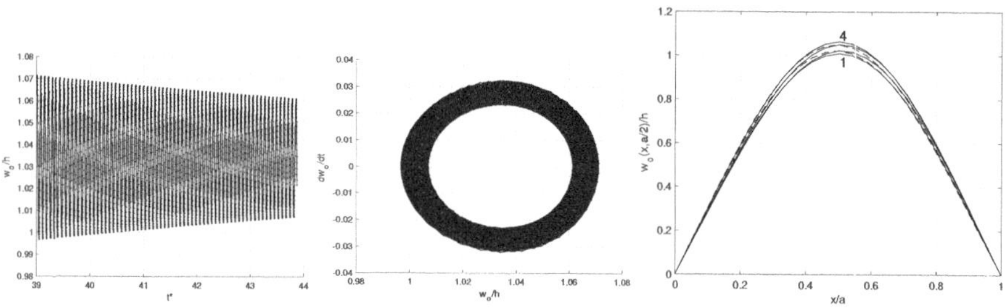

(a) (b) (c)

Figure 10.34 Hydroelastic nonlinear response of a simply supported square plate, $\Lambda = 500$, and initial $w_o/h = 0.20$. (a) Deflection history at $(x/a, y/a) = (0.5, 0.5)$. (b) Phase portrait of time signal at $x/a = 0.5$. (c) Deflection profiles $w_o(x, a/2)$ at different times.

decreases with time as shown 10.34a for the deflection at the panel center, and the phase portrait in Figure 10.34b also indicates convergence to a non-zero deflection. A few deflection profiles around the mean static profile are shown in Figure 10.34c; the time interval between two consecutive profiles is $10\Delta t$. While $\Lambda = 500$ falls in the recovery phase of the linear theory, if the panel is subjected to large disturbance, the nonlinear theory predicts divergence to a static profile. Again, the preferred mode of divergence is the typical first mode in which the deflection is maximum at the center. According to the linear theory, pressure loading of $\Lambda = 700$ is in the post coalescence flutter phase, and growing oscillatory response is expected. Figure 10.35 shows the deflection at the panel center as predicted by time integration of the nonlinear system without structural damping. Initially, the response is erratic followed by oscillations with decreasing amplitude implying slow convergence to a static deflection. The oscillations near the end of the simulation time are depicted in Figure 10.36a for the deflection at the three points $(x/a, y/a) = (0.25, 0.5), (0.5, 0.5), (0.75, 0.5)$; the response is aperiodic. The phase portrait at panel center is shown in Figure 10.36b and it also indicates aperiodic response. Sixteen consecutive deflection profiles are shown

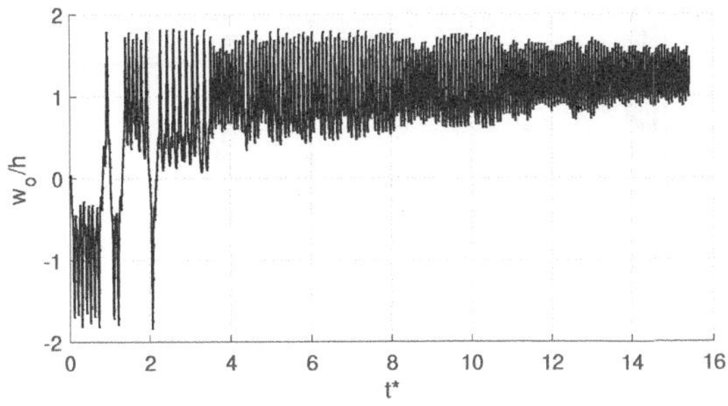

Figure 10.35 Central deflection w_c/h of hydroelastic flutter of a simply supported square plate at $\Lambda = 700$.

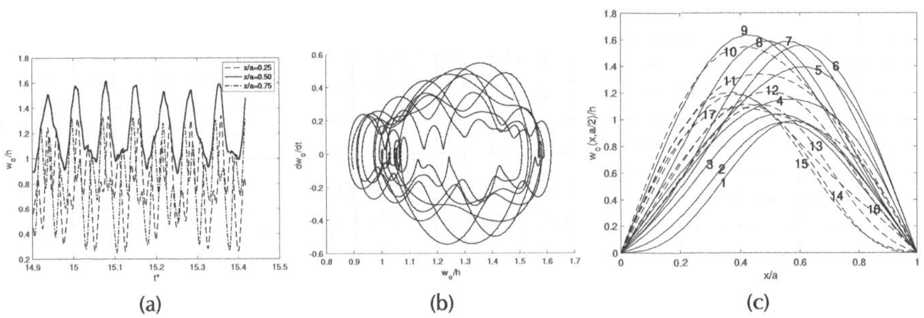

Figure 10.36 Hydroelastic flutter of a simply supported square plate, $\Lambda = 700$, and initial $w_o/h = 0.01$. (a) Deflection history at $y/a = 0.5$ and $x/a = 0.25, 0.50, 0.75$. (b) Phase portrait of time signal at $x/a = 0.5$. (c) Deflection profiles $w_o(x, a/2)$ at different times.

in Figure 10.36c; the time interval between two consecutive profiles is $10\Delta t$. They indicate aperiodic response about a mean static profile similar to the fundamental mode in which the deflection is maximum at the center.

Effects of Structural Damping

Structural damping can significantly affect the transient oscillatory response for hydrodynamic loading greater than that at the onset of divergence. In Equation (10.222), we assume that the structural damping matrix is proportional to the mass matrix, $[C] = 10[M]$, and integrate the equation in time for $\Lambda = 700$, which is post coalescence flutter according to the linear theory. As shown in Figure 10.37, after a few oscillations with large amplitude, damped oscillatory response ensues leading to a static deflection position. The oscillations near the end of the simulation time are depicted in Figure 10.38a for the deflection at the panel center; the response is damped small amplitude oscillations. The phase portrait at panel center is shown in

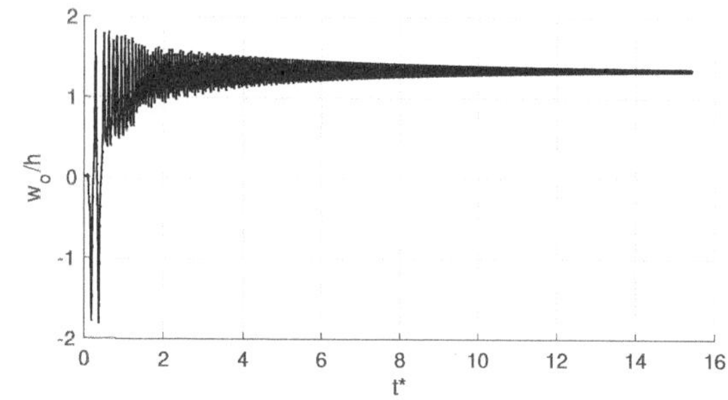

Figure 10.37 Effects of structural damping on central deflection w_c/h for hydroelastic flutter of simply supported square plate at $\Lambda = 700$.

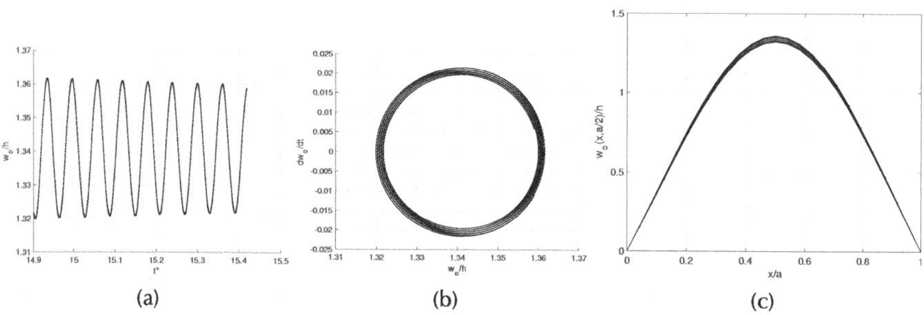

Figure 10.38 Effects of structural damping on hydroelastic flutter of a simply supported square plate, $\Lambda = 700$, and initial $w_o/h = 0.01$. (a) Deflection history at the panel center. (b) Phase portrait of time signal at the panel center. (c) Deflection profiles $w_o(x,a/2)$ at different times.

Figure 10.38b and it also indicates slow convergence to static deflection. A few consecutive deflection profiles are shown in Figure 10.38c; the time interval between two consecutive profiles is $10\Delta t$. They indicate a mean static profile similar to the fundamental mode in which the deflection is maximum at the center.

Nonlinear Static Divergence of Square Panels

Time integration of the nonlinear system for hydrodynamic pressure loading of $\Lambda = 360, 500$, and 700 imply the existence of static equilibrium position. Such a solution is independent of time, and can be determined by solving a nonlinear system obtained by dropping the acceleration and velocity terms in Equation (10.222);

$$\left[[K] - \rho_f V_o^2 \left[H_2^g\right]\right] \{U\} = \{R\} \tag{10.224}$$

where Equation (10.223) gives $\rho_f V_o^2$ in terms Λ. After applying proper boundary conditions, a homogeneous system is obtained. For a prescribed deflection at a point

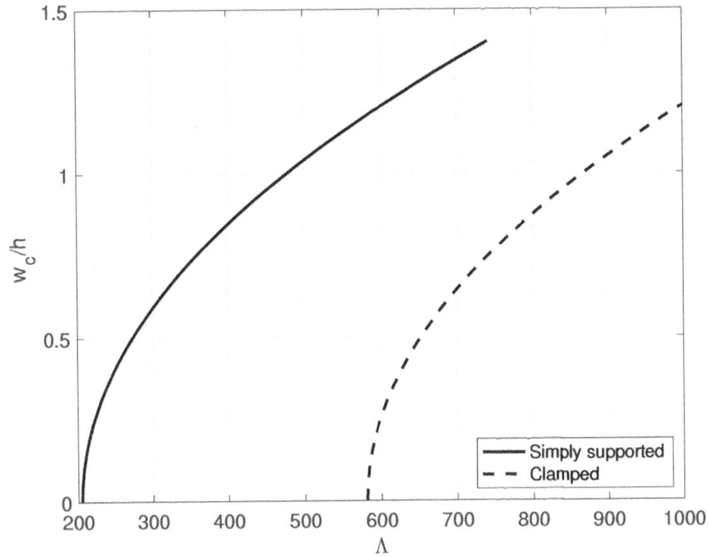

Figure 10.39 Central deflection w_c/h for static hydroelastic divergence of simply supported and clamped square plates.

such as the panel center, an eigenvalue problem can be solved for Λ (or $\rho_f V_o^2$). Alternatively, for an assumed Λ, Newton–Raphson iteration can be used to solve the nonlinear homogeneous system for displacements. We use the eigenvalue solver for Λ near the onset of divergence, and Newton–Raphson iteration for larger Λ. The central deflection is plotted in Figure 10.39 as a function of Λ for both simply supported and clamped square plates. Central deflection values are consistent with values obtained by time integration, and the deflected shape is similar to those in Figure 10.38c. We also remark that the data in Figure 10.39 are independent of the mass ratio parameter μ.

Appendix
LISTS of MATLAB CODES

MATLAB CODES FOR CHAPTER 1

(1) bar_characteristics.m
Method of characteristics for reflection of a right-running wave from a free end of a bar.

(2) string_characteristics.m
Method of characteristics for reflection of a right-running wave from a fixed end of a string.

(3) torsion_rectangular_exact.m
Plots contours of Prandtl stress function and shear stress contours for the exact solution of the torsion of a rectangular section.

MATLAB CODES FOR CHAPTER 2

(1) Example2p1.m
Solve 2nd order ODE by method of weighted residuals; strong form solution.

(2) Example2p2.m
Weak-form solution of 2nd order ODE.

(3) Example2p5.m
Mesh refinement and error analysis of finite-element method for 2nd order ODE.

(4) Krogh1D_MM_exact.m
Comparison between finite element solution and exact analytic solution for oxygen uptake in flat tissues using Mechaelis-Menten kinetics.

(5) KroghSphere.m
Finite-element solution for steady 1D reaction-diffusion problems in cylindrical or spherical domains.

(6) WomersleyPulse.m
Finite-element solution to laminar viscous flow in a tube driven by time-periodic pressure gradient.

MATLAB CODES FOR CHAPTER 3

(1) example3p2.m
Solves the beam of example 3.2.

(2) example3p3.m
Solves the frame of example 3.3.

(3) example3p4.m
Solves the truss of example 3.4.

(4) example3p5.m
Solution for thermoelastic forces in the truss of example 3.5.

MATLAB CODES FOR CHAPTER 4

(1) example4p1.m
Heat conduction on a square domain, exmaple 4.1

(2) example4p2.m
Heat conduction with convection boundary conditions on a square domain, example 4.2

(3) example4p3.m
Poisson's equation on an L-shaphed domain, exmaple 4.3

(4) torsion_semicircle.m
Poisson's equation on a semi-circular domain, example 4.5

(5) Airfoil_C_Kutta_AG.m
Potential flow over 2D airfoils including mesh generation and panel code. Isoparametric quadrilateral bilinear and biquadratic elements.

(6) potential_flow_cylinder_Q4.m
Laplace's equation, potential flow over a cylinder in a channel and comparison with method of images.

(7) shape_2triangle10.m
Torsion of semi-circle section using bicubic isoparametric triangular elements.

(8) Krough_cylinder_Axi_P.m
Oxygen diffusion and consumption in Krogh capillary-tissue cylinder with Michaelis-Menten kinetics

(9) cascade_potential_flow1.m
Potential flow over 2D cascades including mesh generation and panel code. Isoparametric quadrilateral bilinear elements.

MATLAB CODES FOR CHAPTER 5

(1) bioheat_radiation_BC.m
Thermal injuries due to radiation from a hot plate

(2) bioheat_water.m
Thermal injuries due to direct contact with hot water

(3) Bioheat1_single_layer.m
Comparison between exact solution and finite element method for a single-layer skin model

(4) Cooling_fin.m
Transient heat in a cooling fin

(5) Egg_boilling.m
Pasteurization or hard-boiling of a spherical egg

(6) UnsteadyKroghCylinder_first_zero.m
Transient reaction-diffusion in a cylindrical annular region with first- or zero-order reaction rate and comparison with the exact analytic solutions

(7) UnsteadyKroghCylinder.m
Transient reaction-diffusion in a cylindrical annular region with Michaelis-Menten kinetics

(8) UnsteadyKroghSphere_3_Reaction.m
Transient reaction-diffusion in a spherical cell with Michaelis-Menten, first- and zero-order reaction rates

(9) UnsteadyKroghSphere_first_zero.m
Transient reaction-diffusion in a spherical cell

with first- or zero-order reaction rate and
comparison with the exact analytic solutions

(10) UnsteadyKroghSphere.m
Transient reaction-diffusion in a spherical cell
with Michaelis-Menten kinetics

MATLAB CODES FOR CHAPTER 6

(1) string_dynamic.m
Direct time integration using Newmark method for
transverse waves on a taut string, example 6.1.
Comparison with exact solutions.

(2) bar_free_oscil.m
Direct time integration using Newmark method for
longitudinal waves in a bar, example 6.2.
Comparison with exact solutions.

(3) bar_impact_Mo_effects.m
Direct time integration using Newmark method for
longitudinal waves in a bar impacted by a rigid block,
example 6.3. Comparison with exact solutions.

(4) beam_dynamic_1.m
Direct time integration using Newmark method for
free vibration of a cantilever, example 6.4
and mesh refinement

(5) beam_dynamic_2.m
Direct time integration using Newmark method for
forced vibration of a simply-supported fixed beam,
example 6.5, and comparison with exact solution.

MATLAB CODES FOR CHAPTER 7

(1) beam_eigen.m
Natural frequencies of beams using Hermite cubic
and quintic elements. Examples 7.2 and 7.3.

(2) beam_disc_speed.m
Critical speed of a shaft with one disc, example 7.5

(3) frames_eigen_pi.m
Natural frequencies of a pi-frame, example 7.6

(4) beam_eigen_buckling.m
Beam buckling load, example 7.7

(5) beam_SS_frequency_Axial_force.m
Natural frequencies of a beam with axial force, example 7.8

(6) Orr_Summerfeld_H5_V3A_margin.m
Solution to the Orr-Sommerfeld eigenvalue problem
for channel flow, example 7.9

MATLAB CODES FOR CHAPTER 8

(1) example8p2.m
Plane elasticity on one biquadratic element
and comparison with exact solution

(2) Howland_extra.m
Howland problem, stresses in a strip with a central circular
hole

(3) eccentric_cylinders.m
Example 8.4: A circular cylinder pressurized by
an eccentric circular conduit

(4) eccentric_cylinders_Mesh.m
Generate a mesh on the full or half the domain between
exccentric circular cylinders

(5) Florence1.m
Example 8.5: Thermal stresses around an insulated circular hole
in an infinite plate subjected to steady heat flow

(6) Mesh_Culvert.m
Generate a mesh of bilinear or biquadratic quadrilateral
elements for a culvert

MATLAB CODES FOR CHAPTER 9

(1) CPT_Rectangle.m
Example 9.1, static deflection of a rectangular plate
Classical or Kirchhoff-Love plate using conforming
Bogner-Fox-Schmit element

(2) CPT_semi_circ_annular1.m

Example 9.2, static deflection of annular semi-circular plate
Classical or Kirchhoff-Love plate using isoparametric Hermite element

(3) RMPLate_Rectangle.m
Example 9.3, static deflection of a rectangular plate
Reissner-Mindlin plate using isoparametric Lagrange element

(4) CPT_Rectangle_Eigen.m
Example 9.4, natural frequencies of a rectangular plate
Classical or Kirchhoff-Love plate using conforming
Bogner-Fox-Schmit element

(5) RMPlate_circle_eigen.m
Example 9.6, natural frequencies of a circular plate
Reissner-Mindlin plate using isoparametric Lagrange element

(6) RMPlate_circle_linear_hydroelastic.m
Example 9.9, virtual mass effects on natural frequencies,
NAVMI factors of a circular plate
Reissner-Mindlin plate using isoparametric Lagrange element

MATLAB CODES FOR CHAPTER 10

(1) NL_RM_static_square.m
Example 10.1, nonlinear static deflection of a rectangular plate
Reissner-Mindlin plate using isoparametric Lagrange element

(2) NL_RM_static_circle.m
Example 10.2, effects of geometric nonlinearity on static
deflection of a circular Reissner-Mindlin plate

(3) RM_critical_load_uniaxial_square.m
Example 10.3, critical uniaxial loading for a square plate
Reissner-Mindlin plate using isoparametric Lagrange element

(4) RM_buckling_newton.m
Example 10.6, post-buckling of a square plate by uniaxial load
Reissner-Mindlin plate using isoparametric Lagrange element

(5) RM_rectangle_thermal_post_buckling_newton.m
Example 10.11, defelction-temperature path for thermal post
buckling of skewed rectangular plates
Reissner-Mindlin plate using isoparametric Lagrange element

(6) NL_RM_Eigen_23_SS.m
Example 10.12, backbone curve for a square plate
Effects of geometric nonlinearity on natural frequencies of a square Reissner-Mindlin plate using harmonic balance

(7) Linear_supersonic_flutter_rectangle.m
Example 10.14, linear supersonic flutter of a simply supported rectangular Reissner-Mindlin plate

(8) NL_flutter_supersonic_square_SS.m
Example 10.14, nonlinear supersonic flutter of a simply supported rectangular Reissner-Mindlin plate

(9) NL_diverge_hydro_square_Newton.m
Example 10.15: nonlinear hydroelastic divergence of a simply supported square plate using Newton-Raphson method, Reissner-Mindlin plate

References

1. B. Abbasnezhad, N. Hamdami, J. Monteau, and H. Vantankhah. Numerical Modeling of Heat Transfer and Pasteurizing Value During Thermal Processing of Intact Egg. *Food Science & Nutrition*, pages 1–8, 2015. Open Access article published by Wiley Periodicals, Inc.
2. I.H. Abbot and A.E. von Doenhoff. *Theory of Wing Sections*. Dover, New York, NY, 1959.
3. M. Abramowitz and I.A. Stegun. *Handbook of Mathematical Functions*. Dover, New York, NY, 1970.
4. M. Amabili. *Nonlinear Vibrations and Stability of Shells and Plates*. Cambridge University Press, Cambridge, 2004.
5. M. Amabili. Nonlinear Vibrations of Rectangular Plates with Different Boundary Conditions: Theory and Experiments. *Computers and Structures*, 82:2587–2605, 2004.
6. M. Amabili and M.K. Kwak. Free Vibrations of Circular Plates Coupled with Liquids: Revising the Lamb Problem. *Journal of Fluids and Structures*, 10:743–761, 1996.
7. J.H. Argyris and S. Kelsey. *Energy Theorems and Structural Analysis*. Butterworths, 1960.
8. J. Barlow. Optimal Stress Locations in Finite Element Models. *Int. J. Numerical Methods in Egineering*, 10(2):243–251, 1976.
9. J. Barlow. More on Optimal Stress Points—Reduced Integration, Element Distortions and Error Estimation. *International Journal for Numerical Methods in Egineering*, 28(7):1487–1504, 1989.
10. E. Baskharone and A. Hamed. Cascade Flow Analysis with the Finite Element Method. *AIAA Journal*, 19(1):65–71, 1981.
11. F.P. Beer and E.R. Jr. Johnston. *Mechanics of Materials*. McGraw-Hill, New York, NY, second edition, 1992.
12. A. Bhimarasetty and R.N. Govardhan. A Simple Method for Potential Flow Simulation of Cascades. *Sadhana, Indian Institute of Science*, 35(6):649–657, 2010.
13. B.A. Boley and J.H. Weiner. *Theory of Thermal Stresses*. John Wiley & Sons, New York, NY, 1960.
14. A.P. Boresi. *Elasticity in Engineering Mechanics*. John Wiley & Sons, New York, NY, second edition, 1965.
15. A. Boutemedjet, M. Samardžic, L. Rebhi, Z. Rajic, and T. Mouadaa. UAV Aerodynamic Design Involving Genetic Algorithm and Artificial Neural Network for Wing Preliminary Computation. *Aerospace Sciences and Technology*, 84:464–483, 2019.
16. L. Boutet de Monvel and I.D. Chueshov. Oscillations of von Karman's Plate in a Potential Flow of Gas. *Izv. Math.*, 63(2):219–244, 1999.
17. W.E. Boyce and R.C. DiPrima. *Elementary Differential Equations and Boundary Value Problems*. John Wiley & Sons, New York, NY, sixth edition, 1997.
18. A.N. Brooks and T.J.R. Hughes. Streamline Upwind/Petrov-Galerkin Formulations for Convection Dominated Flows with Particular Emphasis on the Incompressible Navier-Stokes Equations. *Computer Methods in Applied Mechanics and Engineering*, 84:199–259, 1982.

19. K. Buettner. Effects of Extreme Heat and Cold on Human Skin. I. Analysis of Temperature Changes Caused by Different Kinds of Heat Application. *Journal of Applied Physiology*, 3(12):691–702, 1951.
20. M.P. Cetingul and C. Herman. A Heat Transfer Model of Skin Tissue for the Detection of Lesions: Sensitivity Analysis. *Physics in Medicine & Biology*, 55:5933–5951, 2010.
21. Y. Chai, F. Li, and Z. Song. Nonlinear Flutter Suppression and Thermal Buckling Elimination for Composite Lattice Sandwich Panels. *AIAA Journal*, 57(11):4863–4872, 2019.
22. I.A. Chang and U.D. Nguyen. Thermal Modeling of Lesion Growth with Radiofrequency Ablation Devices. *BioMedical Engineering OnLine*, 3(27), 2004.
23. C-Y Chia. *Nonlinear Analysis of Plates*. McGraw-Hill, New York, NY, 1980.
24. H-N Chu and G. Herrmann. Influence of Large Amplitudes on Free Flexural Vibrations of Rectangular Elastic Plates. *Journal of Applied Mechanics*, page 532–540, 1956.
25. G. Di, Z. Wu, and D. Huang. The Research on Active Flow Control Method with Vibration Diaphragm on a NACA0012 Airfoil at Different Stalled Angles of Attack. *Aerospace Science and Technology*, 69:532–540, 2017.
26. S.L. Dixon and C. Hall. *Fluid Mechanics and Thermodynamics of Turbomachinery*. Elsevier, New York, seventh edition, 2014.
27. E.H. Dowell. Panel Flutter: A Review of the Aeroelastic Stability of Plates and Shells. *AIAA Journal*, 8(3):385–399, 1970.
28. E.H. Dowell. *Aeroelasticity of Plates and Shells*. Noordhoff International Publishing, Leyden, The Netherlands, 1974.
29. E.H. Dowell. Some Recent Advances in Nonlinear Aeroelasticity. In E.H Dowell, editor, *A Modern Course in Aeroelasticity*. Springer, 2015. Chapter 14.
30. E.H. Dowell and D.B. Bliss. New Look at Unsteady Supersonic Potential Flow Aerodynamics and Piston Theory. *AIAA Journal*, 51(9):2278–2281, 2013.
31. P.G. Drazin and W.H. Reid. *Hydrodynamic Stability*. Cambridge University Press, New York, NY, 1981.
32. J. Dugundji, E.H. Dowell, and B. Perkin. Subsonic Flutter of Panels on Continuous Elastic Foundations. *AIAA Journal*, 1(5):1146–1154, 1963.
33. D.A. Dunavant. High Degree Efficient Symmetrical Gaussian Quadrature Rules for the Triangle. *International Journal for Numerical Methods in Egineering*, 21:1129–1148, 1985.
34. C.L. Dym. *Stability Theory and its Applications to Structural Mechanics*. Noordhoff International Publishing, Leyden, The Netherlands, 1974.
35. G.H. Ellen. The Stability of Simply Supported Rectangular Surfaces in Uniform Subsonic Flow. *Journal of Applied Mechanics*, 68:68–72, 1973.
36. W.C. Elmore and M.A. Heald. *Physics of Waves*. Dover, New York, NY, 1985.
37. A.L. Florence and J.N. Goodier. Thermal Stress at Spherical Cavities and Circular Holes in Uniform Heat Flow. *Journal of Applied Mechanics*, 26:293–294, 1959.
38. R.L. Fournier. *Basic Transport Phenomena in Biomedical Engineering*. CRC Press, Inc. Boca Raton, FL, third edition, 2011.
39. Y.C. Fung. *Foundations of Solid Mechanics*. Prentice-Hall, Inc. Englewood Cliffs, New Jersey, 1965.
40. P.R. Garabedian. *Partial Differential Equations*. Chelsea Publishing Company, New York, NY, second edition, 1976.
41. C. Gonzalez and H. Taha. A Variational Theory of Lift . *Journal of Fluid Mechanics*, 941(A58), 2022.

42. D.J. Gorman. A Highly Accurate Analytical Solution for Free Vibration Analysis of Simply Supported Right Triangular Plates. *Journal of Sound and Vibration*, 89(1):107–118, 1983.
43. D.J. Gorman. Free Vibration Analysis of Right Triangular Plates with Combinations of Clamped-Simply Supported Boundary Conditions. *Journal of Sound and Vibration*, 106(3):419–431, 1986.
44. N. Grover, B.N. Singh, and D.K. Maiti. An Inverse Trigonometric Shear Deformation Theory for Supersonic Flutter Characteristics of Multilayered Composite Plates. *Aerospace Sciences and Technology*, 52:41–51, 2016.
45. X. Guo and C. Mei. Application of Aeroelastic Modes on Nonlinear Supersonic Panel Flutter at Elevated Temperatures. *Computers and Structures*, 84:1619–1628, 2006.
46. H.C. Hardee and D.O. Lee. A Simple Conduction Model for Skin Burns Resulting from Exposure to Chemical Fireball. *Fire Research*, 1:199–205, 1977.
47. S.H. Hashemi and M. Arsanjani. Exact Characteristic Equations for Some of Classical Boundary Conditions of Vibrating Moderately Thick Rectangular Plates. *International Journal of Solids and Structures*, 42:819–853, 2005.
48. S.H. Hashemi, K. Khorshidi, and M. Amabili. Exact Solution for Linear Buckling of Rectangular Mindlin Plates. *Journal of Sound and Vibration*, 315:318–342, 2008.
49. H. Hencky. Über die Berücksichtigung der Schubverzerrung in ebenen Platten. *Ingenieur Archiv.*, 16(1):72–76, 1947.
50. F.C. Jr. Henriques. Studies of Thermal Injury V. The Predictability and the Significance of Thermally Induced Rate Processes Leading to Irreversible Epidermal Injury. *Archives of Pathology*, 43:489–502, 1947.
51. C. Hirsch. *Numerical Computation of Internal and External Flows.* John Wiley & Sons, New York, NY, 1988.
52. R.C.J. Howland. On the Stresses in the Neighbourhood of a Circular Hole in a Strip Under Tension. *Philosophical Transactions of the Royal Society of London A*, 229:49–86, 1930.
53. T.J.R. Hughes. *The Finite Element Method.* Dover, New York, NY, 2000.
54. T.J.R. Hughes, R.L. Taylor, and W. Kanoknukulchai. A Simple and Efficient Finite Element for Plate Bending. *Philosophical Transactions of the Royal Society of London A*, 11:1529–1543, 1977.
55. G.B. Jeffery. Plane Stress and Plane Strain in Bipolar Coordinates. *Philosophical Transactions of the Royal Society of London A*, 221:265–293, 1921.
56. D.J. Johns. Some Panel-Flutter Studies Using Piston Theory. *Journal of the Aerospace Sciences*, 25(11):679–684, 1958.
57. N.N. Johnson, J.P. Abraham, Z.I. Helgeson, W.J. Minkowycz, and E.M. Sparrow. An Archive of Skin-Layer Thicknesses and Properties and Calculations of Scald Burns with Comparisons to Experimental Observations. *Journal of Thermal Science and Engineering Applications*, 3:011003–9, 2011.
58. J. S. Kapor. *Novel Computational Methods for the Study of Compliant-Wall Fluid-Structure Interaction.* 2012. Ph.D. Curtin University, Department of Mechanical Engineering.
59. J. Katz and A. Plotkin. *Low-Speed Aerodynamics.* Cambridge University Press, UK, second edition, 2001.
60. N.P. Kirchner. Computational Aspects of the Spectral Galerkin FEM for the Orr-Sommerfeld Equation. *International Journal for Numerical Methods in Fluids*, 32:119–137, 2000.

61. A. Kornecki, E.H. Dowell, and J. O'Brien. On the Aeroelastic Instability of Two-dimensional Panels in Uniform Incompressible Flow. *Journal of Sound and Vibration*, 47(2):163–178, 1976.
62. T.D. Lagerlund and P.A. Low. Axial Diffusion and Michaelis-Menten Kinetics in Oxygen Delivery in Rat Peripheral Nerve. *American Physiological Society Journal 260 (Regulatory Integrative Comp. Physiol.)*, 29:R430–R440, 1991.
63. T.D. Lagerlund and P.A. Low. Mathematical Modeling of Time-Dependent Oxygen Transport in Rat Peripheral Nerve. *Computers in Biology and Medicine*, 23:29–47, 1993.
64. N.N. Lebedev, I.P. Skalskaya, and Y.S. Uflyand. *Worked Problems in Applied Mathematics*. Dover, New York, NY, 1979.
65. S.H. Lin. Oxygen Diffusion in a Spherical Cell with Nonlinear Oxygen Uptake Kinetics. *J. Theor. Biol.*, 60:449–457, 1976.
66. A.E.H. Love. *A Treatise on the Mathematical Theory of Elasticity*. Dover, New York, NY, fourth edition, 1944.
67. A.D. Lucey, G.J. Cafolla, P.W. Carpenter, and Yang M. The Nonlinear Hydroelastic Behaviour of Flexible Walls. *Journal of Fluids and Structures*, 11:717–744, 1997.
68. A.D. Lucey and P.W. Carpenter. The Hydroelastic Stability of Three-dimensional Disturbances of a Finite Compliant Wall. *Journal of Sound and Vibration*, 165(3):527–552, 1993.
69. K. Makihara and S. Shimose. Supersonic Flutter Utilization for Effective Energy-Harvesting Based on Piezoelectric Switching Control. *Smart Materials Research*, 2012. Article ID 181645, 10 pages doi:10.1155/2012/181645.
70. M. Mamou and M. Khalid. Finite Element Solution of the Orr–Sommerfeld Equation Using High Precision Hermite Elements: Plane Poiseuille Flow. *International Journal for Numerical Methods in Fluids*, 44:721–735, 2004.
71. T.A. Massaro and I. Fatt. Oxygen Diffusion in Large Single-Celled Organisms. *Bulletin of Mathematical Biology*, 31:327–340, 1969.
72. D.A. McDonald. The Relation of Pulsatile Pressure to Flow in Arteries. *Journal of Physiology*, 127:533–552, 1955.
73. D.L.S. McElwain. A Re-examination of Oxygen Diffusion in a Spherical Cell with Michaelis-Menten Oxygen Uptake Kinetics. *Journal of Theoretical Biology*, 71:255–263, 1978.
74. C.G. Merrett and H.H. Hilton. Elastic and Viscoelastic Panel Flutter in Incompressible, Subsonic and Supersonic Flows. *Journal of Aeroelasticity and Structural Dynamics*, 2(1):53–80, 2010.
75. L.M. Milne– Thomson. *Theoretical Hydrodynamics*. The Macmillan Company, New York, NY, fourth edition, 1960.
76. R.D. Mindlin. Influence of Rotatory Inertia and Shear on Flexural Motions of Isotropic, Elastic Plates. *Journal of Applied Mechanics*, 18(1):31–38, 1951.
77. D. Nakos and P. Sclavounos. Ship Motions by a Three-Dimensional Rankine Panel Method. In *18th Naval Hydrodynamic Symposium*, page 21–40, 1990.
78. A.H. Nayfeh and D. T. Mook. *Nonlinear Oscillations*. John Wiley & Sons, New York, NY, 1979.
79. N.M. Newmark. A Method of Computation for Structural Dynamics. *Journal of the Engineering Mechanics Division, ASCE*, page 67–94, 1959.

80. V.V. Novozhilov. *Foundations of the Nonlinear Theory of Elasticity*. Dover, New York, NY, 1953. Dover (1999) republication of the work published by Graylock Press, Rochester, NY, 1953.
81. S.A. Orszag. Accurate Solution of the Orr-Sommerfeld Stability Equation. *Journal of Fluid Mechanics*, 50(4):689–703, 1971.
82. S. Panda and M. Barik. Large Deflection of Arbitrary Thin Plates Using Superparametric Finite Element. *ASIAN Journal of Civil Engineering (BHRC)*, 18(2):207–234, 2017.
83. H.H. Pennes. Analysis of Tissue and Arterial Blood Temperatures in the Resting Human Forearm. *Journal of Applied Physiology*, 1(2):93–122, 1948.
84. J. Petera and J.F.T. Pittman. Isoparametric Hermite Elements. *International Journal for Numerical Methods in Engineering*, 37:3489–3519, 1994.
85. A. Pica, R.D. Wood, and E. Hinton. Finite Element Analysis of Geometrically Nonlinear Plate Behavior Using a Mindlin Formulation. *Computers & Structures*, 11:203–215, 1980.
86. M.W. Pitman and A.D. Lucey. On the Direct Determination of the Eigenmodes of Finite Flow-Structure Systems. *Proceedings of the Royal Society A*, 465:257–281, 2009.
87. M.S.S. Prabhu and S. Durvasula. Thermal Post-Buckling Characteristics of Clamped Skew Plates. *Computers & Structures*, 6:177–185, 1976.
88. T. Prakash, M.K. Singha, and M. Ganapathi. Thermal Post-Buckling of FGM Skew Plates. *Engineering Structures*, 30:22–32, 2008.
89. J.S. Przemieniecki. *Theory of Matrix Structural Analysis*. Dover, New York, NY, 1968.
90. S. A. Ragab. Shape Optimization of Surface Ships in Potential Flow Using an Adjoint Formulation. *AIAA Journal*, 42(2):296–304, 2004.
91. S.A. Ragab and H.E. Fayed. Finite Element Analysis of Aero-Hydroelastic Stability of Arbitrary Shape Panels. *Aerospace Science and Technology*, 90:299–313, 2019.
92. S.A. Ragab and H.E. Fayed. Linear and Nonlinear Flutter of Supersonic Panels of Various Shapes. *Journal of Engineering Mechanics*, 147(3:04020156), 2021. https://doi.org/10.1061/(ASCE)EM.1943-7889.0001899.
93. K.K. Raju and E. Hinton. Nonlinear Vibrations of Thick Plates Using Mindlin Plate Elements. *International Journal for Numerical Methods in Engineering*, 15:249–257, 1980.
94. R.H. Rand. *Lecture Notes on Nonlinear Vibrations*. Cornell University, Ithaca, NY, 2012. https://hdl.handle.net/1813/28989.
95. J.N. Reddy. *Theory and Analysis of Elastic Plates*. Taylor & Francis, Philadelphia, PA, 1999.
96. J.N. Reddy. *Mechanics of Laminates Composite Plates and Shells: Theory and Analysis*. CRC Press, Boca Raton, FL, second edition, 2004.
97. J.N. Reddy. *An Introduction to the Finite Element Method*. McGraw-Hill, New York,NY, third edition, 2006.
98. E. Reissner. On the Theory of Elastic Plates. *Journal of Mathematics and Physics*, 23:184–191, 1944.
99. E. Reissner. The Effect of Transverse Shear Deformation on the Bending of Elastic Plates. *Journal of Applied Mechanics*, 3:69–77, 1945.
100. E. Reissner. On Bending of Elastic Plates. *Quarterly Journal of Applied Mathematics*, 5(1):55–68, 1947.

101. D.D. Jr. Reneau, D.F. Bruley, and M.H. Knisely. A Digital Simulation of Transient Oxygen Transport in Capillary-Tissue Systems, (Cerebral Grey Matter). *AIChE Journal*, 15(6):916–925, 1969.
102. P. Ribeiro. A Hierarchical Finite Element for Geometrically Non-linear Vibration of Thick Plates. *Meccanica*, 38:115–130, 2003.
103. R.D. Richmyer and K.W. Morton. *Difference Methods for Initial-Value Problems*. John Wiley & Sons, New York, NY, second edition, 1967.
104. Z. Rong, B. Cao, and J. Hu. Stability Analysis on an Aeroelastic System for Design of a Flutter Energy Harvester. *Aerospace Science and Technology*, 60:203–209, 2017.
105. R.R. Ryenolds and E.H. Dowell. Nonlinear Aeroelastic Response of Panels. In *34th Structures, Structural Dynamics and Materials Conference*, 1993. AIAA-93-1599-CP.
106. M. H. Sadd. *Elasticity- Theory, Applications, and Numerics*. Academic Press, Oxford, UK, second edition, 2009.
107. M. Sathyamoorthy. Nonlinear Vibration Analysis of Plates: a Review and Survey of Current Developments. *Applied Mechanics Review*, 40:1553–1561, 1987.
108. H. Schlichting. *Boundary-Layer Theory*. McGraw-Hill, New York, NY, seventh edition, 1979.
109. I. Senjanovic, N. Hadzic, N. Vladimir, and D.-S. Cho. Natural Vibrations of Thick Circular Plate Based on the Modified Mindlin Theory. *Archives of Mechanics*, 66(6):389–409, 2014.
110. I.S. Sokolnikoff. *Mathematical Theory of Elasticity*. McGraw-Hill, New York, NY, 1946.
111. S. Srinivas and A.K. Rao. Buckling of Thick Rectangular Plates. *AIAA Journal*, 7(8):1645–1646, 1969.
112. H. Taha and C. Gonzalez. Refining Kutta's Flow over a Flat Plate: Necessary Conditions for Lift . *AIAA Journal*, 61(5), 2023.
113. B.H. Tan, A.D. Lucey, and R.M. Howell. Aero-hydro-elastic Stability of Flexible Panels: Prediction and Control Using Localised Spring Support. *Journal of Sound and Vibration*, 332:7033–7054, 2013.
114. B.H. Tan, A.D. Lucey, and M.W. Pitman. Stability of a structurally inhomogeneous flexible plate in uniform axial flow. In C. Meskell and G. Bennett, editors, *Flow-Induced Vibration*, 2012.
115. J.C. Tannehill, D.A. Anderson, and R.H. Pletcher. *Computational Fluid Mechanics and Heat Transfer*. Taylor & Francis, Philadelphia, PA, second edition, 1997.
116. B. Thwaites. *Incompressible Aerodynamics*. Dover, New York, 1987.
117. S. Timoshenko and J. Gere. *Theory of Elastic Stability*. McGraw-Hill, New York, NY, second edition, 1963.
118. S. Timoshenko and S.W. Krieger. *Theory of Plates and Shells*. McGraw-Hill, New York, NY, second edition, 1959.
119. S. Timoshenko and D.H. Young. *Vibration Problems in Engineering*. D. Van Nostrand Company, Inc., Princeton, New Jersey, third edition, 1955.
120. D.A. Torvi and J.D. Dale. A Finite Element Model of Skin Subjected to a Flash Fire. *Journal of Biomechanical Engineering*, 116:250–255, 1994.
121. N. Tsushima and W. Su. Flutter Suppression for Highly Flexible Wings Using Passive and Active Piezoelectric Effects. *Aerospace Science and Technology*, 65:78–89, 2017.
122. N. Tsushima and W. Su. A Study on Adaptive Vibration Control and Energy Conversion of Highly Flexible Multifunctional Wings. *Aerospace Science and Technology*, 79:297–309, 2018.

123. E. Tubaldi, F. Alijani, and M. Amabili. Non-Linear Vibrations and Stability of a Periodically Supported Rectangular Plate in Axial Flow. *International Journal of Non-Linear Mechanics*, 66:54–65, 2014.
124. K. Washizu. *Variational Methods in Elasticity and Plasticity*. Pergamon Press, New York, NY, third edition, 1982.
125. F.M. White. *Viscous Fluid Flow*. McGraw-Hill, New York, NY, second edition, 1991.
126. J.R. Womersley. Method for the Calculation of Velocity, Rate of Flow and Viscous Drag in Arteries When the Pressure Gradient is Known. *Journal of Physiology*, 127:553–563, 1955.
127. R.C. Zhou, Z. Lai, D.Y. Xue, J-K. Huang, and C. Mei. Suppression of Nonlinear Panel Flutter with Piezoelectric Actuators Using Finite Element Method. *AIAA Journal*, 33(6):1098–1105, 1995.

Index